Environmental Science and Engineering

Environmental Engineering

Series editors

More information about this series at http://www.springer.com/series/3172

Liangtong Zhan · Yunmin Chen
Abdelmalek Bouazza
Editors

Proceedings of the 8th International Congress on Environmental Geotechnics Volume 2

Towards a Sustainable Geoenvironment

Springer

Editors
Liangtong Zhan
MOE Key Laboratory of Soft Soils and Geoenvironmental Engineering, College of Civil Engineering and Architecture
Zhejiang University
Hangzhou, China

Abdelmalek Bouazza
Department of Civil Engineering
Monash University
Clayton, VIC, Australia

Yunmin Chen
MOE Key Laboratory of Soft Soils and Geoenvironmental Engineering, College of Civil Engineering and Architecture
Zhejiang University
Hangzhou, China

ISSN 1863-5520 ISSN 1863-5539 (electronic)
Environmental Science and Engineering
ISSN 1431-2492
Environmental Engineering
ISBN 978-981-13-2223-5 ISBN 978-981-13-2224-2 (eBook)
https://doi.org/10.1007/978-981-13-2224-2

Library of Congress Control Number: 2018957622

This Springer imprint is published by the registered company Springer Nature Singapore Pte Ltd.
The registered company address is: 152 Beach Road, #21-01/04 Gateway East, Singapore 189721, Singapore

Preface

Issues associated with environmental geotechnics continue to be a major preoccupation for governments, public and private organizations and the general community around the world. With the support from the International Society for Soil Mechanics and Geotechnical Engineering (ISSMGE) and the Technical Committee of Environmental Geotechnics (TC215), the Environmental Geotechnics Congress Series has been held regularly since 1994 (Canada 1994; Japan 1996; Portugal 1998; Brazil 2002; UK 2006; India 2010; Australia 2014) and established itself as a highly influential forum for exchange and discussion on the subject. Following the success of 7ICEG in Melbourne, 2014, the 8th International Congress on Environmental Geotechnics (8ICEG) held on 28 October–1 November 2018 in Hangzhou, China, continues tackling challenging issues in the broad range of environmental geotechnics.

The congress theme is "Towards a Sustainable Geoenvironment". "Sustainable Development is to meet the needs of the present without compromising the ability of future generations to meet their own needs". Geoenvironment is a specific compartment of the environment and comprises portions of geosphere, hydrosphere and biosphere. Under this theme, the congress will cover a broad range of topics and will provide an excellent opportunity for academics, engineers, scientists, government officials, regulators and planners to present, discuss and exchange the latest advancements and developments in the research and application of environmental geotechnics.

Out of 340 abstracts received, 8ICEG chose 255 full manuscripts and 6 extended abstracts submitted from 28 countries and regions in 5 continents for publication in this conference proceedings, which provides a platform for scholars and practitioners to share and exchange their experiences with their peers, especially with those from developing countries with impending geoenvironmental issues. There are several features of the proceedings:

- Eleven comprehensive manuscripts, together with 6 extended abstracts, were contributed from the Plenary Lecturers from 11 countries, which includes 8 academicians, 4 editors of major journals and 6 chairpersons of international renowned organizations. Their contributions provided in-depth insights into the chosen topics.

- Emerging topics, such as sustainability, bio-geoengineering and geoenvironmental aspects in energy geotechnics have strong appearance, which suggests developments towards a better living geoenvironment is the focus for future developments.
- Manuscripts from "Belt and Road Countries" emerged and showed strong presence in 8ICEG, including 24 manuscripts from 10 countries. Environmental problems in those countries take high priority for the governments.

Financial supports from the National Natural Science Foundation of China (41842018) and the Chinese Program of Introducing Talents of Discipline to University (the 111 Project, B18047) are acknowledged. The time and efforts of the associate editors and the reviewers for the proceedings are greatly appreciated.

We hope you enjoy 8ICEG in the beautiful city of Hangzhou!

Sincerely yours,

Liangtong Zhan
Yunmin Chen
Abdelmalek Bouazza
Editors of the Proceedings of 8ICEG

The original version of the book was revised: The book was inadvertently published without incorporating the author corrections. The correction to the book is available at https://doi.org/10.1007/978-981-13-2224-2_92

Organization

List of Associate Editors

List of Committees

International Advisory Committee

Craig H. Benson, USA
Malek Bouazza, Australia
Zuyu Chen, China
Raffaello Cossu, Italy
Pierre Delage, France
Roger Frank, France
Delwyn G. Fredlund, Canada
Russell Jones, UK
Edward Kavazanjian, USA
Yuanming Lai, China
Mario Manassero, Italy
Charles W. W. Ng, Hong Kong SAR
William Powrie, UK
R. Kerry Rowe, Canada
Daichao Sheng, Australia
Devendra N. Singh, India
Kenichi Soga, USA
Hywel R. Thomas, UK
Ikuo Towhata, Japan
Ju Wang, China
Weimin Ye, China
Jianmin Zhang, China
Jorge G. Zornberg, USA

Technical Committee

Chair: Yunmin Chen, China
Co-chair: Malek Bouazza, Australia; Liangtong Zhan, China
Secretary: Bate Bate, China; Shuai Zhang, China

Ishtiaque Anwar, Singapore
D. N. Arnepalli, India
Kazem Badv, Iran
Xiaohong Bai, China
Loretta Batali, Romania
Maria Eugenia Boscov, Brazil
Herminia Cano, Spain
Roger Clark, UK
Dimitrios Coumoulos, Greece
John Cowland, Hong Kong SAR
Yujun Cui, France
Manoj Datta, India
Gemmina Di Emidio, Belgium
Andrea Dominijanni, Italy
Abbas El-Zein, Australia
Evelina Fratalocchi, Italy
Eugene Gallagher, UK
Antonio Gens, Spain
Jean-Pierre Gourc, France
Pinjing He, China
Georg Heerten, Germany
Liming Hu, China
Stephan Jefferis, UK
Han-Yong Jeon, South Korea
Takeshi Katsumi, Japan
Edward Kavazanjian, USA
Eugeniusz Koda, Poland
Hideo Komine, Japan
Lingwei Kong, China
Craig Lake, Canada
William Likos, USA
Zhiping Lin, Taiwan
Xianchang Ling, China
Songyu Liu, China
Lurdes Lopes, Portugal
Mario Manassero, Italy
Nicola Moraci, Italy
Catherine Mulligan, Canada
V. G. Ofrikhter, Russia
Ernest Olinic, Romania
Ali Pak, Iran
Marina Pantazidou, Greece
Fernando Pardo, Spain
G. V. Ramana, India

Organizing Committee

Yuemiao Liu	Beijing Research Inst. Uranium Geology, China
Zhihong Zhang	Beijing U. of Tech., China
Fan Lv	Tongji U., China
Weilie Zou	Wuhan U., China
Hefu Pu	Huazhong U. of Sci. and Tech., China

Organized by

Ministry of Education Key Laboratory of Soft Soils and Geoenvironmental Engineering, Zhejiang University

The Chinese Institution of Soil Mechanics and Geotechnical Engineering (CISMGE)

The Hong Kong Geotechnical Society (HKGES)

Under the Auspices of

TC215 Environmental Geotechnics, ISSMGE

Supported by

International Geosynthetics Society (IGS)

International Waste Working Group (IWWG)

Commission on Radioactive Waste Disposal of International Society for Rock Mechanics (ISRM)

Commission on Waste Disposal of the International Association of Engineering Geology and the Environment (IAEG)

List of Plenary Lectures

Speakers	Affiliation	Topic
Distinguished Lecture		
Kerry Rowe	Queens University, Canada	Environmental Geotechnics: Past, Present and Future
Kerry Rowe Lecture		
Mario Manassero	Politecnico di Torino, Italy	On the Fabric and State Parameters of Active Clays for Contaminant Control
Keynote Lectures		
Craig Benson	University of Virginia, USA	Sustainability in Reuse of Solid Wastes
Ningwu Chang	California EPA, USA	Brownfield Redevelopment at Contaminated Landfill Site
Yunmin Chen	Zhejiang University, China	Waste Mechanics and Sustainable Landfilling Technology: Comparison between HFWC and LFWC MSWs
Hywel Thomas	Cardiff University, UK	Deep Ground and Energy: Carbon Sequestration and Coal Gasification
Invited Lectures		
Malek Bouazza	Monash University, Australia	Failures in Containment Systems: Lessons Learned
Michael A. Celia	Hamburg University, Germany	Modeling Geological Storage of Carbon Dioxide with a Focus on Leakage Risk Assessment
Pierre Delage	Ecole des ponts Paris Tech, France	Micro-macro Effects in Bentonite Engineered Barriers for Radioactive Waste Disposal
Nathalie Touze Foltz	Irstea, France	Performance Issues of Barrier Systems for Landfills
Stephan Jeffries	Environmental Geotechnics Limited, UK	Low Permeability Vertical Barriers: The State of the Art and the Research Needs for the Future

Takeshi Katsumi	Kyoto University, Japan	Towards Sustainable Soil Management Reuse of Excavated Soils with Natural Contamination
Ed Kavazanjian	Arizona State University, USA	Biogeotechnical Engineering Applications for Environmental Protection and Restoration
Olaf Kolditz	Dresden University of Technology, Germany	Workflows in Environmental Geotechnics: Status-Quo and Perspectives
William Powrie	University of Southampton, UK	Climate and Vegetation Impacts on Infrastructure Cuttings and Embankments
Krishna Reddy	University of Illinois at Chicago, USA	Risk, Sustainability and Resiliency Considerations in Polluted Site Remediation
Devendra Narain Singh	Indian Institute of Technology, India	Centrifuge Modeling of Contaminant Transport in Geomaterials
Rainer Stegmann	Hamburg University, Germany	Development of Waste Management in The Last 30 Years
Antonio Thome	University of Passo Fundo, Brazil	Remediation Technologies Applied in Polluted Soils: New Perspectives in This Field
Fuming Wang	Zhengzhou University, China	Jet Grouting for Leakage Prevention
General Report		
Charles Shackelford	Colorado State University, USA	Pollutant Transport
William Powrie	University of Southamptom, UK	Waste Mechanics

Contents

Landfills of Solid Wastes

(Edited by Xiaobing Xu, Dongxing Wang, and Chuang Yu)

Pozzolanic Properties of Municipal Solid Waste Incineration (MSWI) Fly Ash Under the Actions of Three Different Activators

Ping Chen, Xiao-qing Ding, Hao Zheng, and Hui Xu(✉)

School of Civil Engineering and Architecture, Zhejiang Sci-Tech University, Hangzhou, China
xuhui@zstu.edu.cn

Abstract. This study investigated the pozzolanic capacity of MSWI fly ash in the solidification process. The MSWI fly ash were solidified by using three kinds of activators, i.e., Na_2CO_3, $CaSO_4$ and Na_2SiO_3, they were all cured for 7 and 28 days. Each of them involved two cases, i.e., the activator mixing proportions were 3 and 6% in dry mass basis, respectively. After that, the unconfined compressive strength (UCS) test and water immersion test were conducted on the solidified samples, to investigate the influence of the kind and proportion of activator on MSWI fly ash solidification. The results showed that all the three kinds of activators had played a role in promoting the solidification process, and the Na_2CO_3 performed best. The UCS value of the samples with the action of Na_2CO_3 enhanced with the increase of activator mixing proportion and curing time. The 28-day UCS value of the sample with Na_2CO_3 mixing proportions of 6% reached 2.924 MPa, and the solidified sample also showed a good water tolerance. The effect of $CaSO_4$ on the MSWI fly ash solidification was complex. The UCS value increased with the curing time when the $CaSO_4$ mixing proportion was 3%, however, it decreased with the curing time when the $CaSO_4$ mixing proportion was 6%.

Keywords: MSWI fly ash · Pozzolanic property · Activator
Unconfined compressive strength · Water tolerance

1 Introduction

Incineration is one of the most widely used methods to treat the MSW. Even though incineration is efficient for treating MSW, one of the shortcomings of this technology is the significant production of solid residues, e.g., fly ash [1]. MSWI fly ash is regarded as hazardous material mainly due to its high content of heavy metals. Therefore, MSWI fly ash needs solidification/stabilization treatment before landfill disposal [2]. A deep understanding of the pozzolanic properties of MSWI fly ash has important theoretical significance in the solidification/stabilization treatment of MSWI fly ash.

At present, many studies have been conducted to investigate the pozzolanic properties of coal fly ash, mineral powder and other materials. Zhang et al. [3] studied the pozzolanic properties of mineral slag, steel slag and coal fly ash under the action of NaOH. The results showed that the mineral slag obtained the best effect, followed by

L. Zhan et al. (Eds.): ICEG 2018, ESE, pp. 3–9, 2019.
https://doi.org/10.1007/978-981-13-2224-2_1

steel slag and coal fly ash. Shi et al. [4] found that 4% $CaCl_2.2H_2O$ can significantly improve the later strength of lime-pozzolan cement. Sivapullaiah et al. [5] found that the adding of $CaSO_4$ accelerated the increase of UCS value of coal fly ash. Sun et al. [6] studied the effect of different alkali activator (NaOH, Na_2CO_3, Na_2SiO_3) on slag and coal fly ash, and the alkali activator constituted of NaOH and Na_2SiO_3 performed the best. It is seen from the above studies that the effects of activators adding on the pozzolanic behaviors of coal fly ash, mineral slag and steel slag have been well studied. However, few researches have been carried out to focus on the pozzolanic properties of MSWI fly ash.

The aim of this study is to investigate the pozzolanic properties of MSWI fly ash in the solidification process. Four groups of solidification tests were conducted on the MSWI fly ash samples, which were solidified by deionized water, Na_2CO_3, $CaSO_4$ and Na_2SiO_3 solutions, respectively. After that, the UCS and water immersion tests were carried out on the solidified samples to investigate the influence of the kind and proportion of activators on the pozzolanic properties.

2 Materials and Methods

2.1 Materials

MSWI fly ash. The MSWI fly ash used in this study was collected from a MSWI plant with the burning environment of circulating fluidized bed in Hangzhou, China. The MSWI fly ash is dark yellow in colour, as shown in Fig. 1. The chemical composition was determined by an energy dispersive X-ray spectrometer. The test result was shown in Table 1. It was observed that the MSWI fly ash contained a certain amount of CaO, SiO_2, Al_2O_3 and Fe_2O_3, and the content of CaO was as high as 44.1%.

Fig. 1. MSWI fly ash

Activator. Three types of activators were used in this study, i.e., Na_2CO_3, $CaSO_4$ and Na_2SiO_3. Na_2CO_3 was selected to take into account the high CaO content in fly ash. Na_2CO_3 not only plays a role in activating the pozzolanic reactivity [5], but also reacts with CaO to generate the precipitate of $CaCO_3$ to produce a function of auxiliary cementation. $CaSO_4$ is a widely used activator [6], which may generate hydration

Table 1. Chemical compositions of MSWI fly ash (%)

Compound	CaO	SiO_2	Fe_2O_3	Al_2O_3	MgO	Na_2O	K_2O	SO_3	P_2O_5	Cl	Others
Percent (%)	44.07	9.82	5.47	9.85	3.13	3.99	3.19	2.89	3.56	10.32	3.71

reaction with pozzolanic materials. Na_2SiO_3 is an activator which is commonly used to promote the solidification process of Portland cement [7].

2.2 Experimental Process

Four groups of solidification tests were conducted on the MSWI fly ash samples, as shown in Table 2. B0 was solidified by water with a water-cement ratio of 0.35, and cured for 7 and 28 days, which served as the reference group. B1-1 and B1-2 was solidified under the similar condition to B0, except that 3% and 6% Na_2CO_3 was added to the fly ash, respectively. 3 and 6% $CaSO_4$ were added in B2-1 and B2-2, respectively, and 3 and 6% Na_2SiO_3 were used in B3-1 and B3-2, respectively. The groups B1, B2 and B3 were performed to investigate the effects of Na_2CO_3, $CaSO_4$ and Na_2SiO_3 adding and their proportions on the solidification capacity of MSWI fly ash.

Table 2. Program of solidification/stabilization treatments

Number	Liquid solid ratio	Activator	Proportion (dry mass basis) (%)
B0	0.35	–	0
B1-1		Na_2CO_3	3
B1-2		Na_2CO_3	6
B2-1		$CaSO_4$	3
B2-2		$CaSO_4$	6
B3-1		Na_2SiO_3	3
B3-2		Na_2SiO_3	6

Taking B2-1 for an example, the experimental processes were presented as follows:

(a) The MSWI fly ash samples were dried at the temperature of 105 ± 5 °C, and mixed with 3% Na_2CO_3, then stirred at a water-cement ratio of 0.35. After that, the mixture was filled into a self-designed moulding cylinder in three lifts. As shown in Fig. 2, the moulding cylinder consists of a split PVC tube with the inner diameter of 36 mm and the height of 80 mm, a layer of geotextile lined between the sample and the PVC tube, and two porous stones placed on the bottom and top of the PVC tube, respectively. After filling and compacting of the mixture, the moulding cylinder was placed in an incubator with a temperature of 20 ± 2 °C and a humidity of $\geq 95\%$ for curing. After curing for 24 h, the moulding cylinder was removed, and the sample was continued to be maintained for 7 and 28 days.

(b) Subsequently, the UCS test was conducted on the solidified sample by using a servo mechanical press (CMT4000, China). The servo mechanical press can supply a maximum force of 30 kN with an accuracy of 1 N and a loading rate of

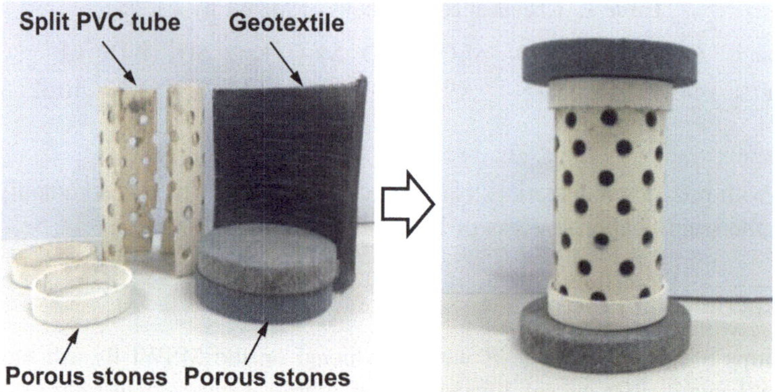

Fig. 2. Cylindrical mold for solidifying fly ash

2 mm/min. The UCS value was determined by using the formula: P = F/A, where P is the compressive strength (MPa), F is the total maximum load recorded at the point of fracture (N), and A is the area of loaded surface (mm^2).

(c) The water immersion test was also conducted on the solidified sample after curing for 7 days. The sample bulk with an area of about 2 cm^2 was immersed by deionized water in a beaker. Then, the sample bulk was kept immersion and observed the shape change carefully. The time durations for recording the observations were 10, 30 min, 1, 2, 4, 12 h, 1, 3, 7, 10, and 14 days.

3 Results and Analysis

3.1 Effect of Activator Proportion on the UCS Value

Figure 3 shows the UCS value of the solidified MSWI fly ash samples. The UCS value of the reference group (0% activator added) was obtained as 0.28 and 0.50 MPa after curing for 7 and 28 days, respectively. It was observed that the UCS value of the solidified samples under the action of the activators significantly improved when compared with the reference group. This indicated that all the three activators were likely to stimulate the pozzolanic capacity of the MSWI fly ash. As seen from Fig. 3(a), the UCS value of the solidified samples after curing for 7 days tended to increase with an increase in activator proportion. Among the three activators, $CaSO_4$ performed best. When the adding proportions of $CaSO_4$ were 3 and 6%, the UCS values of the solidified samples reached 1.18 and 1.96 MPa, respectively, which were 318.02 and 592.58% higher than that of the reference group. As shown in Fig. 3(b), the UCS values of the solidified samples cured for 28 days also remarkably increased with the activator proportions for the cases of adding Na_2CO_3 and Na_2SiO_3. However, the UCS value of the solidified sample achieved the peak value of 2.90 MPa for the case of adding 3% $CaSO_4$, and decreased to 2.23 MPa when the $CaSO_4$ proportion was raised up to 6%. This might be associated with the formation of ettringite (AFt) in the fly ash solidification process under the action of $CaSO_4$. The AFt has the characteristic of swell effect, which tended to cause the solidified samples to slightly break [8]. This phenomenon was also found in the solidification process of sludge by the authors and was verified by scanning electron microscopy (SEM) [9].

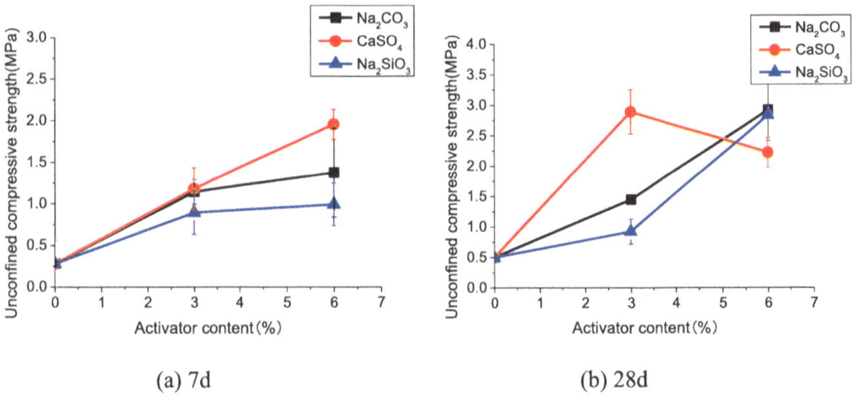

Fig. 3. UCS values of the solidified samples with different activator proportions

3.2 Effect of Curing Age on the UCS Value

The effects of curing age on the UCS value of the solidified MSWI fly ash samples are given in Fig. 4. As the curing age increased, the UCS values of solidified samples increased, which implied that the solidification processes were likely to progress with time. It was observed in Fig. 4(a) that $CaSO_4$ performed the best on stimulating the pozzolanic activity among all the three activators when the adding proportion was 3%. When activated by $CaSO_4$, the UCS values of the solidified samples after curing for 7 and 28 days reached 1.18 and 2.90 MPa, respectively, which were 318.02 and 478.24% higher than that of the reference group. The performance of Na_2CO_3 took the second place, and followed by Na_2SiO_3. As shown in Fig. 4(b), when the adding proportion of activators was 6%, the case of adding $CaSO_4$ again performed the best after curing for 7 days, however, the UCS value of this case increased slightly after curing for a longer time of 28 days. For the curing time increased from 7 days to 28 days, the UCS values of the solidified samples significantly increased from 1.37 to 2.92 MPa for the case activated by Na_2CO_3, and from 0.99 to 2.84 MPa for the case activated by Na_2SiO_3.

3.3 Water Immersion Behavior

The water immersion behaviors of the solidified samples are showed in Table 3. The sample of the reference group gradually collapsed after immersing in the water for 7 days. Similar results were also observed in the cases of adding 3 and 6% $CaSO_4$, and the samples were damaged in 7 and 10 days, respectively. This may be associated with

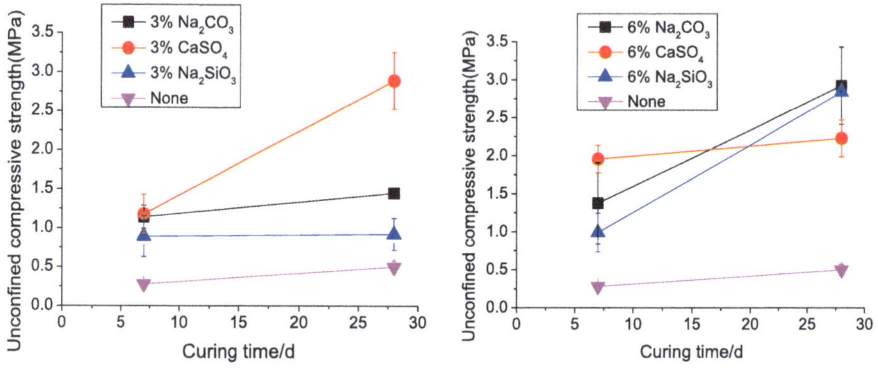

(a) proportion of 3%　　(b) proportion of 6%

Fig. 4. UCS values of the solidified samples under different curing ages

Table 3. Water immersion test results

Number	Activator	Proportion (%)	Damage time
B0	/	/	7 d
B1-1	Na_2CO_3	3	No damage was seen in 14 d
B1-2	Na_2CO_3	6	No damage was seen in 14 d
B2-1	$CaSO_4$	3	7 d
B2-2	$CaSO_4$	6	10 d
B3-1	Na_2SiO_3	3	No damage was seen in 14 d
B3-2	Na_2SiO_3	6	No damage was seen in 14 d

the structural weakness induced by water. However, no visible damage was observed in the samples solidified by adding Na_2CO_3 and Na_2SiO_3 after immersing in the water for 14 days.

4 Summary and Conclusion

The findings from this study are summarized as follows

(1) The reference group, with no activator added, had UCS values of 0.28 and 0.50 MPa after curing for 7 and 28 days, respectively. The solidified sample cured for 7 days collapsed after immersing in the water for 7 days.
(2) When the sample was solidified with the action of the activator Na_2CO_3, its UCS value increased with increasing Na_2CO_3 mixing proportion and curing time. The 28-day UCS values reached 2.92 MPa for the samples with Na_2CO_3 mixing proportion of 6%, which was 483.43% higher than the reference group. No collapse was observed in the 7-day solidified sample even if it was immersed in the water for 14 days. Similar behavior was observed in the solidified sample added

with Na_2SiO_3, however, its UCS value was lower than that added with Na_2SiO_3 under the given mixing proportion and curing time.

(3) When the sample was solidified with the action of $CaSO_4$, its UCS value achieved the peak value of 2.90 MPa for the case of adding 3% $CaSO_4$, and decreased to 2.23 MPa when the $CaSO_4$ proportion was raised up to 6%. The solidified samples cured for 7 days collapsed after immersing in the water for 7–10 days.

From the perspectives of compressive strength and water tolerance of the solidified MSWI fly ash, the optimal activator among the three ones studied in this work was Na_2CO_3, followed by Na_2SiO_3, and the last was $CaSO_4$.

References

1. Padmi T, Tanaka M, Aoyama I (2009) Chemical stabilization of medical waste fly ash using chelating agent and phosphates: heavy metals and ecotoxicity evaluation. Waste Manag 29:2065–2070
2. Song ZX, Wang LA, Lin X, Liu YY, Yuan H et al (2008) Experimental study on properties and cement solidification of municipal solidwaste incineration fly ash. Res Environ Sci 21 (4):163–168
3. Zhang SQ, Huang SY (2008) Study on the hydraulic activity and alkali activation of mineral admixtures. Fly Ash Compr Util 4:6–9
4. Shi C, Day RL (1993) Chemical activation of blended cements made with lime and natural pozzolans. Cem Concr Res 23(6):1389–1396
5. Sivapullaiah PV, Baig MAA (2011) Role of gypsum in the strength development of fly ashes with lime. J Mater Civ Eng 23(2):197–206
6. Bai EL, Xu JY, Li H, Zhang T et al (2014) Experiment study on the stimulating characteristics of slag-fly ash cementitious material system to different alkali-activator. Sci Technol Eng 14 (1):96–99
7. Zhang ML (1999) Effect of water glass on hardening of cement slurry. Jiangxi Coal Technol 2:47–50
8. Jun Y, Oh JE (2015) Use of gypsum as a preventive measure for strength deterioration during curing in class F fly ash geopolymer system. Materials 8(6):3053–3067
9. Chen P, Feng B, Zhang LT (2014) Solidification of dewatered sewage sludge using bottom ash of MSWI as skeleton material. China Environ Sci 34(10):2624–2630

Incorporating Thermal Effects in Modeling of MSW Landfills

Girish Kumar[1(✉)], Kevin Kopp[2], Krishna R. Reddy[1], James L. Hanson[2], and Nazli Yeşiller[2]

[1] Department of Civil and Materials Engineering, University of Illinois at Chicago, 842 West Taylor Street, Chicago, IL 60607, USA
{gkumar6,kreddy}@uic.edu

[2] Department of Civil and Environmental Engineering, California Polytechnic State University, San Luis Obispo, CA 93407, USA
{kkopp,jahanson,nyesille}@calpoly.edu

Abstract. Heat is one of the primary byproducts of biodegradation of municipal solid waste (MSW). Biodegradation of MSW in landfills induces changes in physical properties, mechanical response of MSW, and flow of leachate within the MSW pore spaces. Moreover, biodegradation of MSW in landfills is temperature dependent and consequently the engineering properties of MSW are all influenced by waste temperatures. Thus, landfills are complex systems with interrelated processes and it is crucial to account for all these interdependencies to accurately predict the coupled behavior of MSW. In this study, a coupled-thermo-hydro-bio-mechanical model was formulated that incorporates the effect of temperature on heat generation and biodegradation of MSW. The model integrates a two-phase flow hydraulic model, a plane-strain formulation of the Mohr-Coulomb mechanical model, a first order decay biodegradation model, and a one-dimensional heat conduction model with temperature-dependent heat generation. Numerical simulations were carried out using a typical landfill configuration with leachate injection simulating a bioreactor landfill. The simulations were carried out with and without temperature effects to determine the influence of temperature on MSW behavior. The results indicate a significant influence of temperature on the MSW response (degradation and settlement) and underscore the importance of incorporating thermal effects in numerical modeling of MSW landfill systems.

Keywords: Landfill · Heat generation · Biodegradation · Settlement Coupled processes · Numerical model

1 Introduction

Several studies involving numerical modeling of the coupled behavior of MSW have been reviewed by Reddy et al. (2017a). Most of the coupled models developed for analysis of MSW behavior in landfills do not account for the interdependency between heat generation and temperature distribution effects on biodegradation of waste. In this study, a coupled thermo-hydro-bio-mechanical (CTHBM) modeling framework is

L. Zhan et al. (Eds.): ICEG 2018, ESE, pp. 10–17, 2019.
https://doi.org/10.1007/978-981-13-2224-2_2

presented that combines and simultaneously solves for thermal, hydraulic, biological and mechanical response within MSW.

2 Modeling Framework

The coupled hydro-bio-mechanical model as described in Reddy et al. (2017b) is integrated with a one-dimensional heat conduction model developed by Hanson et al. (2013) which accounts for seasonal temperature changes at the ground surface and temperature dependent heat generation induced by biodegradation. The heat transfer in the MSW landfills was described by the classical 1-D heat conduction formulation as shown in Eq. (1) below:

$$k_t \frac{\partial^2 T}{\partial z^2} + \dot{q} = C_t \frac{\partial T}{\partial t} \tag{1}$$

where k_t is coefficient of thermal conductivity (W/m-K), $\dot{q}$ is rate of heat generation (heat source term) (W/m^3), C_t is volumetric heat capacity (kJ/m^3-K), T is temperature, z is distance, and t is time. The heat generation from biodegradation was represented by heat generation rate functions that were developed empirically from the field investigations of temperature data at different landfills with waste of different ages (Yeşiller et al. 2005; Hanson et al. 2008). The heat generation functions accounted for the normal landfill operations (placement rate, placement moisture and placement waste density) and were based on the net heat gain from the waste decomposition. An exponential growth-decay heat generation rate function was formulated to simulate this heat generation in landfills (Eq. (2)):

$$H = A\left[\frac{Bt}{B^2 + 2Bt + t^2}\right]e^{-\sqrt{\frac{t}{D}}} \tag{2}$$

where H is the heat generation rate (W/m^3), A is the peak heat generation factor (W/m^3), B is the shape factor (days), D is the decay rate factor (days), and t is the time (days). The model also incorporated seasonal temperature variations into the thermal model to realistically predict the spatial and temporal distributions of temperature within the landfills. The seasonal temperature variations are represented by a sinusoidal function and the temperature variation at the ground surface is determined using Eq. (3) below:

$$T(t) = T_m - A_s \cos\left[\frac{2\pi}{365s}(t - t_o)\right] \tag{3}$$

where T_m is mean annual surface temperature (°C), A_s is the amplitude of the temperature wave (°C), s is equal to one day expressed in seconds (86,400), t is time, and t_o is a phase constant (seconds). This thermal model developed by Hanson et al. (2013) was validated with extensive field monitored temperatures in landfills at four different regions in North America. The model is used herein to predict the spatial and temporal distribution of temperatures within the landfill.

The biodegradation model used in this study is based on USEPA's LandGEM model. The biodegradation model quantifies the rate of methane gas produced, q, with time based on the mass of the waste in the landfill as given by Eq. (4):

$$q(t)_{CH_4} = kL_oMe^{-kt} \tag{4}$$

where k is the decay rate constant (yr^{-1}), L_o is the biochemical methane potential (BMP) of the waste (m^3/Mg), M is the mass of MSW (Mg), e is the base of natural logarithms, and t is time (yr). The rate of waste biodegradation is formulated to account for both moisture and waste temperature as given in Eq. (5).

$$k(S,T) = \frac{T(mf_w + n)}{C_N\left[1 + \exp\left(\frac{T}{4} - 18\right)\right]} \tag{5}$$

where $f_w = (S - S_r)/(1 - S_r)$ is the water content factor similar to effective saturation, S is the degree of saturation of the waste, S_r is the residual degree of saturation of the waste, T is the temperature of the waste, m and n are linear constants based on the lower and upper bounds of k which were assumed to be 0.05 and 0.3, respectively, and C_N is a normalization constant equal to 57.33. The relation between the decay rate constant for hydrolysis of the organic compounds and the temperature was adopted from Young (1989). The extent of biodegradation of the waste was represented by the degree of degradation (DOD) which is defined as the ratio of methane gas generated over the maximum methane potential of the waste (calculated based on BMP). The functionality of the biodegradation model and the influence of DOD on the waste properties are provided in Reddy et al. (2017b).

2.1 Modeling Scenarios

A typical landfill cell model with a total width of 15 m was selected for the numerical analyses performed in this study (Fig. 1). The waste was assumed to be placed in ten 3-m-thick lifts, making up a total waste height of 30 m. Each 3-m-thick waste layer was assumed to be placed over approximately 50 days. The waste was placed on native soil (subgrade) consisting of silty clay. Since this study focused mainly on the waste behavior, the bottom liner systems consisting of compacted clay or geosynthetics were not modeled explicitly; however, the silty clay subgrade would have similar properties as compacted clay and the model can be extended to include these components. A 1 m × 1 m horizontal trench (HT) similar to the one presented in Reddy et al. (2017b) was modeled at a depth of 10 m from the waste surface to simulate leachate injection as a part of bioreactor landfill operations. The thermal boundary conditions involved fixing a far field temperature boundary, 75 m below the bottom of the waste mass, to a mean annual earth temperature of 12.7 °C. In addition, a sinusoidal surface temperature boundary condition with a mean surface temperature of 13 °C and a surface temperature amplitude of 19.4 °C was applied to simulate seasonal temperature variations at the ground surface. The initial heat capacity and thermal conductivity of the waste across the landfill depth were assumed to be 2,700 J/kg-K and 1.3 W/m-K, respectively. The mass heat capacity and thermal conductivity of the subgrade soil were assumed to be 1,300 J/kg-K and 2.4 W/m-K, respectively.

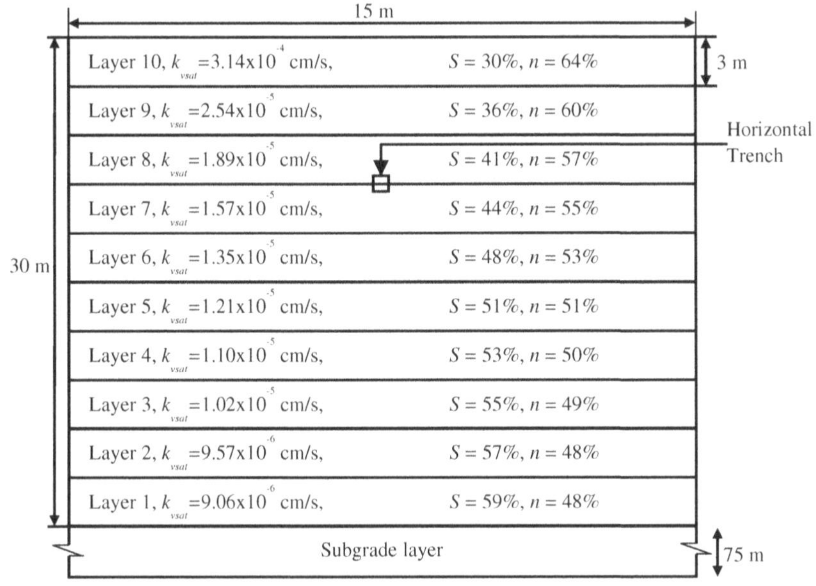

Fig. 1. Landfill model and the initial waste properties used in the study (k_{vsat} = saturated vertical hydraulic conductivity, S = degree of saturation, n = porosity)

Six numerical simulations were performed: one of these was based on the coupled hydro-bio-mechanical (CHBM) framework presented in Reddy et al. (2017b). The study by Reddy et al. (2017b) did not consider any element of temperature in the framework but implicitly assumed optimum temperatures in the waste during simulations. However, the effects of variable waste moisture content on biodegradation of waste were included. Herein, this simulation is denoted as "CHBM". Four of the simulations used different constant temperature (spatially and temporally) of the waste for the entire duration of the simulation. The four constant temperatures used were 20, 30, 40, and 50 °C. These simulations are denoted as "CTHBM T = X °C" with X representing the constant temperature of waste. Finally, one simulation was conducted that incorporated the spatial and temporal temperature variations based on the heat generation from biodegradation of waste and the seasonal temperature changes as described in Sect. 2. This simulation is denoted as "CTHBM T = Variable".

3 Results and Discussion

Figure 2 shows the variation of the DOD with depth for the six simulations carried out. Figure 2(a) shows the results for the simulation with CHBM, Fig. 2(b) to (e) show the results for simulations with constant waste temperatures of 20, 30, 40 and 50 °C, respectively, and Fig. 2(f) shows the results for the simulation that considers spatial temperature variations in the waste.

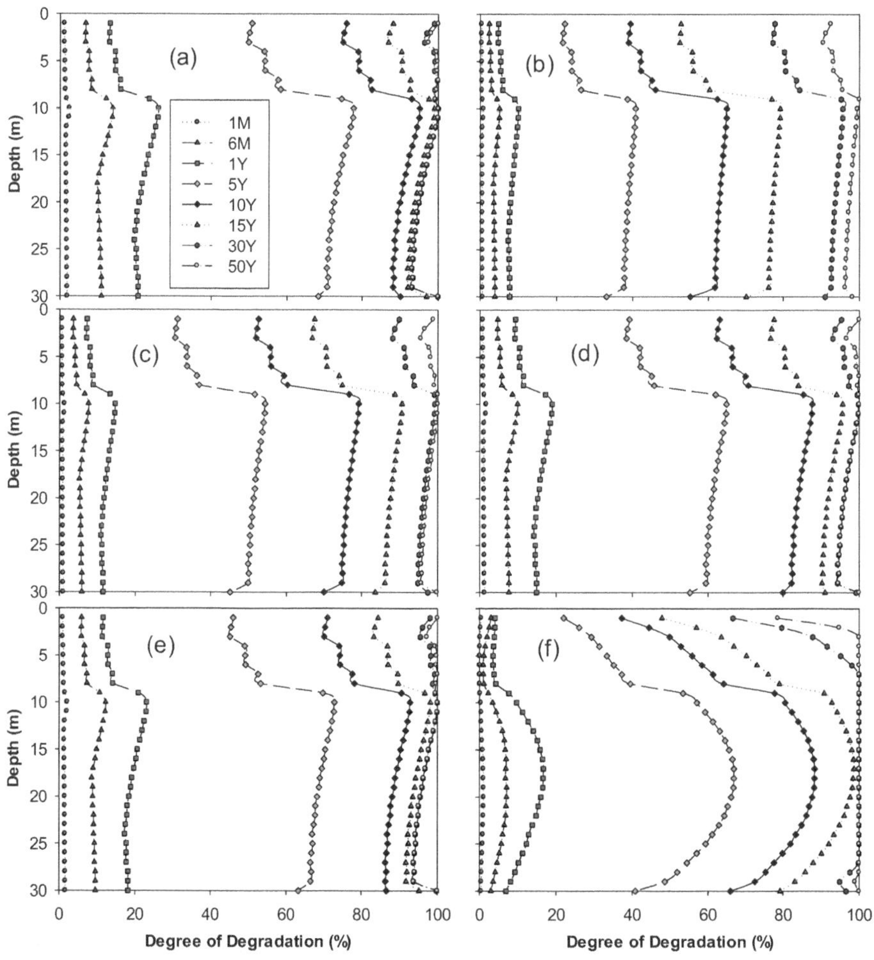

Fig. 2. Variation of DOD along landfill depth for simulation with (a) CHBM, (b) constant waste temperature of 20 °C, (c) constant waste temperature of 30 °C, (d) constant waste temperature of 40 °C, (e) constant temperature of 50 °C, (f) with variable temperature effects

The variation of DOD with depth at eight selected time intervals of 1 month (1M), 6 months (6M), 1 year (1Y), 5 years (5Y), 10 years (10Y), 15 years (15Y), 30 years

(30Y), and 50 years (50Y) after the onset of leachate injection which began after the completion of the placement of all the waste lifts. Due to the high moisture conditions induced by leachate injection, the waste zones around and below the leachate injection location (near HT) undergo enhanced degradation, thereby resulting in higher DOD. As the temperature of the waste increases from 20 to 50 °C, the corresponding rates of waste degradation also increase. The optimum temperature value for peak primary degradation rates (hydrolysis) occurs at approximately 60 °C as reported by Young (1989). The "CHBM" simulation (Fig. 2a) assumes this scenario by default and the results of this simulation are based on the degradation rates corresponding to the optimum temperature value.

Unlike all the other cases, Fig. 2(f) shows a significant difference in the variation of the DOD with depth in addition to the variation with time. This simulation is based on the spatially and temporally varying temperature as predicted by the thermal model. The DOD with depth is represented by a convex profile over the depth of the landfill. The variation in DOD with depth as observed in Fig. 2(f) could be more indicative of the realistic variation of waste degradation with depth as it captures the combined influence of both temperature and moisture content on the biodegradation of waste mass.

Figure 3(a) to (c) show the predicted spatial variation of temperature with time for waste at depths of 2 m (topmost layer), 17 m (middle layer), and 29 m (bottommost layer) from the surface. The variation of temperature with time in the topmost layer is highly influenced by the seasonal surface temperature fluctuations and hence the resulting temperature distribution in the topmost layer follows a sinusoidal variation.

Figure 3(b) shows the waste temperatures in the middle waste layer where the temperature reaches a maximum of 56 °C after about 5 years of leachate injection and thereafter diminishes exponentially due to the decreasing rates of degradation in the long-term. The seasonal temperature changes do not or negligibly influence the temperature at this depth location. Figure 3(c) shows the variation of temperature with time for waste located in the lowermost layer. No influence of the seasonal ground surface temperature fluctuations can be seen at this location. The temperature in this bottom layer of waste is lower than the temperature at the mid-depth due to the underlying subgrade temperatures. The predicted temperatures in the three layers discussed above are well within the typical temperatures reported in field investigations (Yeşiller et al. 2015; Hanson et al. 2010).

Figure 4 shows the settlement response of the waste for the six simulations conducted. The settlement responses for the simulations with different constant temperatures show that as the waste temperature increases towards the optimum temperature for biodegradation, the settlement rates also increase, which corresponds well with higher waste decomposition rates. The settlement rates for the simulation with CHBM were the highest as this simulation assumes optimum temperature of the waste by default and the degradation rate varies solely based on the changes in moisture contents of the waste.

The settlement trend with time for the "CTHBM T = Variable" case was influenced by the spatial and temporal variations of both waste temperature and waste moisture. In each of the other five simulations, temperature remained constant and only the moisture variation occurred. The settlement with time curve for the "CTHBM T = Variable"

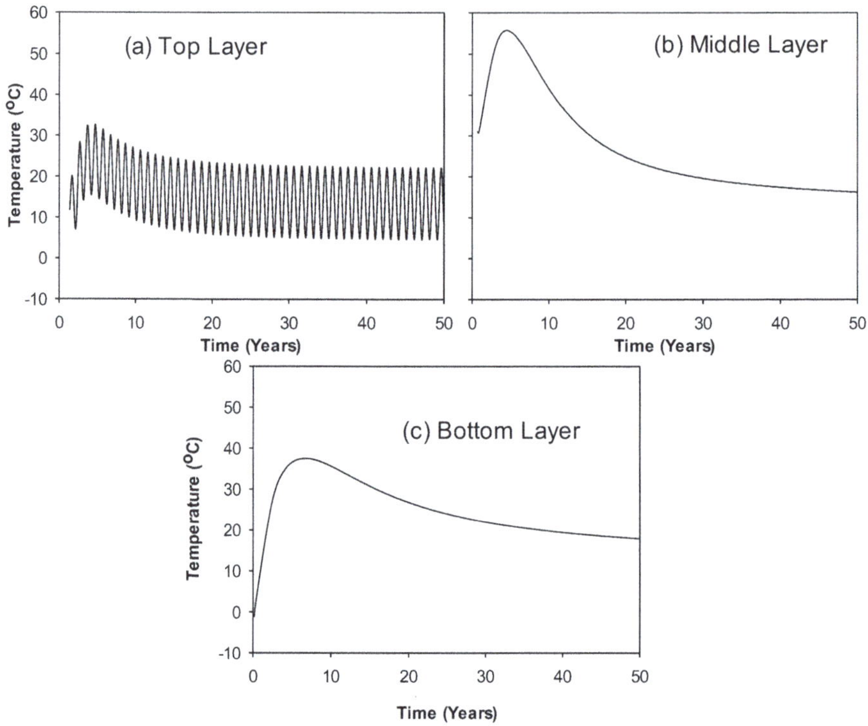

Fig. 3. Temperature variation with time at depth (a) top layer (2 m), (b) middle layer (17 m), (c) bottom layer (29 m).

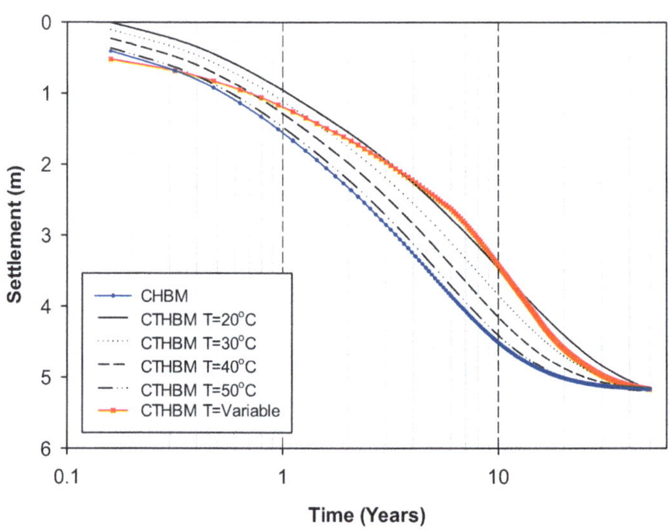

Fig. 4. Settlement response for variable temperature conditions

case shows an initial low settlement rate compared to other cases and in the long-term the settlement response was intermediate of the highest and the lowest settlement rates. This indicates that the different spatial and temporal distributions of temperatures within the landfill could vary the waste degradation rates and thereby influence the mechanical response of the waste mass within the landfill. Overall, the predicted settlement trends in all cases are in good agreement with the typical observed settlement response for bioreactor landfill conditions as reported in Bareither et al. (2012).

4 Conclusions

The model framework proposed in this study integrates the effects of temperature on the biodegradation of waste to realistically predict the waste behavior in landfills. The results show a strong influence of temperature on the biodegradation of waste and consequently on the waste settlement. The modeling framework provides settlement response similar to that observed in laboratory bioreactor cell experiments and in the field.

Acknowledgements. This project is funded by the U.S. National Science Foundation (grant CMMI #1537514), which is gratefully acknowledged.

References

Bareither CA, Benson CH, Edil TB (2012) Compression of municipal solid waste in bioreactor landfills: mechanical creep and biocompression. J Geotech Geoenviron Eng 139(7):1007–1021

Hanson JL, Liu WL, Yeşiller N (2008) Analytical and numerical methodology for modeling temperatures in landfills. In: Khire et al (eds) GeoCongress 2008: geotechnics of waste management and remediation, GSP no. 177. ASCE, Reston, pp 24–31

Hanson JL, Yeşiller N, Oettle NK (2010) Spatial and temporal temperature distributions in municipal solid waste landfills. J Environ Eng 136(8):804–814

Hanson JL, Yeşiller N, Onnen MT, Liu WL, Oettle NK, Marinos JA (2013) Development of numerical model for predicting heat generation and temperatures in MSW landfills. Waste Manag 33(10):1993–2000

Reddy KR, Kumar G, Giri RK (2017a) Modeling coupled processes in municipal solid waste landfills: an overview with key engineering challenges. Int J Geosynth Ground Eng 3(1):6

Reddy KR, Kumar G, Giri RK (2017b) Influence of dynamic coupled hydro-bio-mechanical processes on response of municipal solid waste and liner system in bioreactor landfills. Waste Manag 63:143–160

Yeşiller N, Hanson JL, Liu WL (2005) Heat generation in municipal solid waste landfills. J Geotech Geoenviron Eng 131(11):1330–1344

Yeşiller N, Hanson JL, Yee EH (2015) Waste heat generation: a comprehensive review. Waste Manag 42:166–179

Young A (1989) Mathematical modelling of landfill degradation. J Chem Technol Biotechnol 46(3):189–208

Improving the Slope Stability of Municipal Solid Waste Dumps Using Reinforced Soil Berms: A Case Study

A. Agarwal[1(✉)], M. Datta[1], G. V. Ramana[1], and R. Satyakam[2]

[1] Department of Civil Engineering, Indian Institute of Technology Delhi, Delhi, India
apoorvaagarwal6@gmail.com
[2] NTPC Limited, Greater Noida, India

Abstract. The slope stability of an existing dump at Delhi has been analyzed for (a) dry waste, (b) with pore water pressure, and (c) with earthquake. The results show that existing slopes are just stable in the dry case and instability is triggered by development of pore water pressures due to infiltration in rainy season as well as by seismic forces. In order to stabilize the slope of the dumps, the traditional way is to flatten the slope by extending the toe of the slope outward in horizontal direction or by moving the crest inwards. This is not a feasible solution in Delhi where there is a scarcity of land and waste heights and volume are very large. The alternative considered in the paper is the construction of reinforced soil (RS) berm at the base of Municipal Solid Waste (MSW) dumps. A parametric study has been carried out considering the influence of the following factors (a) height of waste and (b) slope angle of waste for three cases viz., original slope, slope with 10 m high RS berm and slope with 20 m high RS berm. Apart from internal stability, the external stability of the complete system is analyzed using the limit equilibrium method of analysis. An optimum solution is suggested based on the results of the study. The study shows that RS berm offers a more acceptable solution for stabilizing the slopes of waste dumps where limited space is available beyond the toe of the existing dump.

Keywords: MSW dumps · Reinforced soil berms · Slope stability

1 Introduction

Population growth (around 1.13% in year 2017) [7] and rapid urbanization (33.1% in 2016) [8] in India is leading to major generation of Municipal Solid Waste (MSW). With the urbanization and increase in income level, the amount of solid waste generation increases [2]. Delhi being the capital city of India is having largest density of population in India. The population density of the city is around 11320 person/km^2 [1]. On an average Delhi generates about 0.41 kg/day/capita of waste [4]. Major portion of the generated MSW goes into landfill. Delhi has three active MSW waste dumpsites i.e., at Bhalswa, Ghazipur and Okhla. The area covered by these landfills is at least 1% of Delhi's total area [3].

L. Zhan et al. (Eds.): ICEG 2018, ESE, pp. 18–25, 2019.
https://doi.org/10.1007/978-981-13-2224-2_3

Slope stability is a major issue in the existing MSW dumps. These are majorly un-lined dumping grounds in which the waste is tipped from the top of the dump leading to steep slopes. In the beginning, these dumps have low heights and gentle slopes. But due to exponential increase in generation of MSW in recent years, the height of dumps had grown rapidly. Also, there is an increase in slope angle of the dumps as the waste is tipped from the top of the dump in a progressive manner. In September 2017, Delhi has experienced a slope failure at one of the dumping sites which is around 50 m high and the slope angle varies between 35° and 40°. This paper presents the results of a study undertaken for enhancing the stability of the dump by flattening of slopes and provision of reinforced soil (RS) berms.

2 Problem Statement

The waste dump was commissioned in 1984 and is not designed according to the MSW Rules of 2000. It is an un-lined dump in which the waste is tipped from the top of the dump leading to steep side slopes. The site should have shut down in 2006 but continued to be operational due to non-availability of land. The dump has crossed a height of 50 m and spread on an area of 70 acres. The steepest slope angle is 42° (1.1 (H):1.0(V)). About 2000 MT of garbage is dumped at the site every day. For the purpose of analysis, the cross section shown in Fig. 1 was considered. Width of 15 m is available beyond the toe of the dump beyond which an unlined (kutcha) drain is present. Flattening of the slope was the first alternative considered as shown in Fig. 1. However due to limitations in space beyond the toe, provision of berms was also considered (Fig. 2). A RS berm was feasible as the conventional berm would intrude into the drain. Two RS berms, each 10 m high (along with toe strengthening) along with flattening of slope above the berms was considered for analysis as shown in Fig. 3.

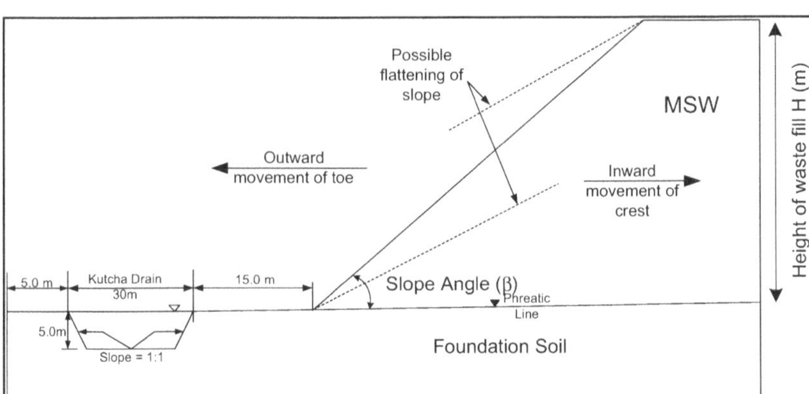

Fig. 1. Section considered for analysis

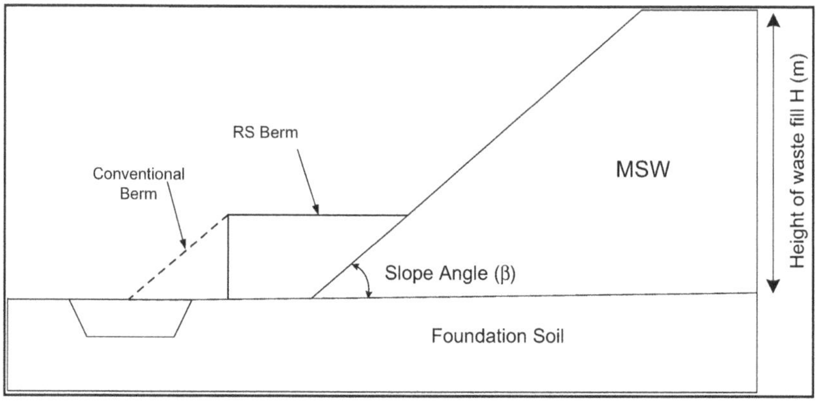

Fig. 2. Section showing RS berm and conventional berm

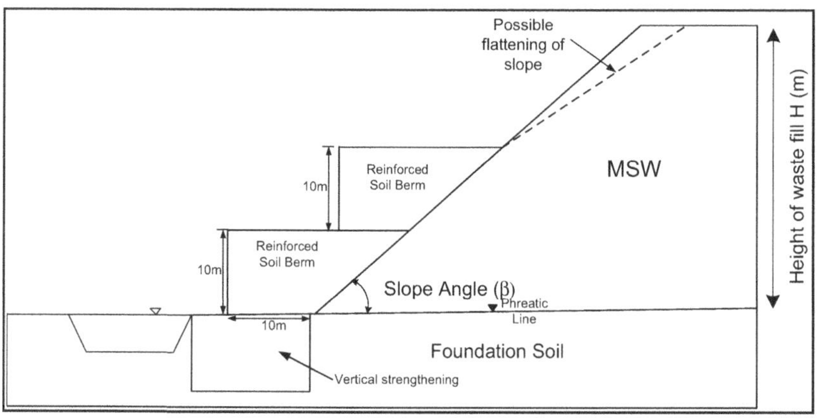

Fig. 3. Section showing RS wall and vertical strengthening of the toe and flattening of slope above RS wall

3 Parameters Considered for Analysis

Waste Parameters: Shear strength parameters of MSW have been reported by Ramaiah et al. [6] for two MSW dumps i.e., Ghazipur and Okhla. Experiments were carried out in the lab on large direct shear test boxes (300 by 300 mm) and shear strength properties for two cases of horizontal displacement (25 and 55 mm) have been given. For the analysis small displacement parameter (which corresponds to 25 mm (8.3%) horizontal displacement in large shear test box) is considered as slope stability is governed by the critical parameters. The parameters considered are cohesion intercept = 13 kN/m^2; angle of shearing resistance = 23°; unit weight of aged MSW = 12 kN/m^3.

Sub-Soil Parameters: The sub-soil (Delhi Silt) parameters considered for analysis are cohesion intercept = 0 kN/m^2; angle of shearing resistance = 30°; unit weight of sub-soil = 19 kN/m^3.

4 Stability Analysis

The analysis is carried out using Morgenstern Price Method. In this method the Factor of Safety (FoS) is computed using both force and moment equilibrium.

The stability analysis is carried out for different conditions:

(a) Height of waste = 60, 50 and 40 m.
(b) Inclination of waste = 42°(1.1:1), 34°(1.5:1), 27°(2:1), 22°(2.5:1), 18°(3:1)

The following assumptions are made for the analysis:

(c) The sub-soil is considered up to depth of approximately 20 m.
(d) The water table is considered at ground level.
(e) Seismic Coefficients: Horizontal seismic coefficient = 0.12 g; Vertical seismic coefficient = 0.08 g.
(f) Pore water pressure: Analysis is done for Ru = 0.1.

For different conditions following Factor of Safety (FoS) are required:

(i) For static stability for long term (dry case) – 1.5
(ii) For seismic stability (pseudo static) – 1.1
(iii) For static stability for short term (with Ru) – 1.3

Stability analysis is carried out highlighting the influence of height and inclination on slope without berm at base of the slope and with two 10 m each high RS berm at the base of the slope. Three conditions are considered under each head. The conditions are:

(i) Dry Slope
(ii) Slope under earthquake loading
(iii) Slope under pore water pressure (Ru = 0.1).

The dimensioning of the berms is as per FHWA-NHI-10-024. The internal design of the berms is as per BS8006:2010. The following properties of the fill is considered while designing the berm; angle of shearing resistance = 32°; unit weight of fill = 19 kN/m^3.

Though the backfill material is cohesionless material, the backfill is considered having equivalent cohesion. The tension in each soil reinforcement layer is converted into equivalent cohesion using the pseudo-cohesion concept [5]. The equivalent cohesion is found to be 130 kN/m^2. For the analysis, cohesion is considered as 200 kN/m^2 as the berm needs more soil reinforcement than the design. Vertical strengthening is also done at the base of RS berm so that adequate bearing capacity will be achieved.

5 Results

The FoS for the existing slope in various conditions are listed in Table 1. The desirable factor of safety for long-term stability of dry slope is 1.5. As observed the existing mound is just stable in dry state but FoS drops below 1.0 with EQ and pore water pressure condition.

Table 1. FoS for existing slope H = 50 m and slope inclination = 42°(1.1(H):1.0(V))

Condition	Dry	EQ	Ru = 0.1
FoS	1.04	0.83	0.94

The variation in FoS for different slope inclinations and different heights is presented in the following subsections.

5.1 Influence of Slope Inclination

In this case, the slope inclination is varied from 42° to 18°. The results shown are considered for three cases:

In Fig. 4a, the slope angle is from the toe of the berm to the crest of the slope. In Fig. 4b and c, the slope inclination is above the RS berm. The desirable FoS for dry case is 1.5. From Fig. 4a it could be concluded that the existing dump slope is stable in between 2.5(H):1.0(V) to 3.0(H):1.0(V). From Fig. 4b and c it can be seen that when RS berm is introduced at the toe of the slope, then the stable slope angle ranges from 2.5(H):1.0(V) to 2.0(H):1.0(V) for the slope above the RS berm.

5.2 Influence of Height

In this case, the slope height is varied from 60 to 40 m. The results shown are considered for three cases:

From Fig. 5a, it is evident that existing slope is stable for slope inclination of 3.0 (H):1.0(V) for all the heights. From Fig. 5b and c, the slope inclination can be made steeper to 2.5(H):1.0(V) to 2.0(H):1.0(V) with the introduction of RS berm at the toe of the existing dump.

5.3 Typical Failure Surfaces

From Fig. 6, it can be seen that the critical failure surface in the existing slope is observed to be the base failure. With the introduction of the RS berm, critical failure surface moves from base failure upwards to a slope failure above the RS berm (Fig. 7).

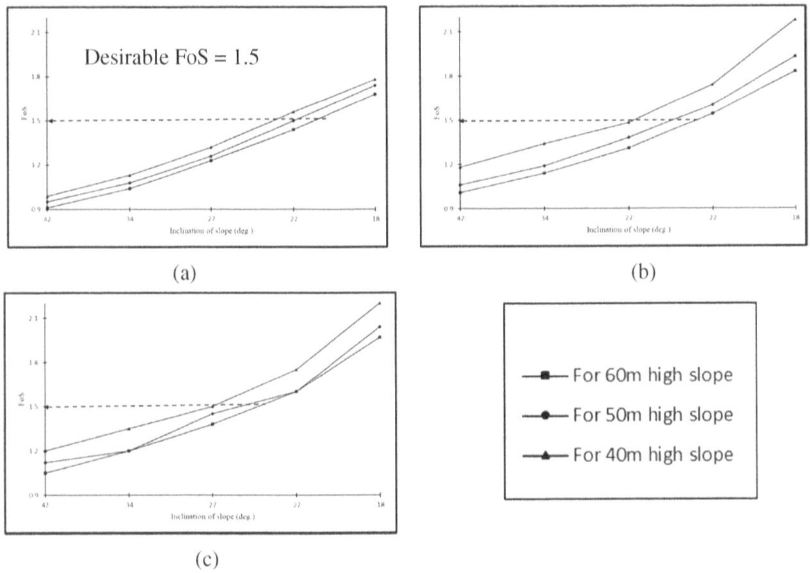

Fig. 4. FoS for variation in slope inclination for dry slope (a) without berm at base (b) with 10 m high RS berm at base (c) with 20 m high RS berm at base

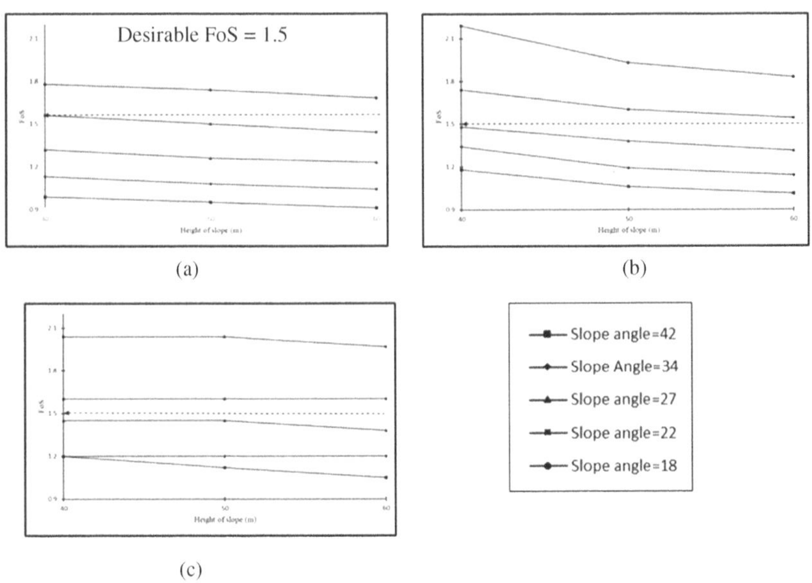

Fig. 5. FoS for variation in slope inclination for dry slope (a) without berm at base (b) with 10 m high RS berm at base (c) with 20 m high RS berm at base

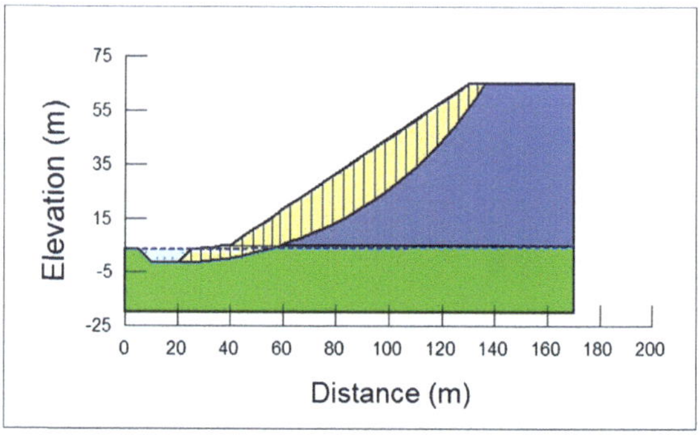

Fig. 6. Critical failure surface for slope angle = 42° and height = 50 m; FoS = 1.04

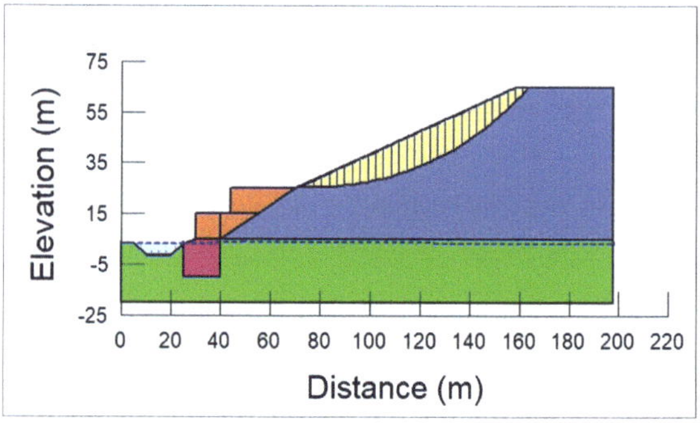

Fig. 7. Critical failure surface for slope angle above the berm = 27° and height = 50 m; FoS = 1.51

6 Discussions

From the results of the analysis as given above, it is evident that with decrease in slope inclination, the FoS increases significantly. Also, with decrease in height, the increase in FoS is not significant.

The FoS for various conditions depicted above, are also checked for pore water pressure conditions and seismic condition. For both the cases FoS increases with decrease in slope inclination and decrease in height as that for dry slope.

From Figs. 4 and 5 it could be concluded that for height range from 40 to 60 m, the safe slope inclination is 22° i.e., 2.5(H):1.0(V). With the introduction of berms up to 20 m high, the safe slope inclination above the RS berm is increased to 27° i.e., 2.0 (H):1.0(V).

RS berm acts like a stabilizing mass at the base of slope. It will help in improving the FoS of the slope. It is better than the other conventional methods of flattening the slope as it takes less space on ground.

7 Conclusions

Following Conclusions Are Arrived at from the Slope Stability Analysis

1. The existing waste dump can undergo sudden failure under rainfall conditions or earthquake loading. The critical failure surface is observed to be base failure.
2. To make the waste dump stable by flattening the slopes, the inclination has to be reduced from 1.1(H):1.0(V) to 2.5(H):1.0(V). Since adequate space is not available beyond the toe, approximately 1.0 million tons has to be excavated to flatten the slope.
3. Provision of reinforced stabilizing berm at the toe helps in improving the stability to some extent because the critical failure surface moves from base failure upwards to a slope failure above the berm.
4. The slope inclination above the berm has to be flattened from 1.1(H):1.0(V) to 2.0 (H):1.0(V). As a consequence, the quantity of waste to be excavated reduces by more than 40% of the original value. This is a more acceptable solution than conventional solution of flattening the slope as limited space is required beyond the toe of the dump.

References

1. Census (2011) Census of India 2011. https://www.census2011.co.in/states.php. Accessed 5 Mar 2018
2. Hoornweg D, Bhada-Tata P (2012) What a waste: a global review of solid waste management
3. Kumar D, Alappat BJ (2003) Monitoring leachate composition at a municipal landfill site in New Delhi, India. Int J Environ Pollut 19(5):454–465
4. Kumar S, Smith SR, Fowler G, Velis C, Kumar SJ, Arya S, Kumar R, Cheeseman C (2017) Challenges and opportunities associated with waste management in India. R Soc Open Sci 4 (3):160764
5. Leshchinsky D, Tatsuoka F (2003) Reinforced soil engineering
6. Ramaiah BJ, Ramana GV, Manoj Datta (2017) Mechanical characterization of municipal solid waste from two waste dumps at Delhi, India. Waste Manag 68:275–291
7. Worldometers (2017) Live population meter. http://www.worldometers.info/world-population/india-population/. Accessed 5 Mar 2018
8. World Development Indicators (2016). http://databank.worldbank.org/data/reports.aspx?source=2&series=SP.URB.TOTL.IN.ZS&country. Accessed 5 Mar 2018

Slope Stabilization and Capacity Expansion at Tianziling Landfill in Hangzhou, China

Liangtong Zhan, Jiwu Lan(✉), Wei Li, and Yunmin Chen

MOE Key Laboratory of Soft Soils and Geoenvironmental Engineering, Zhejiang University, Hangzhou 310058, Zhejiang, China
lanjiwu@zju.edu.cn

Abstract. Slope stability was carried out on the Tianziling landfill to investigate its safety at current and subsequent landfilling stages. The investigation was based on an understanding the high leachate generation rate and waste mechanics of the municipal solid wastes with a high content of kitchen wastes. The slope stability analyses indicate that the potential failure mode at the current state is a shallow slip induced by the high perched leachate mound. With the increase in the height of landfill, the failure mode will change from shallow slip to an overall slip along the bottom composite liner, which is lying on the sloping ground with an inclination of 1:4. The early warning levels for both the perched leachate mound and the main leachate mound in landfill were determined at different stages when the landfill is filled to different elevations. Engineering measures are proposed to enhance the slope stabilization as well as to increase landfill capacity. The measures include improving the configuration of liner system to be lined on the sloping ground, establishing three-dimensional drainage system to drawdown leachate level, raising the height of the downstream retaining dam, applying 10 m thick surcharge layer on the downstream slope, and setting up a safety monitoring system. Landfill stability analyses indicate that with these measures it is able to fill the landfill to the design height of 130 m, which will be the highest landfill in China.

Keywords: Landfill stability · Failure mode · Early warning leachate level
Engineering measures

1 Overview of Project

Tianziling Landfill is located in the north of Hangzhou, Zhejiang Province, China. The landfill consists of the first stage that has been closed and ecologically reclaimed and the second stage that is under operation. The second stage is located on the downstream of the first stage, being extended westward by 440 m. The second stage is lined with a composite liner, and a vertical barrier was set at downstream. It covers an area of 96 hectares and has a maximum design elevation of 165 m and a total landfill capacity of 22.02 million cubic meters. According to the average daily dumping volume of 2671 t/d, the designed service life is 24.5 years.

Figure 1 is engineering geological profile of Tianziling landfill. There is a retaining dam set at the downstream of the second stage of landfill. The upstream part is lying on

L. Zhan et al. (Eds.): ICEG 2018, ESE, pp. 26–34, 2019.
https://doi.org/10.1007/978-981-13-2224-2_4

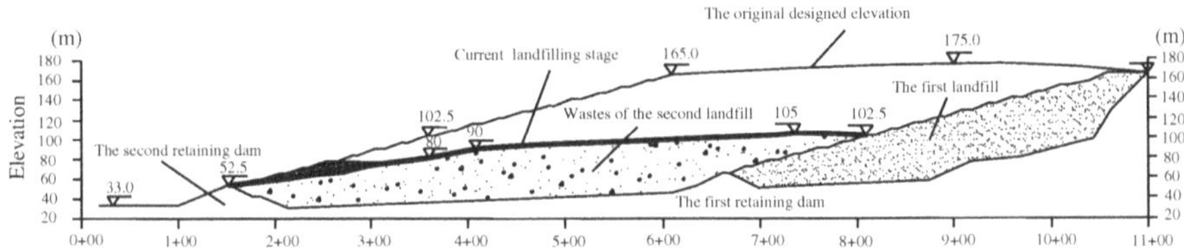

Fig. 1. Engineering geological profile of Tianziling Landfill

the sloping surface of the first stage of landfill. The composite liner system lying on the ground forms a weak interface in the landfill. The second stage of landfill mainly consists of kitchen wastes, and the water content on the basis of wet weight is as high as 60%. With the increase in the height of landfill, the safety operations of the second landfill are faced with enormous challenges. According to operation records, the average daily dumping volume of the second landfill has reached 4,556.83 t, and the maximum daily dumping volume is 6302 t, which greatly exceeds the designed value of 2671 t, so that the second landfill is in an overcapacity operation state. The on-site gas wells reveal that the depth of the perched leachate level is 2–6 m below the landfill surface. The pore-water pressure measurements near the landfill bottom have shown that the head of the main leachate mound has reached 10 m, far exceeding the 0.3 m specified by the national standard. The increase of landfill volume and landfilling rate will increase the difficulty to drawdown leachate level, and will reduce the landfill stability.

The Institute of Geotechnical Engineering of Zhejiang University was commissioned by Hangzhou Environment Group Co., Ltd. to carry out stability analysis for the second landfill to propose engineering measures required for safe control. According to the assessment results, if no engineering measures are taken and the landfill continues to be filled adopting the current technology, the factor of safety for the overall stability of the landfill will be reduced from the current 1.6 to 1.35 when the elevation of the landfill platform reaches 96.25 m. The corresponding capacity of landfill volume is 8.78 million cubic meters, which is significant lower than the pre-designed 22.02 million cubic meters. When the elevation of the landfill platform reaches 133.75 m, the factor of safety for the overall stability of landfill is reduced to 1.0, and the risk of failure is very high. The corresponding capacity of the landfill is 16.55 million cubic meters. The engineering measures are proposed to enhance the slope stabilization as well as to increase landfill capacity.

2 Engineering Geology and Hydrogeology

2.1 Engineering Geological Characteristics

Bedrock Characteristics. The bedrock under the landfill has a stable stratum structure and a large bearing capacity, which basically will not affect the landfill stability.

Configuration of the Composite Liner System. Figure 2 shows the HDPE geomembrane (GM) of the composite liner lying on the ground. The liner system consist of the geotextile, GM, geosynthetics clay liner (GCL). GM is rough or smooth. The composite liner system has a weak interface between geotextile and GM as well as between GM and GCL. There is a risk tha the landfill may be sliding along the interface. Direct shear tests are carried out to obtain the shear strength of the interface. The maximum frictional angle for the interface between geotextile and smooth GM is 13° at the bottom flat ground. The frictional angle of the interface between geotextile and rough GM is 17° at the sloping ground with an inclination greater than 1:4.

Fig. 2. HDPE GM of the basal composite liner

Landfill Characteristics. The landfilled waste body can be divided into three engineering geological layers according to the degree of degradation and engineering characteristics: ① Shallow waste soil: The color is grayish white, and the main components are plastic waste, nylon bags, fibers, cloths and other kitchen waste, and the layer thickness is 0–10 m; ② Middle waste soil: the color is gray-brown and the layer thickness is 8–20 m; ③ bottom waste soil: The color is gray black, compared with shallow and middle wastes, the degree of degradation for paper is relatively high, and the humus content is significantly increased. The layer thickness is 12–20 m. The strength characteristics of various layers are shown in Table 1.

2.2 Hydrogeological Characteristics

Leachate Production. The main waste filled in the second landfill is kitchen waste, and the water content is as high as 60–70% (wet-basis). After landfilling, it is degraded to produce a large amount of leachate. The produced leachate in the upper part under the influence of gravity is moving downward to the drainage layer. With the duration of operation, the void ratio in the low layer of waste will be gradually decreased, and the drainage layer may be gradually clogged. Thus the drainage ability of leachate at the landfill will decrease to be less than the leachate production rate. A large amount of

Table 1. The strength characteristics of various layers

Material	Parameter			
	Saturated weight (kN/m^3)	Natural weight (kN/m^3)	Cohesion (kPa)	Internal friction angle (°)
Shallow waste	11.3	10	20	25
Middle waste	12.5	10.3	5	28
Bottom waste	13.9	12.2	0	36

leachate accumulates in the waste body to form the main leachate mound. There is a difference in leachate leavel at different stages when the landfill is filled to different elevations.

Leachate Level in the Waste Body. There were 633,610 cubic meters of leachate accumulating in the second landfill to form the main leachate mound or the perched leachate mound according to preliminary calculations, as of 2013. The depth of perched leachate depth is 5–8 m. The measurements of pore pressure transducers indicated that the height of the main leachate mound above the drainage layer was up to 9–10 m.

3 Slope Stability Assessment

3.1 Analysis Method

The landfill stability analyses were conducted by using the Morgerstern-Price method in the GEOSLOPE software package. According to "Technical code for geotechnical engineering of municipal solid waste sanitary landfill" (CJJ176-2012) [1], the safety factor for this grade of landfill is greater than 1.35 when the leachate level corresponding to normal meteorological conditions is considered.

3.2 Stability Analysis of Current Landfill Stage

Table 2 shows the calculation results of landfill stability for each section for analysis. The main leachate mound is 10 m in height, the perched leachate mound is 20 m, and the average depth of perched water is 4 m, the bottom surface of the perched leachate mound is 24 m below the landfill surface. The stability of shallow slip, deep slip, and overall slip along the liner are calculated.

Taking the Section 3-3 as an example, the failure mode of landfill is shown in Fig. 3. The current stability of shallow slip cannot meet the regulated requirements, and the current stability of the deep slip and the overall sliding along the liner can meet the regulated requirements. The main potential failure mode is a shallow slip. The high perched leachate level is the main reason of shallow slip, and the perched leachate level needs to be controlled to below the early warning level corresponding to a safety factor of 1.15.

Table 2. The calculation results of landfill stability

Failure mode	Section 1-1	Section 2-2	Section 3-3	Section 4-4
Shallow slip	1.196	1.055	0.966	0.997
Deep slip	2.349	2.196	2.53	2.312
Slip along the liner	1.574	1.957	1.633	2.269

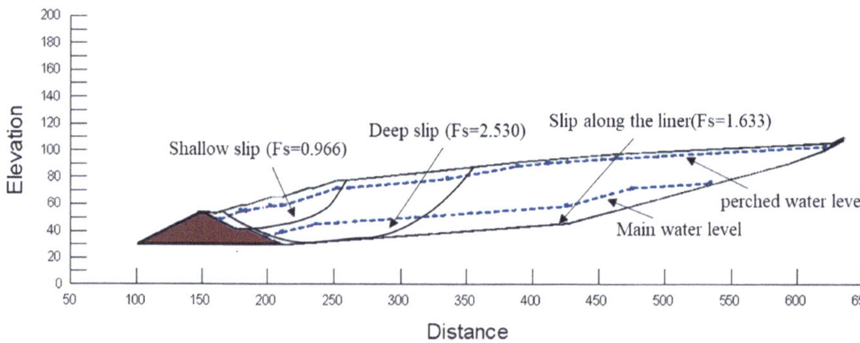

Fig. 3. The failure mode of current landfill stage

3.3 Stability Analysis of Landfill with Subsequent Landfilling Stages

According to the landfilling plan, the landfilling elevation needs to be gradually increased to 96.25 m platform, 115 m platform, 133.75 m platform and 165 platform. At the same time, different daily dumping volume will be considered because of the excessive amount of waste in Hangzhou, and the average daily dumping volume of 3500 t, 4000 t, 5000 t and 6500 t is considered. The total output of leachate can be calculated using the ZJU leachate production model. This calculation method has been adopted by the standard of CJJ176-2012.

The calculated liquid-solid ratios from the ZJU model at different daily dumping volume of 3500 t, 5000 t, and 6500 t are 0.69, 0.82, and 0.91 by 2020, respectively. The height of the main leachate level is 20.1, 30.7, and 40.2 m, and the thickness of the perched mound is 36.4, 56.1, and 68.5 m, respectively. There are significant differences both in the perched leachate mound and the main leachate mound at different stages when the landfill is filled to different elevations. In order to find the variation of the slope stability and the failure mode with the different stages, the daily dumping volume of 5000 t is selected and the Section 3-3 is used as the analysis section. It should be noted that the control requirements of the Section 3-3 for perched leachate level and main leachate level is most critical.

Figure 4 shows that the safety factor of landfill sliding along the bottom liner decreases from the current 1.633 to 1.354, when the elevation of the landfill reaches 96.25 m. When the elevation of the landfill reaches 115 m, the safety factor further decreases to 1.161, so that the safety factor is lower than the control standard. When the elevation of the landfill reaches 133.75 m, the safety factor is further reduced to 0.990, the safety factor is lower than the equilibrium state.

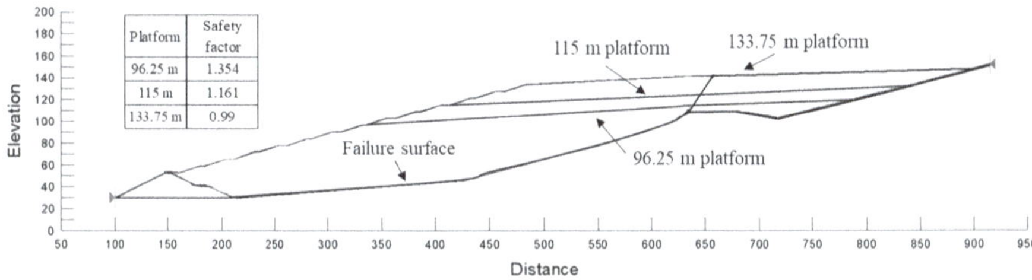

Fig. 4. The landfill stability with subsequent landfilling stages

Figure 5 shows the decrease of safety factor with an increase in the landfill elevation at different daily dumping volume. With the increase of the height of the landfill, the failure mode of landfill gradually shifts from the shallow slip to the overall slip, and the overall stability is decreased. When the elevation of the landfill reaches 100 m, the safety factor can no longer meet the standard of 1.35. The failure mode is along the interface of the composite liner at the bottom of the landfill, which is the weakest. With the increase in the height of landfill, the GCL is prone to hydration, and the peak friction angle at the interface between the GM and the GCL is decreased.

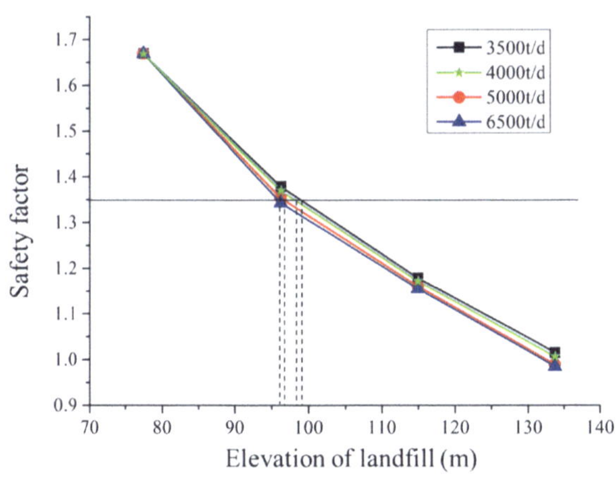

Fig. 5. The safety factor of subsequent landfilling at different daily dumping volume

3.4 Determination of the Early Warning Leachate Level

The stability of the shallow slip at the landfill is mainly controlled by the perched leachate mound. The control standard of the safety factor is 1.35 to determine the early warning levels for both the perched leachate mound and the main leachate mound.

We obtain the early warning levels for the different landfilling stages (current elevation, 96.25, 115, 133.75 m, and 165 min elevation), as shown in Table 3. The full leachate level in Table 3 represents that even if the main leachate level reaches the bottom of the perched leachate mound, the safety factor of the landfill still meets the requirements. From Table 3, it can be seen that the landfill platform reaches 133.75 m, the safety factor cannot meet the requirements even when the main leachate level of Section 1-1 and Section 3-3 is controlled at a 0 m. When the landfill platform reaches final design elevation of 165 m, and the main leachate level of Section 1-1, Section 2-2, and Section 3-3 is 0 m, the safety factor cannot meet the requirements. For the situations when it is impossible to control the water level to meet the requirements, engineering measures need to be taken to enhance the stability of the landfill.

Table 3. The early warning level at different stages

Stage (m)	Section 1-1		Section 2-2		Section 3-3		Section 4-4	
	Perched (m)	Main (m)	Perched (m)	Main (m)	Perched (m)	Main (m)	Perched (m)	Main (m)
Current	7	Full	8	Full	8.6	Full	8.9	Full
96.25	8.4	13.6	10.8	Full	12.3	21.4	12.1	Full
115	8.7	0.2	12.8	22.6	12.6	10.2	12.5	Full
133.75	10.4	0	11	7	12.6	0	12.6	Full
165	10.4	0	11.4	0	15	0	14.2	13.8

4 Engineering Measures of Slope Stabilization and Capacity Expansion

With the increase in the height of the landfill and the leachate levels, the landfill stability at subsequent landfilling stages is obviously lower than the control standard, and it is difficult to reach the original design elevation. Engineering measures are proposed to enhance the slope stabilization as well as to increase landfill capacity.

4.1 Improving the Configuration of Liner System

GCL is prone to hydration at the current and subsequent landfilling stages, which can easily cause the peak friction angle of the GM and GCL interface to be decreased. Once the slip occurs along the interface, the landfill has a greater risk of catastrophic failure. Therefore, it is recommended to remove the GCL layer and use textured GM in the composite liner system to be lined on the sloping ground.

4.2 Controlling the Leachate Levels in the Landfill

It is critical to controll the leachate level in the landfill by setting up a three-dimensional drainage system during the landfilling operation. The system is made up of horizontal drainage trenches under the dumping area and deep vertical wells at the area where the landfill reaches its design elevation. The deep wells could serve for both leachate pumping and landfill gas extractions. It is suggested that at the first state 29 large-diameter deep wells be built at the front sloping area where the design elevation has been reached. The wells drilled down to a depth of 30–40 m. Reducing the leachate level above the liner system will enhance the landfill stability.

4.3 Raising the Height of the Downstream Retaining Dam

Figure 6 shows the profile of the landfill after taking engineering measures, including raising the height of the downstream retaining dam, applying 10 m thick surcharge layer on the downstream slope. The overall stability of the landfill will be improved. For each section, the overall stability of landfill for using MSWs with weight of 10 kN, waste soils with weight of 20 kN, and waste aggregates with weight of 22 kN. The perched leachate level taken the warning level corresponding to the final design elevation, and the main leachate level is set as 10 m by considering the feasibility of the project. The calculation results are shown in Table 4. Based on the analyses, the following measures are proposed to enhance the landfill stability: raising the height of the downstream retaining dam by 10 m, applying 10 m thick surcharge layer on the downstream slope, and control both the perched and main leachate level below the early warning levels. Candidate materials for surcharge may include waste soils, MSW incineration bottom ash, or waste aggregates.

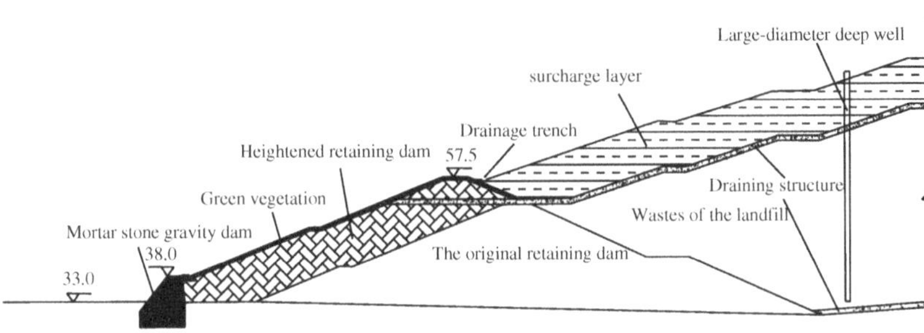

Fig. 6. The profile showing the raising in the height of the downstream retaining dam

4.4 Setting up a Safety Monitoring System

The elevation of the retaining dam of the second landfill is 50 m, and the final elevation is 165 m. According to the standard of CJJ176-2012, the safety level is set as Grade 1 and must be monitored. The main and perched leachate level and horizontal displacement of the landfill should be monitored as a routine monitoring item.

Table 4. The overall stability of landfill for using different materials as surcharge layer

Material	Profile			
	Section 1-1	Section 2-2	Section 3-3	Section 4-4
MSWs (10 kN)	1.502	1.286		1.532
Waste soils (20 kN)		1.435	1.321	
Waste aggregates (22 kN)			1.354	

5 Conclusions and Recommendations

This paper presents the landfill stability analysis at current and subsequent landfilling stages. Based the analyses, engineering measures for enhancing slope stabilization and capacity expansion are proposed. The conclusions are as follows:

(1) The field measured that there exists two kinds of leachate mound in the landfill, i.e., perched and main leachate mound.
(2) The safety factor the shallow slip at the current stage is obviously lower than the control standard. Measures should be taken to drawdown the perched leachate level as soon as possible. It is unlikely for the current landfill to take place a deep slip or a slip along the bottom liner.
(3) The greater the daily dumping volume, the greater the amount of leachate accumulated in the landfill, and the higher the perched and main leachate mound. With the increase in the height of landfill, the potential failure mode gradually shifts from the shallow slip to the deep and overall slip. The GM/GCL interface in the composite liner at the bottom of the landfill is the weakest plane because the GCL is prone to hydration.
(4) Early warning levels of both the perched and main leachate level are determined for different stages of landfilling. Engineering measures are proposed to enhance the slope stabilization as well as to increase landfill capacity. The measures include improving the configuration of liner system to be lined on the sloping ground, establishing three-dimensional drainage system to drawdown leachate level, raising the height of the downstream retaining dam, and setting up a safety monitoring system.

Acknowledgements. The authors acknowledge financial support from Major Science and Technology Projects of Zhejiang Province (No. 2015C03021) and National Science Fund for Distinguished Young Scholars (No. 51625805).

Reference

1. CJJ176-2012 (2012) Technical code for geotechnical engineering of municipal solid waste sanitary landfill. Ministry of Housing and Urban-Rural Development, Beijing

Cyclic Simple Shear Testing of Degraded Municipal Solid Waste from California Under Constant Volume and Constant Load Conditions

Xunchang Fei[1(✉)] and Dimitrios Zekkos[2]

[1] School of Civil and Environmental Engineering, Nanyang Technological University, Singapore, Singapore
xcfei@ntu.edu.sg

[2] Department of Civil and Environmental Engineering, University of Michigan, Ann Arbor, MI, USA
zekkos@geoengineer.org

Abstract. For municipal solid waste (MSW) landfills that are situated in seismically active regions, the response of MSW under dynamic loading is critical to landfill design. In this study, a large-size simple shear device is utilized to perform cyclic testing on MSW under constant load (CL) and constant volume (CV) conditions which are considered equivalent to drained and undrained conditions respectively. The MSW sample was excavated from a California landfill located in a seismically active zone and was fully degraded in the laboratory in controlled conditions. Degraded MSW specimens were compacted and consolidated, and then sheared cyclically using cyclic stress ratios (CSR) of 0.1, 0.2, 0.3 and 0.4 sequentially. For a given number of cycles, higher CSR and higher vertical stress result in higher vertical strain and shear strain under CL conditions, and higher excess pore pressure ratio and shear strain under CV conditions. When subjected to cyclic loading, MSW is found to experience significant vertical strain under CL conditions (i.e., compression), or generate potentially significant pore pressures under CV conditions.

Keywords: Municipal solid waste · Simple shear · Cyclic loading

1 Introduction

A significant amount of the generated municipal solid waste (MSW) is disposed of in landfills near populated areas. Thus, there are thousands of landfills in urban and suburban areas worldwide and many are located in seismically active regions. These landfills are susceptible to damage during earthquakes that may cause financial losses, environmental contamination and even loss of life. Although the performance record of modern MSW landfills during large earthquakes is short, damages in landfills components were reported during the 1994 Northridge earthquake (Augello et al. 1995; Matasovic et al. 1995). Therefore, understanding the dynamic response of disposed MSW remains critical to landfill design in seismic regions.

L. Zhan et al. (Eds.): ICEG 2018, ESE, pp. 35–42, 2019.
https://doi.org/10.1007/978-981-13-2224-2_5

The dynamic response of MSW is most commonly investigated in the laboratory using triaxial, direct shear, and simple shear devices, although recent efforts have successfully measured the dynamic response of MSW in-situ (Sahadewa et al. 2015; Sahadewa et al. 2014). A large specimen size is needed to account for the large particle sizes (up to 10s of cm) and heterogeneity of MSW (Athanasopoulos 2011). Cyclic shearing of MSW has been performed using large-size triaxial devices (e.g., Ramaiah and Ramana 2017; Zekkos et al. 2008). Although cyclic loading under simple shear condition is believed to be more representative of earthquake loading, cyclic simple shear (Matasovic and Kavazanjian 1998; Yuan et al. 2011) and monotonic simple shear (Landva and Clark 1990; Zekkos and Fei 2017) testing of MSW remains limited.

Cyclic testing of MSW is typically conducted under drained conditions as MSW is generally unsaturated. Undrained cyclic behaviour of MSW has not been explored in detail but warrants investigation, because MSW in the lower part of a landfill can become saturated by leachate and the potential for gas generation in leachate-recirculation or wet landfills may result in the occurrence of high gas pressure. In a simple shear device, drained conditions are achieved by maintaining a constant vertical load (CL) on a specimen during shearing, whereas undrained conditions are simulated by shearing the specimen in constant volume (CV), i.e., without volume change.

In this study, MSW is sheared cyclically using a large-size simple shear device in both CL and CV conditions with the intent to compare the differences in dynamic response.

2 Methodology

The MSW sample was excavated from a landfill in California, United States, around 100 km west of Los Angeles and located in a seismically active region. The waste was disposed of less than a year before excavation and was minimally degraded based on visual observation, although no food waste was identified. The sample was shipped in sealed drums to the laboratory. The MSW had an average dry moisture content of 28%. Its composition by dry weight was 68% of <20 mm particles classified as SP-SM (Unified Soil Classification System) and 32% of >20 mm particles consisting of paper, soft plastic, wood, hard plastic, metal and cobbles. The MSW sample was reconstituted and degraded under enhanced biodegradation conditions in a large-size laboratory simulator for around 900 days. The experimental setup and degradation process are reported in Fei and Zekkos (2018) as specimen CA2. The MSW was "soil-rich" and showed the lowest levels of settlement and methane generation among the seven specimens tested. At the end of the experiment the specimen had experienced 16% vertical settlement due to biodegradation and generated 187 L of methane. Additional settlement was deemed low and methane generation stopped despite continued leachate recirculation. The degraded MSW consisted of 72% <20 mm particles and 28% >20 mm particles by dry weight. One noticeable difference after degradation is that >20 mm paper waste was mostly degraded and converted largely to <20 mm particles.

The degraded MSW was carefully retrieved from the simulator and transferred to the simple shear device with minimal disturbance for monotonic shear testing. The sheared MSW was then reconstituted and tested again using the same procedure. The monotonic shear responses between the two specimens were similar, indicating that specimen reconstitution to the same density and moisture ("undisturbed" and reconstituted) as the original degraded specimen was not affecting the results considerably (data now shown, see an example in Fei and Zekkos 2015). For this study, the specimens were reconstituted using the degraded MSW and compacted in four layers by dropping a 98 N hammer from a height of 0.9 m 18 times per layer. The as-compacted dry ensity of the specimens was 1250 ± 20 kg/m^3 and moisture content was maintained at 25 ± 2%.

The large-size cyclic simple shear device was developed and described in detail in Zekkos et al. (2018). Each specimen is contained in a stack of frictionless shear rings and has a diameter of 30 cm and a height between 12 and 13 cm. Once a specimen was placed in the device, a consolidation vertical stress (σ'_{v0}) was applied for 24 ± 1 h prior to cyclic loading. The σ'_{v0} used in this study was 100, 200 and 400 kPa. Vertical displacement of a specimen was maintained near zero to ensure CV conditions, while significant vertical displacement occurred during shearing under CL conditions. Cyclic testing was performed at a frequency of 0.1 Hz using a constant cyclic stress ratio (CSR, applied shear stress divided by σ'_{v0}). For each CSR, the loading stopped after 150 cycles, except for four steps which were manually stopped earlier due to difficulty in maintaining a constant CSR (Table 1). Subsequently, the specimen was re-consolidated at the same σ'_{v0} and subjected again to cyclic testing at a higher CSR. The sequential CSR values were 0.1, 0.2, 0.3 and 0.4 (except 0.35 in one test). Excess pore pressure ratio (r_u), shear strain (ε_h) and single amplitude shear strain ($\varepsilon_{h,single}$) were calculated after each cycle for CV tests (Zekkos et al. 2018), while vertical strain (ε_v), ε_h and $\varepsilon_{h,single}$ were calculated after each cycle for CL tests.

The sheared specimen was retrieved with negligible material loss and manually disassembled to erase fabric structure and stress history and the material was reused for the next test. Deionized water was sprayed on the MSW as necessary to maintain a constant moisture content.

3 Results and Discussion

Testing conditions and cyclic loading responses of the MSW specimens under CV and CL conditions are summarized in Table 1. Two specimens were tested under CV conditions at σ'_{v0} of 100 and 400 kPa and CSR ranging from 0.1 to 0.35. Three specimens were tested under CL conditions at σ'_{v0} of 100, 200 and 400 kPa and CSR ranging from 0.1 to 0.4. The values of r_u, $\varepsilon_{h,single}$ and ε_v after the 15th cycle are reported. These values after the last cycle (up to 150 cycles) are indicative of long-term repetitive loading response of the MSW and are used for further analysis as well.

Table 1. Summary of testing conditions and cyclic loading response of the MSW specimens

σ'_{v0} kPa	γ_t (kg/m^3) prior to each step	CSR	Number of cycles	$\varepsilon_{h,single}$ (%) after 15th/last cycle	ε_v (%, CL) or r_u (CV) after 15th/last cycle
Constant load (CL) condition					
100	1516	0.1	150	0.19/0.18	0.10/0.32
	1521	0.2	150	0.53/0.50	0.17/0.64
	1532	0.3	150	1.03/1.05	0.24/1.05
	1551	0.4	27	1.91/2.00	0.25/0.55
200	1580	0.1	150	0.29/0.27	0.24/0.40
	1588	0.2	150	0.78/0.73	0.25/0.84
	1604	0.3	150	1.55/1.53	0.28/1.26
400	1633	0.1	150	0.44/0.40	0.19/0.60
	1646	0.2	150	1.19/1.11	0.34/1.06
	1667	0.3	150	2.48/3.01	0.33/1.49
Constant volume (CV) condition					
100	1480	0.1	150	0.19/0.19	0.15/0.22
	1482	0.2	150	0.57/0.70	0.27/0.42
	1488	0.3	66	1.36/4.27	0.38/0.69
	1514	0.35	41	3.12/8.55	0.51/0.64
400	1640	0.1	150	0.43/0.45	0.18/0.28
	1645	0.2	150	1.48/2.22	0.36/0.57
	1669	0.3	30	3.19/4.41	0.36/0.47

Cyclic Constant Load Testing: Example CL testing results for the specimen at σ'_{v0} = 100 kPa are shown in Fig. 1. Shear strain remains practically constant at CSR = 0.1 whereas at CSR = 0.4, the shear strain is increasing with more cycles, i.e., the MSW is experiencing shear modulus degradation (Fig. 1a and b). Vertical strain during cyclic shearing increases and approaches a constant value asymptotically at CSR = 0.1 (Fig. 1c), but rapidly increases at CSR = 0.4 with increasing number of cycles.

The vertical strain and shear strain after the 15th and the last cycle are plotted in Fig. 2a and b. Both strains increase with increasing σ'_{v0} and CSR, and the maximum values are 1.5% for vertical strain and 3% for shear strain at 150 cycles. The evolution of ε_v and $\varepsilon_{h,single}$ is plotted in Fig. 2c. The shear strain continues to increase with increasing ε_v at high CSR, otherwise, the shear strain remains constant whereas vertical strain increases with increasing number of cycles, i.e., the specimen densifies during shearing. Noticeably, ε_v exceeds 0.5%, and even reaches 1.5% in one loading step, when subjected to high enough σ'_{v0} and CSR and after around 100 cycles. This finding highlights that CL cyclic loading of MSW could induce permanent settlement that may cause damage to landfill infrastructure components including final cover and gas venting systems.

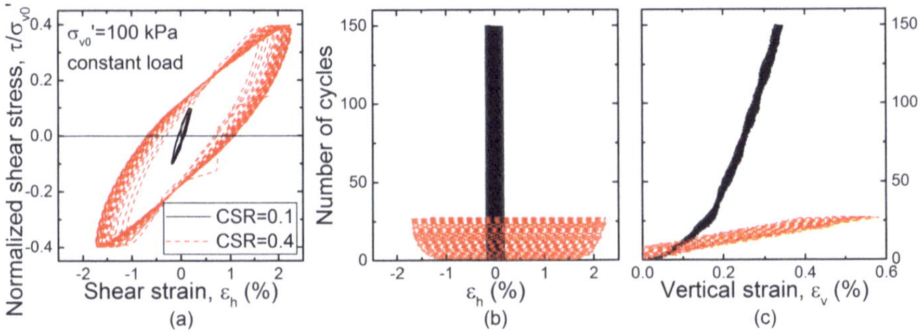

Fig. 1. Example CL testing results at σ'_{v0} = 100 kPa and CSR of 0.1 (black solid line) and 0.4 (red dotted line): (a) normalized shear stress (τ/σ'_{v0}) vs shear strain (ε_h); (b) number of cycles vs ε_h; and (c) number of cycles vs vertical strain (ε_v)

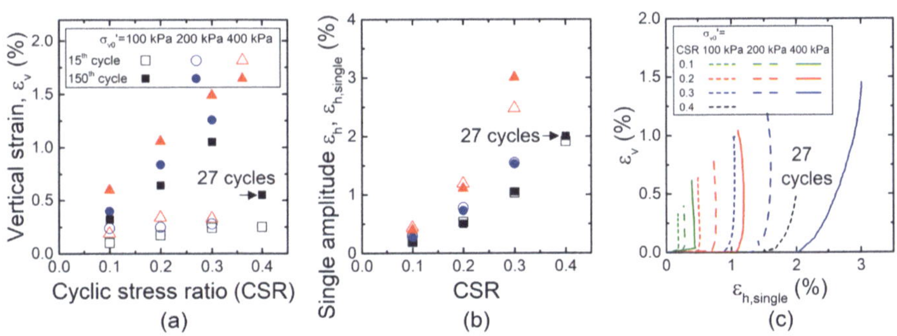

Fig. 2. Cyclic response of MSW under constant load conditions (CL) for different cyclic stress ratio (CSR): (a) relationship with vertical strain (ε_v); (b) relationship with single amplitude shear strain ($\varepsilon_{h,single}$); and (c) evolution of ε_v and ε_h

Cyclic Constant Volume Testing: Example CV cyclic testing results for the specimen at σ'_{v0} = 100 kPa are shown in Fig. 3. The typical "butterfly-loop" relationship between normalized shear stress and normalized effective vertical stress develops during testing for both CSRs (Fig. 3a). The specimen exhibits a stiffer response and lower damping at CSR = 0.1 compared to CSR = 0.35 (Fig. 3b). Note that for the CSR = 0.35 test, the level of shear strains imposed were increasingly higher and thus the device was challenged to maintain the shear stress amplitude at large number of cycles. Nevertheless, constant shear stress amplitude was maintained for many cycles. The evolutions of r_u and ε_h with the number of cycles is shown in Fig. 3c and d. The pore pressure increase is low (r_u = 0.1) for the CSR = 0.1 test, and becomes higher (r_u = 0.5 – 0.6) for the higher CSR indicating the potential for generation of significant pore pressures in CV conditions.

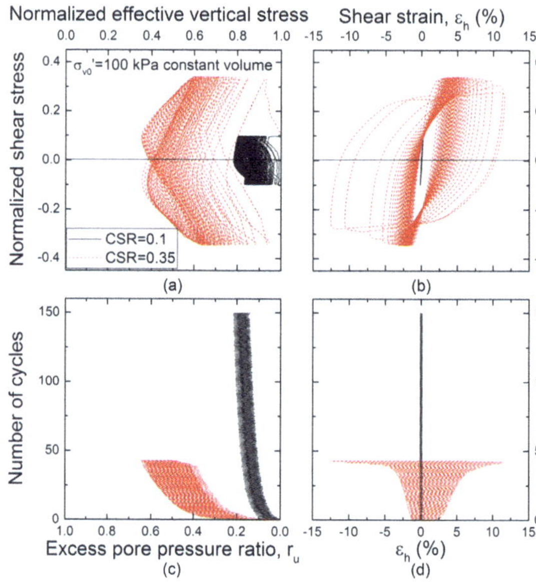

Fig. 3. Example CV testing results at σ'_{v0} = 100 kPa and CSR of 0.1 (black solid line) and 0.35 (red dotted line): (a) normalized shear stress (τ/σ'_{v0}) vs normalized effective vertical stress (σ'_v/σ'_{v0}); (b) τ/σ'_{v0} vs shear strain (ε_h); (c) number of cycles vs excess pore pressure ratio (r_u); and (d) number of cycles vs ε_h

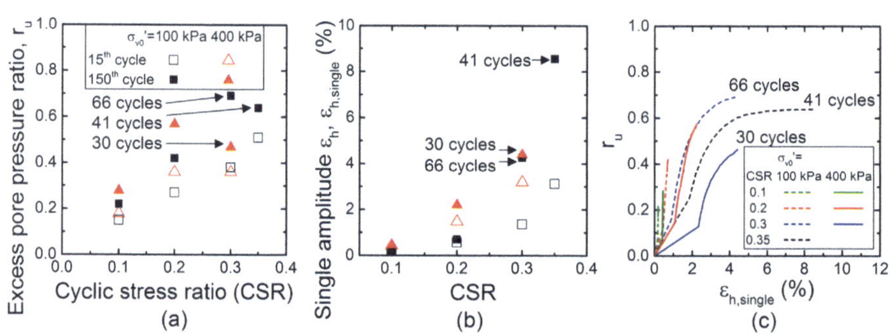

Fig. 4. Cyclic-loading response of MSW under CV conditions for different cyclic stress ratio (CSR): (a) relationship with excess pore pressure ratio (r_u); (b) relationship with single amplitude shear strain ($\varepsilon_{h,single}$); and (c) evolution of r_u and $\varepsilon_{h,single}$

The excess pore pressure after the 15th and the last cycle is found to increase with increasing CSR. For the same CSR, higher σ'_{v0} results in higher r_u (Fig. 4a). Figure 4b shows that $\varepsilon_{h,single}$ increases with increasing σ'_{v0} and CSR after the same number of cycles. The maximum r_u and $\varepsilon_{h,single}$ values approach 0.7 and 8.6%, respectively. Similar to CL condition, the dynamic response of MSW under CV condition is significantly affected by σ'_{v0} and CSR.

As shown in Fig. 4c, considerable excess pore pressure and ε_h are generated after the first loading cycle, indicated by an abrupt change in the slope of the $r_u - \varepsilon_{h,single}$ curve starting from the origin. Subsequently, excess pore pressure continues to accumulate while ε_h increase is much slower. When CSR is 0.3 or higher, with increasing number of cycles, $\varepsilon_{h,single}$ keeps increasing while r_u eventually approaches a maximum value between 0.6 and 0.7. This material behaviour is similar to ratcheting which describes sustained long-term accumulation of shear strain during repetitive loading (Pasten et al. 2014). Overall, mobility failure appears to be a more pronounced failure mechanism for MSW under CV cyclic loading condition adopted in this study, and sudden loss of shear capacity is not observed.

4 Summary and Conclusions

A MSW sample retrieved from a California landfill located in a seismically active region was fully degraded in the laboratory and then tested for its dynamic response using a large-size cyclic simple shear device. The specimens were tested under constant load (considered equivalent to drained) and constant volume (considered equivalent to undrained) conditions at 100 to 400 kPa of σ'_{v0}. Each specimen was sheared at a frequency of 0.1 Hz using cyclic stress ratios of 0.1, 0.2, 0.3 and 0.4 sequentially.

For the same number of loading cycles, increasing CSR and σ'_{v0} results in increasing vertical and shear strain in CL conditions, and higher pore pressures and shear strains in CV conditions. Shear strain is lower under CL condition than CV condition due to continuous vertical compression and densification of MSW during loading. Vertical compression during cyclic loading is of potential concern in CL conditions. In CV conditions, the specimens experienced a pore pressure increase that could approach $r_u = 0.7$ at high CSR, but sudden loss of strength is not observed.

Acknowledgements. This research was supported by the National Science Foundation (NSF) Division of Computer and Communication Foundations under Grant no. 1442773. Any opinions, findings, conclusions and recommendations expressed in this paper are those of the authors and do not necessarily reflect the views of the NSF.

References

Athanasopoulos GA (2011) Laboratory testing of municipal solid waste. In: Zekkos D (ed) Geotechnical characterization, field measurement, and laboratory testing of municipal solid waste, pp 195–205

Augello AJ, Matasovic N, Bray JD, Kavazanjian Jr E, Seed RB (1995) Evaluation of solid waste landfill performance during the Northridge earthquake. In: Proceedings of earthquake design and performance of solid waste landfills. ASCE geotechnical special publication no. 54

Fei X, Zekkos D (2015) Large-size controlled degradation and simple shear testing of municipal solid waste from Michigan. In: Proceedings of XVI European conference on soil mechanics and geotechnical engineering. ISSMGE, Edinburgh, Scotland, pp 2753–2758

Fei X, Zekkos D (2018) Coupled experimental assessment of physico-biochemical characteristics of municipal solid waste undergoing enhanced biodegradation. Geotechnique. https://doi.org/10.1680/jgeot.16.p.253

Landva AO, Clark JI (1990) Geotechnics of waste fill. In: Landva A, Knowles GD (eds) Geotechnics of waste fills—theory and practice, vol 1070. American Society of Testing and Materials, Philadelphia, pp 86–103

Matasovic N, Kavazanjian E Jr (1998) Cyclic characterization of OII landfill solid waste. J Geotech Geoenviron Eng 124(3):197–210

Matasovic N, Kavazanjian Jr E, Augello AJ, Bray JD, Seed RB (1995) Solid waste landfill damage caused by 17 Jan 1994 Northridge earthquake. In: Proceedings of California Department of Conservation, Division of Mines and Geology Special Publication 116. California Department of Conservation, Sacramento, CA

Pasten C, Shin H, Santamarina JC (2014) Long-term foundation response to repetitive loading. J Geotech Geoenviron Eng 140(4):04013036

Ramaiah BJ, Ramana GV (2017) Study of stress–strain and volume change behavior of emplaced municipal solid waste using large-scale triaxial testing. Waste Manag 63:366–379

Sahadewa A, Zekkos D, Woods RD, Kenneth HS (2015) Field testing method for evaluating the small-strain shear modulus and shear modulus nonlinearity of solid waste. Geotech Test J 38 (4):427–441

Sahadewa A, Zekkos D, Woods RD, Stokoe KH, Matasovic N (2014) In-situ assessment of the dynamic properties of municipal solid waste at a landfill in Texas. Soil Dyn Earthq Eng 65:303–313

Yuan P, Kavazanjian E Jr, Chen W, Seo B (2011) Compositional effects on the dynamic properties of municipal solid waste. Waste Manag 31(12):2380–2390

Zekkos D, Athanasopoulos-Zekkos A, Hubler J, Fei X, Zehtab K, Marr W (2018) Development of a large-size cyclic direct simple shear device for characterization of ground materials with oversized particles. Geotech Test J 41(2)

Zekkos D, Bray JD, Riemer MF (2008) Shear modulus and material damping of municipal solid waste based on large-scale cyclic triaxial testing. Can Geotech J 45(1):45–58

Zekkos D, Fei X (2017) Constant load and constant volume response of municipal solid waste in simple shear. Waste Manag 63:380–392

Measuring Soil Strain Using Fibre Optic Sensors

Susanga Costa[1(✉)], Gayan Kahandawa[1], Jian Chen[2], and Jianfeng Xue[3]

[1] School of Science, Engineering and IT, Federation University Australia, Ballarat, VIC, Australia
susanga.costa@federation.edu.au
[2] School of Engineering, University of Melbourne, Melbourne, Australia
[3] School of Engineering and Information Technology, ADFA Campus, University of New South Wales, Kensington, Australia

Abstract. Monitoring subsurface soil movement is important in many geotechnical engineering applications such as stability of slopes, road embankments and settlement in foundations. Soil displacement measurement is also helpful in understanding the formation of shrinkage cracks. Clay soils undergo shrinkage during drying and experience substantial stresses and strains, which results in shrinkage cracks. This paper presents a novel approach to measure soil strain using Fibre Bragg grating (FBG) sensors. In the experiments described, FBG sensors have been used to investigate the strain development in clay during drying.

FBG sensors are fabricated in the core region of specially fabricated single mode low-loss germanium doped silicate optical fibres. The grating is the laser-inscribed region with a periodically varying refractive index, which reflects a specific light wavelength. Due to the applied strain, ε, there is a change in the wavelength which can be measured and is directly proposal to strain.

Kaolin clay, mixed with water close to the liquid limit, was allowed to dry under room temperature. The specimens were prepared in thin, long linear shrinkage moulds. FBG sensors were placed inside soil at the centre of the specimen. The strain development during drying underwent four phases moving from compression to tension. An oscillating nature of strain was also observed throughout the drying process. Results obtained are useful to develop analytical solutions to describe stress-strain behavior of drying soil.

Keywords: Clay · Drying · Fibre optic sensors · Strain

1 Introduction

Expansive soils undergo shrinking and swelling due to change in moisture content. Thus, the characteristics of shrink-swell behaviour is of great importance to serviceability of geotechnical structures. In Australia, about 20% of the total land area is covered by expansive soils [1]. Traditionally, shrink-swell potential is determined using shrinkage limit test and swelling test.

L. Zhan et al. (Eds.): ICEG 2018, ESE, pp. 43–50, 2019.
https://doi.org/10.1007/978-981-13-2224-2_6

Determination of soil displacement and strain with conventional methods such as local displacement transducers (LDT) and strain gauges are challenging due to their limitations with electromagnetic interference and possibility of corrosion in the presence of water. Alternatively, researchers have used particle image velocimetry (PIV) technique [2, 3], laser technology [4] and X-ray tomography [5] to measure soil movement. However, these methods too have limitations. PIV technique can only be used to measure surficial movements. Sub-surface movement can be observed with laser and X-ray technology, however, the measurement of local small strains is still challenging.

This paper investigates the possibility of using optical fibre sensors to measure soil strain during shrinkage. The type of sensors used is fibre Bragg grating (FBG) sensors. The application of FBG sensors to soil strain measurement is novel. Xu et al. [6] reported to have used FBG sensors in a modified triaxial apparatus. They attached the FBG sensors to a metal bender element which translated the soil strain to the optical cable. In the present study, FBG sensors were placed in direct contact with soil which was subjected to drying.

2 Fibre Bragg Grating (FBG) Sensors

Fibre Bragg Grating (FBG) sensors have been used for measuring strain of structures efficiently for more than two decades. The main application area of the sensors was limited to composite structures. Recent advances in FBG sensor technologies have provided great opportunities to develop more sophisticated strain monitoring systems for soil strain measurement and monitoring.

The diminutive FBG sensor, provides the ability to embed the sensors in soil samples without scarifying the material properties of the specimen. With the soil shrinkage, the embedded sensor compresses with soil and the respective strain in the soil is been measured by the sensor. This sensor is suitable for networking since it has a narrow bandwidth and wide wavelength operating range. Hence, it can be highly multiplexed, which means the same fibre can have multiple sensors at different locations along the length of the fibre. Multiplexed sensors can be used to measure strain at multiple locations in a soil sample using a single fibre. Since FBG sensors are non-conductive, they can be operated in electromagnetically noisy and/or wet environments without any effect. FBG sensor is made up of glass which is environmentally more stable and with a long life span. FBG sensor signal can be monitored from longer distances making it suitable for remote sensing due to its low transmission loss [7–9].

2.1 Principle of Bragg Sensor

FBG sensors are fabricated in the core region of specially fabricated single mode low-loss germanium doped silicate optical fibres. The grating is the laser-inscribed region which has a periodically varying refractive index. This region reflects only a narrow band of light corresponding to the Bragg wavelength λ_B, which is related to the grating period Λ_o [10].

$$\lambda_B = \frac{2n_o \Lambda_o}{k} \tag{1}$$

where k is the order of the grating and n_o is the initial refractive index of the core material prior to any applied strain.

Due to the applied strain, ε, there is a change in the wave length, $\Delta\lambda_B$, for the isothermal condition

$$\frac{\Delta\lambda_B}{\lambda_B} = \varepsilon P_e \tag{2}$$

where P_e is the strain optic coefficient and calculated as 0.793.

Figure 1(a) shows reflected spectrums of an FBG sensor. Monitoring the peak wavelength allows the stain at that point to be calculated. Figure 1(b) shows the change of the spectrum due to loading. Wave shift of the FBG sensors under loading is recorded and the readings are converted to strain using Eq. 2.

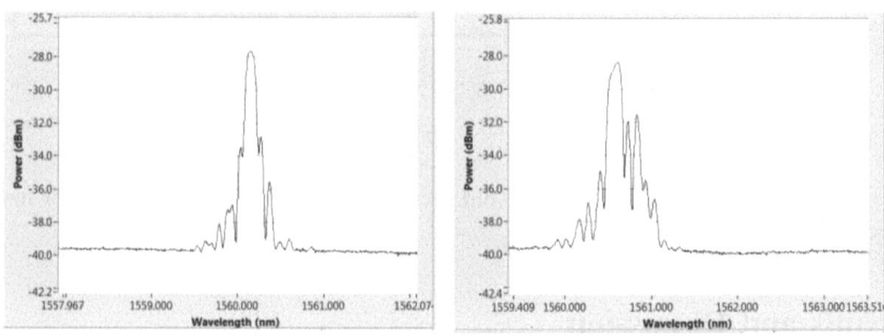

(a) (b)

Fig. 1. Reflected spectrum of an FBG sensor.

3 Materials and Shrinkage Experiments

A series of drying shrinkage tests was conducted with kaoline clay. Consistency limits of kaoline are Liquid Limit (LL) 70% and Plastic Limit (PL) 42%. Specimen preparation was similar to linear shrinkage test. The moulds used in the current study were made of Perspex and rectangular in shape. The internal dimensions were L = 250 mm, W = 25 mm and D = 18 mm. A specimen prepared for testing is shown in Fig. 1(a)

Soil was prepared at a moisture content of 80%, slightly above the LL, in order to obtain a smooth and homogenous clay layer. A thin layer of petroleum gel was applied to the base and the sides of the mould to allow free shrinkage. Specimens were kept on an electronic balance and were dried in the air under two floodlights. Each test was carried out for about 36 h.

It was decided to place the sensor at the centre of the specimen to minimise any boundary effects. The configuration is shown in Fig. 2(b). Approximately, 50 mm long segment of the optical fibre (including the 10 mm gauge length of the sensor) was buried inside clay. The sensor recorded 10 readings per second (10 Hz). The accuracy of FBG analyser used in the experiments is in nanometre scale, which is more than required for soil strain measured in micrometre scale.

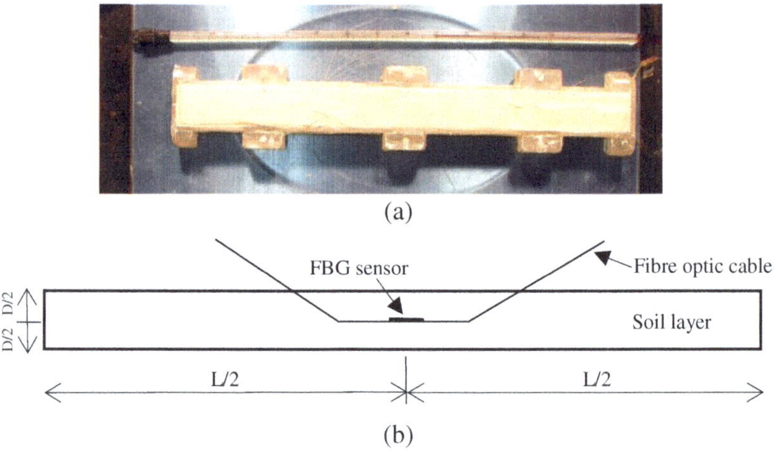

Fig. 2. (a) Specimen prepared for testing (top view) (b) Location of the FBG sensor inside the soil layer (side view)

4 Results and Discussion

Strain measurements obtained during drying is shown in Fig. 3. The curve represents the average strains of all the tests conducted. For brevity, strains measured at 15 min intervals were used for plotting. The centre of the clay layer undergoes compressive strain for most part of the drying period. At the beginning of the tests, there is a short period, up to about 2 h, where the strain is tensile. This is due to the initial heating up of the clay layer. The specimens were prepared at room temperature which was in the range of 22° to 24 °C. However, the temperature in the surrounding of the experimental set up remained steady at 31.5 °C due to the constant heat emitted by the lights. Therefore, the specimens experienced a warming up period resulting an expansion.

The accuracy of measurements depends largely on how well the sensor is attached to clay. In the current study, visual observations indicated a fairly strong adherence between the sensor and soil. Additionally, the sensor recorded data from the very beginning of the test implying there was no slip. As the drying progressed, the adherence of sensor to soil strengthened further.

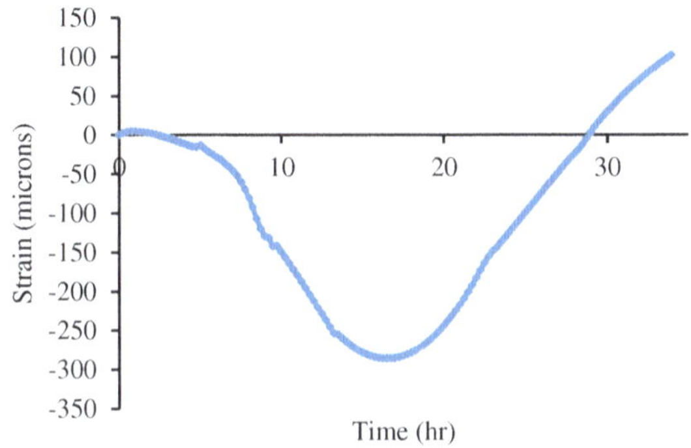

Fig. 3. Development of strain with drying time

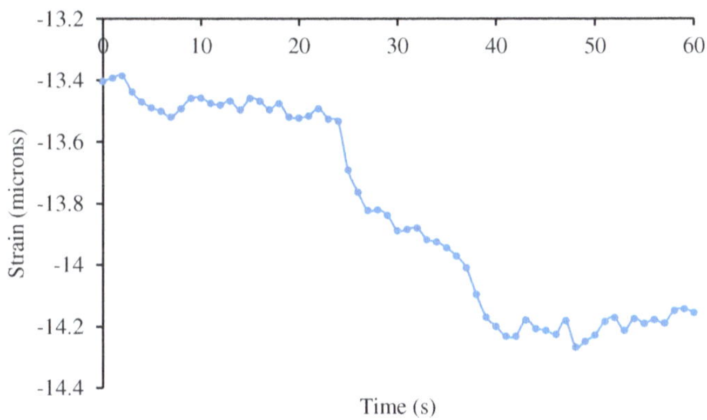

Fig. 4. Strain oscillations observed within the 300th minute (5 h)

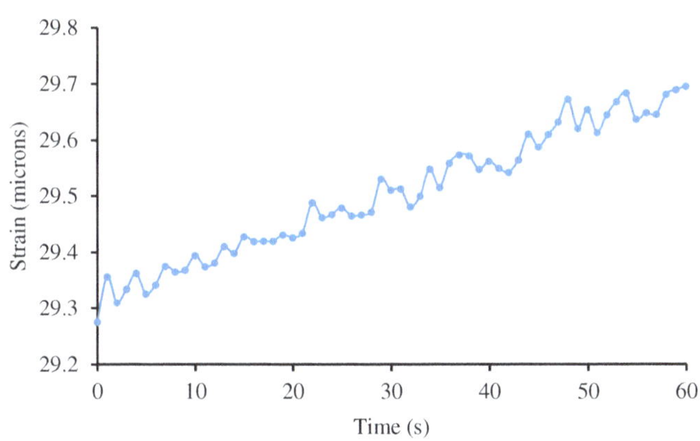

Fig. 5. Strain oscillations observed within the 1800th minute (30 h)

The curve in Fig. 3 suggests a smooth progression in strain development. However, a closer view will show that strains undergo fluctuations. Figures 4 and 5 show magnified sections of Fig. 3 at 300th and 1800th minute respectively. At the 300th minute, the specimen is subjected to increasing compressive strain. Figure 4 shows the fluctuations inside this minute with measurements taken at every 1 s.

By the 1800th minute, the strain at the middle of the specimen has started to undergo tensile strains. Figure 5 shows the fluctuating behaviour within the selected minute. The fluctuations are more prominent at this stage. Similar observations were made by Costa and Kodikara [11] using PIV image analysis under restrained shrinkage conditions prior to desiccation crack initiation.

Two observations can be made from the results shown in Figs. 4 and 5. First, strains of a drying soil element oscillates around a mean value. It is possible that these oscillations are inherent to drying shrinkage. Second, the mean value changes with the loss of water. This is initially compressive indicating a reduction in volume due to removal of water (500th minute). Once the specimen has undergone substantial drying, the mean value takes a positive value (tensile) suggesting an increase in volume due to air entry (1800th minute). However, this volume increase occurs only locally as the overall result would be a reduction in volume for the whole specimen.

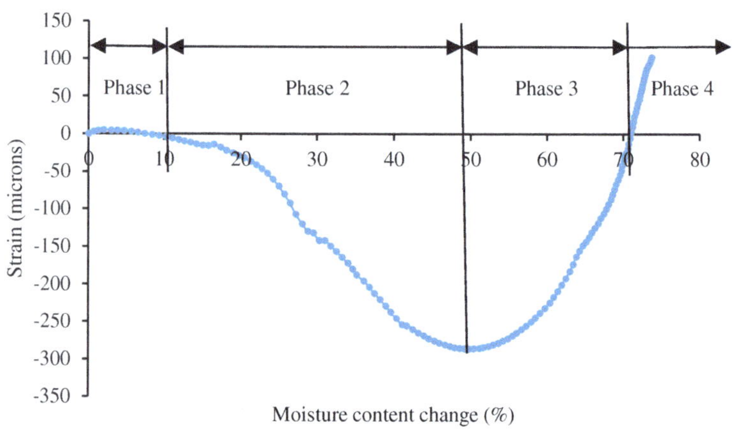

Fig. 6. Variation of strain with the moisture content change

Figure 6 shows the variation of strain as the moisture content reduced. It is possible to identify four phases of drying from this curve. Phase 1 refers to the initial period where the specimens undergo warming up as explained above. No significant development in strain can be seen at this phase. Strain is slightly tensile due to the expansion of pore water. The total moisture content reduces by about 10%.

As the specimen starts to lose water, the specimen undergo compression. This is phase 2. The specimen loses nearly about 40% of its water during this phase. The shrinkage continues until the compressive strain reaches a maximum.

The third phase can be considered a relaxation phase. The middle of the specimen relieves compression. It is important to remember that this is apparent from a soil element at the centre of the specimen. The strain of the whole specimen is still compressive as the soil mass continues to shrink. A reduction of about 20% of moisture is evident during this phase.

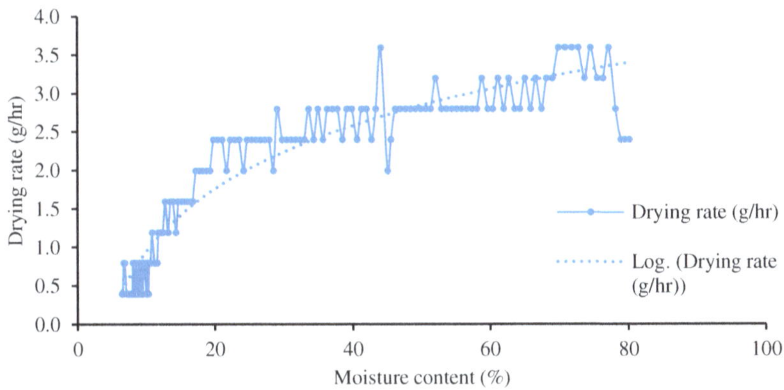

Fig. 7. Variation of drying rate of the specimens with moisture content

The end of the relaxation phase marks the beginning of tensile strain development. By this time, the specimen has lost about 70% of its moisture and the drying process slows down as shown in Fig. 7.

In free shrinkage, tensile strain development (Phase 4) starts at a later stage. The specimen is close to its shrinkage limit by this time and cannot generate sufficient tensile stresses to cause a tensile failure. In other words, strain energy built up is not adequate to produce a fracture surface. Thus, the free shrinkage specimens rarely produce desiccation induced shrinkage cracks. When the shrinkage is restrained at the base (sometimes on sides as well), phase 4 is reached at an early stage while the specimens still have much moisture to lose. Consequently, the strain curve will continue to develop into a parabolic shape with a maximum tensile strain. This will result in a stress profile similar to what was proposed by Kodikara and Choi [12, 13]. If the specimen continue to dry having such a tensile stress profile, it will reach a state where tensile stress exceeds the tensile strength of the material causing a fracture.

5 Conclusions

Fibre optic sensors were successfully implemented to measure sub-surface strain in drying soil specimens. The development of strain at the centre of the specimens was determined. It is possible to identify four phases of strain development during drying: warming up, compression, relaxation and tension. Another interesting behaviour of strain was its oscillating nature. The strains oscillates around a mean value at any given instance.

While the FBG sensors produced reliable measurements, it is necessary to verify them using another method. The current study was limited to test the feasibility of FBG sensors to measure soil strain. Further investigations are needed to study the stress-strain behaviour of drying soil in detail.

References

1. Richards BG, Peter P, Emerson WW (1983) The effects of vegetation on the swelling and shrinking of soils in Australia. Geotechnique 33(2):127–139
2. Costa S, Kodikara J, Thusyanthan NI (2008) Study of desiccation crack evolution using image analysis. In: Toll et al. (eds) Unsaturated soils: advances in geo-engineering. Taylor & Francis Group, London
3. Shannon B, Kodikara J, Rajeev P (2015) The use of restrained ring test method for soil desiccation studies. Goetechn Test J 38(1):98–112
4. Sanchez M, Atique A, Kim S, Romero E, Zielinski M (2003) Exploring desiccation cracks in soils using a 2D profile laser device. Acta Geotech 8:583–596
5. Julina M, Thyagaraj T (2018) Quantification of desiccation cracks using X-ray tomography for tracing shrinkage path of compacted expansive soil. Acta Geotech. https://doi.org/10.1007/s11440-018-0647-4
6. Xu DS, Borana L, Yin JH (2014) Measurement of small strain behaviour of a local soil by fiber Bragg grating-based local displacement transducers. Acta Geotech 9:935–943
7. Hill KO, Meltz G (1997) Fiber Bragg grating technology fundamentals and overview. J Lightwave Technol 15(8):1263–1276
8. Kersey AD, Davis MA, Patrick HJ, LeBlanc M, Koo KP, Askins CG, Putnam MA, Friebele EJ (1997) Fiber grating sensors. J Lightwave Technol 15(8):1442–1463
9. Kahandawa G, Epaarachchi J, Wang H, Lau K (2012) Use of FBG sensors for SHM in aerospace structures. Photonic Sensors 2:203–214
10. Kashyap R (1999) Fibre Bragg gratings. Academic Press Publications, London
11. Costa S, Kodikara J (2010) Shrinkage development during soil desiccation. In: Buzzi et al. (eds) Unsaturated soils: theoretical and numerical advances in unsaturated soil mechanics. Taylor & Francis Group, London
12. Kodikara J, Choi X (2006) A simplified analytical model for desiccation cracking of clay layers in laboratory tests. In. Miller et al. (eds) Unsaturated soils 147(2):2558–2567. ASCE Geotechnical Special Publication
13. Costa S, Kodikara J, Barbour SL, Fredlund DG (2018) Theoretical analysis of desiccation crack spacing of a thin, long soil layer. Acta Geotech 13:39–49

Effect of Landfill Odorous Gas on Surrounding Environment: A Field Investigation and Numerical Analysis in a Large-Scale Landfill in Hangzhou, China

Siliang Shen[1], Qiao Wang[2], Yunmin Chen[1], Xinru Zuo[1], Feiyu He[2], Shuangke Fei[2], and Haijian Xie[1,2](✉)

[1] MOE Key Laboratory of Soft Soils and Geoenvironmental Engineering, Zhejiang University, Hangzhou 310058, China
xiehaijian@zju.edu.cn

[2] College of Civil Engineering and Architecture, Zhejiang University, Hangzhou, China

Abstract. Landfill odorous gas emission has been a serious environmental problem. In this paper, in situ monitoring and numerical analysis model Calpuff were adopted to analyze the odour nuisance problem near a large-scale landfill in Hangzhou, east of China. The static chamber technique was used to measure the odorous gas emission intensity on the working area surface in the landfill during the Sep. 2016 to Apr. 2017, irregularly. The obtained results indicated that the emission rate of H_2S varied from 6.14 mg/m^2/h to 91.7 mg/m^2/h. The average emission rate in each month was then used in the Calpuff model to analyze the emission pattern of the H_2S. The terrain and meteorological conditions at the site were also considered. Concentration of H_2S at the community which is 4.2 km away from the landfill is 15 times greater than its olfactory thresholds (0.62 μg/m^3). In order to keep the odorous influence area within landfill site, proper engineering measures should be taken to ensure the emission rate of H_2S drops to the 15% of its original value.

Keywords: Odorous gas emission · Calpuff · H_2S

1 Introduction

The degradation of municipal solid waste (MSW) in the landfill generates less than 1% v/v non-methane compounds, which was considered to be one of the main sources of odor emissions (Allen et al. 1997). The MSW in China contains much more kitchen wastes than that in the developed countries. This leads to a high leachate level when they degraded. The high leachate level causes relatively low LFG collection efficiency. More than 50% of the total LFG emitted uncontrolled. The odor-related nuisance has become the major causes of residents' complaints to the local authorities (Henshaw et al. 2006). H_2S is the main landfill odor gas and was always chosen as the representative odor gas of landfills (Ding et al. 2012; Duan et al. 2014).

L. Zhan et al. (Eds.): ICEG 2018, ESE, pp. 51–59, 2019.
https://doi.org/10.1007/978-981-13-2224-2_7

The increasing number of complaints around the landfill may attribute to the increase of landfill gas emission rates. It is also directly related to the meteorological conditions on the site (Chemel et al. 2012). Some specific weather conditions including low mixing height or low wind speed could severely affect the odour dispersion process. Landfills are often located in complex terrain area. The meteorological conditions are difficult to predict, which leads to the complexity of odour prediction.

The purpose of evaluating personal exposure in the odours area, the most effective way could be the quantification of compounds received at the receptors. The odour measurement on the field is very complicated (Brandt et al. 2011). Capelli et al. (2011) indicated that dispersion models have become a common tool to simulate the dispersion of pollutants into the air. In China, the related national standard, i.e., Guidelines for environmental impact assessment-Atmospheric environment, which was issued by Ministry of Environmental Protection also recommended the models such as Aermod and Calpuff to analyse the gas migration in the range of tens of meters to tens of kilometres. Recent studies tend to use Calpuff to assess the odor effects caused by the landfill gas due to the limitations of other models (Barclay and Borissova 2013).

This paper presents the results obtained through in-situ monitoring and the numerical analysis by using Calpuff model. Calpuff model, cooperated with the local meteorological data and residents' complaints records, the area that would be influenced by odorous gas was analysed. Proper method should be adopted to mitigate the odour problem to the residential areas.

2 Materials and Methods

2.1 Site Description

The landfill site is located at the north part of Hangzhou City (120°12′E, 30°23′N). It is the first standardized valley-type sanitary landfill in China. The area of the landfill is 700,000 m^2. It was designed to contain 22 million m^3 of MSW. The site does not accept medical, industrial and hazardous waste. The daily amount of landfilled MSW increased to 5000 t/d since 2017.

As shown in Fig. 1, the total surface area of the landfill is 438,000 m^2 and comprises three parts: the working area (A), the temporary cover area (B) and the final cover area (C). The 1.5 mm thick high density polyethylene (HDPE) was used as the temporary cover material for the temporary cover area. Gas wells were installed to collect landfill gas for electricity generation. The landfill is surrounded by many residential areas and more than 10,000 people live within a 5 km radius of the landfill. The size of working area is approximately 6000 m^2.

2.2 Meteorological Information

Meteorological data including air temperature, barometric pressure, and wind velocity and air humidity were collected and stored every 1-h using a fixed meteorological station on the landfill site. During the field measurement, the prevailing wind directions in the whole monitoring periods are all Easterly–North Easterly (56.25°–78.75°). The

A: working area; B: temporary HDPE cover area; C: final cover area

Fig. 1. Satellite image of the landfill.

atmospheric stability on site refers to the date of the city, due to the lacked date on daytime insolation. The atmospheric stability class of the city are dominated by E and F classes, which accounted for 18.51% and 68.34% of that total (Liu 2013). For the stability class of E, the atmospheric condition is slightly stable. In the stability class of F, the atmospheric condition is stable.

2.3 Gas Sampling and Analysis

Sampling points were distributed on the landfill working area. The measurement numbers were arranged in accordance with US. EPA requirements, with a total of 12 sampling points for each campaign.

A static chamber and 100 ml-Tedlar bags were used to collect landfill gas from selected locations. The chamber used in this study, consisted of a glass cylinder and a steel pedestal. The chamber is 550 mm high and has a diameter of 500 mm. Once the chamber was placed on the sampling location, the groove in the steel pedestal was filled with water to keep the chamber sealed. All of the valves were then shut off to allow gas to accumulate in the chamber. Gas was sampled every 1 h for duration of 4 h. A 1-L empty aluminium bag was connected to the chamber to equilibrate inner pressure. Gas emission rate was determined by the following Eqs. (1) and (2) (Senevirathna et al. 2006).

$$VC_{t+\Delta t} = VC_t + j'S\Delta t \tag{1}$$

$$j' = \frac{V}{S}\frac{dC}{dt} \tag{2}$$

where V is the volume of the chamber (m^3); C is the gas concentration (mg/m^3); S is the area of the chamber base (m^2); j′ is the gas flux ($mg/m^2/h$); and t is time (min).

Sulphur-containing gas analysis was carried out by a gas chromatograph with a flame photometric detector (FPD, HC-3, Hubei, china shafts Technology Co., Ltd). FPD conditions were as follows: Nitrogen was used as the carrier gas at the pressure of 0.1 MPa. The pressure of hydrogen and oxygen were adjusted to 0.04 MPa to form hydrogen-rich flame. The temperature of column and detector were held at 125 °C and 100 °C, respectively. 1 mL sample from anaerobic tubes was directly injected into the injection port manually during the measurement. The detection limits is 0.05 ppm for hydrogen sulphide.

2.4 Calpuff Model

CALPUFF model was approved by the US Environmental Protection Agency (EPA) as a useful tool to determine atmospheric dispersion, especially in areas with pollution sources such as landfills. It combines three major components: CALMET, CALPUFF, and CALPOST. A detailed description of Calpuff model could be found in Scire et al. (2000).

2.4.1 Meteorological Data

The weather data used in this case were processed in the Fifth-Generation National Center for Atmospheric Research, NCAR/Penn State Mesoscale Model (MM5), by using the data re-analysis of the U.S. National Centers for Environmental Predictions (NCEP 2016). Using the MM5 as the initial guess wind field has been proved to be trustful in different researches (Wang 2013).

The domains in the MM5 model for the processing of the meteorological parameters of the study area consisted of a main domain (Domain 1) with a resolution of 1 km per point and a nested domain (Domain 2) with a resolution of 1 km per point. The integration step was 9 s.

2.4.2 Odor Threshold Regulations

The odor annoyance criterions of different countries are also rather diverse (Elbir et al. 2007). Mostly are based on 98th percentile method proposed by the Lombardia Region guidelines (Sironi and Capelli 2010). In odor diffusion modelling, very short averaging times (e.g., a few minutes) must be used (Mussio et al. 2001). Odours can be experienced intermittently for short periods. In Calpuff model, the averaged 1-h odour concentrations are converted to shorter averaging time periods based on the power law equation of the form:

$$C_s = C_h \left(\frac{T_h}{T_s}\right)^p \tag{3}$$

where C_s is the required shorter time average concentration; C_h is the hourly average concentration calculated by the Calpuff; T_h is the 1-h period expressed in minutes; T_s is the Short term period in minutes; Exponent (p): the Calpuff manual advises the use of 1/5th power relationship. Nagata and Takeuchi (2003) showed that the olfactory threshold of H_2S was 0.41 ppb (0. 62 ug/m^3). Therefore, the H_2S 1-h average limited concentration was 0.34 ug/m^3.

2.5 Operation of CALPUFF

Figure 2 illustrates this study's diagram for calpuff operation. The dimensions of the simulated area were 10 km $\times$ 10 km, with a receptor every 100 m. The map used in this study was a Universal Transverse Mercator (UTM) Reference Ellipsoid WGS-84, Global Coverage, UTM Zone 11 with coordinates 30.38°N and 120.21°E. The mesh used was 10 $\times$ 10 km and the 11 vertical levels for meteorology. The orography and land use were defined using the pre-processors 'Terrel' and 'CTGPROC'.

3 Results and Discussion

3.1 The Distribution of Complains

According to the official record, a detailed complains inventory (including date, duration and location of olfactory nuisances) is available for the whole 2016. 79 complain cases were received. The locations of the complaint cases were within 5 km away from the landfill. The odour-affected areas were intensively distributed around the landfill (see Fig. 2). The numbers denote different residential areas. In terms of distribution of complains, No. 11 and No. 12 have the largest number of complains. These two parts have 37 and 8 complains, respectively. In terms of the month of complaints, all of the complaints centralized in the second half of 2016, especially in September, October and November (Fig. 3).

3.2 Emission Rate of Odorous Gas

All dispersion models need reasonably accurate emission data to produce reliable results. The H_2S emission rate on site during the monitoring period was shown in Fig. 4. The highest emission rate in each campaign varied from 21.52 mg/m^2/h to 91.7 mg/m^2/h with its peak in September, 2016. The detected mean rates was 29.3 mg/m^2/h, 17.49 mg/m^2/h, 17.87 mg/m^2/h, 14.02 mg/m^2/h and 12.19 mg/m^2/h, respectively. Considering the extreme condition, the average emission rates of each campaign were chosen for the analysis. The size of working area was nearly 6000 m^2.

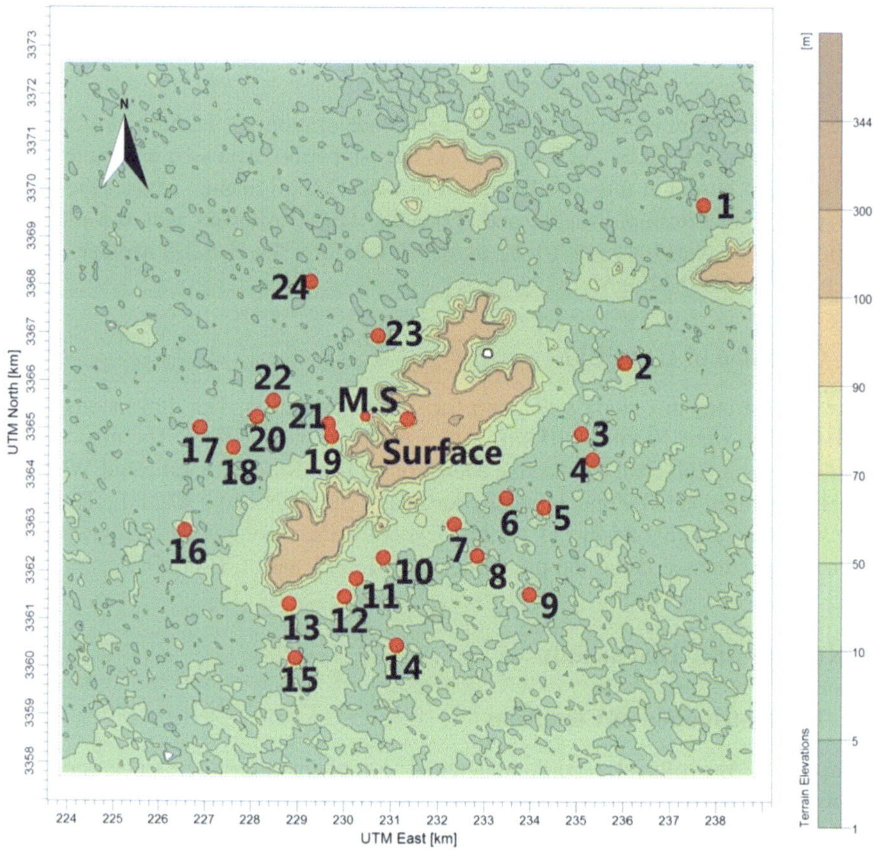

Fig. 2. The distribution of complained residential areas (The numbers denote the different position Surface denotes the landfill working area and the origin of odour emission M.S denotes the meteorological station in the landfill)

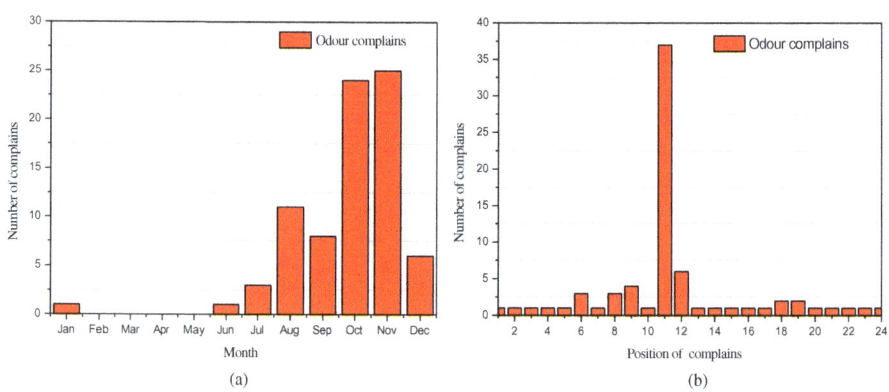

Fig. 3. The distribution of odor-related complains (a) by months (b) by positions

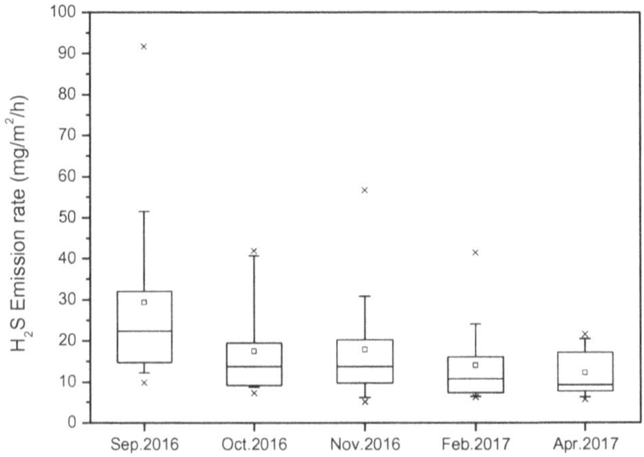

Fig. 4. The H_2S emission rates in each campaign

4 Assessment of Odour Influence by Calpuff

The calculated wind speed and direction were compared with the monitored data during September, to November 2016 (Table 1). The results indicated that MM5 model is able to capture the temporal variations in wind speed and direction at the point of the fixed meteorological station in these months. The analysis was mainly focused on these months.

Table 1. The comparison of calculated and monitored wind parameters

	Monitored prevailing wind direction	Calculated prevailing wind direction	Monitored wind speed range (m/s)	Calculated wind speed range (m/s)
September	Easterly–North Easterly (ENE)	East (E)	1.6–3.3	1.6–3.3
October	ENE	ENE	1.6–3.3	1.6–3.3
November	ENE	ENE	1.6–3.3	1.6–3.3

Assuming that the gas emitted uniformly from the surface, the mainly affected area was No. 11 according to the complaints inventory. There are several housing estates in the No. 11 area. The population there is nearly 2000. This area lies to the southwest to the landfill site and 4.2 km away.

The result indicated that the situation will be deteriorating with the high emission rate. Assuming the emission rate is 29.3 mg/m^2/h, all the selected area will be influenced by the odorous gas. The averaged concentration at No. 11 area reached 9.1 ug/m^3, which is 15 times greater than the olfactory threshold. Some improvement measures such as controlling exposed area or increasing the gas collection should be adopted to mitigate the gas emission. When the gas emission rate is reduced to the 15% of its original emission rate, the high concentration area could be control in the landfill site (Fig. 5).

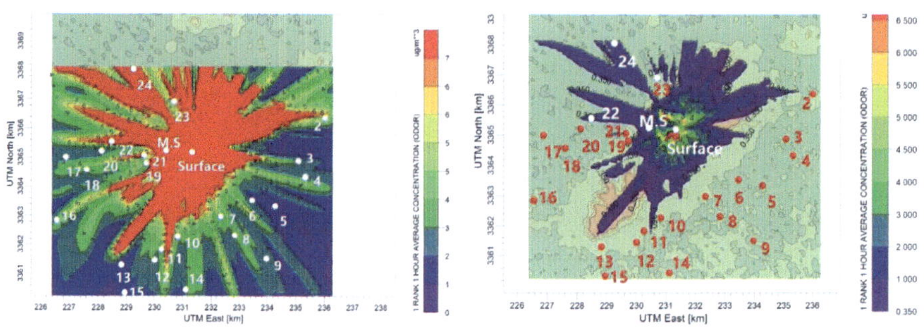

(a)H_2S-influencd area with the maximum emission rate on site

(b) H_2S-influencd area with the reduced emission rate on site

Surface denotes the landfill working area and the origin of odour emission
M.S denotes the meteorological station in the landfill

Fig. 5. The comparison of H_2S-influenced area sizes with different gas emission rates

5 Conclusions

The odorous gas, represented by H_2S was monitored and analysed in a large-scale landfill in Hangzhou, China. Calpuff model was also adopted to evaluate the odor influence to the surrounding residential areas. The main conclusions are as follows:

1. All complains were distributed within 5 km of radius to the landfill and centralized in the second half of 2016, especially in September, October and November.
2. The prevailing wind directions in September, October and November are all Easterly–North Easterly. The atmospheric stability class of city are dominated by E and F class, which accounted for 86.85% in total. It demonstrated that the atmosphere is stable in most of the time.
3. The emission rate of H_2S varied from 6.14 mg/m^2/h to 91.7 mg/m^2/h. Only when the gas emission rate is reduced to the 15% of its origin, the high concentration area could be control in the landfill site. Proper engineering measures should be conducted to limit uncontrolled gas emission. The integrity of the cover system is of great importance for controlling gas emission. It is important to restrict the size of exposure area, which could reduce the water infiltration and to increase the gas collection rate. The pretreatment methods such as MSW source amount reduction, reduction of water content and organic content of the MSW before landfilling would also be effective management methods for the decrease of gas generation rate.

References

Allen MR, Alan Braithwaite A, Hills CC (1997) Trace organic compounds in landfill gas at seven UK waste disposal sites. Environ Sci Technol 31(4):1054–1061

Henshaw P, Nicell J, Sikdar A (2006) Parameters for the assessment of odour impacts on communities. Atmos Environ 40(6):1016–1029

Ding Y, Cai CY, Hu B, Xu YE, Zheng XJ, Chen YX, Wu WX (2012) Characterization and control of odorous gases at a landfill site: a case study in Hangzhou, China. Waste Manag 32 (2):317–326

Duan ZH, Lu WJ, Li D, Wang HT (2014) Temporal variation of trace compound emission on the working surface of a landfill in Beijing. China Atmos Environ 88:230–238

Chemel C, Riesenmey C, Batton-Hubert M, Vaillant H (2012) Odour-impact assessment around a landfill site from weather-type classification, complaint inventory and numerical simulation. J Environ Manage 93(1):85–94

Brandt RC, Elliott HA, Adviento- Borbe MA, Wheeler EF, Kleinman PJ, Beegle DB (2011) Field olfactometry assessment of dairy manure land application methods. J Environ Qual 40 (2):431–437

Capelli L, Sironi S, Rosso RD, Céntola P, Rossi A, Austeri C (2011) Olfactometric approach for the evaluation of citizens' exposure to industrial emissions in the city of Terni, Italy. Sci Total Environ 409(3):595–603

Barclay JJ, Borissova M (2013) Potential problems using AERMOD to implement current odour regulations for WWTPs. Water Sci Technol 225:1932–1946

Senevirathna DGM, Achari G, Tiaratchi JPA (2006) Laboratory evaluation of errors associated with the determination of landfill gas emissions. Can J Civ Eng 33(3):240–244

Scire JS, Strimaitis DG, Yamartino RJ (2000) A user's guide for the CALPUFF dispersion model. J Nonnewton Fluid Mech 223:257

Wang XW (2013) Study on the effect of vehicle exhaust on visibility based on Calpuff model of Hangzhou. ZheJiang University (in Chinese)

Elbir T, Dincer F, Muezzinoglu A (2007) Evaluation of measured and predicted odor concentrations around a meat packaging and rendering plant. Environ Eng Sci 24(3):313–320

Sironi S, Capelli L, Céntola P, Rosso RD, Pierucci S (2010) Odour impact assessment by means of dynamic olfactometry, dispersion modelling and social participation. Atmos Environ 44 (3):354–360

Nagata Y, Takeuchi N (2003) Measurement of odor threshold by triangle odor bag method. Odor Meas Rev 118:118–127

Mussio P, Gnyp AW, Henshaw PF (2001) A fluctuating plume dispersion model for the prediction of odour-impact frequencies from continuous stationary sources. Atmos Environ 35(16):2955–2962

Balance Between Cover Resistance and Pump Capacity for Designing Vertical Gas Wells

Qi-Teng Zheng[1,2], R. Kerry Rowe[2], and Shi-Jin Feng[1(✉)]

[1] Department of Geotechnical Engineering, Tongji University, Shanghai 200092, China
fsjgly@tongji.edu.cn

[2] GeoEngineering Centre at Queen's-RMC, Department of Civil Engineering, Queen's University, Kingston K7L 3N9, Canada

Abstract. Considering exponential decreases in landfill gas (LFG) generation and gas permeability of waste with depth, a design method for vertical gas well spacing to achieve a 90% recovery rate is proposed through a quantitative analysis of over 500,000 scenarios calculated using a normalized analytical solution. Both increasing cover resistance and pump capacity can allow an increase in the well spacing while maintaining a 90% recovery rate but may result in a maximum gas pressure below the cover ranging from −5 kPa to tens of kPa. A balance between cover and pump effects is to be achieved in this paper by constraining a maximum gas pressure of 1 kPa below the cover and a recovery rate of 90%. With this goal, the design cover resistance is shown to be strongly related to LFG generation and vertical gas permeability at the top of the waste and the non-homogeneity of LFG generation with depth, but is independent of vacuum pressure and non-homogeneity of gas permeability with depth. A chart suitable for designing to meet these objectives is proposed and illustrated by a practical application.

Keywords: Landfill · Waste non-homogeneity · Vertical gas well
Maximum cover-gas pressure · Design

1 Introduction

After the landfill closure, vertical wells are the most effective landfill gas (LFG) collection method [1]. The recovery efficiency of a series of gas wells depends primarily on waste properties, cover properties and well design parameters. The waste properties include LFG generation rate, gas permeability and waste non-homogeneity with depth. Waste non-homogeneity plays a particularly important role in LFG recovery [2]. However, it is difficult to consider the waste non-homogeneity in the design and few studies focus on this question.

In [3], a two-dimensional axisymmetric analytical model was developed to describe the LFG flow around a vertical well by defining exponential decreases of gas permeability and LFG generation rate of waste with depth as:

L. Zhan et al. (Eds.): ICEG 2018, ESE, pp. 60–67, 2019.
https://doi.org/10.1007/978-981-13-2224-2_8

$$K_{GV}^{*}(z) = K_{GV0}^{*}10^{-a_1 z^*}, \quad s(z) = s_0 e^{-a_2 z^*} \tag{1}$$

where terms marked with an asterisk are dimensionless, K_{GV} is the vertical gas permeability of waste ($K_{GV}^{*} = K_{GV}/H^2$); s is the LFG generation rate per bulk volume (kg m^{-3} s^{-1}); z is the depth ($z^* = z/H$); and, H is the waste thickness (m). K_{GV0}^{*} and s_0 refer to K_{GV}^{*} and s of the top waste (at $z^* = 0$). a_1 and a_2 represent the waste non-homogeneity in gas permeability and LFG generation with depth. Using this solution, over 560,000 scenarios were modelled combining typical values of waste properties, cover properties and well parameters. Through a quantitative study, Zheng et al. [3] concluded that most of the parameters could be integrated together and the recovery response of a vertical gas well could be captured by:

- *Cover resistance parameter* $NP_1 = h_c^* K_{GV0}^* / K_c^* (0 \leq NP_1 \leq 10^3)$, defining the ability of cover to resist LFG emission;
- *Pump capacity parameter* $NP_2 = -p_w P_{atm}/A_0$ ($0 \leq NP_2 \leq 5 \times 10^4$), defining the capacity of vacuum pressure to collect LFG;
- *Well spacing parameter* $NP_3 = -\left(\log_e(r_w^*) + 1\right)/k^*$ ($0.05 \leq NP_3 \leq 9$), defining the coupled effect of well radius and horizontal spacing;
- *Permeability decay constant* a_1 ($0 \leq a_1 \leq 4$), defining the decrease in gas permeability of waste with depth;
- *Generation decay constant* a_2 ($0 \leq a_2 \leq 40$), defining the decrease in LFG generation of waste with depth;

where h_c is the cover thickness ($h_c^* = h_c/H$); K_c is the gas permeability of cover ($K_c^* = K_c/H^2$); p_w is the vacuum pressure at the well (kPa); P_{atm} is the atmospheric pressure (nominally 101.3 kPa); r_w is the well radius ($r_w^* = r_w/R$); R is half the vertical well spacing (m); and, k is the degree of waste anisotropy ($k^* = kH^2/R^2$). The *top waste parameter* A_0 (kPa^2) is expressed as

$$A_0 = \mu R_G T s_0 / K_{GV0}^{*} \tag{2}$$

where μ is the LFG dynamic viscosity (kg m^{-1} s^{-1}); R_G is the LFG constant (J kg^{-1} K^{-1}); T is the absolute temperature in Kelvin (K).

To conservatively design a vertical well field that will achieve 90% recovery, the *well spacing parameter* (NP_3) was examined as a simple function of *cover resistance* (NP_1) and *pump capacity* (NP_2) *parameters* in [4]. The function is the sum of four parts: minimum well spacing parameter, cover term with passive collection, pump term for an uncovered situation, and term that addresses interaction between cover and pump systems. Five coefficients in the function can be estimated from the *permeability* and *generation decay constants* (a_1 and a_2). The proposed method is a means of designing a vertical well with consideration of the waste non-homogeneity. However, these studies raised another question: how to balance cover (NP_1) and pump effects (NP_2) for optimum design?

The first objective of this paper is to investigate the effects of key variables on the *maximum gas pressure directly below the cover*, called as *maximum cover-gas pressure*. The second objective is to use a critical *maximum cover-gas pressure* to balance cover and pump effects while maintaining 90% recovery. The final objective is to propose a simple method to design the *cover resistance parameters*.

2 Critical Maximum Cover-Gas Pressure

As noted in [4], the *well spacing parameter* NP_3 increases with an increase in *cover resistance* (NP_1) and *pump capacity* (NP_2) *parameters* at a *gas recovery rate* (*GRR*) of 90%. For a specific NP_3, various combinations of NP_1 and NP_2 values can achieve a 90% recovery rate, but the corresponding *maximum cover-gas pressure*, P_{CG} (relative pressure), ranges from approximately −5 kPa to tens of kPa and even hundreds of kPa. A P_{CG} value less than 0 kPa indicates an unreasonably high vacuum pressure and an uneconomical design, while a P_{CG} larger than 10 kPa may cause LFG explosion and decrease landfill cover stability. Thus, the *maximum cover-gas pressure* can be used to further define the relationship between cover resistance and pump capacity.

A slide failure caused by an average gas pressure of 1.7 kPa underneath the final cover was reported by [5], thus the critical *maximum cover-gas pressure* was conservatively assumed as 1 kPa. Above a critical (minimum) value of A_0, the *maximum cover-gas pressure* P_{CG} increases with increasing *top waste parameter* A_0 and decreasing vacuum pressure p_w (Fig. 1). Greater cover resistance (i.e., a higher value of NP_1) generally increases P_{CG}, however, an opposite is true for a relatively high vacuum pressure at low A_0 values ($p_w = -2$ kPa in Fig. 1). The relationship between A_0 and P_{CG} was fitted by a second order polynomial to identify values of A_0 for a combination of a_1, a_2, NP_1, p_w, and NP_3 at $P_{CG} = 1$ kPa (solid circles in Fig. 1). Both a passive and an active collection scenario ($p_w = 0$ and −2 kPa) are studied in the following.

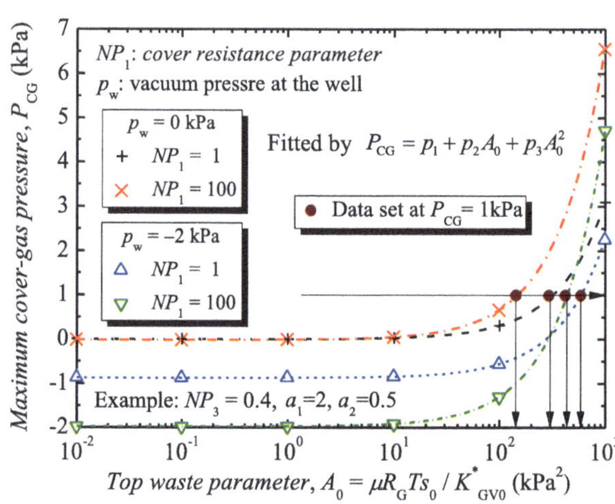

Fig. 1. Relationship between *maximum cover-gas pressure*, P_{CG}, and *top waste parameter*, A_0, identifying values of A_0 for two values of *cover resistance parameter*, NP_1, and vacuum pressure at the well, p_w, ($a_1 = 2$, $a_2 = 0.5$, $NP_3 = 0.4$) for $P_{CG} = 1$ kPa.

3 Result Analysis

3.1 Passive Collection Scenario

For the passive collection ($p_w = 0$ kPa), the *well spacing parameter* NP_3 for 90% recovery ($GRR = 90\%$) linearly increases with an increase in cover resistance (i.e., NP_1) but remains unchanged with *top waste parameter* A_0 (dashed lines in Fig. 2) [4]. Together with the data set at $P_{CG} = 1$ kPa (Fig. 1), a modified data set for both $P_{CG} = 1$ kPa and $GRR = 90\%$ was captured at the interaction points (solid circles in Fig. 2). The new data set, used in Fig. 3, includes four independent variables, i.e., A_0, a_1, a_2 and NP_1, and NP_1 is an important design parameter while NP_3 can be calculated from NP_1 and A_0.

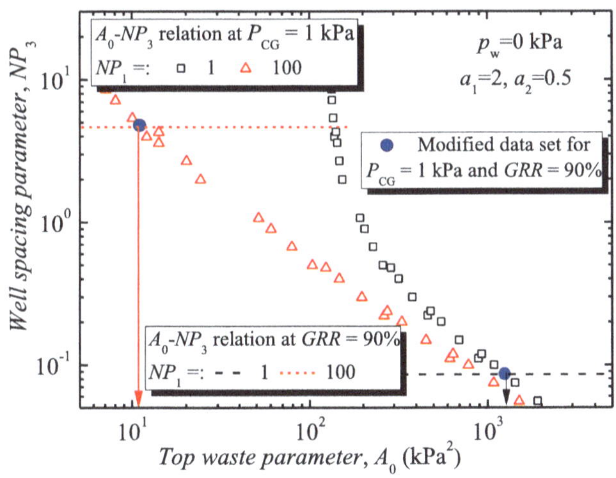

Fig. 2. Relationship between *well spacing parameter*, NP_3, and *top waste parameter*, A_0, identifying values of A_0 for two values of *cover resistance parameter*, NP_1, ($a_1 = 2$, $a_2 = 0.5$, $p_w = 0$ kPa) for $P_{CG} = 1$ kPa and $GRR = 90\%$.

The value of the *cover resistance parameter* NP_1 at $P_{CG} = 1$ kPa and $GRR = 90\%$, designated as the *design cover resistance parameter* NP_1, decreases with an increase in *top waste parameter* A_0 (e.g., with a higher generation rate s_0 or lower vertical gas permeability K^*_{GV0}, at the top of the waste) and a decrease in *generation decay constant* a_2 (Fig. 3). The decrease in the *design* NP_1 is to decrease the gas pressure below a low-permeability cover caused by more generated LFG. The *design* NP_1 was almost independent of the *permeability decay constant* a_1, indicating little effect of the decreasing gas permeability of waste with depth on the *maximum cover-gas pressure*. Therefore, the cover resistance is limited by a *maximum cover-gas pressure* of 1 kPa. A design chart will be provided after discussing the active collection.

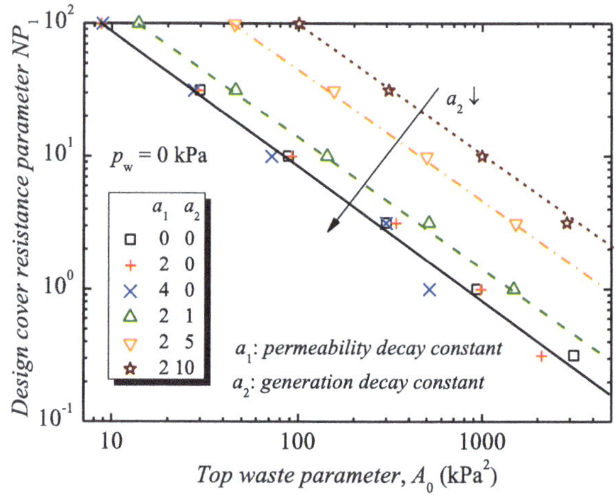

Fig. 3. Relationship between *design cover resistance parameter* NP_1 and *top waste parameter*, A_0, for different values of *permeability* and *generation decay constants*, a_1 and a_2 ($p_w = 0$ kPa).

3.2 Active Collection Scenario

Considering a vacuum pressure at the well of $p_w = -2$ kPa to represent active collection and following the similar steps as in passive collection scenario just examined, a modified data set of [*pump capacity parameter* NP_2 ($NP_2 = -p_w P_{atm}/A_0$), *permeability decay constant* a_1, *generation decay constant* a_2, *cover resistance parameter* NP_1] was captured for $P_{CG} = 1$ kPa and a target 90% collection, like solid circles in Fig. 4. The corresponding *well spacing parameter* NP_3 can be calculated from NP_1 and NP_2.

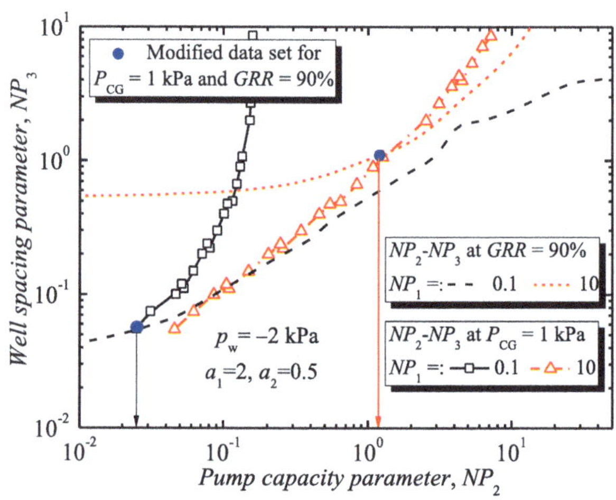

Fig. 4. Relationship between *well spacing parameter*, NP_3, and *pump capacity parameter*, NP_2, identifying values of NP_2 for two values of *cover resistance parameter*, NP_1, ($a_1 = 2$, $a_2 = 0.5$, $p_w = -2$ kPa) for $P_{CG} = 1$ kPa and $GRR = 90\%$.

With the constraints of a *maximum cover-gas pressure* P_{CG} of 1 kPa and *gas recovery rate GRR* of 90%, the *design cover resistance* NP_1 values at $p_w = -2$ kPa almost overlap with those at $p_w = 0$ kPa (Fig. 5) indicating that the *design* NP_1 is independence of vacuum pressure p_w. As the vacuum pressure increases, P_{CG} would decrease if there were no other changes (i.e. all other things being equal) but to maintain $GRR = 90\%$, the well spacing could increase with P_{CG} remaining essentially unchanged.

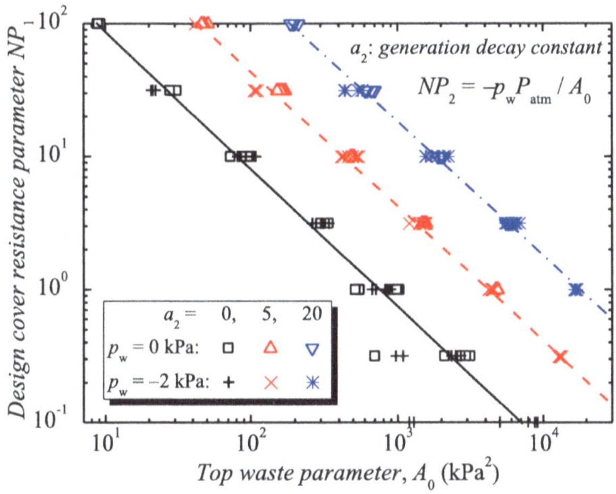

Fig. 5. Comparison of relationships between *design cover resistance parameter* NP_1 and *top waste parameter*, A_0, for three values of *generation decay constant*, a_2, at passive ($p_w = 0$ kPa) and active collection ($p_w = -2$ kPa)

A design chart for both passive and active collection scenarios can be proposed to estimate the upper limit for *cover resistance parameter* NP_1 by constraining $P_{CG} = 1$ kPa and $GRR = 90\%$ (Fig. 6). Curves were fitted to the computer generated data in Fig. 6 with an R-Squared greater than 0.95 and 97% of the fitted data within the range $[\log_{10}(NP_1) - 0.2]$ to $[\log_{10}(NP_1) + 0.2]$.

3.3 Case Application

To illustrate the application of the design chart, consider Area A/B cell at the Central Solid Waste Management Center landfill, Sandtown, Delaware. The landfill construction and parameter estimation were described in detail by [3]. The essential parameters are: waste thickness $H = 20$ m; vertical gas permeability of the top waste $K_{GV0} = 7.9 \times 10^{-12}$ m^2, giving $K^*_{GV0} = K_{GV0}/H^2 = 2 \times 10^{-14}$; LFG generation rate of the top waste $s_0 = 6.3 \times 10^{-7}$ kg m^{-3} s^{-1}; temperature in the landfill $T = 40$ °C (313 K), $\mu R_G T = 1.19$ kg m s^{-3}; *permeability decay constant* $a_1 = 3.0$; *generation decay constant* $a_2 = 0.35$. These lead to a *top waste parameter* $A_0 = \mu R_G T s_0 / K^*_{GV0} = 37.5$ kPa^2.

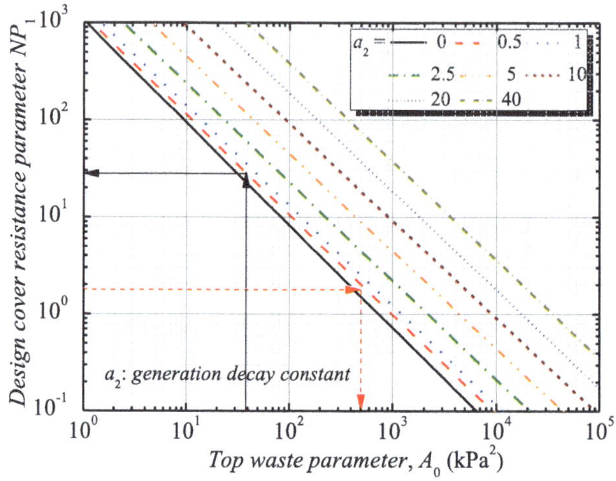

Fig. 6. Design chart for *cover resistance parameter*, NP_1, at P_{CG} = 1 kPa and *GRR* = 90%

With $a_2 = 0.35$ and $A_0 = 37.5$ kPa^2, the *design cover resistance parameter* NP_1 was estimated as 28 following the solid arrow in Fig. 6. With $K_c = 1 \times 10^{-14}$ m^2 this gives a cover thickness $h_c = H \times NP_1 K_c / K_{GV0} = 70$ cm. This high value for NP_1 is caused by the low LFG generation rate and high gas permeability of the waste. In this case, the high NP_1 corresponds to a low risk of high gas pressure beneath the cover. Figure 6 can also be used to estimate the *maximum cover-gas pressure* P_{CG} for different NP_1 values. For example, a minimum cover thickness of about 45 cm ($NP_1 = 1.8$ if $K_c = 1 \times 10^{-13}$ m^2) is required for closure of non-hazardous landfills according to Resource Conservation and Recovery Act (RCRA) Subtitle D. To achieve $P_{CG} = 1$ kPa and following the dashed arrow in Fig. 6, $NP_1 = 1.8$ corresponds to $A_0 = 500$ kPa^2. Based on the second order polynomial for the A_0 - P_{CG} relationship in Fig. 1 ($P_{CG} = p_1 + p_2 A_0 + p_3 A_0^2$), the *maximum cover-gas pressure* P_{CG} at $NP_1 = 1.8$ and $A_0 = 37.5$ kPa^2 lies in the range of $1 \times 37.5^2/500^2 = 0.006$ kPa to $1 \times 37.5/500 = 0.08$ kPa for $p_w = 0$ kPa ($p_1 = 0$). This P_{CG} range would further decrease with an increase in vacuum pressure.

Together with an active collection system, the Area A/B was eventually covered by sandy soil with a thickness of 43 cm [6], approximately equal to a cover resistance of $NP_1 = 1.8$. From the above studies, the *maximum cover-gas pressure* P_{CG} in this situation is close to atmospheric pressure. In this case, with vacuum pressure $p_w = -2$ kPa and well radius $r_w = 0.3$ m, the well spacing for 90% recovery was estimated as 36 m using the function proposed by [4]:

$$NP_3 = cf_1 + cf_2 NP_1 + cf_3 NP_2^{cf_4} + cf_5 NP_1 NP_2^2 \quad (3)$$

where cf_1, cf_2, cf_3, cf_4 and cf_5 are coefficients affected by *permeability* and *generation decay constants*, a_1 and a_2. This design is conservative but uneconomical, and a thicker cover with lower permeability (e.g., $NP_1 = 28$) could be used to increase the well spacing for 90% recovery to about 55 m for $p_w = -2$ kPa and $r_w = 0.3$ m, and to balance the pump and cover effects through having a P_{CG} value of 1 kPa. A value of

$NP_1 = 28$ corresponds to a 70 cm thick cover with an equivalent gas permeability of 1×10^{-14} m^2. Therefore, the design for a cover system can be optimized using Fig. 6, followed by an optimized well spacing for 90% recovery. Note that the *design cover resistance parameter* NP_1 estimated from Fig. 6 is treated as an upper limit with $P_{CG} = 1$ kPa and if a lower NP_1 was preferred, the corresponding *maximum cover-gas pressure* could also be estimated using this approach.

4 Conclusion

This study investigated the changes in maximum gas pressure directly below the cover as a function of cover properties, vacuum pressure, well spacing and waste non-homogeneity. A critical maximum gas pressure of 1 kPa was set below the cover to balance the effects of cover properties and vacuum pressure for 90% LFG recovery. The cover properties to meet these objectives were shown to be strongly influenced by the LFG generation and gas permeability of the top waste, and the non-homogeneity of LFG generation with depth, but to be essentially independent of vacuum pressure or non-homogeneity of gas permeability with depth. A design chart for cover properties was proposed and illustrated by a practical application.

References

1. Townsend TG, Wise WR, Jain P (2005) One-dimensional gas flow model for horizontal gas collection systems at municipal solid waste landfills. J Environ Eng 131:1716–1723
2. Tinet AJ, Oxarango L (2010) Stationary gas flow to a vertical extraction well in MSW landfill considering the effect of mechanical settlement on hydraulic properties. Chem Eng Sci 65:6229–6237
3. Zheng Q-T, Rowe RK, Feng S-J (2018) Recovery response of vertical gas wells in non-homogeneous landfills (under review)
4. Zheng Q-T, Rowe RK, Feng S-J (2018) Design of vertical gas collection wells in non-homogeneous landfills (under review)
5. Thiel R (1999) Design of a gas pressure relief layer below a geomembrane cover to improve slope stability. Proc Geosynth 99:235–251
6. Delaware Solid Waste Authority (2006) Application for major modification to permit SW-03/01 for Central Solid Waste Management Center Sandtown, Delaware

A New Consolidation Model for Unsaturated High-Kitchen-Waste-Content MSW

Ke Li, Wenjie Xu(✉), and Yunmin Chen

MOE Key Laboratory of Soft Soils and Geoenvironmental Engineering, Zhejiang University, Hangzhou 310058, China
wenjiexu@zju.edu.cn

Abstract. Kitchen waste refers to waste which is degradable and with high content of organic matter such as lettuce, flesh, peels, etc. Kitchen waste is more easily degraded than non-kitchen waste and produces a greater amount of leachate and landfill gas. It is vital to study the consolidation of municipal solid waste landfills with high kitchen waste content. A new consolidation model for unsaturated high-kitchen-waste-content MSW is proposed after summarizing the consolidation model for unsaturated porous media and high-kitchen-waste landfill.

The model takes pore gas pressure and matrix suction as variables. The model is based on hydro-mechanics coupled model for unsaturated porous media, taking bulk modulus of solid particles in waste into account. The porosity and relative permeability of waste change during loading process. A typical example is simulated using OpenGeoSys software to study the influence of age, bulk modulus, permeability on consolidation of municipal solid waste. The model is instructive.

Keywords: Consolidation · Municipal solid waste · Unsaturated

1 Introduction

It is vital to study the consolidation of municipal solid waste (MSW) landfills for the following reasons: (1) MSW is a high-water-content and high-organic-content unsaturated soil composed of solid, liquid and gas phases. Its mechanical properties changes overtime. (2) The research is instructive to risk management because seepage process of MSW is complex due to increasing applied load and degradation. (3) Accurate consolidation model is required to calculate storage capacity by calculating settlement.

Kitchen waste refers to waste degradable and with high content of organic matter such as lettuce, flesh, peels, etc. Kitchen waste is more easily degraded than non-kitchen waste so that produces more leachate and landfill gas. Landfills in China have a high content of kitchen waste. In Europe and United States, the content of kitchen waste in landfills is less than 30%. In contrast, in China, this content is about 40% [1] in the north and 70% [2] in the south. There is a great demand for research on consolidation models of high-kitchen-waste-content MSW.

The research on the consolidation of high-kitchen-content landfills is the research on the consolidation of unsaturated porous media. Terzaghi [3] established the theory of one-dimensional consolidation of saturated soils. Biot [4] developed a three-dimensional

L. Zhan et al. (Eds.): ICEG 2018, ESE, pp. 68–76, 2019.
https://doi.org/10.1007/978-981-13-2224-2_9

model. It differs from Terzaghi's model in that pore water pressure in Biot's model is multiplied by Biot coefficient to correct its effect. Biot, Bishop and Skempton [5], and Suklje [6] all defined the express of Biot coefficient. Jennings [7], Croney [8], Bishop [9], Aitchison [10] have improved and extended the concept of Biot effective stress to unsaturated porous media of two-phase or multiphase. Bishop's effective stress is written as $\sigma'_{ij} = (\sigma_{ij} - u_a) + \chi(u_a - u_w)$, where u_a is pore air pressure, u_w is pore water pressure, χ is Bishop coefficient. The more common one is Bolzon's [11] equation as $\sigma'_{ij} = \sigma_{ij} - [S_r u_w + (1 - S_r)u_a]\delta_{ij}$, where S_r is relative saturation.

The hydro-mechanics coupled governing equations in unsaturated soil could be derived involving the expression of effective stress, seepage and the constitutive equation. Based on Richard's equation [12], a fluid-solid coupled model takes the effect of transient porous media deformation on fluid mass into account. Richard equation is written as $\rho_a n \cdot \partial s_a/\partial t + s_a n c_a \cdot \partial p_a/\partial t - \nabla[(\nabla p_a - \rho_a \mathbf{g}) \cdot \rho_a k^a_{rel}\mathbf{k}/\mu_a] = 0$, where subscript a represents a fluid phase, ρ_a is fluid density in porous media, s_a is fluid saturation in porous media, n is porosity of porous media, the first two terms of equation come from the expand of $\partial(\rho_a n s_a)/\partial t$ and $c_a \cdot \partial p_a = \partial \rho_a/\partial t$. The hydro-mechanics coupled models assume $\rho_a s_a \cdot \partial n/\partial t$ as non-zero, and eliminates item with $\partial n/\partial t$ by add the solid deformation equation and Richard's equation. Lewis [13], Schrefler [14], Rutqvist [15] have proposed hydro-mechanics coupled equations.

Models for high-kitchen-waste-content landfills are as follows: Chen's [16] solid-liquid-gas coupled model is for waste with macro pores. Matrix suction is zero due to macro pores. The model incorporated Gibson and Lo's biodegradation and creep model. Guo's [17] model for saturated MSW considers the solid and pore water volume change due to degradation and intracellular water release, and effect of degradation on the compression attenuation and its influence on excess pore pressure. Guo takes Chen's compression model [18] which considers secondary consolidation. Liu's [19] model is gas-mechanical coupled. Dissipation of pore water pressure is much faster than that of pore gas pressure, so the coupling of water is ignored. Liu [20] proposed a coupled model of degradation-consolidation-solute-transport, considering the effect of degradation on compressibility and pore volume and changes in permeability caused by changes in pore volume, without considering changes in pore volume caused by applied loads. None of the model above considers the compression of solid particles. Solid particles in high-kitchen-waste-content MSW have nonnegligible compressibility. Traditional consolidation model would lead to a bias in stress distribution calculation.

In this paper, a new consolidation model for unsaturated high-kitchen-waste-content MSW is proposed. The model is based on consolidation model for unsaturated porous media, hydro-mechanics coupled with considering the compressibility of solid particles. A typical example is simulated using OpenGeoSys software to study the influence of age, bulk modulus, and permeability on consolidation of municipal solid waste.

2 A New Consolidation Model for Unsaturated High-Kitchen-Waste-Content MSW

The hydro-mechanics coupled model proposed by Lewis was applied to the consolidation of unsaturated high-kitchen-waste MSW. The basic assumptions are as follows: (1) Waste is homogeneous, isotropic, and not completely saturated; (2) Solid particles, pore water and pore gas are all compressible; (3) The additional stress in the waste is infinitely evenly distributed along the horizontal plane, so the compression of the soil layer and the water seepage are one-dimensional; (4) Fluid seepage in the waste obeys Darcy's law; (5) Both the relative permeability coefficient and the compression coefficient of the waste change during consolidation process. Pore gas pressure p^g and matrix suction p^c are basic variables.

Governing equation for water:

$$S_w\left(\frac{n}{K_w}+\frac{\alpha-n}{K_s}\right)\frac{\mathrm{d}p^g}{\mathrm{d}t}-\left[\frac{n}{K_w}S_w+\frac{\alpha-n}{K_s}S_w^2+\left(n-\frac{\alpha-n}{K_s}S_wp^c\right)\frac{\mathrm{d}S_w}{\mathrm{d}p^c}\right]\frac{\mathrm{d}p^c}{\mathrm{d}t}$$
$$+\nabla\cdot[-\frac{\mathbf{k}k^{rw}}{\mu^w}(\nabla p^g-\nabla p^c+\rho^w g\nabla z)]+\alpha S_w\nabla\cdot\mathbf{v_s}=0$$

Governing equation for gas:

$$(1-S_w)\left(\frac{n}{K_g}+\frac{\alpha-n}{K_s}\right)\frac{\mathrm{d}p^g}{\mathrm{d}t}+\left\{\frac{\alpha-n}{K_s}(1-S_w)S_w-\left[n+\frac{\alpha-n}{K_s}(1-S_w)p^c\right]\frac{\mathrm{d}S_w}{\mathrm{d}p^c}\right\}\frac{\mathrm{d}p^c}{\mathrm{d}t}$$
$$+\nabla\cdot\left[-\frac{\mathbf{k}k^{rg}}{\mu^g}(\nabla p^g+\rho^g g\nabla z)\right]+\alpha(1-S_w)\nabla\cdot\mathbf{v_s}=0$$

where S_w is saturation of water, n is porosity, α is Biot coefficient, K_w, K_g and K_s are water, gas and solid bulk moduli (kPa), p^g and p^c are pore gas pressure and matrix suction (kPa), $\mathbf{v_s}$ is velocity of solid skeleton ($\mathrm{m\cdot s^{-1}}$), $\mathbf{k}$ is intrinsic permeability ($\mathrm{m^2}$), k^{rw} and k^{rg} are relative water permeabilities, μ^w and μ^g are water and gas viscosities ($\mathrm{kg\cdot m^{-1}s^{-1}}$), ρ^w and ρ^g are water and gas densities ($\mathrm{kg\cdot m^{-3}}$).

The derivation process is as follows: Considering the conversion of the material derivative and time differential of the variable f_π in the π phase reference system as $\mathrm{d}^\pi f_\pi/\mathrm{d}t=\partial f_\pi/\partial t+\nabla f_\pi\cdot\mathbf{v}^\pi$, where $\mathbf{v}^\pi$ is the velocity of π phase. In conjunction with Richard's equations, the continuity equations for solids and fluids (for pore water as an example) can be obtained:

$$\frac{1-n}{\rho^w}\frac{\mathrm{d}^s\rho^s}{\mathrm{d}t}-\frac{\mathrm{d}^s n}{\mathrm{d}t}+(1-n)\nabla\cdot\mathbf{v}^s=0$$
$$\frac{n}{\rho^w}\frac{\mathrm{d}^s\rho^w}{\mathrm{d}t}+\frac{\mathrm{d}^s n}{\mathrm{d}t}+\frac{n}{S_w}\frac{\mathrm{d}^s S_w}{\mathrm{d}t}+\frac{1}{S_w\rho^w}\nabla\cdot(\rho^w\frac{\mathbf{k}k^{rw}}{\mu^w}(-\nabla p^w+\rho^w g))+n\nabla\cdot\mathbf{v}^s=0$$

Add equations above to get:

$$\frac{n}{\rho^w}\frac{\mathrm{d}^s\rho^w}{\mathrm{d}t} + \frac{1-n}{\rho^w}\frac{\mathrm{d}^s\rho^s}{\mathrm{d}t} + \frac{n}{S_w}\frac{\mathrm{d}^s S_w}{\mathrm{d}t} + \frac{1}{S_w\rho^w}\nabla\cdot(nS_w\rho^w\mathbf{v}^{\mathbf{vs}}) + \nabla\cdot\mathbf{v}^{\mathbf{s}} = 0$$

The material derivatives of fluid density, solid density, and saturation were expanded respectively, and the pore water pressure was expressed as the difference between the pore gas pressure and the matrix suction. Finally, the governing equation can be written as:

$$S_w\left(\frac{n}{K_w}+\frac{\alpha-n}{K_s}\right)\frac{\mathrm{d}p^g}{\mathrm{d}t} - \left[\frac{n}{K_w}S_w + \frac{\alpha-n}{K_s}S_w^2 + \left(n - \frac{\alpha-n}{K_s}S_w p^c\right)\frac{\mathrm{d}S_w}{\mathrm{d}p^c}\right]\frac{\mathrm{d}p^c}{\mathrm{d}t}$$
$$+\nabla\cdot\left[-\frac{\mathbf{k}k^{rw}}{\mu^w}(\nabla p^g - \nabla p^c + \rho^w g\nabla z)\right] + \alpha S_w\nabla\cdot\mathbf{v_s} = 0$$

According to Zienkiewicz's theory of effect of hydrostatic pressure on the deformable solids, the material derivative of solid density is expanded to a formula including Biot coefficient and bulk modulus of solid particle.

$$\frac{1}{\rho^s}\frac{\mathrm{d}^s\rho^s}{\mathrm{d}t} = \left[\frac{\alpha-n}{K_s}\frac{d^s p^s}{dt} + (\alpha-1)\nabla\cdot\mathbf{v}^{\mathbf{s}}\right]$$

According to Bolzon's equation of effective stress, equivalent effective stress p^s is written as $p^s = S_w(p^c - p^g) + (1 - S_w)p^g$, thus the above control equation can be obtained.

3 Case Studies

3.1 Model Set Up

The high-kitchen-waste-content MSW column is 10 m high and 0.1 m wide (i.e., one dimension). The initial water level of MSW is 6 m high because it represents the general water level of many landfills. Both the bottom boundary and the lateral boundary are impervious and immovable. The top boundary is free to move. The initial state of the soil column is in static equilibrium. The top boundary of the soil column is undergoing one-year-long uniform loading to 39.5 kPa finally (see Fig. 1). A model was built in OpenGeoSys software (see Fig. 2) and it was divided into 100 cell units that were 0.1 meters high and 0.1 m wide. It contains 202 nodes and 4 boundary lines.

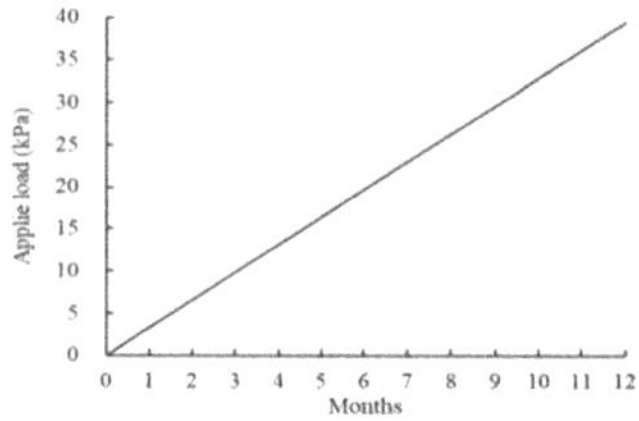

Fig. 1. Load-time curve model of MSW

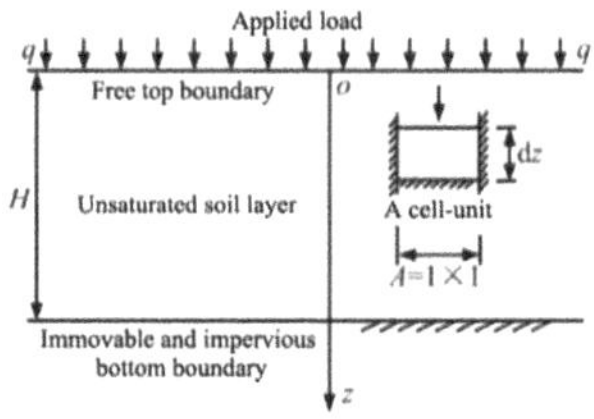

Fig. 2. Model of MSW

3.2 Case Set Up

Five cases were set up. Three cases were normal cases of low, middle and high ages. Two cases were dangerous cases (extreme permeability or bulk modulus of solid particles) based on high age. Parameters of MSW of all cases are shown in Table 1. Case 1–3 are normal MSW of low age, middle age and high age, respectively. Case 4 is MSW of high age with extreme intrinsic permeability. Case 5 is MSW of high age with extreme bulk modulus. E is elastic modulus, α is Biot coefficient, K_s is bulk modulus of solid particles, ϕ is porosity, **k** is intrinsic permeability, $S_{\max}$ and $S_{\min}$ are maximum and minimum saturation, α_v, n_v, and m_v are parameters for vG-M model.

Table 1. Parameters of MSW of all cases

Case 1	Loading rate (Pa/t)		E (kPa)	α	K_s (kPa)	ϕ
	1.25E−3		185.8	0.822	1159.7	0.730
	k (m^2)	$S_{\max}$	$S_{\min}$	α_v	n_v	m_v
	7.84E−11	0.987	0.233	0.88	1.58	0.36
Case 2	Loading rate (Pa/t)		E (kPa)	α	K_s (kPa)	ϕ
	1.25E−3		294.1	0.790	1556.3	0.688
	k (m^2)	$S_{\max}$	$S_{\min}$	α_v	n_v	m_v
	1.11E−13	0.959	0.390	0.45	1.44	0.30
Case3	Loading rate (Pa/t)		E (kPa)	α	K_s (kPa)	ϕ
	1.25E−3		519.2	0.706	1962.1	0.615
	k (m^2)	$S_{\max}$	$S_{\min}$	α_v	n_v	m_v
	1.11E−13	0.959	0.390	0.45	1.44	0.30
Case 4	Loading rate (Pa/t)		E (kPa)	α	K_s (kPa)	ϕ
	1.25E−3		519.2	0.706	1159.7	0.550
	k (m^2)	$S_{\max}$	$S_{\min}$	α_v	n_v	m_v
	1.99E−15	0.959	0.390	0.45	1.44	0.30
Case 5	Loading rate (Pa/t)		E (kPa)	α	K_s (kPa)	ϕ
	1.25E−3		1240.6	0.706	4909.0	0.615
	k (m^2)	$S_{\max}$	$S_{\min}$	α_v	n_v	m_v
	1.11E−13	0.959	0.390	0.45	1.44	0.30

3.3 Numerical

Figures 3, 4 and 5 show the change over time of the excess pore pressure in case 1 to 3. For MSW near the top, the excess pore pressure keeps increasing in all cases. For MSW at the bottom, the excess pore pressure stables at low values in case 1 and case 2, while keep increasing in case 3.

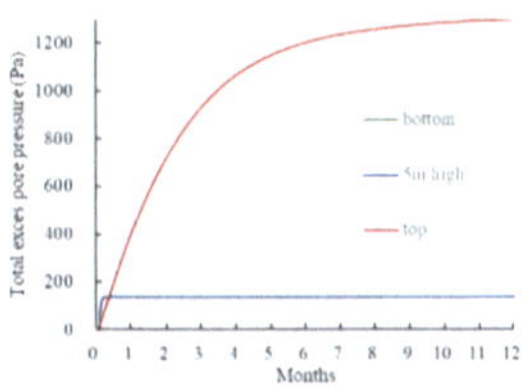

Fig. 3. Total excess pore pressure-time curve of case 1

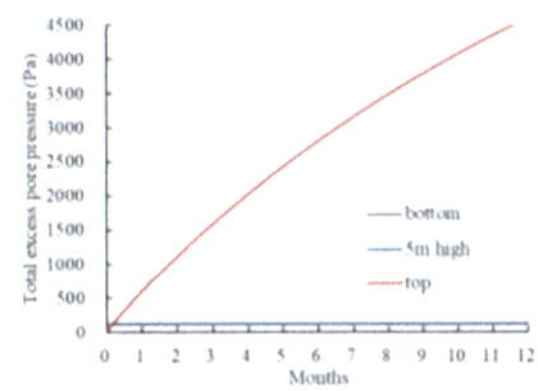

Fig. 4. Total excess pore pressure-time curve of case 2

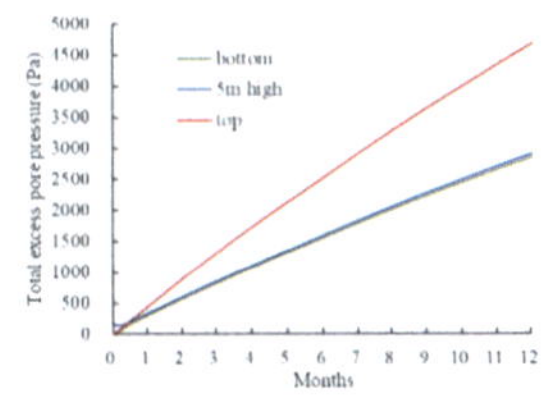

Fig. 5. Total excess pore pressure-time curve of case 3

Figures 6, 7, 8, 9, 10 and 11 show the change over time of the excess pore water and gas pressure in case 1–3. Gas pressure is only shown in the height of 6–10 m for it equals to the water pressure below 6 m. The excess pore water pressure is low in the saturation region and has little correlation with depth but higher in the unsaturated region and the closer to the top boundary the larger. The excess pore gas pressure for 3 cases are similar.

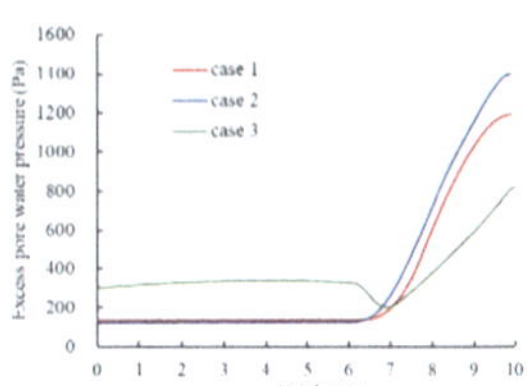

Fig. 6. Excess pore water pressure of case 1 2 3-height curve of day 30

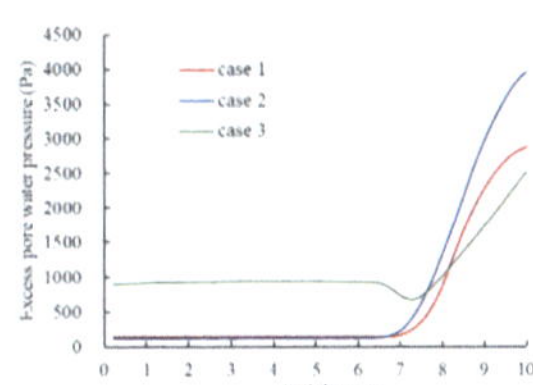

Fig. 7. Excess pore water pressure of case 1 2 3-height curve of day 100

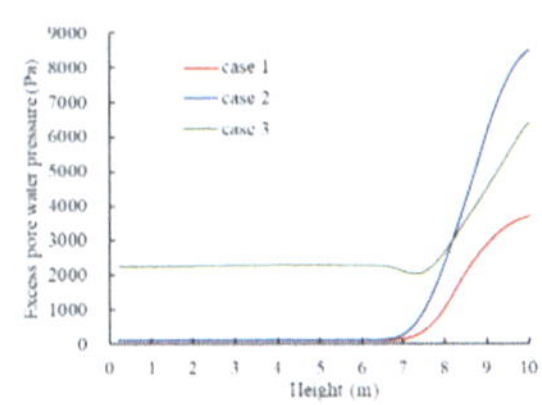

Fig. 8. Excess pore water pressure of case 1 2 3-height curve of day 270

The change in the excess pore pressure with time and space for case 4 is shown in Figs. 12, 13 and 14. The excess pore water pressure distribution is more relative with the depth in the saturated region. The excess pore gas pressure dissipates more slowly and continues to accumulate. Compared to case 3, the excess pore water and gas pressure in the saturated region are smaller; the excess pore water pressure in the unsaturation region has little difference but the excess pore gas pressure has great difference.

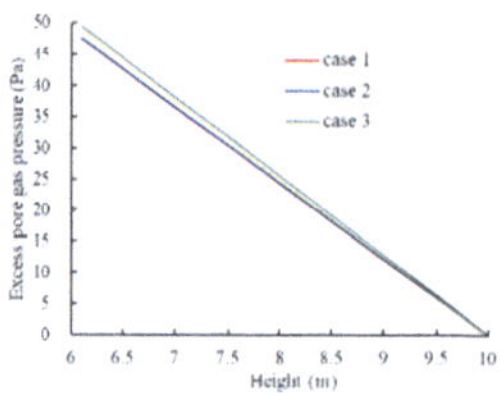

Fig. 9. Excess pore gas pressure of case 1&2&3-height curve of day 30

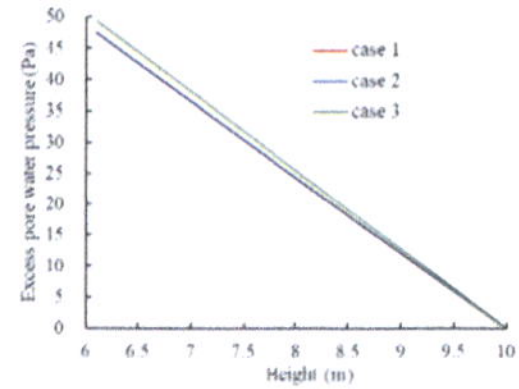

Fig. 10. Excess pore gas pressure of case 1&2&3-height curve of day 100

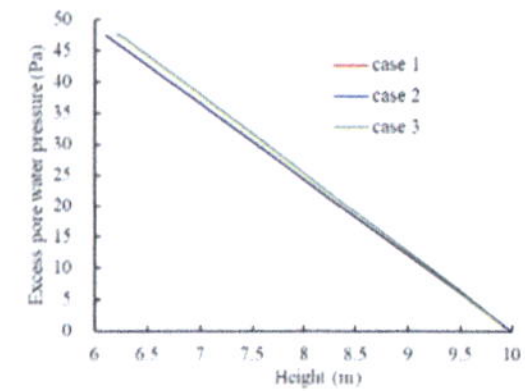

Fig. 11. Excess pore gas pressure of case 1&2&3-height curve of day 270

The change in the excess pore pressure with time and space for case 5 is shown in Figs. 15, 16 and 17. The total pore pressure in the bottom and middle area are similar. The excess pore water pressure increases with time, and the excess pore gas pressure in the unsaturated region decreases with time. In comparison with case 3, the excess pore gas pressure is similar, but the excess pore water pressure is greater.

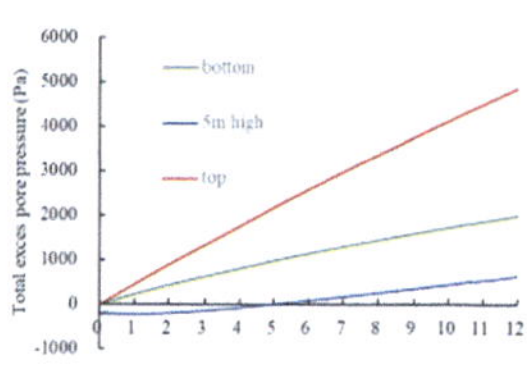

Fig. 12. Total excess pore pressure-time curve of case 4

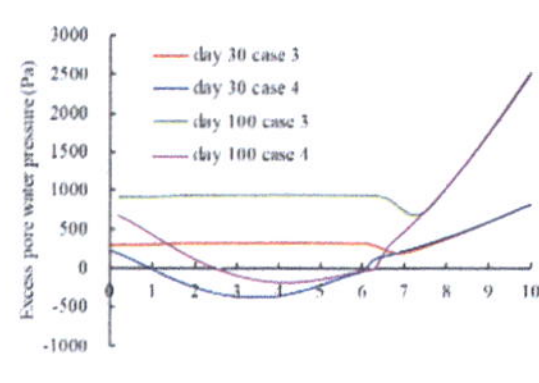

Fig. 13. Excess pore water pressure-height curve of case 4&3

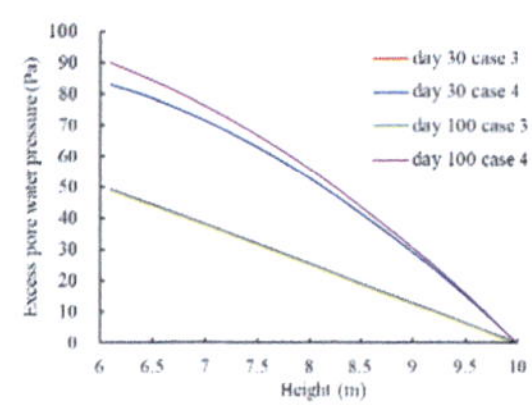

Fig. 14. Excess pore gas pressure-time curve of case 4

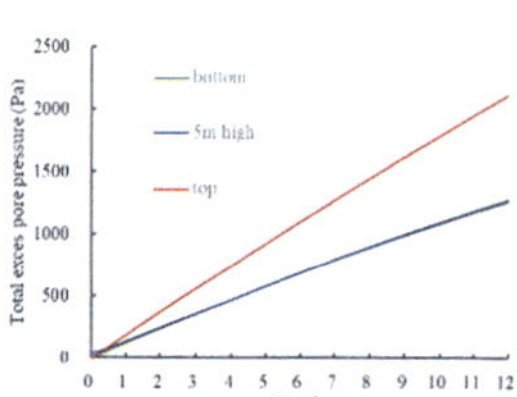

Fig. 15. Total excess pore pressure-time curve of case 5

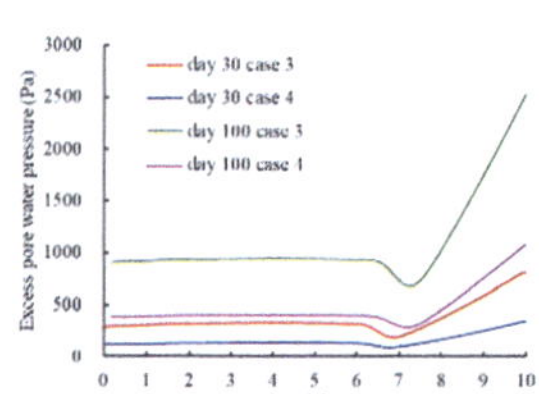

Fig. 16. Excess pore water pressure-height curve of Case 5&3

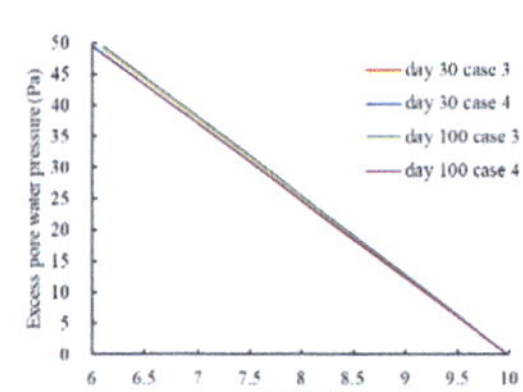

Fig. 17. Excess pore gas pressure-height curve of case 5&3

4 Conclusion

(1) A one-dimensional consolidation model of unsaturated high-kitchen-waste content MSW landfill considering the compressibility of particles was proposed based on consolidation model for unsaturated porous media. A case is studied using OGS software.
(2) For the same age, the total excess pore pressure first increases with time, and then be steady or continue to increase, to be detailed, the pore water pressure continues to increase and the pore gas pressure is equal to pore water stress in the saturated area while dissipates in the unsaturated region.
(3) The rate and trend of change of pore pressure varies according to age. It takes more time to reach stability for high age MSW.
(4) With the decrease of permeability, both the growth rate and value of pore pressure increase. The difference in the distribution of excess pore pressure with depth in saturated region is large.
(5) As the bulk modulus of solid particles increases, both the growth rate and value of pore pressure decrease, where the growth of pore water pressure is smaller while of pore gas pressure is larger.
(6) The model could be further developed considering solute migration and degradation to more accurately predict the pressure in landfills.

References

1. Bi XH, Xu XB (2000) Treatment of municipal solid waste in China and suggestions. J Dialectics Nat (04):43–49+96
2. Gao G, Dong YW, Jin HB, Huang W (2000) Researches on countermeasures of municipal waste disposal and management. Urban Environ Urban Ecol 02:39–41
3. Terzaghi K (1925) Erdbaumechanik auf bodenphysikalischer grundlage. Leipzig Deuticke, Vienna
4. Biot MA (1941) General theory of three-dimensional consolidation. J Appl Phys 12(2):155–164
5. Skempton A, Bishop AW (1954) Soils. North-Holland, Amsterdam
6. Suklje L (1969) Rheological aspects of soil mechanics. Wiley, London
7. Jennings J (1957) Discussion on MS Youssef's paper. In: Proceedings of the 4th international conference on soil mechanics. ISSMFE, vol 3, p 168
8. Corney D, Coleman J, Black W (1958) Movement and distribution of water in soil in relation to highway design and performance. In: Water and its conduction in soils, highway res board, Special Report, Washington, DC, (40):226–252
9. Bishop AW (1959) The principle of effective stress. Teknisk Ukeblad 39:859–863
10. Aitchison G (1960) Relationships of moisture stress and effective stress functions in unsaturated soils. In: Proceedings of the conference of the British national society of the international society for soil mechanics and foundation engineering, London, pp 20–25
11. Bolzon G, Schrefler B, Zienkiewicz O (1996) Elastoplastic soil constitutive laws generalized to partially saturated states. Geotechnique 46(2):279–289
12. Richards LA (1931) Capillary conduction of liquids through porous mediums. J Appl Phys 1 (5):318–333

13. Lewis RW, Schrefler BA (1998) The finite element method in static and dynamic deformation and consolidation of porous media. Wiley, London
14. Schrefler BA, Scotta R (2001) A fully coupled dynamic model for two-phase fluid flow in deformable porous media. Comput Methods Appl Mech Eng 190(24):3223–3246
15. Rutqvist J, Borgesson L, Chijimatsu M, Kobayashi A, Jing L, Nguyen TS, Noorishad J, Tsang C-F (2001) Thermohydromechanics of partially saturated geologic media: governing equations and formulation of four finite element models
16. Chen YM, Xie Y, Zhan LT (2006) One-dimensional consolidation model for landfills considering solid-liquid-gas interaction. Chin J Geotechn Eng 28(02):184–190
17. Guo QG (2016) A theoretical study on one-dimensional degradation-consolidation of saturated municipal solid waste. Zhejiang University
18. Chen YM, Han K, Fredlund DG et al (2010) Secondary compression of municipal solid wastes and a compression model for predicting settlement of municipal solid waste landfills. J Geotech Geoenviron Eng 136(5):706–717
19. Liu XD, Shi JY, Hu YD (2011) Coupled mechanical-gas settlement model and calculation for MSW by considering biodegradation. Chin J Geotech Eng 05:693–699
20. Liu HL (2016) Research on biodegradation-consolidation-solute migration coupled behaviors of municipal solid waste and landfill stabilization. Zhejiang University

Leachate Properties of the Stabilized MSW Incineration APC Residues in Pilot-Scale Mono-landfill

Junjie Qiu, Hua Zhang, Siyuan Yu, Liming Shao, and Pinjing He(✉)

Institute of Waste Treatment and Reclamation, Tongji University,
Shanghai 200092, China
solidwaste@tongji.edu.cn

Abstract. The dramatic development of municipal solid waste (MSW) incineration recently has brought a surge in the generation of air pollution control (APC) residues, a kind of hazardous waste. Landfilling after stabilization of APC residues plays an essential role in the hazardous waste management. Few studies have focused on the leachate quality of mono-landfills for stabilized APC residues in China. This investigation concentrates on simulating mono-landfills by percolation leaching tests in a pilot scale, the leachate quality data from which could offer fundamental basis both for the design of mono-landfill sites (if leachate would be generated) and chemical stabilization processes. Moreover, granulation of the stabilized APC residues could offer a fairly better working condition for transportation and landfilling operation; thus, another purpose of this work is to analyse the leaching behaviour of contaminants and tentatively conclude the effect of granulation on the leachate characteristics of stabilized APC residues in mono-landfill site. Three models were used to describe the leaching rate of the specific contaminants in landfill sites.

Keywords: Municipal solid waste · Air pollution control residues
Landfill leachate · Granulation · Model

1 Introduction

Incineration is a widely used technology to treat municipal solid waste (MSW) (Chiang et al. 2009; Hyks et al. 2009; del Valle-Zermeño et al. 2014), which can recover energy. In decades, the number of MSW incineration plant is three times more than 10 years ago, which caused a seven-fold increase in the amount of MSW transformed by incineration. Nevertheless, as each ton of MSW is incinerated, 15–40 kg APC residues are produced in grate furnace (Quina et al. 2008) or 100–200 kg APC residues are generated in fluidized bed incinerator (Nie 2008), which exert a potential threat to environment.

Characterised by the heavy metal and trace organic pollutants, APC residues are landfilled after solidification/stabilization in many countries (Quina et al. 2008). Solidification/stabilization is a technology that employ additives or binders to immobilize hazardous components initially present in waste (Wiles 1996). In China, commercial chemical agents are used to stabilize APC residues. However, the size of the

L. Zhan et al. (Eds.): ICEG 2018, ESE, pp. 77–84, 2019.
https://doi.org/10.1007/978-981-13-2224-2_10

residues is fine, which likely cause smog in the period of transportation or landfill operation. And, it is showed that a good correlation between increasing fine-particle concentration and higher mortality related to cardiopulmonary disease (Dockery et al. 1993). Granulating the stabilized residues could be a useful method to avoid smog.

In the landfill site, when MSWI residues meet with water, three typical physical or chemical processes(leachate production, gas generation and heat release) could happen. In previous investigation, Xia and He present the calcium concentration of leachate is fairly high in a MSWI button ash and MSW co-landfill site, which could result in clogging of the leachate collection system (Xia et al. 2015; He et al. 2017), and inhibit the bio-treatment of waste leachate (Xia et al. 2017). APC residues also contain about 15% calcium and bulk soluble salt (Quina et al. 2008). According to the source of water in a landfill site (Yang et al. 2015), if stabilized APC residues were mono-landfilled, the generation of leachate would be reduced. To our knowledge, there are limited reports revealing the leachate quality of APC residues in a mono-landfill site. Column test is a feasible method to simulate the leachate quality(Eighmy et al. 1997). In addition, the pilot-scale study is quite essential to design and maintain real projects (del Valle-Zermeño et al. 2014).

Therefore, the object of this paper is to simulate the leachate quality of stabilized APC residues (including stabilized residues which have been granulated and not) in a mono-landfill site by a pilot-scale column test. The law of pollutants release was analyzed by the regression analysis of three models.

2 Material and Methods

2.1 Sample

APC residues came from a MSWI plant in Changzhou, China. The first residues was fly ash stabilized by a commercial agent, which is referred to as the "SFA" for short. The second residues was granulated fly ash after stabilization, which is referred to as the "GFA". SFA were fine particles, while GFA resembles ellipsoid with polar radiuses of (2.3 ± 0.2) cm and (1.3 ± 0.2) cm and equatorial radius of (0.5 ± 0.1) cm (n = 11). Moisture content of SFA and GFA were (3.4 ± 0.8) % and (5.8 ± 0.2) % (n = 3) respectively. Two kinds of resides were crushed smaller than 1 mm, then Cl, Br and S were directly tested by X-ray fluorescence spectrometry (XRF, UniQuant, Thermo, USA). Moreover, after acid digestion by $HCl/HClO_4/HF/HNO_3$ (Zhang et al. 2008), other elements were analyzed by inductively coupled plasma optical emission spectrometry (ICP-OES, ICP-725ES, Agilent, USA). Table 1 shows the element composition of the residues, which are similar to the composition of other residues in a review (Quina et al. 2008).

Table 1. Elemental composition of SFA and GFA in dry basis

Element	Unit	SFA	Deviation	GFA	Deviation
Ca[a]	mg/g	115	± 17	122	± 3
K[a]	mg/g	32	± 5	24	± 1
Na[a]	mg/g	16	± 3	17.2	± 0.3
Al[a]	mg/g	4.0	± 0.7	6.6	± 0.8
Mg[a]	mg/g	4.0	± 0.6	6.2	± 0.2
Zn[a]	mg/g	3.7	± 0.6	3.3	± 0.2
Fe[a]	mg/g	3.6	± 0.6	4.6	± 0.7
Pb[a]	mg/g	1.3	± 0.3	1.3	± 0.1
P[a]	mg/g	1.0	± 0.2	1.4	± 0.1
Cu[a]	mg/g	0.7	± 0.1	0.69	± 0.04
Ba[a]	mg/g	0.4	± 0.1	0.7	± 0.1
Mn[a]	mg/g	0.16	± 0.03	0.22	± 0.02
B[a]	mg/g	0.13	± 0.01	0.15	± 0.01
Cd[a]	mg/kg	89	± 16	76.3	± 3.3
Cr[a]	mg/kg	59	± 10	104	± 17
As[a]	mg/kg	36	± 6	42	± 5
V[a]	mg/kg	7.1	± 1.1	13.3	± 0.7
Se[a]	mg/kg	1.1	± 1.6	417	± 130
Co[a]	mg/kg	1.0	± 0.5	2.6	± 1.5
Cl[b]	mg/g	149		149	
Br[b]	mg/g	5.7		4.8	
S[b]	mg/g	27.8		21.0	

[a]Determined by ICP-OES after acid digestion.
[b]Determined using XRF.

2.2 Column Test

To investigate the leachate quality of SFA and GFA in a mono-landfill site, pilot scale column tests were conducted. The working condition is showed in Table 2. Soft tap water was used as leachant. In addition, the material of the column was high density polyethylene (HDPE), which do not release metals or organic matters in a high salty and alkaline solution. For homogenises of the distribution of the leachant, tap water went through the silica sand (20–50 meshes) before contacting with residues. And, a nylon filter screen (250 meshes) was installed at the end of the column to avoid loss of residues. Leachate samples were taken at different time. The end liquid to solid ratio was 10 L/kg.

Table 2. Working conditions of the column test

	Ash sample	Inner diameter of column (m)	Height of ash (m)	Flow rate (L/d)	Flow regime	Mass of ash in dry basis (kg)
1#	GFA	0.36	1.5	60	DF[a]	136.1
2#	GFA	0.36	1.5	60	UF[b]	120.9
3#	SFA	0.36	1.5	60	UF[b]	103.5
4#	50% GFA + 50% SFA	0.36	1.5	60	UF[b]	134.3

[a]DF: down flow;
[b]UF: up flow.

2.3 Leachate Analysis

pH and conductivity were tested immediately after sampling. Leachates were filtrated through 0.45 μm mixed cellulose acetate filters. Filtrates were saved in polyethylene bottle for other tests. Cl^- was titrated by $AgNO_3$ solution; The concentrations of Ca, K, Na, Ba, Cd, Cr, Cu, Ni and Zn were determined by ICP-OES and ICP-725ES (Agilent, USA); Using an atomic fluorescence spectrometer (Beijingjitian, AFS-8230) to test the concentration of As in the leachate.

2.4 Data Analysis

The accumulated leaching amount of each element was calculated from the leaching curve by cumulative Simpson's integration with uneven spacing (Hyks et al. 2009). Leaching concentration curves were fitted by three models, CSTR model (Guyonnet et al. 2008), advection–dispersion model (Grathwohl and Susset 2009) and mass transfer model (Brouwers 1997), using "curve fitting" tool box (Matlab, R2014b). Correlation analysis of the different elements were calculated by data analysis (add in, Excel 2016).

3 Result and Discussion

3.1 Leachate Quality

The L/S ratio in this paper can be extrapolated to the landfill age of a real landfill site by the Eq. (4.1) (Hjelmar 1990; Kosson et al. 1996; Kosson 2002). t represents the landfill age (year); L/S stands for liquid to solid ratio (L/kg); ρ means the stacking density of the landfill body (kg/m^3); H is the height of the landfill body (m); I is behalf of the anticipated infiltration rate (cm/year).

$$t = (L/S \cdot \rho \cdot H)/(10I) \tag{4.1}$$

Figure 1 shows the leaching curves of pH, Cl^- and Pb. Other concentrations of elements in the leachate were lower than the limit of standard for pollution control on the landfill site of municipal solid waste (GB 16889-2008) in China.

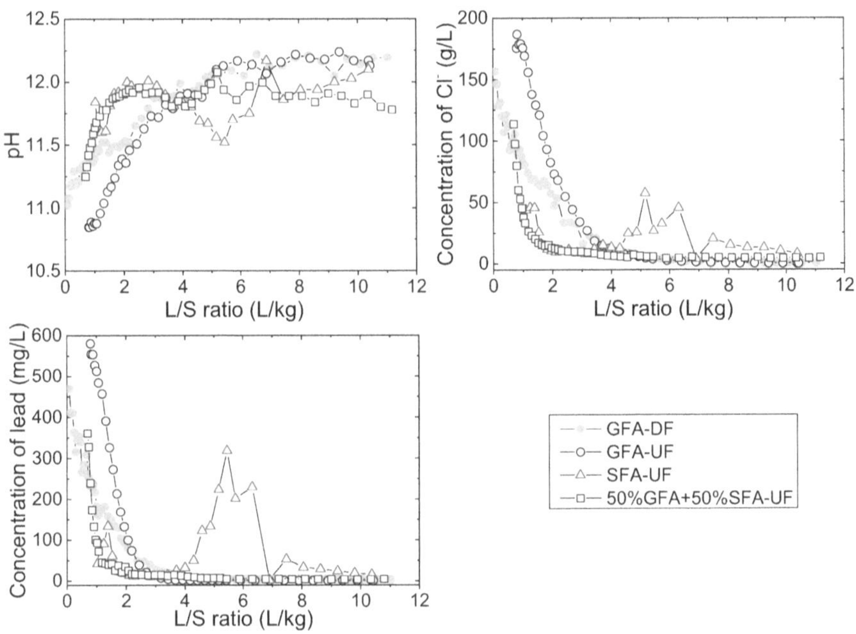

Fig. 1. Leachate pH, Cl^- and Pb concentrations in the leachates varying with L/S ratio

pH increased with *L/S*, which is similar to other researches (Quina et al. 2011; Guyonnet et al. 2008). The control solid of pH were probably $CaSO_4$, CaOHCl, $CaCl_2$ or other calcium salts that influence ionization balance of calcium hydroxide (Astrup et al. 2006). The concentration of Cl^- and Pb declined in exponential decay curve. When *L/S* < 2 L/kg (means landfill age of less than 10 years), the concentration of Cl^- was about 50–200 g/L. Meanwhile, the concentration of Pb was up-to 600 mg/L. When *L/S* = 2 L/kg, approximately 80% of Cl^- and 70% of Pb were released. For Cl^-, Na and K, such elements were controlled by availability (Hyks et al. 2009). It is generally recognized that amphoteric metal Pb was controlled by pH during the leaching process (Quina et al. 2009). However, our results were fairly different, which may stem from that Pb was not stabilized by chemical agent and existed in $PbCl_2$ or other soluble lead salts. And, it was also indicated by the similarity of the leaching curves of Pb and Cl^-.

3.2 Leaching Dynamics

The Pearson correlation coefficients of the concentrations of K, Na, Ca and Cl^- were around 0.980–0.999. The perfectly excellent correlation suggest that these elements were released in same mode. Moreover, concentrations of Pb and Cl^- also show relatively good correlation with Pearson correlation coefficients of 0.95. Equations (4.2) to (4.4) represent the CSTR model (Guyonnet et al. 2008), advection–dispersion model (Grathwohl and Susset 2009) and mass transfer model respectively (Brouwers 1997). *a*, *b* and *c* are the constant of model.

$$C = a \cdot e^{-b \cdot L/S} \tag{4.2}$$

$$C = c\left(a - 0.5\text{erfc}\left((1 - a \cdot L/S)\Big/\left(2\sqrt{b \cdot L/S}\right)\right)\right) \tag{4.3}$$

$$C = a \cdot (L/S)^b \tag{4.4}$$

For “SFA”, when the L/S ratio was raising from 4 to 5.5 L/kg, the concentrations of Pb and Cl^- went up in a sudden. It may be caused by preferential flow. Nevertheless, the three models did not consider the preferential flow, so it makes no sense to conduct regression analysis. Table 3 shows the regression coefficient R^2 of the three models, which indicates fitting results: CSTR model ≈ advection–dispersion model > mass transfer model. It reveals that the leaching rate is controlled by solute diffusion rate in bulk liquid other than mass transfer rates between solid and liquid. Thus, the similarity of leaching curves of “GFA-UF”, “GFA-DF” and “50%GFA + 50%SFA-UF” could be explained. What’s more, CSTR model and advection–dispersion model shared similar fitting results, for advection–dispersion model can be transformed into CSTR model with larger axial diffusion (Levenspiel 1999).

Table 3. Regression coefficient and parameters of the three models

		CTSR model	Advection-dispersion model	Mass transfer model
Cl^-	1#	0.9859	0.9330	0.8103
		148.9[a], 0.5721[b]	1.033[a], 0.4297[b], 185.6[c]	63.19[a], − 0.4684[b]
	2#	0.9927	0.9937	0.9413
		358.9[a], 0.7597[b]	59.04[a], 0.00005[b], 1316[c]	161.2[a], − 1.252[b]
	3#	–	–	–
	4#	0.9220	0.8844	0.9609
		66.18[a], 2.658[b]	278.5[a], 0.00004[b], 1345[c]	4.582[a], − 2.342[b]
Pb	1#	0.9881	0.9607	0.8324
		465.1[a], 0.8339[b]	303.8[a], 0.000005[b], 746[c]	157.7[a], − 0.5361[b]
	2#	0.9822	0.9990	0.9407
		358.9[a], 0.7597[b]	1.092[a], 1.104[b], 1078[c]	161.2[a], − 1.252[b]
	3#	–	–	–
	4#	0.9658	0.9519	0.9789
		7612[a], 4.349[b]	386.3[a], 0.00004[b], 2414[c]	103.2[a], − 3.56[b]

[a,b,c]Mean the regression parameters of each three models.

To sum up, several conclusions are as follows: (1) In the beginning of the landfill site (less than 10 years), stabilized residues could release about 80% of Cl^- and other major elements, and generate high salty leachate. (2) Granulation has few effect on the leaching rate of major elements, but it reduces the preferential flow in the landfill site, which could slow down the release of pollutants. (3) For some specific elements, such as Pb, even though stabilized by chemical agent, the release risk seems not to go down.

Acknowledgement. We thank Haohao Zhang, Yuan Gu, Jiawei Zeng, Xiangxiang Cheng, Jianchen Li, Yang Liu and Junda Ren for help in the laboratory. We'd also like to thank Changzhou Domestic Waste Treatment Center for the experimental site. This study was financially supported by scientific and technological projects of Urban-Rural Development of the People's Republic of China (No. 2016-K4-001).

References

Astrup T, Dijkstra JJ, Comans RNJ, Sloot HAVD, Christensen TH (2006) Geochemical modeling of leaching from MSWI air-pollution-control residues. Environ Sci Technol 40:3551–3557

Brouwers HJH (1997) Leaching models for multiple immersed materials and for granular materials flushed in a column. J Hazard Mater 53:1–17

Chiang KY, Tsai CC, Wang KS (2009) Comparison of leaching characteristics of heavy metals in APC residue from an MSW incinerator using various extraction methods. Waste Manag 29:277–284

Del Valle-Zermeño R, Formosa J, Prieto M, Nadal R, Niubó M, ChimeNos JM (2014) Pilot-scale road subbase made with granular material formulated with MSWI bottom ash and stabilized APC fly ash: environmental impact assessment. J Hazard Mater 266:132–140

Dockery DW, Xu X, Spengler JD, Ware JH, Fay ME, Jar FB, Speizer FE (1993) An association between air pollution and mortality in six U.S. cities. New Engl J Med 329:1753

Eighmy TT, Chandler AJ, Hjelmar O, Kosson DS, Sawell SE, Vehlow J, Sloot HAVD, Hartlen J (1997) Municipal solid waste incineration residues

Grathwohl P, Susset B (2009) Comparison of percolation to batch and sequential leaching tests: theory and data. Waste Manag 29:2681–2688

Guyonnet D, Bodénan F, Bronslaot G, Burnol A, Chateau L, Crest M, Méhu J, Moszkowicz P, Piantone P (2008) Multiple-scale dynamic leaching of a municipal solid waste incineration ash. Waste Manag 28:1963–1976

He P-J, Pu H-X, Shao L-M, Zhang H (2017) Impact of co-landfill proportion of bottom ash and municipal solid waste composition on the leachate characteristics during the acidogenesis phase. Waste Manag 69:232–241

Hjelmar O (1990) Leachate from land disposal of coal fly ash. Waste Manag Res 8:429–449

Hyks J, Astrup T, Christensen TH (2009) Long-term leaching from MSWI air-pollution-control residues: leaching characterization and modeling. J Hazard Mater 162:80–91

Kosson DS (2002) An integrated framework for evaluating leaching in waste management and utilization of secondary materials. Environ Eng Sci 19:159–204

Kosson DS, Sloot HAVD, Eighmy TT (1996) An approach for estimation of contaminant release during utilization and disposal of municipal waste combustion residues. J Hazard Mater 47:43–75

Levenspiel O (1999) Chemical reaction engineering. Ind Eng Chem Res 38:1055–1076

Nie Y (2008) Development and prospects of municipal solid waste (MSW) incineration in China. Front Environ Sci Eng China 2:1–7

Quina MJ, Bordado JC, Quinta-Ferreira RM (2008) Treatment and use of air pollution control residues from MSW incineration: an overview. Waste Manag 28:2097–2121

Quina MJ, Bordado JC, Quinta-Ferreira RM (2009) The influence of pH on the leaching behaviour of inorganic components from municipal solid waste APC residues. Waste Manag 29:2483–2493

Quina MJ, Bordado JC, Quinta-Ferreira RM (2011) Percolation and batch leaching tests to assess release of inorganic pollutants from municipal solid waste incinerator residues. Waste Manag 31:236–245

Wiles CC (1996) Municipal solid waste combustion ash: state-of-the-knowledge. J Hazard Mater 47:325–344

Xia Y, He PJ, Pu HX, Lü F, Shao LM, Zhang H (2017) Inhibitory effect of high calcium concentration on municipal solid waste leachate treatment by the activated sludge process. Waste Manag Res 35:508–514

Xia Y, Zhang H, Phoungthong K, Shi DX, Shen WH, Shao LM, He PJ (2015) Leaching characteristics of calcium-based compounds in MSWI Residues: From the viewpoint of clogging risk. Waste Manag 42:93

Yang N, Damgaard A, Kjeldsen P, Shao L-M, He P-J (2015) Quantification of regional leachate variance from municipal solid waste landfills in China. Waste Manag 46:362–372

Zhang H, He PJ, Shao LM (2008) Fate of heavy metals during municipal solid waste incineration in Shanghai. J Hazard Mater 156:365

Influence of Atmospheric Pressure on Methane Emissions from Earthen Landfill Covers

T. Wu(✉) and L. T. Zhan

MOE Key Laboratory of Soft Soils and Geoenvironmental Engineering, Zhejiang University, Hangzhou 310058, China
chinawt@zju.edu.cn

Abstract. An earthen landfill cover test site, consisted of a 0.9 m thick compacted loess layer underlain by a 0.3 m thick gas distribution layer, was constructed at the Jiangcungou landfill in Xi'an, China. Nine methane emission test points were evenly arranged in the test area, and methane emissions were measured at each test point for approximately 35 days. The atmospheric pressure was recorded by the weather station installed near the test site. During 2015, the atmospheric pressure decreased first and then increased with time, and was basically symmetrical about the middle of the year. Methane emission from the earthen landfill cover could be directly related to atmospheric pressure changes, and a significantly negative relationship was found between the measured methane emissions and surface atmospheric pressure. An improved understanding of the atmospheric pressure variations and the influence of atmospheric pressure on methane emission from earthen covers are meaningful for developing a methane emission model and for mitigating methane emissions.

Keywords: Earthen landfill cover · Methane emission · Atmospheric pressure

1 Introduction

Methane (CH_4) was the second important greenhouse gas (GHG) after carbon dioxide (CO_2). Methane contributed approximately 0.5 Wm^{-2} to total radiative forcing by long-lived GHGs, which made it a powerful GHG with a global warming potential (GWP) of 28 over a period of 100 years (Hofmann et al. 2006; IPCC 2014). In addition, the atmospheric concentration of CH_4 had increased 150% since 1750 and landfills were ranked as the third largest anthropogenic CH_4 emission source (IPCC 2014). As a consequence, mitigating the amount of CH_4 released from landfills becomes an urgent project to be solved.

The passive methane emission from landfills generally follows the pressure gradient, and thus is affected by atmospheric pressure. Atmospheric pressure was defined as a key parameter in gas flux model developed by Young (1990). Czepiel et al. (2003) used atmospheric tracer methods to investigate the influence of atmospheric pressure on the whole landfill methane emissions, and a negative relationship was found. Gebert et al. (2006) demonstrated that the direction of advective gas flow reversed more than once a day at a passively vented landfill site, and this was mainly driven by atmospheric pressure variation.

L. Zhan et al. (Eds.): ICEG 2018, ESE, pp. 85–92, 2019.
https://doi.org/10.1007/978-981-13-2224-2_11

Although earthen landfill cover systems were increasingly being used to reduce deep drainage (Scanlon et al. 2006), the field experimental study on methane emissions from the earthen landfill covers and the influence of atmospheric pressure have not been well investigated. A full-scale earthen cover was constructed at the Jiangcungou landfill in Xi'an, China. The objective of this paper is to investigate the effects of atmospheric pressure on methane emissions from the earthen cover. An improved understanding of the effects of atmospheric pressure on methane emissions will be meaningful for developing a methane emission model in landfills.

2 Materials and Methods

2.1 Site Description and Construction

The earthen landfill cover test site investigated in this paper is located at the Jiangcungou landfill, 16 km east of Xi'an city (34°14′34″N–109°05′5″E). The landfill has been in operation since 1994 and is the only sanitary landfill in Xi'an, servicing a population of millions of people. Currently, the landfill occupies an area of 73 ha, and accepts all the municipal solid waste (MSW) generated in Xi'an city (approximately 8500 t/d). The MSW is deposited in a number of 10 m thick filled platforms that are set back at embankment slopes of 3H: 1V (where H and V mean horizontal and vertical, respectively). The test site was constructed on the slope between the 7th and 8th platform, and was completed in January 2014. The fill age of the MSW under the test site was about 3–5 years at the construction time.

The test site measures 20 m (W) × 30 m (L) and follows a slope of 14.5° (Fig. 1). The core test facility was a lysimeter (10 × 24 m) which was located in the center of the test site with a 5 m wide buffer area on both sides. And it was hydraulically isolated from the surrounding environment (i.e., base and sidewalls) by HDPE geomembrane. The cover mainly consisted of two parts, namely, 0.3 m thick gas distribution layer and 0.9 m thick compacted loess layer. The landfill gas (LFG) was supplied by a gas collection well, and was pumped into the gas distribution layer via an induced draft fan. As shown in Fig. 1, the earthen cover was divided into three sections (i.e., upper part, middle part and lower part), and three methane emission test points were evenly arranged in each section.

2.2 Experimental Operations

Tests on CH_4 emissions from the earthen landfill cover were carried out mainly in November 2015. Methane emissions were measured by static chamber techniques (Park et al. 2016). A total of nine methane emission test points were set up in the experimental area, and numbered 1 to 9, respectively (Fig. 1).

The static chambers (0.5-m in diameter and 0.5-m high) used in this paper were made of plexiglass. A thermometer was fixed inside the chamber to obtain the temperature, and a fan was used to mix the gas sufficiently. A fixed steel base was arranged at each test point, and a water filled U-groove was added for taking measurements. The CH_4 concentration in the chamber was measured by a laser methane detector (TGE-SA3C32A). According to Park et al. (2016), the LMD should be placed at the

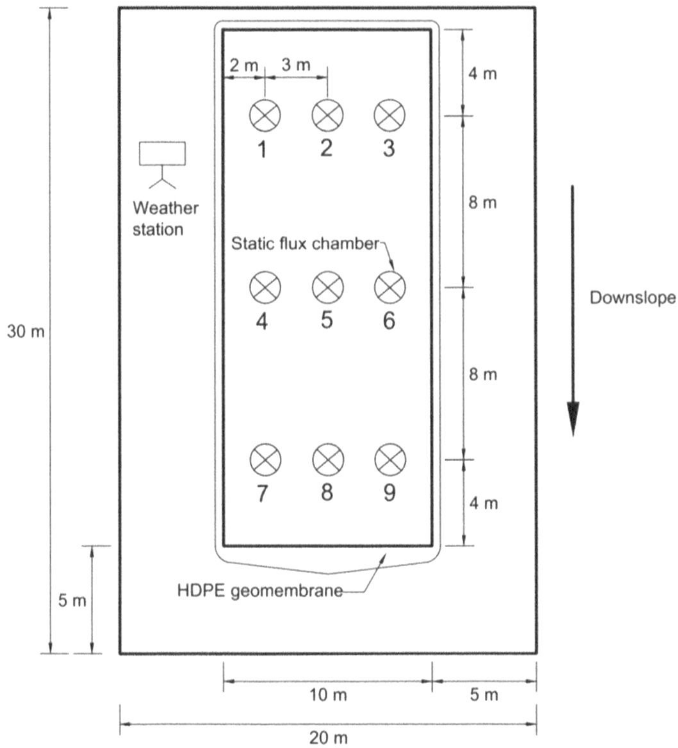

Fig. 1. Earthen landfill cover test facility – Plan view

top of the chamber. However, it was found that the test results of the methane concentration had considerable error when test areas encountered cracks. In order to avoid the influence of predominant outflow and obtain stable test results, the LMD was placed on the side of the chamber at the time of testing. The methane emission flux was calculated by

$$J = \frac{V}{A}\left(\frac{dC}{dt}\right)\frac{P_a \mathrm{M}}{\mathrm{R}T_a} \times 1440 \tag{1}$$

where J is the methane emission (g m^{-2} d^{-1}), V is the total volume of the chamber and frame (m^3), A is the surface area covered by the chamber (m^2), dC/dt is the slope of CH_4 concentration versus time curve (m^3 CH_4 m^{-3} air min^{-1}), P_a is the atmospheric pressure (Pa), R is the gas constant (J mol^{-1} K^{-1}), T_a is the Kelvin temperature of the gas in the chamber (K), and M is the mole mass of methane (g mol^{-1}).

2.3 Evaluation Model for Methane Emissions

Gas migration in the earthen landfill covers was mainly dependent on diffusive and advective transport process. For simplicity, the gas inflow was assumed under steady state, and the experimental study was focused on the advective transport process. The methane emission could be calculated as:

$$J = F_{in} - R_{ox} \tag{2}$$

where F_{in} is the methane inflow flux (g m^{-2} d^{-1}), and R_{ox} is the methane oxidation capacity of the landfill cover (g m^{-2} d^{-1}).

Methane emissions from landfills could be oxidized by methanotrophic bacteria in soils or compost utilizing oxygen (Dunfield et al. 1993). Scheutz et al. (2004) demonstrated that the methane oxidation rate of the soil cover could be calculated as:

$$R_{ox} = r_{ox} d_a \rho_b \tag{3}$$

where r_{ox} is the methane oxidation rate per unit mass of soil (μg g^{-1} d^{-1})), d_a is the oxidation zone (m), and ρ_b is the soil bulk density (Mg m^{-3}).

According to the Darcy's law, the methane inflow flux could be determined by

$$F_{in} = \frac{P_b}{RT_b} M\left(-\frac{k_g}{\mu_g}\frac{\partial u_g}{\partial z} C_b\right) \times 86400 = \frac{P_b}{RT_b}\frac{k_g}{\mu_g}\frac{MC_b}{L}(P_b - P_a) \times 86400 \tag{4}$$

where P_b is the absolute gas pressure in the gas distribution layer (Pa), k_g is the gas permeability (m^2), μ_g is the viscosity coefficient of gas (Pa s), $\partial u_g/\partial z$ is the gas pressure gradient (Pa m^{-1}), C_b is the methane volume fraction in LFG, L is the thickness of the loess cover (m), and T_a is the Kelvin temperature of the LFG in the gas distribution layer (K).

Substituting Eqs. (3) and (4) into Eq. (2), we could obtain:

$$J = -\left(86400\frac{P_b}{RT_b}\frac{k_g}{\mu_g}\frac{MC_b}{L}\right)P_a + \left(86400\frac{P_b^2}{RT_b}\frac{k_g}{\mu_g}\frac{MC_b}{L} - r_{ox} d_a \rho_b\right) = -AP_a + B \tag{5}$$

where A and B refer to the corresponding polynomials, respectively.

3 Results and Discussion

3.1 Atmospheric Pressure Variation

The one-year variation of the surface atmospheric pressure in Xi'an area is shown in Fig. 2. In 2015, the atmospheric pressure was in the range from 950 to 990 mbar. Although the diurnal mean values of atmospheric pressure fluctuated greatly over time, the distribution of surface atmospheric pressure within one year still showed a clear pattern. The atmospheric pressure decreased first and then rose with time, and the atmospheric pressure value distribution showed symmetry about the middle of the year. The minimum values of atmospheric pressure were obtained near July 1, while the maximum values were acquired around January 1 and December 1. In other words, the maximum and minimum atmospheric pressure appeared in winter and summer, respectively. The results indicated that methane emissions in summer might be higher than that in winter, and more attention should be paid to the control of methane emission from the earthen landfill cover in summer.

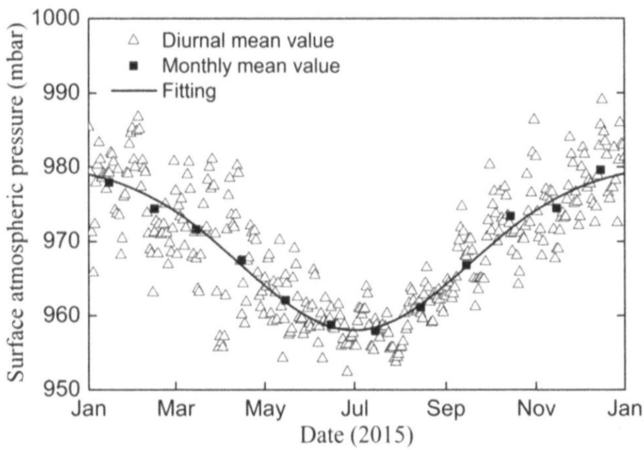

Fig. 2. The change curve of atmospheric pressure in 2015

Figure 3 illustrates the representative variation curves of atmospheric pressure within 24 h. The four curves in the figure represent typical atmospheric pressure distributions for the four seasons in spring, summer, autumn and winter, respectively. Although the distribution range of atmospheric pressure values varied at different test times, the variation law in 24 h was consistent. Atmospheric pressure at the test site generally reached its maximum in 08:00–12:00, then fell steadily and reached its minimum in 16:00–20:00. Furthermore, the diurnal atmospheric pressure difference at the test site was generally around 5 mbar. In fact, unpleasant odors near the landfill tended to be most pronounced in the evening. Mastering the variation law of atmospheric pressure was of great significance to regulation of odor emissions from landfills. In a landfill with LFG collection system, the odor emissions might be effectively mitigated by increasing the pumping rate when the atmospheric pressure was low.

3.2 Methane Emission

Atmospheric pressure was an important factor affecting methane emissions from landfills. Figure 4 shows the relationship between the methane emissions measured at 9 test points and atmospheric pressure. The atmospheric pressure was generally in the range of 950–970 mbar during the test period. The measured maximum methane emission is approximately 78.68 g CH_4 m^{-2} day^{-1}. Due to the heterogeneity of the earthen cover, its gas permeability has a large spatial variability, resulting in a large variation in the measured methane emissions under the same atmospheric pressure. However, the range of methane emissions measured at each test point shows a trend of steady decrease with the increase of atmospheric pressure. The results indicated that the

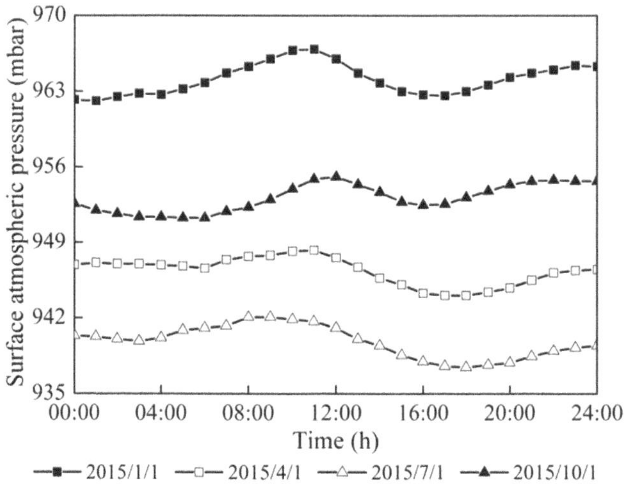

Fig. 3. Curves of surface atmosphere pressure with time in 24 h at different measuring times

methane emissions decreased with increasing atmospheric pressure, as described earlier (Poulsen et al. 2003).

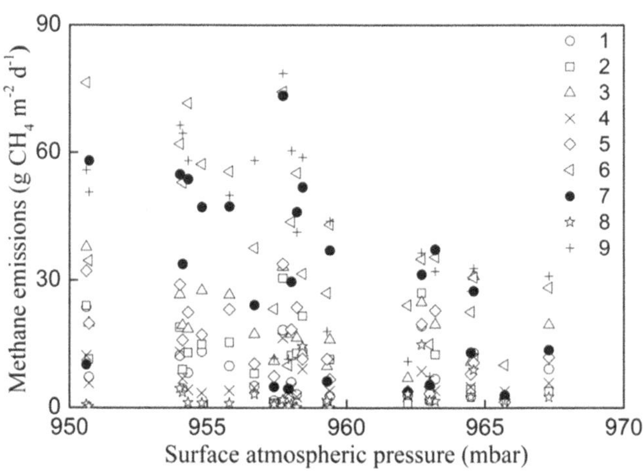

Fig. 4. Effect of surface atmosphere pressure on methane emissions

The test point 6 was chosen as the representative test point for the convenience of analysis. Figure 5 presents the measured methane emissions at test point 6 as a function of surface atmospheric pressure, which is consistent with the results of Czepiel et al. (2003). When other parameters remain constant, Eq. (5) demonstrates a strict linear relationship between methane emission and atmospheric pressure. However, the measured methane emissions fluctuated dramatically, and the resulting correlation coefficient, R^2, of the linear regression was 0.39. The methane emissions were closely related to the gas permeability and methane oxidation capacity of the earthen cover on the basis of Eq. (5). The uncontrollable ambient conditions (e.g., moisture content and temperature) changed significantly over time, leading to variations of the gas permeation rate and methane oxidation rate of the loess cover (Boeckx and Van Cleemput 1996; Zhan et al. 2014). As a consequence, the measured methane emissions might fluctuate considerably under the same atmospheric pressure due to changes in temperature and soil moisture content.

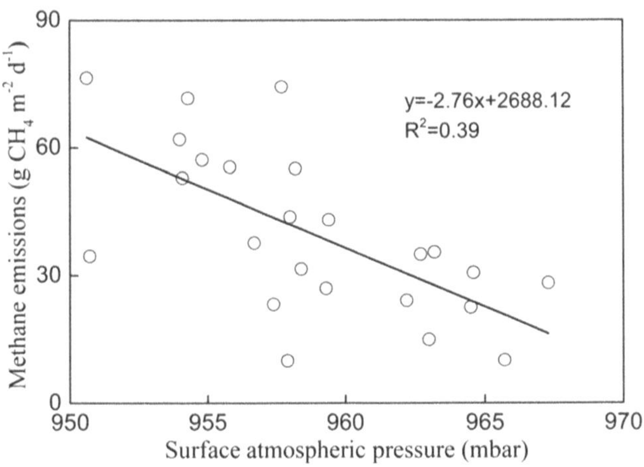

Fig. 5. Relationship between surface atmosphere pressure and methane emissions measured at test point 6

4 Conclusion

A full-scale testing facility was constructed at the Jiangcungou landfill in Xi'an, China. The test results show a significant negative correlation between methane emissions from the earthen landfill cover and atmospheric pressure. The variation of atmospheric pressure in Xi'an during the year of 2015 showed obvious regularity. The atmospheric pressure decreased first and then increased with time, and was basically symmetrical about the middle of the year. The diurnal atmospheric pressure difference was generally around 5 mbar. Atmospheric pressure and its seasonal changes should be taken into account when considering the long-term control of methane emissions from earthen landfill covers. In the daily control of methane emissions, detailed adjustments could be made based on changes in atmospheric pressure within 24 h.

Acknowledgements. The authors are very grateful for the financial support of the National Basic Research Program of China (973) via Grant No. 2012CB719805 and the National Natural Science Foundation of China via Grant No. 51625805.

References

Boeckx P, Van Cleemput O (1996) Methane oxidation in a neutral landfill cover soil: influence of moisture content, temperature, and nitrogen-turnover. J Environ Qual 25(1):178–183

Czepiel PM, Shorter JH, Mosher B, Allwine E, McManus JB, Harriss RC, Kolb CE, Lamb BK (2003) The influence of atmospheric pressure on landfill methane emissions. Waste Manag 23 (7):593–598

Dunfield P, Knowles R, Dumont R, Moore TR (1993) Methane production and consumption in temperate and subarctic peat soils: response to temperature and ph. Soil Biol Biochem 25 (3):321–326

Hofmann DJ, Butler JH, Dlugokencky EJ, Elkins JW, Masarie K, Montzka SA, Tans P (2006) The role of carbon dioxide in climate forcing from 1979 to 2004: introduction of the Annual Greenhouse Gas Index. Tellus B 58(5):614–619

IPCC (2014) Climate Change 2014: Synthesis Report. Contribution of Working Groups I, II and III to the Fifth Assessment Report of the Intergovernmental Panel on Climate Change [Core Writing Team, Pachauri RK, Meyer LA (eds)]. IPCC, Geneva, Switzerland

Park JK, Kang JY, Lee NH (2016) Estimation of methane emission flux at landfill surface using laser methane detector: influence of gauge pressure. Waste Manag Res 34(8):784–792

Poulsen TG, Christophersen M, Moldrup P, Kjeldsen P (2003) Relating landfill gas emissions to atmospheric pressure using numerical modelling and state-space analysis. Waste Manag Res J Int Solid Wastes Public Clean Assoc Iswa 21(4):356

Scanlon BR, Reedy RC, Keese KE, Dwyer SF (2006) Evaluation of evapotran- spirative covers for waste containment in arid and semiarid regions in the southwestern usa. Vadose Zone J 5 (2):55–71

Scheutz C, Mosbaek H, Kjeldsen P (2004) Attenuation of methane and volatile organic compounds in landfill soil covers. J Environ Qual 33(1):61–71

Zhan TLT, Yang YB, Chen R, Ng CWW, Chen YM (2014) Influence of clod size and water content on gas permeability of a compacted loess. Can Geotech J 51(12):1468–1474

An Analytical Model for Methane Oxidation and Migration in Double-Layer Landfill Biocover

Shi-Jin Feng, Zhang-Wen Zhu, and Hong-Xin Chen(✉)

Key Laboratory of Geotechnical and Underground Engineering of the Ministry of Education, Department of Geotechnical Engineering, Tongji University, Shanghai 200092, China
chenhongxin@tongji.edu.cn

Abstract. Landfill has become one of the main atmosphere methane sources. Biocover systems were proposed to mitigate methane emission by utilizing the biodegradation existing in landfill cover soils. The methane emission from landfill should be estimated by considering the methane oxidation and migration process in the biocover system. For the typical biocover system consisting of a gas distribution layer and a methane oxidation layer, an analytical model is proposed to investigate the mechanism of methane emission. The model is effectively verified against the published experiment results. The influence of effective diffusivity of each layer and advection is investigated. The results reveal that the existing single-layer analytical model cannot be applied to double-layer biocover. Putting soil with high effective diffusivity on that with lower value can better mitigate methane emission from cover. The upward advection in cover soils has significant negative effect on mitigating methane emission and should be strictly controlled.

Keywords: Biocover · Analytical model · Methane oxidation
Double-layer soil

1 Introduction

Methane (CH_4) is one of the major greenhouse gases and is more powerful than carbon dioxide (CO_2) to aggravate the greenhouse effect. Due to the anaerobic degradation of organic waste in landfill, CH_4, as one of the major compositions of landfill gas (LFG), is produced continuously [1]. The biocover was proposed as a cost-effective measure to mitigate CH_4 emission [2]. When CH_4 migrates through the biocover system via diffusion and advection, the methanotrophic bacteria in cover soils can utilize CH_4 as their carbon and energy source [1], mitigating CH_4 emission efficiently. Therefore, the migration of CH_4 through biocover system is necessary to be researched.

Analytical models provide basic insight into the mechanism of CH_4 migration through cover soils and are convenient to be used in practical engineering. However, only a few analytical models were developed to simulate CH_4 migration in landfill cover systems considering diffusion, advection and CH_4 oxidation. Recently, Yao et al. [3] proposed a single-layer analytical model to estimate the CH_4 reduction in cover

L. Zhan et al. (Eds.): ICEG 2018, ESE, pp. 93–100, 2019.
https://doi.org/10.1007/978-981-13-2224-2_12

soils. Although diffusion, advection and CH_4 oxidation are all considered in this model, it cannot be applied in the typical double-layer biocover system which consists of a gas distribution layer and a methane oxidation layer [2]. To the best of our knowledge, there is no analytical model for CH_4 migration through double-layer biocover system considering diffusion, advection and CH_4 oxidation.

In this study, an analytical model is developed to describe CH_4 migration through double-layer biocover. This model takes diffusion, advection and CH_4 oxidation into consideration. The solution is verified against the published results of a soil column experiment. Then the mechanism of CH_4 migration is investigated.

2 Mathematical Model

2.1 Governing Equation

In cover soils, the area where CH_4 oxidation occurs is the aerobic zone while the rest area is the anaerobic zone (Fig. 1). The governing equation of this model can be built as follows:

$$D_i \frac{d^2 c_{j,i}^{\mathrm{aer}}}{dz^2} + v \frac{d c_{j,i}^{\mathrm{aer}}}{dz} = R_{j,i}, z \le L_{\mathrm{b}}, \quad i = 1, 2\,(\text{Aerobic zone}) \tag{1}$$

$$D_i \frac{d^2 c_{j,i}^{\mathrm{ana}}}{dz^2} + v \frac{d c_{j,i}^{\mathrm{ana}}}{dz} = 0, z > L_{\mathrm{b}}, \quad i = 1, 2\,(\text{Anaerobic zone}) \tag{2}$$

where D_i (m^2/s), v (m/s), $R_{j,i}$ ($mol/m^3/s$) are the effective diffusivity, gaseous Darcy flow velocity and consumption rate of chemical j due to oxidation in the i-th layer, respectively; L_{b} is the depth of the aerobic/anaerobic interface where O_2 concentration reaches the threshold $c_{\mathrm{O}}^{\mathrm{cr}}$, below which no oxidation happens; $c_{j,i}^{\mathrm{aer}}$ and $c_{j,i}^{\mathrm{ana}}$ are the concentrations of chemical j in the aerobic zone and anaerobic zone, respectively. As for chemical j, m represents CH_4 and O represents O_2 in the following part. The effective diffusivity of CH_4 is close to that of O_2. The oxidation terms in Eqs. (1) and (2) are related to the first-order reaction rate constant in each layer, k_i (s^{-1}):

$$R_{\mathrm{m},i} = k_i c_{\mathrm{m},i} \text{ and } R_{\mathrm{O},i} = \beta k_i c_{\mathrm{m},i} \tag{3}$$

where β is the stoichiometric conversion factor of this oxidation reaction.

For the advection part, the gaseous Darcy flow velocity v (m/s) in cover soils can be calculated by

$$v = \bar{k}_{\mathrm{p}} \Delta P / (\mu L) \tag{4}$$

where μ (kg/m/s) is the dynamic viscosity of gas; ΔP (Pa) is the pressure difference between atmosphere and landfill; L (m) is the total thickness of the biocover; $\bar{k}_{\mathrm{p}}$ (m^2) is the mean intrinsic permeability of the biocover and can be expressed as

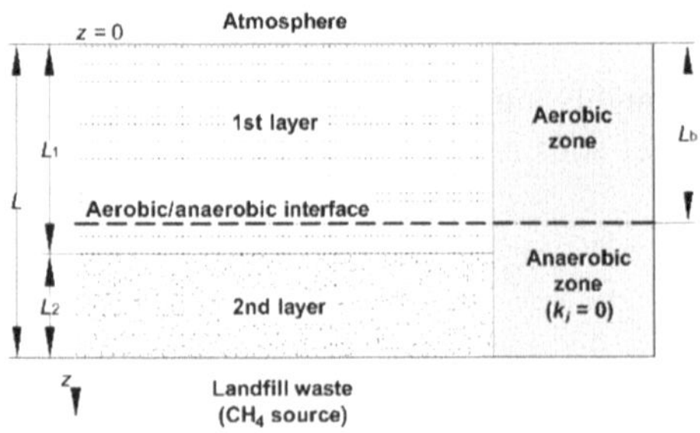

Fig. 1. Mathematical model for CH_4 and O_2 transport through the double-layer biocover.

$$\bar{k}_{\mathrm{p}}=L/(L_1/k_{\mathrm{p},1}+L_2/k_{\mathrm{p},2}) \tag{5}$$

where L_1 (m) and L_2 (m) are the thicknesses of first layer and second layer, respectively; $k_{\mathrm{p},1}$ (m^2) and $k_{\mathrm{p},2}$ (m^2) are the intrinsic permeabilities of each layer.

The boundary conditions are as follows:

$$c_{\mathrm{m},1}^{\mathrm{aer}}(0)=0, \quad c_{\mathrm{m},2}^{\mathrm{ana}}(L)=c_{\mathrm{m}}^{\mathrm{s}} \tag{6}$$

$$c_{\mathrm{O},1}^{\mathrm{aer}}(0)=c_{\mathrm{O}}^{\mathrm{atm}}, \quad D_2\frac{dc_{\mathrm{O},2}^{\mathrm{ana}}(L)}{dz}+vc_{\mathrm{O},2}^{\mathrm{ana}}(L)=0 \tag{7}$$

where $c_{\mathrm{m}}^{\mathrm{s}}$ is the CH_4 concentration at the bottom of cover system and $c_{\mathrm{O}}^{\mathrm{atm}}$ is the O_2 concentration in the atmosphere.

In this model, the continuity of chemical concentration and flux between two layers should be satisfied and can be expressed as

$$c_{j,1}^{(\mathrm{aer\ orana})}(L_1)=c_{j,2}^{(\mathrm{aerorana})}(L_1) \tag{8}$$

$$D_1\frac{dc_{j,1}^{(\mathrm{aerorana})}(L_1)}{dz}+vc_{j,1}^{(\mathrm{aerorana})}(L_1)=D_2\frac{dc_{j,2}^{(\mathrm{aerorana})}(L_1)}{dz}+vc_{j,2}^{(\mathrm{aerorana})}(L_1) \tag{9}$$

Similar continuity should also be satisfied at the aerobic/anaerobic interface:

$$c_{j,\mathrm{r}}^{\mathrm{aer}}(L_{\mathrm{b}})=c_{j,\mathrm{r}}^{\mathrm{ana}}(L_{\mathrm{b}}) \tag{10}$$

$$D_{\mathrm{r}}\frac{dc_{j,\mathrm{r}}^{\mathrm{aer}}(L_{\mathrm{b}})}{dz}+vc_{j,\mathrm{r}}^{\mathrm{aer}}(L_{\mathrm{b}})=D_{\mathrm{r}}\frac{dc_{j,\mathrm{r}}^{\mathrm{ana}}(L_{\mathrm{b}})}{dz}+vc_{j,\mathrm{r}}^{\mathrm{ana}}(L_{\mathrm{b}}) \tag{11}$$

where r equals 1 or 2 because the aerobic/anaerobic interface may appear in either of the two layers.

2.2 Analytical Solution of the Model

L_b should be estimated roughly initially. The general solution to the governing equation of CH_4 migration in cover soils can be formulated as

$$c_{m,i}^{(aer,ana)} = K_{1,i}^{(aer,ana)} e^{t_i^{(aer,ana)} z} + K_{2,i}^{(aer,ana)} e^{r_i^{(aer,ana)} z} \tag{12}$$

where

$$t_i^{(aer,ana)} = \left(-v + \sqrt{v^2 + 4D_i k_i}\right) \Big/ (2D_i) \text{ and } r_i^{(aer,ana)} = \left(-v - \sqrt{v^2 + 4D_i k_i}\right) \Big/ (2D_i) \tag{13}$$

In the anaerobic zone, the reaction rate constant, k_i, is regarded as zero and $r_i = 0$. According to the continuity equation (Eqs. (8)–(11)) and the boundary conditions (Eqs. (6) and (7)), the relationship of CH_4 concentration and its gradient between the top boundary and the bottom can be formulated as

$$\begin{aligned} \begin{bmatrix} c_{m,2}(L) \\ c'_{m,2}(L) \end{bmatrix} &= \begin{bmatrix} e^{-vL/D_2} & 1 \\ -ve^{-vL/D_2}/D_2 & 0 \end{bmatrix} \begin{bmatrix} K_{1,2}^{ana} \\ K_{2,2}^{ana} \end{bmatrix} \\ &= \begin{bmatrix} e^{-vL/D_2} & 1 \\ -ve^{-vL/D_2}/D_2 & 0 \end{bmatrix} \mathbf{A} \begin{bmatrix} 1 & 1 \\ t_1 & r_1 \end{bmatrix}^{-1} \begin{bmatrix} c_{m,1}(0) \\ c'_{m,1}(0) \end{bmatrix} \end{aligned} \tag{14}$$

The matrix $\mathbf{A}$ in Eq. (14) is the product of transfer matrixes representing the continuity relationship on the interfaces. If the aerobic/anaerobic interface is located in the first layer, the matrix $\mathbf{A}$ should be formulated as

$$\mathbf{A} = \begin{bmatrix} e^{-vL_1/D_2} & 1 \\ 0 & -v \end{bmatrix}^{-1} \begin{bmatrix} e^{-vL_1/D_1} & 1 \\ 0 & -v \end{bmatrix} \begin{bmatrix} e^{-vL_b/D_1} & 1 \\ 0 & -v \end{bmatrix}^{-1} \begin{bmatrix} e^{t_1 L_b} & e^{r_1 L_b} \\ D_1 r_1 e^{t_1 L_b} & D_1 t_1 e^{r_1 L_b} \end{bmatrix} \tag{15}$$

otherwise,

$$\mathbf{A} = \begin{bmatrix} e^{-vL_b/D_2} & 1 \\ 0 & -v \end{bmatrix}^{-1} \begin{bmatrix} e^{t_2 L_b} & e^{r_2 L_b} \\ D_2 r_2 e^{t_2 L_b} & D_2 t_2 e^{r_2 L_b} \end{bmatrix} \begin{bmatrix} e^{t_2 L_1} & e^{r_2 L_1} \\ D_2 r_2 e^{t_2 L_1} & D_2 t_2 e^{r_2 L_1} \end{bmatrix}^{-1} \begin{bmatrix} e^{t_1 L_1} & e^{r_1 L_1} \\ D_1 r_1 e^{t_1 L_1} & D_1 t_1 e^{r_1 L_1} \end{bmatrix} \tag{16}$$

Then the coefficients in the general solution for each layer, $K_{1,i}^{(aer,ana)}$ and $K_{2,i}^{(aer,ana)}$, can be calculated according to Eqs. (14)–(16). Based on the solution to CH_4 concentration, the solutions to CH_4 flux in the aerobic zone and the anaerobic zone can be derived. The CH_4 flux at the top of biocover is the CH_4 emission rate.

Additionally, the general solution to the governing equation for O_2 migration in cover soils can be formulated as

$$c_{\mathrm{O},i}^{(\mathrm{aer,ana})} = \begin{cases} \beta c_{\mathrm{m},i}^{\mathrm{aer}} + H_{1,i}^{\mathrm{aer}} \mathrm{e}^{-vz/D_i} + H_{2,i}^{\mathrm{aer}} & \text{(aerobic zone)} \\ H_{1,i}^{\mathrm{ana}} \mathrm{e}^{-vz/D_i} + H_{2,i}^{\mathrm{ana}} & \text{(anaerobic zone)} \end{cases} \quad (17)$$

The solution procedure for O_2 concentration in soils is similar to that for CH_4. The depth of the aerobic/anaerobic interface should be adjusted according to the comparison between the calculated O_2 concentration at the aerobic/anaerobic interface and the O_2 concentration threshold $c_{\mathrm{O}}^{\mathrm{cr}}$. If the calculated O_2 concentration at L_b is less than $c_{\mathrm{O}}^{\mathrm{cr}}$, L_b should be decreased; otherwise increased. This procedure will be repeated several times. Finally, the appropriate L_b can be determined by means of dichotomy.

3 Results and Discussion

3.1 Model Verification

In this study, a column experiment conducted by De Visscher et al. [4] are used to verify the analytical solution. The input parameters are as follows: $D_1 = 3.55$ 10^{-6} m^2/s, $v = 6.95 \times 10^{-6}$ m/s, $k_1 = 9.12 \times 10^{-4}$ s^{-1}, $L = 0.4$ m, and $\beta = 2$. The O_2 concentration limited threshold, $c_{\mathrm{O}}^{\mathrm{cr}}$, is 0.01. The boundary conditions are $c_{\mathrm{m}}^{\mathrm{s}} = 0.23$ and $c_{\mathrm{o}}^{\mathrm{atm}} = 0.21$.

The comparison of the variations of CH_4 and O_2 concentration with depth is shown in Fig. 2. The results of the present method match the experimental results reported by De Visscher et al. [4] reasonably well. Although the O_2 concentration at shallow depth has a little deviation, the CH_4 emission rate is close to the measured results (see Fig. 2), which means that the proposed model can well estimate CH_4 emission from cover soils.

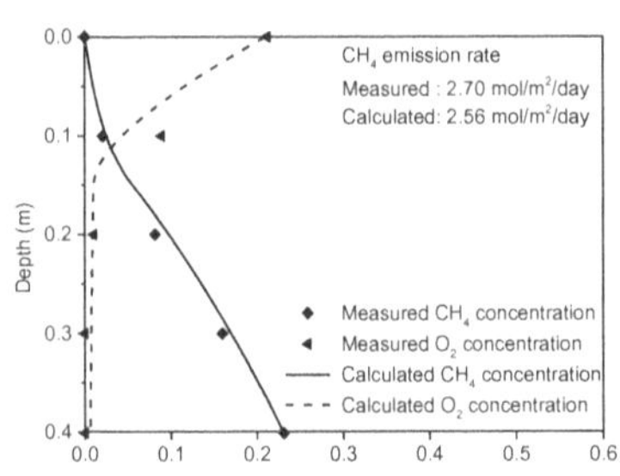

Fig. 2. Comparison between the proposed analytical solution and the reported measured results [4].

3.2 Influence of Effective Diffusivity

Effective diffusivity governs the diffusion process and are generally different in different layers. In this part, the influence of effective diffusivity of each layer is discussed. The comparison between homogeneous ($D_1 = D_2$) cover and double-layer biocovers are shown in Fig. 3. For the homogeneous cover, three values of effective diffusivity are adopted; namely, 2.90×10^{-6}, 1.14×10^{-6}, 2.02×10^{-6} m^2/s (the average of the first two values). Other input parameters are listed in Table 1. The results of homogeneous cases are greatly different from those of the double-layer case. Therefore, simplifying the double-layer biocover as a homogenous cover to simulate CH_4 migration may deviate from the real scenario significantly.

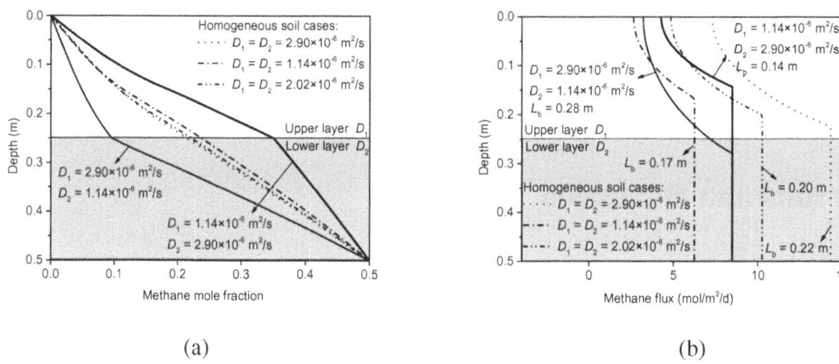

Fig. 3. Methane concentration (a) and flux (b) profiles for different effective diffusivities ($L_1 = L_2$).

In the above double-layer biocover, the effective diffusivity of the lower layer is smaller than that of the upper layer. In Fig. 3, another scenario is also simulated, in which the lower layer has higher effective diffusivity (2.90×10^{-6} m^2/s). The results show that this kind of biocover has poorer capacity of impeding CH_4 emission. The reason is as follows. If the upper layer has a larger effective diffusivity, the lower effective diffusivity of the lower layer will impede CH_4 entering the cover from landfill. On the other hand, the higher effective diffusivity of the upper layer facilitates the entering of O_2 into the cover from the atmosphere, which enlarges the aerobic zone in the cover and can consume more CH_4. Therefore, material with low effective diffusivity is suggested to be used at the lower part of biocover while that with high effective diffusivity is used at the upper part.

3.3 Influence of Advection

In order to investigate the influence of advection on the CH_4 migration, different pressure differences are adopted. The other input parameters are also listed in Table 1. The gaseous Darcy flow velocity can be calculated by Eq. (4). The CH_4 and O_2 concentration profiles given different gaseous Darcy flow velocities are shown in Fig. 4. Larger gaseous Darcy flow velocity leads to larger CH_4 concentration in shallow area, indicating that upward advection impedes O_2 entering the biocover from atmosphere and promotes CH_4 emission from the landfill. The aerobic zone is so thin with a larger gaseous Darcy flow velocity that the CH_4 oxidation is insufficient. The gas distributions are similar when the gaseous Darcy flow velocity is low, showing that the influence of advection can be neglected.

Table 1. Basic input parameters for methane migration analysis in double-layer biocover [5].

Parameters	Value	Parameters	Value	Parameters	Value
$\varphi_{g,1}$	0.27	L (m)	0.5	$\bar{k}_p$ (m^2)	10^{-13}
D_1 (m^2/s)	2.90×10^{-6}	L_1 (m)	0.25	μ (kg/m/s)	1.8×10^{-5}
$\varphi_{g,2}$	0.26	k (s^{-1})	10^{-4}	ΔP (Pa)	100
D_2 (m^2/s)	1.14×10^{-6}	B	2	c_O^{cr}	0.01

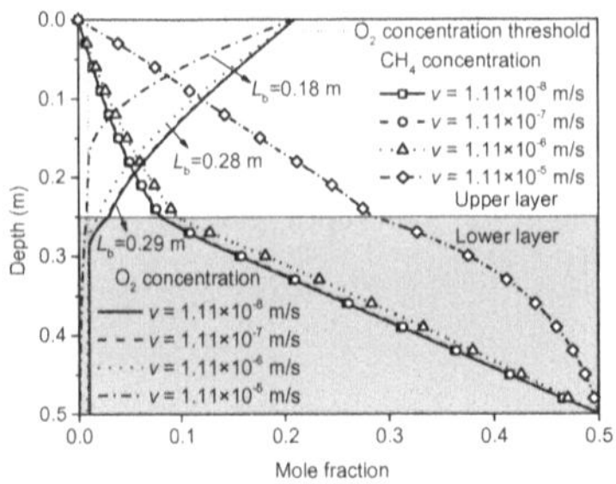

Fig. 4. Variation of gas concentration with gaseous Darcy flow velocity in biocover ($L_1 = L_2$).

4 Conclusions

An analytical model is developed to describe CH_4 migration through the typical double-layer biocover in this study. The model is effectively verified against the published experimental results. The influence of gaseous Darcy flow velocity and effective diffusivity is investigated. The upward advection has significantly negative impact on mitigating CH_4 emission through biocover and should be strictly controlled. It is suggested that the soil with low effective diffusivity is employed at the lower part of the biocover while that with high effective diffusivity is employed at the upper part. This double-layer biocover is beneficial to promote O_2 entering the cover soils and enlarge the aerobic zone in biocover so that more CH_4 can be consumed in cover soils.

Acknowledgments. Much of the work described in this paper was supported by the National Natural Science Foundation of China under Grant Nos. 41572265 and 41661130153, the Shanghai Shuguang Program under Grant No. 16SG19, the National Program for Support of Top-Notch Young Professionals, and the Newton Advanced Fellowship of the Royal Society under Grant No. NA150466. The writers would like to greatly acknowledge all these financial supports and express their most sincere gratitude.

References

1. Scheutz C, Kjeldsen P, Bogner JE, De Visscher A, Gebert J, Hilger HA, Huber-Humer M, Spokas K (2009) Microbial methane oxidation processes and technologies for mitigation of landfill gas emissions. Waste Manag Res 27:409–455
2. Huber-Humer M, Gebert J, Hilger H (2008) Biotic systems to mitigate landfill methane emissions. Waste Manag Res 26:33–46
3. Yao Y, Su Y, Wu Y, Liu W, He R (2015) An analytical model for estimating the reduction of methane emission through landfill cover soils by methane oxidation. J Hazard Mater 283:871–879
4. De Visscher A, Thomas D, Boeckx P, Van Cleemput O (1999) Methane oxidation in simulated landfill cover soil environments. Environ Sci Technol 33:1854–1859
5. Yao Y, Wu Y, Tang M, Wang Y, Wang J, Suuberg EM, Jiang L, Liu J (2015) Evaluation of site-specific lateral inclusion zone for vapor intrusion based on an analytical approach. J Hazard Mater 298:221–231

Effects of Leachate Concentration (Na^+, Pb^{2+}, COD) on Non-Darcy Flow of Compacted Clay

Shengwei Wang(✉), Wei Zhu, Haoqing Xu, Shi Shu, and Xihui Fan

Institute of Geotechnical Engineering,
Yangzhou University, Yangzhou 225127, China
dvlove2007@126.com

Abstract. Waste placed in landfills may produce pollutants with a high concentration of complex components from chemical and biological degradations. The hydraulic conductivity and threshold gradient of the clay liner are affected by the high concentration of complex components. In order to study the influence of leachate concentration on permeability of compacted clay, three pollutants namely Na^+, Pb^{2+} and COD were selected to conduct permeability study for compacted clay. COD is abbreviation of chemical oxygen demand. COD solution was prepared from dissolved glucose for the experiments in the paper. Flexible wall permeameter was adopted to perform a series of laboratory experiments. The results showed that Na^+, Pb^{2+} and COD concentration increased, the hydraulic conductivity increased and the threshold gradient decreased. When three solutions were provided with the same concentration, Pb^{2+} had the greatest influence on hydraulic conductivity and threshold gradient of saturated compacted clay, followed by Na^+ and COD solution. In the specific environment of landfills, COD had the greatest influence on hydraulic conductivity and threshold gradient of saturated compacted clay, followed by Na^+ and Pb^{2+} solution, which meant it was necessary to conduct further studies on the effect of COD solution on hydraulic conductivity and threshold gradient of saturated compaction clay.

Keywords: MSW landfill · Compacted clay
Leachate concentration (Na^+, Pb^{2+}, COD) · Hydraulic conductivity
Threshold gradient

1 Introduction

Due to the low permeability characteristic, compacted clay is always used as impermeable liners in landfills. Darcy's Law is a classic theory which has been widely used for seepage analyses in porous medium. However, Darcy's Law is not suitable for some porous medium, such as compacted clay. In 1869, King Hagen first discovered that water in porous media seepage process presented non-Darcy phenomenon (Longmuir 2004). Thereafter, a number of researchers also found that the seepage

---

Project: Natural Science Foundation of Jiangsu Province (No. BK20170501).

L. Zhan et al. (Eds.): ICEG 2018, ESE, pp. 101–108, 2019.
https://doi.org/10.1007/978-981-13-2224-2_13

process in compacted clay deviated from Darcy's law in the case of low seepage velocity and there exists a threshold gradient i_0 (Wang et al. 2017; Wang et al. 2018). Miller and Low (1963) thought that there was a "quasicrystal" structure at the interface between clay particles and water. A shear stress is required to deform the water-membrane structure to initiate seepage.

Waste in landfills will degrade over time and produce a large amount of high concentrations and complex compositions of contaminants. These contaminants can pass through the landfill liner by advection and diffusion and thereby pollute the underground environment. The COD can be selected as the key indicator for pollution alerts and be used to assess the environmental risk that is posed by municipal-solid-waste landfill sites (Shu et al. 2018). In order to prevent contamination of groundwater, the construction requirement for compacted clay liner is that the hydraulic conductivity must be no more than 1.0×10^{-7} cm/s (Qian et al. 2002). The leachate levels in these landfills tend to be very high and even reach more than 10 m (Chen and Zhan 2007). When leachate level is more than 10 m, Luo (2004) put forward that the hydraulic conductivity is no more than 1.0×10^{-8} cm/s, liner will still be unable to meet the whole service life. Wang et al. (2016) suggested that a threshold gradient for compacted clay liners be included to prevent high leachate leakage. Thus, the study of the high concentration effects on the hydraulic conductivity and threshold gradient of clay liner is very necessary.

The work related to non-Darcy flow permeation is still scarce. Zhu et al. (2016) conducted permeability experiments on four different sand-clay mixtures, among which dialysate was permeated by deionized water and $CaCl_2$ solution with a concentration of 0.2 mol/L. It is revealed that if the $CaCl_2$ solution was put into the mixed samples composed of of the four mixtures, the hydraulic conductivity was higher than those put by deionized water. Miller and Low (1963) used Wyoming Bentonite to conduct permeability experiments. It is found that under different ion concentration conditions, when lithium concentration dropped from 45.9% to 27.5%, threshold gradient decreased from 60 to 0, $i_0 = 6.7 - 10$ when $T = 20$ °C.

The hydraulic conductivity k and threshold gradient i_0 of compacted clay have a significant effect on leachate seepage and containment transport in waste containment facilities. Landfill leachate was composed of different sorts of pollutants with a high concentration. Previous scholars found that changes of leachate concentration could exert great influence over k and i_0 but were unaware of its internal mechanism. In order to solve the above-mentioned problems, this research studied clay permeability law of pollutants dialysate under three different concentrations. After analyzing the influence of Na^+, Pb^{2+} and COD concentration on dialysate viscosity and the interaction between particles and water, and why the influence of Na^+, Pb^{2+} and COD concentration on hydraulic conductivity and threshold gradient of compacted clay. The effect of three pollutions on the compacted clay liner was carried out, which has important guiding significance for the selection of the clay liner and the pollution control.

2 Experimental Studies

2.1 Materials

The clay used in this study was taken from Nanjing, Jiangsu province, China. The mineral components were tested by X-ray diffraction (XRD) atlas analysis. X-ray diffraction analysis of clay was conducted by a Rigaku D/max-rC rotating anode X-ray powder diffractometer. Air-dried powdered sample (particle size < 75 μm) of clay was used. The X-ray source was a Cu anode operating at 40 kV and 100 mA using Cu Kα radiation (λ = 1.5406 Å). The run speed was 3°/min. Data was collected between 5° and 40° in 2θ increments (Mitchell and Soga 2005). The mineralogical characterizations by X-ray diffraction of the clay is shown in Fig. 1. The particle distribution was tested by using Malvern 2000 laser particle analyzer. The grain size curves is presented in Fig. 2.

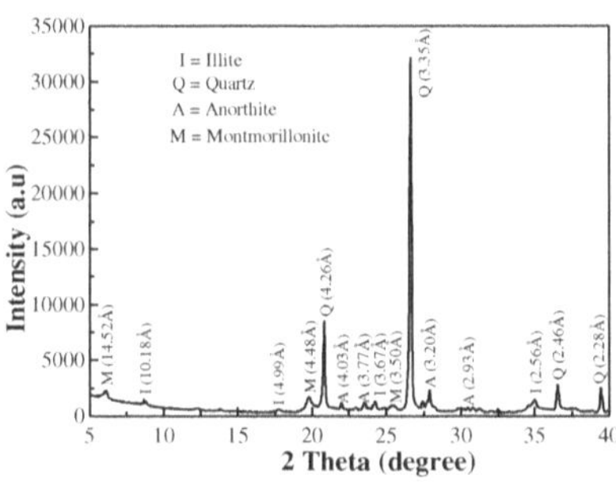

Fig. 1. X-ray diffraction pattern

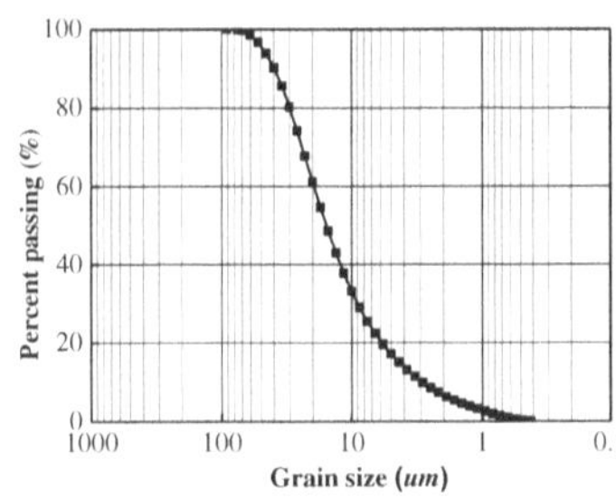

Fig. 2. Grain size distribution curve

Liquid limit(LL), Plastic limit(PL) and Specific gravity(G_s) tests were conducted per GB/T 50123-1999 method (China MOC 1999). From the relationship established between water content of soil and cone penetration depth, the water content corresponding to a depth of 17 mm is the liquid limit, and that corresponding to a depth of 2 mm is the plastic limit. The particle distribution was tested by using Malvern 2000 laser particle analyzer. Based on Usual Soil Classification System (USCS), it was low plasticity clay (CL). The cation exchange capacities (CEC) of the materials were determined using the JC/T593-1995 method (State Construction Materials Industry Administration 2004). The properties of the clay are presented in Table 1.

2.2 Leachate

In order to study the effect of solution concentration on permeability law, leachate were sampled in six typical Chinese landfills, which are Tianziling Landfill in Hangzhou, Shuige Landfill in Nanjing, Qizishan Landfill in Suzhou, Laogang Landfill in Shanghai, Jiangcungou Landfill in Xi'an as well as Xiaping Landfill in Shenzhen. Measured indicators included TN, TP, COD, Pb, Na, etc. The results of leachate component measurement regarding six typical Chinese landfills are shown in Table 2. It is indicated that COD content reached highest in the leachate, as high as 20000 mg/L. In terms of positive ion, Na^+ had the highest content, reaching 3500 mg/L.

Table 1. Basic properties of tested clay

Specific gravity, G_s	Liquid index, w_L (%)	Plastic limit, w_P (%)	Plasticity index, I_P (%)
2.72	48	24	24

Table 2. Analysis of leachates from six landfills in China

Concentration	Tianziling Landfill	Shuige Landfill	Qizishan Landfill	Laogang Landfill	Jiangcungou Landfill	Xiaping landfill
TN (mg/L)	1928.41	1858.41	1061.91	2317.91	2296.41	1998.41
TP (mg/L)	63.02	47.43	47.43	9.36	17.61	85.6
COD_{cr} (mg/L)	2762.79	16725.91	15442.64	19830.58	14635.97	20573.9
BOD_5 (mg/L)	2534	2750	3210	2329	1994	2351
Pb (mg/L)	0.27	0.11	0.05	0.21	0.03	0.98
Cr (mg/L)	0.47	0.3	0.23	0.4	0.42	1.31
Cd (mg/L)	0.041	0.017	0.017	0.017	0.004	0.342
Na (mg/L)	2425.84	1901.17	2556.3	1917.05	1765.02	3459.43

Two kinds of electrolyte solutions (NaCl, $PbCl_2$) and COD (glucose configuration) were used as the target dialysate in this study. The concentration of each component is listed in Table 2. Dialysate configuration was completed based on the corresponding leachate concentration from six measured landfills. Sodion (Na^+) solutions concentration were 0 mg/L, 1000 mg/L, 2000 mg/L and 4000 mg/L; Lead ion (Pb^{2+}) solutions concentration were 0 mg/L, 1 mg/L, 2 mg/L, 3 mg/L and 4 mg/L; COD (glucose) solutions concentration were 0 mg/L, 10000 mg/L, 30000 mg/L and 60000 mg/L.

2.3 Specimen Preparation

The soils were first air-dried, crushed, and sieved. Soils with particle size less than 1 mm were used for the laboratory experiments. Corresponds to the concentration of penetrants was added with stirring to make the moisture content of 22%. The soil was compacted in three equal layers into a cutting ring with a diameter of 7 cm and a height of 4 cm. Before the next layer was filled, the previous layer must be scarified to make seepage-proofing of the interface between two contact layers. The prepared soil samples were then transferred into the vacuum saturating chamber for 2 h and immersed in the corresponding concentration of penetrants for 24 h. It is noted that saturation degree of the samples can reach over 98%.

2.4 Permeability Test

In order to more accurately measure the seepage curve, permeability test should be conducted under a relatively large hydraulic gradient. Meanwhile, for the purpose of preventing lateral seepage in the process, flexible wall permeameter with constant head was used. First, the sample was loaded into a pressure chamber after saturation. Each sample was experienced three processes which are venting, consolidation, and permeation. It is assumed that 20 m solid waste with density of 1 g/cm^3 was loaded over the landfill liner. Thus, 200 kPa of consolidation pressure was adopted in the experiments. If the water level in the drainpipe did not change under a certain consolidation pressure, it means that the primary consolidation was finished. Then, the seepage pressure was applied on the sample. In order to prevent change of the porosity of the sample under high seepage pressure, the seepage pressure was applied using the way from high to low. At each level of seepage pressure, the water level in the drainpipe was read at intervals of two hours. The next level of seepage pressure was then applied until the water level in the drainpipe became stable.

3 Results and Discussion

3.1 Permeation Rules of Clay with Three Kinds of Pollutants

The porosities of all the samples under a confining pressure of 200 kPa are between 41% to 42%. This shows that all of the samples have the same porosity during the tests. At different leachate concentrations, seepage velocity v increased with the increase of hydraulic gradient i and exhibited good linear relationship. However, its reverse extension cord did not pass through the origin point, which showed a significant non-Darcy phenomenon, as shown in Fig. 3. The intersection of the $v \sim i$ line and i axis was defined as threshold gradient i_0. The slope of the $v \sim i$ line was hydraulic conductivity k.

3.2 Leachate Concentration Effects on k and i_0

Based on the seepage curves shown in Fig. 3, hydraulic conductivity k and threshold gradient i_0 can be obtained at different leachate concentration, as shown in Fig. 4. The values of hydraulic conductivity change in the 1.0×10^{-9} cm/s magnitude at different leachate concentration. All hydraulic conductivity values are less than 1.0×10^{-7} cm/s, which satisfy the requirement for landfill clay liner. Figure 4 shows that the values of the threshold gradient i_0 decease with increasing the Na^+, Pb^{2+} and COD concentration, which range from 58.5–65, 61–65 and 5–65 for compacted clay, respectively.

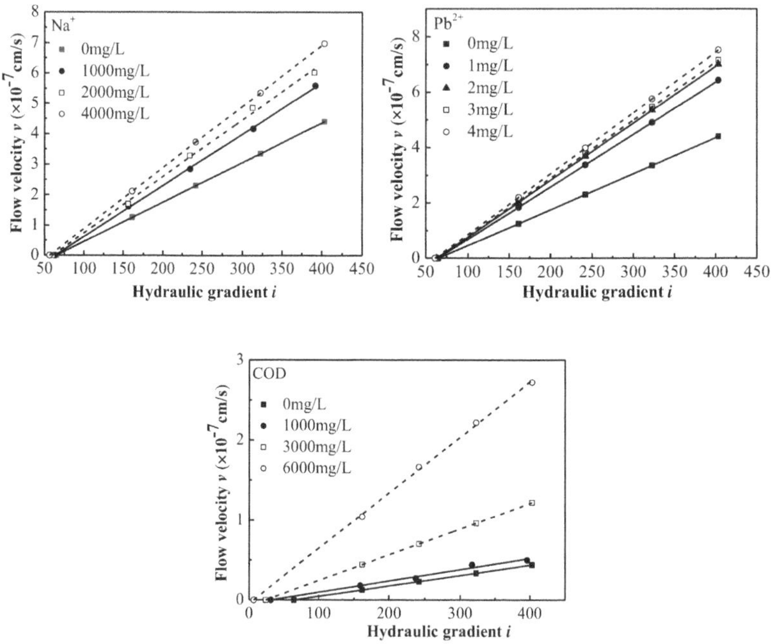

Fig. 3. Seepage curves of clay with three kinds of leachate

As leachate concentration increases, hydraulic conductivity increases and threshold gradient decreases. Based on seepage theory, the relationship between hydraulic conductivity k and intrinsic permeability K is as follows:

$$k = K \times \rho_w \times g/\mu \tag{1}$$

Based on previous studies, the relationship between the threshold gradient and apparent fluidity (K/μ) under normal temperature conditions is as follows(Prada and Civan 1999; Zeng et al. 2010):

$$i_0 = \mathrm{C} \times (K/\mu)^{\mathrm{D}} \tag{2}$$

where C and D are constants, K is intrinsic permeability, cm^2, μ is water dynamic viscosity, mpa s, ρ_w is fluid density, g/cm^3, g is gravitational acceleration, m/s^2.

On the hand, three kinds of solutions may affect water dynamic viscosity. As leachate concentration increases, water dynamic viscosity increases. On the other hand, three kinds of solutions may affect water morphology among particles, which would further influence permeability of saturated compacted clay.

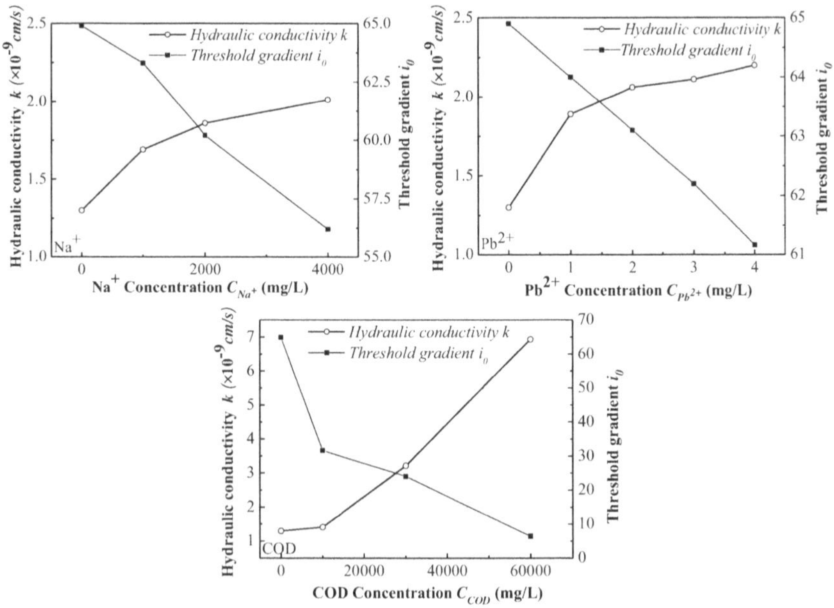

Fig. 4. Concentration vs. hydraulic conductivity, threshold gradient

3.3 Analyze the Permeability of Clay for Three Kinds of Leachate

When three solutions were provided with the same concentration, it is illustrated that Pb^{2+} had the greatest influence on hydraulic conductivity and threshold gradient of saturated compacted clay, followed by Na^+ and COD solution. However, due to the special environment of landfills, by measuring leachate component from six landfills, it is found that average concentration of COD reached 18000 mg/L in leachate, average concentration of sodium ion (Na^+) was 2000 mg/L while average concentration of lead ion (Pb^{2+}) solution was only 2 mg/L. In this sense, in the specific environment of landfills, COD had the greatest influence on hydraulic conductivity and threshold gradient of saturated compacted clay, followed by Na^+ and Pb^{2+} solution. So it was necessary to conduct further studies on the effect of COD solution on hydraulic conductivity and threshold gradient of saturated compaction clay.

4 Conclusions

From our study, the following conclusions can be drawn.

1. Under the same consolidation pressure (σ = 200 kPa), permeability tests were conducted on compacted clay at different concentration of Na^+, Pb^{2+} and COD. It is concluded that as the concentration of three pollutants increased, the hydraulic conductivity was increasing, but the threshold gradient was decreasing.

2. Effects of leachate concentration(Na^+, Pb^{2+}, COD) on non-darcy flow of compacted clay was studied. On the one hand, there existed certain differences with regards to solution viscosity. On the other hand, three kinds of solutions may affect water morphology among particles, which would further influence permeability of saturated compacted clay.
3. When three solutions were provided with the same concentration, it is illustrated that Pb^{2+} had the greatest influence on hydraulic conductivity and threshold gradient of saturated compacted clay, followed by Na^+ and COD solution. However, in the special environment of landfills, COD concentration had the greatest influence on hydraulic conductivity and threshold gradient of saturated compacted clay.

References

Chen Y, Zhan L (2007) Geo-environmental issues associated with landfills of municipal solid wastes. In: Proceedings of the 13rd Asian regional conference on soil mechanics and geotechnical engineering, pp 169–192

China MOC (1999) Standard for soil test method. GB/T 50123. Ministry of Construction of People's Republic of China, Beijing, China

Longmuir G (2004) Pre-Darcy flow: a missing piece of the improved oil recovery puzzle? SPE. https://doi.org/10.2118/89433-MS

Luo C (2004) Study on geoenvironmental characteristics of clay and the behavior of clay liner in MSW landfills. Zhejiang University, Hangzhou

Miller R, Low P (1963) Threshold gradient for water flow in clay systems. Soil Sci Soc Am J 27 (6):605–609

Mitchell J, Soga K (2005) Fundamentals of soil behavior, 3rd edn. Wiley, New York

Prada A, Civan F (1999) Modification of Darcy's law for the threshold pressure gradient. J Petrol Sci Eng 22(4):237–240

Qian X, Koerner R, Gray D (2002) Geotechnical aspects of landfill design and construction. Prentice Hall Inc., Upper Saddle River chap 18

Wang S, Zhu W, Qian X, Xu H, Fan X (2017) Temperature effects on non-Darcian flow of compacted clay. Appl Clay Sci 135:521–525

Wang S, Zhu W, Qian X, Xu H, Fan X (2016) Study of threshold gradient for compacted clays based on effective aperture. Environ Earth Sci 75(8):693

Wang S, Zhu W, Fei K, Xu C, Zhang N (2018) Study on non-Darcian flow sand-clay mixtures. Appl Clay Sci 151:102–108

Shu S, Zhu W, Wang S, Ng C, Chen Y, Abraham C (2018) Leachate breakthrough mechanism and key pollutant indicator of municipal solid waste landfill barrier systems: centrifuge and numerical modeling approach. Sci Total Environ 612:1123–1131

State Construction Materials Industry Administration (2004). The test method of bentonite. JC/T 593 - 1995. China Standards Press, Beijing, China

Zeng J, Cheng S, Kong X, Kong X, Guo K (2010) Non-Darcy flow in oil accumulation (oil displacing water) and relative permeability and oil saturation characteristics of low-permeability sandstones. Pet Sci 7(1):20–30

Zhu W, Xu H, Wang S, Fan X (2016) Influence of $CaCl_2$ solution on the permeability of different clay-based cutoff walls. Rock Soil Mech 37(5):1224–1230

Opening of an Environmentally Friendly Landfill of Non-hazardous Industrial Waste as Support for Closing an Uncontrolled Waste Landfill

Ernest Olinic[1(✉)] and Tatiana Olinic[2]

[1] Department of Geotechnical and Foundation Engineering, Technical University of Civil Engineering Bucharest, Bucharest, Romania
ernest.olinic@utcb.ro

[2] Department of Environmental Engineering and Land Reclamation, University of Agronomic Sciences and Veterinary Medicine Bucharest, Bucharest, Romania

Abstract. The paper refers to the closure of two old dumps, one for hazardous waste and one for non-hazardous waste. The hazardous waste landfill is bordered by perimeter dykes with dimensions of 80 × 80 m. Its closure was achieved by the execution of a closure system in accordance with the European legislation in force. In order to design and execute the closure of the non-hazardous waste landfill it is necessary to relocate waste in order to obtain 1: 3 slopes. Under these conditions, the maximum capacity of the closed landfill is 89500 m^3, resulting a deficit of 191100 m^3 of waste for which there is no possibility of relocation to the site. An alternative solution to waste relocation is to build a support structure on the eastern side of the existing landfill to ensure waste stability. A classical support structure (retaining wall, piles, etc.) will have very large dimensions and will involve significant economic efforts. An optimal solution is to build a filling up to the level of the existing deposit, i.e. to make an ecological landfill of non-hazardous industrial waste. The paper deals also with the regulations that had to be followed and the difficulties to obtain construction authorization.

Keywords: Old dump · Landfills closure · Waste stability

1 Introduction

Closure of a landfill should be done when it has reached its maximum storage capacity or when it no longer complies with the environmental and technical requirements.

The paper refers to the closure of two old dumps, one for hazardous waste and one for non-hazardous waste.

These landfills were built in 1970 by the Romanian public authorities and then, in 1991, were taken over by a private company. In the hazardous waste landfill were stored liquid waste until 1996, and after 2001 only basalt sludge and waste from the dyestuff industry were discharged. Since 2005, waste disposal was stopped in this landfill. In the non-hazardous landfill were stored municipal solid waste, non-hazardous

L. Zhan et al. (Eds.): ICEG 2018, ESE, pp. 109–117, 2019.
https://doi.org/10.1007/978-981-13-2224-2_14

sludge, industrial solid waste and construction and demolition debris until 2010, when waste disposal was stopped.

2 Topographical Conditions

The landfills site is located in the southwest part of city of Medias, at a distance of about 5 km from the city center and at an altitude of 360…390 m.

The access road to the site is the western border of the landfills. The landfills are bordered to the south, east and north part by the countryside (Fig. 1).

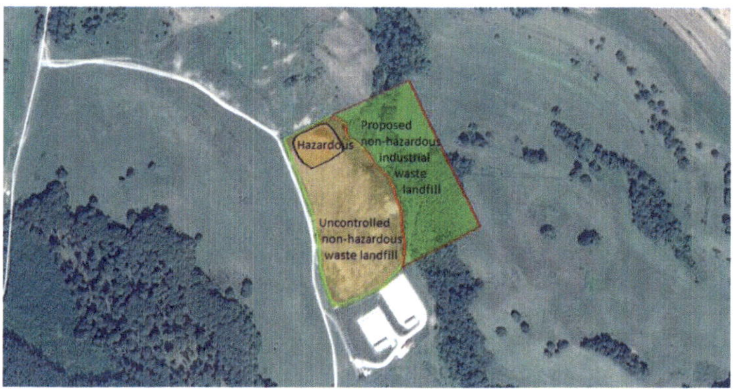

Fig. 1. Plan view of the landfills

The landfill of non-hazardous waste covers an area of approx. 38000 m^2, on a land with a general slope of approx. 1: 5, and natural ground level are descending in the west-east direction. The access in the site is on the west side, where the final level of waste is approximately equal to the road level.

The hazardous waste landfill has a rectangular shape, with dimensions of approx. 80 × 80 m and an area of 0.64 ha. It is bordered on all sides by a perimetral dike with a width of 2 to 3 m, a height up to 6 m and with slopes of approx. 1:2.

On the eastern side of the site there is a level difference of approx. 17 m, where the current slope of the waste is about 1:1. The eastern side of the site coincides with a stream, which is a tributary of the Târnava Mare River and has a small flow, which varies according to the quantity of precipitation.

3 Solutions for the Closure System

In Romania, the closure system of a landfill must comply with the requirements of OM 757/2004 "Technical Norm for Waste Storage". Depending on the type of waste, the surface occupied and the volume of waste stored, the closure solutions are different.

3.1 The Closure Solution for Non-hazardous Waste Landfill

According to the legislation in force, the solutions for the closure system of a non-hazardous waste landfill must complain the following requirements [1]:

- The maximum slope of the storage areas must be 1: 3;
- Landfills with slopes between of 1: 3 and 1: 5 must provide special berms in order to achieve roads for access. The berms are designed and constructed every 10 m high.
- The top landfill lining system must be composed of the layers shown shown below.
- Gas extraction wells must be uniformly distributed in the mass of gas-generating waste. The gas extraction wells are placed symmetrically and at equal distances between them (recommended, about 50 m). The gases will be recovered or burned.
- To collect rainwater falling on landfill site, it is necessary to construct a perimeter ditches and then to discharge the conventional clean waters into an emissary.

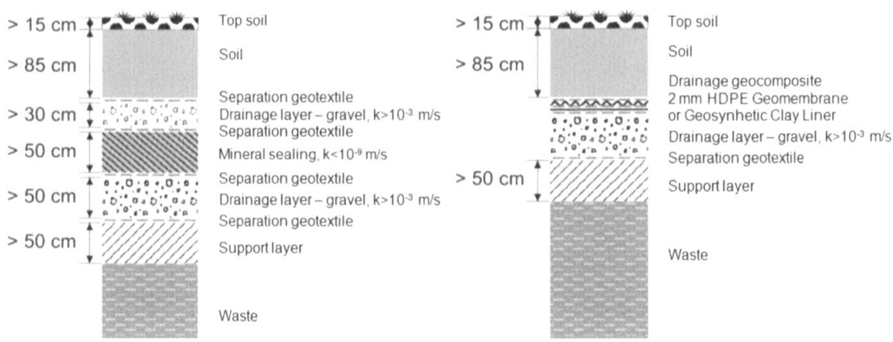

Fig. 2. Top landfill lining systems for non-hazardous landfills (required solution/alternative solution [1])

The article analyzes the conditions imposed by the national legislation in force regarding the closure of the landfills.

The solution for the closure of the landfill is presented in Fig. 3, where the existing profile of the landfill and the designed solution are shown, solution which provides a slope of 1:3 and 5 m wide berms for every 10 m on the vertical. It is noted that there is a quantity of excess waste that can not be relocated because the storage capacity is insufficient.

3.2 The Closure Solution for the Hazardous Waste Landfill

During the operation of the hazardous waste landfill, liquid waste was also stored. At the time of the closure project, a small amount of liquid waste was stored in the landfill, covering approx. 50% of the total landfill area. Considering the low amount of liquid

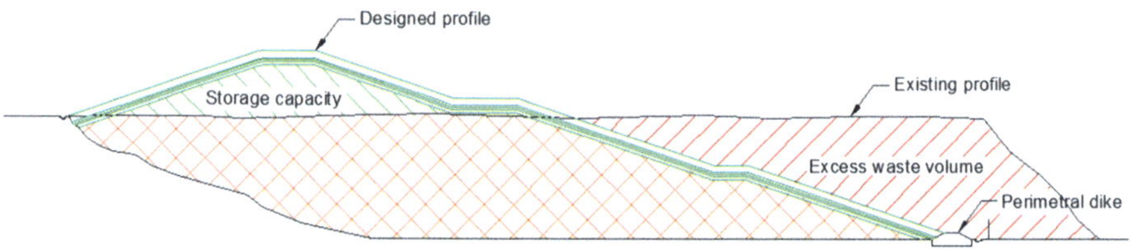

Fig. 3. Cross section through landfill (East-West direction) indicating the existing profile and the design solution

waste, it is assumed that they will be assimilated by the filler materials, and will partially fill the pores of the material.

The proposed works for the closure of the hazardous waste landfill fully reproduce the requirements of OM 757/2004 (Fig. 4).

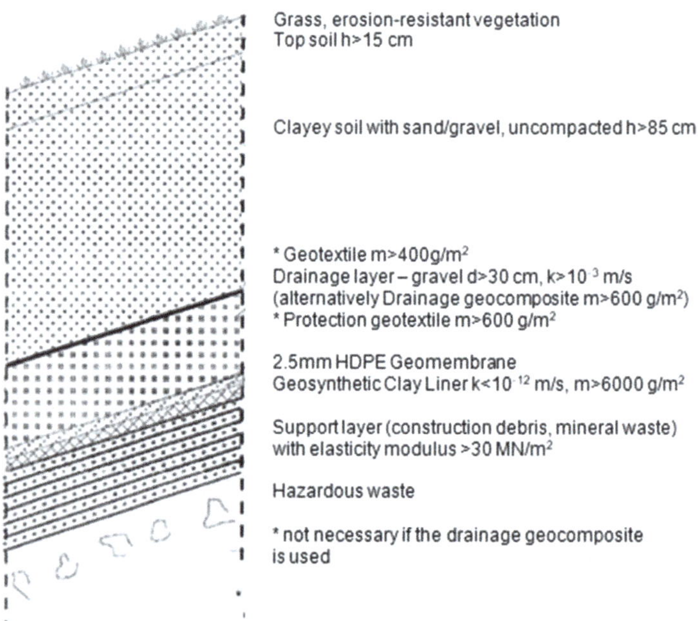

Fig. 4. The proposed closure system, identical to that required by OM 757/2004

The entire sealing system is designed with slopes of 1:10 to the perimeter dyke of the deposit. In order to collect runoff, a trapezoidal trench will be executed on the perimeter of the landfill. The trapezoidal trench will collect rainwater that falls on the surface of the landfill; water considered as conventional clean water that will be discharged into the stream that flow in the eastern side of the site.

4 Environmental Analysis for the Closure of the Non-hazardous Waste Landfill

The topography of the site and the operation of the landfill prior to the entry into force of the current waste disposal legislation, have led to excessive waste disposal and overcapacity of the landfill. To comply with current legislation, it would be necessary to excavate 70% of the landfilled waste and relocate it to another non-hazardous landfill site.

4.1 Solution 1 – Relocation of 191.000 m^3 of Waste

In order to have a quantitative image of the volume of waste which should be relocated, a topographical survey was made in May 2014 (Fig. 5). Based on an estimate shape of the initial ground profile and on the actual topographical survey, it was possible to calculate the approximate volume of land filled waste, respectively 280.600 m^3.

Fig. 5. 3D model of the current situation of the landfill

For the execution of the closure system of the non-hazardous waste landfill it should be necessary to relocate the waste in order to obtain 1: 3 slopes. Figure 6 shows the designed solution, where the maximum capacity of the closed landfill is 89.500 m^3 of waste, resulting a deficit of 191.100 m^3 of waste that can not be relocated within this non-hazardous waste landfill.

To comply with the legislation in force it would be necessary to excavate 70% of the stored waste and relocate it to an ecological landfill of municipal solid waste. The closest ecological landfills to the non-hazardous landfill site, are:

- Sighişoara Municipal Solid Waste Landfill at a distance of 44 km
- Cristian Municipal Solid Waste Landfill (CMID Sibiu) at a distance of 58 km
- Sânpaul Municipal Solid Waste Landfill (CMID Mureş) at a distance of 60 km

Currently, the Sighisoara municipal solid waste landfill has no extra capacity to store 191.000 m^3 of waste, and waste disposal in landfills from Cristian or Sânpaul,

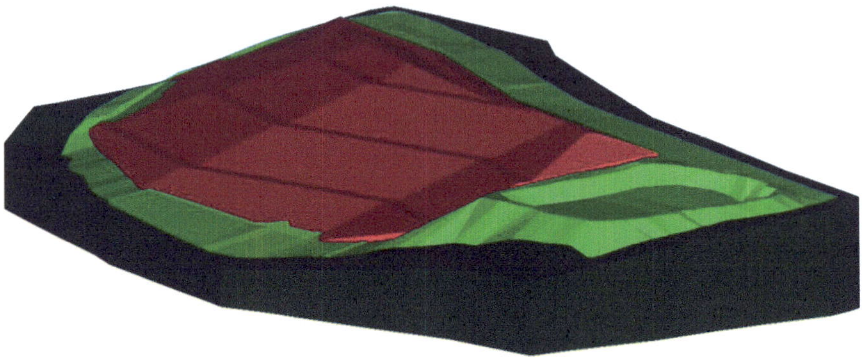

Fig. 6. 3D model of the designed shape of the landfill

beyond the very high costs, would bring significant changes to the national waste management strategy.

4.2 Solution 2 – Construction of an Ecological Landfill of Non-hazardous Industrial Waste

The need for the relocation of 191.000 m^3 of non-hazardous municipal waste arises from the obligation imposed by the legislation to achieve a slope of 1:3. Beside legislative imposition [6], engineering practice showed that, in the medium and long term, the stability of waste can be ensured on slopes less steep than 1:2.5…1:3.

An alternative solution to waste relocation is to establish a support structure on the eastern side of the existing landfill to ensure the stability of the waste. A classical support structure (retaining walls, pilots, etc. …) will have very large dimensions and will involve significant economic efforts. An optimal solution is to build a filling up to the level of the existing deposit by designing and creating a non-hazardous industrial waste landfill.

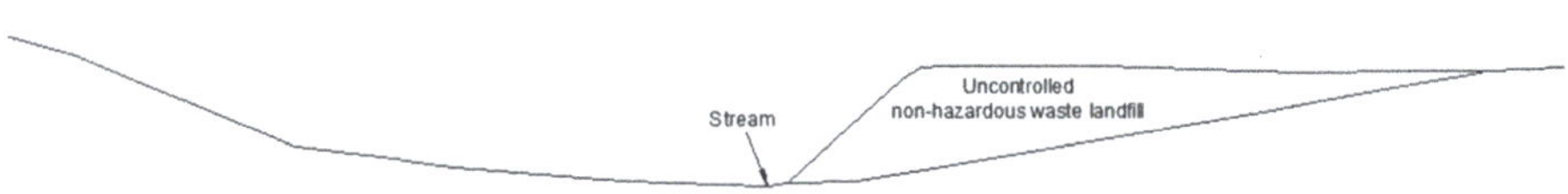

Fig. 7. The current situation

By designing a new ecological landfill of non-hazardous industrial waste on the eastern side of the existing landfill, it should be obtained a filling with the role of supporting the waste from the non-ecological landfill. This structure will ensure the long-term stability of the existing landfill, will reduce the volume of waste which must be relocated and will ensure the storage capacity.

To ensure the transport of the water (including waters from heavy rain periods) at the base of the existing landfill there will be placed a large diameter drain on the line of the existing stream.

At the base of the existing landfill, on the eastern side, a compacted earth prism will be constructed, prism which will ensure the short-term stability of the waste in the existing landfill.

In the eastern area of the existing landfill the bottom lining system of the ecological landfill of non-hazardous industrial waste will be made in accordance with the national legislation in force.

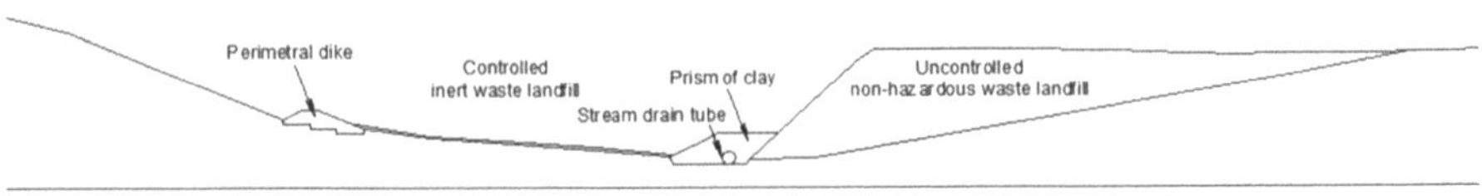

Fig. 8. Construction of the new ecological landfill bottom lining system

After the construction of the new ecological landfill of non-hazardous industrial waste, the waste will be relocated from the existing landfill in order to provide a slope that allows the construction of a double landfill sealing system:

- A landfill top lining system for the initial non-hazardous landfill
- A landfill bottom lining system for the new ecological landfill of non-hazardous industrial waste

The construction of the new ecological landfill will have a slopes of 1:2 … 1:2.5 in order to ensure the waste stability for the short term. On the long term, the stability will be ensured by the industrial waste that will be disposed in this new ecological landfill. In this case, the volume of relocated waste is reduced to about 15.000 m^3 which can be partially relocated even in the plateau area of the existing landfill.

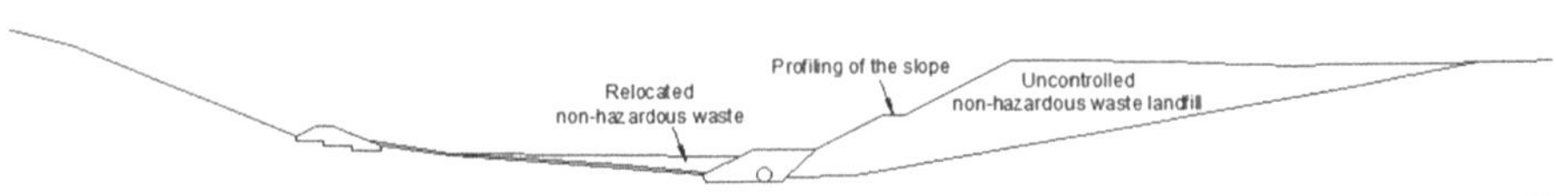

Fig. 9. Relocation of non-hazardous waste to the new ecological landfill

After the relocation of the waste from the existing landfill, it will be possible to build the closure system according to the national legislation in force.

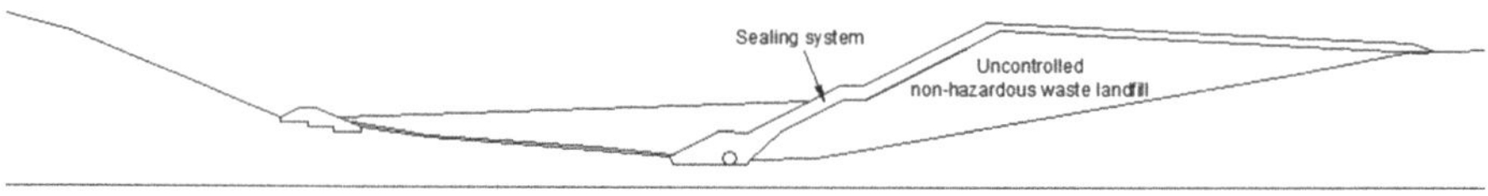

Fig. 10. The closure of non-hazardous waste landfill

After the exploitation of the new ecological landfill of non-hazardous industrial waste, this can be closed according to the national legislation in force.

Fig. 11. The closure of the new ecological landfill of non-hazardous industrial waste

4.3 Comparative Analysis of Proposed Solutions

Solution 1 – relocation of 191.000 m^3 of waste	Solution 2 – construction of an ecological landfill of non-hazardous industrial waste
The solution has only technical and economical disadvantages: Waste relocation can be done in two ecological landfills situated at distances of approx. 60 km The above-mentioned landfills are part of the national non-hazardous municipal waste management plan and have been designed without taking into account the storage of such a large amount of waste The storage capacity of each of the two ecological landfills is approx. 1.5 mil. m^3, respectively, 20 years after commissioning. The volume of waste that needs to be relocated from the Medias landfill is approx. 13% of storage capacity, which would reduce the lifetime by about 2.5 years In Romania, the new legislation on waste disposal came into force in 2004, and the economic operator had only 6 years to collect the funds necessary to close and monitor the landfill. Basically, the amounts collected during this period of six years (2004–2010) far exceed the funds required for closing and monitoring the landfill	The solution has the following advantages: The costs for closing the non-compliant landfill are much lower due to: -The waste transport distance is much smaller; -The volume of waste to be relocated is much smaller; -Storage costs much lower The new ecological landfill of non-hazardous industrial waste will ensure the long-term stability of the non-compliant landfill The solution does not influence or overlap with the regional and national strategy for municipal waste management The opening of an ecological landfill of non-hazardous industrial waste in Medias will serve economic operators near the site The opening of an ecological landfill of non-hazardous industrial waste would provide to the operator the necessary funds for the closure and monitoring of the non-hazardous landfill site

5 Conclusions

The Romanian legislation on Waste Management Strategy is harmonized with the European Union Directives. Recently, it was reviewed and approved a new version of The National Waste Management Plan [5] which adopt European policy objectives on WMP Guide 2012: Preparing a Waste Management Plan - A methodological guidance note and WPP Guide 2012: Preparing a Waste Prevention Programme – Guidance document.

An important aspect for alignment with EU requirements is the closure of non-compliant waste landfill sites. This is a significant problem for many countries in eastern Europe and around the world.

In order to increase the waste storage capacity, the achievement of steep slopes that exceed stable slope is a practice made known in several scientific papers, and moreover, there are presented many case studies in which waste slope failures have occured (Athanasopoulos et al., 2013, [3]).

The closure of non-compliant landfills that were opened before the entry into force of legislation on the design of ecological landfills represents an ecological problem that belongs both to economic operators and local authorities.

The lack of legislation on waste disposal in Romania before 2003 led to an uncontrolled exploitation of the landfills site by reaching volumes that exceeds the storage capacity and very steep slopes that are stable only on the short term.

Usually, non-compliant landfills are located in areas with steep slopes, near watercourses or areas with difficult foundation conditions and their closure solution requires extensive investigations and detailed technical projects.

References

1. Athanasopoulos G, Vlachakis V, Zekkos D, Spiliotopoulos G (2013) The December 29th 2010 Xerolakka Municipal Solid Waste landfill failure. In: Proceedings of the 18th international conference on soil mechanics and geotechnical engineering, Paris, France, pp 302–312
2. Feodorov V, Manea S, Batali L, Sofrone D (2000) Use of geosynthetics in landfills. Case studies from Romania. In: Proceedings of the second European geosynthetics conference. Patron Editore, Bologna, Italy, pp 509–514
3. Reddy KR, Munwar Basha B (2014) Slope stability of waste dumps and landfills: state-of-the-art and future challenges. In: Proceedings of Indian geotechnical conference ICG-2014, Kakinada, India, pp 2311–2337
4. Optimum Consulting SRL (2014) Technical project for closure of non-hazardous municipal solid waste landfill and hazardous industrial waste landfill in Medias. Beneficiary: Prestsal SA Medias, Tarnava, Sibiu County
5. HG 942/2017: Approval of the national waste management plan
6. OM 757/2004: Technical norm on waste storage

Installation and Performance of Horizontal Wells for Leachate Level Control in Tianziling MSW Landfill, China

Jie Hu[1(✉)], Han Ke[1], Zu Yu Chen[2], Ji Wu Lan[1], Liang Tong Zhan[1], and Yun Min Chen[1]

[1] MOE Key Laboratory of Soft Soils and Geoenvironmental Engineering, Institute of Geotechnical Engineering, Zhejiang University, Hangzhou 310058, China
hujie1993@zju.edu.cn

[2] China Institute of Water Resources and Hydropower Research, Beijing 100044, China

Abstract. This report describes a pilot test at Tianziling MSW landfill to assess the feasibility of lowering the leachate level using horizontal wells. Three horizontal wells S01, S02 and S03 (length 50 m, 56 m and 55 m, respectively) were successfully installed using the improved directional pipe-protected drilling method. The average leachate flow rate of S01 and S03 were 10.66 m^3/day and 3.93 m^3/day, respectively after installation of the pipes. After 94 days draining of S01, the maximum and average leachate level drawdowns around S01 were 2.7 m and 1.7 m, respectively. After 42 days draining of S03, the maximum and average leachate level drawdowns around S03 were 1.5 m and 1.1 m, respectively. The radius of influence for S01 and S03 were in the range of 40–50 m. This test in general demonstrates the effectiveness of horizontal wells in lowering leachate level of MSW landfills. The obtained experience can be useful for civil projects that have similar concerns of draining the ground water.

Keywords: MSW · Landfill · Horizontal well · Drilling · Leachate Flow rate

1 Introduction

In the process of landfilling, a large amount of leachate is produced due to the rainfall infiltration and degradation of the waste. Due to the low hydraulic conductivity of waste in deep layer and the gradually clogging of leachate drainage system, leachate cannot be effectively removed from landfills [1]. The development of high leachate level would bring ground water pollution, reduce landfill gas recovery efficiency and give rise to landslide disasters [2]. To ensure the safe operation of landfills, vertical wells are conventionally used as a contingency or long-term measure to lower the high leachate level for its construction convenience and good performance [3]. However, it may be insufficient to meet the leachate level control requirement if only vertical wells are installed [4]. Horizontal wells, with their likely greater ratio of submerged screen to

L. Zhan et al. (Eds.): ICEG 2018, ESE, pp. 118–126, 2019.
https://doi.org/10.1007/978-981-13-2224-2_15

total installed length, represent a potential alternative to lower leachate level. Cox et al. [5] and Beaven et al. [6] from University of Southampton took likely the first reported horizontal wells dewatering field test in Rainham landfill site in Essex, United Kingdom. Three horizontal wells HW1, HW2 and HW3 (length 256 m, 236 m and 250 m, respectively) were installed using the directional drilling method. The average leachate flow rates of HW1, HW2 and HW3 were 7.00, 4.84 and 2.42 m^3/day, respectively. The pore pressures in the waste around the wells fell down by 80 kPa after 595 days draining. It indicated that horizontal wells could be an effective means of controlling leachate level in landfills.

The aim of this work was to assess the feasibility of using horizontal wells to lower leachate level in Chinese MSW landfill. It firstly focused particularly on the difficulties of installing the horizontal wells in these heterogeneous materials lack of sufficient geological information such as the landfilled waste. This paper illustrated and discussed the technology and design details to facilitate a successful installation of the horizontal well. The paper also reviews the monitored leachate flow rate and pore pressures in the waste around the wells to evaluate the performance of the project. The engineering signification of the horizontal wells has been discussed.

2 Materials and Drilling Method

2.1 Site Description

Tianziling landfill is located in Gongshu District of Hangzhou, China. It is a valley infill landfill that was put into operation in 1991. Its original capacity and design service life were 6 million m^3 and 16 years, respectively. The lateral and vertical expansion project has been in progress since 2007. Its original capacity increased to 22 million m^3 since the expansion project started. Currently, approximately 8000 tons of MSW is disposed in this landfill daily. The total thickness of the waste body was approximately 50 m. As shown in Fig. 1, the landfill consisted of a number of filled platforms that were set back at an embankment slope and a rock-fill dam retains the lowest platform. A flat region that was at least 60 m away from the dam of the landfill was chosen for the tests. Three horizontal wells S01, S02 and S03 were installed to lower the high leachate level which was nearly 5 m beneath the surface.

The design length of the horizontal wells S01, S02 and S03 were 50 m, 100 m and 50 m, respectively. The space between each wells was 25–30 m. As shown in Fig. 1, the well casing starts with a non-opening section in front with a length of 10 m, followed by an opening section with the screened pipe. The wells were made up of galvanized steel which has good performance of resistance against corrosion and deformation. The angle of inclination to horizontal of the pipe was 5% (height/length). The diameter and thickness of the pipe were 133 mm and 6 mm, respectively. The slot diameter was 12 mm accounting for 6% of the total area.

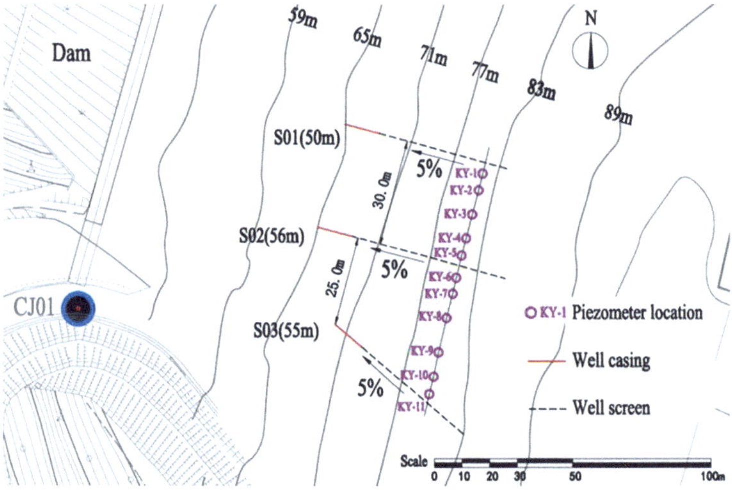

Fig. 1. Layout of horizontal wells in Tianziling MSW landfill

2.2 Installation Method and Equipment

A method called directional pipe-protected drilling which was similar with that of Cox et al. [5] was adopted. Its construction procedures are as follows:

(1) The directional drill rod was firstly drilled into the formation at the designed angle of inclination to create a pilot bore.
(2) Once the pilot bore was completed, an overwash casing was drilled into the formation to create the space for placing the well screen.
(3) After the overwash casing had been installed, drilling fluid (e.g. soap liquid) was pumped into the overwash casing to keep it free of waste.
(4) The well screen was subsequently pushed into the overwash casing. Finally, the overwash casing was pulled out of the borehole, leaving the well screen in place.

The installation machine was DDW-20 horizontal directional drilling rig which was produced by Sinohydro Foundation Engineering Co., Ltd. The length and width of the drilling rig were 4 m and 2 m, respectively. Its relatively small size made it easy to move at the landfill surface. The torque capacity, pullback capacity and rotational speed of drilling rig were 6000 N.m, 200 kN and 10–160 r/min, respectively. The torque capacity is the key parameter that determines the maximum length of horizontal wells when using directional pipe-protected drilling method.

2.3 Monitoring Methods

To evaluate the performance of the horizontal wells, the leachate and gas flow rates of wells and pore pressures in the waste around the wells were monitored. The monitoring method of leachate and gas flow rates was similar to that used in Cox et al. [5]. Each well was capped by a well head connected to the separation tank with 110 mm diameter high density polyethylene (HDPE) pipe. The volume of separation tank was

$3\ m^3$ which was large enough to allow the gas to separate from the leachate. Leachate flowed from the hole that is opened at the bottom of the tank, then passed through an electromagnetic leachate flowmeter to measure the leachate flow rates. Meanwhile, the gas was vented to atmosphere through a separate impeller gas flowmeter for monitoring its flow rates.

Piezometers were installed within the landfill so that the pore pressure response to dewatering from the wells could be monitored. The piezometers were installed using a 110 mm diameter rotary barrel auger and the boreholes were sealed by bentonite. Figure 1 showed the plan view of the piezometers. The piezometers were arranged to be perpendicular to the trend of the wells. As pore pressures within landfills should not in general be expected to vary hydrostatically with depth [6], 16 piezometers were installed at different depth. As shown in Fig. 2, the response zone of the piezometers (KY01L-KY05L and KY06-KY11) is expected to be level to the horizontal wells. The elevation of response zone in the upper horizon (KY01U-KY05U) is 3 m higher than the horizontal wells.

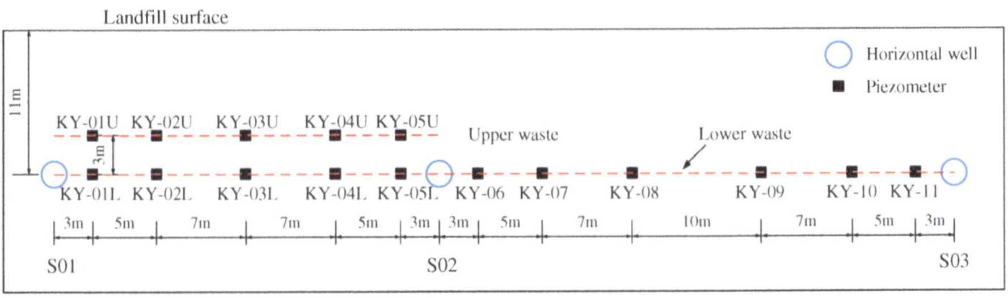

Fig. 2. Cross sectional view of the horizontal wells and piezometers

3 Difficulties Encountered and Solutions

In the first trial of S01, the directional drill rod was easy up to 50 m into the formation. After the pilot bore was complete, an overwash casing was drilled over the drill rod which acted as a guide. In the process of overwash casing drilling, flushing fluid was pumped through the drill rod to

- lubricate and cool the drilling assembly,
- convey cuttings in suspension from the borehole,
- provide a fluid pressure to prevent collapse of the borehole wall.

The permeability of the waste around the borehole should not be disturbed, therefore, the drilling fluid used there was clean water. Figure 3 showed the torque varying with the casing drilling length in the first trial of S01. When the overwash casing was drilled into 20 m, the torque was increased to 6000 N.m, at which the drilling could not be performed further. Thus the casing was pulled out. The over-torque was mainly due to:

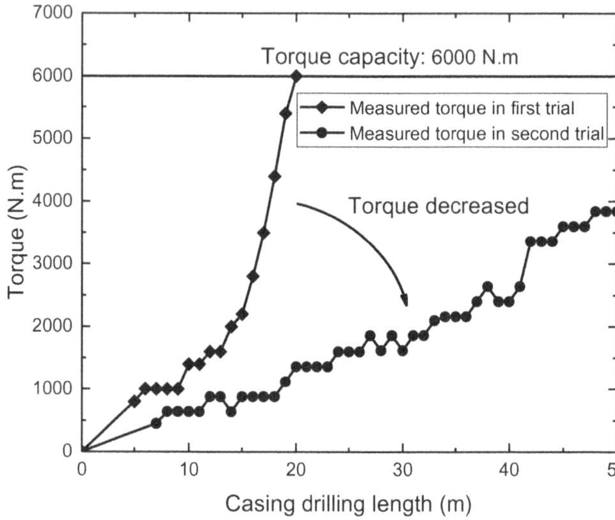

Fig. 3. Variation of torque with casing drilling length in the trial of S01

(1) Improper drilling method. The overwash casing was drilled after the rod had been completely penetrated into the formation. If the overwash casing could not catch the process of the pilot drilling, the borehole would collapse. The collapsed borehole would make the drilling of casing more difficult.
(2) Improper drilling bits. The diameter of casing drilling bit was same with that of casing, which would make the casing attached tightly with the surrounding waste. The drill rod bit was a flat plate, and it drilled forward through pressing the waste. The strength of the waste after being pressed would increase, making the waste harder be cut off.
(3) Improper drilling fluid. Selection of proper drilling fluid is important. Possible fluid used in normal drilling includes bentonite slurry, clean water and saponifier. Saponifier is only used in case when the difficult geological formation is encountered. However, the drilling was occasionally hampered due to a poor judgment on whether the saponifier should be adopted, so that the difficult ground cannot be lubricated and passed smoothly.

To accomplish the design length (i.e., 50 m) of S01, some improvements were made as following:

(1) The overwash casing was drilled simultaneously with that of the rod. The drill rod was only 2 m ahead of the overwash casing.
(2) The diameter of casing bit was changed to be larger (shown in Fig. 4).
(3) The drill rod bit was changed to a cutting bit (shown in Fig. 5), which cut off the waste more efficiently.
(4) Saponifier was frequently added to the clean water to lubricate the casing.

Fig. 4. Improvement of the casing bit

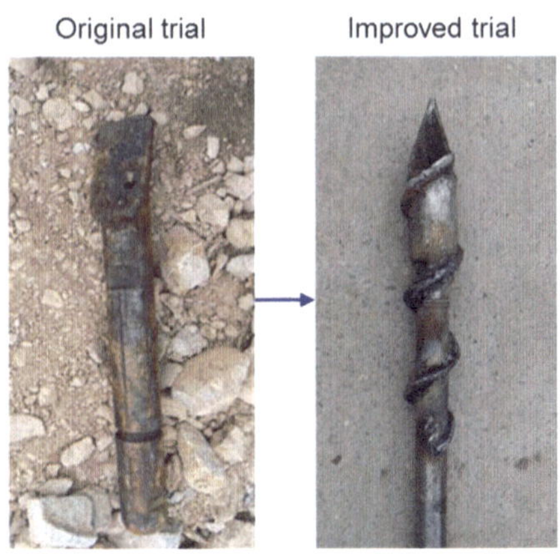

Fig. 5. Improvement of the drill rod bit

Figure 3 showed the torque varying with the casing drilling length in the second trial of S01. The torque decreased remarkably to only 4000 N.m when casing drilling length reached 50 m with the above mentioned improvements. The improvements mainly decreased the friction between the casing and waste, and then reduced the torque. Through the repeated drilling of horizontal well S01, the improved drilling methods and installation experience were obtained. The installation process of the horizontal wells S02 and S03 were relatively smooth except a breakdown of the 16 m-long casing in S02. Finally, the installed length of S01, S02 and S03 were 50 m, 56 m and 55 m, respectively.

4 Monitoring and Discussion

4.1 Leachate Flow Rates

Figure 6 showed the leachate flow rate and cumulative leachate flow of S01 varying with the draining time. The leachate flow was high in the early stage with a rate of 75 m^3/day. During the drilling process, a large amount of fluid was injected into the borehole. Part of leachate in the early stage might come from the injected drilling fluid. The leachate flow rate declined quickly to 20 m^3/day at day 3. Then, the leachate flow rate ranged from 20.5 to 34.5 m^3/day from day 3 to 14. It was mainly due to the release of the landfill gas. As the draining continued, the leachate level above the horizontal well decreased gradually. Thus, the hydraulic gradient of leachate flow to the well would decrease. Then, the leachate flow rate decreased gradually to 3 m^3/day at day 90. The cumulative leachate flow of S01 increased to 959 m^3 at day 90 and the average leachate flow rate was 10.66 m^3/day.

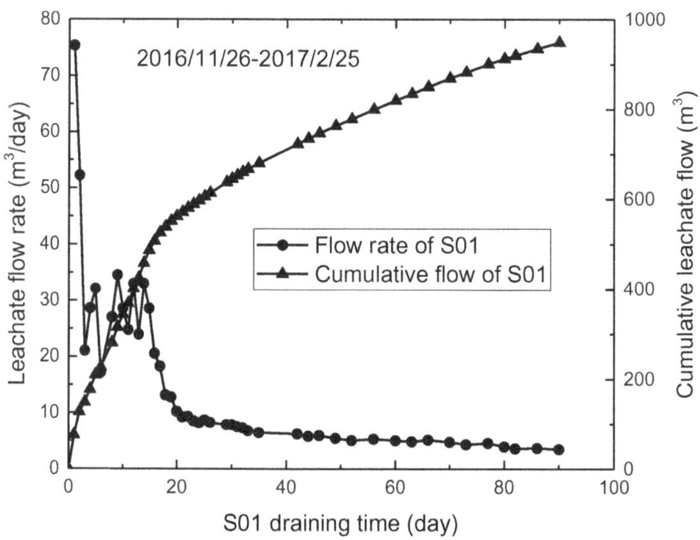

Fig. 6. Leachate flow rate and cumulative leachate flow from S01

As the 16 m-long casing was broken down at the borehole during the installation of S02, the leachate and gas flow rates were very low (<1 m^3/day). The flow rates of S02 had not been measured. Similar to S01, the leachate flow rate of S03 was also high in the early stage and then decreased quickly. The leachate flow rate of S03 was 19.6 m^3/day at day 1 and decreased to 0.5 m^3/day at day 40. The average leachate flow rate of S03 was 3.93 m^3/day.

4.2 Pore Pressures

Figure 7 showed the initial (i.e., day 20) and final (i.e., day 94) pore pressures in different horizontal distance from S01. Obvious decrease of pore pressures in the upper and lower sections could be observed. With the horizontal distance from S01 increased, the decreased value of pore pressures reduced. For the upper and lower sections, the decreased values of pore pressures at distance 3 m were 20.34 kPa and 26.95 kPa, respectively. And the decreased values of pore pressures at distance 27 m were 10.03 kPa and 10.84 kPa, respectively. If it was assumed that the pore pressure gradient between 3 and 27 m continued linearly to the original piezometric level, the radius of influence for S01 might be estimated as 50.3 m for the upper section and 43.1 m for the lower section.

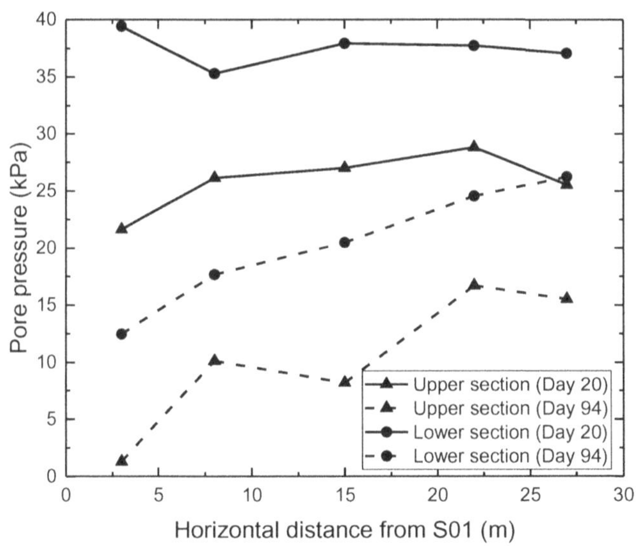

Fig. 7. Cross section of initial and final pore pressures across S01

Similar to S01, the pore pressures around S03 decreased faster in the early stage. The average pore pressures around S03 decreased from 22.13 kPa at day 1 to 11.05 kPa at day 42. After 42 days draining of S03, the decreased values of pore pressures at distance 32 m and 37 m from S03 were 18 kPa and 7 kPa, respectively. According to the pore pressure gradient, the radius of influence for S03 was estimated as 40.2 m. All the measurement indicated the pore pressures had been reduced irreversibly.

5 Conclusions

The horizontal wells field tests have been carried out at Tianziling landfill to assess the feasibility of using horizontal wells to lower leachate level in Chinese MSW landfill. The conclusions drawn from the field test are as follows:

(1) The pilot tests have demonstrated the feasibility of installing horizontal wells in MSW landfills using the directional pipe-following drilling method. The difficulties in drilling through complicated landfills can be overcome by the technology that includes improved casing bit and drill rod bit and proper use of lubricate fluid.
(2) The leachate flow rate of the horizontal wells decreased quickly with the draining time. For S01 and S03, the initial (i.e., day 0) leachate flow rates are 75.0 m^3/day and 19.6 m^3/day, respectively. The average leachate flow rate of S01 and S03 are 10.66 m^3/day and 3.93 m^3/day, respectively.
(3) After 94 days draining of S01, the maximum and average leachate level drawdowns around S01 are 2.7 m and 1.7 m, respectively. After 42 days draining of S03, the maximum and average leachate level drawdowns around S03 are 1.5 m and 1.1 m, respectively. The influence radius for S01 and S03 are in the range of 40–50 m.

Acknowledgements. The work was supported by the National Natural Science Foundation of China (Grant Nos. 51578503 and 41502276), and the Major Science and Technology Projects of Zhejiang Province (Grant No. 2015C03021). Great thanks to Hangzhou Environment Group Co., Ltd and Sinohydro Foundation Engineering Co., Ltd.

References

1. Zhan TLT, Xu XB, Chen YM, Ma XF, Lan JW (2015) Dependence of gas collection efficiency on leachate level at wet landfills of municipal solid wastes and its improvement methods in China. J Geotech Geoenvironmental Eng 141(4):1–11
2. Ke H, Hu J, Xu XB, Wu XW, Li YC, Lan JW (2018) Analytical solution of leachate flow to vertical wells in municipal solid waste landfills using a dual-porosity model. Eng Geol 239:27–40
3. Slimani R, Oxarango L, Sbartai B, Tinet AJ, Olivier F, Dias D (2017) Leachate flow around a well in MSW landfill: analysis of field tests using Richards model. Waste Manag 63:122–130
4. Al-Thani AA, Beaven RP, White JK (2004) Modelling flow to leachate wells in landfills. Waste Manag 24(3):271–276
5. Cox SE, Beaven RP, Powrie W, Cole DJ (2006) Installation of horizontal wells in landfilled waste using directional drilling. J Geotech Geoenvironmental Eng 132(7):869–878
6. Beaven RP, Cox SE, Powrie W (2007) Operation and performance of horizontal wells for leachate control in a waste Landfill. J Geotech Geoenvironmental Eng 133(8):1040–1047

Model Uncertainties in Long-Term Settlement Prediction of Landfill Waste

Yuekai Xie(✉), Jianfeng Xue(✉), and Carthigesu T. Gnanendran(✉)

School of Engineering and IT, UNSW Canberra at ADFA, Canberra, ACT, Australia
yuekai.xie@student.adfa.edu.au,
{jianfeng.xue, r.gnanendran}@adfa.edu.au

Abstract. Settlement in municipal solid waste (MSW) in landfills and its change with time is a complex process, because of the break-down potential of solids, heterogeneity of the material, variable particle size, and decomposition processes of organics induced by biodegradation. Waste properties, composition, distribution, and load applied vary significantly with depth and location. Various models have been proposed to predict the long-term settlement of landfill. This paper investigates the variation of settlement predictions of a landfill site using 11 different models proposed in the literature. Time required for 50% and 90% (t_{50} and t_{90}) settlement to finish is also investigated. The coefficient of variation of t_{50} and t_{90} can be as high as 0.8 and 0.6 respectively. This investigation demonstrates the necessity of using probability-based method for long term settlement analysis and landfill design.

Keywords: Uncertainty · Waste settlement · Model variation

1 Introduction

The production of the waste, especially of municipal solid waste, has significantly increased in the past few decades. The settlement of landfill waste refers to the vertical deformation of the waste material, which is a direct consequence of the mechanical compression, particles rearrangement and biodegradation of the organic fraction of MSW (Melo et al. 2017). The deformations induced by settlement of waste can contribute to negative effects in the landfill structure, such as damages in the leachate and biogas collection system, cracks over the landfill cover soil, and collapsing of the waste body into the landfill (Ling et al. 1998; Durmusoglu et al. 2005).

There are mainly five stages in the waste settlement process: Stage I is defined as the initial settlement, which results from instantaneous mechanical compression of waste (Grisolia and Napoleoni 1995). Stage II, the primary settlement, is caused by the continuous compression and movement of waste, including crushing, distortion, reorientation, bending and/or breaking of waste particles (Ivanova et al. 2008). Stage III is generally defined as the intermediate settlement, resulting from the combination of creep and initial biodegradation of organics or the secondary settlement, which includes intermediate settlement and the early stage of biodegradation. Stage IV refers to the biodegradation of organics, which generally contributes to the high settlement in the

L. Zhan et al. (Eds.): ICEG 2018, ESE, pp. 127–134, 2019.
https://doi.org/10.1007/978-981-13-2224-2_16

landfills (Fei and Zekkos 2013). Stage V, the residual settlement, which is small, is induced by both mechanical settlement and decomposition of organics, and this lasts up to a couple of decades (Leonard et al. 2000). Settlements in stages I and II are considered as short-term settlement. They are mainly due to the change of particle orientation, fine material movement into larger voids, and collapse of void spaces or distortion, bend, break and redirection of waste materials (Sharma and Reddy 2004). Settlements in stages III to V are mainly caused by the physical-chemical changes, consolidation, biodegradation and interaction, the processes except the other four (Li 1990).

Based on various monitoring data, typical settlement in a landfill may range from 15% to 50% of its initial height, depending the properties of the waste materials and the construction procedure (e.g. thickness of the cover soil, degree of compaction). Park et al. (2007) analyzed the monitoring data from 9 landfill sites (different aged landfills) and found that the rate of mechanical settlement caused by compression is relatively constant and rapid, which can be over 10% of initial height in the first month while the rate of biodegradation settlement is much slow, especially during the early stage of decomposition. The rate can rise steadily, reaching the peak after few months. In the first 2 years, the settlement can reach about 15% of the landfill thickness and can further increase to around 20% to 40% in 6 years, accounting for over 60% of the total settlement. Due to the long-term effect of the biodegradation, even 25 years after the completion of a landfill, the additional settlement is estimated to be 1% to 8% of the initial height (Park et al. 2007). Other factors such as the ratio between external load (drainage gravel and cover soil) and self-weight of the waste materials can directly affect the load-induced settlement (short-term), and mechanical creep and biodegradation (long-term). Higher load applied can reduce the void ratio in waste materials, decreasing gas and leachate transportation, and the growth of the microorganisms.

Various models have been developed to estimate the long-term settlement of landfill waste. However, large variation in the predicted settlement can be obtained using different models (Simões and Catapreta 2013). The objective of this paper is to review and investigate the relative merits of a number of existing models for predicting the settlement of waste in a landfill with the emphasis of the uncertainties in the estimation of settlement at a particular time.

2 Models Investigated

There are mainly four types of models used for the prediction of landfill settlement: the rheological models, regression models, consolidation models, and biodegradation models. These models take the sum or part of the immediate, primary and long-term settlement as the total settlement (Reddy et al. 2017). Below are the models to be investigated in this paper.

Source	Models for settlement (ΔH) prediction	
Gibson and Lo (1961)	$H\Delta P\left[a+b\left(1-e^{-(\lambda/b)t}\right)\right]$	(1-1)
Marques (2001)	$H\left(C_{ce}log\left(\frac{P_0+\Delta P}{P_0}\right)+b\Delta P(1-e^{-ct})+E_{dg}(1-e^{-kt})\right)$	(1-2)
Oweis and Khera (1986)	$H(C_{ce}log(\frac{P_0}{} + \Delta PP_0)+C_{\alpha}log\left(\frac{t}{t_1}\right))$	(1-3)
Bjarngard and Edgers (1990)	$H\left(C_{ce}log\left(\frac{P_0+\Delta P}{P_0}\right)+C_{\alpha 1}log\left(\frac{t_2}{t_1}\right)+C_{\alpha 2}log\left(\frac{t}{t_2}\right)\right)$	(1-4)
Hossain and Gabr (2005)	$H\left(C_{\alpha}log\left(\frac{t_2}{t_1}\right)+C_{\beta}log\left(\frac{t_3}{t_2}\right)+C_{\alpha f}log\left(\frac{t}{t_3}\right)\right)$	(1-5)
Park and Lee (1997)	$H\left(C_{\alpha}log\frac{t}{t_1}+E_{dg}\left(1-e^{-kt}\right)\right)$	(1-6)
Sowers (1973)	$HC_{ce}\log\left(\frac{P_0+\Delta P}{P_0}\right)+H_1C_{\alpha}\log\left(\frac{t}{t_1}\right)$	(1-7)
Ling et al. (1998)	$t/(1/\rho_0+\mathrm{t}/\Delta H_{ult})$	(1-8)
Marques et al. (2003)	$HC_{ce}\log\left(\frac{\frac{1}{2}\gamma_iH_i+\sum_{j=i+1}^{N}\Delta P_{ij}}{\frac{1}{2}\gamma_iH_i}\right)$ $+Hb[\frac{1}{2}\gamma_iH_i\left(1-e^{-c(t-t_i)}\right)+H\sum_{j=i+1}^{N}b\Delta P_{ij}\left(1-e^{-c(t-t_j)}\right)]$ $+HE_{dg}\left(1-e^{-k(t-t_i)}\right)$	(1-9)
Hettiarachchi et al. (2005)	$HC_{ce}\log\left(\frac{P_0+\Delta P}{P_0}\right)+H(1-n_i)\sum_{j=1}^{4}f_{sj}\left(\frac{G_{si}}{G_{sj}}\right)\left(1+w_{j(t)}G_{sj}\right)\left(1-e^{-kt}\right).$	(1-10)
Hettiarachchi et al. (2009)	$HC_{ce}log\left(\frac{P_0+\Delta P}{P_0}\right)+H\frac{M_{si}}{\rho_w}\sum_{j=1}^{4}\left(\frac{f_{sj}}{G_{sj}}\right)\left(1-e^{-kt}\right)$	(1-11)

In the equations, H is the initial height of waste; P_0 is the initial effective stress; ΔP is the change in effective stress; C_{ce} is the primary compression ratio; C_{α} is the secondary compression index; $C_{\alpha 1}$ is the intermediate secondary compression ratio; $C_{\alpha 2}$ is the long-term secondary compression ratio; a is the primary compression parameter; b is the coefficient of secondary mechanical compression; c is the secondary mechanical compression rate; λ/b is the rate of secondary compression; ρ_0 is the settlement rate; ΔH_{ult} is the ultimate settlement; t_1 is the time for primary compression; t_2 is the time for secondary compression; t_3 is the time for completion of biological compression; t is the time of interest; H_1 is the waste thickness at the start of secondary settlement; k is the first-order decomposition strain rate; $C_{\alpha f}$ is the creep index; C_{β} is the biodegradation index; E_{dg} is the total strain due to waste degradation; N is number of lifts; M_{si} is the weight of solids; n is the porosity; G_{si} is the initial overall specific gravity of waste; G_{sj} is the specific gravity of jth waste; $w_{j(t)}$ is the gravimetric moisture content; t_i (t_j) is the time when i^{th} (j^{th}) lift is filled; f_{sj} is the waste solids fraction.

3 Landfill Waste Properties

To study the variation of settlement predicted a given landfill site, the settlement data and experimental results from Bandeirantes Landfill, Brazil in Table 1 are used (Marques et al. 2003). The compositions of the waste solids and corresponding specific gravity of each type are based on the recommendations by Hettiarachchi et al. (2009).

The assumed landfill waste properties and characteristics are listed in Table 2. The regression results for required parameters of some of the models using two-year monitoring data are listed in Table 3.

Table 1. Reference data from Marques et al. (2003)

Parameters	Range	Average (applied value)
C_{ce}	0.073–0.122	0.106 (0.106)
b	$2.92*10^{-4}$–$7.26*10^{-4}$	$5.27*10^{-4}$ ($5*10^{-4}$)
c	$9.69*10^{-4}$–$2.55*10^{-3}$	$1.79*10^{-3}$ (0.0015)
k	$6.77*10^{-4}$–$2.51*10^{-3}$	$1.14*10^{-3}$ (0.001)
E_{dg}	0.131–0.207	0.159 (0.18)

Table 2. Assumed landfill waste and soil properties

Waste thickness	30 m	Time to fill one lift	200 d
Thickness of cover soil	2 m	Moisture content	30%
Thickness of drainage gravel	0.4 m	Non-biodegradable (G_s = 3)	35%
γ_w (waste)	7.5 kN/m^3	Slowly biodegradable (G_s = 2)	25%
γ_c (cover soil)	19.5 kN/m^3	Moderately biodegradable (G_s = 1.2)	35%
γ_d (drainage gravel)	18 kN/m^3	Highly biodegradable (G_s = 1)	15%

Table 3. Reference data

Reference	C_α	
	Recommended	Adopted
Sowers (1973), Hossain et al. (2005)	0.015–0.07	0.03
Oweis and Khera (1986)	0.0138–0.035 for $C_{\alpha1}$	0.002 for $C_{\alpha1}$
	0.132–0.321 for $C_{\alpha2}$	0.020 for $C_{\alpha2}$

4 Results and Discussion

Figure 1 summarizes the total strain calculated using the models listed in Table 1, which ranges from 13.6% to 51.5%. The rate for the mechanical settlement is rapid in first few months; the settlement is significantly slower during the biodegradation stages and decreases with time. Over half of the settlement occurs after 2 years, and the rate of settlement slows down after around 5 years in most models. For some of the models (1-3, 1-8, 1-9, and 1-10), settlement continues to grow significantly even after 5 years, and the biodegradation processes could last for a long time, i.e. more than 30 years.

Except the models by Gibson and Lo (1961), Ling et al. (1998) and Hossain and Gabr (2005), all other models are also applied using the maximum and minimum values defined in Tables 1 and 3 respectively and the results are shown in Figs. 2 and 3 respectively. The simulation results using different parameters are similar in the first few months, that is, the mechanical settlement stages. The rate of settlement can be

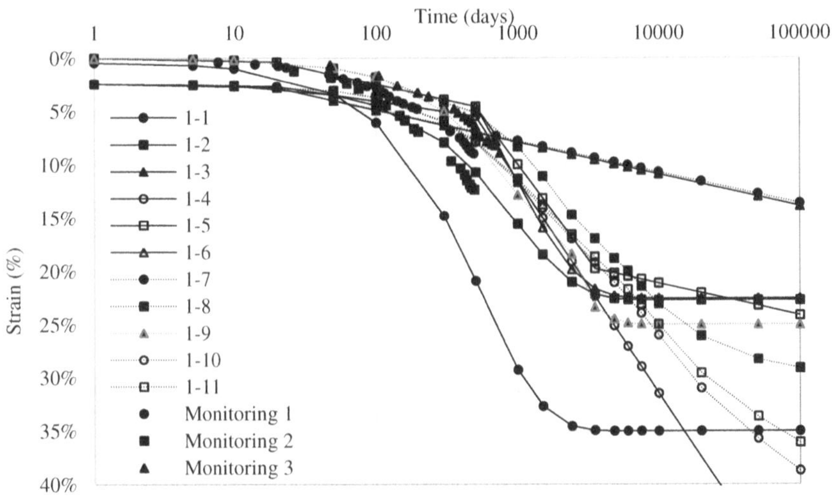

Fig. 1. Simulation results of models selected

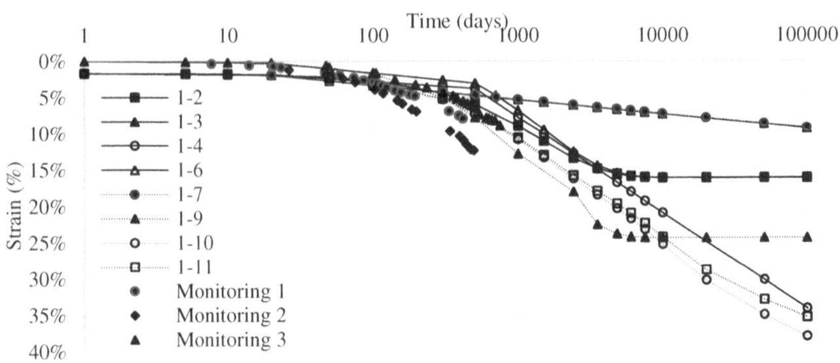

Fig. 2. Model simulation results using the minimum values defined in Table 1

significantly different during the biodegradation stage (or secondary settlement) due to the material properties variation and distribution of the materials.

As can be seen in Fig. 4, there is a significant variation between the model predictions when using different parameters (the maximum, minimum and average values). In models 1-10 to 1-11, once the waste component is determined, the total settlement is not very sensitive to the rate of biodegradation and compressibility variations. Other models, such as models 1-4 are sensitive to the variation of the parameters, e.g. compression parameters. The variation between the long-term settlement is much more significant using maximum parameters than that of average and minimum values (among different locations).

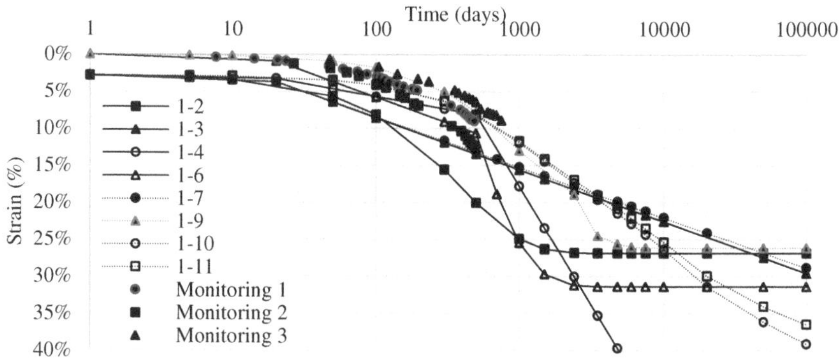

Fig. 3. Model simulation results using the maximum values defined in Table 1

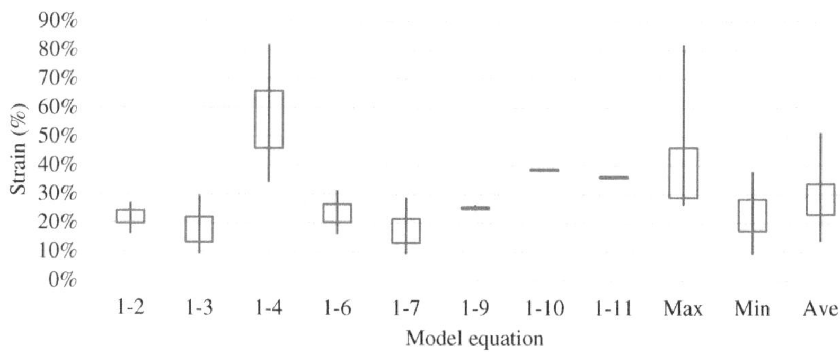

Fig. 4. Variation of simulation results with parameters listed in Table 1

The time required to reach 50% and 90% of total settlement is calculated using all three types of parameters (the minimum, the average and maximum). The results are compared in Fig. 5. Large variation in the time to reach 50% and 90% settlement can be found in equations 1-2 and 1-6. The mean t_{50} and t_{90} are 2047 days and 29270 days

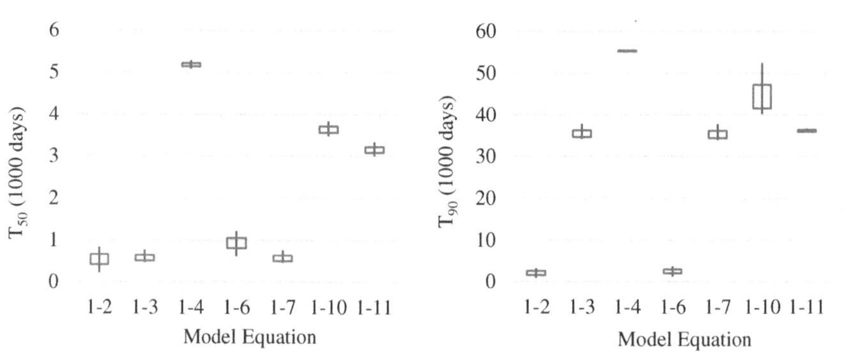

Fig. 5. Time for 50% and 90% settlement to finish predicted with the models

respectively when using average values listed in the tables. The coefficient of variation of t_{50} and t_{90} are 0.85 and 0.61 respectively, which is very high. The time to reach 90% settlement can exceed 100 years based on several models (e.g. models 1-4, 1-10 and 1-11). This suggests that large model uncertainty exists in the prediction of landfill settlement.

5 Conclusion

The variation of the settlement of a landfill and time for the settlement to finish predicted with eleven models have been investigated. A historical data of a landfill site reported in the literature has been used to examine the models. The long-term settlement calculated from the models ranges from 13.6% to 51.5% of initial landfill thickness, which is within the typical range (15% to 50%). The average total settlement is about 28.39%, with coefficient of variation of 10.74%. The time required for 50% and 90% total settlement ranges from 226 to 5283 days and 820 to 55534 days respectively, with mean values of 2047 days and 29270 days. The coefficient of variation of t_{50} and t_{90} are 0.85 and 0.61 respectively. This investigation shows that that large variation and uncertainty exist in the prediction of long term settlement of a landfill site. It is suggested that a probability-based method is required for the prediction of landfill waste settlement.

Acknowledgement. The first author received scholarship from UNSW for his PhD study. The project also received funding from ACT NoWaste, ACT government, Australia.

References

Bjarngard A, Edgers L (1990) Settlement of municipal solid waste landfills. In: The annual Madison waste conference

Durmusoglu E, Corapcioglu M, Tuncay K (2005) Landfill settlement with decomposition and gas generation. J Environ Eng 131(9):1311–1321

Fei X, Zekkos D (2013) Factors influencing long-term settlement of municipal solid waste in laboratory bioreactor landfill simulators. J Hazard Toxic Radioact Waste 17(4):259–271

Gibson R, Lo K (1961) A theory of consolidation for soils exhibiting secondary compression. Norges Tekniske Vitenskapsakad, Trondheim

Hettiarachchi C, Meegoda J, Hettiaratchi J (2005) Towards a fundamental model to predict the settlements in bioreactor landfills. In: Waste containment and remediation, pp 1–15

Hettiarachchi H, Meegoda J, Hettiaratchi P (2009) Effects of gas and moisture on modeling of bioreactor landfill settlement. Waste Manag 29(3):1018–1025

Hossain M, Gabr M (2005) Prediction of municipal solid waste landfill settlement with leachate recirculation. In: Waste containment and remediation

Ivanova L, Richards D, Smallman D (2008) The long-term settlement of landfill waste. Proc Inst Civ Eng Waste Resour Manag 161(3):121–133

Li X (1990) Finite-element analysis for immiscible two-phase fluid flow in deforming porous media and an unconditionally stable staggered solution. Commun Appl Numer Methods 6 (2):125–135

Ling H, Leshchinsky D, Mohri Y, Kawabata T (1998) Estimation of municipal solid waste landfill settlement. J Geotech Geoenvironmental Eng 124(1):21–28

Marques A (2001) Compaction and compressibility of municipal solid waste. Ph.D. thesis, Sao Paulo University, Sao Carlos, Brazil

Marques A, Filz G, Vilar O (2003) Composite compressibility model for municipal solid waste. J Geotech Geoenviron Eng 129(4):372–378

Melo M, Caribé R, Ribeiro L, Sousa R, Monteiro V, de Paiva W (2017) Settlement behavior of municipal solid waste due to internal and external environmental factors in a lysimeter. J Environ Sci Health A Toxic/Hazard Subst Environ Eng 51(14):1205–1215

Oweis I, Khera R (1986) Criteria for geotechnical construction on sanitary landfills. In: International symposium on environmental geotechnology, USA, pp 205–222

Park H, Lee S (1997) Long-term settlement behavior of landfills with refuse decomposition. J Resour Manag Technol 24(4):159–165

Park H, Park B, Lee S (2007) Analysis of long-term settlement of municipal solid waste landfills as determined by various settlement estimation methods. J Air Waste Manag Assoc 57 (2):243–251

Reddy K, Kumar G, Giri R (2017) Modeling coupled processes in municipal solid waste landfills: an overview with key engineering challenges. Int J Geosynth Ground Eng 3(1):6

Sharma H, Reddy K (2004) Geoenvironmental engineering. Wiley, New York

Simões G, Catapreta C (2013) Monitoring and modeling of long-term settlements of an experimental landfill in Brazil. Waste Manag 33(2):420–430

Sowers G (1973) Settlement of waste disposal fills. In: 8th international conference on soil mechanics and foundation engineering, Moscow, pp 207–210

Temperature Variation Under the Consideration of Convection and Heat Generation in Landfills

Hao Lei(✉), Jianyong Shi, and Xun Wu

Key Laboratory of Ministry of Education for Geomechanics and Embankment Engineering, Hohai University, Nanjing, China
leihao2015@hhu.edu.cn

Abstract. Heat generation occurs in municipal solid waste (MSW) landfills due to degradation of the organic fraction of the MSW, which results in elevated temperature. There is a great influence on the temperature distribution from leachate migration in landfills. Therefore, it is beneficial to the operation and management of the landfill to study the thermal evolution in a landfill under the effect of heat generation and heat convection. In this paper, the heat convection-diffusion model of landfill is established, and the analytical solution of the equation is obtained. The solution is used to analyze the variation of temperature under the effect of heat generation and heat convection in a landfill. The existing theory is verified by the measured data, and the influencing factors are analyzed. The calculation results show that convection in the landfill and heat generation parameters all have great influence on the temperature of the landfill. When the convection velocity becomes larger, the influence of depth of the atmospheric temperature increases. The temperature of the landfill increases with the increase of peak heat generation rate factor *A*.

Keywords: Landfill · Temperature distribution · Convection-diffusion Heat generation

1 Introduction

Heat, gas and leachate are the main by-products of organic matter degradation in landfill. Heat which is released during the degradation process, increase the temperature of the landfill. High temperatures reduce the service life of the geomembrane in the liner system (Rowe and Islam 2009). High temperatures change the mechanical properties of the pipes in the landfill (Krushelnitzky and Brachman 2013). High temperatures also result in the drying of the GCL, which reduce the impervious performance of the liner system (Southen and Rowe 2011). Furthermore, the microorganisms involved in the degradation of the landfill are sensitive to temperature. The change of temperature affects the degradation of the waste. At present, many scholars have developed models to predict the temperature distribution in landfills. Yesiller et al. (2005) analyzed the change of the temperature in a landfill and the thermal performance of the MSW through field test. Hanson et al. (2008) and Hanson et al. (2013) proposed an experiential heat generation rate function to describe the heat generation of landfill,

L. Zhan et al. (Eds.): ICEG 2018, ESE, pp. 135–144, 2019.
https://doi.org/10.1007/978-981-13-2224-2_17

and simulated the distribution of temperature in landfill site by numerical method. A comprehensive summary of the thermal performance, temperature distribution and physical properties of waste in landfill sites was made by Yesiller et al. (2015). However, the influence of leachate migration on the temperature distribution was not considered in their researches. Yoshida and Rowe (2003) established the heat conduction equation of a landfill, which considered the effect of convection on the heat transfer of the landfill, with convection velocity of 2.0 mm/d. Gholamifard et al. (2008) studied the change of temperature in landfill leachate recirculation by field test and numerical simulation. Leachate recirculation can change the temperature distribution in landfill site in a short time. Numerical methods were adopted in the study of Yoshida and Rowe (2003) and Gholamifard et al. (2008). The analytical solutions can provide deep physical insight.

At present, an analytical model is necessary to consider the influence of leachate migration on the temperature distribution in landfills. In this paper, a heat convection-diffusion equation in the landfills considering the leachate migration is established, and the analytical solution of the equation is obtained. This solution can be used to analyze the change of the temperature during moisture movement in landfill, and it is of major significance to the operation and management of landfills.

2 Equation Development and Analytical Solution

2.1 Equation Development

Heat generation occurs in municipal solid waste (MSW) landfills because of decomposition of the organic fraction of the MSW. The heat generated results in elevated temperatures in a landfill compared to atmospheric temperature and ground temperature. Leachate migration takes place during rainfall, leachate recirculation and leachate drainage. There is a great influence on the temperature distribution from leachate migration in landfills. In order to study the variation of temperature distribution in the landfill site under the conditions of heat convection and heat generation, a convection-diffusion equation for heat transfer in a landfill is established.

As shown in Fig. 1, a MSW layer with a thickness of H represents sections of the landfill and a one-dimensional approximation is justified. We adopt a coordinate system, z, whose positive direction is downward, with the top chosen as the origin.

The proposed heat conduction model in the landfill has the following assumptions: ① The velocity of seepage at each depth is uniform and constant; ② Heat transfer only includes thermal diffusion and heat convection. The heat convention-diffusion equation of a landfill can be obtained as Eq. (1) (Yoshida and Rowe 2003).

$$[n\rho_w c_w + (1-n)\rho_s c_s]\frac{\partial T}{\partial t} = \lambda\frac{\partial^2 T}{\partial z^2} - \rho_w c_w u_w\frac{\partial T}{\partial z} + f(t) \tag{1}$$

where, T is the temperature of MSW (°C), n is void ratio, ρ_w is density of water (kg/m^3), c_w is the specific heat capacity of water (J/(kg · °C)), ρ_s is density of MSW (kg/m^3), c_s is the specific heat capacity of MSW (J/(kg · °C)), u_w is the heat convection

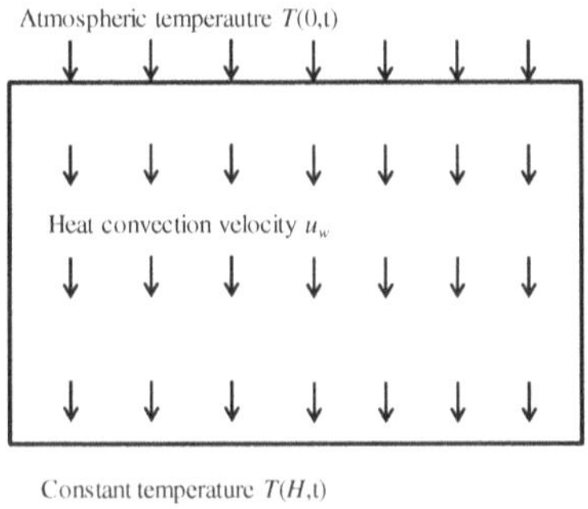

Fig. 1. Schematic diagram of the convective diffusion in landfills.

velocity (m/s), $f(t)$ is heat generation rate function (J/(m^3·s)), λ is heat conduction coefficient (W/(m·K)).

It is assumed that the temperature of the landfill is uniformly distributed at the initial time, and the initial condition for the equation is as follows:

$$T(z,0) = T_0 \tag{2}$$

where T_0 is the initial temperature of MSW.

Due to the heat exchange between the surface of the landfill and the atmosphere, the temperature of the upper surface of the landfill is close to the temperature of atmosphere. Therefore, the temperature of the upper surface of the landfill can be expressed as a function of sinusoidal form. The upper boundary conditions of the model are:

$$T(0,t) = T_m + A_S \sin\left[\frac{2\pi(t+t_0)}{365}\right] \tag{3}$$

where T_{m} is the average temperature (°C), A_{s} is the amplitude of atmosphere temperature (°C), t_0 is the phase difference of temperature (day). The temperature of the lower boundary is relatively constant.

The lower boundary is set as the average temperature of the landfill base, as Eq. (4).

$$T(H,t) = T_m \tag{4}$$

2.2 Analytical Solution

The analytic solution of the equation is obtained by using the superposition principle and the separation variable method. Equation (5) is the analytic solution.

$$T(z,t) = \left[\sum_{n=1}^{\infty} \Phi_n(t)\sin\left(\frac{p_n z}{H}\right) + w(z,t)\right] e^{\alpha z + \beta t} \tag{5}$$

where $p_{\mathrm{n}} = \pi, 2\pi, 3\pi, \ldots$

$$\alpha = \frac{v}{2k}, k = \frac{\lambda}{n\rho_w c_w + (1-n)\rho_s c_s} \tag{6}$$

$$\beta = -\frac{v^2}{4k}, v = \frac{\rho_w c_w u_w}{n\rho_w c_w + (1-n)\rho_s c_s} \tag{7}$$

$$w(z,t) = C(t)z + D(t) \tag{8}$$

$$\Phi_n(t) = \frac{2}{H}\int_0^t e^{-\frac{kp_n^2}{H^2}(t-\tau)}\int_0^H f^*(z,\tau)\sin\left(\frac{p_n z}{H}\right)\mathrm{d}z\mathrm{d}\tau + F_n e^{-\frac{kp_n^2 t}{H^2}} \tag{9}$$

$C(t)$ and $D(t)$ in Eq. (8) are given as follows:

$$C(t) = \frac{\mathrm{e}^{-\alpha H - \beta t}T_m}{H} - \frac{\mathrm{e}^{-\beta t}}{H}\left\{T_m + A_s \sin\left[\frac{2}{365}\pi(t+t_0)\right]\right\} \tag{10}$$

$$D(t) = \left\{T_m + A_s \sin\left[\frac{2}{365}\pi(t+t_0)\right]\right\}\mathrm{e}^{-\beta t} \tag{11}$$

$f^*(z,\tau)$ and F_n in Eq. (9) are given as follows:

$$f^*(z,\tau) = \frac{f(t)e^{-\alpha z - \beta\tau}}{n\rho_w c_w + (1-n)\rho_s c_s} - \frac{dC(\tau)}{d\tau}z - \frac{dD(\tau)}{d\tau} \tag{12}$$

$$F_n = \frac{2}{H}\int_0^H [T_0 e^{-\alpha z} - w(z,0)]\sin\left(\frac{p_n z}{H}\right)\mathrm{d}z \tag{13}$$

where $C(\tau)$ and $D(\tau)$ in Eq. (12) are given in Eq. (10).

3 Solution Validation

3.1 Verification of Thermal Evolution with Leachate Flow

Gholamifard et al. (2008) established the coupling equation of heat generation – heat conduction – gas diffusion - liquid migration in a landfill. The changes in the temperature of the landfill in a long term and the temperature changes in the recirculation of leachate in short term are analyzed. The experiment on the temperature change of the landfill site under the recirculation of leachate was carried out, and the test results were simulated by the coupling Equation. The heat convection-diffusion Equation development by this paper is used to simulate the experiment on the influence of temperature by leachate recirculation in Gholamifard et al. (2008). The convection velocity is 57 m/d respectively, other parameters are shown in Table 1. The parameters are same as Gholamifard et al. (2008). The results of the temperature at a depth of 2 m are shown in Fig. 2. It can be seen from the Fig. 2 that the consistency between calculated temperature and the field data is good. Leachate recirculation can quickly reduce the temperature of waste.

Table 1. Model parameters

Parameters	Value	References
n	0.3	Gholamifard et al. (2008)
λ (W/m K)	0.6/0.8	Gholamifard et al. (2008)
ρ_w (kg/m^3)	1000	Haynes (2011)
c_w (J/(kg °C))	4200	Haynes (2011)
ρ_s (kg/m^3)	1000	Gholamifard et al. (2008)
c_s (J/(kg °C))	1000	Gholamifard et al. (2008)
H (m)	10	Gholamifard et al. (2008)

3.2 Verification of Thermal Evolution with Heat Generation

The heat generated results in long-term elevated waste temperatures with respect to local air temperature and landfill base temperature. In order to analyze the influence of leachate migration on temperature distribution in a landfill under the condition of heat generation, it is needed to be taken into account the impact of the heat generation source item. The degradation test by Liu et al. (2012) shows that the biodegradation rate of MSW can be described by a function with a single peak value. In this paper, it is considered that the rate of heat generation of MSW is related to the degradation rate of MSW. The heat generation function used in this paper is similar to the degradation rate function developed by Liu et al. (2012). The heat generation function is shown in Eq. (14).

$$f(t)=A\frac{t}{B}e^{-\frac{t}{B}} \tag{14}$$

where A is peak heat generation rate factor(peak heat generation rate = A/e) (W/(m K)), B is the time of maximum heat generation rate (day). The values of A and B are derived by fitting the heat generation rate function of the Michigan landfill in Hanson et al. (2013), A = 2.87 W/(m^3·s), B = 540.5 day. The upper boundary temperature is set as a sinusoidal function with the change of the external temperature. The average temperature of the Michigan landfill is 14.7 °C, and the amplitude of the temperature is 17.3 °C. The other parameters are shown in Table 2.

The Michigan landfill is located in the Michigan state of the United States of America, located in the cold area, and has a humid climate. The average annual rainfall is 1046 mm. In order to study the temperature change and heat generation property in a landfill, many sensors for measuring temperature were installed in landfill site, and temperature sensors were arranged horizontally and vertically. The sensor has measured a large number of temperature data for 5 years, which gives a better understanding of thermal evolution inside the landfill. The heat convection speed is neglected, the equation is simplified into a heat conduction equation. This equation is

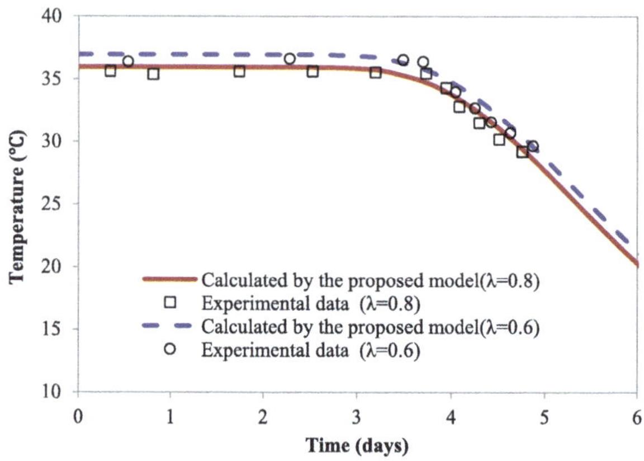

Fig. 2. The simulation of experimental data at depth of 2 m

Table 2. The model parameters of Michigan landfill

Parameters	Value	References
n	0.6	Feng et al. (2014)
λ (W/m K)	1	Hanson et al. (2013)
ρ_w (kg/m^3)	1000	Haynes (2011)
c_w (J/(kg °C))	4200	Haynes (2011)
ρ_s (kg/m^3)	1000	Hanson et al. (2013)
c_s (J/(kg °C))	2000	Hanson et al. (2013)
H (m)	33	Hanson et al. (2013)

used to simulate the temperature change of Michigan landfill, and the simulation results are shown in Fig. 3. It can be seen that the equation can better simulate the temperature change of the Michigan landfill. The temperature in the landfill is greatly influenced by atmosphere temperature in the shallow layer. At a depth of 7 m, the waste temperature in the landfill changes lightly with atmospheric temperature. The temperature in the middle layer of the landfill is generally higher than that in the upper portion.

4 Parameter Analysis

4.1 Influence of Convection Velocity

From Fig. 4, it can be seen that the waste temperature at a depth of 5 m is greatly influenced by the external temperature. The temperature of the MSW at a depth of 5 m fluctuates with the atmospheric temperature. When the depth is 10 m, the temperature fluctuates greater when the convection velocity is 0.01 m/d, than when the convection velocity is 0.001 m/d where there is almost no temperature fluctuation. When the convection velocity becomes larger, the influence of depth of the atmospheric

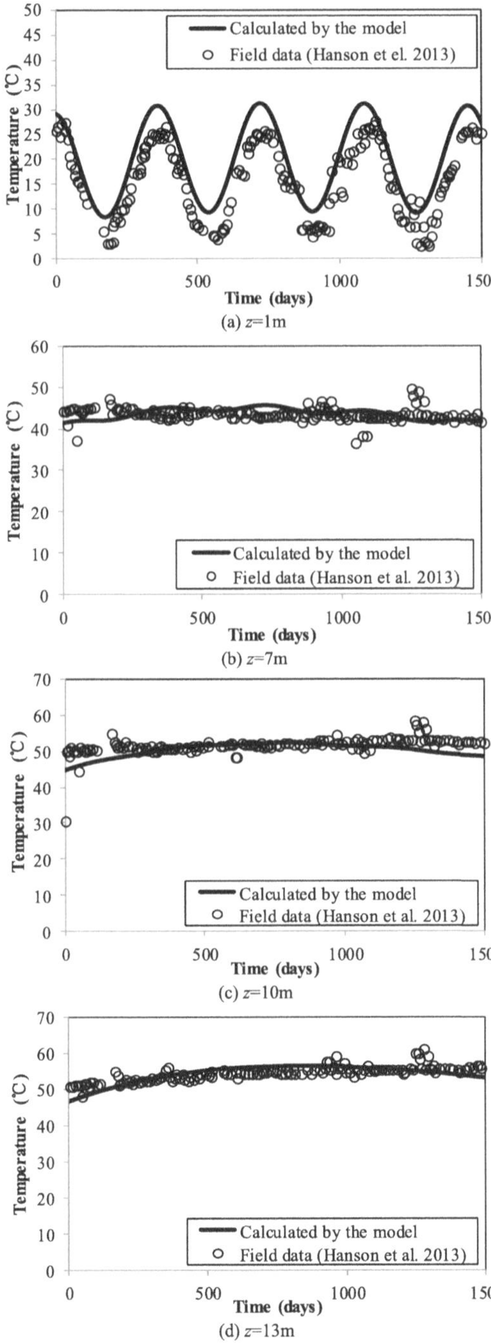

Fig. 3. Temperature simulation results of Michigan landfill

temperature increases. At the depth of 5 m, the waste temperature is lower than that of 0.001 m/d where the flow velocity of leachate is 0.01 m/d. This is because the greater the flow velocity, the more cold liquid flows into the upper boundary. The influence of the atmosphere temperature on the shallow MSW is even greater. The maximum temperature difference is about 10 °C at the depth of 5 m. The highest temperature appeared at about 20 m, the maximum temperature difference was 2 °C. The time required to reach the highest temperature was about 1750 days when $u_w = 0.001$ m/day, whereas the time required to reach the highest temperature was about 1570 days when $u_w = 0.01$ m/day. When the heat convection velocity is small, the maximum temperature is reached early. In conclusion, the heat convection in a landfill has a great influence on the temperature distribution in the landfill.

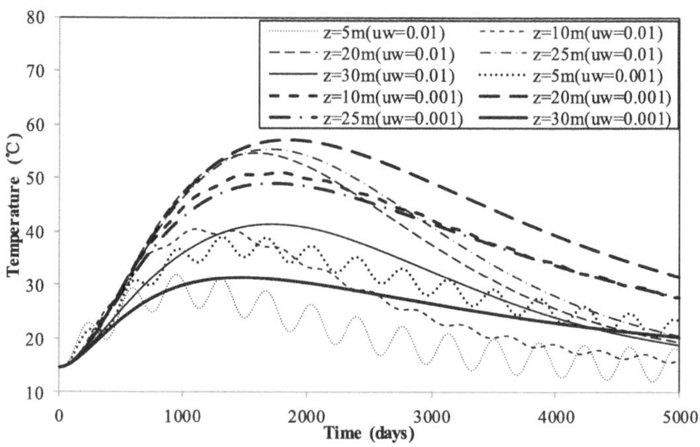

Fig. 4. The influence of the heat convection on temperature distribution

4.2 Influence of Heat Generation Parameters

At present, there are some researches on the heat generation capacity of MSW. The heat generation capacity of MSW is mainly related to the composition of the organic fraction, the temperature, pH, water content and other factors of MSW. The heat generation capacities of landfills are generally different, with some landfills varying greatly. There is a wide range of heat generation capacity. It is necessary to analyze the sensitivity of the thermal parameters in the landfill. The parameters of the single-peak heat generation rate function include peak heat generation rate factor *A* and the time of maximum heat generation rate *B*. The variation of the waste temperature with time in different depths of the landfill under different heat generation parameters is shown in Fig. 5. It can be seen that when the peak heat generation rate factors *A* are 2.87 W/m^3, 0.5*2.87 W/m^3 and 0.1*2.87 W/m^3 and the time of maximum heat generation rate *B* is kept constant, the temperature distribution in the landfill is greatly different. The highest temperature appears at a depth of 15 m, and the highest temperatures with a different *A* value are 58 °C, 36 °C and 19 °C, respectively. When the heat generation parameter *A* is 0.1*2.87 W/m^3, the heat migration from the upper and lower boundaries of a landfill is similar to the amount of heat generated by the MSW. Therefore, there is slight change in temperature. It can be seen that the heat generation parameters of MSW are very important for predicting the temperature inside the landfill. The heat generation rule of different landfills is main factor used in the study of the temperature changes inside the landfill.

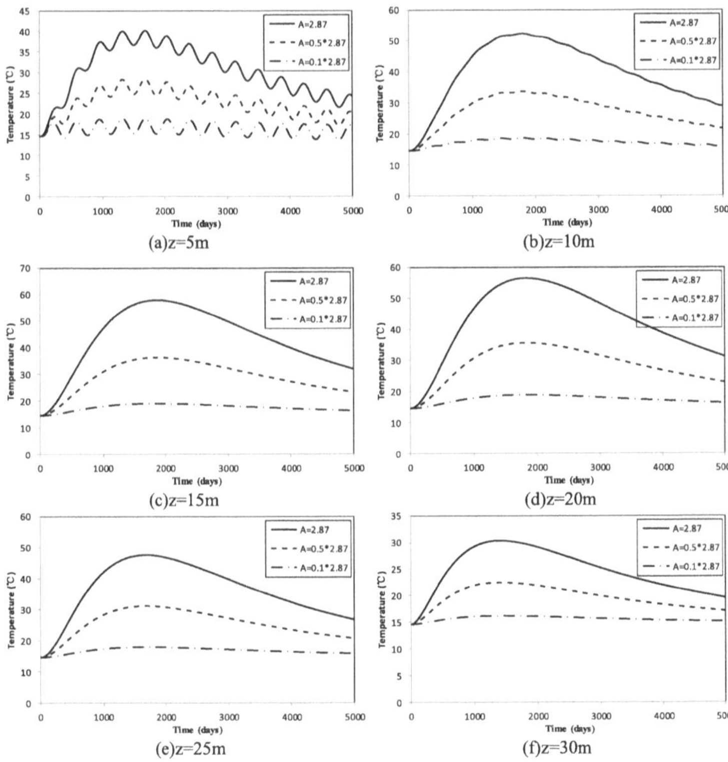

Fig. 5. The variation of the temperature with time at different depths.

5 Conclusion

Elevated temperatures occur in municipal solid waste (MSW) landfills due to the heat generation from the degradation of MSW. Leachate migration takes place during rainfall, leachate recirculation and leachate drainage, which causes heat convection. There is a great influence on the temperature distribution from heat convection in a landfill. In this paper, a heat convection-diffusion equation considering heat generation is developed, and the analytical solution of this equation is obtained. It is concluded that heat convection has a great influence on the temperature change of landfill, and the value of heat generation parameters also is a main factor for predicting temperature distribution in landfills. Therefore, when predicting the temperature distribution in landfill, the influence of convection on heat conduction should be considered and the appropriate heat production parameters should be selected.

Acknowledgments. The current study was financially supported by the key fund of China Natural Science Foundation (41530637) and the general fund of China Natural Science Foundation (41372268).

References

Feng SJ, Jiao Y, Zheng QT (2014) Research on leachate recirculation in landfills with considering MSW stratification. Chin J Undergr Space Eng 10(6):1263–1269

Gholamifard S, Eymard R, Duquennoi C (2008) Modeling anaerobic bioreactor landfills in methanogenic phase: long term and short term behaviors. Water Res 42(20):5061–5071

Hanson JL, Liu WL, Yesiller N (2008) Analytical and numerical methodology for modeling temperatures in landfills. Geotech Spec Publ 78(177):2599–2604

Hanson JL, Yeşiller N, Onnen MT et al (2013) Development of numerical model for predicting heat generation and temperatures in msw landfills. Waste Manag 33(10):1993–2000

Haynes WM (2011) CRC handbook of chemistry and physics, 95th edn. CRC Press, Boca Raton, pp 882–920

Krushelnitzky RP, Brachman RWI (2013) Buried high-density polyethylene pipe deflections at elevated temperatures. Geotext Geomembr 40(5):69–77

Liu XD, Shi JY, Qian XD et al (2012) Biodegradation behavior of municipal solid waste with liquid aspects: experiment and verification. J Environ Eng 139(12):1488–1496

Rowe RK, Islam MZ (2009) Impact of landfill liner time-temperature history on the service life of hdpe geomembranes. Waste Manag 29(10):2689–2699

Southen JM, Rowe RK (2011) Numerical modelling of thermally induced desiccation of geosynthetic clay liners observed in laboratory experiments. Geosynth Int 18(5):289–303

Yesiller N, Hanson JL, Liu WL (2005) Heat generation in municipal solid waste landfills. J Geotech Geoenviron 131(11):1330–1344

Yeşiller N, Hanson JL, Yee EH (2015) Waste heat generation: a comprehensive review. Waste Manag 42:166–179

Yoshida H, Rowe RK (2003) Consideration of landfill liner temperature. In: Proceedings Sardinia

Experimental Study on the Compression and Pore Characteristics of Undisturbed Loess Polluted by Landfill Leachate

Shicheng Xu, Haijun Lu(✉), Dinggang Li, Jixiang Li, and Weiwei Wang

School of Civil Engineering and Architecture, Wuhan Polytechnic University, Wuhan 430023, China
lhj_whpu@163.com

Abstract. This paper aims to investigate the compression and pore characteristics of underlying stratum in landfill. Batch tests were conducted to measure the compressibility as well as the pore, mineral, and microstructure characteristics of the undisturbed loess polluted by landfill leachate under different concentrations. Results show that the compressibility and compressive deformation of undisturbed soil increased as the landfill leachate concentrations increased. The reduction range of the void ratio was from 22.1% to 30.8%. The leachate pollution caused a decrease in the content of illite and montmorillonite, which reduced by 4.59% and 3.8%, respectively. The NMR signal of undisturbed soil increased as the landfill leachate concentrations increased. The pore radius of the undisturbed soil was mainly distributed in the range of 0.01–1 μm and can be divided into two ranges. The inter-particle pores and intra-aggregate pores mainly existed in undisturbed soil polluted by landfill leachate. Leachate caused an increase in inter-particle pore, whereas a decrease in intra-aggregate pore radius.

Keywords: Undisturbed loess · Landfill leachate · Compressibility Pore

1 Introduction

A considerable amount of landfill leachate is produced in landfills due to rainfall eluviation, microbiological degradation and so on, which contains high concentrations of heavy metals and organic pollutants [1]. Leachate leaking caused by the destruction of anti-seepage system changed the physical and chemical properties, the mineral composition and microstructure of underlying stratum [2, 3]. Eventually permeability and instability of underlying stratum affected by uneven settlement and decline in carrying capacity. Therefore, the evolution law of pore structure of underlying stratum needs to be got urgently in landfills.

H. Lu—Contributed equally to this work and should be considered co-first author.

L. Zhan et al. (Eds.): ICEG 2018, ESE, pp. 145–153, 2019.
https://doi.org/10.1007/978-981-13-2224-2_18

Amadi et al. [4] found that higher bentonite content result in a decrease in hydraulic conductivity. Rowe et al. [5] discovered that the particle size has an effect on the degree of logging in a granular media. Cao et al. [6] found that the amount of ions adsorbed by soil particles increased with the increase of ion concentration. Lu et al. [7] discovered that at increased adsorption capacity, the pore volume decreased and the specific surface area increased. Khodary and Du et al. [8, 9] found that landfill leachate reduced the thickness of the diffuse double layer (DDL), and resulted in soil shrinkage and caused a decrease in repulsive forces. Nayak et al. [10] discovered that the maximum dry density and the hydraulic conductivity of soils polluted by NaCl increased, the optimum moisture content (OMC) decreased with increasing sodium chloride concentrations. Hou et al. [11] found that pore size polluted by landfill leachate was decreased while average pore radius was increased.

In this paper, the compression and pore characteristics of undisturbed soil polluted by landfill leachate were examined using methods such as consolidation, nuclear magnetic resonance, X-ray diffraction, and scanning electron microscope tests. This study aims to reveal the law of pore for undisturbed soil polluted by different concentrations of landfill leachate. Results from this study can provide theoretical and technical support for assessing the underlying stratum in landfill.

2 Testing Materials and Methods

2.1 Testing Materials

The undisturbed loess was collected from a construction site in Huangpi District, Wuhan City and drilled by SD type of undisturbed soil sampler (Beijing Tethys Technology Co., Ltd.). The physical characteristics and mineral compositions measured by X-ray diffraction are given in Tables 1 and 2.

Table 1. Basic physical characteristics of undisturbed loess

w/(%)	ρ/(g/cm^3)	S_r/(%)	W_L/(%)	W_P/(%)	I_p/(%)	e	G_s/(g/cm^3)
21.7	2.10	87.3	34.4	18.8	15.6	0.63	2.68

The fresh landfill leachate was obtained from Chenjiachong landfills in Wuhan City, and its chemical compositions are listed in Table 3. Three kinds of solutions with different concentrations were used in the test by diluting with distilled water: 0% (distilled water only), 50% (volume ratio of distilled water and leachate is 1:1), 100% (leachate only).

2.2 Testing Methods

2.2.1 Consolidation Test

The undisturbed loess specimens (diameter = 20 mm and height = 60.8 mm) were presaturated with different concentrations of leachate (0%, 50%, and 100%) for 24 h in a BHG vacuum saturating cylinder (Nanjing Soil Instrument Company). A WG single

Table 2. Basic chemical and mineral composition of undisturbed loess

pH	Soluble salt (%)	Organic (%)	Mineral composition/%					
			Quartz	Illite	Montmorillonite	Muscovite	Albite	Kaolinte
5.5	0.6	1.8	23.41	23.10	20.94	21.09	6.69	4.76

Table 3. Chemical characteristics of landfill leachate

Contaminants	Parameter values/ ($mg{\cdot}L^{-1}$)	Determination method	Contaminants	Parameter values/ ($mg{\cdot}L^{-1}$)	Determination method
DO	3.7	GB/T 7489-1987	TOC	692.5	HJ/T 501-2009
N-NO_3	16.57	HJ/T 346-2007	COD	4107	HJ/T 399-2007
N-NO_2	0.50	HJ/T 197-2005	Cl^-	2022.9	GB/T 11896-1989
N-NH_4^+	2779.9	HJ/T 666-2013	Na^+	1855.6	GB/T 11904-1989
P-PO_4	21.2	HJ/T 669-2013	Fe	193	HJ/T 345-2007
TP	25.5	HJ/T 671-2013	Zn	20	GB/T 7472-1987
BOD_5	1457.1	HJ/T 505-2009	Cu	0.32	HJ/T 485-2009
$SO4^{2-}$	5.8	GB/T 13196-91	Ni	2.55	GB/T 11912-1989

lever consolidation instrument (NSIC) was used in consolidation tests according to the Chinese "Specification of soil test (SL237-1999) [12]".

2.2.2 X-ray Diffraction (XRD) Tests

The EMPYREAN X-ray diffraction (Holland PANalytical Co. Ltd) was used to determine the mineral component of undisturbed soil. According to "Analysis method for clay minerals and ordinary non-clay minerals in sedimentary rocks by the X-ray diffraction (SY/T 5163-2010) [13]", approximately 300 g samples polluted by landfill leachate were used in XRD test. Samples were air-dried and smashed to pass through a screen of 0.075 mm.

2.2.3 Microstructure Test

S-3000N scanning electron microscope (Hitachi, Japan) and the CPMG technique with a PQ-001 MiniMR NMR setup (the Institute of Rock and Soil Mechanics, Chinese Academy of Sciences) were used to determine the micromorphology, pore characteristics of undisturbed soil, respectively. The samples (about approximately 1 mm × 1 mm × 1 mm) and the samples (about approximately 2 mm × 2 mm × 2 mm) taken in undisturbed soil polluted by 3 kinds of landfill leachate were used in SEM and NMR tests. Theses samples were preprocessed using vacuum freeze-drying technology. Pore size was calculated according to the soil-water characteristic curve [14].

3 Results and Discussion

3.1 Compressibility Property

Figure 1 shows consolidation curves for undisturbed soil polluted by landfill leachate. The compressibility and compressive deformation of undisturbed soil increased as the landfill leachate concentrations increased. The reduction range of the void ratio was from 22.1% to 30.8%. The compressibility of undisturbed soil decreased significantly when the leachate concentrations was 100%, which indicated deterioration of structural characteristics for undisturbed soil.

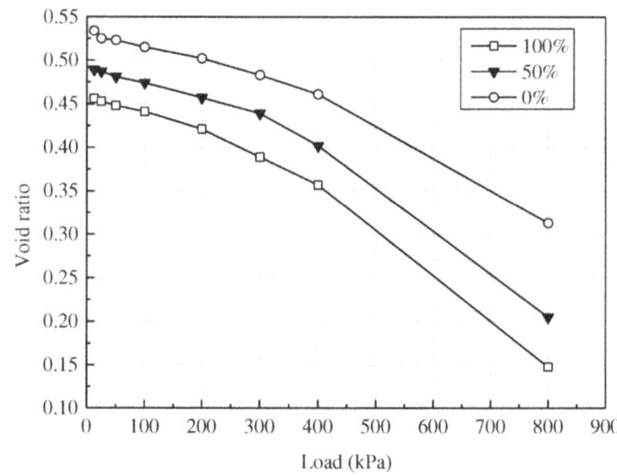

Fig. 1. Consolidation curves for undisturbed soil with different concentrations of leachate.

The change of pore was attributed to the structural damage caused by leachate. At increased leachate concentrations, coarse particles change into fine particles, the arrangement of soil particles changed. The external loading changes the equilibrium osmotic pressure of the system. The compressional deformation of undisturbed soil polluted by landfill leachate under loadings was mainly plastic deformation caused by soil structure destruction.

3.2 Mineral Composition

Figure 2 shows the X-ray diffraction of undisturbed soil with different concentrations. Montmorillonite peaks at $2\theta = 5.05°$ and illite peaks at $2\theta = 35.07°$ and 62.01° became smaller, indicating that the pollution of leachate caused a decrease in the content of montmorillonite and illite.

Table 4 shows main mineral compositions of undisturbed soil polluted by leachate. The illite and montmorillonite content decreased from 23.10% to 18.51% and from 20.94% to 17.14%, respectively. The albite content increased from 6.69% to 11.76%. The change of other mineral content was more stable after leachate pollution.

The properties of non-clay minerals such as quartz and muscovite remain relatively stable. These minerals are less affected and can be ignored when soil are eroded by

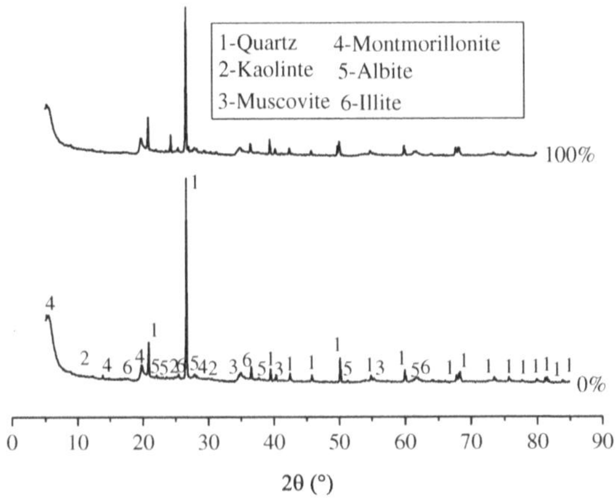

Fig. 2. X-ray diffraction of undisturbed soil with different concentrations.

Table 4. Main mineral compositions of undisturbed soil polluted by leachate

Concentrations	Mineral					
	Quartz	Illite	Montmorillonite	Muscovite	Albite	Kaolinte
0%	23.41	23.10	20.94	21.09	6.69	4.76
100%	24.88	18.54	17.10	22.33	11.76	5.39

leachate. Clay minerals have small particle size, which leads to the strong interaction between the physical-chemical effects. The montmorillonite is a dioctahedron structure, and the illite is a scaly structure. When the soil was eroded by leachate, the water molecules, metal ions and organic pollutants in leachate were easy to enter the crystal layer, and the corresponding cation exchange capacity (C.E.C) is stronger [15], which also makes montmorillonite and illite more hydrophilic and active in chemical properties. The organic matter and metal ions in the leachate reacted with the montmorillonite and illite in the clay minerals and form many compounds. The pore volume increased, which was consistent of the consolidation and NMR test.

3.3 Pore Characteristics

T_2 distribution of undisturbed soil polluted by landfill leachate are shown in Fig. 3. The area between the T_2 distribution curve and lateral axis was defined as the peak area, which represents water content in the soil sample. The maximum NMR signal corresponding T_2 represented the optimal radius of pores saturated with water [16]. There were two peaks of undisturbed soil polluted by landfill leachate at 0.01–2 ms and 10–300 ms, respectively. The NMR signal of undisturbed soil increased as the landfill leachate concentrations increased, indicating that leachate caused an increase in the amount of pores in this range. The NMR signal of the first peak was obviously larger than that of the second peak.

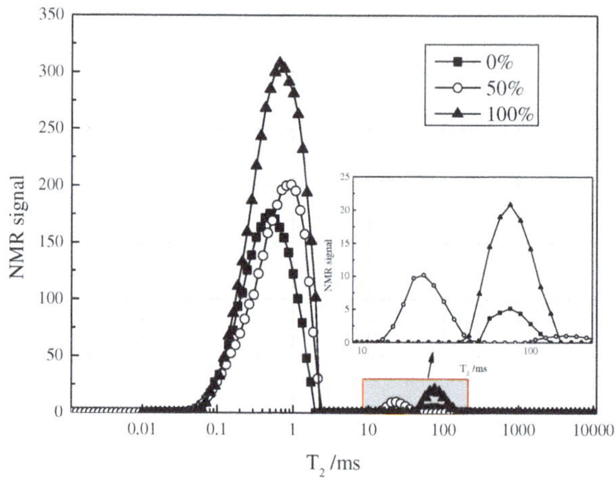

Fig. 3. T_2 distribution of undisturbed soil polluted by landfill leachate.

Pore radius of undisturbed soil polluted by landfill leachate are shown in Fig. 4. Pore structure changed after being polluted. It indicated that the pore radius of the undisturbed soil was mainly distributed in the range of 0.01–1 μm and can be divided into two ranges. The second range was 9–70 μm. The pore distribution density increased with the increased in leachate concentrations.

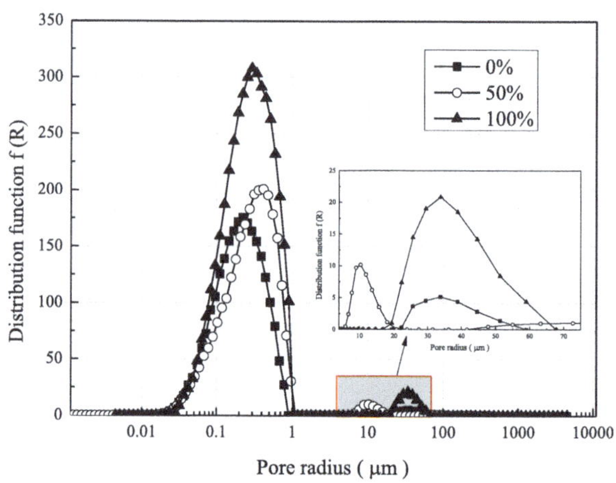

Fig. 4. Pore size distribution of undisturbed soil polluted by landfill leachate.

An accurate classification for pore diameter was the key to the study of the evolution law of soil microstructure. The ranging from 0.014 μm to 1.800 μm was defined as inter-particle pore. The ranging from 1.800 to 70 μm was defined as intra-aggregate pore. The ranging from 70 to 4000 μm was defined as inter-aggregate pore. Generally, the ranging exceeding 600 μm was defined as macroscopic pores [17]. The inter-particle pores and intra-aggregate pores mainly existed in undisturbed soil polluted by landfill leachate. The number of pores and pore radius increased with the increase of leachate concentrations, which indicates leachate result in a large internal space of undisturbed soil and compressibility of soil changed, and the structure of undisturbed soil was destroyed. This result consists with the following consolidation test.

The pore radius of the corresponding peak values increased with the increased in leachate concentrations ranging from 0.01–1 μm, indicating that leachate caused an increase in inter-particle pore radius. However, the location of the corresponding peak values decreased with the increased in leachate concentrations ranging from 9–70 μm, indicating that leachate caused a decrease in intra-aggregate pore radius. The organic pollutant and microorganisms in leachate were adsorbed in the soil surfaces, thereby blocking the interconnected pores and causing a decrease in inter-aggregate pore [18].

3.4 Surface Micromorphology

Figure 5 shows the SEM of undisturbed soil with leachate concentrations of 0% and 100%. In Fig. 5(a), the undisturbed soil not eroded by leachate has not seen more pores, and the aggregates were larger. In Fig. 5(b), soil structure polluted by leachate was destroyed seriously, the pores and cracks in the undisturbed soil polluted by leachate increased, which be attributed to the following two reasons: (a) binding material of soil particles was corroded and dissolved by landfill leachate; and (b) some mineral compositions (illite and montmorillonite) decreased and whose mineral crystal lattice collapsed [19] based on the results of XRD test.

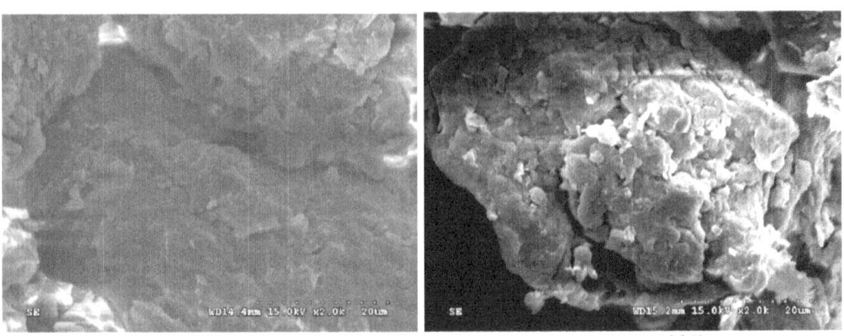

(a) 0% (b) 100%

Fig. 5. SEM of undisturbed soil with different leachate concentrations. (a) 0%, (b) 100%.

4 Conclusions

The intent of this paper is to explore the compression and pore characteristics of undisturbed soil polluted by landfill leachate. Consolidation, nuclear magnetic resonance, X-ray diffraction, and scanning electron microscope tests were performed to evaluate compression characteristics, pore parameter, mineral composition, and surface morphology of undisturbed soil polluted by landfill leachate. The main conclusions are as follows.

(1) The compressibility and compressive deformation of undisturbed soil increased as the landfill leachate concentrations increased. The reduction range of the void ratio was from 22.1% to 30.8%.
(2) The illite and montmorillonite content decreased from 23.10% to 18.51% and from 20.94% to 17.14%, respectively. The soil structure polluted by leachate was destroyed seriously, pores and cracks in undisturbed soil polluted by leachate increased.
(3) The NMR signal of undisturbed soil increased as the landfill leachate concentrations increased. The pore radius of the undisturbed soil was mainly distributed in the range of 0.01–1 μm and can be divided into two ranges. The pore distribution density increased with the increased in leachate concentrations. The inter-particle pores and intra-aggregate pores mainly existed in undisturbed soil polluted by landfill leachate. Leachate caused an increase in inter-particle pore, whereas a decrease in intra-aggregate pore radius.

Acknowledgements. The authors would like to express their great appreciation for funding provided by the "National Natural Science Foundation of China (11672216, 11602183).

References

1. Xue Q, Li JS, Liu L (2013) Experimental study on anti-seepage grout made of leachate contaminated clay in landfill. Appl Clay Sci 81(4):438–442. https://doi.org/10.1016/j.clay.2013.06.026
2. Nayak S, Sunil BM, Shrihari S (2007) Hydraulic and compaction characteristics of leachate-contaminated lateritic soil. Eng Geol 94(3–4):137–144. https://doi.org/10.1016/j.enggeo.2007.05.002
3. Li JS, Xue Q, Wang P, Liu L (2013) Influence of leachate pollution on mechanical properties of compacted clay: a case study on behaviors and mechanisms. Eng Geol 167(18):128–133. https://doi.org/10.1016/j.enggeo.2013.10.013
4. Amadi AA, Eberemu AO (2012) Delineation of compaction criteria for acceptable hydraulic conductivity of lateritic soil-bentonite mixtures designed as landfill liners. Environ Earth Sci 67(4):999–1006. https://doi.org/10.1007/s12665-012-1544-z
5. Rowe RK, Armstrong MD, Cullimore DR (2000) Particle size and clogging of granular media permeated with leachate. J Geotech Geoenviron Eng 126(9):775–786. https://doi.org/10.1061/(ASCE)1090-0241(2000)126:9(775)
6. Cao LW, Zhang J, Wang Y, Lian XY, Li H (2009) Experimental research on geotechnical behaviors of compacted clay influenced by metal cation. Procedia Earth Planet Sci 1 (1):1016–1023. https://doi.org/10.1016/j.proeps.2009.09.157
7. Lu HJ, Zhang Q, Dong YQ, Li JX, Zhang X (2016) The adsorption capacity, pore structure, and thermal behavior of the modified clay containing SSA. Adv Mater Sci Eng 2:1–7. https://doi.org/10.1155/2016/9894657
8. Khodary SM, Negm AM, Tawfik A (2018) Geotechnical properties of the soils contaminated with oils, landfill leachate, and fertilizers. Arab J Geosci 11(2):1–17. https://doi.org/10.1007/s12517-017-3372-7

9. Du YJ, Shen SL, Liu SY, Hayashi S (2009) Contaminant mitigating performance of Chinese standard municipal solid waste landfill liner systems. Geotext Geomembr 27(3):232–239. https://doi.org/10.1016/j.geotexmem.2008.11.007
10. Nayak S, Sunil BM, Shrihari S, Sivapullaiah PV (2010) Interactions between soils and laboratory simulated electrolyte solution. Geotech Geol Eng 28(6):899–906. https://doi.org/10.1007/s10706-010-9333-5
11. Hou J, Li JZ, Chen YJ (2017) Coupling effect of landfill leachate and temperature on the microstructure of stabilized clay. Bull Eng Geol Env 5:1–12. https://doi.org/10.1007/s10064-017-1099-z
12. SL 237-1999 (1999) Specification of soil test. China Water Power Press, Beijing
13. SY/T 5163-2010 (2010) Analysis method for clay minerals and ordinary non-clay minerals in sedimentary rocks by the X-ray diffraction. Petroleum Industry Press, Beijing (in Chinese)
14. Tian H, Wei CF, Wei HZ, Yan RT, Chen P (2014) An NMR-based analysis of soil–water characteristics. Appl Magn Reson 45(1):49–61. https://doi.org/10.1007/s00723-013-0496-0
15. Sunil BM, Shrihari S, Nayak S (2009) Shear strength characteristics and chemical characteristics of leachate-contaminated lateritic soil. Eng Geol 106(1):20–25. https://doi.org/10.1016/j.enggeo.2008.12.011
16. Gao SQ, Chapman G, House W (2009) Application of low field NMR T_2 measurements to clathrate hydrates. J Magn Reson 197(2):208–212. https://doi.org/10.1016/j.jmr.2008.12.022
17. Shear DL, Olsen HW, Nelson KR (1993) Effects of desiccation on the hydraulic conductivity versus void ratio relationship for a natural clay. National Academy Press, Washington, DC
18. Mehmet SO, Nejat C, Suna OC, Guler TD, Vildan C, Nese T (2013) Hydraulic conductivity and removal rate of compacted clays permeated with landfill leachate. Desalin Water Treat 51(31–33):6148–6157. https://doi.org/10.1080/19443994.2013.769662
19. Oztoprak S, Pisirici B (2011) Effects of micro structure changes on the macro behaviour of Istanbul (Turkey) clays exposed to landfill leachate. Eng Geol 121(3–4):110–122. https://doi.org/10.1016/j.enggeo.2011.05.005

Landfill Storage Capacity Analysis Method by Considering Foundation Settlement and Its Application

Hai-Long Liu[1(✉)], Zhuo-Fei Wu[1], and Yu-Chao Li[2]

[1] Dalian Maritime University, Dalian, China
liuhailong@dlmu.edu.cn
[2] Zhejiang University, Hangzhou, China

Abstract. The increase of municipal solid waste (MSW) and the reduction of available urban land lead to the development of large-scale landfill. The landfill foundation is assumed to be rigid in existing models for calculating landfill compression. The self-weight of MSW increases with landfill expansion, and the landfill foundation settlement should not be ignored anymore. The estimation of landfill foundation settlement is of great importance in predicting landfill storage capacity and for the design of effective long-term landfill facilities. The implemented model for calculating landfill compression displacement and storage capacity is able to consider increase in foundation settlement induced by landfilling process. A calculation program is developed base on this proposed model, which can estimate landfill storage capacity. The quantitative effects of initial compaction on MSW landfill settlement and storage capacity are investigated by using the program via consideration of a hypothetical case. According to the investigation, a significant increase in storage capacity can be achieved by rational use of landfill foundation compression and intensive initial compaction. The quantitative investigation presented aims to encourage landfill operators to improve management to enhance storage capacity. Furthermore, prediction of landfill foundation settlement is helpful to protect leachate and gas management infrastructure at bottom of modern landfills.

Keywords: Landfill · Compression · Foundation settlement · Storage capacity

1 Introduction

Municipal solid waste (MSW) is a highly compressible material, and the compression of landfills resulting from MSW displacement can be as large as 25% ~ 50% of the initial filling height (Ling et al. 1998). The height of MSW bodies in modern sanitary landfills has exceeded 100 m. The overburden load on foundation of landfills increases significantly, and compression of landfill foundation should not be ignored, especially when there are soft soil layers in the foundation. Foundation settlement increases storage capacity of a landfill and extend its service life. On the other hand, foundation settlement has detrimental influences on the integrity of the structures at the bottom of a landfill, such as the leachate drainage and collection system. Failed leachate drainage system increases the risk of instability (Jang 2013). Most studies focus on the vertical

L. Zhan et al. (Eds.): ICEG 2018, ESE, pp. 154–161, 2019.
https://doi.org/10.1007/978-981-13-2224-2_19

displacement of MSW bodies and the foundation of landfills is generally assumed rigid (Gao et al. 2017), while the estimation method of landfill capacity considering compression of landfill foundation is rarely reported.

This article proposes a revised approach for calculating settlement and storage capacity of landfills by considering the compression of landfill foundation. In order to improve the understanding of the effect of foundation settlement on landfill storage capacity, the quantitative influences of different initial compaction are investigated.

2 Calculation of the Settlement and Storage Capacity of the Landfill

2.1 General Assumptions

The settlement and storage of a landfill can be transformed into a calculation of the settlement and storage of a series of independent column units, neglecting the interaction between column units. Some assumptions were made to calculate settlement of a column unit:

(1) When a MSW landfill is in normal operation, the leachate drainage system should keep the main leachate level in MSW body at a very low level, and the bottom barrier can effectively prevent leachate leakage. Leachate generated during biodegradation can be assumed to be drained immediately, and the main leachate level in MSW body has no effect on the foundation displacement of landfill;
(2) The overburden load on MSW and foundation is continuously changed with MSW placement process and solid mass loss of MSW. In order to simplify the calculation, MSW placement process is approximated by a serial of instant layers. The overburden load on every MSW layer and foundation is only renewed when a new MSW layer is disposed, and the solid mass loss due to biodegradation of MSW is instantaneously generated at the same time;
(3) Compared with degradation compression, vertical displacement due to the mechanical compression complete quickly. It can be assumed that the primary compression of MSW body and the displacement of foundation complete instantly during the filling period.

2.2 Settlement of the Landfill

The compressive volumetric strain of landfill is divided into two parts: the compressive strain of MSW body and foundation,

$$\varepsilon(\sigma, t) = \varepsilon_{\mathrm{w}}(\sigma, t) + \varepsilon_{\mathrm{f}}(\sigma, t) \tag{1}$$

where $\varepsilon(\sigma, t)$ is the volumetric strain of landfill, $\varepsilon_{\mathrm{w}}(\sigma, t)$ and $\varepsilon_{\mathrm{f}}(\sigma, t)$ are volumetric strain of MSW body and foundation soil with a fill age of t under the overburden stress σ, respectively.

The compressive volumetric strain of MSW body is expressed in the implemented model (Chen et al. 2010) as:

$$\varepsilon_{\mathrm{w}}(\sigma,t)=C'_{\mathrm{c0}}\log\left(\frac{\sigma}{\sigma_0}\right)+\left[\left(C'_{\mathrm{c\infty}}-C'_{\mathrm{c0}}\right)\log\left(\frac{\sigma}{\sigma_0}\right)+\varepsilon_{\mathrm{dc}}(\sigma_0)\right][1-\exp(-ct)] \tag{2}$$

where C'_{c0} and $C'_{\mathrm{c\infty}}$ are compression ratios for placed fresh MSW and fully decomposed MSW, respectively, and can be obtained from compression tests of fresh and fully decomposed MSWs; σ_0 is pre-consolidation pressure, which is dependent on initial compaction; $\varepsilon_{\mathrm{dc}}(\sigma_0)$ is the sum of ultimate volumetric strains of decomposition compression and mechanical creep under pre-consolidation pressure, c is the MSW decomposition rate, which can be obtained from degradation compression tests. The calculation of MSW body compression strain has been described in another paper by the authors (Li et al. 2013).

The settlement of landfill foundation is calculated by using layerwise summation method. The accumulated compression deformation from bottom of landfill to underground at any depth is expressed as:

$$S(z)=\int_0^z \varepsilon_{\mathrm{f}}(\sigma,t)\mathrm{d}z=\int_0^z \frac{K(z)\sigma_{\mathrm{f}}(t)}{E_s}\mathrm{d}z \tag{3}$$

where $S(z)$ is accumulated compression of landfill foundation; E_s is compressive modulus of soil, which can be obtained according to the $e \sim p$ curve of soil; $K(z)$ is the average additional stress coefficient, which is affected by calculation depth; $\sigma_{\mathrm{f}}(t)$ is addition stress at fill age t.

The overburden load on landfill foundation increases with filling process and decreases due to biodegradation of MSWs. According to the general assumption above, $\sigma_{\mathrm{f}}(t)$ is updated when a new layer of MSW is added to the column unit:

$$\int_0^z \sigma_{\mathrm{f}}(t_i\rightarrow t_{i+1})\mathrm{d}z=\int_0^z [\sigma_{\mathrm{f}}(t_i)+\Delta\sigma_1(t_{i+1})-\Delta\sigma_2(t_{i+1})]\mathrm{d}z \tag{4}$$

$$\Delta\sigma_2(t_{i+1})=\frac{\Delta h}{1-w}\frac{d_{\mathrm{s}}\gamma_{\mathrm{w}}}{1+e_0}\{1-\theta_0[1-\exp(-ct_{i+1})]\} \tag{5}$$

where $\Delta\sigma_1(t_{i+1})$ is self-weight of new MSW layer; $\Delta\sigma_2(t_{i+1})$ is solid mass loss; d_{s} is specific gravity of the solid particles; γ_{w} is unit weight of water; w is water content of the MSW by mass; θ_0 is degradable solid content, which is the ratio of the mass of degradable components to that of total solid for the fresh MSW. d_{s} and w are assumed to be constant in the following investigation for the sake of simplification.

For a given fill volume of the landfill, the storage capacity can be represented by the average placed MSW weight in unit fill volume at closure, which can be expressed as follows:

$$\overline{\gamma}=\frac{\sum_{j=1}^{J}\gamma_0 A_j H'_j}{\sum_{j=1}^{J} A_j H_j} \tag{6}$$

where γ_0 is total unit weight of placed fresh MSW after initial compaction; A_j, H'_j and H_j are the area, accumulative filling height and the design height of the jth column unit, respectively; and J is the total number of column units of the landfill.

A calculation program was developed based on MATLAB, which can be used to estimate the settlement of column units. The framework of this program is shown in Fig. 1: material parameters, including characteristic properties of MSWs and $e \sim p$ curve of foundation soil, are input and fill schedule of MSWs is also need to be set up at the beginning. The overburden load of every layer and the addition stress of foundation are renewed after every fill operation. The settlement and the accumulated height of landfill are updated, and the results are output when the accumulated height of landfill is equal to the design height.

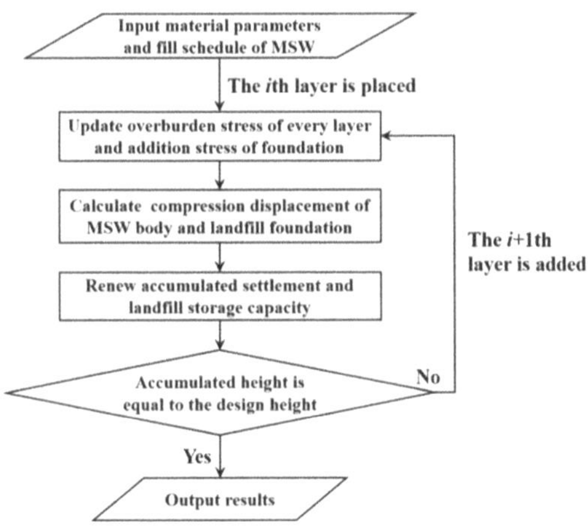

Fig. 1. Framework of landfill settlement calculation program.

3 Case Study

3.1 Case Profile

In this section, the model described above is applied to investigate the influences of scale and initial compaction on settlement and storage capacity of a hypothetical landfill, which is shown in Fig. 2. Property parameters of MSW are list in Table 1, and $e \sim p$ curves of foundation soil are given in Fig. 3. The parameters selected for the MSW are based on Hanson et al. (2010), Stoltz et al. (2010), and Chen et al. (2009) (see Table 1). To investigate the influence of initial compaction on settlement and storage capacity the value of the initial void ratio is varied, while the other parameters remain at the original value. While it is recognized that this approach does not fully consider the complete multivariate nature of the problem it does offer a clear assessment of the influence of these various factors in isolation.

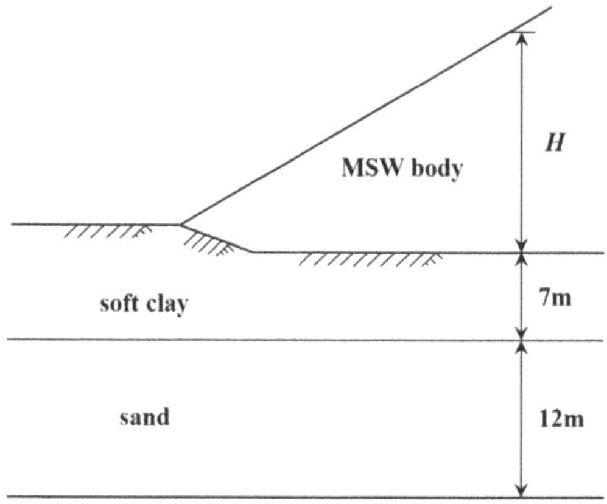

Fig. 2. A hypothetical landfill case.

3.2 Results

The initial void ratio of MSW is affected by compaction operation when MSW is placed in a landfill. The typical initial void ratio values are reported in the range between 1.0 and 3.0 in most literatures (Zhan et al. 2008; Park and Park 2009). So two different degrees of compaction, corresponding to e_0 = 1.0 and 3.0, are considered to investigate the influence of initial compaction on landfill compression displacement and storage capacity. e_0 = 1.0 indicates the scenario having a good initial compaction; while e_0 = 3.0 is close to that of having poor or no initial compaction, which is common in developing countries, such as China.

As shown in Fig. 4, for a given design height, a higher degree of compaction, which corresponds to a lower initial void ratio, results in a greater foundation settlement and a smaller accumulated compression of MSW body, as expected. For instance, S_{MSW} = 3.73 m and $S_{foundation}$ = 1.87 m for e_0 = 1.0, while the calculation results are 18.46 m and 0.83 m respectively, for e_0 = 3.0 when H = 60 m. The compression of foundation is 66% of total settlement for e_0 = 1.0, while it is only 4% for e_0 = 3.0. The compressibility of MSW is much greater than that of foundation soil, which results significant vertical displacement in MSW body. The settlement calculation results are close to the methods without considering the compression of foundation when MSW have poor initial compaction. This response is accords with field measured record (Xu et al. 2015). Figure 4 also show that the proportion of foundation settlement to total displacement decreases with design height. For example, the proportion value is 63% for H = 20 m, while it is 24% for H = 100 m, when e_0 = 1.0.

Table 1. Property parameters of MSW.

Parameter	Value	Parameter	Value
d_s	1.7	$C'_{c\infty}$	0.17
w	70%	$\varepsilon_{dc}(\sigma_0)$	0.25
θ_0	20%	c	0.01/month
σ_0	30 kPa	r (filling rate)	1 m/month
C'_{c0}	0.3		

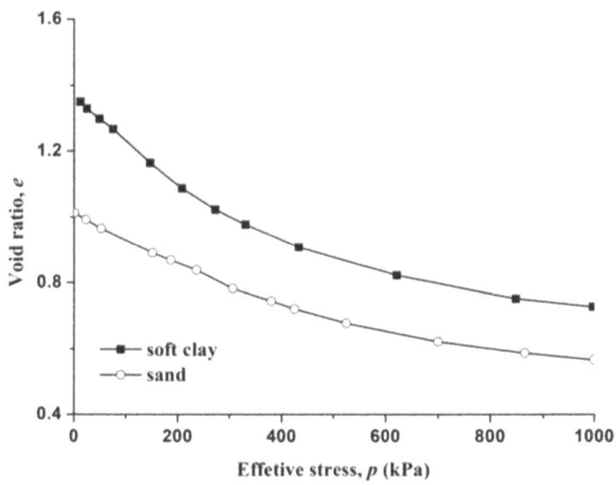

Fig. 3. The *e*–*p* curves of foundation soil.

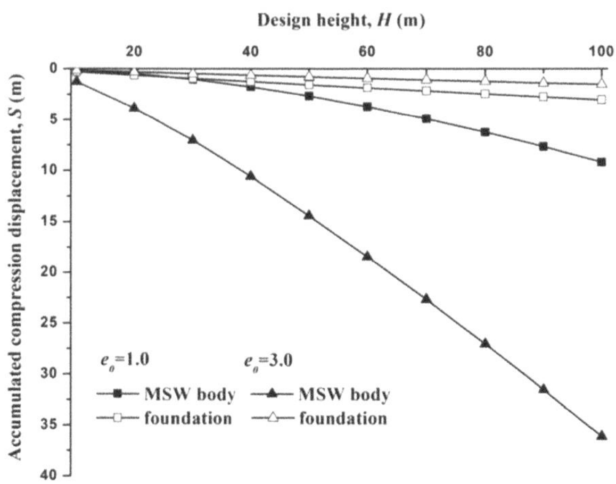

Fig. 4. Accumulated compression displacement at closure with respect to design height

Figure 5 shows the influence of foundation settlement on landfill capacity storage. The landfill capacity storage calculation results without considering foundation deformation are less than the implemented method. Foundation compression displacements result in a greater $\overline{\gamma}$ that means a greater landfill capacity storage. For the scenario of $e_0 = 1.0$, $\overline{\gamma}$ is 14.8 kN/m^3 by neglecting the deformation of foundation for $H = 50$ m, while it is 16.5 kN/m^3 when foundation compression is considered. The difference of calculation results between the two methods decreases with an increment of design height. The reason for this response is that the compression of MSW in the deeper layers is much greater than foundation soil when landfill has a large design height.

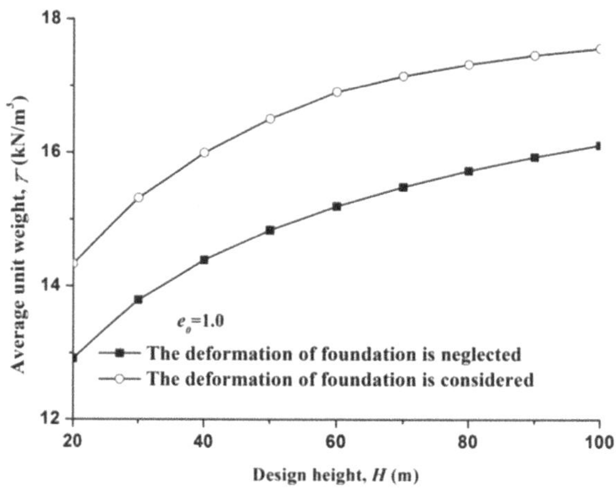

Fig. 5. Comparison of different methods for estimating landfill capacity storage

4 Summary

A revised approach for estimating the settlement and storage capacity of landfills was proposed. The settlement and storage of a landfill could be achieved via a calculation of the settlement and storage of a series of independent column units, and the compression displacement of foundation was considered by using layerwise summation method. A calculation program was developed based on the implemented approach.

The quantitative influences of different initial compaction were investigated via a hypothetical landfill case. The results show that improvement of initial compaction can increase the operational compression displacement of foundation, which is close to accumulated compression of MSW body for the landfills that have a low design height. Estimate results of the settlement and storage capacity of general landfills are more accurate by considering the compression displacement of landfill foundation, however, it has limited influence on calculation results of settlement and storage capacity for the landfills with poor initial compaction.

Acknowledgements. The comments and suggestions of the reviewers are appreciated. The financial support received from the Fundamental Research Funds for the Central Universities of China (Grant No. 3132016342) are gratefully acknowledged.

References

Chen YM, Zhan LT, Wei HY, Ke H (2009) Aging and compressibility of municipal solid wastes. Waste Manag 29:86–95

Chen YM, Ke H, Fredlund DG, Zhan LT, Xie Y (2010) Secondary compression of municipal solid wastes and a compression model for predicting settlement of municipal solid waste landfills. J Geotech Geoenviron Eng 136:706–717

Hanson JL, Yesiller N, von Stockhausen SA, Wong WW (2010) Compaction characteristics of municipal solid waste. J Geotech Geoenviron Eng 136:1095–1102

Gao W, Xu WJ, Bian XC, Chen YM (2017) A practical approach for calculating the settlement and storage capacity of landfills based on the space and time discretization of the landfilling process. Waste Manag 69:202–214

Jang YS (2013) Field-monitored settlement and other behavior of a multi-stage municipal waste landfill Korea. Environ Earth Sci 69(3):987–997

Ling HI, Leshchinsky D, Mohri Y, Kawabata T (1998) Estimation of municipal solid waste landfill settlement. J Geotech Geoenviron Eng 124:21–28

Li YC, Liu HL, Cleall PJ, Ke H, Bian XC (2013) Influences of operational practices on municipal solid waste landfill storage capacity. Waste Manage Res 31(3):273–282

Park HI, Park B (2009) Prediction of MSW long-term settlement induced by mechanical and decomposition-based compressions. Int J Environ Res 3:335–348

Stoltz G, Gourc JP, Oxarango L (2010) Characterisation of the physico-mechanical parameters of MSW. Waste Manag 30:1439–1449

Xu XB, Zhan LT, Chen YM, Guo QG (2015) Parameter determination of a compression model for landfilled municipal solid waste: an experimental study. Waste Manage Res 33(2):199–210

Zhan LT, Chen YM, Ling WA (2008) Shear strength characterization of municipal solid waste at the Suzhou landfill, China. Eng Geol 97:97–111

Experiment on Monitoring Leakage of Landfill Leachate Through Electrical Resistivity Tomography

Ping Yang and Yao-hui Liu(✉)

Tongji University, Siping Road, 200039 Shanghai, China
liuliu8525@163.com

Abstract. Leachate leakage has become a widespread problem in most landfills due to the increasing service time. The soil property, water quality and residents' safety around the landfill was seriously affected by the leachate leakage of landfill. A physical simulating set-up was created to simulate the leakage of leachate in the layer under the influence of flowing water. The monitoring of leachate leakage was conducted using improved electrical resistivity tomography (ERT). The result showed that the accuracy of improved ERT was more than 85% and the laboratory test relying on the simulating set-up and improved ERT was feasible. The laboratory test provides material basis for investigate the leachate diffusion quantitatively.

Keywords: Leachate leakage · Electrical resistivity tomography
Real-time monitoring · Physical simulation

1 Introduction

Landfill leachate is one of the major concerns during the operation and post-closure period of landfills [1]. Landfill leachate contains a variety of pollutants that may potentially contaminate the groundwater and affect the quality of surface water [2]. Leachate will leak in the stratum with the rainfall and underground water. It may cause harm to human health if the leachate infiltrates into the drinking water and agricultural water. Therefore it is significant to monitor the leakage of leachate.

The complexity of underground medium and the shortage of direct measurement were main difficulties for the monitoring of leachate leakage. Electrical resistivity tomography (ERT) was an effective method to detect underground mediums through the difference of electrical resistivity. Artificially generated electric currents were supplied to the soil and the resulting potential differences were measured [3]. Compared with other methods, electrical resistivity tomography had some unique advantages: this method was highly automated which allows for quick and timely field inspection; and was non-destructive and sensitive [1, 4–8]. Therefore, ERT was used to detect the leachate distribution [9–13].

ERT has been used to detect the distribution of leachate in different landfills which the leakage had finished for a long time [9, 14–16]. Then some authors monitored the leachate leakage in real time through ERT and the results showed that this method was

L. Zhan et al. (Eds.): ICEG 2018, ESE, pp. 162–169, 2019.
https://doi.org/10.1007/978-981-13-2224-2_20

effective [17–19]. Many experiments have showed that ERT has been one of the most effective methods to monitor the leachate leakage.

Although it has been mature to detect the leakage of leachate, most researches were based on field tests. Many uncontrollable factors, such as complex underground structure, uncontrollable groundwater, overlying plants and climate, affect results of field tests. In addition, filed tests may result in the second pollution during experiment process. The influence of different factors on leachate leakage could be quantitative through indoor simulating experiments. However, few physical simulating experiments were conducted due to the limitation of experimental condition [1, 20].

To monitor the process of leachate leakage in the underground in real time, a physical simulating experiment was conducted and improved ERT equipment was created.

2 Materials and Methods

2.1 Flowing Water Simulating System

The flowing water simulating system was used to simulate underflow in soil. The system consisted of water injection device and a simulation box. The simulation box (200 cm * 100 cm * 98 cm) consisted of injection tank, soil tank and drainage tank, as shown in Fig. 1.

Fig. 1. Flowing water simulating system

A pump was used to inject water to the injection tank (100 cm * 10 cm * 98 cm). Some drain pipes were arranged in front of injection tank with the height of 40 cm, 50 cm, 60 cm, 70 cm, 80 cm and 90 cm. The water head height was controlled through opening different drain pipes.

Water entered the soil through a water pipe in past simulating experiments [1]. This method would result in uneven distribution of groundwater. Hence, some mini drain holes were arranged in the board between injection tank and soil tank. The diameter of mini drain hole was 10 cm and the distance between these holes was 40 cm. Water could enter soil evenly through these holes.

Soil box (180 cm * 100 cm * 98 cm) was used to place soil. Nonwovens were arranged at the interface of injection tank and drainage tank to prevent soil particles leaking.

The structure of drainage tank was similar to injection tank. Mini drain holes were distributed in the front board and drainage pipes with different height were arranged in another side to control the water head height in drainage tank.

The flowing rate was controlled by adjusting the water head height of injection tank and drainage tank. This controlling method could keep the flow rate in a constant level.

2.2 Improved ERT Equipment

Small-Scale Electrode

The original ERT equipment was WGMD-9. The routine electrode of ERT was a copper bar which diameter was 1.2 cm and the length was 15 cm. Routine electrodes were widely used in the field measurement where the length of profiles were hundreds meters. However, routine electrodes would result in the deformation of soil and alteration of flowing direction in the physical simulating experiment due to its great scale (Fig. 2). The distance between electrodes in physical experiments was usually 0.05–0.1 m. Large-scale electrode affected the measurement accuracy and small-scale electrodes adapting to the physical experiment were created. The radius of improved electrode was 0.3 cm and the length was 5 cm. The measurement accuracy of small-scale electrodes was above 95% in the calibration test.

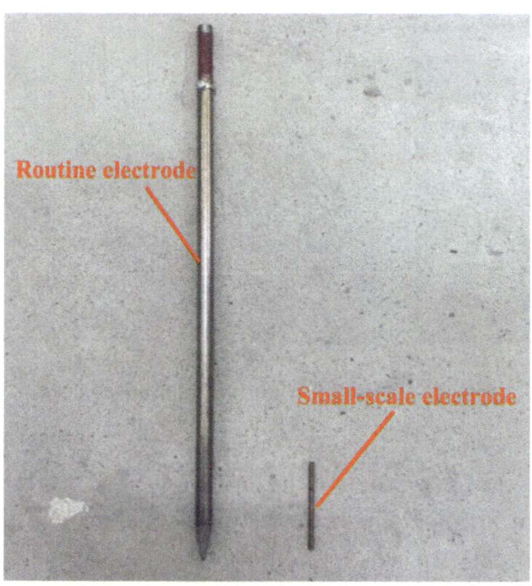

Fig. 2. Routine electrode and small-scale electrode

Quickly Connecting Device

A lot of time was wasted to change the profile during the process of ERT measuring. The long time for changing profiles resulted in a long gap between each measure. It will also affect the monitoring result when the flowing rate was great.

A quickly connecting device was designed to change profiles in a short time. The quickly connecting device consisted of two parts. One part was connected with electrodes in turns. Another part was connected with the cable. The measurement could start only connecting two parts of connecting device, as shown in Fig. 3. This method shortened the time of changing profiles from 10 min to 10 s.

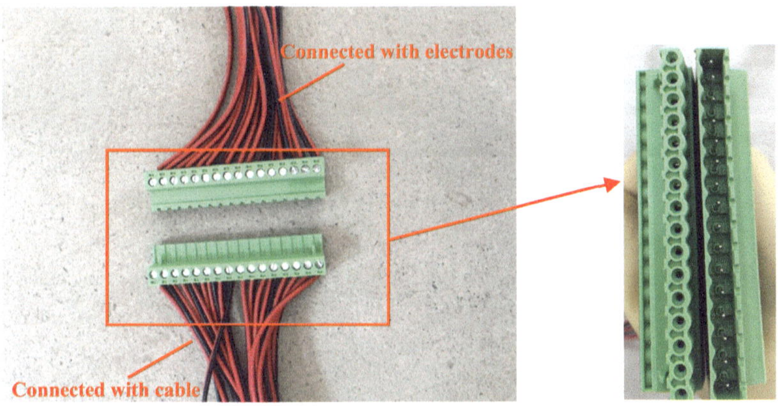

Fig. 3. Quickly connecting device

2.3 Physical Simulating Experiments

To monitor the leachate migrate in real time, a physical simulating experiment was conducted. The procedures of experiment were as follows:

1. Layout sand in the soil box. The thickness of sand layer was 45 cm and the permeability coefficient was 1.82 * 10–3 cm/s. The grain diameter of sand ranged from 0.30–0.42 mm. The sand layer was overlaid with 2 cm thick soft clay to fix electrodes and diminish the ground resistance.
2. Four ERT profiles were layout in the soil. Each profiles consisted of 30 electrodes. The electrode distance was 0.05 m and the profile distance was 0.15 m.
3. Keep the water head height in injection tank was 50 cm and the water head height in drainage tank was 40 cm.
4. Leachate was injected in the soil through a pipe. The initial leakage depth was 0.1 m and the location was shown in Fig. 4. The dropping rate was controlled by a valve in 30 ml/h.
5. Monitor the diffusion using the arrangement modes of δA and AMN. Measure the leakage one time each half-hour.

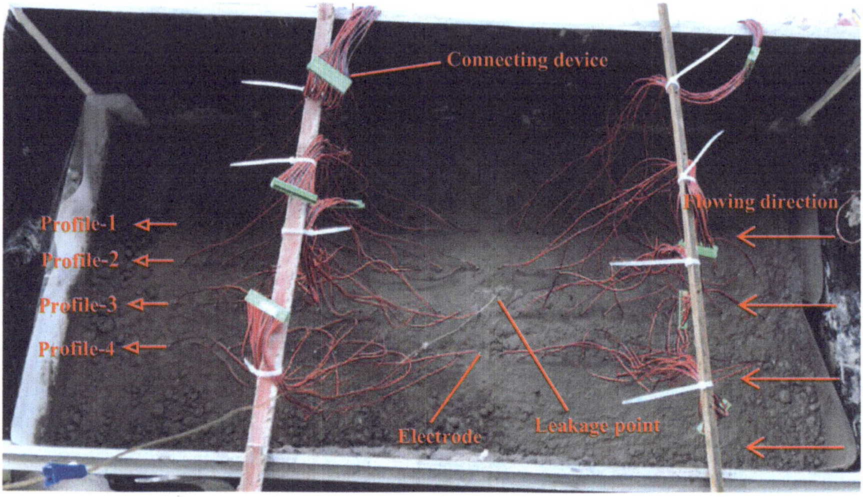

Fig. 4. Physical simulating experiment

6. Stop the experiment when the leakage lasted for 6 h. Clean equipment and process data.

3 Results and Analysis

Only the results of profile-3 were analysed in this paper due to the length limitation. The measured electrical resistivity was inversed through RES2D and results were shown in Fig. 5. The region with lower electrical resistivity correspond to the leakage area due to the electrical resistivity of leachate was much lower than sand. The lower electrical resistivity represented the higher concentration of leachate.

The electrical resistivity increased with the increase of depth due to the influence of soil box and edge effect. However, difference between leakage scope and background was enough distinct to be distinguished.

When the experiment time increased to 1 h, the range of leakage could be distinguished through ERT. It was obvious that leakage range enlarged and the concentration kept increasing with the increase of time.

As shown in Fig. 5, leachate migrated quickly along the flowing due to the high permeability of sand layer. At the same time, leachate migrated along the vertical direction due to the influence of gravity. A fraction of leakage migrated along the reverse flowing direction in the Fig. 5(d).

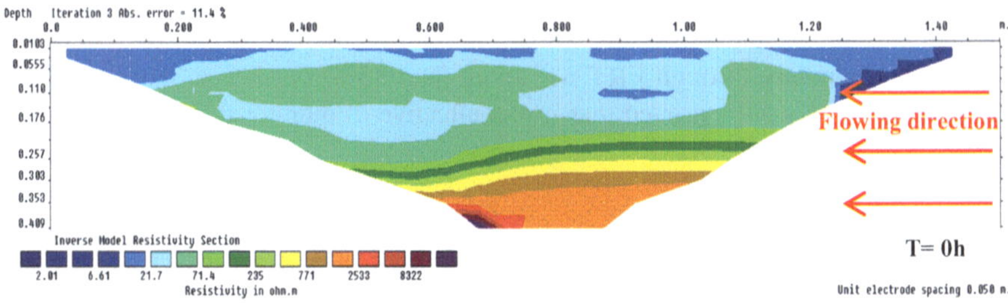

(a) Inversion result of profile-3 before leakage

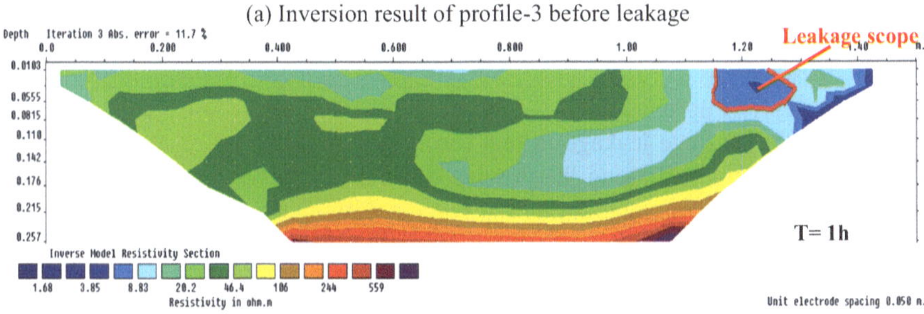

(b) Inversion result of profile-3 at 1 hour

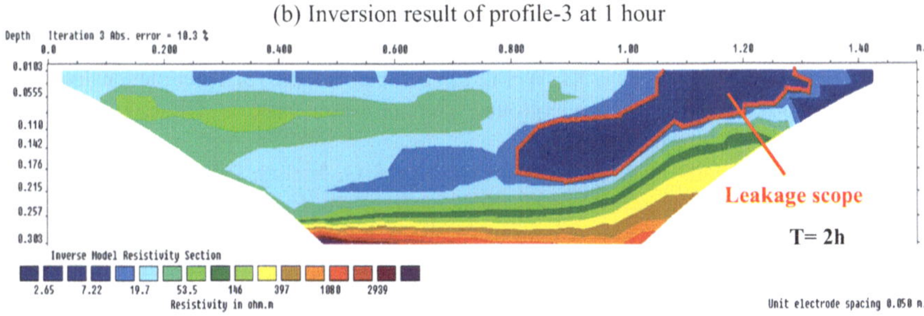

(c) Inversion result of profile-3 at 2 hour

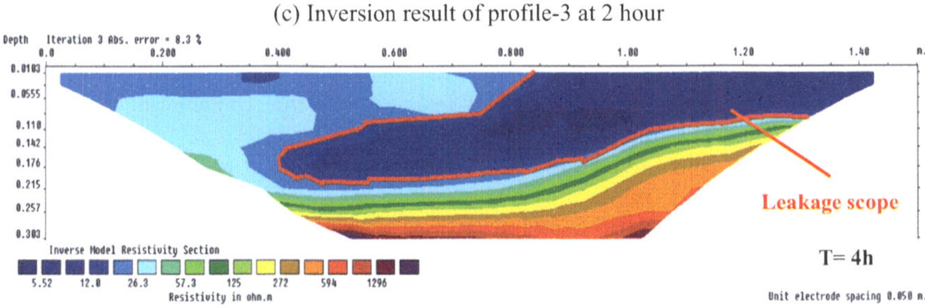

(d) Inversion result of profile-3 at 4 hour

Fig. 5. Inversion results of profile-3

The contained soil was dig up after experiment and its leachate concentration was examined through chemical methods. The results showed that the accuracy of detecting the pollution boundary through ERT was more than 85%.

The experiment results proved that it was effective to monitor the leachate leakage in real time through ERT.

4 Results and Analysis

1. Improved ERT equipment with small-scale electrodes and quickly connecting devices was designed. The improved ERT had the characteristic of high resolution and high efficiency.
2. The accuracy of Improved ERT was more than 85%. Improved ERT was effective to monitor the leachate leakage in real time.
3. Most leachate migrated along the flowing direction in the sand layer. Only a small fraction of leachate migrated along the reverse flowing direction.

References

1. Jia YG, Pan YY, Shang H, Guo XJ, Wei L, Li HJ, Shan HX (2012) Three-dimensional (3D) dynamic monitoring for an underground contamination process induced by Landfill Leakage. Environ Forensics 13:68–81
2. Christensen TH, Kjeldsen P, Albrechtsen HJ, Heron G, Nielsen PH, Bjerg PL, Holm PE (2009) Attenuation of landfill leachate pollutants in aquifers. Critical Rev Environ Sci Technol 24(2):119–202
3. Samouelian A, Tabbagh CA, Bruand A, Richard G (2004) Electrical resistivity survey in soil science: a review. Soil Tillage Res 83:173–193
4. Buselli G, Davis GB, Barber C, Height MI, Howard SHD (1992) The application of electromagnetic and electrical methods to groundwater problems in urban environments. Explor Geophys 23(4):543–555
5. Lima OAL, Sato HK, Porsani MJ (1995) Imaging industrial contaminant plumes with resistivity techniques. J Appl Geophys 34(2):93–108
6. Dahlin T (1996) 2D resistivity surveying for environmental and engineering application. First Break 14(7):275–283
7. Aristodemou E, Thomas-Betts A (2000) DC resistivity and induced polarization investigations at a waste disposal site and its environments. J Appl Geophys 44(2–3):275–302
8. Ogilvy R, Meldrum P, Chambers J, Williams G (2002) The use of 3D electrical resistivity tomography to characterise waste and leachate distribution within a closed landfill, Thriplow, UK. J Environ Eng Geophys 7(1):11–18
9. Maurya PK, Ronde VK, Fiandaca G, Balbarini N, Auken E, Bjerg PL, Christiansen AV (2017) Detailed landfill leachate plume mapping using 2D and 3D electrical resistivity tomography - with correlation to ionic strength measured in screens. J Appl Geophys 138:1–8
10. Casado I, Mahjoub H, Lovera R, Fernandez J, Casas A (2015) Use of electrical tomography methods to determinate the extension and main migration routes of uncontrolled landfill leachates in fractured areas. Total Environ 506:546–555

11. Konstantaki LA, Ghose R, Draganov D, Diaferia G, Heimovaara T (2015) Characterization of a heterogeneous landfill using seismic and electrical resistivity data. Geophysics 80, EN13–EN25
12. Vargemezis G, Tsourlos P, Giannopoulos A, Trilyrakis P (2015) 3D electrical resistivity tomography technique for the investigation of a construction and demolition waste landfill site. Geophys. Geod 59:461–476
13. Wang TP, Chen CC, Tong LT, Chang PY, Chen YC, Dong TH, Liu HC, Lin CP, Yang KH, Ho CJ, Cheng SN (2015) Applying FDEM, ERT and GPR at a site with soil contamination: a case study. Appl Geophys 121:21–30
14. Feng SJ, Bai ZB, Cao BY, Lu SF, Ai SG (2017) The use of electrical resistivity tomography and borehole to characterize leachate distribution in Laogang landfill. China. Environmental Science and Pollution Rearch 24(25):20811–20817
15. Bichet V, Grisey E, Aleya L (2016) Spatial characterization of leachate plume using electrical resistivity tomography in a landfill composed of old and new cells (Belfort, France). Eng Geol 211:61–73
16. Belmonte-Jimenez SI, Jimenez-Castaneda ME, Perez-Flores MA, Campos-Enriquez JO, Reyes-Lopez JA, Salazar-Pena L (2012) Characterization of a leachate contaminated site integrating geophysical and hydrogeological information. Geofis Int 51(4):309–321
17. Clement R, Oxarango L, Descloitres M (2011) Contribution of 3-D time-lapse ERT to the study of leachate recirculation in a landfill. Waste Manag 31:457–467
18. Audebert M, Clement R, Duquennoi MC, Loisel S, Touze-Foltz N (2016) Understanding leachate flow in municipal solid waste landfills by combining time-lapse ERT and subsurface flow modelling - Part I: Analysis of infiltration shape on two different waste deposit cells. Waste Manag 55:165–175
19. Audebert M, Oxarango L, Duquennoi C, Touze-Foltz N, Forquet N, Clement R (2016) Understanding leachate flow in municipal solid waste landfills by combining time-lapse ERT and subsurface flow modelling - Part II: Constraint methodology of hydrodynamic models. Waste Manag 55:176–190
20. Xiao K, Xue Q, Liu K (2015) Influence factor analysis of the electricity-dipole method for landfill leakage detection on HDPE membrane. Environ Earth Sci 74(7):6225–6232

Monitoring of Biogas from Two Covers of Lysimeters Filled with Municipal Solid Waste

Caio Satoshi Misawa, Marina Mostiack Pomaleski, Rafael Souza Rodrigues dos Santos, and Miriam Gonçalves Miguel(✉)

University of Campinas, Campinas 13083-889, Brazil
mgmiguel@fec.unicamp.br

Abstract. In Brazil, conventional covers of sanitary landfill are consisted of fine soils compacted with tractor treadmill. Nonetheless, some clays have swelling behavior and present shrinkage cracking when they are dry, impairing their performance with relation to control the flow of water into the waste mass and the flow of gases out of the waste mass. This control is necessary to minimize the leachate and gas volume generated into de waste mass. This research aimed to evaluate the performance of two conventional covers of two lysimeters filled with municipal solid waste (MSW) of Campinas city, Brazil. For this, the concentration of gases (CO_2 and CH_4) that was coming out of covers was obtained using flux chamber and correlate with the concentration of gases that were drained by central pipe. Disturbed soil samples from two covers were characterized by geotechnical tests such as grain-size distribution, consistent limits, unit weight and moisture content. Permeability testes were also performed with specimens trimmed of undisturbed soil samples collected of two covers. The soil samples were classified as silt clayed-sandy and they presented micro aggregation. Saturated permeability coefficients presented orders of magnitude of 10^{-4} and 10^{-5} cm/s. Gas concentrations from central drain were greater than gas concentrations from the cover, twice for CO_2 and 30% for CH_4, approximately. The precipitation influenced the gas concentrations from central drain and from covers differently and this fact is discussed.

Keywords: Biogas · Municipal solid waste · Cover

1 Introduction

Biogas has a biogenic origin and so it can be used as a substitute for fossil fuels in energy production. It is the result of chemical reactions carried out by strictly anaerobic bacteria during the decomposition of organic matter, and therefore can to reduce greenhouse gas emissions [1].

Municipal solid waste disposed in sanitary landfills provides the generation of biogas which is composed mainly of carbon dioxide (CO_2) and methane (CH_4), contributing to the greenhouse effect. Methane's Global Warming Potential (GWP) is 25 (based on carbon dioxide) for a 100-year horizon and 72 for the 20-year horizon [1]. In

L. Zhan et al. (Eds.): ICEG 2018, ESE, pp. 170–177, 2019.
https://doi.org/10.1007/978-981-13-2224-2_21

Brazil, CH_4 emissions from the Waste Treatment Sector were 1,060 and 1,104 million Mg^2 in 2000 and 2005, respectively, representing an increase of approximately 4% in the sector's CH_4 emission [2]. In addition, activities related to solid waste management emit approximately 14% of the total methane produced by anthropogenic activities [3].

Methods have been developed in order to monitor the amount of gas emission through the landfill cover, such as flux chamber. The flux chamber is an accurate and nonintrusive method for measuring emission rates from individual locations [4].

The flux chambers have been used to evaluate the concentration of the contained gases in the chamber in relation to the time to determine the flow of gases. The use of the flux chamber has some advantages, such as the low cost, ease of handling and the simultaneous evaluation of more than one gas. However, several long-term tests are required to obtain the emission of gases.

There are two types of the flux chambers: dynamic and static chambers (see Fig. 1). In the dynamic chamber, the gases are diluted in an airflow, analyzed and directed to the atmosphere (open cycle). In the static chamber, the gases are confined, analyzed and returned to the chamber (closed cycle).

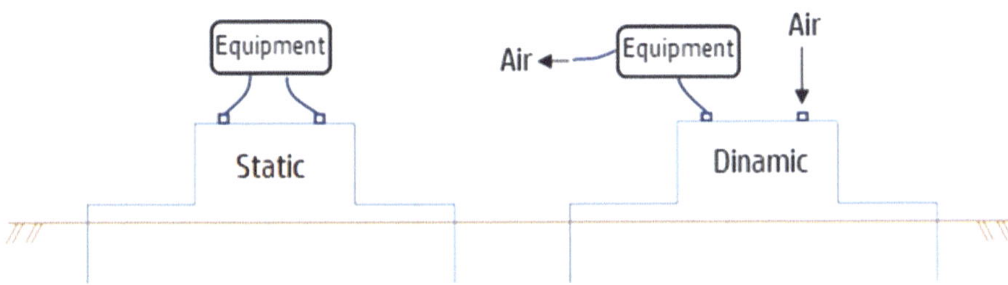

Fig. 1. Static and dynamic flux chamber

This research aimed to monitor and analyze the biogas (methane and carbonic gas) that escapes the covers of two lysimeters filled with municipal solid waste from the city of Campinas, state of São Paulo, Southeastern Brazil, using a static flux chamber. Data are compared with the concentration of the gases from central drain of lysimeters and precipitations during the monitoring period.

2 Material and Method

2.1 Lysimeters

Two lysimeters (L4 and L5) were filled with MSW from the city of Campinas, state of São Paulo, Southeastern Brazil. The MSW gravimetric composition was: 41.9% organics, 14.6% paper, 12.5% plastic, 7.9% pruning, 4.6% fabric, 3.6% diapers and sanitary pads, 2.0% wood, 1.9% metal, 1.8% glass, 1.1% construction waste, and 8.1% miscellaneous.

The cross-sections of lysimeters were 1.43 m × 1.43 m and 1.38 m × 1.43 m, for L4 and L5, respectively. The lysimeters were built in 2015 [4]. Currently, MSW is in the methanogenic phase of the anaerobic biodegradation [5].

Both lysimeters have a concrete base with 2% slope superimposed by a drainage layer composed by stone for the leachate of 20 cm thickness (lower stone layer). A horizontal tube of PVC of 3.81 cm in diameter is connected to a recipient of 35 L of capacity to collect the leachate. Above the stone layer, a compacted MSW was disposed with 190 cm thickness, approximately. The MSW unit weight values were 8.8 $kN.m^{-3}$ and 6.1 $kN.m^{-3}$ for L4 and L5, respectively. In each of the lysimeters was also installed a vertical PCV tube (central drain) with 5.08 cm in diameter from the lower stone layer to the outlet to atmosphere, at which its tip has a gas metering register. The central drain has small 8 mm holes distributed along its length to capture the gases produced by MSW biodegradation. Another stone layer (upper stone layer) was disposed on MSW layer with 20 cm thickness. Finally, a mineral cover of about 50 cm thickness was disposed on the upper stone layer. The longitudinal profile of lysimeters can be seen in Fig. 2.

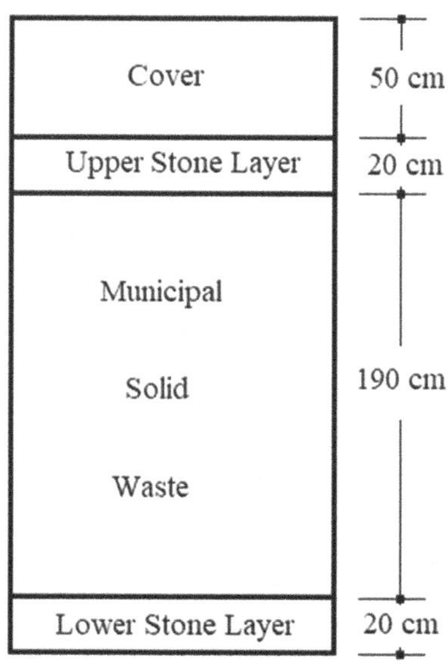

Fig. 2. Longitudinal profile of lysimeters

The soils of the covers was compacted until they reached the unit weight of 18 kN/m^3, approximately. Two kinds of soils were used: soil 1 for L5 cover and soil 2 for L4 cover [5].

2.2 Sampling and Soil Characterization

Disturbed soil samples 1 and 2 were subjected to the following geotechnical tests: (a) grain size analysis [7], particle density [8], liquid and plastic limits [9] and shrinkage limit [10].

Undisturbed soil samples 1 and 2 were collected by crimping of metal rings (5 cm of diameter and 2 cm of height) in the covers. After the leveling of the faces of the soil samples inside the ring, the bulk density was determined by ratio between mass and volume of specimen inside the metal rings. During the leveling, the detached soil portions were used to determine the moisture content [11].

The disturbed soil samples 1 and 2 were compacted into 1,000 L volume metal cylinders with unit weight and moisture content characteristic of each cover. Specimens were prepared trimming the compacted soil samples to obtain a cylindrical specimen (10 cm diameter and 12 cm height) to be submitted to permeability tests with variable hydraulic load [12].

2.3 Biogas and Precipitation Data

A static flux chamber was used to confine the gases escaped from the covers of lysimeters. The Eagle equipment, produced by RKI Instruments, was used to measure the concentrations of CH_4 and CO_2 gases contained in the static flux chamber and in the central drain. The monitoring were performed weekly. Seventeen measurements were obtained during four months between July and October 2017 for the central drain and sixty four measurements for the cover.

The precipitation data were provided by a station, distant 500 m, approximately, from the location of the lysimeters, in the period between June 30, 2017 and September 30, 2017. Data were accumulated in weeks.

2.4 Statistical Method

The T test are used to compare whether the means of the two data sets were significantly different. For this, a confidence level of 95% was adopted, that is, $P = 0.05$. However, if the calculated P value is less than 0.05, the null hypothesis will be rejected, that is, the means of the two sample spaces are different and consequently there is difference in the means of the spaces. Two hypothesis were considerate in order to verify if the soils influenced the emission of the gases from the cover: (a) Concentrations of CO_2 in L4 and L5 are equal, and (b) Concentrations of CH_4 in L4 and L5 are equal.

3 Results

Results of the characterization tests of the soil samples 1 and 2 were presented in Table 1. Results of the T test were shown in Table 2. The average concentrations of the gases from the central drain and from the covers were compared (see Table 3).

Table 1. Results of tests characterization

	Soil sample 1	Soil sample 2
Sandy (%)	20	7
Silt (%)	50	71
Clay (%)	30	22
Moisture content (%)	11.34	21.15
Bulk density (g/cm^3)	1.55	1.44
Particle density (g/cm^3)	2.67	2.71
Shrinkage limit (%)	11.9	15.7
Plastic limit (%)	21.5	25.6
Limit liquid (%)	37.2	40.2
Plasticity index (%)	15.7	14.5
Dry density (g/cm^3)	1.39	1.19
Void ratio	0.92	1.29
Saturated coefficient of permeability (cm/s)	2.2×10^{-5}	3.5×10^{-4}

Table 2. Results of the T test

Hypothesis	P value	Confidence index	Result
a	3.47E − 07	0.05	Rejected
b	0.055	0.05	Not rejected

Table 3. The average concentrations of the gases from the central drain and from the covers

	L4		L5	
	Cover	Drain	Cover	Drain
CO_2 (mg/L)	9.34	17.04	8.53	19.94
CH_4 (mg/L)	11.77	16.13	13.97	19.50
Cover/drain CO_2 (%)	54.81		42.78	
Cover/drain CH_4 (%)	72.97		71.64	

Both soils were classified as silt clayed-sandy. The saturated coefficient of permeability values were high due to micro aggregation. Both soils are classified as low compressibility clay (CL) by the Unified Soil Classification System (USCS) [13].

The soils of L4 and L5 covers differ in the emission of CO_2 but they do not differ in the emission of CH_4, according to statistical test (see Table 2).

The L5 cover retained approximately the same CH_4 gas percentage as the L4 cover, evidencing the hypothesis (b) of statistical analysis. For CO_2 gas, the L5 cover retained considerably more gas than the L4 cover, confirming the hypothesis (a) of statistical analysis (see Table 3).

The shrinkage behavior of soils 1 and 2, which can lead to cracks, influences the escape of gases. Soil 1, presents in the L5 cover, is in the semi-solid consistency state, which allows the soil to shrinkage during the drying process (with cracking), between moisture content values 11.9% and 21.5% (Table 1), that is, between the shrinkage limit (SL) and the plastic limit (PL). Soil 2 is in the same state of consistency between higher values of moisture content, which are, 15.7% and 25.6% (SL and PL, respectively). This means that soil 1, under a drying process, needs to reach lower moisture contents to shrink and to cease shrinkages, when compared to soil 2. Under the wetting process, the opposite occurs, that is, soil 1 needs of a moisture content lower than that of soil 2 to reach its SL and transpose from the solid consistency to the semi-solid, in which the cracks with increasing moisture content gradually occur. In addition, the physical characteristics of the two soils 1 and 2 corroborate to the above behavior. Soil 1, after compaction, had lower void ratio than soil 2 (see Table 1).

Average concentrations of CO_2 e CH_4 from central drain and cover of the L4 and the L5 were compared with precipitation during the monitoring (see Figs. 3 and 4, respectively).

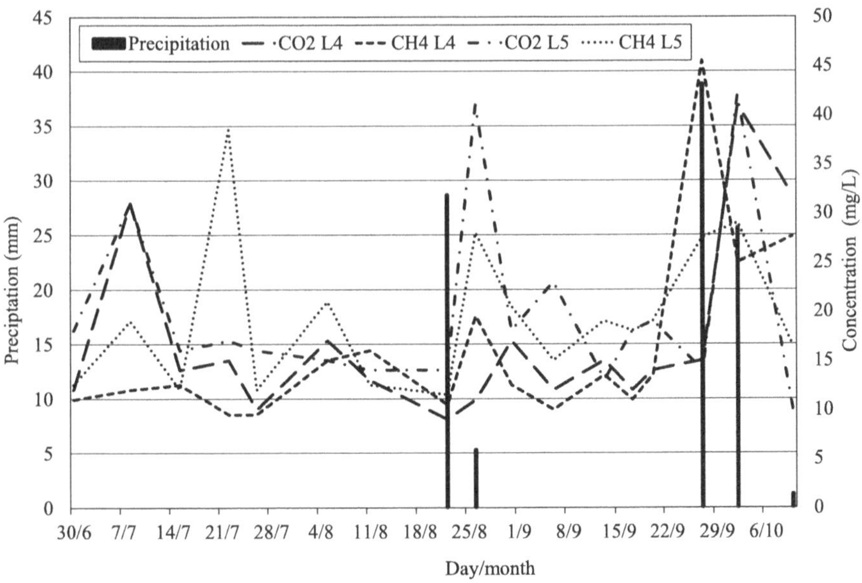

Fig. 3. Concentration of CO_2 e CH_4 from central drain and precipitation

There were three concentration peaks of L4 and L5 gases from central drain in the first weeks, between the last week of August and the first week of September, and between September 29 and October 6 (Fig. 3). In the last two precipitation periods, the peaks occur soon after the precipitations. As the water infiltrates through the cover and reaches the MSW layer, the MSW biodegradation is accelerated, increasing the gas production. At the same time the precipitations raised the soil moisture content, reduces its permeability to the air in the covers, and consequently, the escape of the gas (Fig. 4). Therefore, the gas drains mainly through the central drain and not through the cover. However, after the precipitations, concentrations of the gases from the central

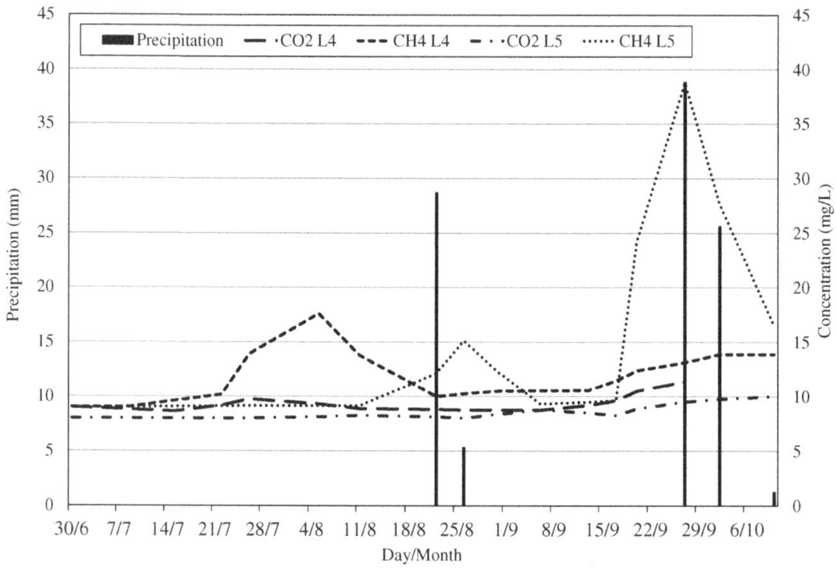

Fig. 4. Concentration of CO_2 e CH_4 from the covers and precipitation

drain and the cover increase, because the soil suffers shrinkage with draying process, cracks and so allows the escape of the gases. In addition, the lower moisture content increases its air permeability, also facilitating the escape of gases.

The precipitations did not influence the concentrations in either of the two gases for L4 cover, but for L5 cover, CH_4 concentrations were reduced after precipitations, and CO_2 concentration remained practically constant for the 4 months of monitoring.

4 Conclusion

The statistical analysis showed that the different soils of covers influenced the emission of CO_2, but did not influence the emission of CH_4. The L5 cover retained more gas, both CO_2 and CH_4, than the L4 cover. However, for the CH_4 gas, the L5 cover retained approximately the same percentage of gas as the L4 cover. For CO_2, the L5 cover retained much more gas than the L4 cover, corroborating with the statistical analysis.

After the precipitations, there was a reduction in the concentrations of gases from the covers due to moisture content increased, making it difficult to escape the gases through the covers. However, precipitations influenced the increase of the biogas production with the biodegradation of the municipal solid waste, through the infiltration of the rainwater in the soil of the covers. However, after the precipitations, the concentrations of the gases from the central drain and the covers increase, due to the decrease in moisture content and, consequently, cracking and increase in air permeability.

References

1. UNEP. https://wedocs.unep.org/bitstream/handle/20.500.11822/8648/Waste&ClimateChange.pdf?sequence=3. Accessed 30 Mar 2018
2. MCT (2010) Ministério de Ciência e Tecnologia. Emissões de gases de efeito estufa no tratamento e disposição de resíduos. Segundo Inventário Brasileiro de Emissões Antrópicas de Gases de Efeito Estufa. Brasília – DF
3. IEA. https://www.iea.org/publications/freepublications/publication/India_methane.pdf. Accessed 30 Mar 2018
4. Reinhart DR, Cooper DC, Walker BL (1992) Flux chamber design and operation for the measurement of municipal solid waste landfill gas emission rates. J Air Waste Manag Assoc 42(8):1067–1070
5. Favery RLT, Manzatto MP, Moretto RL, De Almeida GHTDD, Miguel MG, Teixeira EN (2016) Study of influential factors on the compaction of municipal solid waste in lysimeters. In: Proceedings of 31st international conference on solid waste technology and management, ICSW2016, Philadelphia, Widener University, pp 548–558, 3–6 April 2016
6. Barlaz MA, Staley BF, de los Reyes FL (2010) Anaerobic biodegradation of solid waste. In: Mitchell R, Gu J (eds) Environmental microbiology. Wiley, Hoboken, pp 281–299
7. ASTM D 422-63 (2007) Standard test method for particle-size analysis of soils. American Society for Testing and Materials, West Conshohocken, PA
8. ASTM D 854-14 (2014) Standard test methods for specific gravity of soil solids by water pycnometer. American Society for Testing and Materials, West Conshohocken, PA
9. ASTM D 4318-10 (2010) Standard test method for liquid limit, plastic limit and plasticity index of soils. American Society for Testing and Materials, West Conshohocken, PA
10. ASTM D 427-04 (2004) Test method for shrinkage factors of soils by the mercury method (Withdrawn 2008), ASTM International, West Conshohocken, PA
11. ASTM D2216-10 (2010) Standard test methods for laboratory determination of water (Moisture) content of soil and rock by mass, ASTM International, West Conshohocken, PA
12. ASTM D 5856-15 (2015) Standard test method for measurement of hydraulic conductivity of porous material using a rigid-wall, compaction-mold permeameter. American Society for Testing and Materials, West Conshohocken, PA
13. ASTM D2487 (2011) Standard practice for classification of soils for engineering purposes (unified soil classification system). Standard language: English

Landfilling of Oil and Gas Exploration and Production Wastes: Geotechnical and Environmental Considerations

Ricardo C. de Abreu(✉) and Jonathan E. Fourrier

Fourier & de Abreu Engineers, L.L.C., Baton Rouge, USA
ricardo@fdaengineers.com

Abstract. The United States Environmental Protection Agency (EPA) characterizes the wastes generated during the exploration, development, and production of crude oil, natural gas, and geothermal energy as "special wastes" that are exempt from federal hazardous waste regulations. This study is based on a landfill in the state of Louisiana permitted to dispose four types of exploration and production (E&P) wastes. Due to the nature of some E&P wastes received at the landfill, a solidification agent, usually sawdust, is added and mixed prior to final disposal. E&P wastes can have an impact on the slope stability of landfills due to their distinct shear strengths and the concern of development of additional pore pressures due to the reduction of the hydraulic conductivities caused by the utilization of some solidification agents (e.g., sawdust). From this study, the authors developed the necessary geotechnical parameters to conduct a slope stability analysis for the design of future disposal cells and proposed general operational and monitoring guidance concerning the expansion of the landfill. One of the main conclusions of the study was that even though the utilization of sawdust/E&P waste (water-based) may improve the shear strength of the total waste mass, the development of inexorable pore pressures may reduce the safety factor of the landfill slopes.

Keywords: Oil · Gas · Exploration · Production · Slope stability
Shear strength

1 Introduction

1.1 A Historical Background of E&P Wastes in the United States

Along with its 1978 proposed hazardous waste management requirements, the United States Environmental Protection Agency (EPA) also proposed less stringent standards for several categories of high volume wastes. The EPA's position was that these high volume "special wastes" had lower levels of contaminants than other hazardous wastes being regulated under the Resource Conservation and Recovery Act (RCRA), the United States public law that creates the framework for the proper management of hazardous and non-hazardous solid waste [1]. Based on the EPA's position, the United States Congress exempted the special wastes from the RCRA Subtitle C hazardous waste regulations until the EPA could conduct a more detailed study and regulatory

L. Zhan et al. (Eds.): ICEG 2018, ESE, pp. 178–185, 2019.
https://doi.org/10.1007/978-981-13-2224-2_22

evaluation. The EPA issued a regulatory determination in 1998 which declared that management of exploration and production (E&P) wastes under RCRA Subtitle C (hazardous waste regulations) was not required. As a result of this action, E&P wastes are exempt from Subtitle C regulations. Despite this exemption, there is no prohibition against management of E&P wastes under state regulations, or under the less strict non-hazardous solid waste regulations (RCRA Subtitle D). Therefore, under this framework, E&P wastes are able to be managed at non-hazardous (Subtitle D) solid waste landfills such as the subject facility.

1.2 Description of the Facility

The subject facility is a non-hazardous solid waste landfill located in the state of Louisiana in the southeastern United States. The landfill has been permitted since 2003 by the state regulatory authority to accept industrial wastes, MSW, and construction and demolition (C&D) debris. Since receiving authority from the state to accept E&P wastes in 2009, the facility's total annual solid waste intake rate has averaged approximately 191,500 metric tons per year, of which approximately 21% has constituted E&P wastes [2]. The types of E&P wastes that the facility is allowed to receive comprise water-based drilling wastes, oil-based drilling wastes, and contaminated soils.

The mean daily maximum temperature for the facility's region is 24.6 °C. The mean daily minimum temperature for the facility's region is 12.8 °C. The average annual rainfall is 131 cm [3].

Although the State of Louisiana allows co-disposal of certain E&P wastes in industrial non-hazardous solid waste landfills, these types of waste can have different mechanical and physical properties when compared to Municipal Solid Waste (MSW) and typical industrial wastes. The general literature is very scarce about the geotechnical characteristics of E&P wastes; therefore, the authors conducted a study to investigate E&P waste properties and their impact on the slope stability of the landfill. At the time of the study, the landfill primarily stabilized water-based E&P drilling wastes with sawdust. An alternative method for stabilization, as proposed by the facility, was to utilize selected MSW and absorbent industrial wastes as a solidification agent.

2 Study Program

2.1 Scope of Study

The analysis of E&P wastes in landfill stability cannot be done through the routine geotechnical methodology employed to other waste materials that have been more extensively researched (e.g., MSW), unless a very conservative approach is considered. The authors believed that the use of a very conservative approach could result in a significantly reduction of the landfill's capacity, therefore it seemed appropriate to consider at least an expedited testing program to evaluate the effects of E&P wastes on the landfill's geotechnical stability. A plan for E&P sampling, testing, and analysis was then developed by the authors.

Although the geotechnical properties of MSW are relatively known, not much information is available in the general literature regarding the properties of E&P wastes. The landfill intended to mix stabilized E&P wastes with MSW/industrial wastes for disposal, however it would be very difficult to collect and test representative and undisturbed samples of mixed E&P and MSW/industrial wastes. Moreover, the parameters generated by such tests would hardly be representative of landfill conditions, where pore-pressures develop inside the waste mass, therefore reducing the effective stresses. The approach employed by the authors was to evaluate the characteristics and mechanical properties of pure sawdust-stabilized E&P waste and generate appropriate parameters for the mix of E&P/MSW/industrial wastes based on these tests and on MSW's available literature for evaluation of slope stability.

After the sampling and laboratory phase was completed, all data were compiled and analyzed in order to characterize the E&P wastes received at the facility regarding the geotechnical and mechanical properties tested. Then, different scenarios, including the use of sensitivity analysis, were utilized to interpret the geotechnical characteristics of the waste mass (municipal and industrial solid wastes + E&P waste + soil). A slope stability analysis was then conducted to evaluate the slopes and maximum elevation of the landfill.

2.2 Testing Program Description (Phases I & II)

A plan for E&P wastes sampling, testing, and analysis was developed by the authors specifically for this project. In accordance with the plan, testing would be conducted in two phases. Phase I of the study consisted of an expedited testing program performed to initially characterize E&P wastes; the Phase I results were used to recommend a more comprehensive batch of tests to be conducted during Phase II. Phase II consisted of more specific testing selected to acquire additional data which would complement the Phase I testing.

For Phase I, water-based drilling wastes were collected in buckets and solidified with sawdust and MSW. The tests performed consisted of particle size analysis, Atterberg limits, field capacity, organic content, compaction, direct shear strength, and hydraulic conductivity.

For Phase II, water-based drilling wastes were solidified with sawdust and tested for hydraulic conductivity and triaxial compression (consolidated-undrained) under different moisture contents.

Testing of oil-based drilling wastes was not performed, since these types of wastes were not available at the landfill at the time the study was conducted.

3 Results

3.1 Phase I Testing

Particle Size Analysis and Atterberg Limits. The particle size analysis of the tested sawdust/E&P waste sample (sawdust and water-based drilling waste) presented 0.5% of gravel-size particles, 36.3% of sand-size particles (32.8% of total being fine sand), 33.3% of silt-size particles, and 29.9% of clay-size particles. Figure 1 shows the grain size distribution of the sample.

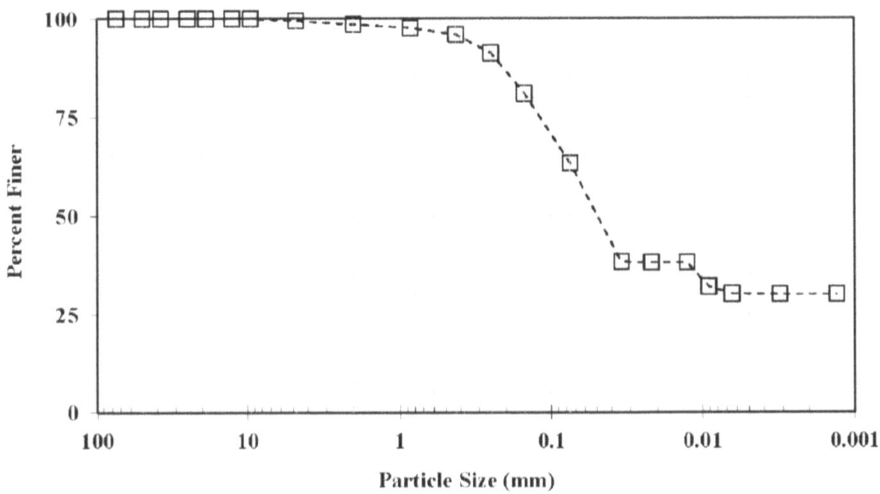

Fig. 1. Particle size analysis of sawdust/E&P waste sample

The Atterberg limits for the tested sawdust/E&P waste sample were observed to be: LL = 311, PL = 219, and PI = 92. Those values are significantly higher than for ordinary soils due to the fact that soil particles are generally considered "impermeable", which cannot be assumed for sawdust. If plotted on a plasticity chart, this material would be defined as MH/OH material and the simplified UCSC classification for this material would be sandy silt.

Field Capacity and Organic Content. The tested sawdust/E&P waste sample presented very high field capacity (471.6%), due to the absorption characteristics of the sawdust material. The field capacity of the samples was determined by saturation of the samples using a paint filter mesh and allowing the samples to drain by gravity for a period of three days.

The organic content of the tested sawdust/E&P waste sample was also observed to be high (87.7%), mainly due to the material's sawdust composition. This value is similar to the organic content of the tested MSW (79.6%) and MSW/E&P waste (79.8%).

The tested MSW/E&P waste sample presented high field capacity (121%) when compared to soils, but it was much smaller than the field capacity of the tested sawdust/E&P waste. This value was also higher than the values reported in the general literature for MSW, usually between 30 and 70%.

Compaction. The tested sawdust/E&P waste sample presented a maximum dry unit weight equal to 10.6 kN/m^3 and an optimum water content equal to 197.7% when tested according to Standard Proctor conditions (water content defined as the ratio between the mass of water and the dry mass of the material). Those values are very distinct from the ones generally obtained for soils. However, it may indicate that the sawdust/E&P waste can be significantly compacted using standard Proctor effort if the dry unit weight of the compacted sample is a "real" dry unit weight (i.e., if the interior of the particles were also dry). It must be noted that the sawdust unit weight under dry and loose state is significantly lower (generally around 2.2 kN/m^3).

The Standard Proctor compaction test curve for the MSW/E&P waste sample did not result in an optimum moisture content or a maximum dry unit weight associated with an optimum moisture content, which is common for MSW. The maximum observed dry unit weight was 14.7 kN/m^3 and the minimum was 9.8 kN/m^3 at the moisture content range tested (between 35% and 145%, approximately).

Direct Shear. The direct shear test conducted in a small specimen of sawdust/E&P waste (6.4 cm in diameter by 2.5 cm in height) compacted to approximately 90% of the maximum dry unit weight and at optimum water content presented no peak in shear stresses at all three normal stresses analyzed (152 kPa, 290 kPa, and 496 kPa) to a maximum displacement of 0.5 cm. At 0.5-cm displacement, the linear Mohr- Coulomb parameters are $c' = 14.7$ kPa and $F' = 33.1^{\circ}$. This test lead to the conclusion that although the parameters of the direct shear test could not be utilized as final, they indicated that strength testing of sawdust/E&P waste could produce useable results if consolidated-undrained (CU) triaxial compression shear tests would be performed. These CU tests were performed as part of Phase II.

The direct shear test conducted in a specimen of MSW/E&P waste using a box of dimensions 30.5 cm × 30.5 cm × 10.2 cm presented no peak in shear stresses at all three normal stresses analyzed (152 kPa, 290 kPa, and 496 kPa) to a maximum displacement of 7.5 cm. At 2.5-cm displacement, the linear Mohr-Coulomb parameters are $c' = 28.9$ kPa and $F' = 14^{\circ}$. At 7.5-cm displacement, the linear Mohr-Coulomb parameters are $c' = 0$ kPa and $F' = 28.5^{\circ}$. Those shear strength parameters are in-line with the low-boundary values observed in the general literature regarding all types of MSW. However, for shredded MSW using a direct shear box, the literature parameters are generally higher than the ones observed.

Hydraulic Conductivity. The hydraulic conductivity of a specimen of sawdust/E&P waste compacted to approximately 90% of the maximum dry unit weight and at optimum water content was 2.7×10^{-5} cm/s for a confining pressure equal to 34.5 kPa. According to de Abreu [4], this hydraulic conductivity is lower than for MSW of developed countries and is comparable to the hydraulic conductivity of MSW of developing countries and to the hydraulic conductivities of sludges, ashes, and silts.

Since the testing was performed at a relatively low confining pressure, hydraulic conductivities at higher pressures (e.g., considering the waste mass internal pressures in a landfill) could be even lower. Additional hydraulic conductivity tests were performed during Phase II.

3.2 Phase II Testing (Water-Based Drilling Wastes Solidified with Sawdust)

Hydraulic Conductivity. For Phase II, samples of sawdust/E&P waste were compacted to approximately 90% of the maximum dry unit weight and at optimum water content, and tested under two distinct confining pressures (290 kPa, and 496 kPa). For the sample tested at the confining pressure of 290 kPa, the observed hydraulic conductivity was equal to 3.2×10^{-6} cm/s. For the confining pressure equal to 496 kPa, the hydraulic conductivity was equal to 1.3×10^{-6} cm/s.

Those hydraulic conductivities are lower than for MSW of developed countries and are similar to the hydraulic conductivities of MSW of developing countries, sludges, ashes, and silts. They are also in line with the results of the Particle Size Analysis conducted during Phase I for a sample of sawdust/E&P waste (0.5% of gravel-size particles, 36.3% of sand-size particles (32.8% of total being fine sand), 33.3% of silt-size particles, and 29.9% of clay-size particles).

It is important to note that under those low hydraulic conductivities, the development of pore pressures due to the gas generated from the waste mass cannot be ignored and may be significant to the stability of the landfill [5]. Under this condition, the waste mass also cannot be considered as being freely drained as typically observed in MSW landfills at the United States and other developed countries [4].

Consolidated-Undrained (CU) Triaxial Compression Tests. Three triaxial shear tests were recommended and conducted for Phase II in samples of sawdust/E&P waste. CU Triaxial Shear tests back-pressured saturated with pore pressure measurements were conducted in samples compacted to 90% of the standard Proctor maximum dry unit weight at target moisture contents of 200%, 300%, and 400%. Confining pressures at each one of the 3 tests were set at 152 kPa, 290 kPa, and 496 kPa. Figure 2 presents the Mohr-Coulomb diagram for the test performed at the target moisture content of 200%. Table 1 represents a summary of the linear Mohr-Coulomb parameters at 15% strain.

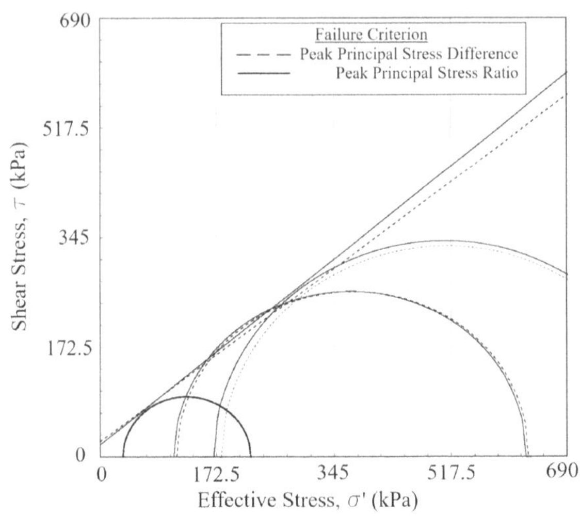

Fig. 2. Mohr Coulomb diagram for CU triaxial test performed on sawdust/E&P waste at 200% moisture content target

Table 1. CU shear strength parameters for sawdust/E&P waste at 15% strain

Moisture content	Cohesion intercept (kPa)	Friction angle
200%	24.1	38.4°
300%	40.0	29.7°
400%	42.7	27.0°

Although these parameters should not be used directly in a slope stability analysis, they do indicate that the shear strengths of sawdust/E&P waste are relatively high when compared to the parameters typically utilized for MSW in slope stability analyses in Louisiana ($c' = 14.4$ kPa and $F' = 23^{o}$). It must be considered that in a mix of MSW and sawdust/E&P waste, it is likely that the sawdust/E&P waste will cause some significant increase in the MSW's shear strength when above a certain percentage. It was also observed from the CU Triaxial Shear tests that the sawdust/E&P waste material developed internal pore pressures throughout the performance of the tests, which is typical of low permeability materials (e.g., clays).

4 Application of the Testing Results of Sawdust/ E&P Waste in a Slope Stability Analysis

As it can be seen from the previous sections of this article, the presence of sawdust/E&P waste may have double effect in the geotechnical stability of a MSW landfill. For one side, it may increase the shear strength of the waste mass and, by the other, it may introduce significant pore pressures.

It is virtually impossible to develop exact parameters of shear strength and pore pressure for the mix of MSW and sawdust/E&P waste that may exist in the landfill, therefore, a sensitivity study was developed in order to conduct the slope stability analysis. Two sets of shear strength parameters were analyzed. For the first one, the conventional MSW parameters were utilized (c′ = 14.4 kPa and F′ = 23°). On the second one, enhanced MSW parameters were adopted (c′ = 23.9 kPa and F′ = 27°), which consider that the sawdust/E&P waste will have some beneficial effect in shear strength of the MSW.

Distinct pore pressure parameters (r_u) were adopted for the analyses, varying between 0 and 0.3. According to Boscov et al. [6], in their study about pore pressure development in landfills, "pore pressure represented by r_u parameter varying from 0.1 to 0.3 yield safety factors similar to those obtained by more realistic assumptions, i.e. independent leachate piezometric levels in each cell." The slope stability model for this project was analyzed with the various pore pressure parameters ($r_u = 0.0$, 0.1, 0.2, and 0.3) which were applied to the two sets of shear strength parameters (conventional MSW and enhanced MSW/E&P/sawdust). The minimum safety factors for the landfill, according to the assumptions of this study, were approximately 1.3 to 1.4, even when these very conservative parameters were utilized.

5 Conclusions and Recommendations

One of the main conclusions of the study was that even though the utilization of sawdust/E&P waste (water-based) may improve the shear strength of the total waste mass, the development of inexorable pore pressures may reduce the safety factor of the landfill slopes. The 1.3 to 1.4 safety factors resulting from the slope stability analysis even when utilizing the more conservative parameter combination (conventional MSW parameters and r_u's equal to 0.2 and 0.3) may be considered very conservative since they assume the development of high pore pressures and relatively low strength parameters. Despite these findings, adoption of a rigorous geotechnical monitoring program to mitigate the effects of the possible development of such pressures in the waste mass of the landfill, was still recommended by the authors. It is important to note that it is possible that in cases where a landfill does not have adequate drainage, and depending on the quantities of gas generated, pore pressures can be even higher than the ones assumed in this study.

Regarding the use of selected MSW as stabilizing agent for E&P wastes, the increase in moisture content in a landfill may have two negative effects: possible accumulation of liquids inside "trapped" areas of the waste mass, and possibly a reduction on shear strength due to more efficient waste decomposition. The study concluded that the increase of the internal pore-pressures is a main concern in this case, as in the case of bioreactor landfills. The authors suggest that the utilization of an active gas system helps to reduce the internal pore pressures of the waste mass, and the adequate design, installation, and operation of such system is highly recommended if E&P waste is planned to be disposed of at a MSW landfill. Based on the knowledge of the authors, the adequate operation of such system without interruptions lasting more than 48 h is fundamental to the stability of the slopes.

Finally, due to the characteristics of E&P wastes observed in this study, it is also recommended to avoid placing the E&P waste material near the surface of final slopes or on preferential planes parallel to the slopes.

References

1. U.S. EPA (2002) Exemption of oil and gas exploration production waste from Federal Hazardous Waste Regulations. EPA publication EPA530-K-01-004
2. Louisiana Department of Environmental Quality: Annual Disposer Reports (2009–2016)
3. NOAA—National Centers for Environmental Information. Climate normals, means and extremes. Shreveport, LA. https://www.ncdc.noaa.gov/. Accessed 23 May 2018
4. de Abreu R (2000) Compressibility of sanitary waste masses (in Portuguese), M.Sc. Dissertation, Universidade de Sao Paulo, Brazil
5. Merry S, Fritz W, Budhu M, Jesionek K (2006) Effect of gas on pore pressures in wet landfills. J Geotech Geoenviron Eng ASCE 132(5):553–561
6. Boscov M, Campi T, Afonso B, Tavares J, Tercini, J, Bergamo L, Ribeiro M, Juliano R (2011) Effects of pore pressure on slope stability of sanitary landfills. In: VII international conference on engineering and computer education, pp 64–68

Dependency of Landfill Gas Generation Parameters on Waste Composition Based on Large-Size Laboratory Degradation Experiments

Sampurna Datta(✉) and Dimitrios Zekkos

University of Michigan, Ann Arbor, MI 48109, USA
sampurna@umich.edu

Abstract. Landfill gas (LFG) is a product of the biodegradation of municipal solid waste (MSW) under anaerobic conditions. LFG primarily consists of methane (CH_4) (40–60%) and carbon dioxide (CO_2) (40–60%), both greenhouse gases. Methane has high energy potential that remains largely untapped as a national energy source. In order to recover LFG for energy generation purposes, a reliable estimate of gas generation at landfill sites is necessary. To that end, numerous LFG generation models have been developed with different assumptions made. In this study, three gas generation models – two first order decay (LandGEM and IPCC) and one sigmoidal model (Modified-Gompertz) are considered. The cumulative methane yield prediction of these models is fitted against three degradation experiments on well-characterized MSW specimens with significantly different waste composition ranging from "waste-rich" to "soil-rich". The results indicate that the sigmoidal model better captures the evolution of methane yield compared to first order decay model and majority of the model parameters follow a systematic trend as a function of waste composition.

Keywords: Municipal solid waste · Biodegradation · LFG generation model

1 Introduction

MSW disposed in landfills decompose in an anaerobic environment leading to the production of landfill gas (LFG). LFG is primarily composed of methane (CH_4) ($\sim$50%), and carbon dioxide (CO_2) ($\sim$50%). CH_4 is a potent greenhouse gas when released but it is also a sustainable source of energy when harvested. As a result, various models have been developed to provide a better estimation of methane generation at landfill sites such as zero order models where the biogas generation rate produced in landfill sites remain steady with time (Triangular [7]); first order kinetic models where the effect of carbon depletion in the waste over time is considered (LANDGEM [8], IPCC 2006 [5], SWANA [6]) and growth models where the biogas generation kinetics is compared to a bacterial growth process (Modified-Gompertz). A primary output of such landfill gas generation models is cumulative methane production curve, where the cumulative methane production is plotted against time. The

L. Zhan et al. (Eds.): ICEG 2018, ESE, pp. 186–193, 2019.
https://doi.org/10.1007/978-981-13-2224-2_23

patterns these curves follow have implications on the degradation of MSW. The kinetics of the process and ultimately the shape of the methane production curves is primarily controlled by the biodegradability of MSW and the performance of the microbial population inside the waste mass.

Most of the available models commonly used follow first order kinetics. However, significant commonality has been found between bacterial growth curves and the cumulative gas production curve [9]. In this study, three gas generation models – two first order decay (LandGEM and IPCC) and one growth model (Modified-Gompertz) are selected to fit against a dataset of degradation experiments conducted at the University of Michigan on three well-characterized large-size MSW specimens and the results are investigated with respect to waste composition.

2 Experimental Setup

2.1 Waste Characterization

The MSW specimens used in this study were collected from Los Reales Landfill in Arizona (AZ), Austin Community landfill in Texas (TX) and Lamb Canyon Landfill in California (CA) and transported in sealed drums to the laboratory. Field composition was characterized according to the procedure described by [10]. Waste was segregated into finer and coarser fraction using a 20-mm sieve. The initial total mass, mass percentage on a wet basis, moisture content, and volatile solids (VS) in <20 mm fraction were determined by [4] and are provided in Table 1. The biodegradability of each specimen prior to degradation is evaluated using density of biodegradable waste ($\gamma_{B,I}$) [3] defined by Eqs. 1 and 2.

$$B_0(\%) = \frac{\mathit{Dry\,mass\,of\,food,\,yard\,waste,\,paper\,and\,mass\,of\,VS\,in}\,<20\,\mathit{mm\,fraction}}{\mathit{Total\,initial\,dry\,waste\,mass}} \tag{1}$$

$$\gamma_{B,I}\left(\frac{kg}{m^3}\right) = \frac{B_0(\%)}{100} \times \gamma_{d,I}\left(\frac{kg}{m^3}\right) \tag{2}$$

where, $\gamma_{d,I}$ is the dry unit weight of the specimen after immediate compression was practically completed. Briefly, AZ waste is more "waste-rich" containing the maximum amount of biodegradable material per unit volume ($\gamma_{B,I}$), while CA waste is the least biodegradable per unit volume and is "soil-rich"; TX waste is intermediately biodegradable.

2.2 Degradation Testing of MSW

The specimens were reconstituted based on their field waste composition and placed in 42-L columns for degradation testing. Three columns (d = 300 mm, h = 500 mm) accompanied by systems to recirculate and store leachate, and systems to monitor the long-term waste degradation process including biogas, changes in solid weight and

Table 1. Composition, moisture content, volatile solids and biodegradability of three MSW specimens

Parameter	Value		
	AZ	TX	CA
Paper (%)	21.4	10.7	4.6
Wood (%)	2.1	4.3	5.8
Soft plastic (%)	8.5	5.9	3.9
<20 mm particles (fine particles) (%)	67.9	79.1	68.3
Hard plastic, cobbles and metals (%)	0	0	16.2
Initial total weight (kg)	21.69	29.31	29.02
Initial average moisture content (%)	32.7	37.7	28.1
Volatile solids in <20 mm fraction (g/g dry)	0.113	0.07	0.086
Initial percentage of biodegradable waste, B_O (% dry)	30.6	16.4	10.3
Initial density of biodegradable waste, $\gamma_{B,I}$ (kg/m^3)	143	98	66

volume were constructed and are described in detail by [2]. The materials were placed at a loose state, i.e., without significant compaction. Negligible vertical stress (<1 kPa) was applied to the specimens. The temperature of each column remained constant at 40 ± 3 °C using a heating blanket. Leachate was recirculated three times a week and involved short-term (~15 min) submergence of the specimens followed by gravity drainage of the liquids through the base of the column. Biogas composition and volume were continuously measured to assess the progress of waste degradation. The experiments were considered complete after 900 days, 1500 days and 850 days of operation for AZ, TX and CA specimens, respectively, several months after observing no additional biogas generation, and the setups were disassembled.

3 Landfill Gas Generation Models

3.1 First Order Decay (FOD) Models

LandGEM. LandGEM was developed to predict LFG and methane generation from MSW landfills in the United States. The model requires two methane generation parameter inputs: L_O and k. All landfilled waste is considered a single phase within the model. This model estimates the annual gas generation based on the first-order degradation kinetic over a specified period [8]. In this study, this model is represented as

$$V_{CH_4} = \sum L_0 M_i (1 - e^{-kt}) \tag{3a}$$

where V_{CH4} is the cumulative methane generation (L CH_4); k is the decay constant (day^{-1}); L_O is methane generation potential (L CH_4/kg wet waste); M_i is the mass of waste placed in the column (kg); and t is the age of the waste.

LandGEM (modified). In a simple first order decay model it is assumed that methane generation starts immediately after deposition of the waste. However, the generation of gas in the field is often delayed perhaps as a result of acclimation of microorganisms, but also, non-optimal gas recovery, or the lack of landfill cover [1]. Therefore, LandGEM has been modified to account for this lag by introducing an additional parameter t_{lag} (days) in Eq. 3b.

$$V_{CH_4} = \sum L_0 M_i (1 - e^{-k(t - t_{lag})}) \tag{3b}$$

IPCC Waste Model (modified). The IPCC model is equally capable of predicting methane generation from landfills because it only requires readily available data on methane generation inputs and annual tonnages [5]. The model uses parameters of *DDOCm* and *k*, as shown in Eqs. 4 and 5, which incorporate several additional parameters to transform the carbon source to annual methane generation. The lag phase (t_{lag}), which is identified by IPCC is incorporated in this study.

$$V_{CH_4} = \frac{\sum DDOCm_i \times (1 - e^{-k(t - t_{lag})}) \times 16}{12} \times F \tag{4}$$

where V_{CH4} is the mass of methane generated which is converted into volume by dividing by density of methane (ρ_{CH4} = 0.656 kg/m^3 at STP); $DDOCm_i$ is the mass of degraded decomposable organic carbon of waste component i and F is the fraction of methane in biogas.

$$DDOCm_i = \sum M_i \times DOC_i \times DOC_f \times MCF \tag{5}$$

where M_i is the mass of waste component i; DOC_i is degradable organic carbon concentration of waste component i; DOC_f is the fraction of anaerobically degradable carbon in landfill; and *MCF* is the methane correction factor that may be used to reduce the predicted methane generation of landfills that are partially aerobic. IPCC waste model has been used to apply component-specific DOC_i values to user-defined waste composition data as a multiphase model.

3.2 Sigmoidal or Growth Model

Modified Gompertz Function. Growth curves typically demonstrate a phase in which the specific growth rate begins at a value of zero (minimum asymptote) and then accelerates to a maximum growth rate in a certain period denoted as the lag phase. In addition, the growth curves contain a final phase in which the growth rate decreases and finally reaches zero, so that a point of saturation or the maximum asymptote is reached [9].

A kinetic model modified by [11] from the Gompertz growth equation is implemented to determine the methane yield potential as follows:

$$BG = BGP \exp\left\{-\exp\left[\frac{R_m e}{BGP}(\lambda - t) + 1\right]\right\} \tag{6}$$

where, BG is the cumulative methane yield, BGP is the methane yield, R_m is the maximal daily methane yield, λ is the lag time and e is the mathematical constant equal to 2.718.

4 Analysis

The three models are fitted against each of the three MSW specimens (AZ, TX and CA) using non-linear regression. The regression has been carried out for 95% confidence interval of the parameters. The goodness of fit is assessed using two parameters: adjusted R^2 (ranges from 0 to 1) which is a relative measure of fit and Root Mean Square Error (RMSE) which is an absolute measure of fit. Higher values of adjusted R^2 and lower values of RMSE indicate better fit.

The DOC values for wood, paper and soft plastic have been adopted as 0.43, 0.40 and 0 kg C/kg waste, respectively, the default values recommended in IPCC 2006. DOC for fine particles has been assumed to be the initial volatile solids content measured for the three specimens as indicated in Table 1, in terms of wet weight of specimen. MCF has been considered as 1.0 since the degradation experiments were conducted under well-controlled anaerobic conditions. The fraction of methane, F, is assumed to be 0.5 which is reasonable considering methane concentrations varied from 40–60% in the three MSW specimens. However, the DOC_f values which represent the anaerobic biodegradability of carbon for each component has been determined by trial and error from the regression results ($DDOCm$ values) using guidelines from literature.

5 Results

Figure 1a–c show a typical example of the regression analysis results for AZ specimen. Similar behavior has been observed for the other two specimens. LandGEM and IPCC waste models are collectively designated as FOD (modified) models since they follow the same behavior, despite having distinct methodologies. Adjusted R^2 values close to 1 and low RMSE values indicate that both the modified first order models and growth model fit well the experimental data. For the FOD model, methane generation begins immediately although there is a time lag in the experiment (Fig. 1a). The modified FOD model completely offsets the lag phase in the methane generation (Fig. 1b), while Modified-Gompertz successfully captures the behavior (Fig. 1c) with a lag phase in the beginning followed by exponential phase and then by a stationary phase. The kinetics of the anaerobic degradation process is better captured by a growth curve. Hence, while first-order models are most widely used and are easier to implement, it is a sigmoidal function that fits better LFG generation, especially at the early stages of the process. AZ, the "waste rich" specimen (with the highest $\gamma_{B,I}$ of all the three specimens), has the maximum methane yield of 1100 L while CA, the "soil rich" specimen, has the lowest yield of 180 L in 550 d (Fig. 1e); TX generated 475 L of methane in 550 d (Fig. 1d).

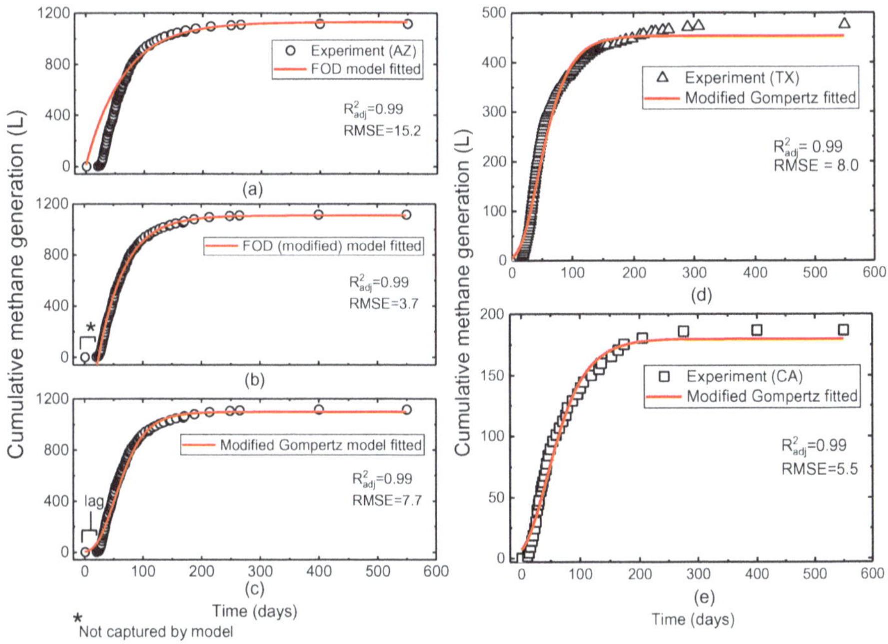

Fig. 1. Non-linear regression fitting of (a) FOD model (AZ), (b) Modified FOD model (AZ), (c) Modified Gompertz model (AZ), (d) Modified Gompertz model (TX), (e) Modified Gompertz model (CA), on cumulative methane generation curves from large-scale experiments.

From Table 2, it is observed that as the biodegradability of the specimen increases, i.e. as we move from "soil-rich" CA to "waste-rich" AZ, most of the parameter values show an increasing trend. It is reasonable to expect that with higher biodegradability, a faster methanogenic daily yield (higher R_m) occurs. The methane yield potential (L_0 or *BGP*) or the amount of decomposable carbon (*DDOCm*) is higher in AZ than in TX and CA specimens. It is interesting to note that a significant lag phase of about 25 days is observed in AZ specimen, while TX and CA have a shorter time lag ranging from 10 to 16 days. As indicated in [9], during the lag phase, the initial breakdown of insoluble substrates occurs through hydrolytic bacteria. Thus, it can be suggested that with higher biodegradability or greater availability of biodegradable material, it takes more time for hydrolysis to occur, hence, the higher lag time observed in AZ specimen. However, the decay rate (k) for all the three specimens are comparable indicating it is independent of waste composition.

When comparing the default parameters recommended by US EPA for LandGEM with the regression results of the degradation experiments, it is observed that the decay rate, k, is an order higher for the experiments (k = 0.015 d^{-1} to 0.022 d^{-1}) than default LandGEM values (k = 0.7 yr^{-1} or 0.002 d^{-1} for wet bioreactor), which emphasizes that the experiments are well-controlled with negligible losses ensuring enhanced biodegradation throughout. On the other hand, default L_0 provided in LandGEM for wet bioreactors is 96 L CH_4/kg while the experiments exhibited values ranging from

Table 2. Regression analysis results - model parameters with 95% confidence interval

Specimen	First order decay model				Growth model		
	LandGEM				Modified Gompertz		
	L_0 (LCH_4/kg)	IPCC					
		k (d^{-1})	t_{lag} (d)	$DDOCm$ (kgC)	BGP (L CH_4)	R_m (LCH_4/d)	λ (d)
AZ	51.3 ± 0.2	0.022 ± 7e−4	25 ± 1	1.1 ± 4e−3	1100 ± 6	14 ± 0.8	22 ± 3
TX	15.9 ± 0.1	0.021 ± 9e−4	16 ± 2	0.5 ± 3e−3	456 ± 4	6 ± 0.4	14 ± 3
CA	6.5 ± 0.1	0.015 ± 8e−4	10 ± 2	0.2 ± 3e−3	180 ± 5	2 ± 0.2	7 ± 5

6.5 to 51 L CH_4/kg. This lower L_0 may be attributed to the absence of food waste (that is highly biodegradable) in the MSW specimens used for degradation testing, which would otherwise be available in landfilled waste.

The anaerobic biodegradability or DOC_f values (as referred by IPCC) obtained using degraded decomposable organic carbon content ($DDOCm$) from regression analysis range from 0–0.20 for wood, 0.2–0.60 for paper and 0–0.43 for fine particles for the conditions and assumptions considered in this study (Table 3), as opposed to a single DOC_f value of 0.5 for all waste components as recommended by IPCC 2006, which would overestimate methane yield.

Table 3. DOC_f values (IPCC waste model parameter) obtained in this study.

Specimen	DOC_f (wood)	DOC_f (paper)	DOC_f (fine particle, <20 mm)
AZ	0–0.20	0.2–0.60	0–0.43
TX	0–0.20	0.2–0.37	0–0.12
CA	0–0.10	0.2–0.36	0–0.05

6 Conclusion

Three LFG generation models have been fitted against a dataset of large scale laboratory biodegradation experiments. The influence of waste composition on model behavior has been investigated through three distinct MSW specimens with a range of waste composition. The most significant findings from this work are as follows:

- Sigmoidal growth models better represent the characteristic shape of cumulative methane generation curve, than first order decay models. The kinetics of the anaerobic process is better captured by growth curves.
- With an increase in biodegradability, parameters methane yield potential, mass of degraded decomposable organic carbon, daily maximum yield and lag phase all increase. However, the decay constant for first-order LFG models shows no correlation with waste composition.

- IPCC DOC_f values obtained range from 0–0.2 for wood, 0.2–0.6 for paper and 0–0.43 for fine particles, as opposed to a single value of 0.5 recommended by IPCC for all waste components.

Acknowledgement. This research was supported by the National Science Foundation (NSF) Division of Computer and Communication Foundations under Grant No. 1442773 and Environmental Research and Education Foundation (EREF) Scholarship. ConeTec Investigations Ltd. and the ConeTec Education Foundation are acknowledged for their support to the Geotechnical Engineering Laboratories at the University of Michigan. Any opinions, findings, conclusions and recommendations expressed in this paper are those of the authors and do not necessarily reflect the views of the NSF or ConeTec.

References

1. Faour A, Reinhart D, You H (2007) First-order kinetic gas generation model parameters for wet landfills. Waste Manage 27:946–953
2. Fei X, Zekkos D, Raskin L (2014) An experimental setup for simultaneous physical, geotechnical, and biochemical characterization of municipal solid waste undergoing biodegradation in the laboratory. Geotech Test J 37(1):1–12
3. Fei X (2016) Experimental assessment of coupled physical-biochemical-mechanical-hydraulic processes of municipal solid waste undergoing biodegradation. Ph.D. thesis, University of Michigan, Ann Arbor
4. Fei X, Zekkos D (2018) Coupled experimental assessment of physico-biochemical characteristics of municipal solid waste undergoing enhanced biodegradation. Geotechnique, p 253. https://doi.org/10.1680/jgeot.16
5. Intergovernmental panel on climate change (2006) Solid waste disposal. In: Pipatti R, Svardal P (eds) 2006 IPCC guidelines for national greenhouse gas inventories, vol 5, waste. Intergovernmental Panel on Climate Change, Geneva, pp 1–40
6. SWANA (1998) Comparison of models for predicting landfill methane recovery. Publication no. GR-LG 0075. The Solid Waste Association of the North America (SWANA), Dallas
7. Tchobanoglous G, Theisen H, Vigil S (1993) Integrated solid waste management. McGraw-Hill, New York
8. US EPA (2016) Inventory of US greenhouse gas emissions and sinks: 1990–2014, EPA 430-R-16-002, Washington
9. Ware A, Power N (2017) Modelling methane production kinetics of complex poultry slaughterhouse wastes using sigmoidal growth functions. Renew Energy 104:50–59
10. Zekkos D, Kavazanjian E, Bray JD, Matasovic N, Riemer MF (2010) Physical characterization of municipal solid waste for geotechnical purposes. J Geotech Geoenviron Eng 136:1231–1241
11. Zwietering MH, Jongenburger I, Rombouts FM, Riet KV (1990) Modelling of the bacterial growth curve. Appl Environ Microbiol 56:1875–1881

Laboratory Measurements of K_0 of Municipal Solid Waste

Pengbo Yuan[1,3(✉)], Edward Kvazanjian Jr.[2], and Wenwu Chen[1,3]

[1] School of Civil Engineering and Mechanics, Lanzhou University, Lanzhou 730000, Gansu, China
yuanpb@lzu.edu.cn

[2] The Center for Bio-Inspired and Bio-Mediated Geotechnics, Arizona State University, P.O. Box 873005, Tempe, AZ 85287-3005, USA

[3] Key Laboratory of Mechanics on Disaster and Environment in Western China, The Ministry of Education of China, Lanzhou 730000, Gansu, China

Abstract. Laboratory measurements of the coefficient of earth pressure at rest, $K_{0,}$ of municipal solid waste were made in a large-scale testing device using tactile pressure sensors. The effects of waste composition, compaction and overconsolidation ratio (OCR) on the coefficient earth pressure at rest, K_0 of municipal solid waste were evaluated. The tests results showed that K_0 of municipal solid waste decreased with increasing fiber content and OCR of the waste. K0 decreased at first with increasing of applied vertical stress and then became approximately constant value at 0.48, 0.35, and 0.23 for MSW 100% < 20 mm, MSW 65% < 20 mm and MSW 35% < 20 mm respectively. Compaction effort also affected the initial value of K_0. However, with increasing vertical stress the effect of compaction decreased and was finally eliminated.

Keywords: Municipal solid waste · Landfill · Tactile pressure sensor OCR

1 Introduction

Lateral stresses are very important considerations in the design of a variety of landfill components, including side slope liners, vertical and side slope risers, vaults and conduits, retaining wall, and deep foundations installed for post-closure development (Kavazanjian 2006). Knowledge of in situ lateral stresses is also important in interpreting laboratory test data, particularly with respect to reconciling results from different types of tests (e.g. from triaxial and simple shear tests) and from isotopically consolidated laboratory tests with field behavior. It is also crucial for numerical modeling of geotechnical engineering problems of landfill involving non-linear waste behavior. Despite its importance to this broad range of landfill engineering problems, little attention has been paid to evaluation of lateral stresses and earth pressure in municipal solid waste (MSW). This paper presents laboratory measurements of K_0 of municipal solid waste made in large scale laboratory testing a using tactile pressure sensor system. The tests investigated the effects of waste composition, compaction and overconsolidation ratio (OCR) on K_0 of municipal solid waste.

L. Zhan et al. (Eds.): ICEG 2018, ESE, pp. 194–201, 2019.
https://doi.org/10.1007/978-981-13-2224-2_24

2 K_0 Values from Previous Studies

The only published data on in situ lateral stresses in MSW landfills are measurements of K_0 made at a landfill in Calvert, England, using a pressuremeter by Dixon et al. (1999). These measurements, showed a wide scatter for K_0, from less than 0.2 to 1.0. Most of measurements were made in relatively young (1 to 3 year old) waste. The one measurement made in older (11 year old) waste yielded the maximum value of 1.0, leading the investigators to suggest that K_0 may increase with time and degradation. Landva et al. (2000) measured K_0 on reconstituted waste specimens in a special "split-ring" consolidator. Tests were conducted using vertical normal stresses of up to 275 kPa. Values of K_0 from 0.26 to 0.4 were measured by these investigators on five different specimens with split-ring device, with the measured value of K_0 inversely related to the fiber content of the waste. Towhata et al. (2004) measured K_0 in triaxial compression tests by first isotropically consolidating the specimen at a very low stress (approximately 10 kPa) and then increasing the cell pressure to maintain zero lateral strain while the specimens were axially loaded. The lateral stress coefficient values leveled off and became approximately constant at between 0.25 and 0.35 for vertical normal stresses greater than approximately 250 kPa.

Using elasticity theory, K_0 can be derived from the measurements of the Poisson's ratio. Sharma et al. (1990) performed in-situ downhole seismic velocity measurements in the Richmond landfill in California. Both shear and compression wave velocities were measured and the resulting Poisson's ratio was estimated to be 0.49, resulting in a value of K_0 of 0.96 using elasticity theory. Houston et al. (1995) reported MSW Poisson's ratio values decreasing from 0.25 at a depth of 3 m to 0.11 at a depth of 10 m based upon compressional and shear wave velocity measurements at the Northwest Regional Landfill in Maricopa County, Arizona. These values correspond to a K_0 of 0.12–0.33 based upon elasticity theory. Matasovic and Kavazanjian (1998) presented Poisson's ratio data calculated from shear wave and compressional wave velocity measurements at the OII Landfill in Los Angeles County, California. Values ranged significantly but the majority of the Poisson's ratio ranged between 0.25 and 0.4, corresponding to elasticity theory-based values of K_0 of 0.33 to 0.67.

Zekkos (2005) measured Poisson's ratio in monotonic and cyclic triaxial tests conducted at the University of California at Berkeley (UCB). Poisson's ratio was measured on two different groups of specimens: one group with 100% of the particles less than 20 mm in dimension and a second group with between 62% and 76% of the particles smaller than 20 mm. Typical values of Poisson's ratio varied from 0.2 to 0.4 (though one value as low as 0.05 was measured), with the larger values associated with the specimens with the larger percentage of smaller particles (i.e. with 100% smaller than 20 mm). These values are equivalent to a K_0 of 0.25–0.67. It should be pointed out that it may not appropriate, especially at large strains, to estimate the coefficient of earth pressure at rest, K_0 by using Poisson's ratio. Zekkos (2013) discussed the anisotropy in municipal solid waste based on the experimental results. Cai et al. (2015) evaluated the feasibility of predicting the K0 value of natural soil deposits on the basis of the measured shear wave velocities from the seismic piezocone penetration tests (SCPTU).

3 Tactile Pressure Sensors

Among the innovative technologies used by geotechnical engineering in recent years are tactile pressure sensors. Grid-based tactile sensor technology was first developed in response to a need identified in dentistry in 1987 (Podoloff and Benjamin 1989). Paikowsky et al. (1997) were the first to report on grid-based tactile pressure sensor application for geotechnical purposes. Wilson et al. (2003) assessed the repeatability and accuracy of grid-based tactile pressure sensors and compared the results from grid-based tactile sensors to results from a pressure film. Brimacombe et al. (2009) discussed the effect of calibration techniques on the accuracy and repeatability of grid-based sensors and concluded that their user-defined calibration algorithms provide more accurate force measurements than the manufacturers built-in calibration routines. Palmer et al. (2009) used grid-based pressure sensors to measure the soil-structure interaction forces.

The manufacturer's instructions for the grid-based tactile sensors used in this study notes that the sensors require conditioning, equilibration, and calibration before using. The instructions caution that the repeatability and accuracy of the grid-based tactile pressure sensors used in this study can be affected by load rate, time, shear stress, contact properties, sensor range in use and hysteresis upon unloading. Therefore, it was very important to develop a proper calibration and conditioning procedure for the sensors used in this study to mitigate these effects.

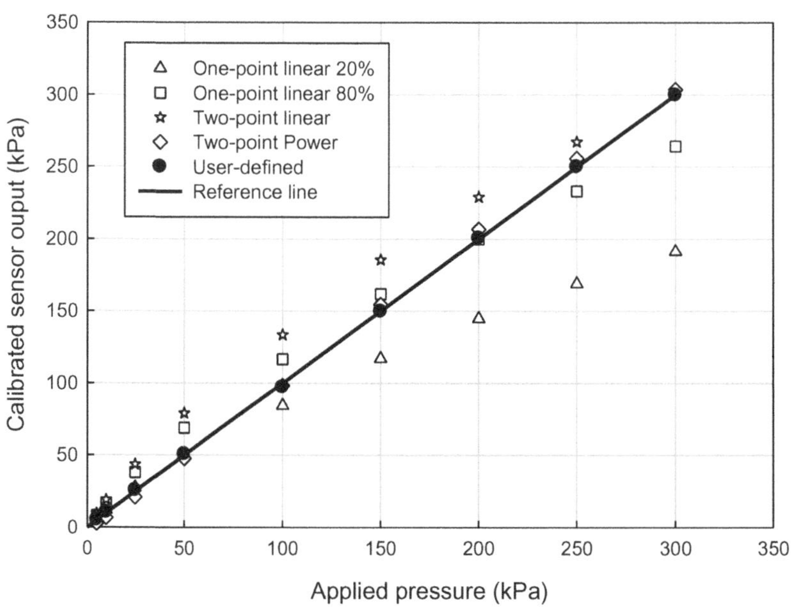

Fig. 1. Comparison of Tekscan pressure sensor different calibration methods

Model 5101 grid-based tactile pressure sensors manufactured by Tekscan, Inc. were used in this research Detailed information about Model 5101 pressure sensors can be found at www.tekscan.com. In accordance with manufacturer recommendations, the pressure sensors were conditioned, equilibrated, and calibrated before use. Calibration was performed after conditioning and equilibration. The same load rate, load range, and holding time were used in the calibration as used in the subsequent testing. Separate calibration files were developed for loading and unloading in order to account for hysteresis of the pressure sensors. Sensors are often calibrated using a one-point or two-point method. However, for the tests described herein, the sensors were calibrated using all 9 steps in the planned loading sequence. At the beginning of the testing program, two one-point calibration methods (one using 20% 0f the maximum load and one using 80% of the maximum load), a two-point linear calibration method, and a two-point power law calibration method were evaluated. However, none of these methods proved satisfactory. Therefore, a piece-wise linear user-defined calibration method was developed. Figure 1 illustrates the accuracy of these five different calibration methods.

4 Materials and Testing Program

The MSW used in this research was collected from the Tri-Cities landfill in the San Francisco bay area of California. Details of the associated field investigation, including field borehole logs, SASW testing, in situ unit weight tests, and waste characterization are reported by Zekkos (2005). The MSW tested at ASU came from the A-3 sample group. First, the waste was screened to separate the >20 mm material from the <20 mm material. Then, the size of the >20 mm particles was reduced to a maximum dimension of approximately 100 mm. The screened and reduced-size MSW was then used to create three different MSW specimens: one in which 100% of the constituents passed through a 20 mm sieve, one in which only 65% of its constituents passes through a 20 mm sieve 65%, and one in which only 35% of its constituents passed through a 20 mm sieve.

Specimens were compacted into a 152 mm-tall stack of Teflon coated rectangular steel frames with an annular opening 406 mm × 304 mm in dimension. The rings were aligned and then fixed against translation, providing a rigid steel box for the specimens. Specimens were prepared using 100%, 65% and 35% of the <20 mm material, by weight. Prior to preparing the specimens, the MSW was mixed thoroughly in a small mechanical mixer. Before compacting the MSW into the steel box, the grid-based pressure sensor pads were fixed by strong tape vertically against the short side of the rectangular frame stack. Kevlar textile was then fixed with tape on top of the pressure sensors to protect them during test from sharp material in the MSW. The waste was compacted 3 layers of equal mass placed such that the long axis of the large particles had a preferred horizontal orientation. Compaction was accomplished using a 13.6 kg hammer that was dropped repeatedly on a rectangular steel plate 114.3 mm × 114.3 mm in dimension from a constant height of 34 cm. Compaction was applied in a pattern of 12 spots designed to provide uniform coverage in each layer. Three (3) blows were applied to each spot. Information on the reconstituted MSW specimens prepared in this manner, including total unit weight, compaction moisture content, and waste composition, is presented in Table 1.

Table 1. Summary of ASU K_0 test specimens of municipal solid waste

Specimen	Moisture content (%)	Composition (% by weight)					Compaction style (blow/spot)[b]	Total unit weight (kN/m³)[a]
		<20 mm	Paper	Soft plastics	Wood	Gravel		
20 mm < 100%	16.5	100	0	0	0	0	3	11.6
20 mm < 65%	22.7	65	15	3	7	10	3	9.3
20 mm < 35%	21.4	35	45	5	5	10	3	6.4

[a]Unit weight after compaction.
[b]Every specimen has 3 layers and each layer had 12 compaction spots.

Specimens were loaded vertically via a rigid steel plate a few millimeters smaller on each side than the annular opening to avoid interference with the sides. A seating vertical stress of 25 kPa was applied followed by vertical stresses of 50 kPa, 100 kPa, 150 kPa, 200 kPa, 250 kPa, and 300 kPa respectively. After the 300 kPa pressure level, the vertical pressure was decreased sequentially to 250 kPa, 200 kPa, 150 kPa, 100 kPa, 50 kPa, and 25 kPa respectively. Every applied pressure level was held constant for 30 min prior to moving to the next vertical pressure level. Lateral pressure as recorded by the tactile pressure sensor system was captured once every minute.

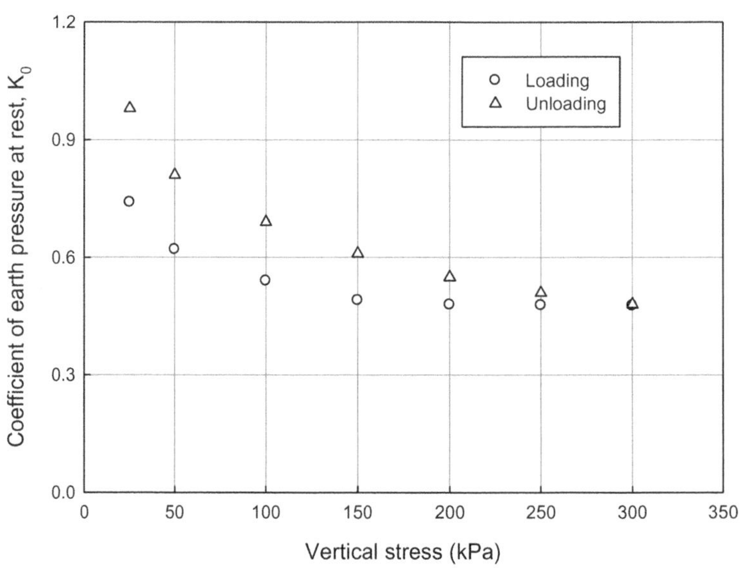

Fig. 2. ASU K_0 test on reconstituted MSW 100% < 20 mm specimen

5 Tests Results and Discussion

Figures 2, 3 and 4 present plots of K_0 for the three of MSW compositions during loading and unloading versus applied vertical stress. Figures 2, 3 and 4 showed that K_0 decreased at first with increasing of applied vertical stress and then became approximately constant value at 0.48, 0.35, and 0.23 for MSW with 100% < 20 mm, 65% < 20 mm and 35% < 20 mm, respectively, when the applied vertical normal stresses was greater than approximately 150 kPa. Figures 2, 3 and 4 also show that K_0 decreased with increasing percentage of >20 mm material. This observation is concordant with Landva et al. (2000), noted that their measured K_0 values were inversely related to the fiber content of their MSW specimens. The value of K_0 also increased systematically during unloading, showing that MSW K_0 increases with increasing OCR like in soils. The MSW specimens experienced a pre-consolidation pressure due to sampling compaction. At first, the applied vertical stress is smaller than the compaction-induced pre-consolidation pressure until the applied vertical stress eliminated the effect of compaction.

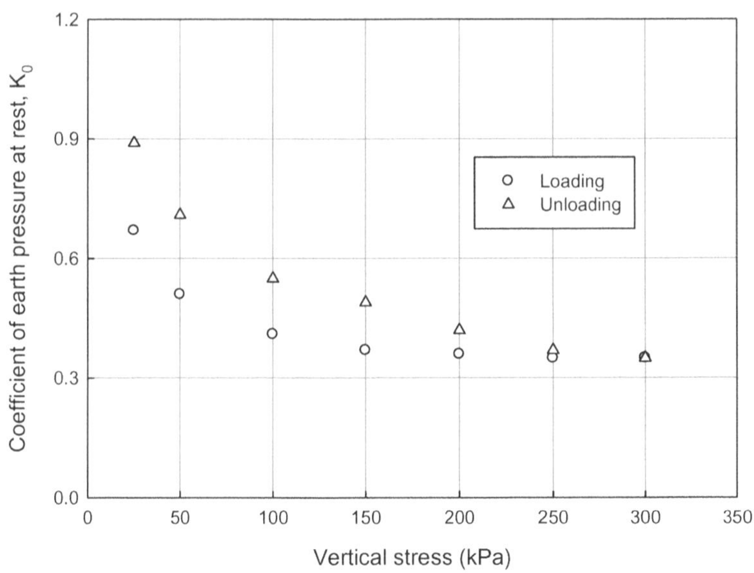

Fig. 3. ASU K_0 test on reconstituted MSW 65% < 20 mm specimen

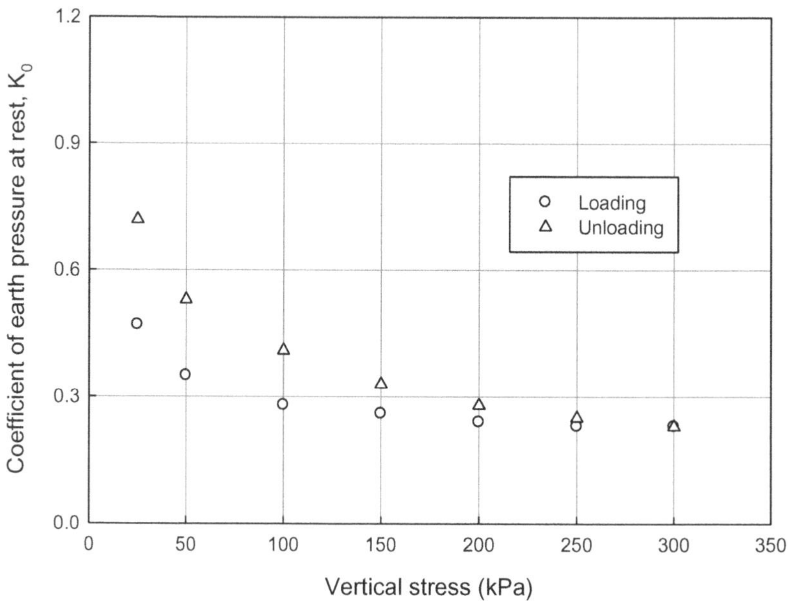

Fig. 4. ASU K_0 test on reconstituted MSW 35% < 20 mm specimen

6 Conclusions

The results of this study demonstrated that compaction of MSW resulted in a coefficient of earth pressure at rest, K_0, consistent with that of an overconsolidated material. The value of K_0 following compaction decreased consistently with increasing vertical pressure until it became relatively constant. Both the initial and ultimate values of K_0 increased with increasing percentage of <20 mm material. This observation is consistent with Landva et al's conclusion regarding the influence of fiber content on K_0 (Landva et al. 2000). The value of K_0 was also shown to increase with increasing OCR.

Acknowledgments. The work described in this paper was funded by the National Science Foundation Division of Civil, Mechanical, and Manufacturing Innovation under Grant CMMI-0635435 and by the China Scholarship Council. Any opinions, findings, conclusions, and recommendations expressed herein are those of the authors and do not reflect the views of the National Science Foundation. The authors also wish to acknowledge the contribution of John Kondziolka and Zachery Shafer, undergraduate laboratory assistants at Arizona State University, to this work, who helped conduct the tests and reduce the data reported herein.

References

Kavazanjian E (2006) Waste mechanics: recent findings and unanswered questions. In: Geoshanghai international conference, pp 34–54

Dixon N, Jones DRV, Whittle RW (1999) Mechanical properties of household waste: In: Situ assessment using pressuremeters. In: Proceedings of seventh international waste management and landfill symposium, vol 3, pp 453–460

Landva AO, Valsangkar AJ, Pelkey SG (2000) Lateral earth pressure at rest and compressibility of municipal solid. Can Geotech J 37(6):1157–1165

Towhata I, Kawano Y, Yonai Y, Koelsh F (2004) Laboratory tests on dynamic properties of municipal wastes. In: Proceedings of the 11th conference in soil dynamics and earthquake engineering and 3rd international conference on earthquake geotechnical engineering, vol 1, pp 688–693

Sharma HD, Dukes MT, Olsen DM (1990) Field measurements of dynamic moduli and Poisson's ratios of refuse and underlying soils at a landfill site. In: Proceedings of the symposium on geotechnics of waste fills—theory and practice, Pittsburg, pp 57–70, 10–13 September 1989. Publ Philadelphia: ASTM, 1990 (ASTM. International Journal of Rock Mechanics & Mining Sciences & Geomechanics Abstracts, 28(6):A356 (1991)

Houston WN, Houston SL, Liu JW et al (1995) In-situ testing methods for dynamic properties of MSW landfills. In: Earthquake design and performance of solid waste landfills. ASCE, pp 73–82

Matasović N Jr, Kavazanjian E (1998) Cyclic characterization of OII landfill solid waste. J Geotech Geoenviron Eng 124(3):197–210

Zekkos DP (2005) Evaluation of static and dynamic properties of municipal solid-waste. University of California, Berkeley

Zekkos D (2013) Experimental evidence of anisotropy in municipal solid waste. In: Coupled phenomena in environmental geotechnics

Cai G, Liu S, Puppala AJ (2015) Assessment of the coefficient of lateral earth pressure at rest (K_0) from seismic piezocone tests (SCPTU). In: Geo-Frontiers congress, pp 2278–2287

Podoloff R, Benjamin M (1989) Tactile sensor for analyzing dental occlusion. SOMA Eng Hum Body 3(3):1–6

Paikowsky SG, Hajduk EL (1997) Calibration and use of grid-based tactile pressure sensors in granular material. Geotech Test J 20(2):218–241

Wilson DR, Apreleva MV, Eichler MJ et al (2003) Accuracy and repeatability of a pressure measurement system in the patellofemoral joint. J Biomech 36(12):1909

Brimacombe JM, Wilson DR, Hodgson AJ et al (2009) Effect of calibration method on Tekscan sensor accuracy. J Biomech Eng 131(3):034503

Palmer MC, O'Rourke TD, Olson NA et al (2009) Tactile pressure sensors for soil-structure interaction assessment. J Geotech Geoenviron Eng 135(11):1638–1645

Triaxial Tests to Characterize the Shear Strength Behavior of Mechanical Biological Treatment Waste in Hangzhou, China

Zhenying Zhang(✉), Wenqiang Guo, Yingfeng Wang, Yuxiang Zhang, Hui Xu, Dazhi Wu, and Kaifu Liu

Zhejiang Sci-Tech University, Hangzhou 310018, China
zhangzhenyinga@163.com

Abstract. Mechanical biological treatment (MBT) waste from the Tianziling landfill in Hangzhou, China, was collected for composition analysis, sample preparation, and conduction of triaxial unconsolidated-undrained and consolidated-undrained tests in an environmental geotechnical laboratory. The principal findings were as follows. (1) The component analysis indicated that the MBT waste comprised plastics, textiles, glass, lime soil, unidentified materials, and so on. (2) The triaxial compression test results showed that the deviatoric stress gradually increased with an increasing axial strain. The relationship between the deviatoric stress and axial strain took the form of a strain hardening curve. (3) The stress–strain relationship of the MBT waste conforms to the Duncan–Chang model. The ranges of the model parameters a and b were obtained. A power function relationship between the model parameters a and b and the confining pressure was established. (4) The values of the cohesion, internal friction angle, effective cohesion, and effective internal friction angle of the shear strength parameters were obtained. The cohesion c was 28.6 kPa, the internal friction angle φ was 23.3°, the effective cohesion c' was 33.5 kPa, and the effective internal friction angle φ' was 32.2°. These results can provide a theoretical basis for the stability analysis of MBT landfills.

Keywords: MBT · Triaxial test · Strain hardening · Shear strength parameter Duncan–Chang model

1 Introduction

The rapid development of China's economy and continuous increase in urbanization mean that more people are flocking to cities. Thus, such cities generate more municipal solid waste (MSW) each year. To maximize the reduction of MSW, Hangzhou introduced advanced MSW treatment technology from overseas: mechanical biological treatment (MBT) waste disposal. Although MSW is reduced, the final product after MBT still needs to be landfilled [1]. In China, there has been no research on the strength characteristics of MBT waste, and there are no design specifications and standards for MBT landfills [2, 3]. In order to provide reasonable advice and guidance for the design and construction of MBT landfills and prevent their collapse and destruction, the strength characteristics of MBT waste need to be studied.

L. Zhan et al. (Eds.): ICEG 2018, ESE, pp. 202–209, 2019.
https://doi.org/10.1007/978-981-13-2224-2_25

Powrie and Beaven [4] assumed that the cohesion is zero and carried out large-scale direct shear tests on MBT waste with particle sizes of 0–20 mm. They obtained internal friction angles of 45.8°–47°. Kühle-Weidemeier [5] studied the shear strength characteristics of MBT waste in Germany and found that the cohesion varied in the range of 10–62 kPa. He concluded that the cohesion is difficult to determine from MSW research and has limited reliability. Stela Fucale [6] conducted a direct test on German MBT waste and acquired that cohesion is 16 kPa and internal friction is 40.1°.

Most of the above researchers used a direct shear apparatus, but such equipment has some defects; for example, the shear stress distribution in the shear plane is not uniform, and stress concentrations occur at the edges. In this study, component analysis was carried out on MBT waste from Hangzhou, China, and the triaxial apparatus was used to study the strength characteristics. The results can lay a foundation for the stability analysis of MBT landfills in China.

2 Test Materials and Sample Preparation

MBT waste from the Hangzhou Tianziling Domestic Waste Treatment Plant was transported to the environmental geotechnical laboratory of Zhejiang Sci-Tech University in a tightly sealed plastic bucket. The MBT waste is shown in Fig. 1a. The manual sorting method was used to obtain the components and contents of the MBT waste. Table 1 presents the composition and wet weight content of the MBT waste.

Table 1. Content of MBT waste components (wet weight content %).

Component	Plastic	Rubber	Textile	Wood	Stones	Ceramics	Glass	Metal	Fine	Unidentified
Content	25.1	5.0	10.7	6.4	4.1	3.0	14.2	2.3	16.4	12.8

To reduce the influence of the dimensional effect on the test results, the MBT waste first needed to be disposed of. Larger waste such as plastic bottles and large pieces of metal, glass, wood, and silk were picked out. Then, the rest of the waste was put in an oven, and dried at 65 °C to a constant weight. As per the principle that the particle size of MBT waste be no more than one-eighth of the diameter of the sample [8], the dried MBT waste was passed through a sieve with round apertures of 62.8 mm/8 = 7.725 mm. The bottom sieved material was used as a test material and is shown in Fig. 1b. As per the domestic waste soil test procedure [8], the sieve residue with a particle size of no more than 7.725 mm was remolded into the samples required for the triaxial test. First, 212.36 g of the sorted MBT waste was weighed and evenly spread on a geomembrane. Then, 106.18 g of pure water was weighed and evenly sprayed on the MBT waste. This was evenly mixed, covered with plastic wrap, and allowed to stand for 24 h. The MBT waste was used to fill the sample cylinder with five equal layers. After each layer was placed in the sample cylinder, it was compacted with the same force. To avoid stratification of the MBT waste, the surface was gently rubbed after compaction. The next layer was then placed in the cylinder and so on until it was full. In order to prevent the samples from rebounding, the prepared samples were sealed for a week before the mold was removed. Figure 1c shows a sample after the mold was removed was shown in Fig. 1c. Each sample had a diameter of 61.8 mm and height of 125 mm. The basic physical parameters are presented in Table 2.

Fig. 1. From a to c is original waste, prepared waste and sample

Table 2. Basic physical indicators.

Density (g/cm^3)	Water content (%)	Specific weight	Void ratio
0.85	50	1.71	2.02

3 Test Methods and Procedures

The unconsolidated-undrained (UU) and consolidated-undrained (CU) tests were carried out on samples prepared according to the parameters presented in Table 2. The test was conducted under four different confining pressures of 50, 100, 200 and 300 kPa. As per the test procedures, the samples were placed in a pressure chamber [8]. The samples were saturated to the water head saturation and back pressure saturation. For the UU test, the drain valve was closed after saturation. A confining pressure of 50 kPa was applied first, followed by the vertical deviatoric stress. The test was conducted at a displacement rate of 0.625 mm/min. After the confining pressure test was completed at this level, tests were performed at confining pressures of 100, 200, and 300 kPa according to the above method until all of the tests were completed and all test data were collected. For the CU test, the sample was saturated first. A confining pressure of 50 kPa was applied, and the drainage valve was opened to consolidate the sample. After consolidation, the confining pressure was kept constant, and a vertical deviatoric stress was imposed. The test was performed at a displacement rate of 0.625 mm/min. After the test was performed at this confining pressure, further tests were performed at confining pressures of 100, 200, and 300 kPa according to the above method. During the test, the displacement meter, measuring ring, and hole pressure meter were used to take measurements. Meanwhile, the deformation of the sample was observed, and the test was stopped when the sample broke.

4 Test Results and Analyses

4.1 Results of the Unconsolidated-Undrained Shear Test

Figure 2 shows the relationship between the deviatoric stress and axial strain. At different confining pressures under UU conditions, this relationship was relatively close and basically independent of the confining pressure. This is consistent with the uniqueness of the stress–strain curves of saturated soils that coincide with each other under different confining pressures. Figure 2 is not completely coincident because the MBT components were complex and varied. Some components were covered with impervious materials, and it was difficult to achieve full saturation during the test. As the axial strain was increased, the deviatoric stress gradually increased. When the axial strain exceeded 15%, the trend of the stress–strain curve rose upward, and the deviatoric stress growth rate became faster. This may be because MBT waste contains a large amount of fibrous material, which exhibited reinforcing properties when the axial strain exceeded 15%.

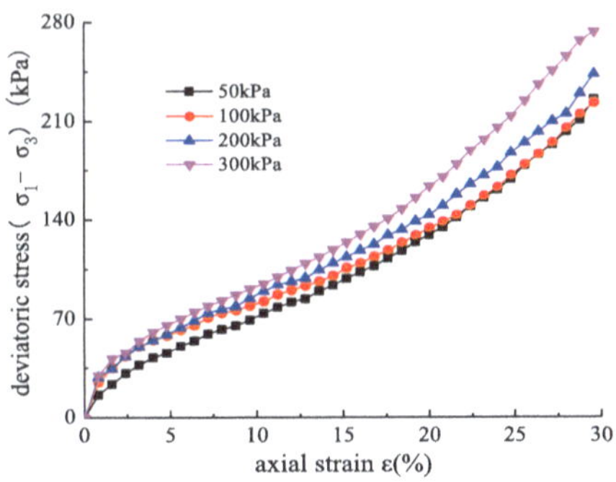

Fig. 2. Relationship between the deviatoric stress and axial strain (UU test).

In analyses of the landfill stability, the shear strength is usually measured at 15% axial strain. Figure 3 shows the strength envelope at 15% axial strain. Under UU conditions, the shear strength parameters of the MBT waste were as follows: cohesion of around 40.8 kPa and internal friction angle of around 4.8°.

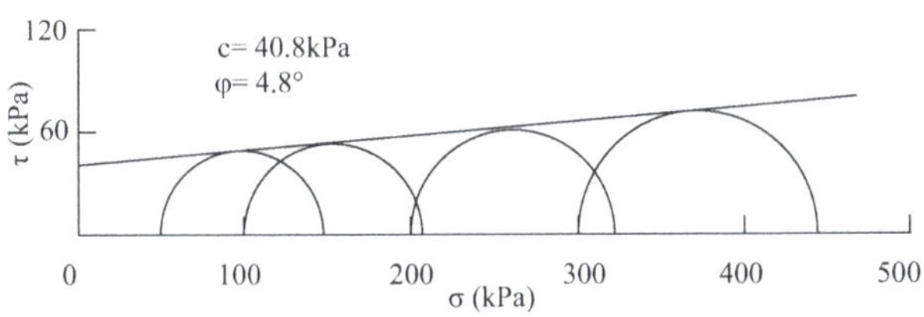

Fig. 3. Strength envelope at 15% axial strain (UU test).

4.2 Results of the Consolidated-Undrained Shear Test

Figure 4 shows the samples before and after failure in the CU test. The samples showed significant bulging and breaking. Figure 5 shows the relationship between the deviatoric stress and axial strain.

Fig. 4. During the test (left) and at the end (right).

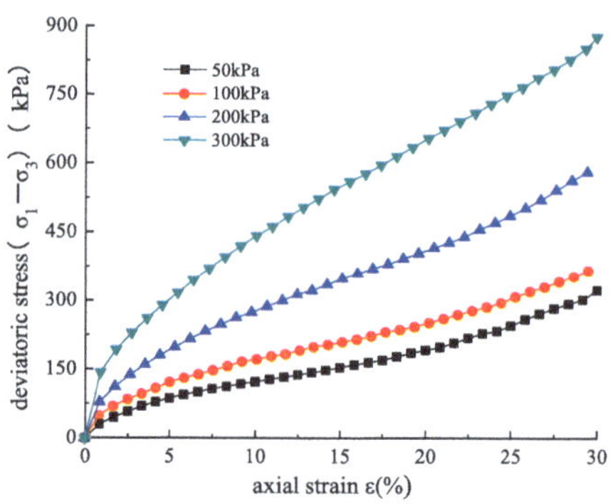

Fig. 5. Relationship between the deviatoric stress and axial strain (CU test).

The main difference in stress under different confining pressures was when the axial strain was the same; then, the deviatoric stress increased with the confining pressure. This may be because increasing the confining pressure decreased the void ratio of the sample, so the structure of each component became more compact. Mutual bite and cooperation increased, the friction increased, and the sample's ability to resist deformation increased. Figure 5 also shows that the deviatoric stress increased with the axial strain, which embodies a strain hardening curve. When axial deformation began to occur, the deviatoric stress increased rapidly with a slight increase in the axial strain, and the growth rate of the deviatoric stress increased faster. This is because the void ratio of the sample after consolidation was greatly reduced. This greatly increased the density and reduced the distance between components. The structure was denser, and the friction increased rapidly under the vertical load condition. At this time, the shear strength was mainly offered by friction. As the axial strain continued to increase, the deviatoric stress also increased gradually, but the growth rate slowed down. As the axial strain gradually increased, the fine particles between the component materials moved toward the internal pores of the porous material due to shearing.

When the axial strain exceeded 15%, the rate of increase in the deviatoric stress starts to quickly accelerate again. This is because the continuous accumulation of axial strain caused the fine particles in the sample components to move completely into the internal pores of the loose porous materials. The entire sample was now very dense. At this time, filament-like, strip-like, block-like, and other similar reinforcing materials began to exert reinforcing properties. The shear strength was given not only by the friction between components but also by the fibrous reinforcement. When these reinforced materials were stretched, they resisted the tensile deformation because of their own elasticity. This trend caused a reaction in the surrounding materials. Under this reaction force, the slip displacement of the surrounding soil was restricted, so the shear strength of the whole soil sample was improved. The growth rate of the deviatoric stress thus accelerated along the stress–strain curve.

Because of the complexity and high compressibility of MBT waste, the deviatoric stress does not have a peak or near-peak point as the axial strain increases. Based on the allowable deformation capacity of the liner system, pipe system, etc. in an actual landfill, the shear stress at an axial strain of 15% should be taken as the shear strength. The strength envelope is shown in Fig. 6.

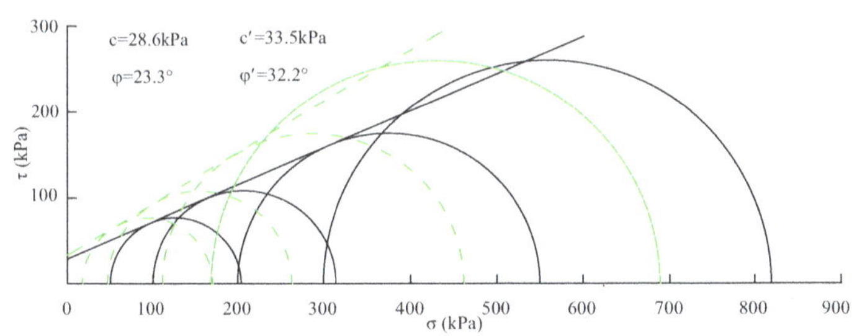

Fig. 6. Strength envelope at 15% axial strain (CU test).

At a 15% axial strain, MBT waste has a cohesion c of 28.6 kPa, internal friction angle φ of 23.3°, effective cohesion force of c' of 33.5 kPa, and effective internal friction angle φ' of 32.2°. According to statistics in the relevant literature [8, 9], cohesion and internal friction angle values have been obtained in the ranges of 0–80 kPa and 5°–40°, respectively.

4.3 Constitutive Model

As shown in Fig. 5, when the axial strain does not exceed 15%, the stress–strain relationship of MBT waste can be fitted to a hyperbola, as shown in Fig. 7. This is consistent with the Duncan–Chang model. The hyperbolic stress–strain relationship can be expressed as follows:

$$\frac{\varepsilon}{\sigma_1 - \sigma_3} = a + b\varepsilon \tag{1}$$

where σ_1 is the major principal stress (kPa), σ_3 is the confining pressure (kPa), ε is the axial strain, and a and b are the model parameters.

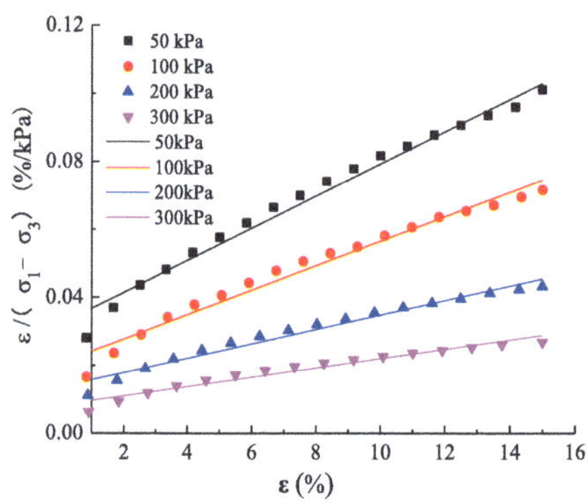

Fig. 7. The curve of $\frac{\varepsilon}{\sigma_1 - \sigma_3} \sim \varepsilon$

The relationship in Fig. 7 can be fitted to a straight line. The fitting correlation coefficient R^2 is greater than 0.94. Figure 8 shows the relationship between the model parameters a and b and the confining pressure. The model parameters a and b decrease as the confining pressure increases. Under the same confining pressure, a is greater than b. The relationship between the model parameters a and b and the confining pressure can be fitted to a power function with a fitting correlation coefficient R^2 of greater than 0.94.

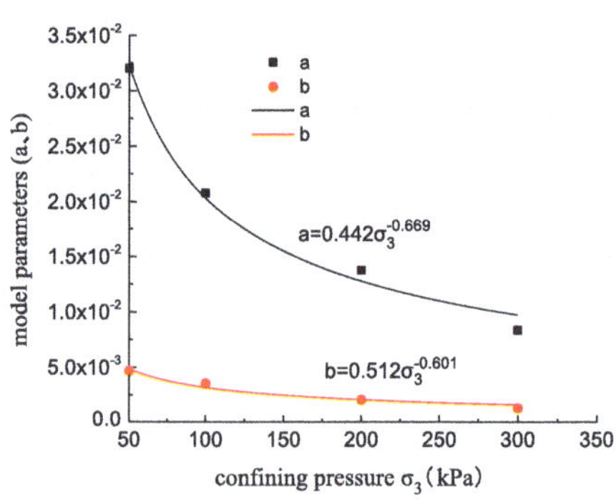

Fig. 8. Curve of the model parameters (a, b) and the confining pressure.

5 Conclusions

1. The components of MBT waste are extremely complicated, and the compressibility is very high. Even if the axial strain reaches 30%, the deviatoric stress does not reach a peak value. The stress–strain behavior of MBT waste is similar to that of conventional MSW and represents a kind of strain-hardening material.
2. The values of the shear strength parameters, cohesion, internal friction angle, effective cohesion, and effective internal friction angle of the MBT waste were obtained, which were 28.6 kPa, 23.3°, 33.5 kPa, and 32.2°.
3. The MBT waste conforms to the Duncan–Chang model. The relationship between the model parameters (*a*, *b*) and the confining pressure can be fitted to a power function.

Acknowledgments. This research was funded by the National Natural Science Foundation of China (Contract Nos. 51478436 and 51678532).

References

1. Qu Y, Zhang J, Zhu W, Pan T, Wang Q, Jin H (2015) Effect of biological treatment of municipal solid wastes. J Environ Eng 9(1):401–406
2. Wu W, Zaixing Y, Meng F (2013) Land arrangement and utilization of closured waste landfill sites. Environ Sanit Eng 4(21):16–18
3. Xu S, Wang G, Wang Z (2010) Evaluation of tensile forces of geomembrane placed on waste landfill slope due to temperature variation and filling height. Rock and Soil Mechanics 31 (10):3120–3124
4. Powrie W, Beaven RP (1999) Hydraulic properties of household waste and implications for landfills. Geotech Eng 137(4):235–247
5. Kühle-Weidemeier M (2006) Geotechnical properties and emission behaviour of mechanically and biologically treated municipal solid waste landfill properties of MBT waste. Waste Consult International, Langenhagen
6. Fucale S, Jucá JFT, Muennich K (2015) The mechanical behavior of MBT-waste. Electron J Geotech Eng 20(13):5927–5937
7. Institute of Rock and Soil Mechanics (2014) CJJ/T 204-2013 technical specifications for soil and geotechnical test of domestic rubbish. China Architecture & Building Press, Beijing
8. Feng S, Zhou Z, Chen Y, Zhan L (2005) Study on shear strength parameters of municipal solid waste. J Zhejiang Univ (Eng Sci) 39(7):987–991
9. Sivakumar Babu GL, Lakshmikanthan P, Santhosh LG (2015) Shear strength characteristics of mechanically biologically treated municipal solid waste (MBT-MSW) from Bangalore. Waste Manag 39:63–70

Research on Water Retention Characteristics and Pore Size Distribution of Landfilled Municipal Solid Waste

Wenjie Zhang(✉) and Lu Lv

Department of Civil Engineering, Shanghai University, Shanghai 200444, China
wjzhang2008@gmail.com

Abstract. Soil-water characteristic curve (SWCC) is used to describe the relation between water content and matric suction in porous media, and reflects to some extent the pore size distribution. However, a complete SWCC of municipal solid waste (MSW) is difficult to achieve by direct measurement because of the existence of macropores. For this reason, SWCCs of the matrix region in MSW were measured by a pressure plate extractor, and SWCCs of the macropore region were determined from water breakthrough tests in this paper. Water retention characteristics and pore size distribution of MSW were analyzed based on the complete SWCC. The results show that the residual water contents of municipal solid waste were high. Field capacity of shallow, middle and deep waste were 38.5%, 42.2% and 46.8%, respectively, corresponding to a matric suction range of 3–8 kPa. As the depth and age increase, the effective water content interval and specific yield decrease. Taking 1-kPa matric suction as a limit, pores in MSW can be divided into macropores and micropores. The probability density function of pore radius is bimodal. As the depth increase, the amount of macropores and their radius decrease, and the amount of micropores increases with the average pore size becomes smaller.

Keywords: Landfilled municipal solid waste · Water retention characteristics Dual-porosity · Macropore · Soil-water characteristic curve

1 Introduction

Along with the rapid increase of municipal solid waste (MSW), the number of landfills in China increases continuously. Migration of contaminant in landfills may cause pollution of soil and groundwater. Study on water retention and hydraulic conductivity characteristics of MSW is important for solving the contamination problems related with landfills. MSW is characterized by complex composition and inhomogeneity. There exists both matrix component such as soil and humus, and large-sized particles such as wood, rocks and a lot of plastics. Usually the matrix component content is small, resulting a lot of macropores, such that there exist preferential flow path in MSW. Preferential flow was identified as an important flow pattern in MSW (Ehrig 1983; Uguccioni and Zeiss 1997; Fellner and Brunner 2010). Water retention characteristic and pore-size distribution of MSW is fundamental to understand preferential flow in MSW.

L. Zhan et al. (Eds.): ICEG 2018, ESE, pp. 210–216, 2019.
https://doi.org/10.1007/978-981-13-2224-2_26

Field capacity is the amount of water content held after excess water has drained under gravity. Field capacity of MSW lies between 22% and 55% (Qian et al. 2002). Chen et al. (2011) suggested field capacity of MSW corresponded to a matric suction of 10 kPa. At present, water retention characteristics of MSW are usually investigated by Soil-water characteristic curve (SWCC). Kazimoglu studied SWCC of artificial MSW by a pressure plate extractor, and found that the SWCC curve could be well fitted by the Van Genuchten equation. Zhang et al. (2013) measured SWCC of MSW in different depths of a landfill and found that saturated and residual water contents of MSW were high, the air-entry value was low, and slope of SWCC was very steep at low suction range. Wu et al. (2012) found that as depth increases, saturated water content of MSW decreased and air-entry value and residual water content increased. SWCC can reflect the water retention characteristics of MSW, however, measuring accurately the SWCC of NSW is difficult because of the existence of a lot of large pores. Water will issued immediately from macropores without an applied air pressure in the extractor, making the SWCC measurement challenging.

There are both macropores and micropores in MSW. The pore size distribution in MSW is similar to that of gap-graded soils. Burger and Shackelford (2001) found that there were microscopic and macroscopic porosity regions in pelletized diatomaceous earth, and the SWCC was bimodal. Zhang et al. (2005) also found the bimodal or multi-modal character of soils composed of two or more pore systems. Bimodal SWCC can be used to investigate water retention characteristic and pore-size distribution of MSW. However, bimodal SWCC of MSW has not been reported till now.

This paper aims to evaluate the water retention characteristic and pore-size distribution of MSW. A complete SWCC of MSW was obtained by pressure plate extractor test combined with the water breakthrough test. Water retention characteristics and pore-size distribution of MSW were analyzed based on the complete SWCC. The quantitative results would be beneficial to simulate water and solute migration in MSW under the conception of preferential flow.

2 Materials and Methodology

2.1 MSW Samples

MSW specimens were taken from shallow, middle, and deep depths in Suzhou landfill, with the MSW age of 0–3, 3–6, and 6–10 years, respectively. Dry densities of these MSW specimens are 0.41, 0.60, and 0.81 g/cm^3, respectively. As depth and age increases, organic and plastic contents decrease and soil content increases, due to degradation. The particle-size distribution obtained by sieving and density-hydrometer method showed that the content of fine-grained particles increased with depth, while the content of large particles decreased. The uniformity coefficients C_u are 173, 121 and 21, and the curvature coefficients C_c are 0.54, 0.17 and 0.73 for shallow, middle and deep MSWs, respectively, illustrating the gap-graded characteristic.

2.2 Measuring SWCC by Pressure Plate Extractor Test

After saturation, MSW samples were put into a volumetric pressure plate extractor to get the SWCCs. Water issued immediately from macropores as the MSW sample was taken out from the saturator. As air pressure in the extractor was applied, this part of water discharged, such that the SWCC at the low suction range takes the form of a straight line (Fig. 1) and is not accurate. This part of SWCC cannot reflect the pore size distribution.

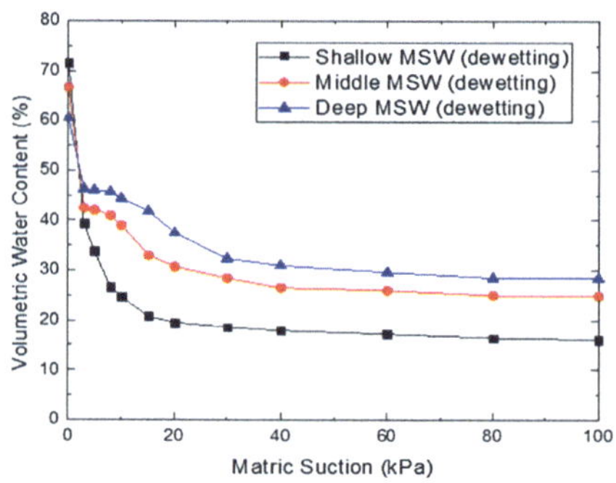

Fig. 1. SWCCs of MSW measured by volumetric pressure plate extractor

2.3 Measuring Field Capacity

Saturated shallow, middle and deep MSW columns were allowed to drain under gravity for 24 h. The corresponding field capacity was measured to be 38.45, 42.15, and 45.80%. In this paper, macropores are defined as those pores that cannot hold water when the soil reaches field capacity. Under this definition, the volume of macropores in MSW was measured according to the amount of drained water. The volume of macropores accounts for 46.23, 36.81 and 24.35% of the total pore volume for shallow, middle, and deep MSW, respectively.

2.4 Adjustment of SWCC by Water Breakthrough Test

Water was added on the surface of the drained column and the effluent was recorded at a interval of 5 s. In this process, water first drained from the largest macropore, then drained from smaller macropores, and finally the drainage became steady. Based on the water breakthrough curves, macropore radii were calculated under the assumption of circular pore shapes. By combination of the flow equation (Radulovich et al. 1989)

$$Q = Av = \pi r^2 \tau L/t \tag{1.1}$$

with the Poiseulle equation

$$Q = \pi r^4 \Delta P/(8\eta\tau L) \tag{1.2}$$

the pore radius is solved to be

$$r = \tau L[8\eta/(t\Delta P)]^{1/2} \tag{1.3}$$

where Q is flow, A is pore area, r is pore radius, v is velocity, τ is tortuosity, L is sample length, t is time, ΔP is pressure, and η is water viscosity.

A matric suction of 1 kPa, corresponding to the equivalent pore radius of 0.145 mm is used as a limit to separate micropore and macropore in this paper. Then τ is calculated to be 7.7, 8.2 and 9.0 for shallow, middle and deep MSW. The total effluent time was partitioned into many time intervals. Each radius calculated from Eq. (1.3) represents the average pore radius corresponding to the time interval. Based on the average radius and column height, the pore volume can be determined. Proportion of pores in each interval is calculated as the ratio of the pore volume in each interval to the total MSW volume.

The matric suctions and the corresponding volumetric water contents of the macropore region were calculated from proportion of pores in each interval based on the Young-Laplace equation:

$$\psi = \frac{2T\cos\alpha}{r} \tag{1.4}$$

where ψ is matric suction, T is surface tension, and α is contact angle. The matric suctions and the water contents of the macropore region were coupled to the SWCC of the matrix region to yield a complete SWCC (Fig. 2).

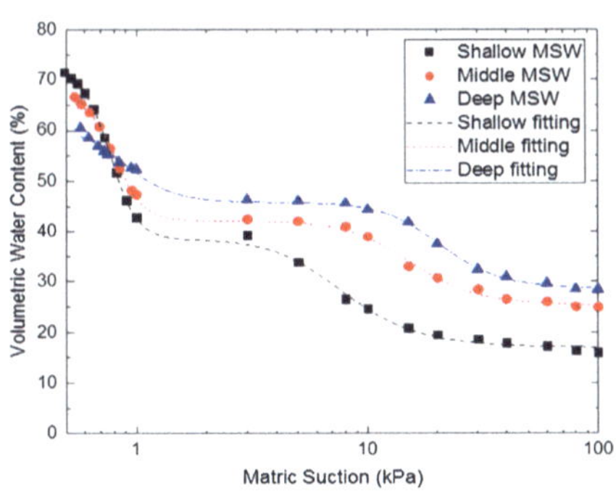

Fig. 2. The complete SWCCs of landfilled waste and the regression

2.5 Fitting of the Complete SWCCs

As the dual porosity of MSW is concerned, the general form of SWCC can be expressed as

$$\theta(\psi) = (\theta_s - \theta_{fc})\theta_1(\psi) + (\theta_{fc} - \theta_r)\theta_2(\psi) + \theta_r \tag{1.5}$$

where $\theta(\psi)$ denotes the volumetric water content under which all pores with matrci suction more than or equal to ψ are filled with water, 1 and 2 represent macropore and micropore, θ_s is the saturated volumetric water content, θ_r are residual water content, and θ_{fc} is field capacity.

Combining Van Genuchten equation (Van Genuchten 1980) with Eq. (1.5), a modified SWCC is obtained

$$\theta_e = \sum_{i=1}^{2} w_i \left[\frac{1}{1+(\alpha_i \psi)^{n_i}}\right]^{m_i} \tag{1.6}$$

where θ_e is effective volumetric water content, α_i, n_i and m_i are Van Genuchten parameters. The complete SWCCs are fitted using Eq. (1.6) with results listed in Table 1.

Table 1. Parameters for fitting of the bimodal SWCCs

MSW	w_1	θ_{s1}	α_1	n_1	θ_{s2}	α_2	n_2	θ_r	θ_s	R^2
Shallow	0.6062	0.3307	1.3270	7.9480	0.3845	0.1514	3.1210	0.1697	0.715	0.999
Middle	0.5932	0.2455	1.2646	7.5960	0.4215	0.0755	3.4030	0.2532	0.667	0.999
Deep	0.4607	0.1474	1.1906	4.6430	0.4580	0.0544	3.7060	0.2855	0.605	0.998

The pore size probability density function can be written as Eq. (1.7). Substituting the fitted parameters (Table 1) into Eq. (1.7), the pore size probability density was obtained.

$$f(r) = \sum_{i=1}^{2} w_i m_i n_i \frac{(\alpha_i \cdot 0.1455/r)^{n_i}}{r} \cdot \left[\frac{1}{1+(\alpha_i \cdot 0.1455/r)^{n_i}}\right]^{m_i+1} \tag{1.7}$$

3 Results and Conclusions

3.1 Water Retention Characteristics

Field capacity of shallow, middle, and deep MSW is 38.45, 42.15, and 45.80%, respectively. It can be seen from Fig. 2 that these volumetric water contents correspond to a horizontal segment on SWCC, with the matric suction ranging from 3 to 8 kPa. The residual water content of MSW was obtained by curve fiting (Table 1), to be 16.97, 25.32, and 28.55 for shallow, middle, and deep MSW, respectively. These values all correspond to a matric suction of about 100 kPa on the SWCC. Water Retention parameters of MSW are shown in Fig. 3. As depth increases, saturated water content decreases, whereas field capacity and residual water content increase, such that the effective water content interval (the difference between the saturated and the residual water content) and the specific yield (the difference between the saturated water content and the field capacity) decrease.

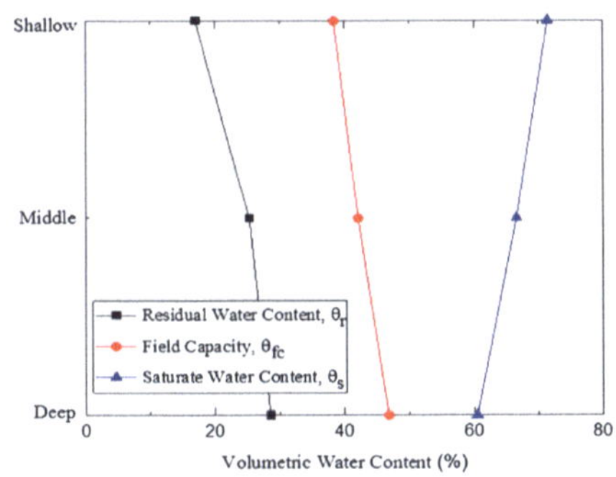

Fig. 3. Retention parameters of MSW

It should be noted that the fitted residual water content of MSW is much higher than common soils such as clay and sand (usually ranges from 2% to 5%, Fredlund and Rahardjo 1993). There are so many macropores in MSW that the matrix regions were separated into many individual parts. As air pressure in the extractor increased, water in macropores was drained and finally micropores became unconnected to each other. The water retained in these micropores could not be drained and were regarded as a part of the residual water content. This reflects a special pore size distribution character in MSW. The residual water content of deep MSW is higher because there are more micropores in deep MSW than in shallow MSW.

3.2 Pore Size Distribution Determined from Complete SWCCs

The volume ratio curves were obtained from analysis of the pore size probability density, and is shown in Fig. 4. The Y-axis represents the ratio of pore volume corresponding to certain pore radius to the total pore volume. The curves are obviously bimodal and can be divided into macropore parts and micropore parts. The average macropore radii are 0.193, 0.184 and 0.173 mm for shallow, middle and deep MSW, and the average micropore radii are 0.022, 0.011 and 0.008 mm, respectively.

It should be noted that the radii in Fig. 4 are equivalent radii which were calculated based on the assumption of equivalent flow capability. Though the shapes of pores in MSW may be far from circular, the results here can still be used to investigate the water flow pattern in MSW.

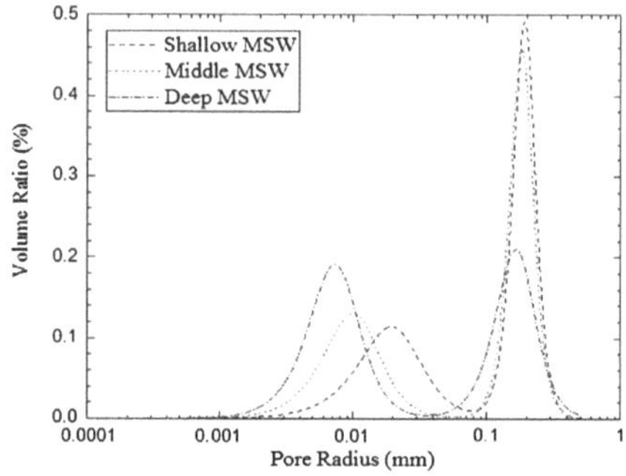

Fig. 4. Volume ratio distribution of pores determined from SWCC tests

Acknowledgments. This research was supported by National Natural Science Foundation of China (Nos. 51478256 and 41772300).

References

Burger CA, Shackelford CD (2001) Evaluating dual porosity of pelletized diatomaceous earth using bimodal soil water characteristic cure functions. Can Geotech J 38:53–66

Chen Y-m et al (2011) Ministry of Housing and Urban-Rural Development of the People's Republic of China. Technical code for geotechnical engineering of municipal solid waste sanitary landfills (CJJ176-2011). China Architecture and Building Press, Beijing (in Chinese)

Ehrig H-J (1983) Quality and quantity of sanitary landfill leachate. Waste Manag Res 1:53–68

Fellner J, Brunner PH (2010) Modeling of leachate generation from MSW landfills by a 2-dimensional 2-domain approach. Waste Manag 30:2084–2095

Fredlund DG, Rahardjo H (1993) Soil mechanics for unsaturated soils. Wiley, New York

Qian XD, Koerner RM, Gray DH (2002) Geotechnical aspects of landfill design and construction. Prentice-Hall Inc, Upper Saddle River

Radulovich R, Solorzano E, Sollins P (1989) Soil macropore size distribution from water breakthrough curves. Soil Sci Soc Am J 53:556–559

Uguccioni M, Zeiss C (1997) Comparison of two approaches to modeling moisture movement through municipal solid waste. J Environ Sys 25:4–63

Van Genuchten MT (1980) A closed-form equation for predicting the hydraulic conductivity of unsaturated soils. Soil Sci Soc Am J 4:892–898

Wu H, Wang H, Zhao Y et al (2012) Evolution of unsaturated hydraulic properties of municipal solid waste with landfill depth and age. Waste Manag 32:463–470

Zhang L, Chen Q (2005) Predicting bimodal soil water characteristic curves. J Geotech Geoenviron 131:666–670

Zhang W-J, Zhang G-g, Chen Y-m (2013) Analyses on a high leachate mound in a landfill of municipal solid waste in China. Environ Earth Sci 70:1747–1752

Numerical Simulation of Failure Process of Buildings Under the Impact of Flow Slide

Shi-Jin Feng, Hong-Yu Gao, and Hong-Xin Chen(✉)

Key Laboratory of Geotechnical and Underground Engineering of the Ministry of Education, Department of Geotechnical Engineering, Tongji University, Shanghai 200092, China
chenhongxin@tongji.edu.cn

Abstract. Flow slide is a form of landslide that can pose a great threat to the surrounding population and buildings. The collapse of buildings is an important cause of injury and death, so it is highly important to understand the failure process of buildings under the impact of flow slide and put forward effective methods for the hazard mitigation. In this paper, complex fluid-structure interaction between flow slide and buildings is simulated using SPH-FEM method. Flow material is modeled by SPH method which is a powerful tool for handling fluid problems, while the buildings are modeled by FEM elements with erosion algorithm. The runout characteristics and failure process of buildings under the impact of flow slide are both simulated. Different failure mechanisms are identified at different locations. The results indicate that this method is an applicable tool for analyzing the failure process of buildings and is promising to be used to assess the risk of buildings to flow slide.

Keywords: Flow slide · Building failure · Fluid-structure interaction Numerical simulation

1 Introduction

With the urbanization process, landfills have been widely constructed to dispose construction waste, municipal solid waste, hospital waste and so on (Ghobadi et al. 2013). Due to the high pore pressure in the landfills, slope failures may occur and evolve into high speed destructive flow slide, posing a great danger to people and buildings in the impact area. For example, on December 20, 2015, a catastrophic flow slide occurred at a construction waste landfill in Guangming New District of Shenzhen, China, located at (E 114°18'48", N 22°40'5"). Nearly 1.8 million cubic meters of waste traveled 1.2 km down the hill, covering an area of about 350×10^3 m^2 and causing 33 buildings to be destroyed or damaged and 76 persons dead or missing (Ouyang et al. 2016; Yin et al. 2016). Therefore, it is greatly important to understand the runout characteristics of flow slide and the failure mechanisms of buildings.

Fluid-structure interaction plays a key role in the design of mitigation measures. However, it is still a challenge for researchers. It is normally too complex to solve analytically, so numerical simulation is often employed to model the interaction between fluid and structures, and some methods have been applied successfully.

L. Zhan et al. (Eds.): ICEG 2018, ESE, pp. 217–223, 2019.
https://doi.org/10.1007/978-981-13-2224-2_27

For example, Huang et al. (2014) utilized Arbitrary Lagrangian-Eulerian (ALE) method to model the interaction between debris flow and flexible barriers; Shan and Zhao (2014) adopted CFD-DEM method to simulate the impact of granular flow on a water reservoir; Dai et al. (2016) applied Smoothed Particle Hydrodynamics (SPH) method (Lucy 1977; Gingold and Monaghan 1977; Liu and Liu 2010) to simulate the impact of debris flow on check dams. However, ALE mesh has to cover the whole computational domain which is high-cost for modeling a large volume of flow slide. CFD-DEM method is also limited by its computational efficiency. Moreover, none of these methods can take into account the large deformation and even the failure of buildings.

In this paper, SPH-FEM method in LS-DYNA is presented to investigate the interaction between flow slide and buildings. FEM is employed to form the extensive boundary and flow material is simulated by SPH method. Erosion algorithm is applied for the building elements to capture the large deformation and the failure process of buildings.

2 Methodology

2.1 Framework of the Method

The digital terrain data are converted and imported into LS-DYNA to form the digital elevation model at first. The ground is modeled by rigid shell elements fixed in space. The flow material is modeled as SPH particles to accommodate high mobility, free surface and large deformation, while the buildings are simulated by traditional FEM elements with erosion to reflect the failure process (Anghileri et al. 2005; Zhang et al. 2011). There are two stages in the numerical analysis, namely flow stage and impact stage. At the flow stage, the flow material comprising a huge number of particles flows down by gravity. At the impact stage, the buildings fail under the impact of flow slide.

2.2 Modeling of Flow Slide

The motion of each SPH particle is calculated on the basis of the interaction with other neighboring particles weighted by a kernel function, as shown in Eq. (2.1)

$$\langle f(x_i) \rangle = \sum_{j=1}^{N} \frac{m_j}{\rho_j} f(x_i) W_{ij} \tag{2.1}$$

where x_i represents the concerning particle, and x_j is a neighboring particle in the kernel area; W is a smooth function.

The flow slide has complex rheological properties and there are so many parameters that can influence the results. To simplify the problem, the flow material is modeled by the elastic-plastic hydrodynamic model (Hallquist 2006) to simulate the runout characteristics of flow slide. In order to make the problem solvable, the equation of state (EOS) must be specified to describe the relationship between the pressure and the volume change (Ma et al. 2008). The widely used Gruneisen equation is adopted and the formulas are shown as follows:

$$p = \frac{\rho_0 C^2 \mu\left[1 + \left(1 - \frac{\gamma_0}{2}\right)\mu - \frac{a}{2}\mu^2\right]}{\left[1 - (S_1 - 1)\mu - S_2\frac{\mu^2}{\mu+1} - S_3\frac{\mu^3}{(\mu+1)^2}\right]} + (\gamma_0 + a\mu)E \tag{2.2}$$

$$p = \rho_0 C^2 \mu + (\gamma_0 + a\mu)E \tag{2.3}$$

where C is the intercept of the v_s-v_p curve, v_s and v_p are the shock velocity and particle velocity, respectively; S_1, S_2, S_3 are the coefficients of the slope of the v_s–v_p curve; γ_0 is the Gruneisen gamma; a is the first order volume correction to γ_0; $\mu = \rho/\rho_0 - 1$. Equation (2.2) is for the compressed materials and Eq. (2.3) for the expanded materials.

2.3 Interaction Between Flow Slide and Buildings

As a Lagrangian method, SPH method can be processed by the penalty-based contact algorithms to couple with FEM elements. The set of SPH particles is defined as a slave while the finite element surface is defined as a master to achieve the coupling between these two solvers. Contact is checked between nodes and building surfaces in every circle. The contact algorithm uses a Coulomb friction model with an exponential function to reflect the transition from static to dynamic friction.

For modeling the material failure, an element erosion algorithm is applied which has been used with satisfactory results for the collapse process of buildings (Shi et al. 2010; Wang et al. 2010). The plastic kinematic model is taken to model the deformable building elements. When the plastic strain exceeds failure strain, building elements will fail and then the failed elements will be deleted. Failure mechanisms of buildings can also be identified according to the flow behavior and deformation of the buildings.

3 Simulation of the Failure Process of Buildings Impacted by Flow Slide

3.1 Input Information

In this study, the failure process of buildings during the flow slide of waste landfill in Guangming New District of Shenzhen, China is simulated. The computational domain is 1603.8 m in length and 1036.8 m in width covering an area of 1.66×10^6 m^2. The ground is formed by 227,119 shell elements with resolution of 2.7 m. Figure 1 shows the top view of the topography. As shown in Fig. 1, there is a stone pit surrounded by mountains that looks like an armchair. The only exit is a saddle back in the north which is so narrow that the rainfall was largely accumulated.

The initial source of the flow slide is reconstructed to fit the surrounding topography. The SPH node spacing is set as 5 m so the flow body comprises a total of 59,269 SPH particles. All the SPH particles have a uniform density of 1820 kg/m^3, which is determined by the in-situ test. The relevant parameters of the flow slide are summarized in Table 1.

(a) (b) (c)

Fig. 1. Topography in Guangming new district: (a) pre-event topography, (b) post-event topography, (c) topography in LS-DYNA before the event

Table 1. Parameters of the material and EOS adopted for the flow slide

Density	ρ (kg/m^3)	1820	C	1484
Shear modulus	G (Pa)	1.0×10^6	S_1	1.92
SPH particle mass	m (kg)	2.275×10^5	S_2	0
Gravity	g (m/s^2)	9.81	S_3	0
Time step size	Δt (s)	7.0×10^{-4}	γ_0	0.1

In the study area, 39 buildings concerned are built to investigate the impact process of buildings with severe failure. The height of the buildings ranges from 8 to 16 m, and the location of the buildings is determined by satellite image as shown in Fig. 2(a). The material parameters for the concrete are assigned to the buildings as shown in Table 2. A friction coefficient of $\mu = 0.1$ is chosen to reflect the basal friction. The simulation time is set as 60 s, which is close to that of the sliding process at the site.

3.2 Analysis of Flow Slide

Figure 2 shows the plan view of snapshots during the flow process recorded in the numerical analysis. The time difference between each snapshot is 20 s. In Fig. 2(a), the particles in the original slope are created with a uniform space of 5 m at $t = 0$ s. All the SPH particles are accelerated by gravity. In Fig. 2(b), the waste flushes down around 500 m during the first 20 s, indicating that the flow slide moves very fast at the early stages. The flow slide passes the saddle back at a high speed and the shacks in the flow path are destroyed immediately.

In Fig. 2(c), at $t = 40$ s, the waste reaches the first row of buildings at a speed of around 15 m/s. The flow depth near the buildings reaches 8 m, which is almost half the height of buildings. The collapse of seven buildings are observed and the presence of

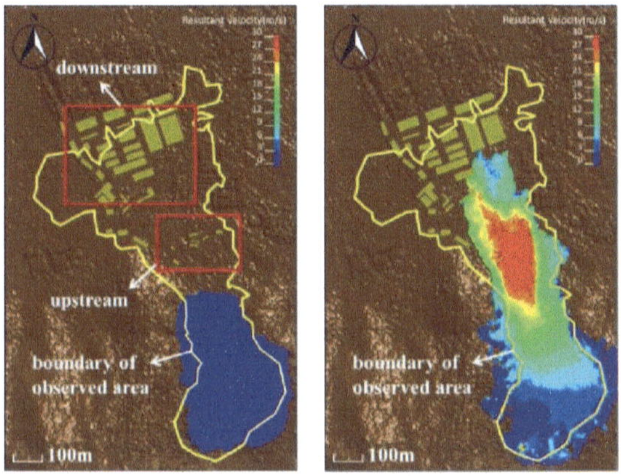

(a) (b)

Fig. 2. Snapshots of velocity of the flow slide: (a) t = 0 s, (b) t = 20 s, (c) t = 40 s, d t = 60 s

Table 2. Parameters of the material adopted for the buildings

Density	ρ (kg/m^3)	2000
Young's modulus	E (GPa)	20
Poisson's ratio	ν	0.2
Yield stress	σ_0 (MPa)	10

the buildings causes the flow to branch into lateral directions. In Fig. 2(d), the flow slide mostly deposits in the downstream region and the velocity reduction of the particles at the building zone is obviously observed at t = 60 s. The flow mass continues to spread at the relatively flat area. Although there is a small deviation in the right region where the waste flows a longer distance, the disposition area overall matches the real scenario reasonably well.

3.3 Failure Process of Buildings

In the analysis domain, 28 buildings are destroyed or damaged in one minute. Two regions in Fig. 2(a) are adopted to investigate the failure process of buildings, as shown in Fig. 3.

For buildings near the landfill, the strong impact of flow slide quickly shatters the buildings into pieces as shown in Fig. 2(a). The first floor of the buildings collapses at first and leads to the collapse of the upper floors. For example, the 1# building in the traversing path is flushed away and the elements of the first floor are removed with only the roof left; the side walls of the 2# building fail due to the strong bending of the frontal elements; similar to these buildings, the 3# building on the left side loses frontal walls and side walls at first because of the impact.

As shown in Fig. 3(b), the failure process of buildings at the downstream differs in some ways. A falling trend of velocity is observed, making the flow slide not flush

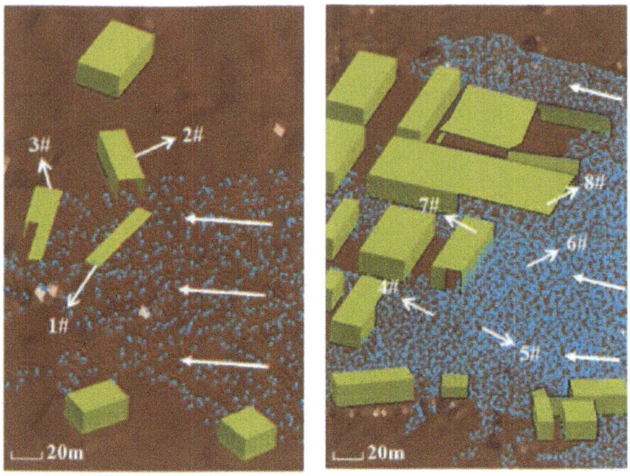

(a) (b)

Fig. 3. Failure process of buildings at different sites: (a) upstream, (b) downstream

down the walls instantly. The duration of the flow slide acting on the buildings ranges from 1 to 18 s before the buildings collapse. The buildings in the mainstream path including 4#, 5# and 6# collapse within several seconds, indicating that the momentum is still localized in the mainstream. While the buildings such as 7# and 8# at the margin of the deposition area maintain for a longer time due to the low momentum of flow slide. When the flow slide approaches the buildings, the deformation and crack develop gradually in the elements on account of the pressure of the flow slide on the buildings.

The results suggest that the building are highly vulnerable to the destructive flow slides. The buildings should be built far enough from the landfill. At the downstream, it is noteworthy that the distribution of momentum is different throughout the deposition region. If hazard mitigation measures can redirect the mainstream flow or shunt the flow into several branches, people will have more time to escape from the hazard.

4 Conclusions

In this study, the three-dimensional SPH-FEM method is used to analyze the failure process of buildings under the impact of flow slide in Guangming New District of Shenzhen, China. SPH method is used to model the flow slide to deal with large deformations. Buildings are set up by shell elements in which erosion algorithm is applied to investigate the failure process of buildings. The results reveal that the numerical results agree with the field data reasonably well. The rapid collapse of buildings is observed at the upstream which indicates that the construction of buildings at these locations should be avoided. As the velocity of the flow slide falls, the failure mechanisms of buildings at the downstream are different. The buildings in the mainstream collapse much faster than those at the marginal area. It is suggested that buildings should not be built near the landfill and that measures should be taken to shunt the flow. Therefore, SPH-FEM method is an applicable tool for studying the interaction between flow slide and buildings, and it is promising to be used for assessing the risk of buildings to flow slide. Further researches will be carried out to comprehensively study the failure mechanisms of buildings and propose effective mitigation measures.

Acknowledgements. The majority of the work described in this paper was supported by the National Natural Science Foundation of China under Grant No. 41602288, the Fundamental Research Funds for the Central Universities, and the Shanghai Chenguang Scheme. The authors would like to express the sincerest gratitude.

References

Anghileri M, Castelletti LML, Tirelli M (2005) Fluid-structure interaction of water filled tanks during the impact with the ground. Int J Impact Eng 31:235–254

Dai Z, Huang Y, Cheng H (2016) SPH model for fluid-structure interaction and its application to debris flow impact estimation. Landslides 14:1–12

Ghobadi MH, Babazadeh R, Bagheri V (2013) Siting MSW landfills by combining AHP with GIS in Hamedan province, western Iran. Environ Earth Sci 70:1823–1840

Gingold RA, Monaghan JJ (1977) Smoothed particle hydrodynamics-theory and application to non-spherical stars. Mon Not R Astron Soc 181:375–389

Hallquist JO (2006) LS-DYNA Keyword User's Manual. Livermore Software Technology Corporation

Huang Y, Yiu J, Pappin J (2014) Numerical investigation of landslide mobility and debris-flow resistant flexible barrier with LS-DYNA. In: Proceedings of the 13th international LS-DYNA users conference

Liu MB, Liu GR (2010) Smoothed particle hydrodynamics(SPH): An overview and recent developments. Arch Comput Methods Eng 17:25–76

Lucy LB (1977) A numerical approach to the testing of the fission hypothesis. Astron J 82:1013–1024

Ma L, Bao RH, Guo YM (2008) Waterjet penetration simulation by hybrid code of SPH and FEA. Int J Impact Eng 35:1035–1042

Ouyang C, Zhou K, Xu Q (2016) Dynamic analysis and numerical modeling of the 2015 catastrophic landslide of the construction waste landfill at Guangming, Shenzhen, China. Landslides 14(2):1–14

Shan T, Zhao J (2014) A coupled CFD-DEM analysis of granular flow impacting on a water reservoir. Acta Mech 225:2449–2470

Shi Y, Li ZX, Hao H (2010) A new method for progressive collapse analysis of RC frames under blast loading. Eng Struct 32:1691–1703

Wang ZL, Wu J, Wang JG (2010) Experimental and numerical analysis on effect of fibre aspect ratio on mechanical properties of SRFC. Constr Build Mater 24:559–565

Xu Q, Peng D, Li W (2017) The catastrophic landfill flowslide at Hongao dumpsite on 20 December 2015 in Shenzhen, China. Nat Hazards Earth Syst Sci 17(17):1–19

Yin YP, Li B, Wang WP (2016) Mechanism of the December 2015 catastrophic landslide at the Shenzhen landfill and controlling geotechnical risks of urbanization. Engineering 2(2):230–249

Zhang Z, Qiang H, Gao W (2011) Coupling of smoothed particle hydrodynamics and finite element method for impact dynamics simulation. Eng Struct 33:255–264

Numerical Simulation of Bioreactor Landfills Subjected to Aeration Using CFD

An-Zheng Li(✉), Shi-Jin Feng, and Ben-Yi Cao

Department of Geotechnical Engineering, Tongji University, Shanghai 200092, China
1350793@tongji.edu.cn

Abstract. Aeration is an effective method to repair old landfills and reduce the post-closure charge. Before designing the aeration systems, reliable prediction of aeration systems performance is important because operators could optimize the system based on it. This study proposes a three-dimensional CFD model of vertical aeration systems using ANSYS Fluent platform. Multiphase flow, multicomponent transports and biochemical reaction are considered to simulate the hydro-biochemical process when aeration occurs in aerobic landfills. This model was verified by experimental data successfully. The results focus on the influence of aeration frequency and well depth in terms of intermittent aeration in landfills.

Keywords: MSW · Bioreactor landfill · CFD simulation Aerobic biodegradation

1 Introduction

The demand of old landfill repair is increasing and aeration is gradually applied to landfill operation, especially for landfills which have stopped landfilling for many years and have almost used up the formation potential of methane.

Since waste in landfills contains leachate and landfill gas, which are both multi-component fluid, numerical models usually treat the leachate and gas transports process as a typical problem of multiphase fluid flow in porous media. The bio-chemical process is another key point when formulating aeration in landfills. There are lots of work about this multi-field coupling problems. However, most models consider only one aspect of them, such as multiphase model without considering biochemical reaction, aerobic biochemical models only considering multi-component [1] or three-dimensional model which assumes the amount of organic matter remains constant [2]. These researches simulate the mechanism of aerobic landfills but are less consistent with reality.

In this study, a three-dimensional CFD model is proposed to simulate the bio-chemical and hydrodynamic behaviour of aeration which couples multiphase flow, multicomponent transports and biochemical reaction simultaneously. Then the appropriate aeration frequency and vertical depth are discussed, which is meaningful for in situ projects.

L. Zhan et al. (Eds.): ICEG 2018, ESE, pp. 224–230, 2019.
https://doi.org/10.1007/978-981-13-2224-2_28

2 Model Development

2.1 Multiphase Flow Equations

The multiphase flow usually was treated as interpenetrating continuum which formulated by the concept of phasic volume fraction α_q (dimensionless). Phasic volume fraction means the ratio of volume occupied by phase q to the volume of void. Each phase (leachate and gas) satisfies the conservation equations of mass and momentum individually. The volume of phase q, V_q is described as follows:

$$V_q = \int_V \alpha_q dV \tag{1}$$

The continuity equation for phase q (either leachate or gas phase) is

$$\frac{\partial}{\partial t}\left(\varepsilon\alpha_q\rho_q\right) + \nabla\cdot\left(\varepsilon\alpha_q\rho_q\vec{v_q}\right) = \varepsilon S_q \tag{2}$$

where ε is the total porosity of MSW (dimensionless), ρ_q is the density of phase q (kg/m^3), $\vec{v_q}$ is the velocity of phase q (m/s), and S_q is the source/sink term of phase q (kg/m^3/s).

The momentum conservation is described by Navier-Stokes-Brinkman equation as follows

$$\frac{\partial}{\partial t}\left(\varepsilon\alpha_q\rho_q\vec{v_q}\right) + \nabla\cdot\left(\varepsilon\alpha_q\rho_q\vec{v_q}\vec{v_q}\right) = -\varepsilon\alpha_q\nabla p + \nabla\varepsilon\bar{\bar{\tau}} + \varepsilon\alpha_q\rho_q\vec{g} - \alpha_q^2\frac{\mu_q}{k_ik_r}\vec{v_q} - \varepsilon\alpha\nabla p_c \tag{3}$$

where p is the liquid/gas pressure (Pa), p_c is the capillary pressure (Pa), $\bar{\bar{\tau}}$ is the shear stress (Pa), g is the acceleration of gravity (m/s^2), μ_q is the dynamic viscosity of phase q (kg m^{-1} s^{-1}), k_i is the intrinsic permeability (m^2), and k_r is the relative permeability (dimensionless).

2.2 Multicomponent Transport Equations

Leachate and gas phases both are mixtures that contain different components. Leachate phase contains O_2, H_2O, CO_2, VFA, and MB, and gas phase contains O_2, CO_2, and N_2. The convection–diffusion equation is adopted to calculate the species transport between leachate and gas:

$$\frac{\partial}{\partial_t}\left(\rho Y_i\right) + \nabla\cdot\left(\rho\vec{v}Y_i\right) = -\nabla\cdot\vec{J_i} + R_i \tag{4}$$

where Y_i represents the mass fraction of species i in each phase, R_i is the mass source/sink term of species i in biochemical reactions, $\vec{J_i}$ is the diffusion flux of species *i* due to the gradient of concentration. This model takes account of the interface mass transfer of oxygen and carbon dioxide by Herny's law.

2.3 Aerobic Biodegradation

This study adopts a two-stage aerobic biochemical model and divides the aerobic biodegradation process to hydrolysis and oxidation reaction to simulate the biochemical process in aerobic landfills.

Hydrolysis Reaction. Selects cellulose ($C_6H_{10}O_5$) was set as biodegradable substance and lignin was non-degradable substance in landfills. The hydrolysis rate obeys the first-order kinetics and considers the impact of moisture and the proportion of non-degradable substances [3]:

$$r_h = r_{h\max} f_w(\alpha_l)\varphi \tag{5}$$

where r_h is the rate of hydrolysis reaction (kg/m^3 s), r_{hmax} is the maximum rate of hydrolysis reaction (kg/m^3 s) in the most favorable environment, f_w is the factor of moisture content, and φ is relative digestibility.

Oxidation Reaction. VFA in leachate is oxidized by oxygen injected from outside in second stage. The consumption rate of VFA in oxidation reaction is calculated via Monod kinetics:

$$S_s = \frac{dC_v}{dt} = -k_m \frac{C_v}{K_s + C_v} C_B K_{o_2} \tag{6}$$

where S_s is the VFA consumption rate (kg/m^3 s), t is time (s), km is the maximum biodegradation rate in high VFA concentration (s^{-1}), Ks is the half saturation constant (kg/m^3), C_v is the concentration of VFA in leachate (kg/m^3), CB is the concentration of microbe in leachate (kg/m^3), and K_{o_2} is the oxygen concentration correction function (dimensionless).

The reproduction rate of aerobic microbe is calculated using Monod kinetics related to the consumption rate of VFA

$$S_s = \frac{dC_s}{dt} = \frac{S_B}{Y_s} = \frac{1}{Y_s}\frac{dC}{dt} \tag{7}$$

where S_B is the VFA consumption rate (kg/m^3·s), and Y_s is the yield coefficient which connects kinetics of VFA and microbe.

The decay rate is calculated using first order kinetics

$$S_D = k_d k_{temp} C_B \tag{1.8}$$

where S_D is the first order kinetics of microbe decay rate (kg/m^3 s), k_d is the microbe decay rate (day^{-1}), k_{temp} is the temperature correction function (dimensionless), and *T* is the temperature (°C).

3 Numerical Solution

3.1 Boundary Conditions

This model simulate multiphase flow and multicomponent transports based on ANSYS Fluent platform. Aerobic biodegradation model is defined by User Defined Functions (UDFs) which is programmed in C language and inserted into the model to calculate aerobic processes.

As shown in Fig. 1, the top surface of landfills is cover system and is set as a zero flux boundary. Leachate collection system (LCS) is set as 0 Pa (free-draining). VWs of aeration systems usually inject air with a specific pressure. The screen of air inlet is the part of VWs sidewall, where H_s is the length of it, H_w is the length of VWs and H_0 is the height of landfill. The model is 30 m high and 60 in diameter. When the ratio of screen length to VWs length is not the object to study, VWs length is equal to 0.5 H_0 and screen length is 1.5 m.

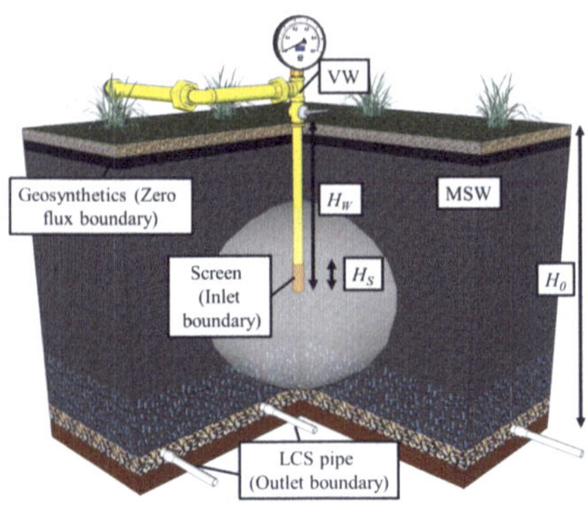

Fig. 1. Geometry model of aerobic landfills.

3.2 Parameter Selection

The hydraulic and waste property parameters of this model were selected according to Table 1. Intrinsic permeability k_i (m^2) is the most important parameter that affects fluid flow, which is obtained from in situ or laboratory tests.

The values of initial biodegradation parameters were selected according to McDougall [1], and the parameters of kinetic were set based on the work of Fytanidis and Voudrias [3]. Table 2 displays default biodegradation parameters in this study.

3.3 Model Validation

A lysimeter test (L-1) from Lee et al. which monitored the oxidation process was simulated by this model [8]. Figure 2 shows that the results of this model correspond to the results of experiment and the proposed model simulates the aerobic process well.

Table 1. Hydraulic parameters and values used in this study.

Parameter	Values	Reference
ρ_d (kg/m^3)	600	Powrie and Beaven [6]
ρ_l (kg/m^3)	1000	
k_i (m^2)	$1.0 \times 10^{-11} \sim 1.5 \times 10^{-13}$	
n_O	$k_V = 1.55 \times 10^{-11} \frac{n^3}{(1-n)^3}$	
θ_L	0.2	
n_{vg}	2.4	White et al. [2]
m_{vg}	0.58	
α_{vg} (kPa^{-1})	1.5	
α_{lr}	0.15	

Table 2. Biochemical parameters and values used in this study.

Parameter	Values	Reference
r_{hmax} (kg m^{-3} s^{-1})	2.89×10^{-6}	McDougall [1]
t	0.7	
c_{VFA0} (kg m^{-3})	1.00	
c_{MB0} (kg m^{-3})	0.25	
C_{CO_20} (%)	38.0	
k_m (kg m^{-3})	4.0	Kim et al. [7]
k_d (day^{-1})	0.05	

The reason of slight discrepancy is that the waste used for lysimeter test may is nonhomogeneous in degradable matter concentration.

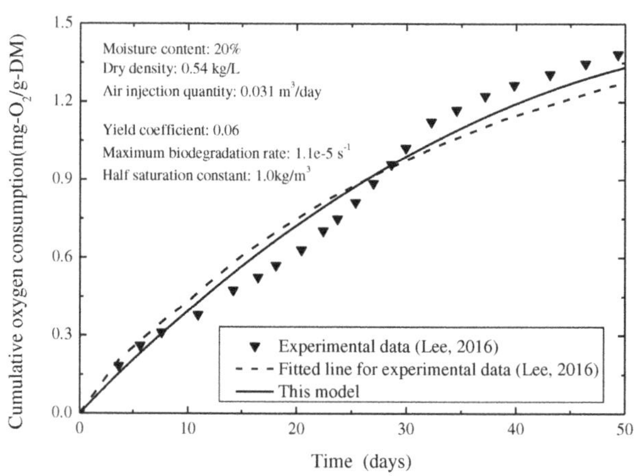

Fig. 2. Comparison of aerobic behavior between this model and previous research.

4 Results and Discussion

4.1 Intrinsic Permeability and Aeration Frequency

Intermittent aeration was focused in past decade and was proved that it is an efficient and low-cost method to establish aerobic bioreactor landfill. In addition, operators need pay attention to choose appropriate frequency of intermittent aeration which decides system efficiency. This study set the aeration time to constant (50% of operating time) and quantified the impact of leachate backflow. Leachate backflow occurs when the aeration pressure source is closed and leachate runs back to vertical wells. In addition, leachate backflow obstructs the oxygen flow in landfills. As shown in Fig. 3, the cumulative oxygen consumption of 3 h-aeration/6 h-total time is 5.3×10^6 L/per-well in the third hundred days, and the value decreases with the decreasing of aeration frequency. Leachate permeates along the reverse direction of VWs when aeration system is launched, and leachate flows back along the direction of VWs to achieve pore pressure balance during the interval when pressure source is closed in landfills.

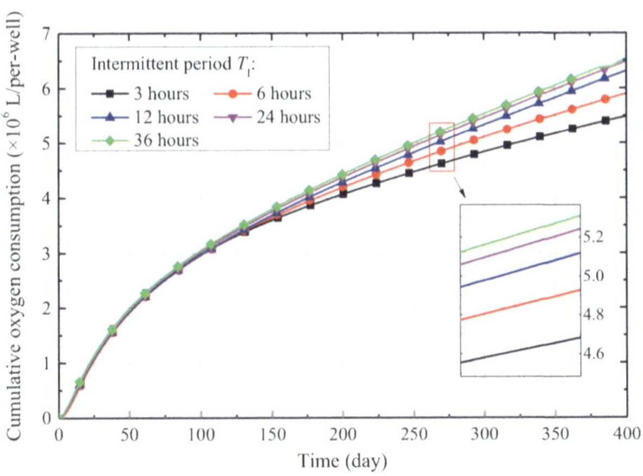

Fig. 3. Variation of cumulative oxygen consumption with time under different intermittent period.

4.2 Ratio of Screen Length to VWs Length

In in situ landfills, VWs is usually conducted by inserting PVC pipe after drilling. As the inlet of air, the bottom part of VWs is perforated. The ratio of screen length to VWs length (H_S/H_W) is a key parameter for aeration systems. Figure 4 shows the variation of influence radius under different H_S/H_W and the pressure at the influence radius distance. The influence radius is about 25 m and increases slowly, but the pressure at the influence radius distance is increase greatly with H_S/H_W. Therefore, when designing aeration system, increasing the screen length is useless and the variety of pressure makes the landfill more unsafe.

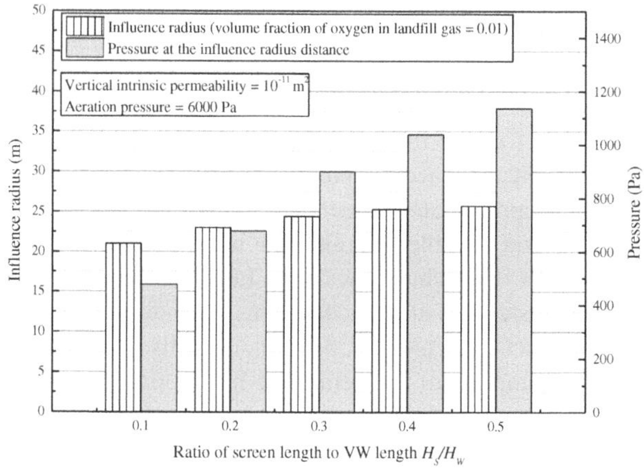

Fig. 4. Variation of influence radius under different HS/HW and the pressure at the influence radius distance.

5 Conclusion

A 3D multiphase flow and multicomponent transport model was used for modeling of aerobic hybrid bioreactor landfills, and the results obtained were in good agreement with previously published data.

References

1. Reddy KR, Kumar G, Giri RK (2017) Influence of dynamic coupled hydro-bio-mechanical processes on response of municipal solid waste and liner system in bioreactor landfills. Waste Manag 63:143–160
2. Fytanidis DK, Voudrias EA (2014) Numerical simulation of landfill aeration using computational fluid dynamics. Waste Manag 34(4):804–816
3. McDougall J (2007) A hydro-bio-mechanical model for settlement and other behaviour in landfilled waste. Comput Geotech 34(4):229–246
4. McDougall JR, Pyrah IC (1999) Moisture effects in a biodegradation model for waste refuse. In: The seventh international waste management and landfill symposium, pp 59–66
5. Reichel T, Haarstrick A, Hempel DC (2005) Modeling long-term landfill emission–a segregated landfill model. In: Proceedings Sardinia 2005, tenth international waste management and landfill symposium S. Margherita di Pula, Cagliari, Italy; 3–7 October 2005
6. Powrie W, Beaven RP (1999) Hydraulic properties of household waste and implications for landfills. Proc Inst Civil Eng-Geotech Eng 137(4):235–237
7. Kim SY, Tojo Y, Matsuto T (2007) Compartment model of aerobic and anaerobic biodegradation in a municipal solid waste landfil. Waste Manag Res 25(6):524–537
8. Lee NH, Park JK, Kang JY, Kang JH (2016) Method to estimate the required oxygen amount and aeration period for the completion of landfill aeration. J Mater Cycles Waste Manag 18 (4):695–702

Impact of Moisture Mobility on Compression Characteristics of MSW Through Soil Structure Measurements

M. V. Shah(✉) and Abhijit J. Brahmbhatt

Department of Applied Mechanics,
L. D. College of Engineering, Ahmedabad, India
mvs2212@yahoo.co.in, brahmbhatt.abhijit@gmail.com

Abstract. The aim of the present research work is to characterize in situ municipal solid waste (MSW) of Pirana landfill, Ahmedabad and study the flow of water through MSW. A series of compaction test were carried out on dry and wet side of optimum and its structural mobility was determined. Suction characteristics, compressibility characteristics and soil microscopy under constant mobility of moisture and its variational flow path into solid surface and sub-surface was evaluated. The results indicate that structural compactness of solid particles which are covered with moisture and chemical diffraction of partial organic matter and its overall effect can be well studied through this type of approach. Suction of MSW is affecting the compression characteristics of MSW and soil microscopy helped in understanding role of drying and wetting under constant degree of compactness and its effect on MSW as hydrodynamic force.

Keywords: Municipal solid waste · Soil suction · Compressibility
Soil microscopy

1 Introduction

Unsaturated soils are often found in numerous portions of the world, mainly at shallow depths from the ground surface and in arid and semi-arid provinces where the ground water table noticeably is at a greater depth (Fredlund and Rahardjo 1993). Suction is one of the most significant parameter relating the moisture condition of unsaturated soils (Bulut et al. 2001). The presence of water and air within the openings between the soil atoms produces capillarity effects which formed suction. There are very few literatures presented the classification of municipal solid waste (MSW) as waste has dissimilar properties at each place and very diverse minerals were present in it. Also, very few were described its suction and compressibility characteristics at different moisture content.

The moisture content plays a noteworthy role in properties and particle arrangement in soils. Thus, it is very much dynamic to understand the performance of MSW compressed at various moisture content for better understanding of its suction as well as compressibility parameters.

The work previously carried out on these waste shows that it fits to nearly silty sand (SM) or poorly graded sand (SP) group of soil as per the IS: 1498-1970 and the waste

L. Zhan et al. (Eds.): ICEG 2018, ESE, pp. 231–238, 2019.
https://doi.org/10.1007/978-981-13-2224-2_29

comprises 70–80% of granulated particles. The MSW used here is segregated at site and particles more than 10 mm size were removed.

In this study the compaction parameters of waste were carried out. Further, the suction and compressibility of solid waste were measured at Optimum moisture content (OMC) and also on dry side and wet side of optimum to imitate its behavior underneath various climatic variations. The total, matric as well as osmotic suction of MSW was measured using filter paper method using ASTM D5298-10 for both soaked as well as un-soaked condition. The sample was soaked into water for 24 h to reflect the behavior of compacted mass of MSW under submerge condition. Further, the void formation and void space in the compressed sample was observed by means of microscope of large magnification factor to observe the impact of moisture in it and the substances present in it.

The gained parameters (suction and compressibility) were compared with silty sand having low plasticity and low FSI (Free swell index) and typically encountered in field to appreciate the influence of organic and inorganic substances present in MSW and to evaluate its suitability as a filler material.

2 Experimental Programs

Municipal solid waste was first characterized by sieve analysis and the results shows that it belongs to SP group of soil as per the IS: 1498-1970. Also, the soil is identified as Silty sand and it found out as non plastic and of low FSI.

Table 1 shows the index and engineering properties of silty sand and MSW, it shows that the Optimum Moisture Content of MSW is 18.5% and that is for silty sand is 12.5%. Thus, to obtained the compressibility and suction parameters and for preparing the samples for soil microscopy, total five different moisture content were selected: two on dry side (−4% OMC, −2% OMC), one at OMC and two on wet side (+2% OMC, +4% OMC) of optimum. The MSW considered here is 10 years old it undergoes several physico-chemical variations along with disintegration over climatic dissimilarity and contains organic and in organic matters in it.

Table 1. Index and engineering properties of MSW and silty sand

Sr. No.	Description	Indian standards	MSW	Silty sand
1	Specific gravity	IS:2720-3	G = 2.25	G = 2.64
2	Grain size distribution	IS:2720-4	cu = 3.428, cc = 1.166	cu = 4.65, cc = 2.13
3	Soil type	IS:1498-1970	Poorly graded sand (SP)	Silty sand (SM)
4	Standard proctor test	IS:2720-7	OMC = 18.5%, MDD = 15.75 kN/m^3	OMC = 12.5%, MDD = 18.75 kN/m^3
5	Direct shear test	IS:2720-13	C = 5.88 kN/m^2, ø = 33°	C = 19.62 kN/m^2, ø = 31°

The proximate analysis and EDAX analysis is carried out on MSW sample for determining the Ash content and elements present in it. The results of proximate analysis and EDAX analysis were shown in Tables 2 and 3 respectively.

Table 2. Proximate analysis of MSW

Sr. No.	Test parameters	Test method	Test results (%)
1	Moisture	1350 (Part 1) 1984	0.80
2	Volatile matters		11.60
3	Ash		86.30
4	Fixed carbon		1.30

Table 3. Elements presented in MSW

Elements	Weight (%)	Elements	Weight (%)
Oxygen (O)	60.71	Chlorine (Cl)	1.02
Magnesium (Mg)	1.94	Potassium (K)	1.49
Aluminium (Al)	4.55	Calcium (Ca)	9.13
Silicon (Si)	15.84	Iron (Fe)	4.42
Sulphur (S)	0.90	**Total**	**100**

2.1 Suction Characteristics of MSW

Mobility of moisture through the solid particles required to be observed thus, suction becomes important phenomena for the soils located above ground water table. In the present study the suction characteristics at different moisture content were measured for both soaked and un-soaked condition and the values obtained for MSW and Silty sand were shown in Tables 4 and 5 respectively.

Table 4. Suction parameters for MSW

Soil type	Water content (%)	Un-soaked condition			Soaked condition		
		Ψ (kPa)	$u_a - u_w$ (kPa)	π (kPa)	Ψ (kPa)	$u_a - u_w$ (kPa)	π (kPa)
MSW	14.5	537.03	134.90	402.13	128.82	26.92	101.90
	16.5	512.86	120.22	392.64	102.33	22.91	79.42
	18.5	488.65	109.65	379.00	91.20	19.50	71.70
	20.5	407.38	107.15	300.23	75.86	16.60	59.26
	22.5	311.13	93.33	237.80	58.88	13.80	45.08

Table 5. Suction parameters for Silty sand

Soil type	Water content (%)	Un-soaked condition			Soaked condition		
		Ψ (kPa)	$u_a - u_w$ (kPa)	π (kPa)	Ψ (kPa)	$u_a - u_w$ (kPa)	π (kPa)
Silty sand	8.5	472.45	385.65	86.80	123.34	94.44	28.90
	10.5	448.37	369.78	78.59	98.78	74.45	24.33
	12.5	426.58	354.81	71.77	87.10	64.57	22.53
	14.5	386.85	324.95	61.90	70.32	51.52	18.80
	16.5	344.58	288.05	56.53	53.31	35.67	17.64

Total suction (Ψ) comprise two parts: matric suction ($u_a - u_w$) and osmotic suction (π). Results clearly shows that the suction reduces with increment in moisture content. There is a greater decrease in suction values of soaked samples as it is subjected to 24 h of soaking to imitate the submergence condition of compressed mass. Suction characteristics helps in understanding capillary moisture retention capacity and its overall influence on compressibility parameters.

Figures 1 and 2 shows the nature of suction parameters over the short range of moisture for MSW for un-soaked condition and soaked condition respectively. It demonstrate that the osmotic suction is greater than matric suction for MSW due to the organic matters present in it.

Figures 3 and 4 demonstrates the nature of suction parameters for silty sand for un-soaked and soaked condition respectively. Total suction for silty sand and MSW were nearly similar. The suction parameters for soaked condition were smaller than that for un-soaked condition because of the increment in moisture during 24 h of soaking.

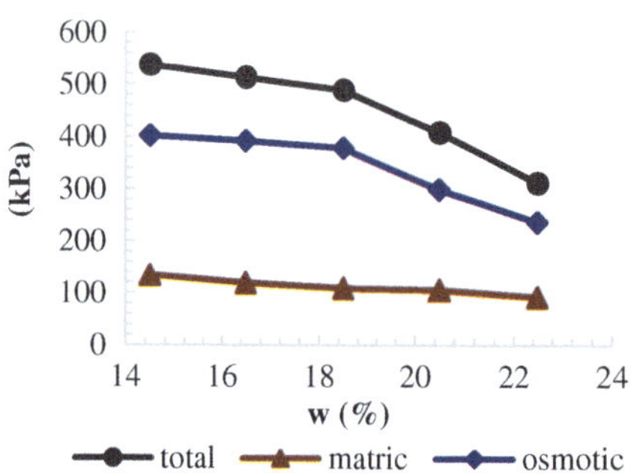

Fig. 1. Ψ vs. w of MSW for un-soaked condition

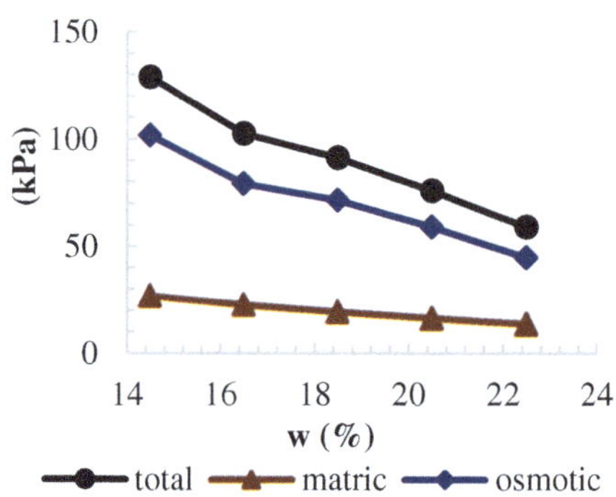

Fig. 2. Ψ vs. w of MSW for soaked condition

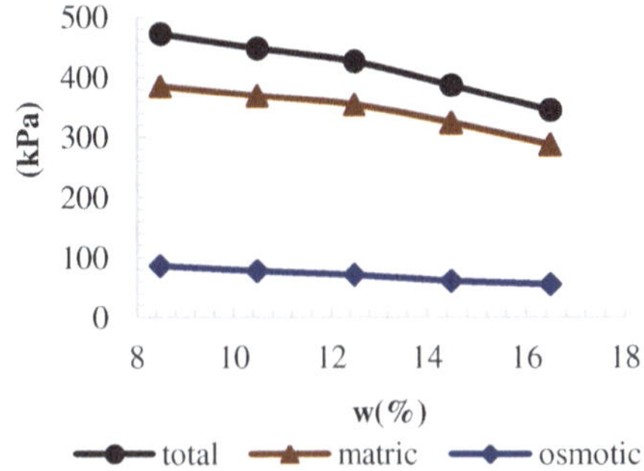

Fig. 3. Ψ vs. w of Silty sand for un-soaked condition

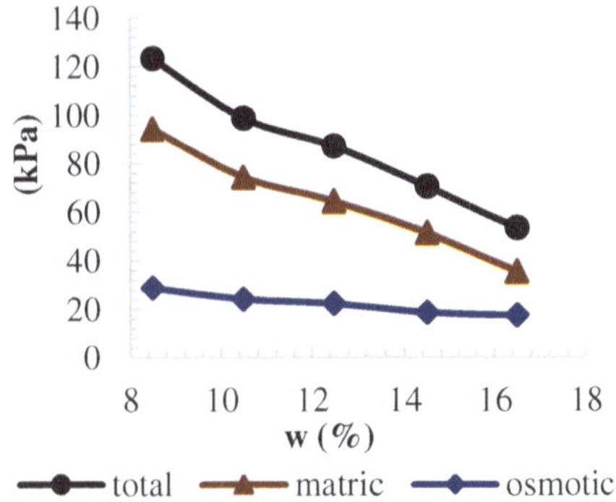

Fig. 4. Ψ vs. w Silty sand for un-soaked condition

2.2 Compressibility Characteristics of MSW

Compressibility of soil depends on compaction of mass and that is affected by moisture dissimilarity, thus moisture difference into compressed mass affects the compressibility of MSW as the particles of MSW involved with each other due to existence of organic matter and moisture in it. To analyse the impact of moisture variation on compressibility, the compression index (Cc) for each moisture variation was determined and compared the characteristics with soil at optimum moisture content.

Figure 5 shows the effect of moisture variation on the compression characteristics of MSW and that values of compression index (Cc) were compared with silty sand. The Cc values were obtained using consolidation test procedure as per IS: 2720-15 and these values were found suitable as a fill material.

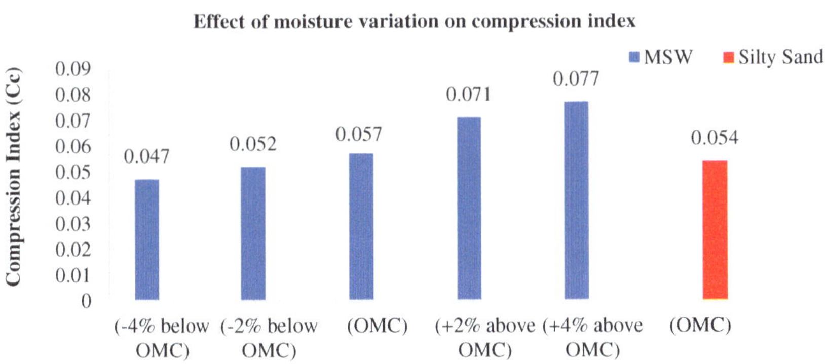

Fig. 5. Effect of moisture variation on compression index

2.3 Soil Microscopy

The effect of moisture movement and organic substance existing in MSW on compressibility parameters essential to be examine and for that the soil microscopy was carried out on various samples of MSW and silty sand. The soil microscopy provides better information concerning segregation of solid particles and void space in compressed mass. Also, the impact of moisture movement was observed as the samples of different moisture content were observed.

The optical microscope used for the observation of compressed structure was of magnification factor 50X. 38 mm dia. samples were prepared at variable moisture content and analysed under microscope. The 2D images of soil samples were rehabilitated to 3D images using Helicon focus software.

Figures 6 and 7 shows the 2D and 3D images of MSW sample respectively and Figs. 8 and 9 demonstrates the 2D and 3D images for silty sand respectively. The microscopical images were judgmentally observed for moisture mobility in vacuums.

Fig. 6. 2D image of MSW

Fig. 7. 3D image of MSW

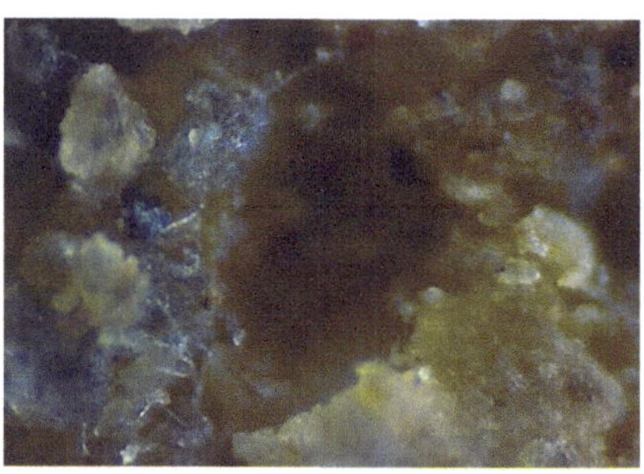

Fig. 8. 2D image of Silty sand

Fig. 9. 3D image of Silty sand

As far as the voids construction is concern there were comparatively more voids present in compressed sample of MSW than that of silty sand at OMC. However, there is decrement in void size as drifting towards wet side and more constant structural arrangement of elements were observed. After the wet samples observed under microscope, these samples were dry overnight for moisture removal and analysed for the cracking pattern developed due to drying of samples. The results shows that the cracks formed and widen as the moisture was removed.

3 Conclusion

The study exposes the importance of suction characteristics computed for MSW at OMC and dry and wet side of optimum. The unconventionality observed in compressibility characteristics and suction parameters were due to mobility of moisture into the solid particles, which is playing dynamic role and generating viscous and non-viscous layer on atoms due to the organic matters present in this MSW. As the MSW considered was about 10 years old it undergoes several physico-chemical variations along with disintegration over climatic dissimilarity thus, suction characteristics becomes identical. The following conclusions were made:

- The suction of MSW and silty sand reduces as moisture content increases and the same nature of suction parameters observed for soaked samples. Thus, MSW shows similar suction–moisture relationship as standard soils (sand, silt, clay, etc.).
- Total suction of MSW is comparable to silty sand. However, the osmotic suction is greater than matric suction for MSW due to the organic matter present in it.
- At OMC, the values of compression index (Cc) for MSW and silty sand were comparable and the influence of moisture mobility was observed by determining the compressibility characteristics at variable moisture content and the increment is observed in Cc values as the moisture content increases.
- The MSW sample comprises larger void compared to Silty sand and as moving towards wet side the void size was reduced.
- The dry samples analysed using microscope displays that cracks become more broader as the samples were dry overnight.
- Soil microscopy facilitated in understanding role of wetting and drying under constant degree of compression and its effect on MSW.
- The movement of moisture plays actual role in the arrangement of elements in the compressed structure. The cracking pattern was varied consequently as the particle separation takes place due to mobility of moisture.
- The results indicate that structural compactness of solid particles which are layered with moisture and chemical diffraction of partial organic matter and its overall effect can be well studied through this type of approach.

The above study exposes the engineering performance of Pirana landfill material under variable moisture content. The study of suction and compressibility characteristics of MSW demonstrate that its performance would be capable as a fill material if the material is converted to inert material after addition of either fly ash, lime, cement or construction debris, so that the reactive effects can be invalidate.

Acknowledgements. The authors are thankful to Dr. G. P. Vadodaria, principal of L D College of Engineering for providing the facilities of institution and motivation towards study. The authors also wish to acknowledge Dr. Kannan Iyer, Assistant Professor of Institute of Infrastructure Technology Research and Management for his esteemed guidance. Image courtesy: R&D department, Dharmsinh Desai University, Nadiad.

References

1. ASTM D5298-10 Standard test method for measurement of soil potential (suction) using filter paper, American Society of Testing and Materials, Annual Book of ASTM Standards
2. Bulut R, Lytton RL, Wray WK (2001) Soil suction measurements by filter paper. ASCE Civil Eng Conf 39:243–261
3. Çokça E, Tilgen HP (2010) Shear strength-suction relationship of compacted Ankara clay. Appl Clay Sci 49:400–404
4. Fredlund DG, Rahardjo H (1993) Soil mechanics for unsaturated soils. Wiley-Interscience Publications, Hoboken
5. Hanson JL, Yesiller N, Stockhausen SAV, Wong WW (2010) Compaction characteristics of municipal solid waste. J Geotech Geoenviron Eng 136:1095–1102
6. IS:1498-1970, Classification and determination of soils for general engineering purposes
7. IS:2720, Methods of test for soils (Part 3, Part 4, Part 7, Part 13 and Part 15)
8. IS:1350 (Part 1) 1984, Methods of test for coal and cock
9. Naveen BP, Sivapullaiah PV, Sitharam TG (2014) Compressibility and shear strength of dumped municipal solid waste. J Solid Waste Technol Manag 40–4:327–334
10. Rahardjo H, Leong EC (2006) Suction measurements. In: Fourth international conference on unsaturated soils, pp 81–104
11. Sreedeep S, Singh DN (2005) Methodology for determination of osmotic suction of soils. Geotech Geol Eng 24:1469–1479

A Constitutive Model for the Stress-Strain-Time Behavior of Municipal Solid Waste

Wu Gao[1(✉)] and Edward Kavazanjian Jr.[2]

[1] MOE Key Laboratory of Soft Soils and Geoenvironmental Engineering, Zhejiang University, Hangzhou, China
gaowu_ed@163.com

[2] Center for Bio-Mediated and Bio-Inspired Geotechnics, School of Sustainable Engineering and the Built Environment, Arizona State University, Tempe, USA

Abstract. A constitutive model for the stress-strain-time behavior of municipal solid waste (MSW) that accounts for waste degradation has been developed. The model is developed within the critical state soil mechanics framework. The model decomposes the total strain into a time-independent elasto-plastic strain and a time-dependent plastic strain. The time-independent strain, including both the elastic part and the plastic part, is characterized by the ellipsoidal yield surface of the Modified Cam-Clay model using the normality condition and the law of associated flow. The time-dependent strain consists of a mechanical creep part and a biodegradation part. The mechanical creep part is evaluated by Taylor's secondary compression equation, the normality rule, and the Modified Cam-Clay yield surface. Biodegradation strain is assumed to be isotropic and is characterized by first order kinetics and mass that decays loss exponentially with time. The decrease in the compressibility of MSW as it biodegrades is accounted for by changing the slope of normal consolidation line and the value of the specific volume at the reference pressure based upon the mass loss due to decomposition. A method for obtaining the model parameters is provided.

Keywords: Municipal solid waste · Constitutive model · Creep
Biodegradation · Compression

1 Introduction

The settlement of the waste mass plays an important role in the performance of many elements of a landfill containment system. Due to the high compressibility of municipal solid waste (MSW), the final vertical displacement (settlement) of a MSW landfill can reach 25–50% of the initial thickness of the MSW body. While this settlement can increase landfill storage capacity and extend its service life, it may damage many of the landfill subsystems, including side slope liner system. Significant tensile loads may be applied to geosynthetic components of a side slope liner in a landfill by MSW settlement. Initial mechanical compression, mechanical creep, and decomposition of organic waste constituents all contribute to this settlement over the life of the landfill.

Practical one-dimensional MSW compression models are available for predicting waste settlement. However, one-dimensional models cannot calculate the horizontal

L. Zhan et al. (Eds.): ICEG 2018, ESE, pp. 239–247, 2019.
https://doi.org/10.1007/978-981-13-2224-2_30

deformation of the MSW body that accompanies vertical settlement or be used to compute strains induced in a side slope liner system by waste settlement. A two-dimensional constitutive model that can be implemented in a finite element or finite difference numerical model is required to calculate loads on landfill subsystems, e.g., to calculate the load induced by settlement on side slope liners. To be realistic, the constitutive model should consider the effect of waste degradation. This is of particularly importance when evaluating the performance of high kitchen waste landfills considering the contribution of degradation to the total settlement. Therefore, a two-dimensional constitutive model, suitable for implementation in a finite element or finite difference model was developed to describe the stress-strain-time behavior of MSW including the effect of degradation.

2 Model Development

The constitutive model developed herein is an extension of the Modified Cam-Clay model [1] and includes consideration of the time-dependent strain caused by both mechanical creep and biodegradation.

In the constitutive model, the mean effective stress p' and the deviatoric stress q state parameters are defined as follows:

$$p' = \frac{1}{3}(\sigma_1' + \sigma_2' + \sigma_3') \tag{1}$$

$$q = \frac{1}{\sqrt{2}}[(\sigma_1' - \sigma_2')^2 + (\sigma_2' - \sigma_3')^2 + (\sigma_1' - \sigma_3')^2]^{\frac{1}{2}} \tag{2}$$

where σ_1', σ_2', and σ_3' are the effective principal stresses.

The model uses the Modified Cam-Clay ellipsoidal yield surface of the form:

$$F = \frac{q^2}{M^2} + p'(p' - p_c') = 0 \tag{3.1}$$

where p_c' is the effective pre-consolidation stress, and M is a constant. The value of M is related to the effective friction angle φ' by the expression:

$$M = \frac{6 \sin \varphi'}{3 - \sin \varphi'} \tag{3.2}$$

The evaluation of deviatoric strains follows the conventional Modified Cam-Clay model. The irrecoverable (plastic) deviatoric strain is obtained based upon the plastic volumetric strain and the normality condition following the law of associated flow.

The total volumetric strain increment $d\varepsilon_v$ may be decomposed as follows:

$$d\varepsilon_v = d\varepsilon_v^e + d\varepsilon_v^p + d\varepsilon_v^{tp} \quad (4.1)$$

$$d\varepsilon_v^{tp} = d\varepsilon_c^t + d\varepsilon_d^t \quad (4.2)$$

where $d\varepsilon_v^e$ is the time-independent elastic volumetric strain increment, $d\varepsilon_v^p$ is the time-independent plastic volumetric strain increment, and $d\varepsilon_v^{tp}$ is the time-dependent plastic volumetric strain increment. The time-dependent plastic volumetric strain $d\varepsilon_v^{tp}$ consists of two parts: the volumetric mechanical creep strain increment $d\varepsilon_c^t$ and the volumetric biodegradation strain increment $d\varepsilon_d^t$.

Figure 1 illustrates how the time-independent compression behavior of MSW is influenced by waste degradation in the model. Biodegradation over time t reduces the specific volume at a unit effective consolidation pressure, p_l', from its value at point R, v_{li}, to the value at point R', v_{lt}. Assuming the ultimate specific volume v_u at an ultimate effective mean normal stress, p_u' (point U on Fig. 1) is unchanged, the change in the specific volume at a unit effective consolidation pressure causes the slope of normal consolidation line to decreases from the initial value λ_i to λ_t. This decrease in slope represents the decrease in the compressibility of MSW due to biodegradation. Biodegradation is assumed not to effect the recoverable elastic strain and thus the slope of elastic rebound line κ is assumed to be unchanged.

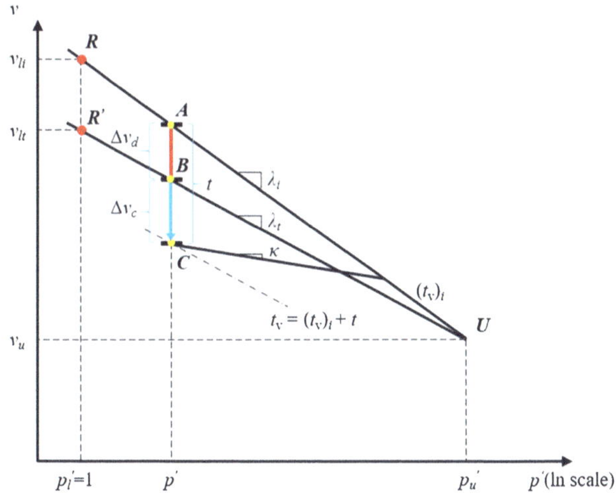

Fig. 1. Compression behavior of MSW (after Schofield and Wroth [1])

An element of MSW that starts at point A on Fig. 1 will move to point C over time t due to mechanical creep and biodegradation under a constant normal effective stress, p', due to the combined effect of mechanical creep and biodegradation. During time t, the normal consolidation line changes from a line that passes through point A to a line that passes through point B and the slope of the normal consolidation line changes to λ_t. The location of point B can be established based upon the change in specific volume due to biodegradation, Δv_d. After point B is established, point C can be found based upon the change in volume due to mechanical creep, Δv_c.

Equations (5) and (6) describe the transition in specific volumes from v_{li} to v_{lt} and the transition in the slope of the normal consolidation line from λ_i to λ_t.

$$v_{lt} = v_{li} - \frac{\Delta v_d (p'_u - 1)}{p'_u - p'} \tag{5}$$

$$\lambda_t = (v_{lt} - v_u)/\ln p'_u = \left[v_{li} - \frac{\Delta v_d (p'_u - 1)}{p'_u - p'} - v_u\right] \Big/ \ln p'_u \tag{6}$$

3 Volumetric Strains

3.1 Time-Independent Elastic Volumetric Strain

Based on the Modified Cam-Clay model, the recoverable time-independent elastic volumetric strain accompanying any changes in p' can be found according to the following expressions:

$$d\varepsilon_v^e = \frac{\kappa}{v_0}\frac{dp'}{p'} = \frac{dp'}{K} \tag{7.1}$$

$$K = \frac{v_0}{\kappa} p' \tag{7.2}$$

where K is the elastic bulk modulus of MSW and v_0 is the initial specific volume of MSW.

It should be mentioned that the recoverable time-independent deviatoric strain that accompanies any change in q can be found according to the following expression:

$$d\varepsilon_s^e = \frac{dq}{3G} \tag{8}$$

where G is the shear modulus of MSW and can be found as:

$$G = \frac{3(1-2\mu)}{2(1+\mu)} K = \frac{3(1-2\mu)}{2(1+\mu)} \frac{v_0}{\kappa} p' \tag{9}$$

in which μ is Poisson's ratio for MSW.

3.2 Time-Independent Plastic Volumetric Strain

The time-independent plastic volumetric strain increment $d\varepsilon_{\mathrm{v}}^{p}$ can be obtained by:

$$d\varepsilon_{\mathrm{v}}^{p}=\frac{\lambda_t-\kappa}{v_0}\frac{dp_c'}{p_c'} \tag{10}$$

3.3 Time-Dependent Plastic Volumetric Strains

Volumetric Mechanical Creep Strain. The volumetric mechanical creep strain increment $d\varepsilon_{c}^{t}$ is calculated based on Taylor's secondary compression equation [2]

$$d\varepsilon_{c}^{t}=\frac{\psi}{v_0 t_{\mathrm{v}}}dt \tag{11.1}$$

$$t_{\mathrm{v}}=(t_{\mathrm{v}})_i+t \tag{11.2}$$

where ψ is the secondary compression coefficient on a ln scale, $(t_{\mathrm{v}})_i$ is the reference time for the normal consolidation line (assumed to be unchanged by biodegradation), and t_{v} is the volumetric time.

Volumetric Biodegradation Creep Strain. The time-dependent volumetric biodegradation strain increment $d\varepsilon_{d}^{t}$ is assumed isotropic and calculated by the function proposed by Park and Lee [3]:

$$d\varepsilon_{d}^{t}=kE_{dg}e^{-kt}dt \tag{12}$$

where k is the rate constant for biodegradation of MSW and E_{dg} is the total amount of strain that can occur because of biodegradation.

4 Hardening Rule

As described by Eq. (13), there are three sources that contribute to the parameter p_c' that describes the hardening of MSW (i.e., the size of the yield surface): the time-independent plastic volumetric strain $\varepsilon_{\mathrm{v}}^{p}$, the time-dependent volumetric mechanical creep strain ε_{c}^{t}, and the volumetric strain due to the biodegradation ε_{d}^{t} [4]:

$$p_c'=p_c'(\varepsilon_{\mathrm{v}}^{p},t_{\mathrm{v}},V_{\mathrm{S}}) \tag{13}$$

where V_{S} is the total volume of the solids, including both the decomposable solids and the inert solids [5].

The increment of p_c' is given by:

$$dp_c'=dp_c'^{1}+dp_c'^{2}+dp_c'^{3} \tag{14.1}$$

where $dp_c'^1$ is the increment due to initial mechanical compression, $dp_c'^2$ is the contribution due to mechanical creep, and $dp_c'^3$ is the increment due to biodegradation. The values of $dp_c'^1$ and $dp_c'^2$ can be found as [4]:

$$dp_c'^1 = \frac{\partial p_c'}{\partial \varepsilon_v^p}\Big|_{(t_v, V_S)} d\varepsilon_v^p = \frac{v_0 p_c'}{\lambda_t - \kappa} d\varepsilon_v^p \tag{14.2}$$

$$dp_c'^2 = \frac{\partial p_c'}{\partial t_v}\Big|_{(\varepsilon_v^p, V_S)} dt = \frac{\psi p_c'}{(\lambda_t - \kappa) t_v} dt \tag{14.3}$$

Based on McDougall [6], the increment in the p_c' due to the biodegradation, $dp_c'^3$, can be defined by:

$$dp_c'^3 = \Omega\,(e_0 - \Lambda)\frac{dV_s}{V_s} \tag{14.4}$$

where e_0 is the initial void ratio upon placement, Ω is a decomposition hardening multiplier that relates magnitude of $dp_c'^3$ to the increment of volumetric strain, and Λ is the decomposition-induced void change parameter. Machado et al. [7] assumed that Λ is proportional to the specific total amount of solid already decomposed as follows:

$$\Lambda = -\Lambda^* \frac{\Delta m_S}{m_{Si}} \tag{15}$$

where m_{Si} is the initial dry mass of the solid in a unit mass of MSW upon placement, Δm_S is the change in dry mass, and Λ^* is the decomposition-induced void ratio change rate.

Decomposition of the decomposable solids is assumed to follow a first order kinetic model with rate constant k as follows:

$$m_D = m_{Di} e^{-kt} \tag{16}$$

where m_D is the dry mass of the decomposable solid in a unit mass of MSW and m_{Di} is the initial dry mass of the decomposable solid in a unit mass of MSW, i.e., the dry mass of decomposable solid upon placement.

Assuming, for simplicity, that the density of the decomposable solid ρ_D and the density of the inert solid ρ_I remain constant during biodegradation, Λ can be related to Λ^* as follows:

$$\Lambda = -\Lambda^* \frac{\Delta m_S}{m_{Si}} = \frac{\Lambda^* m_{Di}(1 - e^{-kt})}{m_{Si}} \tag{17}$$

The change in the volume of solids, dV_S, can be found as:

$$dV_S = \frac{dm_D}{\rho_D} = \frac{-k m_{Di} e^{-kt} dt}{\rho_D} \tag{18}$$

Therefore, $dp_c^{\prime 3}$ can be written as:

$$dp_c^{\prime 3} = \Omega \left[e_0 - \frac{\Lambda^* m_{Di}(1 - e^{-kt})}{m_{Di} + m_{Ii}} \right] \frac{\frac{-k m_{Di} e^{-kt} dt}{\rho_D}}{\frac{m_{Di} e^{-kt}}{\rho_D} + \frac{m_{Ii}}{\rho_I}} \tag{19}$$

where m_{Ii} is the initial dry mass of the inert solid in a unit mass of MSW upon placement.

In the end, the hardening rule is given by:

$$dp_c' = \frac{v_0 p_c'}{\lambda_t - \kappa} d\varepsilon_v^p + \frac{\psi p_c'}{(\lambda_t - \kappa) t_v} dt + \Omega \left[e_0 - \frac{\Lambda^* m_{Di}(1 - e^{-kt})}{m_{Di} + m_{Ii}} \right] \frac{\frac{-k m_{Di} e^{-kt} dt}{\rho_D}}{\frac{m_{Di} e^{-kt}}{\rho_D} + \frac{m_{Ii}}{\rho_I}} \tag{20}$$

5 Parameter determination

In addition to the parameters from the Modified Cam-Clay model as listed in Table 1, there are ten more parameters in the model proposed herein as presented in Table 2. The model parameters can be obtained from laboratory and/or field tests on MSW, MSW composition analysis, or based upon values reported in the literature.

Table 1. Model parameters from the Modified Cam-Clay model.

Parameter	Physical meaning	Method of evaluation
ρ	Dry density of MSW	MSW composition analysis
μ	Poisson's ratio for MSW	Value reported in the literature: 0.2–0.4 [8]; 0.33 [9]
M	Frictional constant	Estimated by Eq. (3.2)
λ_i	Slope of initial normal consolidation line of MSW	One-dimensional compression test of fresh MSW with an unload and reload cycle
λ_t	Slope of normal consolidation line of MSW at biodegradation time t	One-dimensional compression test of MSW at biodegradation time t, or estimated by Eq. (6)
κ	Slope of elastic rebound line	One-dimensional compression test of fresh MSW with an unload and reload cycle
p_l'	Reference pressure	Usually taken as 1 kPa
p_c'	Effective pre-consolidation	Estimated by the stress history
v_{li}	Specific volume at reference pressure on initial normal consolidation line	One-dimensional compression test of fresh MSW with an unload and reload cycle
v_{lt}	Specific volume at reference pressure on normal consolidation line at biodegradation time t	One-dimensional compression test of MSW at biodegradation time t, or estimated by Eq. (5)

Table 2. Model parameters introduced in this paper.

Parameter	Physical meaning	Method of evaluation
$(t_v)_i$	Initial volumetric time	Usually taken as 1 day
ψ	Secondary compression coefficient on a ln scale	Long-term compression tests
k	Rate constant for biodegradation of MSW	Long-term compression tests
E_{dg}	Total amount of strain that can occur because of biodegradation	Long-term compression tests
Ω	A decomposition hardening multiplier	Usually taken as 2.0 kPa [6]
Λ^*	A decomposition-induced void change parameter	Usually taken as 20.0 [7]
m_{Ii}	Initial dry mass of the inert solid in a unit wet mass of MSW	MSW composition analysis
m_{Di}	Initial dry mass of the decomposable solid in a unit wet mass of MSW	MSW composition analysis
ρ_I	Dry density of the inert solid of MSW	MSW composition analysis
ρ_D	Dry density of the decomposable solid of MSW	MSW composition analysis

Due to limited space in this paper, validation of the model using laboratory and/or field data is not given in this paper. However, validation of a simplified version of the model that does not consider the decrease of the compressibility of MSW due to biodegradation, i.e., the change in the slope of the normal consolidation line, has been conducted using experimental data from the literature [10].

6 Conclusion

A new constitutive model has been developed to describe the stress-strain-time behavior of MSW. The model is an extension of the Modified Cam-Clay model. The model includes long-term compression due to both mechanical creep and biodegradation. The decrease in the compressibility of MSW due to biodegradation is accounted for in the model. The model parameters can be obtained from laboratory and/or filed tests on MSW, MSW composition analysis, or values reported in the literature. This model is suitable for use in two-dimensional finite element or finite difference models for analysis of the impacts of the settlement of MSW on the performance of landfill subsystems, including its effect of the integrity of side slope liner system.

Acknowledgements. The National Basic Research Program of China (973 Program) (No. 2012CB719806) and the China Scholarship Council are gratefully acknowledged.

References

1. Schofield AN, Wroth CP (1968) Critical state soil mechanics. McGraw-Hill, London
2. Taylor DW (1948) Fundamental of soil mechanics. Wiley, New York
3. Park HI, Lee SR (2002) Long-term settlement behaviour of MSW landfills with various fill ages. Waste Manag Res 20(3):259–268
4. Borja RI, Kavazanjian E Jr (1985) A constitutive model for the stress–strain–time behaviour of 'wet' clays. Geotechnique 35(3):283–298
5. McDougall JR, Pyrah IC (2004) Phase relations for decomposable soils. Geotechnique 54 (7):487–493
6. McDougall JR (2007) A hydro-bio-mechanical model for settlement and other behaviour in landfilled waste. Comput Geotech 34(4):229–246
7. Machado SL, Vilar OM, Carvalho MF (2008) Constitutive model for long term municipal solid waste mechanical behavior. Comput Geotech 35(5):775–790
8. Sun XL (2007) Characterization of deformation and strength for municipal solid waste. Ph. D. thesis, Dalian University of Technology, Dalian, China (**in Chinese**)
9. Matasovic N, Kavazanjian E Jr (1998) Cyclic characterization of OII landfill solid waste. J Geotech Geoenviron Eng 124(3):197–210
10. Gao W, Kavazanjian E Jr (2018) A constitutive model for municipal solid waste considering mechanical creep and biodegradation-induced compression. Can Geotech J (under review)

Estimation of Strength Parameters of Solid Waste Materials by Corn Penetration and Spiral Pile Pull Out Tests

Shimon Ideguchi, Kiyoshi Omine(✉), and Satoshi Sugimoto

Nagasaki University, 1-14 Bunkyo-Machi, Nagasaki 852-8521, Japan
omine@nagasaki-u.ac.jp

Abstract. There are many kinds of geomaterials such as mixtures with clay, silt, sand or gravel. Solid waste materials are also classified as a geomaterial. Generally, strength parameters of cohesion and frictional angle on the geomaterials are obtained from direct shear test or triaxial compression test. It is difficult to perform with ordinary laboratory tests on samples including a large size of soil particles or waste materials with different size and shape, especially on in-situ condition. The authors developed a large size direct shear test for solid waste material. However, it takes a time for preparing specimen and performing the in-situ test. In this study, the corn penetration test and the spiral pile pull out test were performed for estimating strength constants of various kinds of geomaterials as a simple method. Cohesion of geomaterial is obtained from spiral pile pull out test with different length of the pile. Internal frictional angle is calculated from the cohesion and test result of corn penetration test with Terzaghi bearing capacity formula. Estimated strength parameters by proposed method were compared with the values obtained from the direct shear tests. The applicability of this method is confirmed based on the test results of various geomaterials such as sand, decomposed granite soil, and solid waste materials in laboratory. This proposed method is also applied on in-situ ground.

Keywords: Spiral pile pull out · Corn penetration · In-situ direct shear test
Strength constant

1 Introduction

There are places where waste materials have been heaped up in a steep slope or illegally in Japan, and there is a possibility of collapsing. Generally, stability of solid waste ground has been analyzed by strength parameters on soil mechanics in past study (Zekkos et al., 2010; Towhata et al., 2010). Almost these experiments were performed in laboratory on specific condition. There are few data from in-situ test of solid waste ground. In-situ shear property of solid waste materials has not been clarified. Therefore, it is necessary to establish in-situ evaluation method of solid waste materials. Direct shear test or triaxial compression test is used as a conventional method for determining strength constants of cohesion and internal friction angle. The authors have developed in-situ direct shear apparatus in dimensions of 300 mm × 300 mm × 150 mm (Omine et al. 2014). However, it takes a time for cutting out a sample from the solid waste

L. Zhan et al. (Eds.): ICEG 2018, ESE, pp. 248–254, 2019.
https://doi.org/10.1007/978-981-13-2224-2_31

deposit without disturbance and 2–3 days for obtaining strength parameters of the ground.

In this study, a small size of spiral pile is used as well as a portable cone for a simple test for estimating cohesion and frictional angle of the geomaterials [2]. Spiral pile pull-out and cone penetration tests are performed on several type of soils including solid waste materials. Applicability of the proposed method is confirmed by comparison with test results of the direct shear test.

2 Test Method

2.1 Spiral Pile Pull-Out Test

In this study, a metal pile with a spiral tip is used. Since contact area between soil and metal part of the spiral pile is small, it is assumed that friction of the spiral pile for the pull-out test is correspond to friction of the soil. The dimensions of the spiral pile are shown in Fig. 1. In the test method, embedded depth is changed and peak value of each pull-out force is measured. Relationship between friction divided by the surface area of the pile and embedded depth is obtained. Since the frictional force is influenced by the embedded depth, a value corresponding to zero depth is taken as cohesion of the soil.

Fig. 1. Schematic diagram of spiral pile

2.2 Cone Penetration Test

Portable cone in test method for cone index is used and an electronic load meter is connected at the top. Figure 2 shows the dimensions of the cone. For the test method, cone penetration resistance at 50 mm in depth is measured and the resistance divided by the cone cross sectional area is regarded as bearing capacity of soil. Considering scatter of ground, the tests are performed several times and average resistance is obtained.

Fig. 2. Schematic diagram of portable cone

2.3 Strength Estimation Method

Strength constants are calculated by bearing capacity, q_c, obtained from the cone penetration test and friction obtained from the spiral pile pull-out test. As described in Sect. 2.1, cohesion is obtained as a value of friction at zero depth in the spiral pull-out test. Internal friction angle is obtained from the inverse calculation from cohesion and Terzagi's shallow bearing capacity formulas (1) with the embedded depth of zero.

$$q = \alpha c N_c + \beta \gamma_t B N_r + \gamma_t D_f N_q \tag{1}$$

q: bearing capacity, B: minimum width of foundation, D_f: embedded depth, c: cohesion, γ_t: unit weight of soil, N_c, N_r, N_q: coefficients of bearing capacity, α, β: shape coefficients

3 Direct Shear Test

Shear property of the waste was evaluated using direct shear apparatus developed by Omine et al. (2014). Direct shear apparatus for in-situ testing is shown in Fig. 3. This testing apparatus was developed as in-situ test for ground of solid waste material. It can be also used as direct shear test in laboratory. In the laboratory test, sample is put inside a shear box (width 300 mm, depth 300 mm, height 150 mm) and compacted into three layers using a rammer by adjusting to a predetermined density of the specimen. Shearing is done automatically using a motor, and the shear rate is about 1.67 mm/min. In the case that shear stress does not show a peak, shear displacement of 35 mm is regarded as a failure in accordance with JIS A 1216: 2009.

Fig. 3. In-situ direct shear apparatus

4 Test Result

The target materials were Toyoura Sand, Decomposed granite (density: 3 cases) and waste (laboratory test: 2 cases, in-situ: 4 cases). Composition of wastes is shown in Table 1. The cone penetration test was conducted five or six times for each sample and the results are shown in Table 2. Resistance of decomposed granite soil increases as the density increases. Resistance of the waste materials depends on the location of the samples. Some waste materials show large resistance regardless of small density. The reason is considered that waste materials contain massive material such as a rock.

Table 1. Composition of wastes

	Plastic (%)	Other fabric (%)	Glass, pottery (%)	Rock (%)	Wood (%)	Metal (%)	20 mm under (%)
Chiba lab test	2.66	2.94	10.8	7	0	0	76.6
Sendai lab test	3.12	2.88	10.7	7.3	0	0	76
Chiba 2 section	10.37	6.90	6.29	6.50	2.86	0.29	66.79
Chiba 6 section	2.68	3.04	15.01	14.79	0.06	0.19	64.24
Sendai A section	13.68	17.64	9.14	6.85	0.05	0.63	52.01
Sendai B section	13.47	13.68	7.38	7.54	0.16	0.10	57.67

Table 2. Result of cone penetration test

Samples	Wet density (g/cm^3)	Resistance (kN/m^2)
Toyoura sand	1.41	28.26
Decomposed granite soil (1.4 g/cm^3)	1.40	285.99
Decomposed granite soil (1.5 g/cm^3)	1.51	355.10
Decomposed granite soil (1.6 g/cm)	1.60	404.99
Waste (Chiba lab test)	1.46	403.00
Waste (Sendai lab test)	1.31	476.40
Waste (Chiba 2 section)	1.46	832.48
Waste (Chiba 6 section)	1.35	731.85
Waste (Sendai A section)	1.08	912.10
Waste (Sendai B section)	1.19	809.83

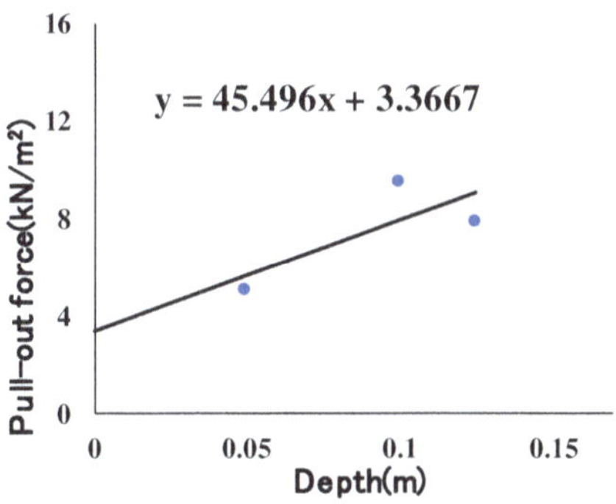

Fig. 4. Result of pull-out test (Chiba lab test)

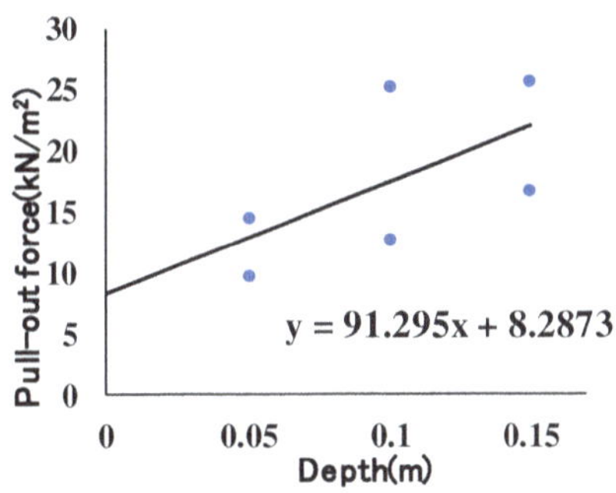

Fig. 5. Result of pull-out test (Sendai B section)

Figures 4 and 5 show typical results of pull-out tests on waste materials (Chiba laboratory, Sendai B section). The pull-out force increases as the depth increases. Although there is a scatter, the linear line is assumed. In this case, intercept of the line is assumed as cohesion of the materials.

Figure 6 shows the results of direct shear test of waste materials (Sendai laboratory). Shear stress of the waste showed no peak value and it tended to increase. In addition, the change in the volume showed a change like a negative dilatancy of loose sand as vertical pressure increases. Strength constants are estimated using the cohesion and internal friction angle calculated from the results of spiral pile pull-out test and cone penetration test.

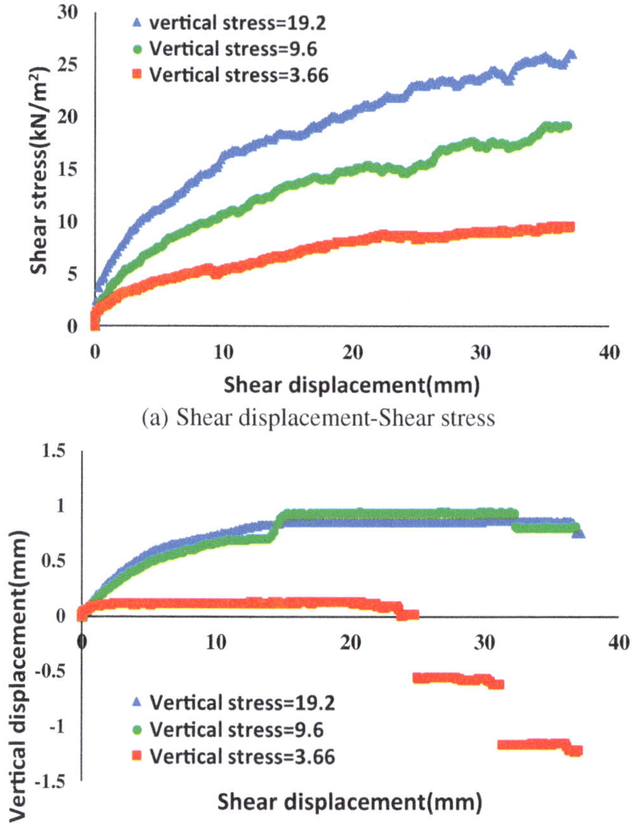

(a) Shear displacement-Shear stress

(b) Shear displacement-vertical displacement

Fig. 6. Relationship among shear displacement, shear stress and vertical displacement (Sendai lab test)

Figure 7 shows a comparison of the cohesion and internal friction angle obtained from the corn penetration/spiral pile pull-out test and the direct shear test. The strength parameters estimated from the simple tests have a relatively good agreement with those of direct shear test. It is considered that the proposed method is applicable to the various types of geomaterials as a rough estimation of the strength constants.

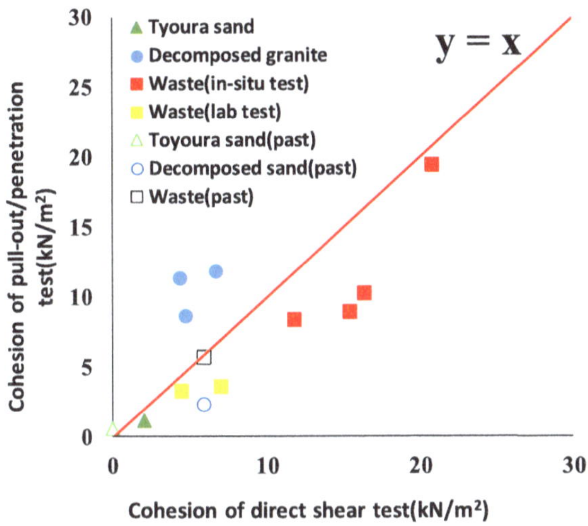

(a) Cohesion

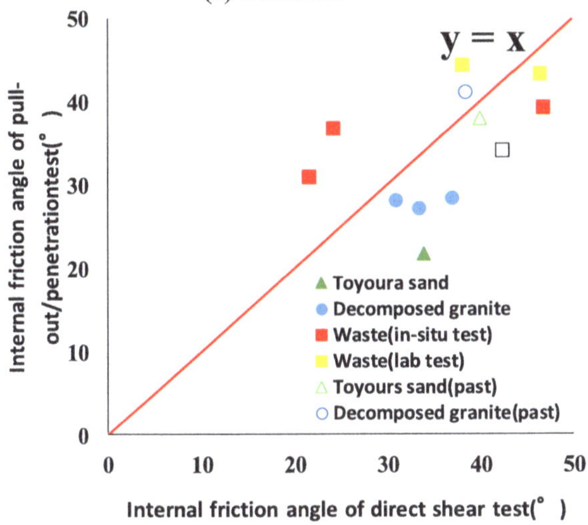

(b) Internal friction angle

Fig. 7. Comparison of intensity constants of direct shear test and spiral pile pull-out test/corn penetration test[1)]

5 Conclusion

In this study, we proposed a simple estimation method of strength constants (cohesion and internal friction angle) by spiral pile pull-out and cone penetration tests. Several types of soils and waste materials are used in a laboratory and in-situ tests. Validity of the proposed method is confirmed by comparison with test results of the direct shear test. It is considered that the proposed method is applicable for simple estimation of strength parameters. In addition, this method is suitable to apply to in-situ ground.

Acknowledgement. This research was supported by the Environment Research and Technology Development Fund (3-1707) of the Ministry of the Environment, Japan.

References

Zekkos Dimitrios, George A et al (2010) Large-scale direct shear testing of municipal solid waste. Waste Manag 30:1544–1555

Ideguchi S, Omine K, Sugimoto S (2018) In-situ strength estimation of slopes collapsed by heavy rain in northern part of Kyushu on July 2017. In: Proceedingd of the Japan national conference on geotechnical engineering, printing (in Japanese)

Omine K, Yamawaki A, Kawasaki M, Doi Y, Sugimoto S, Nagaura B (2014) Estimation of strength of solid waste materials by in-situ direct shear and Earth pressure tests. In: Proceedings of 7th international congress on environmental geotechnics, pp 1227–1233

Towhata I et al (2010) Laboratory test on mechanical properties of municipal solid waste. J Jpn Soc Civil Eng, pp 631–644 (in Japanese)

Geotechnical Engineering Properties of Solidified Sludge in Qizishan Landfill Site, China

Weian Lin(✉), Kaixi An, Yunmin Chen, and Liangtong Zhan

Key Laboratory of Soft Soils and Geoenvironmental Engineering, MOE
Department of Civil Engineering, Zhejiang University, Hangzhou 310027, China
linweian@zju.edu.cn

Abstract. In order to have a better understanding of solidified sludge in a sewage sludge disposal pit, field and laboratory investigation were conducted on the geotechnical properties of the solidified sludge at the Qizishan Landfill in Suzhou, China. The field investigation included borehole sampling. The laboratory test involved the measurements of water content, organic matter content and shear strength for the borehole samples taken from Qizishan landfill sludge pit. The tests results demonstrated that the cohesive intercept (c) of the solidified sludge mostly range from 15 to 60 kPa; friction angle (φ) mostly range from 6° to 40°. The unconfined compressive strength for sludge normally about 180 kPa, but the unconfined compressive strength for some cement-like samples up to 250–300 kPa. Above analysis are useful for the in-situ sludge reinforcement methods, such as first in-situ vacuum preloading combined with subsequent solidification treatment. And it can provide the necessary geotechnical parameter and partly basis for sludge disposal design.

Keywords: Solidified sludge · Consolidation · Geotechnical properties

1 Introduction

With the Chinese continuous progress of economy, there are rapid increasement in the production of sewage sludge. Sewage sludge is the residue of sewage treatment. There is a problem for advanced sewage treatment will lead the amount of sludge increased by 0.5–1 times. Thus, the improvement of sewage treatment efficiency inevitably leads to the increase of sludge production [1]. There are five principal methods for disposal of residual sludge after sewage treatment: to store sludge in vacant pits to form sludge pits [2–5]; mechanical reduction or heat treatment in a dedicated sludge pit landfill or mixed residual sludge with municipal solid waste in a dedicated sanitary landfill [6–10]; as building materials utilization [11]; as agricultural compost [12]; incineration treatment [13]. Due to significant amount of disposal and ideal investment cost, sludge landfill site becomes the main method of sludge disposal. With the development of sludge treatment and disposal technology, the landfill site still remains the main mean of sludge treatment and disposal. Considering the stability of landfill, it is important to learn solidified sludge. The solidification treatment is mixing sludge with cementitious

L. Zhan et al. (Eds.): ICEG 2018, ESE, pp. 255–262, 2019.
https://doi.org/10.1007/978-981-13-2224-2_32

materials. So, the engineering properties of the solidified sludge were quite different from that of fresh sludge, i.e., the solidified sludge had much low organic matter content and water content.

So far as is known to the writer, in-situ reinforcement treatment and stability analysis of the sludge pit lack the necessary geotechnical parameters at present. This paper involved sewage sludge taken from Suzhou Qizishan landfill, conducted a series of field and laboratory tests, and obtained the relationship between physical properties, mechanical properties of its distribution of depth. These tests results involved water content, organic matter content and shear strength. These geotechnical parameters can guide some in-situ reinforcement methods, like sludge vacuum preloading, and provide theoretical basis for disposal scheme design.

2 Sludge Pit Site Investigation

2.1 Engineering Background

The fresh sludge and solidified sludge involved in this article comes from Qizishan landfill sludge pit, which locates in the southwest side of the landfill valley. The sludge pit covers an area of more than 20,000 m^2, the pit includes more than ten tons municipal sewage sludge, and accepted the leachate from factory in a short-term, thus it stays high water content (above 90%) over the years. Preliminary investigation before solidification demonstrated that the sludge pit had a large amount of debris floating on the surface. According to the preliminary investigation, the sludge stays fluidized state, and keeps extremely low strength. As can be seen from the Fig. 1, the leachate water level is high.

Fig. 1. Sewage sludge pit site

In order to figure out the physical and mechanical properties of sludge at different positions and depths, there were samples taken from different depths of 24 boreholes.

2.2 Physical Characteristics of Sludge Without Chemical Treatment

The laboratory test for fresh sludge involved water content and organic matter content. So far as is known to the writer, the water content and organic matter content associate with mechanical characteristics. Firstly, the water content and the internal friction angle, the compressive strength, the vane shear strength are all linearly negative correlation. The influence of water content on the strength of sludge mainly lies in high organic matter content. The surface tension and van der Waals force of the solid particles can absorb water and form a water film on the particle surface. The higher the water content, the greater the water film thickness. Lower water content can reduce the water film thickness. Without the lubrication and barrier of the water film, the cohesion between the sludge particles increases and the strength increases too [14]. Secondly, high organic matter content is the essential reason for high water content and low shear strength [15]. Therefore, the laboratory tests mainly include water content test and organic matter content test.

Water Content

Water content is defined as ratio of mass of water to the mass of dry solid. It was basically determined using the oven-drying method according to ASTM procedure of D2974-07a (2007). And the drying temperature was controlled as 60 °C.

It was found that the water content of sludge has certain volatility with depth. The water content of sludge is generally between rang from 83% to 97%, with an average about 86% (Fig. 2).

Organic matter content

Organic content of sludge was determined by igniting the oven-dried solid matters in a muffle furnace at a temperature of 440 °C according to ASTM procedure of D2974-07a (2007). It was found that the organic matter content was generally range from 30.56% to 46.66%, and the average was 40.56% (Fig. 3).

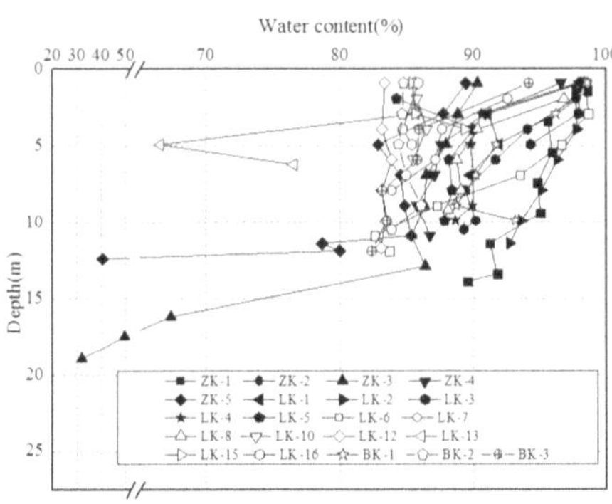

Fig. 2. The water content varies with the depth distribution

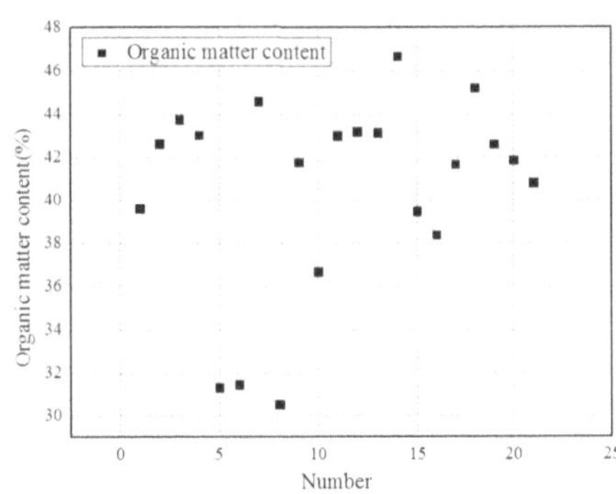

Fig. 3. The organic matter content

3 Solidified Sludge Technology

During the curing process in sludge pit, the crane arm was equipped with the curing agent pipe, and the curing agent pipe inserted the sludge pit. Meantime constant pressure makes sludge mixing with cementitious materials completely. The curing agent injects into the sludge pit for sludge solidification (Fig. 4). The strength in some location within the sludge pit cannot meet the requirements after self-examination. So, subsequent strengthen methods had to be considered. The high pressure swirl technology was used in the reinforcing location to solidify the sludge below 10 m (Figs. 4, 5). The diameter of the high-pressure rotary jet pile is 1 m, and depth for the top entering the sludge

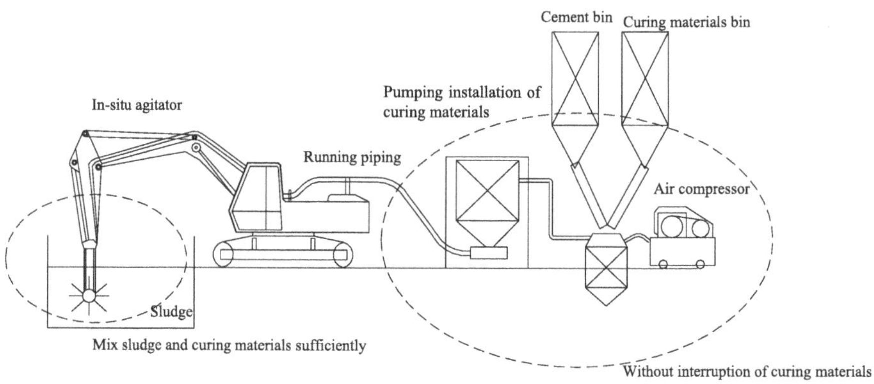

Fig. 4. In-situ curing

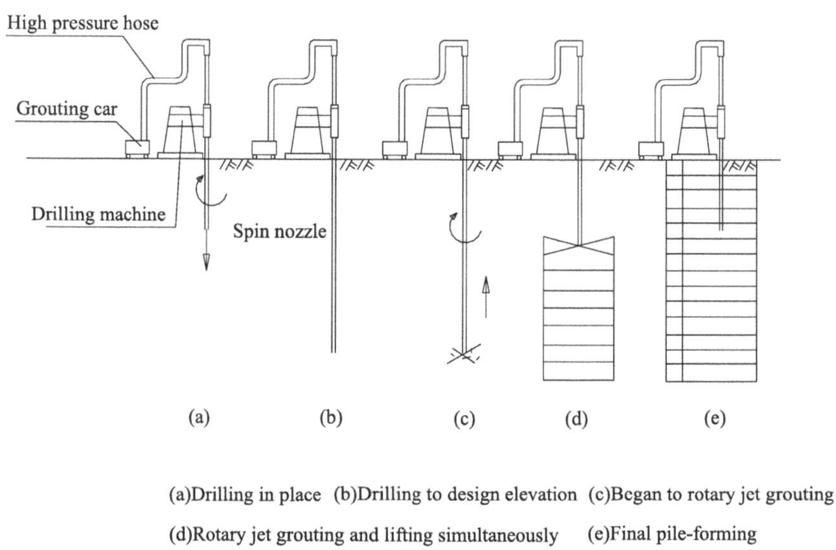

Fig. 5. Jet grouting process

solidification layer is 1 m. There are positive triangular piles set by high-pressure rotary jet pile, and the pile spacing is 3D (replacement rate m = 10%). After the construction, the pressure roller is used to crush the surface (1–2 m), and the compaction degree is not less than 0.9.

4 Sludge with Chemical Treatment

The solidified sludge strength measurement was conducted to parameterize the solidification effect. Based on the field samples, laboratory tests involved the physical properties such as water content, organic matter content. Besides, the direct shear test and unconfined compressive test were conducted for testing the mechanical properties.

4.1 Physical Characteristics of Sludge with Chemical Treatment

Water content

The water content of solidified sludge has a tendency of decreasing with depth (Fig. 6). The water content of solidified sludge generally ranges from 15% to 60%, with an average about 42%. As shown in Fig. 2, the water content of sludge without chemical treatment ranges from 83% to 97%, with an average about 86%. The variation in the water content indicated a significant drop of solidification and reflected the state of sludge correlated to water content. Since local climate and leachate concentration, deep layer of sludge stays in high water content. For further sludge layer, leachate is discharged from the percolation layer.

Organic matter content

The test results indicated that organic matter content generally range from 10% to 26%, the overall average about 18%. Compare with the early fresh sludge on average 40.56% of organic matter, organic matter content of the solidified sludge significantly reduced [16].

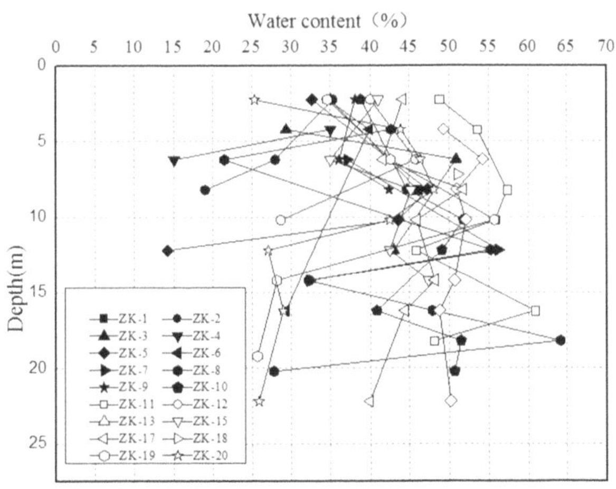

Fig. 6. The water content varies with the depth

4.2 Mechanical Characteristics of Sludge with Chemical Treatment

Direct shear test

Direct shear tests were conducted basically according to ASTM procedure of D3080-04 (2004). Since part of the solidified sludge samples was high in water content, it is difficult to use ring knife molding. There are three vertical pressure level (50 kPa, 100 kPa and 200 kPa) for three specimens from each set of shear test. And it is necessary to adjust pressure during the test process according to the sample hardness.

From the test results, it can be found that the cohesive intercept (c) of the solidified sludge mostly range from 15 kPa to 60 kPa. And the amount of solidified sludge samples of which cohesive intercept more than 10 kPa about 93%. Most of friction angle (φ) range from 6° to 40°. And the amount of solidified sludge samples of which friction angle greater than 10° about 81%.

Since the sample generally contains much rubbish, the direct shear strength is calculated as weighted average (the strength of sludge and the strength of rubbish were weighted according to the percentage of content). MOE Key Laboratory of Soft Soils and Geoenvironmental Engineering researched different ages of rubbish from Suzhou Qizishan landfill by triaxial shear tests. Choosing the strength of 3.5–6 years' rubbish, and taking the cohesive intercept (20.8 kPa) and the friction angle (21.4°) into weighted calculation.

4.3 Unconfined Compression Test

Since unshaped sludge samples are unable to complete sample cutting, the universal testing machine was used for the unconfined compression test. When the computer shows strain reached its peak, the test will stop. Or when the reading is not at the peak, the test will stop when the strain reached 20%.

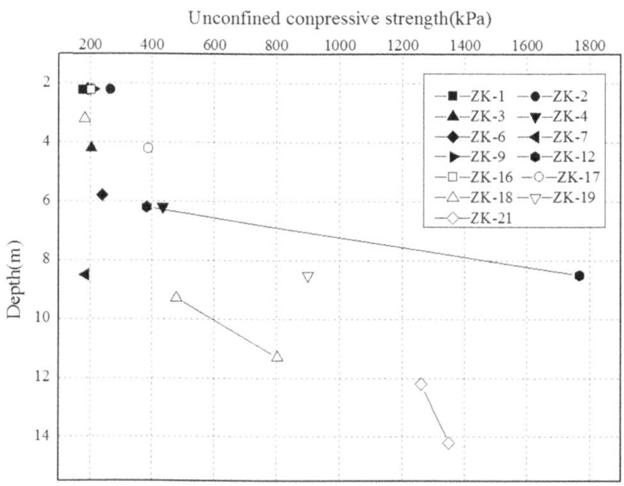

Fig. 7. The unconfined compression strength varies with the depth

The results showed that the strength of the solidified sludge (2 to 8 m) range from 180 to 1500 kPa, and the depth for subsequent strengthen method range from 12 to 15 m. The unconfined compressive strength of solidified sludge is about 180 kPa, while the unconfined compressive strength of cement-like sample reaches 250–300 kPa (Fig. 7).

5 Conclusion

Solidified sludge formed by mixing sludge with cementitious materials. Above tests support the view of high organic matter content is the essential reason of high water content and low shear strength. As can be seen from the previous test results, the solidified sludge from Suzhou Qizishan sludge pit has many special properties. And these properties can provide the necessary geotechnical parameter and partly basis for disposal design.

(1) After solidifying, the water content decreased, the organic matter content reduced for the reaction of sludge with cementitious materials and the strength of the sludge increased.
(2) The weighted calculation makes up for the strength error of the disturbed sludge sample.
(3) Based on the direct shear strength, the cohesive intercept (c) of solidified sludge mostly range from 15 to 60 kPa. Most of friction angle (φ) range from 6° to 40°.
(4) The unconfined compressive strength is about 180 kPa, while the cement sample reaches 250–300 kPa.

References

1. Shen L (2012) Analysis on sludge disposal status and treatment methods in sewage treatment plant. China Sci Technol Overv 15:18–19
2. Claydon JR, Eadie HS, Harding C (1997) Deighton tip—failure, investigation and remedial works. In: Proceedings of the 19th Congress of the International Commission on Large Dams (ICOLD), Florence, Italy, vol 4. ICOLD Press, Paris, pp 233–245
3. Klein A, Sarsby RW (2000) Problems in defining the geotechnical behaviour of wastewater sludges 2000:74–87
4. Lin W, Zhan X, Zhan TL et al (2014) Effect of $FeCl_3$-conditioning on consolidation property of sewage sludge and vacuum preloading test with integrated PVDs at the Changan landfill, China. Geotext Geomembr 42(3):181–190
5. Zhan TL, Zhan X, Lin W, Luo X, Chen Y (2014) Field and laboratory investigation on geotechnical properties of sewage sludge disposed in a pit at Changan landfill, Chengdu, China. Eng Geol 170(4):24–32
6. O'Kelly BC (2004) Geotechnical aspects of sewage sludge monofills. Munic Eng 157 (3):193–197
7. O'Kelly BC (2005) Sewage sludge to landfill: some pertinent engineering properties. J Air Waste Manag Assoc 55(6):765

8. O'Kelly BC (2010) Landfill disposal of alum water treatment residues: some pertinent geoengineering properties. J Residuals Sci Technol 7(7):95–113
9. O'Kelly BC (2016) Assessing the shear strength of municipal sludges and residues for landfill disposal. In: Symposium on urban mining and circular economy
10. Wang MC, Hull JQ, Jao M et al (1992) Engineering behavior of water treatment sludge. J Environ Eng 118(6):848–864
11. Zabaniotou A, Theofilou C (2008) Green energy at cement kiln in Cyprus—use of sewage sludge as a conventional fuel substitute. Renew Sustain Energy Rev 12(2):531–541
12. Elliott HA, Dempsey BA (1991) Agronomic effects of land application of water treatment sludges. J Am Water Works Assoc 83(4):126–131
13. Li J, Li Y, Hohecker HG (2004) Study on the characteristics of sludge incineration technology in fluidized bed incinerator. Environ Eng 22(3):76–79
14. Zhang H, Fan J, Zhao Y (2008) Geotechnical characterization of dewatered sewage sludge for landfilling. J Tongji Univ (Nat Sci) 36(3):361–365
15. Feng Y, Luo X, Lin W (2013) Geotechnical properties measurement of sewage sludge at a disposal site. Rock Soil Mech 34(1):000115–122
16. O'Kelly BC (2016) Geotechnics of municipal sludges and residues for landfilling. Geotech Res 3(4):1–32

Investigation of Construction and Demolition Waste, A Case Study in Deqing, China

Min Xia, Shuai Zhang, and Liangtong Zhan(✉)

Key Laboratory of Soft Soils and Geoenvironmental Engineering, Ministry of Education, Zhejiang University, Hangzhou 310013, China
zhanlt@zju.edu.cn

Abstract. The amount of construction and demolition waste is enormous and still in rapid growth especially in the mega-city like Hangzhou, PR China. However, effective management and technical regulations for dealing with the construction waste are still not available, simple landfill is the main disposal method of C&D waste. In order to reduce the consumption of land spaces by simple landfills, this paper explores the possibility of reclaiming landfills into forest land or farmland. After analyzing the concentration and pollution degree of six heavy metals and two pesticides, the C&D waste landfill can fully satisfy the soil application function corresponding to a third-level standard of Soil Environmental Quality Standard (GB15618-1995). Thus, it can be used for forest land and farmland (except planting vegetables).

Keywords: C&D waste · Heavy metal · Pollution risk assessment

1 Introduction

A huge amount of construction and demolition (C&D) waste has been generated, with the acceleration of urbanization and urban renewal in China. China has become the largest producer of C&D waste in the world [1]. It is estimate that the proportion of C&D waste has reached to 30–40% of the total amount of municipal solid waste in China [2], and continues to increase and estimated to reach 90% in the near future [3]. While at present, most C&D waste is delivered to rural areas for landfill or directly sent to dumpling sites without systemic management and treatment [4], and it is reported that around 75% of Chinese cities are facing the dilemma of "garbage siege" [5]. Due to its consumption of limited landfill spaces and potential environmental pollution, C&D waste have become a great challenge to sustainable urban development [6].

This paper aims to investigate whether C&D waste landfills have caused pollution to the surrounding environment and to explore whether landfills can be reclaimed as forest land or farmland. The results are obtained based on a series of lab tests of C&D waste landfill in Deqing China, 42 soil samples have been taken and tested, further pollution risk assessment was conducted.

L. Zhan et al. (Eds.): ICEG 2018, ESE, pp. 263–270, 2019.
https://doi.org/10.1007/978-981-13-2224-2_33

2 Study Area and Sampling

2.1 Study Area

The study area is a mine pit formed after mining, and the C&D waste is mainly excavated surface soil of the construction site. In order to know whether the C&D waste landfill is fit to be converted into farmland, a pollutant evaluation is needed.

2.2 Sampling

A 9-point (N1–N9) diagonal line sampling scheme was made refer to the "Technical Specification for Soil Environmental Monitoring" (HJ/T 166-2004). Several sampling point branches were set using the plum blossom method around each sampling point, then using quartering method to mix soil samples from one sampling point. At last, 1 kg representative soil samples were selected for storage and analysis. The sampling depth of the surface soil was 20 cm, meanwhile 3 out of 9 sampling points were selected for 2 deep samples (20–60 cm and 60–100 cm), respectively. Besides in order to conduct a comparative test, two background sampling points were selected on the mountain beside the dumps, and deep sampling was conducted (0–20 cm, 20–60 cm, and 60–100 cm). A total of 21 soil samples were collected from the 2# and 3# dumps respectively. The layout of sampling points in 2# and 3# construction waste dump were shown in Figs. 1 and 2.

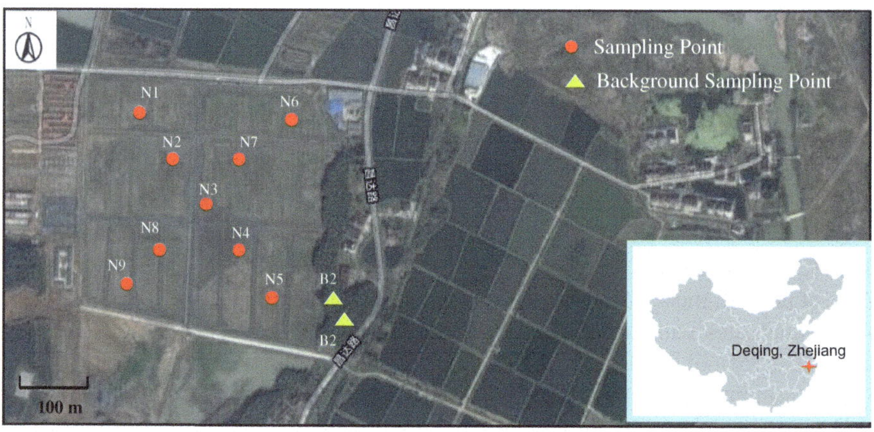

Fig. 1. Layout of sampling points in 2# construction waste dump

Fig. 2. Layout of sampling points in No. 3 construction waste dump

3 Description of Experiments

According to the Soil Environmental Quality Standard (GB15618-1995) and the natural conditions of the C&D waste landfill, six heavy metals (Cadmium, Chromium, Nickel, Copper, Zinc and Lead) and two pesticides (Benzex and DDT) were detected by atomic absorption spectrophotometer (Fig. 3) and gas chromatograph-mass spectrometer (Fig. 4), respectively. In Soil Environmental Quality Standard (GB15618-1995), soil environmental quality is classified into three categories based on soil application functions and conservation. The first-level standard is applicable to national nature reserves, the second-level standard is the soil limit value for human health and agricultural production, and the third-level standard is the soil critical value for normal growth of plants and the protection of agricultural forestry products. According to the aim of this research, the second-level and the third-level standards are selected as evaluation standard.

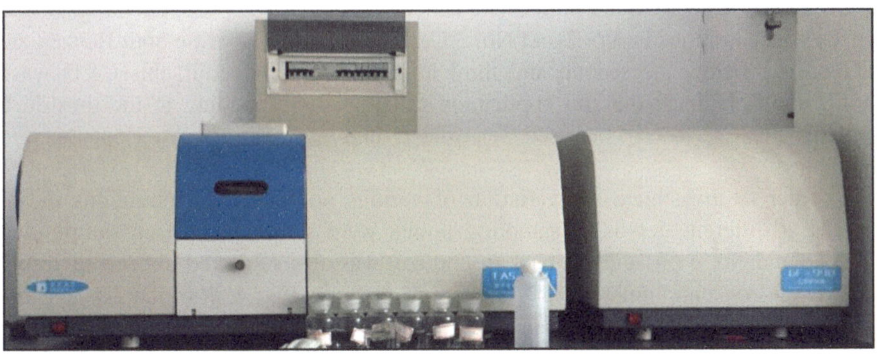

Fig. 3. Atomic absorption spectrophotometer

4 Results and Analysis

4.1 Pollutant Concentration

The test results of No. 2 and No. 3 C&D waste landfill are listed in Table 1. The average concentration of Cd, Pb, Ni and Zn in the soil samples taken from No. 2 and No. 3 pits are larger than that of the background soil, but it should be noted that the soil in the No. 2 and No. 3 pits is alkaline and the background soil is acidic. The second standard for acidic soils is much lower than the alkaline soils.

Table 1. The concentration of pollutants and the standard values

Pollutant	2#	3#	Background soil	2nd standard		3rd standard
				pH < 6.5	pH > 6.5	pH > 6.5
pH	7.82	7.87	4.14	/	/	/
Cd (mg/kg)	0.74	0.71	0.44	<0.3	<0.6	1.0
Cr (mg/kg)	50.59	57.13	53.21	<150	<250	300
Cu (mg/kg)	26.99	23.64	32.33	<50	<100	400
Pb (mg/kg)	11.44	10.95	6.66	<250	<350	500
Ni (mg/kg)	53.15	51.58	46.60	<40	<60	200
Zn (mg/kg)	21.69	21.25	8.37	<200	<300	500
Benzex (mg/kg)	0.001	0.001	/	<0.5	<0.5	1.0
DDT (mg/kg)	0.101	0.011	/	<0.5	<0.5	1.0

As indicated in Table 1, the average concentrations of Cr, Cu, Pb and Zn are much smaller than the second standard, the Ni concentrations are slightly lower than the standard, while the Cd concentrations of No. 2 and No. 3 landfills exceed the second standard by 23% and 18% respectively. While for the background values, the concentration of Cd and Ni exceed the second standard by 42% and 16% respectively. For the third-level standard, the soils in two C&D waste landfill as well as background soils are within the standard value limits. The concentrations of DDT are much higher than Benzex concentrations in No. 2 and No. 3 C&D waste landfill, while both Benzex and DDT are far below the second and third level standards. Overall, this C&D waste landfill can fully meet the soil application function corresponding to the third-level standard, that is, it can be used for forest land and farmland (except planting vegetables).

In order to investigate the variation of various pollutants concentrations in soil samples at different depths, 3 sampling points were selected for deep sampling in No. 2 pit and No. 3 pit, and the background soil was also subjected to deep sampling. The average concentrations of heavy metals are shown in Fig. 4, in which 2-1, 3-1, B-1 represent the surface soil samples in No. 2 pit, No. 3 pit and background sampling point, respectively. Similarly, 2-2, 3-2, B-2 represent soil samples with depth of 20–60 cm, and 2-3,3-3, B-3 represent soil samples with a depth of 60–100 cm.

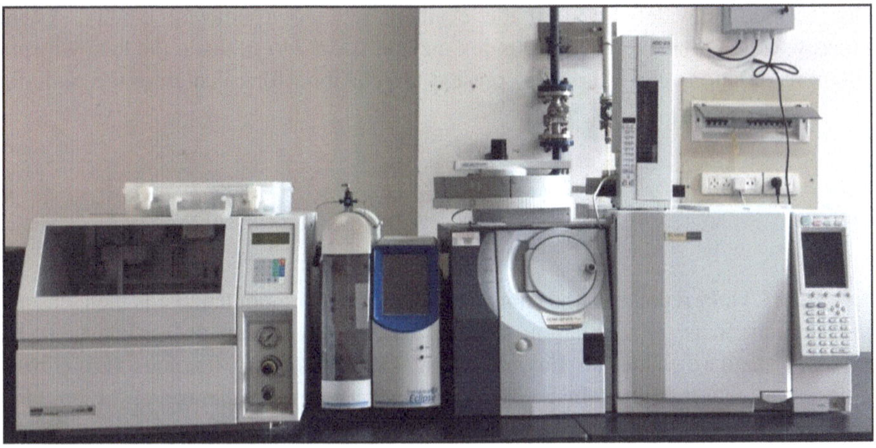

Fig. 4. Gas chromatograph-mass spectrometer

As Fig. 5 shows, the concentrations of heavy metals in soils at different depths did not change significantly.

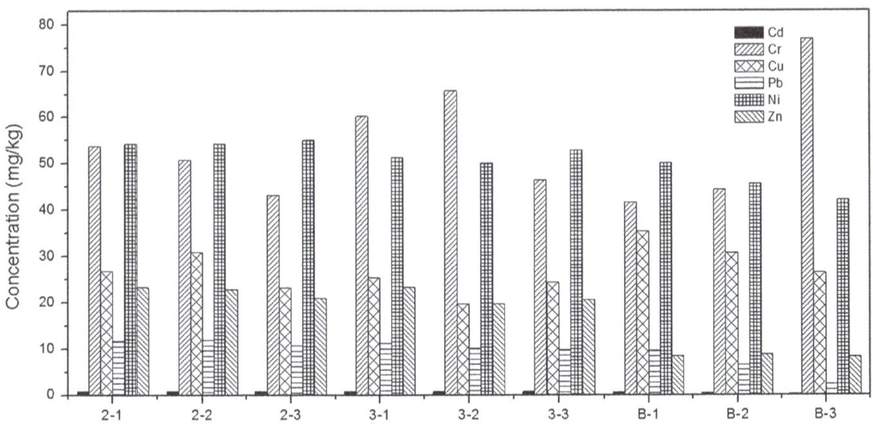

Fig. 5. Heavy metal concentrations in soil samples at different depths

4.2 Pollution Assessment

For evaluation of soil environmental quality, soil heavy metal pollution assessment is indispensable. A variety of soil heavy metal pollution assessment methods at home and abroad have been development, including Muller index method, fuzzy mathematics, gray clustering method, sediment enrichment coefficient method, pollution load index method, index method, etc. Index method is mainly based on the actual measurement values and evaluation criteria to obtain pollution index. Because it is easy to calculate,

Index method is widely used in the evaluation of environmental quality system, Nemerow comprehensive pollution index method is used to evaluate the heavy mental pollution. Equations (1.1) and (1.2) present the equation for calculating the pollution index.

$$\mathrm{Pi} = \mathrm{Ci}/\mathrm{Si} \tag{1.1}$$

$$\mathrm{P_{total}} = \left[\left(\mathrm{P}_{aver}^2 + \mathrm{P}_{max}^2\right)/2\right]^{1/2} \tag{1.2}$$

In Eq. (1.1), Pi represents the single pollution index of pollutant, Ci is the measured concentration, Si is the evaluation standard value of pollutant, this article takes the secondary standard values in Environmental quality standard for soils (GB 15618-1995) as evaluation criteria. In Eq. (1.2), P_{aver} represents the average of the pollution index Pi, P_{max} is the maximum value of the single pollution index, and P_{total} represents the final comprehensive evaluation index.

According to the Technical Specifications for Soil Environmental Monitoring of Farmland (NY/T 395-200), the degree of soil pollution was classified into five levels, security level, alert level, light pollution level, moderate pollution level and heavy pollution level by four Pi values (0.7, 1.0, 2.0, and 3.0). According to the Fig. 6, the Cr Cu Pb Zn pollution index of C&D waste dump sample is 0.21, 0.25, 0.03, and 0.07, respectively, so they are all within safety level. The Ni pollution index is 0.87, which is on an alert level. And the Cd pollution index is 1.21, which belongs to the light pollution level. Cd makes a great contribution to the comprehensive pollution, so it can be considered as the main heavy metal pollutants in the soil. The comprehensive pollution index P_{total} calculated from the six heavy metals is 0.91, it is on the alert level. Similarly the Cr Cu Pb Zn pollution index of background sample is also in security

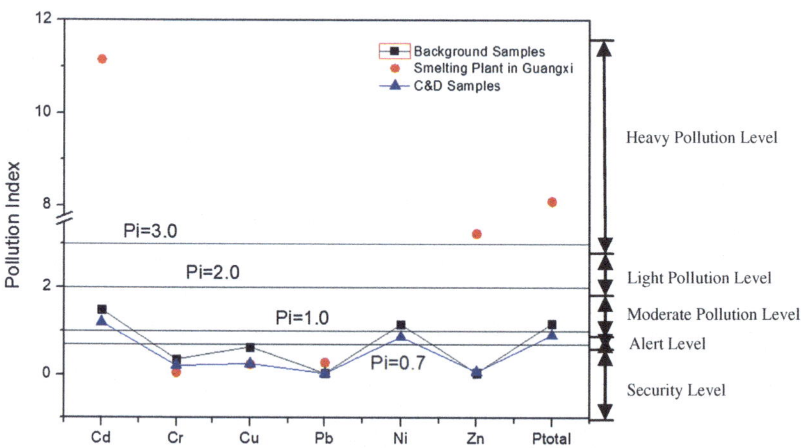

Fig. 6. The pollution index of heavy metals of construction waste dump

level, besides the pollution index of Cd and Ni are in light pollution level, and the comprehensive pollution index of background samples belongs to light pollution levels. Surprisingly the heavy metal pollution index of background samples are generally higher than the C&D waste dump samples. This is not because the heavy metal concentrations of background soil are higher than the concentration of the C&D waste, but the standard value of the acidic soil is much lower than that of the alkaline soil. It shows that this C&D waste landfill does not cause significant pollution and harm to the surrounding environment, it is able to be reclaimed into forest land and farmland (except planting vegetables).

Besides, Wu et al. [7] tested the heavy metals in the construction waste of an abandoned lead-zinc smelting plant in Guangxi. The pollution index of Cr Cu Pb in their research are very similar to this paper, while Cd and Zn belong to heavy pollution level, their pollution index are 11 and 3 times of the results in this paper, respectively.

4.3 Conclusions

(1) Six heavy metals and two pesticides were selected and tested, pollution risk assessment of two C&D waste dumps was conducted in this paper. The results showed that excepting Cd and Ni, the other contaminants concentration are much lower than the secondary standards of Soil Environmental Quality Standard (GB15618-1995), while the third standard values are much higher than all of the 8 contaminants. In addition, the contaminates concentration in soil samples collected from hills next to the pits also have the same variation.

(2) Three sampling points were selected for deep sampling in No. 2 pit and No. 3 pit, and the referenced soil was also subjected to deep sampling. It showed that the concentrations of heavy metals in soils at different depths did not change significantly.

(3) Taking the secondary standard values as the evaluation criteria, pollution index assessment method was used to evaluate the heavy mental pollution in C&D waste dumps and background samples. For C&D waste samples, the comprehensive pollution index P_{total} calculated from the six heavy metals is at the alert level. Cd makes a great contribution to the comprehensive pollution index, so it can be considered as the main heavy metal pollutants in the soil. While the comprehensive pollution index of background samples can be classified into the light pollution level.

(4) Due to the high pollutant concentration in the background soil around the C&D waste dumps, the 8 main pollutant indicators in the C&D waste soil samples are much lower than the third level standard in Soil Environmental Quality Standard (GB15618-1995), it is initially considered that this two C&D waste dumps can be reclaimed into woodland and farmland (except planting vegetables).

References

1. Xiao J, Xie H, Zhang C (2012) Investigation on building waste and reclaim in Wenchuan earthquake disaster area. Resour Conserv Recycl 61(5):109–117
2. Tam V, Lu W (2016) Construction waste management profiles, practices, and performance: a cross-jurisdictional analysis in four countries. Sustainability 8(2):190
3. Yuan H, Shen L (2011) Trend of the research on construction and demolition waste management. Waste Manag 31(4):670–679
4. Zheng L et al (2017) Characterizing the generation and flows of construction and demolition waste in China. Constr Build Mater 136:405–413
5. Huang B et al (2018) Construction and demolition waste management in China through the 3R principle. Resour Conserv Recycl 129:36–44
6. Ding Z, Wang Y, Wu J (2017) ABM based simulation research on construction waste management. In: Proceedings of the 20th international symposium on advancement of construction management and real estate. Springer, Singapore (2017)
7. Wu L, Mo X, Guomei Z et al (2014) Health risk assessment of heavy metal pollution of abandoned lead-zinc smelting plant. Asian J Ecotoxicol 9(3):603–608

Stability Enhancement of Landfills on Sloping Ground Using Earthen Berms at the Toe

Abinash Mahanta(✉), Manoj Datta, and Gunturi Venkata Ramana

Indian Institute of Technology Delhi, Hauz Khas, New Delhi, India
abinashmahanta.am@gmail.com

Abstract. This paper presents the results of a study undertaken to enhance the stability of a landfill on sloping ground of a hilly region. The base of the landfill has a significant elevation difference from one end to the other causing an overall inclination of the base with the horizontal. The study highlights the influence of the following factors on the base sliding stability along the geomembrane interface as: (a) leachate head; (b) pore water pressures in wet waste; (c) seismic forces and (d) smooth versus textured geomembrane (GM). Limit equilibrium methodology is adopted for analyses and three types of failure surfaces are analysed – circular, planar (single straight line) and planar (two straight lines). The study reveals that stability against sliding along liner is low, whenever (a) pore pressures/leachate head are high; (b) earthquake forces are large and (c) smooth GM are used. Usually planar failure surface (two straight lines) is observed as the critical surface.

Furthermore, it is also observed that the stability of the sloping liner can be enhanced significantly by providing adequate size of earthen berm at the toe. After application of the berm at the toe, the possible critical failure surface may pass over the back slope of the berm or pass under the bottom of the berm. This depends on the geometric dimensions (height & base-width/back slope) of the berm. Planar failure surfaces (two and three straight lines) are analysed to check the stability. The study reveals that by varying the berm height as well as the base width, one can arrive at an optimal solution for stability.

Keywords: Landfills · Stability enhancement · Sloping ground
Earthen berms

1 Introduction

1.1 Literature Study

From various documented landfill failure cases, [2, 4, 5] it is observed that stability of landfills is one of the prime factors for ensuring safety of habitat in their immediate vicinity. In hilly regions, waste disposal is very common on sloping ground, sides of hills, in valleys etc. because of paucity of horizontal ground surfaces. Unlike in plains, landfills on sloping ground experience instability along weak interfaces in the basal lining systems [1, 4]. Such failures are reported to involve contributory factors or triggers such as high pore pressures due to rainfall, earthquakes, smooth geomembranes etc. Hence stabilizing measures such as berms at the toe of landfills are often adopted.

L. Zhan et al. (Eds.): ICEG 2018, ESE, pp. 271–278, 2019.
https://doi.org/10.1007/978-981-13-2224-2_34

1.2 Background

In a hilly region of India, an engineered Municipal Solid Waste (MSW) landfill is designed to accommodate 150 metric ton of waste per day for a design life period of 16 years. For this, capacity of an engineered landfill on a particular site is calculated based on the MSW generation data collected from the local municipal corporation of that region. The base of the particular landfill site has a significant elevation difference at the ends due to the shallow bedrock level causing a gradual base inclination, (β_b) of 6° with the horizontal. A two dimensional geometrical prototype as shown in Fig. 1 is modeled closely in accordance to the capacity calculated above with base length, (L_{ws}) 300 m and height of waste, (H_w) 20 m. The waste cover slope is considered to be 1V:2.5H with the bed profile inclination. The stability of the landfill is investigated for various influencing factors on three different modes of failures a, b and c as shown in Fig. 1.

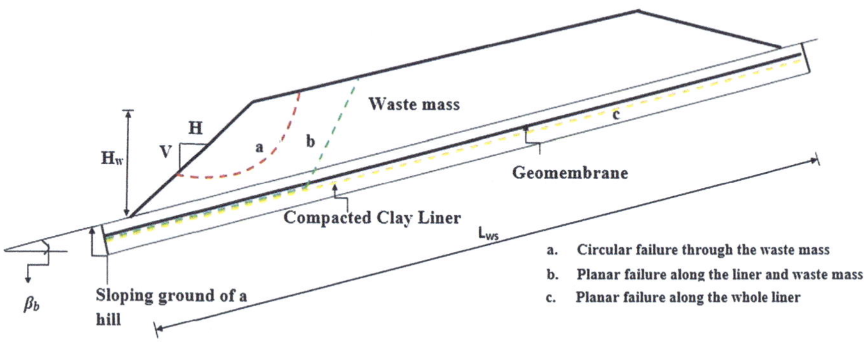

Fig. 1. Typical geometrical prototype of an engineered landfill on a sloping ground.

The stability above landfill is further enhanced by using earthen berm at the toe and a two dimensional geometrical prototype is shown in Fig. 2. The influence of geometric dimensions of the berm such as height of berm (H_b) and width of berm (L_{bs}) parallel to the base slope by varying the back slope (α_b) of the berm are analysed. The stability of the landfill with earthen toe berm is investigated for different modes of failure a, b and c as shown in Fig. 2 for all the critical conditions observed without the toe berm.

1.3 Objective

The objective of the present study is to identify the influence of various factors on stability of an engineered landfill on a sloping ground and presents the details and consequent findings. Furthermore, the influence of earthen toe berm in stability of the landfill is to investigate for different dimensional variability of the berm and an adequate size of earthen berm is to be adopted for final overall stabilization of the landfill.

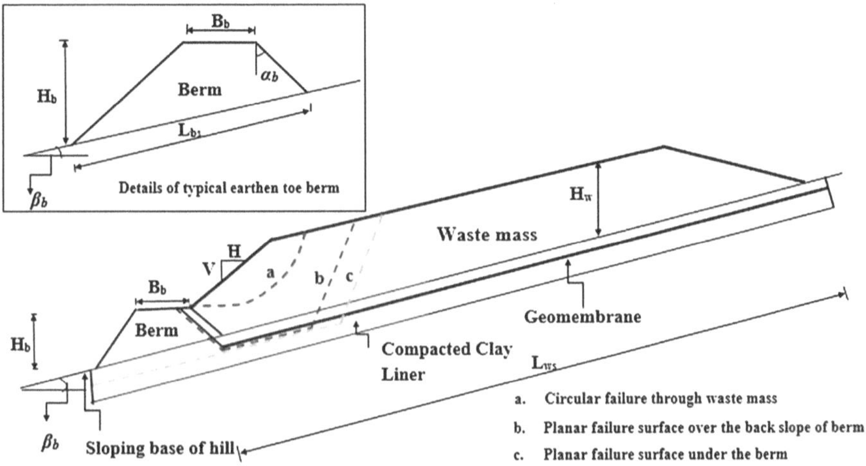

Fig. 2. Typical geometrical prototype of an engineered landfill on a sloping ground with earthen berm at the toe.

2 Landfill on Sloping Ground

2.1 Methods of Stability Analysis

The stability of the engineered landfill on the sloping ground is investigated against failure for three modes of failure: (a) circular failure through the waste mass, (b) planar failure through waste mass and sliding along weakest interface of GM and compacted clay liner (CCL) (two straight lines), and (c) sliding along complete lining system along the interface of GM and CCL (one straight line) under various influencing factors such as pore water pressure in wet waste mass and due to leachate head, earthquakes forces and microscopic roughness of GM. Morgenstern and Price's (M-P) method is found to be more suitable [3] as it satisfies both force and moment equilibrium equations to determine critical failure surface in calculating minimum value of Factor of Safety (FoS). Failure through the waste mass is analysed using circular slip surfaces, failure due to sliding along weak interface of GM and CCL and planar failure through waste mass and sliding along interface of GM and CCL are analysed using straight-line slip surfaces. The seismic condition is analysed by pseudo-static method ($\alpha_h = 0.1$, $\alpha_v = 0.05$).

2.2 Conditions of Analysis

The base case considered for analysis of a landfill on the sloping ground is comprised of height of waste, H_w 20 m, cover slope of 1V:2.5H resting on an inclined base having β_b of 6° with smooth geomembrane in the liner system. This case was analysed for dry and wet waste for static conditions. The influence of leachate head (LH) on the stability is investigated by varying the values of LH from 1 m to 3 m above and parallel to the base. The influence of pore water pressures in waste mass was investigated by varying the values of pore-water pressure ratio (r_u) from 0 to 0.2. The effect of earthquake forces are considered by varying the horizontal earthquake force, α_h from 0.05 to 0.1 and vertical earthquake force, from α_v 0.03 to 0.07. The effect of strengthening the weak interface between GM and CCL is also investigated. For this, the interface between GM and underlying CCL in basal lining system is simulated with interface shearing resistance (δ) value 10° (smooth geomembrane) and 18° (textured geomembrane).

2.3 Materials and Properties

Various geotechnical properties [1] considered in modelling the landfill components are tabulated in Table 1.

Table 1. Material properties used in the stability analyses

Material	Unit weight (kN/m^3)	Cohesion intercept (kPa)	Angle of shearing resistance (°)
Waste	10	13	23
Subgrade soil		—Bedrock—	
Toe berm	20	3	30
		Adhesion intercept (kPa)	Interface shearing angle (δ in °)
Smooth GM-Clay interface	–	0	10
Textured GM-Clay interface	–	0	18

Table 2. Minimum acceptable values of FoS for different conditions of analysis

Conditions of analysis	Acceptable FoS
Dry waste (ru = 0)	1.5
Temporary clogging (leachate head)	1.3
Wet waste/rain/seepage (ru > 0)	1.3
Earthquake (pseudo-static)	1.1

2.4 Results

Effect of Leachate Head

Figure 3 depicts that the effect of LH cause significant change in FoS for all the three modes of failures. The failure along waste mass and sliding along liner and is observed to be the most critical and the landfill in unstable under this mode of failure. The circular failure also shows critical FoS due to application of LH.

Effect of Pore Water Pressure Ratio (r_u)

It is evident from Fig. 4 that effect of r_u causes drastic change in FoS in all the three modes of failures. The planar failure through along liner and waste mass is observed to be the most critical failure mode.

Effect of Earthquake

It is observed from the Fig. 5 that FoS values in all the three failure modes reduce significantly in seismic condition; and for $\alpha_h = 0.1$ and $\alpha_v = 0.07$, both the planar failure modes become critical at the ultimate state of earthquake forces. However two straight line failure mode is the most critical one.

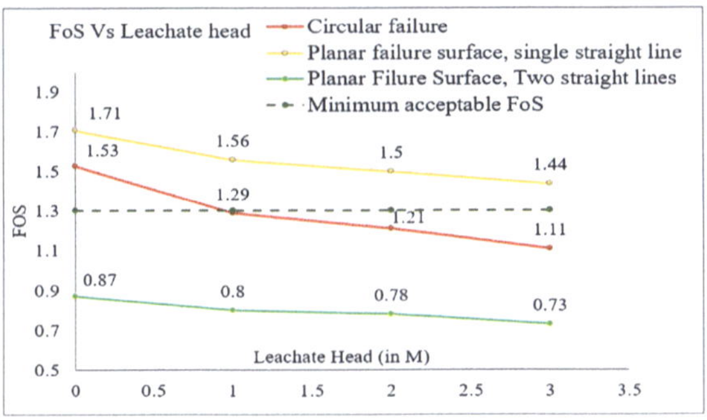

Fig. 3. Effect of leachate head on FoS for a landfill on sloping ground with $\beta_b = 6°$, $H_w = 20$ m and δ =10°.

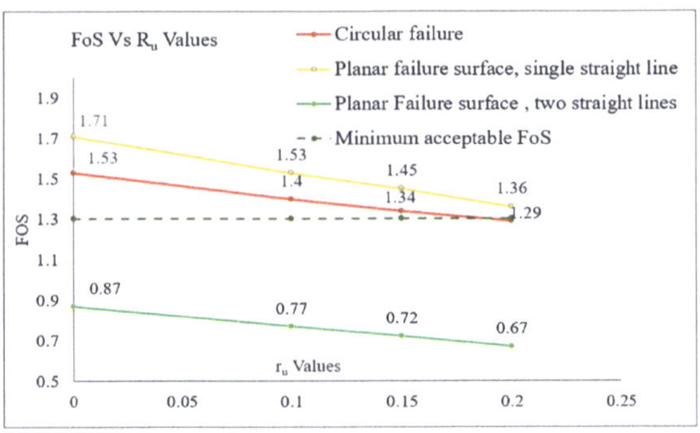

Fig. 4. Effect of r_u on FoS for a landfill on sloping ground with $\beta_b = 6°$, $H_w = 20$ m and $\delta = 10°$.

2.5 Necessity of Stability Enhancement of Landfill on Sloping Ground

It is observed from in the above results that, the landfill is unstable on the sloping ground with smooth GM. Textured GM with $\delta = 18°$ is used to enhance the stability of the landfill on the sloping ground. The FoS increases while using the textured GM.

However it is also investigated that the landfill is still unstable after use of textured GM even if in some dry waste conditions as illustrated in Table 3. It thus necessitates the stability enhancement of the landfill by using toe berm.

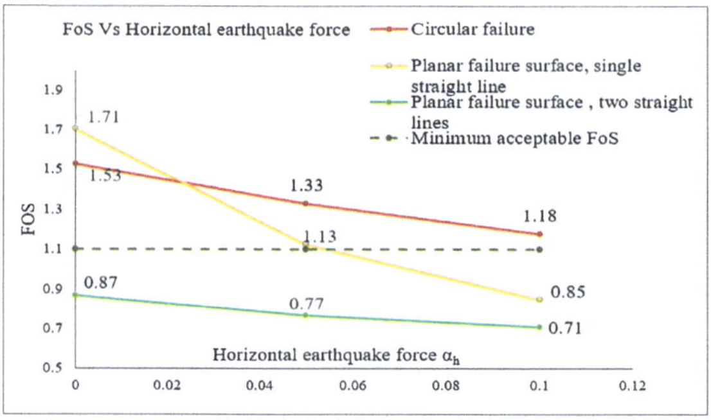

Fig. 5. Effect of horizontal earthquake forces α_h on FoS for a landfill on sloping ground with $\beta_b = 6°$, $H_w = 20$ m and $\delta = 10°$.

Table 3. FoS for smooth and textured GM

Interface shearing angle (δ in °)	FoS		
	Circular failure	Planar, single straight line	Planar, double straight lines
10	1.53	1.71	0.87 (<1.5)
18	1.53	3.14	**1.19** (<1.5)

3 Landfill on Sloping Ground with Earthen Toe Berm

3.1 Stability Analysis

For the landfill with earthen toe berm, stability is investigated for (a) circular failure through the waste mass, (b) planar failure through waste mass and sliding along interface of GM and CCL which passes over the back slope of the berm (three straight lines), (c) planar failure through waste mass and sliding along interface of GM and CCL which passes under the berm (two straight lines). The base case considered for analysis of a landfill on the sloping ground with earthen toe berm is comprised of the base case considered under Sect. 2.2 alongwith textured geomembrane ($\delta = 18°$) in the liner system. Height of berm, H_b is 10 m and back slope, α_b of the berm is 45°. The influence of height of berm on FoS is investigated by varying it from 5 M to 12.5 M. The influence of α_b of the berm is investigated by varying it from 1V:1H (45°) to 1V:2.5H (22°). Material properties are considered from Table 1.

3.2 Results

Effect of Height of Toe Berm

Effect of height of the berm has significant role in FoS under all modes of failure reflecting the highest in case of planar failure over the back slope of the berm. Failure under the berm is critical for all the different heights of berm. For this particular case, for a berm of $H_b = 10$ m and $\alpha_b = 45°$, the FoS for all the three modes of failure is above the minimum acceptable value.

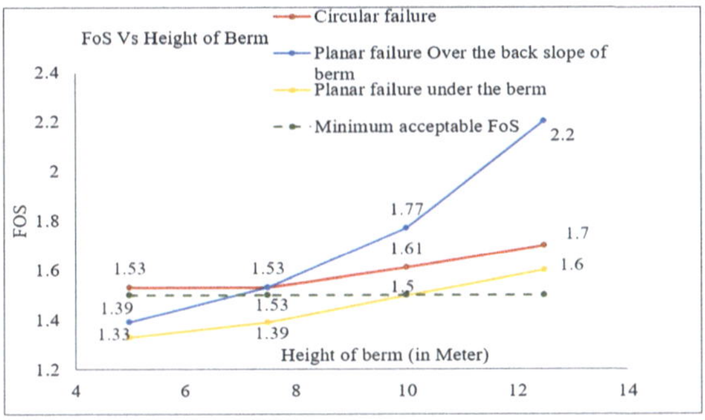

Fig. 6. Effect of height of berm on a landfill on sloping ground with $\beta_b = 6°$, $H_w = 20$ m, $\delta = 18°$ and $\alpha_b = (45°)$.

Effect of Back Slope of the Berm

From Fig. 7 it is evident that the change in back slope has mild effect on FoS for planar failures whereas there is no effect on FoS for the circular failure. An increase in the steepness of the back slope of the berm causes less FoS for the planar failure. However, for this particular case, for a berm of $H_b = 10$ m and $\alpha_b = 22°$ to $45°$, the FoS for all the three modes of failure is above the minimum acceptable value.

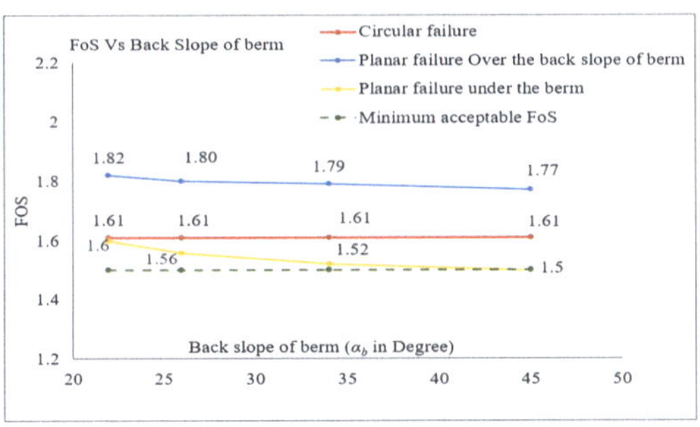

Fig. 7. Effect of back slope of berm on a landfill on sloping ground with $\beta_b = 6°$, $H_w = 20$ m, $\delta = 18°$ and $H_b = 10$ M.

4 Concluding Remarks

This paper presents study of assessment of stability of a MSW landfill on a sloping ground in a hilly region against sliding along interfaces of geosynthetic in the base liner systems as well as stability of waste mass under the influence of various factors. The results highlight the fact that for a base inclination of 6°, stability against sliding along base liner is low when pore water pressures or leachate heads are high, earthquake forces are high and smooth geomembrane is used. Use of textured geomembrane is found to overcome the stability problem upto certain extent. However, in case of critical state of the influencing factors, landfill on the sloping ground is analyzed as unstable even if textured GM is used. In the present study, in addition to the textured GM, to enhance the stability toe berm of different height and back slope angle are used. Thus, it is concluded that factors of safety are high when textured geomembranes are used, and when leachate head and pore water pressures are low and toe berms of adequate size are provided.

References

1. Bergado DT, Ramana GV, Sia HI, Varun H (2006) Evaluation of interface shear strength of composite liner system and stability analysis for a landfill lining system in Thailand. Geotext Geomembr 24(6):371–393
2. Blight GE (2007) Interpretations of surface movement of a landfill built on steepy sloping ground. A cautionary case history. Waste Manag Res 8(25):572–584
3. Fredlund DG, Krahn J (1976) The relationship between limit equilibrium slope stability methods. In: Geotechnical conference proceedings, Vancover, British Coluombia
4. Koerner RM, Soong T-Y (2000) Leachates in landfills—the stability issues. Geotext Geomembr 18(8):293–309
5. Seed RB, Mitchell JK, Bolton H (1990) Kettleman hills waste landfill slope failure I: liner system properties. J Geotech Eng 116(4):447–668

Engineering Barriers

(Edited by Yuchao Li, Hui Xu,
Wenjie Zhang, and Zhihong Zhang)

Equivalent Design and Evaluation of the Liner System for a Hazardous Waste Landfill Vertical Expansion

Xuede Qian[1(✉)], Te-Yang Soong[2], Xianda Zhao[2], and Hang Shi[2]

[1] Michigan Department of Environmental Quality, 525 W Allegan Street, Lansing, MI 48909, USA
qianx@michigan.gov
[2] CTI and Associates, Inc., 28001 Cabot Drive, Novi, MI 48377, USA

Abstract. An alternative liner system utilizing geosynthetic clay liner (GCL) in lieu of the permitted compacted clay liner (CCL) was proposed for a vertical expansion at a hazardous waste landfill sitting over an existing unregulated landfill. An assessment was conducted to demonstrate that the proposed system is technically-sound and effective and meets all federal and state rule requirements. Two critical equivalency assessment items including (i) steady state solute flux and (ii) chemical adsorptive capacity and solute breakthrough time were evaluated and are discussed in this paper. Based on the evaluations and discussions, the proposed alternative GCL system is equivalent or superior to the currently permitted CCL system and is capable of preventing the migration of hazardous constituents into the groundwater or surface water at least as effectively as the CCL system.

Keywords: Equivalency · Liner system · Hazardous waste landfill

1 Introduction

Per Michigan Hazardous Waste Management Rules (Part 111 Rules), the liner system for a hazardous waste landfill must meet the following design requirements (top to bottom):

- 30-cm of protective layer,
- Primary leachate collection/drainage layer (sand or drainage geocomposite),
- 2-mm primary HDPE geomembrane,
- 1.5-m of compacted clay liner (CCL), $k \leq 1 \times 10^{-7}$ cm/s,
- Leak detection layer (sand or drainage geocomposite),
- 2-mm secondary HDPE geomembrane,
- 0.9-m of compacted clay liner (CCL), $k \leq 1 \times 10^{-7}$ cm/s,
- 6-m of subsoil with $k \leq 1 \times 10^{-7}$ cm/s.

In the U.S., both federal and state regulations allow alternative and equivalent liner designs but a detailed "equivalency" demonstration must be provided [1]. In Michigan, an alternate design may be approved if the designer can demonstrate the design will

L. Zhan et al. (Eds.): ICEG 2018, ESE, pp. 281–290, 2019.
https://doi.org/10.1007/978-981-13-2224-2_35

prevent the migration of any hazardous constituent into the groundwater or surface water at least as effectively as the design requirements specified in the Rules. The following sections discuss how the proposed design satisfies this requirement.

2 Proposed Alternative Liner System

The alternative liner system proposes using geosynthetic clay liner (GCL), in lieu of the currently permitted CCL, as an alternative soil component of the liner system for the vertical expansion over an existing unregulated hazardous waste landfill (Fig. 1) in southeastern Michigan. GCL products are factory-manufactured hydraulic barriers consisting of a layer of sodium bentonite supported by geotextiles (woven and/or non-woven) and, in some cases, an additional film of flexible membrane liner (FML) for enhanced barrier performance. These components (sodium bentonite, geotextiles, and FML) are mechanically held together by either needling or chemical adhesive.

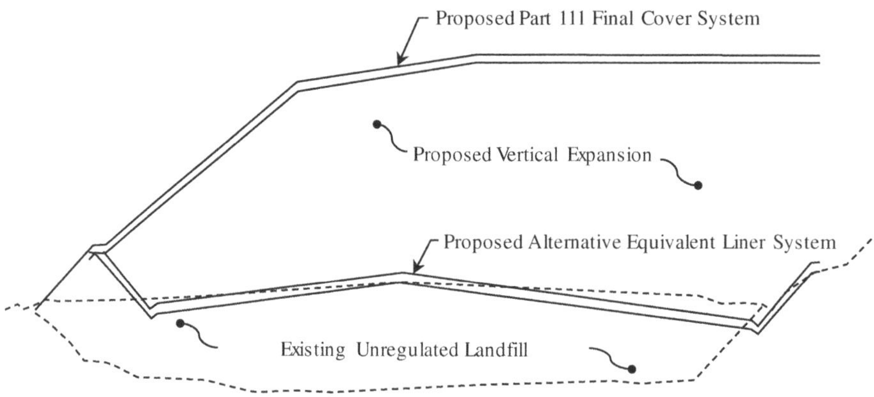

Fig. 1. Cross-section of a new vertical expanded landfill over an existing landfill

Sodium bentonite (the interlayer of GCL) is an effective barrier primarily because it can absorb moisture (i.e., hydrate and swell) producing a dense, uniform layer with extremely low hydraulic conductivity (on the order of 10^{-9} cm/s). Sodium bentonite's exceptional hydraulic properties make GCL superior to CCL with respect to a steady state of water flux even though the thickness of GCL is less than CCL.

Two layers of GCL are proposed to be installed immediately beneath the primary HDPE geomembrane liner. Figure 2 shows the proposed liner construction details. As shown in Fig. 2, the proposed liner system consists of multiple layers of geosynthetic and earthen materials to optimize the performance of the base liner system of the vertical expansion. These layers, along with their respective functions, are tabulated in Table 1 for a direct comparison between the permitted and proposed base liner systems (from top to bottom).

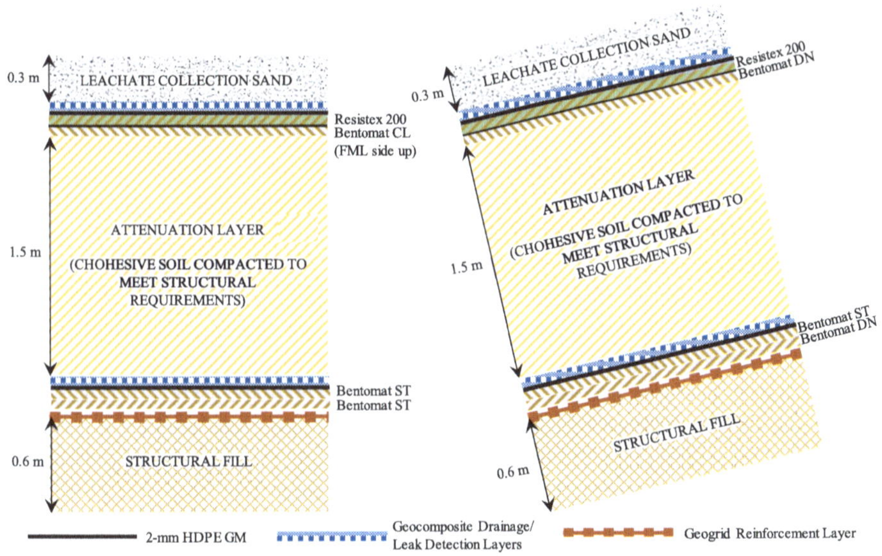

(a) For cell floor and slopes $\leqslant$ 4H:1V (b) For slopes $\geqslant$ 4H:1V

Fig. 2. Proposed alternative equivalent liner system

Table 1. Comparison between permitted and proposed alternative liners (from top to bottom)

Component	Permitted liner system	Proposed alternative liner system
Primary leachate collection	30-cm drainage sand	
	Double-sided drainage geocomposite	
Primary geomembrane liner	2-mm textured HDPE geomembrane	
Primary clay liner	1.5-m CCL ($k \leq 1.0 \times 10^{-7}$ cm/s)	Resistex® 200, by CETCO
		Bentomat® CL, by CETCO
		1.5-m cohesive soil attenuation layer
Leak detection	Double-sided drainage geocomposite	
Secondary geomembrane liner	2-mm textured HDPE geomembrane	
Secondary clay liner	0.9-m CCL ($k \leq 1.0 \times 10^{-7}$ cm/s)	Bentomat® ST, by CETCO
		Bentomat® ST, by CETCO
Base reinforcement	Bi-axial geogrid	
Liner subbase	0.6-c structural fill	

As indicated in Table 1, the main difference between the permitted and proposed liner systems is the use of GCL, including specialty GCL selected for enhanced performance, in lieu of CCL. Other liner components will remain unchanged.

To quantify equivalency between the proposed liner system (GCL) and the permitted liner system (CCL), site-specific leachate test data was provided to the GCL manufacturer for a conservative evaluation of GCL chemical compatibility. After 243 h of permeation, an average permeability of 1.5×10^{-9} cm/s with 0.7 pore volumes of leachate passing through the specimen was measured. This means that the bentonite/polymer blend in the GCL product (Resistex® 200) was hydrating and cutting off flow as designed. For the equivalency demonstration calculations (specifically, the steady-state solute flux), a conservative permeability of 1.0×10^{-8} cm/s was used.

In addition to installing polymer-treated GCL (Resistex® 200) immediately beneath the primary 2-mm HDPE liner on the cell floor, another specialty GCL (Bentomat® CL) is proposed for enhanced protection. Bentomat® CL has an additional flexible membrane liner (FML) laminated on one side of the GCL to offer the highest level of hydraulic barrier performance. By installing this product with the FML side "facing up" towards the cell as indicated in Fig. 2, Bentomat® CL provides another impervious layer to isolate its own bentonite layer from contacting moisture, if any, that may migrate through the primary HDPE geomembrane liner and the overlying GCL (Resistex® 200).

For slopes that are steeper than 4H-to-1V, a standard GCL product (Bentomat® DN) is proposed to replace the FML-laminated GCL (Bentomat® CL) for enhanced slope stability. Bentomat® DN consists of needle-punched, non-woven geotextiles on both sides of the bentonite interlayer. This configuration provides superior shear resistance. The FML-laminated GCL (Bentomat® CL) to be installed on the cell floor will be extended 1.5-m vertically above the toe of the sideslopes for optimized performance.

3 Technical Equivalency Assessment

An equivalency assessment was conducted according to the following steps allowing for a technically-sound, effective and project-focused equivalency demonstration.

1. Identify various technical criterion that are relevant to the proposed liner system.
2. Divide the identified criterion into distinct categories to facilitate a direct technical comparison between GCL (the proposed liner) and CCL (the permitted liner).
3. Identify criterion where technical equivalency between GCL and CCL has already been well-studied, demonstrated and documented by the lining industry (e.g., landfills, surface impoundments, mining, water-proofing of hydraulic structures, etc.), based on past tests and project experiences.
4. Identify criteria which are mainly site-, project-, or product-specific items, and demonstrate equivalency.

The results of the above steps are summarized in Table 2. Both the format and content shown in the table is largely adapted from the well-referenced papers by Koerner and Daniel (1993) [2] and Bonaparte et al. (2002) [3], as well as from general liner engineering practice over the past two decades with some site-specific modifications that are considered appropriate for the construction of the proposed liner.

Table 2. Generalized technical equivalency assessment for liners beneath landfills

Category	Criterion for Evaluation	Equivalency of GCL to CCL		
		GCL is superior	GCL is equivalent	Project-specific
Hydraulic	Steady state water flux	X		
	Breakthrough time - water	X		
	Horizontal flow in seams or lifts		X	
	Horizontal flow beneath FML	X		
	Steady state solute flux			X
	Chemical adsorptive capacity and solute breakthrough time			X
	Permeability to gases (non-issue when installed under FML)	-	-	-
	Generation of consolidation water	X		
Physical/ Mechanical	Freeze-thaw behavior	X		
	Wet-dry behavior	X		
	Vulnerability to erosion (non-issue)	-	-	-
	Total settlement		X	
	Differential settlement	X		
Construction	Ease of placement	X		
	Speed of construction	X		
	Availability of material	X		
	Requirements of water	X		
	Air pollution concerns	X		
	Quality assurance considerations		X	

As shown in Table 2, two criteria ("Steady State Solute Flux" and "Chemical Adsorptive Capacity and Solute Breakthrough Time") will be subjected to detailed project specific comparison between GCL and CCL as presented in the following sections.

4 Steady State Solute Flux

In certain chemical environments, the sodium ions in bentonite can be replaced with cations dissolved in the water that comes in contact with GCL, a process referred to as "cation exchange". This type of exchange reaction can reduce the amount of water that can be held in the interlayer, resulting in decreased swell and increased permeability.

With the design and installation configuration shown in Fig. 2 in mind, the steady state solute flux equivalency demonstration was prepared and presented in Tables 3(a) and (b). Please note that the following assumptions were adopted for additional conservatism:

1. Comparisons are made as if the 2-mm HDPE primary geomembrane liner does not exist. In other words, the superior swelling capability of GCL to "seal" or "plug" holes or imperfections in the overlying HDPE liner is completely ignored.

Table 3. Steady state solute flux equivalency demonstration

(a) for liner on cell floor and slopes ≤ 4(H):1(V)

GCL Layer	Thickness (cm)	*k* (water) (cm/sec)	*k* (leachate) (cm/sec)	Additional adjustment	Adjusted *k* (cm/sec)	Thickness/ Adjusted *k*
Resistex 200	0.95	3E-09	5E-09	2.0	1E-08	9.53E+07
Bentomat CL	0.95	5E-10	5E-10	1.0	5E-10	1.91E+09

k equivalent (k_{eq})* = 9.5E-10 cm/sec

Comparing the steady-state flux (*Q*'s) using Darcy's Law $Q = k\,i\,A$ (assuming no FML)

CCL Option	k_{eq} (cm/sec)	head (cm)	thickness (cm)	gradient, *i*	Flux, *Q* ($cm^3/sec/cm^2$)
1.5-m of CCL	1.0E-07	30.48	152.4	1.20	1.20E-07
Resistex 200 / Bentomat CL	9.5E-10	30.48	1.91	17.0	1.62E-08

Q_{GCL}/Q_{CCL} = **13.5%**

(b) for liner on slopes ≥ 4(H):1(V)

GCL Layer	Thickness (cm)	*k* (water) (cm/sec)	*k* (leachate) (cm/sec)	Additional adjustment	Adjusted *k* (cm/sec)	Thickness/ Adjusted *k*
Resistex 200	0.95	3E-09	5E-09	2.0	1E-08	9.53E+07
Bentomat DN	0.95	5E-09	5E-09	1.0	5E-09	1.91E+08

k equivalent (k_{eq})* = 6.7E-09 cm/sec

Comparing the steady-state flux (*Q*'s) using Darcy's Law $Q = k\,i\,A$ (assuming no FML)

CCL Option	k_{eq} (cm/sec)	head (cm)	thickness (cm)	gradient, *i*	Flux, *Q* ($cm^3/sec/cm^2$)
1.5-m of CCL	1.0E-07	15.24	152.4	1.10	1.10E-07
Resistex 200 / Bentomat DN	6.7E-09	15.24	1.91	9.0	6.00E-08

Q_{GCL}/Q_{CCL} = **54.5%**

* $k_{vertical\ equivalent} = t_{total} / [(t_1/k_1) + (t_2/k_2)]$

2. Per the chemical compatibility evaluation performed using site-specific leachate data, the hydraulic conductivity of the upper GCL (Resistex® 200) is assumed to be 1.0×10^{-8} cm/s, which is more conservative than the tested results suggesting a permeability of 1.5×10^{-9} cm/s. As discussed previously, this adjustment conservatively addresses the concern of chemical compatibility associated with site-specific leachate. This adjustment is also extremely conservative since in reality this GCL layer will be completely covered by a layer of 2-mm HDPE geomembrane liner and hydration of GCL by leachate can only take place only if there is leachate leakage through liner imperfections. The chance of this assumed scenario (i.e., the entire GCL layer is exposed to leachate with an increased hydraulic conductivity) does not practically exist.
3. Values of head-on-liner used in the evaluation were selected as 30 cm for the cell floor (per regulation) and 15 cm for sideslopes steeper than 4(H):1(V). Moreover, while only the standard GCL product (Bentomat® DN) is used in the flux calculation, the calculated maximum head-on-liner will theoretically occur near the toe of

the sideslope where the specialty FML-laminated GCL (Bentomat® CL) will be installed. This presents an additional conservative factor of safety.

4. Technically, an "apples-to-apples" comparison of steady state solute flux should be made by comparing flux that comes from the bottom of the 1.5-m attenuation layer (in the proposed design case) and from the bottom of the 1.5-m CCL layer (in the permitted design case). However, the equivalency evaluation was conservatively conducted by determining the flux that flows through the two layers of GCL and comes out the bottom of the lower GCL layer (Bentomat® CL). In other words, any flow retardation capacity that could be provided by the underlying 1.50-m thick cohesive attenuation layer is completely ignored in this evaluation.
5. Consequent to assumptions 3 and 4 discussed above, the hydraulic gradient (the driving force that causes flow to take place) selected for the proposed liner case is 14 times and 8 times greater than that selected for the permitted liner case for floor and sideslope liners, respectively. This represents another very conservative assumption.

The evaluation of the steady state solute flux criteria is made by dividing the calculated steady state solute flux of the proposed liner (GCL) by the number associated with the permitted liner (CCL). The resulting "ratio", if it is less than or equal to 100%, would indicate that the performance of the proposed liner system is equal or superior to the permitted liner system and technical equivalency is therefore demonstrated.

Input parameters, assumptions, and results of the steady state solute flux evaluation are shown in Tables 3(a) and (b) for cell floor and slopes steeper than 4(H):1(V), respectively.

As shown in Tables 3(a) and (b), the steady state solute flux "ratios" are 14% and 55% for the cell floor and sideslope, respectively. Both numbers are significantly less than 100% indicating the performance of the proposed liner system is superior. Therefore, technical equivalency for the criterion of steady state solute flux has been demonstrated.

5 Chemical Adsorptive Capacity and Solute Breakthrough Time

Federal and State regulations focus on preventing contamination of groundwater (CFR 40 Part 264.301(b) and Michigan Part 111 R299.9620(4)(a)). Therefore, selecting a "point of reference" in the subsoil that has the same hydrogeological characteristics and distance to groundwater and using that point as a reference for both liner systems would be an appropriate approach in demonstrating equivalency.

Two models were established per the concept described above: (a) permitted and constructed MC 6G Phase 1 liner located over natural soil subbase and (b) proposed alternative liner located over an existing landfill (i.e., MC 6G Phase 2). As shown in Fig. 3, the thickness of in-situ clayey subsoils under the existing waste where the proposed vertical expended landfill cell (i.e., MC 6G Phase 2) will be constructed, is approximately the same as the combined thickness of MC 6G Phase 1 liner and its in-situ clayey soil.

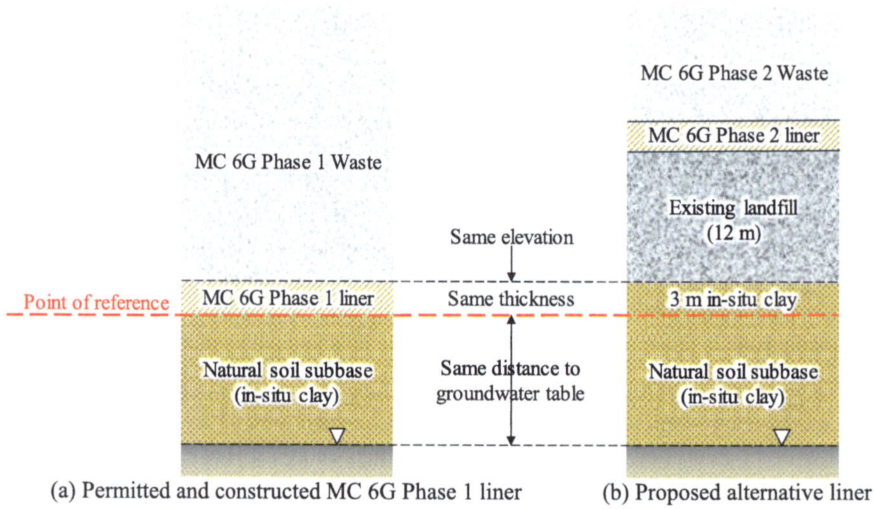

Fig. 3. Conceptual model for chemical adsorptive capacity and breakthrough time comparison

This is an important fact since numerical equivalency, in terms of chemical adsorptive capacity and solute breakthrough time, can already be demonstrated by the 3-m thick in-situ clay present in the proposed vertical expansion (i.e., MC 6G Phase 2) subsoils since all clayey soils (e.g., CCL or in-situ clay) exhibit a similar diffusion coefficient [4].

In addition, as shown in Fig. 3, the proposed alternative liner system contains 2.1-m of cohesive soil layers (1.5-m attenuation layer and 0.6-m structural fill). Since the distance between the potential contaminant source (leachate above the primary liner) and the point of reference is significantly thicker for the proposed MC 6G Phase 2 compared to MC 6G Phase 1, the breakthrough time will be significantly increased in the proposed system.

Another factor impacting the breakthrough time is the steady state flux passing through the liner system (higher flux would lead to shorter breakthrough time). Since it has already been demonstrated (see Tables 3(a) and (b)) that the proposed GCL liner system will significantly reduce the steady state flux, the GCL liner system should also significantly increase the advective breakthrough time.

Moreover, as shown in Fig. 3(b), approximately 12-m of existing waste further separates the new waste in MC 6G Phase 2 from the natural soil and groundwater. This additional waste layer provides additional chemical adsorptive capacity due to the following properties:

- Its anaerobic natural and high sulfide condition could bond heavy metals [5, 6].
- Non-degradable organic material and other materials provide additional adsorption and/or absorption capabilities for organic contaminants [7, 8].

- Additional biological activity reduces the half-life of organic pollutants and reduces potential breakthrough [9, 10].
- Increases the mass transport distance and further reduces the concentration gradient [11, 12].
- Reduces the "concentration gradient" with the contaminants in the existing waste.

Based on the above discussions, the performance of the proposed alternative liner system is clearly superior in the criterion of chemical adsorptive capacity/solute breakthrough time than the permitted reference case (MC 6G Phase 1 liner system). Therefore, technical equivalency has been demonstrated.

6 Conclusions

In the U.S., both federal and state regulations allow alternative and equivalent liner designs as long as detailed equivalency demonstration is provided. An alternative liner system, using GCL in lieu of the permitted CCL, was proposed as an alternative soil component of the liner system for the vertical expansion over an existing unregulated hazardous waste landfill in southeastern Michigan. Based on the above discussions and evaluations, the process can be summarized and concluded as follows:

- An equivalency assessment should be conducted according to the categories of hydraulic, physical/mechanical, and construction allowing for a technically-sound, effective and project-focused equivalency demonstration.
- The evaluation of the steady state solute flux is demonstrated by dividing the calculated steady state solute flux of the proposed liner (GCL) by the corresponding value associated with the permitted liner (CCL). As shown in Tables 3(a) and (b), such steady state solute flux "ratios" are 14% and 55% for the cell floor and sideslope, respectively. Both numbers are significantly less than 100% indicating the performance of the proposed liner system is superior. Therefore, technical equivalency for the steady state solute flux is demonstrated.
- The evaluation of the chemical adsorptive capacity/solute breakthrough time first requires the establishment of a "point of reference". The authors selected a point in the subsoil that has the same hydrogeological characteristics and distance to groundwater. Using that point as a reference for both liner systems supports a rational and appropriate approach in demonstrating equivalency of chemical adsorptive capacity and solute breakthrough time. Based on the above discussions, the performance of the proposed alternative liner system is clearly superior in the criterion of chemical adsorptive capacity/solute breakthrough time than the permitted reference case (MC 6G Phase 1 liner system).

The information presented above demonstrates that the proposed liner system is equivalent or superior to the currently permitted liner system and is capable of preventing the migration of hazardous constituents into the groundwater or surface water at least as effectively as the permitted liner system. Therefore, the proposed liner system is acceptable.

References

1. Qian X, Koerner RM, Grey DH (2001) Geotechnical aspects of landfill design and construction. Prentice Hall Inc., Upper Saddle River
2. Koerner RM, Daniel DE (1993) Technical equivalency assessment of GCLs to CCLs. In: Proceedings of the 7th GRI Seminar, Philadelphia, PA
3. Bonaparte R, Daniel DE, Koerner RM (2002) Assessment and recommendations for improving the performance of waste containment systems. EPA/600/R-02/099
4. Lake CB, Rowe RK (2005) A comparative assessment of volatile organic compound (VOC) sorption to various types of potential GCL bentonites. Geotext Geomembr 23:323–347
5. Bhattacharyya D, Jumawan AB, Grieves RB (2006) Separation of toxic heavy metals by sulfide precipitation. Sep Sci Technol 14(5):441–452
6. Robinson AK, Sum JC (1980) Sulfide precipitation of heavy metals. EPA-600/2-80-139
7. De Gisi S, Lofrano G, Grassi M, Notarnicola M (2016) Characteristics and adsorption capacities of low-cost sorbents for wastewater treatment: a review. Sustain Mater Technol 9:10–40
8. Erses AS, Fazal MA, Onay TT, Craig WH (2005) Determination of solid waste sorption capacity for selected heavy metals in landfills. J Hazard Mater 121:223–232
9. Christensen TH, Kjeldsen P, Albrechtsen HJ, Heron G, Nielsen PH, Bjerg PL, Holm PE (1994) Attenuation of landfill leachate pollutants in aquifers. Crit Rev Environ Sci Technol 24(2):119–202
10. Guan C, Xie H, Wang Y, Chen Y, Jiang Y, Tang X (2014) An analytical model for solute transport through a GCL-based two-layered liner considering biodegradation. Sci Total Environ 466:221–231
11. Shackelford C (2013) Rowe lecture: the role of diffusion in environmental geotechnics. In: Proceedings of the 18th international conference on soil mechanics and geotechnical engineering, Paris, France
12. Xie H, Thomas HR, Chen Y, Sedighi M, Zhan L, Tang X (2015) Diffusion of organic contaminants in triple-layer composite liners: an analytical modeling approach. Acta Geotech 10:255–262

Assessment of Backfill Hydraulic Conductivity in an Instrumented Soil-Bentonite Cutoff Wall

Landon C. Barlow(✉) and Michael A. Malusis

Bucknell University, Lewisburg, PA 17837, USA
lcb020@bucknell.edu

Abstract. The objective of this paper is to present a comparison of measured hydraulic conductivities (*k*) for soil-bentonite (SB) backfill within a 60-m-long section of a 200-m-long, 7-m-deep cutoff wall constructed and instrumented for studying SB backfill properties and variability at the field scale. Backfill *k* was measured using flexible-wall tests (70-mm diameter) on remolded specimens prepared from surface grab samples collected during construction; flexible-wall tests on undisturbed specimens collected from the wall; larger-scale rigid-wall tests (150-mm diameter) on remolded specimens prepared from grab samples; and slug tests conducted within the wall. Applied effective stresses in the laboratory tests ranged from 4–35 kPa, encompassing the range of *in-situ* stresses measured in the backfill after load transfer and consolidation (8–13 kPa). The results indicate low spatial variability in *k* for a given test type, consistent with the observed homogeneity of the backfill. Modest variability in *k* was observed among the different test types, with the slug tests and rigid-wall tests generally yielding slightly higher *k* relative to the flexible-wall tests at field-representative stresses.

Keywords: Cutoff wall · Hydraulic conductivity · Slug test · Soil-bentonite

1 Introduction

Soil-bentonite (SB) cutoff walls, or vertical barriers backfilled with bentonite-admixed soil, are widely used in the US for long-term hydraulic and geoenvironmental containment applications. In these applications, the effectiveness of the wall is governed largely by the hydraulic conductivity (k) of the SB backfill, and designs generally specify a low backfill *k* (e.g., $k < 10^{-6}$ or 10^{-7} cm/s). During construction, backfill *k* typically is verified by means of a quality control/quality assurance (QC/QA) testing program in which small-scale (70–100 mm diameter) laboratory *k* tests (usually flexible-wall tests via ASTM D5084) are conducted on remolded specimens prepared from surface grab samples of field-mixed backfill. The results of these tests may not be representative of the *in-situ* backfill *k,* particularly if the applied stress state on the specimens is not representative of the *in-situ* stress state. The *in-situ* stress distribution within an SB wall is rarely measured, difficult to predict, and impacted by complex load transfer mechanisms. Also, laboratory tests on remolded specimens are inadequate for verifying the absence of defects created while backfilling, or for determining post-construction changes in *k*.

L. Zhan et al. (Eds.): ICEG 2018, ESE, pp. 291–299, 2019.
https://doi.org/10.1007/978-981-13-2224-2_36

These limitations of conventional QC/QA programs are compelling reasons for conducting post-construction assessments of the *in-situ* k of SB cutoff walls based on field testing and/or laboratory testing of undisturbed specimens recovered from the wall. Such assessments are rarely performed in engineering practice and have been conducted only to a limited extent in research studies. A notable study was conducted by Britton et al. [1], who evaluated field and laboratory methods to measure k of a small (20.4-m-long, 2.8-m-deep, 0.6-m-wide) pilot-scale SB cutoff wall. They found that laboratory k values obtained from remolded specimens were consistently lower than laboratory k values obtained from undisturbed specimens and lower than k values obtained from *in-situ* measurements (i.e., slug and pumping tests). The results of the study [1], although limited in size and scope, indicate that a more comprehensive investigation of field k assessment methods for full-scale SB cutoff wall installations is warranted.

Researchers at Bucknell University have constructed and instrumented a 200-m-long, 7-m-deep, and 0.9-m-wide SB cutoff wall near Lewisburg, PA for the purpose of studying the SB properties, behavior, and variability at the field scale. The primary objective of this paper is to present the results of a post-construction assessment of backfill k within a 60-m-long section of the wall and to compare field (*in situ*) measurements of k based on slug tests with laboratory measurements of k performed on both remolded and undisturbed specimens. Lessons learned from this study pertaining to the use of slug tests to measure backfill k in full-scale SB cutoff wall installations are discussed.

2 Background

The SB cutoff wall site is located in Montandon, PA, approximately 3 km east of the Bucknell campus (see Fig. 1a and b). The wall was installed in an alluvial deposit on the property of a local sand and gravel quarry, adjacent to a natural wetland known as the Montandon Marsh (Fig. 1c). The wall was constructed using conventional slurry trenching, with sodium bentonite slurry (5–6 wt% bentonite) used for hydraulic shoring and backfill mixing. The backfill was prepared by mixing imported base soil (stockpiled clayey sand excavated from other areas of the site) with slurry to achieve a slump of 75–150 mm prior to backfilling [2]. Quality control testing included daily soundings of the trench bottom, slurry quality testing (viscosity, filtrate loss, density, and pH), and backfill slump. Also, grab samples (10–20 L) of the field-mixed backfill were collected at every 10 m along the length of the wall, for a total of 20 grab samples.

Stationing was used to delineate position along the wall alignment, with the southern end of the wall located at Station 0 + 06 m and the northern end at Station 2 + 07 m. This study considers the 60-m portion of the wall extending from Station 0 + 50 m to Station 1 + 10 m (Fig. 1d). In this section, monitoring wells and inclinometers were installed adjacent to the wall at Stations 0 + 60 m and 0 + 90 m prior to wall construction, and earth pressure cages (RST Instruments, Maple Ridge, BC, Canada) were embedded in the backfill at Stations 0 + 75 m and 0 + 84 m to monitor the three-dimensional stress state in the backfill. The sensors on each cage include three pressure sensors to measure vertical and horizontal (longitudinal and transverse) total stresses, a vibrating-wire piezometer to measure pore pressure, a biaxial tiltmeter and magnetic compass to monitor the cage orientation, and a thermocouple. Three earth pressure cages were embedded in the backfill at different depths (2.4, 4.4, and 6.4 m) at Station 0 + 84 m, and one cage was embedded at a depth of 6.2 m at Station 0 + 75 to provide replicate measurements of stress near the trench bottom. Additional details are provided by Malusis et al. [2].

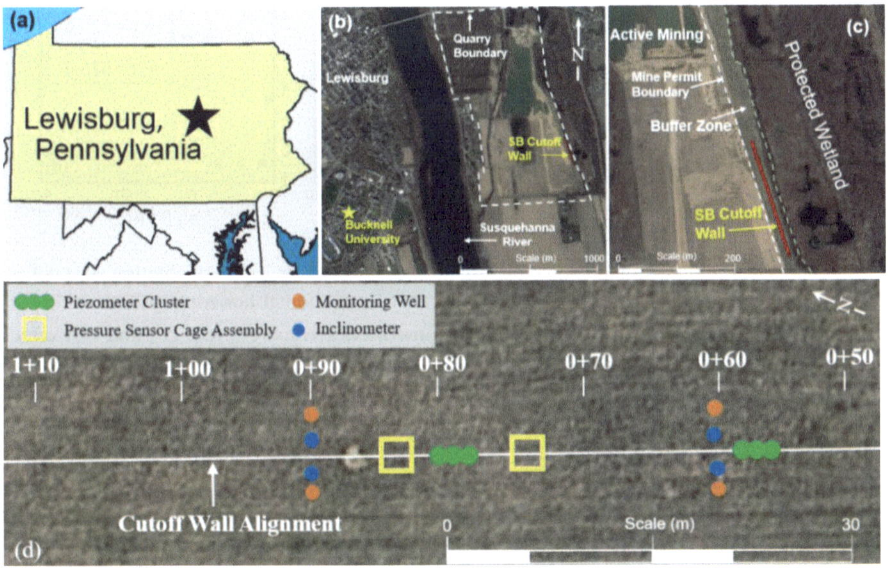

Fig. 1. (a, b) General site location; (c) aerial photo showing where the cutoff wall was constructed; (d) close-up plan view of study area with instrumentation.

The total vertical stress (CT$_z$), horizontal stresses (a* = transverse; *ay* = longitudinal), and pore pressure (u) measured in the wall at Station 0 + 84 m (depth *z* = 6.4 m) over ~19 months (575 days) are shown in Fig. 2a. At completion of backfilling, the total stresses were approximately equal to the weight of the overlying backfill and were nearly isotropic, consistent with the fact that the backfill is placed as a thick, viscous liquid. Over time, as the backfill underwent primary and secondary consolidation, total stresses declined as the load was transferred through shear to the sidewalls of the trench. Likewise, pore pressures dissipated along a similar timeline as the total stresses, yielding low effective stresses that increased slightly during the first year after construction but have now stabilized at 8–13 kPa (see Fig. 2b–d). These effective stresses are considerably lower than would be expected by assuming a geostatic stress distribution in the backfill, but are reasonably consistent with stress estimates given by predictive stress models developed for SB cutoff walls in which load transfer to the trench sidewalls is taken into account [e.g., 3].

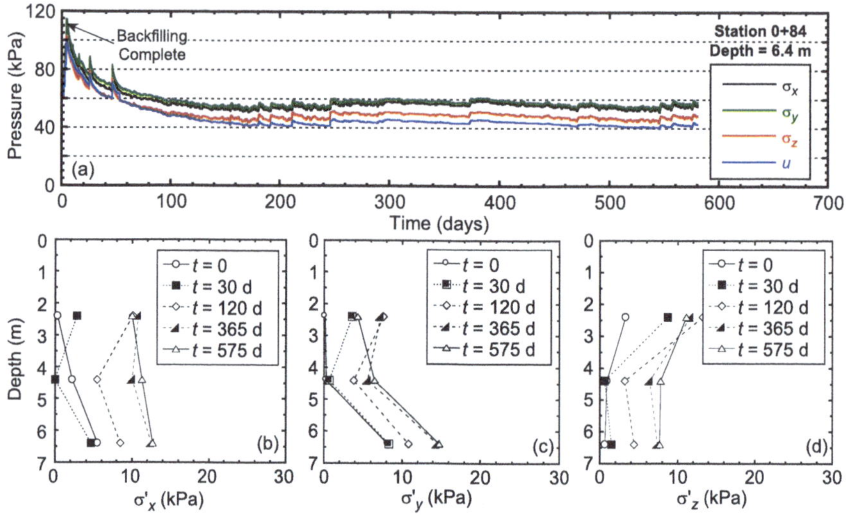

Fig. 2. (a) Total stresses and pore pressures measured over time at Station 0 + 84 m, depth = 6.4 m (σ_x = transverse horizontal stress; σ_y = longitudinal horizontal stress; σ_z = vertical stress; u = pore pressure; (b, c, d) distributions of effective stress (σ_x', σ_y', and σ_z') versus depth (t = time after completion of backfilling).

3 Materials and Methods

3.1 Laboratory Tests

For this study, seven flexible-wall k tests (ASTM D5084 Method C) were performed on remolded specimens (diameter = 70 mm, length = 71–85 mm) prepared from surface grab samples of the field-mixed backfill collected at each 10-m interval along the test section shown in Fig. 1c (i.e., Stations 0 + 50 through 1 + 10 m). The specimen preparation procedures are the same as those described elsewhere [4]. Also, undisturbed backfill samples were recovered from the wall at Stations 0 + 67 m (z = 2.1 m), 0 + 90 m (z = 0.75 m), and 1 + 1.0 m (z = 2.0–3.5 m) using Shelby tubes with a piston sampler. These specimens were tested in a similar manner as the remolded specimens. The loading sequence for each specimen included four stages of isotropic consolidation at average effective stresses ranging from 6.9 to 34.5 kPa. The specimens were permeated with tap water at each loading stage until inflow/outflow balance and steady k were obtained, as required by ASTM D5084.

In addition, two larger-scale (diameter = 150 mm, height = 240–300 mm) falling-head k tests with tap water were conducted on remolded specimens prepared from grab samples collected at Stations 0 + 70 m and 0 + 90 m. These tests were conducted in rigid-wall cells made from plastic pipe. Pea gravel, wire mesh, and filter paper were placed above and below the specimens, and pressures of 3.5–21.7 kPa were applied in stages using weights placed directly on the upper gravel layer.

3.2 In Situ (Slug) Tests

Slug tests were performed to obtain postconstruction measurements of the *in-situ* backfill *k*. The slug test wells consisted of prepacked wells (GeoProbe, Salina, KS) installed by first pushing steel casing (diameter = 7.6 cm) with an expendable cone tip into the wall to the target depth, then inserting the well (a 61-cm-long section of slotted PVC encased in filter sand and steel mesh screen and attached to PVC standpipe; see Fig. 3) into the casing. The casing was then removed, allowing the soft backfill to seal against the well. Six test wells were installed in the test section (Fig. 1c) at depths ranging from 3 to 6 m.

The slug tests were conducted by placing a vibrating wire piezometer into the well, then adding a small volume ("slug") of water to the well to raise the water level from the initial static level H_{wt} to a new level H_p (see Fig. 3), and tracking the decrease in slug height H_s as the water level returned to equilibrium (see Fig. 4a). For this study, small slugs of water were added (initial slug height H_{so} = 25–69 cm) to avoid hydraulic fracturing of the backfill, given the low *in-situ* effective stresses (see Fig. 2). Normalized head dissipation curves (i.e., H_sH_{So}) were plotted on a semi-log scale versus time, and *k* was computed using the following expression:

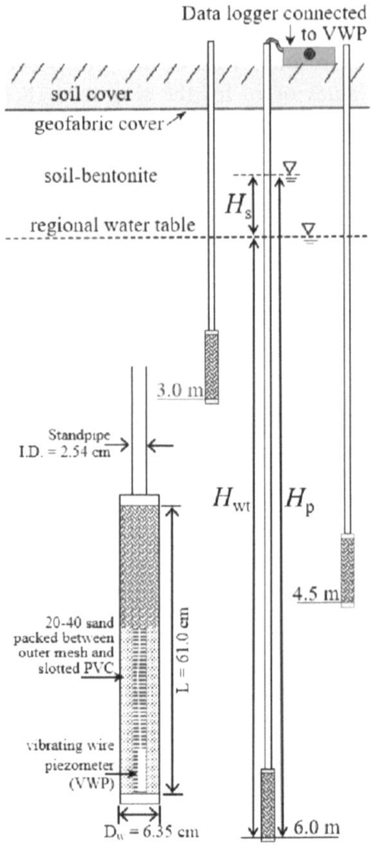

Fig. 3. Schematic of prepacked slug test well.

$$k = \frac{A}{M}m = -\frac{A}{F\Delta t}\ln\left(\frac{H_s}{H_{so}}\right) \tag{1}$$

where A is the cross-sectional area of the standpipe (see Fig. 3), F is a shape factor, and m [= $\ln(H_s/H_{so})/\Delta t$] is the slope of the normalized head dissipation curve (Fig. 4b). The shape factor, F, is based on the dimensions and depth of the well, and different methods for estimating F have been proposed in previous studies [5–10]. The shape factors proposed for SB walls by Britton et al. [8] were used in this study.

4 Results and Discussion

The results of all of the k tests performed for this study are summarized in Table 1. Values of σ' for the slug tests (8.7–11.2 kPa) represent the average *in-situ* effective stresses (mean of σ_x', σ_y', and σ_z') estimated for each well based on Fig. 2. The volumes are the laboratory specimen volumes and the approximate backfill volume represented in the slug tests based on estimated effective radii of 29.7 cm [8]. Narrow ranges of k were obtained for a given test type and stress, indicating a high degree of homogeneity in the backfill. Most of the flexible-wall k values measured at the lower stresses (6.9–13.8 kPa), encompassing the range of *in-situ* stresses inferred at the slug test locations, ranged from 10^{-7} to 2×10^{-7} cm/s, whereas lower k values ($<10^{-7}$ cm/s in most cases) were measured at the higher stresses (20.7–34.5 kPa). The k values from the slug tests ranged from 1.3×10^{-7} to 3.3×10^{-7} cm/s, a range comparable to (though slightly wider and higher than) the range of flexible-wall k at the lower stresses. These results illustrate the importance of conducting lab k tests on SB backfill at stresses representative of the field stresses.

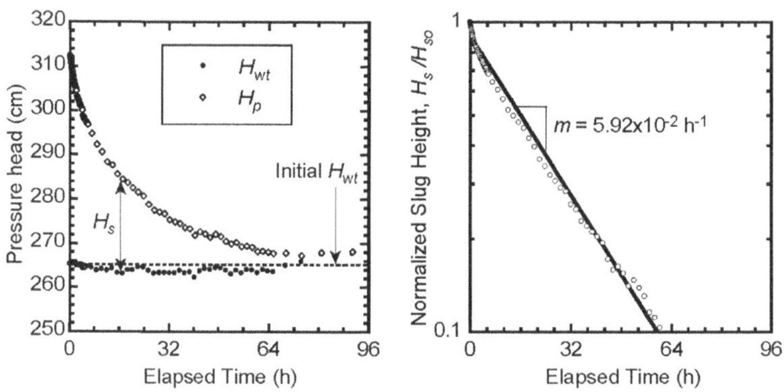

Fig. 4. Example slug test data for test well at Station 0 + 56.2 (depth = 4.5 m): (a) measured pressure heads versus time; (b) semi-log plot of normalized slug height (H_s/H_{so}) versus time.

Measured k values for each test type are plotted as a function of average σ' in Fig. 5. The flexible-wall k results for the remolded specimens are shown as averages with error bars denoting the minimum and maximum k. The undisturbed flexible-wall tests generally yielded k within but at the lower end of the ranges for the remolded specimens, whereas most of the k values from the slug tests and the rigid-wall tests lie above the flexible-wall k ranges. The slightly higher k values for the slug and rigid-wall tests suggest a possible scale effect, as the test volumes associated with these tests were considerably larger than for the flexible-wall tests. A similar effect was observed in the previous study by Britton et al. [1].

Table 1. Summary of hydraulic conductivity test results

Test type	No. of tests	Sample type	Volume (cm^3)	Station (m)	Depth (m)	a′ (kPa)	k ($\times 10^{-7}$ cm/s)
FWR	7	Grab	280–330	0 + 50 to 1 + 10	N/A	6.9	1.2–2.1
						13.8	0.93–1.6
						20.7	0.83–1.4
						34.5	0.70–1.1
RWR	2	Grab	4400–5000		N/A	3.5	5.1
						6.9	2.5–4.1
						13.8	2.1
						21.7	1.7–2.1
FWU		Tube	280–330			6.9	1.1–1.3
	1			0 + 67	2.1	10.2	2.5
	1			0 + 90	0.75	13.8	0.92–1.4
	3			1 + 10	2.0–3.5	20.7	0.80–2.3
						34.5	0.72–2.0
Slug	3	In Situ	83,200	0 + 56.3	4.5	8.7	2.1–3.0
	2			0 + 57.0	6.0	11.2	1.3–1.5
	1			0 + 57.6	3.0	11.2	2.7
	1			0 + 78.9	3.0	11.2	1.9
	1			0 + 79.3	4.0	9.4	2.0–2.1
	3			0 + 80	5.0	9.7	2.4–3.3

FWR flexible-wall tests on remolded specimens, *RWR* rigid-wall tests on remolded specimens, *FWU* flexible-wall tests on undisturbed specimens

Overall, the results indicate that slug testing is a useful method to determine the post-construction k of SB backfill. However, slug tests in SB walls require more time (e.g., 3–5 days) relative to a typical high-k aquifer (e.g., <<1 day). The longer durations introduce potential sources of error in the pressure data caused by changes in barometric pressure and fluctuations in the local groundwater table during the test. These sources of error were minimized in this study by (1) adjusting the pressure heads to account for barometric pressure changes based on hourly barometric pressure readings taken at an on-site weather station, and (2) adjusting the initial static water level (H_{wt}) in the test well based on measured changes in the water levels within the monitoring wells installed adjacent to the wall. Changes in H_{wt} should be minor if no precipitation events occur during a test (e.g., see Fig. 4a).

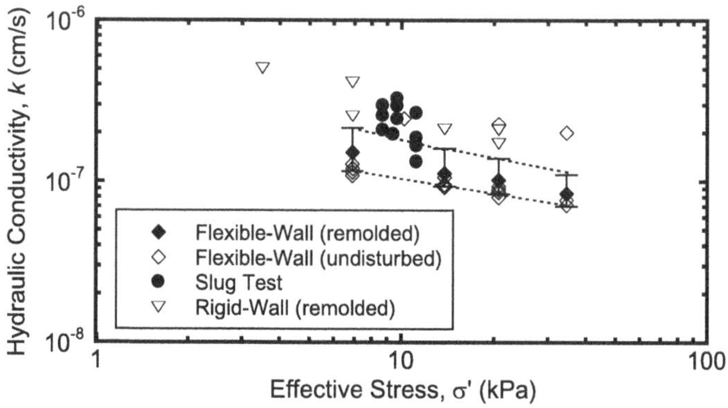

Fig. 5. Measured hydraulic conductivities plotted as a function of average effective stress.

5 Conclusions

The results of this study show that *in-situ* measurements of backfill hydraulic conductivity (k) obtained from slug tests in a field-scale soil-bentonite (SB) cutoff wall compare well with flexible-wall and rigid-wall laboratory k tests conducted on remolded or undisturbed backfill specimens at field representative stresses. The k values from the slug tests and the larger-scale rigid-wall laboratory tests generally were slightly higher than the flexible-wall k values obtained at similar stresses, indicating a slight scale effect associated with the larger test volumes in the slug tests and rigid-wall tests. Also, flexible-wall k values obtained at effective stresses greater than the measured field stresses generally were lower than the k values from the slug tests. This latter finding underscores the importance of conducting laboratory k tests at representative field stresses for SB walls, which tend to be lower than the stresses that would be predicted by assuming a geostatic stress distribution.

Overall, slug testing appears to be a viable method for determining the postconstruction *in-situ* k of SB backfill, provided that the measured pressure heads are adjusted to account for fluctuations in barometric pressure during the test. Fluctuations in the local groundwater table during the test also can be problematic if not accounted for, but can be minimized by avoiding precipitation events.

Acknowledgments. Financial support for this work was provided by the National Science Foundation (NSF Grant 1463198). The opinions and recommendations in this paper are solely those of the authors and are not necessarily consistent with the policies of NSF. The authors thank Central Builders Supply (Montandon, PA) and James Gutelius, Bucknell Director of Civil Engineering Laboratories, for their invaluable support.

References

1. Britton J, Filz G, Herring W (2004) Measuring the hydraulic conductivity of soil-bentonite backfill. J Geotech Geoenviron Eng 130(12):1250–1258
2. Malusis M, Evans J, Jacob R, Ruffing D, Barlow L, Marchiori A (2017) Construction and monitoring of an instrumented soil-bentonite cutoff wall: field research case study. In: Proceedings of 29th central Pennsylvania geotechnical conference
3. Ruffing D, Evans J, Malusis M (2010) Prediction of earth pressures in soil-bentonite cutoff walls. GeoFlorida 2010, ASCE, Reston, VA, pp 2416–2425
4. Malusis M, Barben E, Evans J (2009) Hydraulic conductivity and compressibility of a model soil-bentonite backfill amended with activated carbon. J Geotech Geoenviron Eng 135 (5):664–672
5. Hvorslev M (1951) Time lag and soil permeability in groundwater observations. Bulletin No. 36, U.S. Army Corps of Engineers Vicksburg, MS
6. Bouwer H, Rice R (1976) A slug test for determining hydraulic conductivity of unconfined aquifers with completely or partially penetrating wells. Water Resour Res 12(3):423–428
7. Teeter R, Clemence S (1986) In-place permeability measurement of slurry trench cutoff walls. In: Clemence SP (ed) Use of in-situ tests in geotechnical engineering. Geotechnical Special Publication No. 6, ASCE, Reston, VA, pp 1049–1061
8. Britton J, Filz G, Herring W (2005) Slug test in soil-bentonite cutoff walls using a push-in piezometer tip. In: Waste containment and remediation, GSP 142
9. Choi H, Daniel D (2006) Slug test analysis in vertical cutoff walls. I: analysis methods. J Geotech Geoenviron Eng 132(4):429–438
10. Lim J, Lee D, Zlotnik V, Choi H (2013) Analytical interpretation of slug test in a vertical cutoff wall. Ground Water. https://doi.org/10.1111/gwat.12033

Estimation of Vertical Barrier Performance Based on Microbial Improvement

Yu Zhang[1], Lingling Pan[1], Fei Wang[2], Ning Zhu[2], and Qiang Tang[1(✉)]

[1] School of Urban Rail Transportation, Soochow University, Yangchenghu Campus, Xiangcheng District, Suzhou 215131, China
zhangyu164@foxmail.com, linglingpan1121@foxmail.com, tangqiang@suda.edu.cn

[2] Suzhou Rail Transit Group Co., Ltd, No. 668 Ganjian Road, Suzhou 215004, China
wf5264@163.com, 18249743@qq.com

Abstract. Sanitary landfilling is nowadays the most common way to eliminate municipal solid wastes. However, the durability of landfill is mainly determined by the anti-seepage property of vertical cutoff walls. As microorganism has proved effective in plugging micro-pores, and greatly enhance the barrier performance of cutoff wall for MSW landfills. In this study, Escherichia coli. was selected for preconditioning towards compacted soil specimens and the hydraulic behaviours of the specimens were evaluated. According to the results, the huge amounts of microbial exopolysaccharides produced by Escherichia coli. encountered in the soil pores and formed a pore plugging, eventually caused 81–95% decrease in the hydraulic conductivity, from the initial value 5.3×10^{-5}–5.6×10^{-6} cm/s to the stable value 3.1×10^{-6}–2.9×10^{-7} cm/s (dropped by 1–2 orders of magnitude). The introduction of clay leads to the increase of surface roughness inside the internal pores as well as the diminution of shear forces, resulted in more effective adhesion between microorganism colonization and the surface of the particles. Subsequently, an extensive parametric analysis was conducted by numerical simulation to investigate the transport behaviour of contaminant through vertical cutoff walls. Several decrease of orders of magnitude in the hydraulic conductivity from 1×10^{-8} to 1×10^{-10} m/s resulted in the dramatical increase of the breakthrough time by 431.2% which revealed that a low hydraulic conductivity was of significance for vertical cutoff walls to achieve desirable barrier performance.

Keywords: Landfill · Cutoff wall · Barrier performance
Hydraulic conductivity · Contaminant transport · Bioclogging

1 Introduction

Currently, landfilling is the most common and essential management strategy for the disposal of municipal solid waste worldwide [1, 2]. However, the simple landfills and open dumping sites have pose a security threat on the groundwater environment due to leachate infiltration, especially in developing countries [3]. Vertical cut-off walls are

L. Zhan et al. (Eds.): ICEG 2018, ESE, pp. 300–307, 2019.
https://doi.org/10.1007/978-981-13-2224-2_37

constructed at contaminated sites to contain pollutant in an aquifer and restrict wider proliferation of contaminants in groundwater. Use of natural clay soil in landfills is preferred due to it high sorption capacity, long-term structural stability as well as extremely low hydraulic conductivity [4, 5, 25, 26]. However, clay minerals are natural un-renewable resources, but construction of the barrier wall always consumes a large amount of clay minerals to meet the anti-seepage requirements and contaminant blocking capability [6, 7]. Hence, it is essential to preprocess local materials to meet the hydraulic conductivity standard, which can save a large amount of bentonite soil for landfill cutoff walls construction.

Microbes can produce biofilms in many natural as well as engineered porous medium. An excessive bacterial growth in soil contributes to the biofilm-induced pore block, which thereby dramatically decrease the permeability and groundwater flow [8]. According to Rowe and Booker, leachate flowing through the drainage layer of landfills can cause plugging, thus promoting the microbial growth [9]. Thus, biological clogging can effectively form grout curtains to lessen the spread of heavy metals and organic contamination, and thus avoiding the leakage of construction pit, landfill, or dike. While, the microbiological species used in the earlier researches were not the main bacteria in the landfill leachate. It is generally known that the hypersaline and hypoxic environment of landfill leachate highly restrict the growth of most microbes. Grisey et al. reported the presence of microorganisms in leachate, and found that the Escherichia coli population was extremely high in the leachate [10].

The design of landfill cutoff walls depended almost entirely on the permeability of the material. Nevertheless, these barriers are not impermeable and some solute passes through them, although at rates that are considerably lower than rates of solute transport in the aquifer [11]. This is because fluid flow will still exist from molecular diffusion of pollutants across the barrier [12]. Once the leachate penetrate to the ground water and the soil, it will interact physically and chemically with both the surrounding water and the soil matrix. Consequently, accurate numerical modelling of breakthrough time of contaminant through a barrier wall is important for evaluating remedial activities.

In order to fulfill the above research requirements, this study evaluated the influence of biological clogging on the hydraulic properties of vertical cutoff wall material. Subsequently, numerical models were presented to investigate the migration of pollutants through the cutoff wall. The last section gives a sum-up about the major findings of this research.

2 Materials and Methods

2.1 Bacterium and Permeant Solutions

The type of bacteria used as an inoculum in this study was Escherichia coli (E. coli), which was the most popular bacteria existing in landfill leachate [10]. E. coli was regarded as one type of indigenous micro-organism in soil, while the environmental impact of E. coli will be strictly controlled through migration-cutting-off method proposed by Tang et al. [24]. This bacterium rapidly reproduces and is amenable to laboratory cultured, and more importantly, it is non-pathogen. The growth temperature (i.e., from 7 to 49.5 °C) and pH value (i.e., from 4.5 to 8.0) of E. coli can be adjusted in wide range. Additionally, E. coli can survive in an oxygen-deficient environment (i.e., a nM level) [13]. These features make E. coli an appealing candidate for the environmental application of biofilm.

Soil specimens under treatments with inoculating microbes were permeated with nutrient liquid to promote microbial growth and biofilms production. The culture solution used for E. coli was Luria–Butani (LB) medium which is of the following composition: tryptone, 10 g/L; yeast extract, 5 g/L; and NaCl, 10 g/L [14].

2.2 Soil

The soils used in this study include clay and sand. The clay was one type of locally available clay which collected from Suzhou, China. The sand was quartz sand in which the quartz content was about 99.3%, collected from Fengyang, Anhui, China. The relevant properties and the particle size distributions of soils are presented in Table 1.

Table 1. Physical and chemical properties of studied soils

Properties	Standard	Value	
		Clay	Sand
Physical-chemical properties			
Natural moisture content (%)	JIS A 1203	23.78	0.39
Swelling index (mL/2 g-solid)	ASTM D 5890-06	2	-
pH	JGS 0211	6.6	7.50
EC (mS/cm)	JGS 0212	0.05	0.06
N_2-BET			
Correlation coefficient (R^2)		0.999	-
Specific surface area (m^2/g)		24.49	-
Average pore Diameter (nm)		8.23	-
Total pore volume (cm^3/g)		5.04×10^{-2}	-

2.3 Sample Preparation and Hydraulic Conductivity Tests

Both clay and quartz sand were dried at 105 °C in a vacuum oven for 24 h. The clay was then mixed with quartz sand in different proportion (clay/sand = 2:8, 4:6, and 8:2). After that, the mixtures were compacted in modified rigid-wall compaction-mold permeameters (TST-55, China) following JGS 0162. Finally, use the DIW (De-ionized water) to saturate all specimens for at least 36 h.

Permeation tests were conducted by the constant-head technique according to JIS A 1218. Based on constant-head method, other influence factors will be kept stable, including hydraulic flow rate and osmotic pressure, which was mainly depend on water head. Different hydraulic flow rate and osmotic pressure play important role in the bacteria attaching and biofilms formation. At the first stage, the soil specimens were permeated with DIW to obtain initial stable hydraulic conductivity. Subsequently, the E. coli bacteria suspension cultured in a LB medium at 37 °C for 24 h was used to replace DIW, and then percolate through the soil with a certain water head. The equivalent bacteria concentration between the inlet and the outlet indicated the bacteria was uniformly distributed in the soils and the inoculation process was almost finished. Finally, the CM was used as the permeant solution to flush the inoculated specimens. The hydraulic conductivity was measured every 12 h by monitoring the volume of liquid permeated through the specimens.

3 Results and Discussions

3.1 Experimental Results and Discussions

Table 2 shows the numerical changes in hydraulic conductivity (k) of specimens permeated by DIW. It can be seen that during the permeation process, the hydraulic conductivity of the three mixtures (specimen #1–3) exhibited respective reduction about 74%, 78% and 71%. This time-resolved decrease in k permeated by DIW can be due to the dilution effect of the leached out salts. With the permeation process carried out, the leached out soluble salts became less, which brought about an increase in the thickness of the diffusion double layer and a reduction in the concentration of the outflow in response [15, 16]. Then, the lower concentration of outflow led to lower k. From Table 2, it is also found that the k decreased significantly accompanied with the clay content when permeated with DIW and the k of specimen #3 was lowest. When the sand content of the specimens increased, its gradation was harmful. A little amount of clay particles was inadequate to fill the sand pores and resulted in the great k.

Table 2. Changes in permeability of specimens permeated by DIW (Unit: cm/s)

Specimens	Initial, k	Final, k	Minimum, k	Maximum, k	Stable, k
20%	7.2×10^{-5}	5.2×10^{-5}	5.1×10^{-5}	7.8×10^{-5}	5.3×10^{-5}
40%	5.5×10^{-6}	7.5×10^{-6}	7.4×10^{-6}	9.7×10^{-6}	7.6×10^{-6}
80%	7.9×10^{-6}	5.4×10^{-6}	5.3×10^{-6}	1.0×10^{-5}	5.6×10^{-6}

When live bacterial suspensions permeated through these soil specimens, the hydraulic conductivity of which dropped to about 2.8×10^{-5} cm/s, 6.8×10^{-6} cm/s and 4.8×10^{-6} cm/s only a few hours. After CM was substituted for live bacterial cultures, all the specimens presented time dependent behaviours as shown in Fig. 1. Generally, k reduced by almost one order of magnitude after 12 days of penetration experiments with CM, and the specimen #3 achieved the lowest and the most stable value k. Compared with the original value, the final stable k for specimen #1 (3.1×10^{-6} cm/s) dropped by 94%. In terms of specimen #2 and #3, the k of clay-sand mixture decreased by 81% and 95%. The reduction of k can be related to the biofilms formation which attributed to the nutrients in CM [17]. Biofilms are communities of bacteria attaching and developing on surfaces embedded in a matrix of extracellular polymeric substances (EPS) [18]. The formation of biomass is a main product of the microbial growth and the EPS excretion [19]. Ryder et al. found that EPS provide an extensive surface area per unit volume for microorganism binding and have easier access to attach the soil surface and can lead to floc formation by agglomeration of bacteria that partially blocked the soil pores [20]. With permeation tests going, in soils containing higher quantities of biofilms there is a larger biomass of bacteria contained within those biofilms [21]. Inside the soil, the inoculated bacteria formed several layers of mesh forms of biofilm between soil particles as well as on the soil surface. The formation mesh layers effectively plugged the soil pore and led to the reduction of *k* [22].

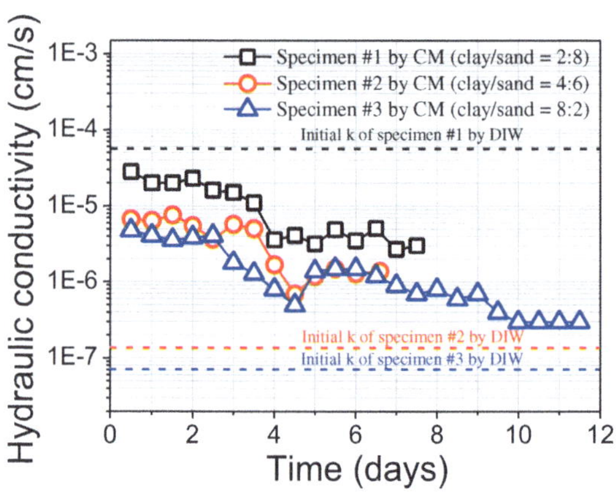

Fig. 1. The effect of microbes on permeability of soil specimens

3.2 Numerical Analysis

For evaluating the effectiveness of the vertical barrier on minimizing the migration of contaminant, a one-dimensional contaminant transport model was established in the COMSOL software. A sketch of contaminant advection-dispersion through the barrier system is presented in Fig. 2. The parametric analysis in this study focused on investigating the effect of variation of selected model parameters upon the contaminant breakthrough time, which was defined as the time required for a pollutant at the barrier exterior to reach 10% of the steady concentration of the leachate [23]. The parameters used in the breakthrough time analysis were summarized in Table 3.

In Table 3, a set of 6 different cases of contaminant diffusion results were listed. The first three cases revealed the variation of breakthrough time corresponding to different hydraulic conductivity in cutoff walls. It shows that, the drop of the *k* from 1×10^{-8} m/s to 1×10^{-10} m/s caused a significant increase in the breakthrough time from 6.0 years to 32.0 years. From these results it may be observed that the soil with low permeability were critical to restrict the transport of leachate, and thereby prevented water resource from being polluted.

Cases 4, 5 and 6 were designed to evaluate the influence of leachate head changes on the breakthrough time. As illustrated in Table 3, the dramatically decrease in the breakthrough time from 24.0 years to 2.1 years as a result of the increase of 32.3% of

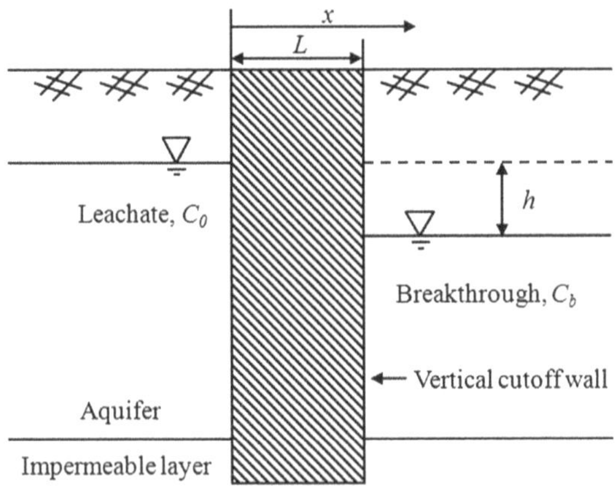

Fig. 2. Sketch of the vertical cutoff wall

Table 3. Comparison of the breakthrough time of pollutants

Cases	K (m/s)	h (m)	L (m)	D (m^2/s)	R_d	T (years)
1	1×10^{-8}	0.3	1	2×10^{-10}	1.1	6.0
2	1×10^{-9}					24.0
3	1×10^{-10}					32.0
4	1×10^{-9}	0.3	1	2×10^{-10}	1.1	24.0
5		3				6.0
6		10				2.1

leachate head. The results indicated that the higher leachate head had a significant adverse impact on the antifouling effect of the cutoff wall. Hence, it was of great significance to control the leachate head at a lower value, and further to reduce the proliferation of contaminant into the surrounding environment.

4 Conclusions

This paper investigated the ability of microbes to plug porous medium. The bacteria Escherichia coli. was selected for preconditioning towards compacted soil specimens. After that, the transport behaviours of leachate contaminants through vertical landfill barrier systems were investigated by numerical modelling. The principal conclusions can be drawn:

1. The large amounts of microbial cells present in the soil specimens apparently reduced the permeability. The effective porosity, which initially received most of the fluid flow, was significantly reduced as a result of the biofilms in the soil pores.

2. The increase of the clay content in the samples promoted the surface roughness inside the internal pores, and thereby enhanced the adhesion of microbes to soil particle surface.
3. Significant increase of breakthrough time by 431.2% as a result of the reduction of several orders of magnitude in hydraulic conductivity from 1×10^{-8} m/s to 1×10^{-10} m/s. Therefore, a low hydraulic conductivity was essential for vertical cutoff walls to effectively restrict the proliferation of pollutants.

References

1. Tang Q, Gu F, Gao YF et al (2018) Desorption characteristics of Cr(III), Mn(II) and Ni(II) in contaminated soil using citric acid and citric acid containing wastewater. Soils Found 58:50–64
2. Tang Q, Zhang Y, Gao YF et al (2017) Use of cement-chelated solidified MSWI fly ash for pavement material: mechanical and environmental evaluations. Can Geotech J 54(11):1553–1566
3. Tang Q, Kim HJ, Endo K et al (2015) Size effect on lysimeter test evaluating the properties of construction and demolition waste leachate. Soils Found 55(4):720–736
4. Tang Q, Liu W, Wang HY et al (2016) Membrane behavior of bentonite-amended Fukakusa clay under K, Na and Ca solutions. J Cent South Univ 23:3122–3131
5. Tang Q, Katsumi T, Inui T et al (2015) Influence of pH on the membrane behavior of bentonite amended Fukakusa clay. Sep Purif Technol 141:132–142
6. Parastar F, Hejazi SM, Sheikhzadeh M et al (2017) A parametric study on hydraulic conductivity and self-healing properties of geotextile clay liners used in landfills. J Environ Manag 202:29
7. Tang Q, Liu W, Li ZZ et al (2018) Removal of aqueous Cu(II) with natural kaolin: kinetics and equilibrium studies. Environ Eng Manag J 17(2):467–476
8. Rowe RK, Yu Y (2013) Modelling of leachate collection systems with filter separators in MSW landfills. ASCE J Environ Eng 139(8):1042–1052
9. Rowe RK, Booker JR (1985) 1-D pollutant migration in soils of finite depth. ASCE J Geotech Eng 111:479–499
10. Grisey E, Belle E, Dat J et al (2010) Survival of pathogenic and indicator organisms in groundwater and landfill leachate through coupling bacterial enumeration with tracer tests. Desalination 261:162–168
11. Harte PT, Konikow LF, Hornberger GZ (2006) Simulation of solute transport across low-permeability barrier walls. J Contam Hydrol 85(3):247–270
12. Daniel DE, Shackelford CD (1988) Disposal barriers that release contaminants only by molecular diffusion. Nucl Chem Waste Manag 8:299–305
13. Sugnaux M, Mermoud S, Da CA et al (2013) Probing electron transfer with Escherichia coli: a method to examine exoelectronics in microbial fuel cell type systems. Bioresour Technol 148:567–573
14. Gao M, Song HY, Zheng HZ et al (2016) Culture of low density E. coli cells in alginate–chitosan microcapsules facilitates stress resistance by up-regulating luxS/AI-2 system. Carbohyd Polym 14:160–165
15. Tang Q, Katsumi T, Inui T et al (2014) Membrane behavior of bentonite amended compacted clay. Soils Found 54(3):329–344

16. Tang Q, Katsumi T, Inui T et al (2015) Membrane behavior of bentonite-amended compacted clay towards Zn(II) and Pb(II). Membr Water Treat 6(5):393–409
17. Carter MQ, Louie JW, Feng D et al (2016) Curli fimbriae are conditionally required in Escherichia coli O157:H7 for initial attachment and biofilm formation. Food Microbiol 57:81–89
18. Carrel M, Morales VL, Beltran MA et al (2018) Biofilms in 3d porous media: delineating the influence of the pore network geometry, flow and mass transfer on biofilm development. Water Res 134:280–291
19. Bottero S, Storck T, Heimovaara TJ et al (2013) Biofilm development and the dynamics of preferential flow paths in porous media. Biofouling 29(9):1069–1086
20. Ryder C, Byrd M, Wozniak DJ (2007) Role of polysaccharides in Pseudomonas aeruginosa biofilm development. Curr Opin Microbiol 10:644–648
21. Phillips MC, Solo-Gabriele HM, Reniers AJHM et al (2011) Pore water transport of enterococci out of beach sediments. Mar Pollut Bull 62:2293–2298
22. Kim G (2004) Hydraulic conductivity change of bio-barrier formed in the subsurface by the adverse conditions including freeze-thaw cycles. Cold Reg Sci Technol 38:153–164
23. Peters GP, Smith DW (2001) Solute transport through a deforming porous medium. Int J Numer Anal Methods 26(7):783–788
24. Tang Q, Gu F, Zhang Y, Zhang YQ, Mo JL (2018) Impact of biological clogging on the barrier performance of landfill liners. J Environ Manag 222:44–53
25. Tang Q, Tang XW, Hu MM, Li ZZ, Chen YM, Lou P (2010) Removal of Cd(II) from aqueous solution with activated Firmiana Simplex Leaf: behaviors and affecting factors. J Hazard Mater 179:95–103
26. Tang Q, Tang XW, Li ZZ, Chen YM, Kou NY, Sun ZF (2009) Adsorption and desorption behaviour of Pb(II) on a natural kaolin: equilibrium, kinetic and thermodynamic studies. J Chem Technol Biotechnol 84:1371–1380

Identifying Active Methanotrophs and Mitigation of CH_4 Emissions in Landfill Cover Soil

Raksha K. Rai[1], Jyoti K. Chetri[1], Stefan J. Green[2], and Krishna R. Reddy[1(✉)]

[1] Department of Civil and Materials Engineering, University of Illinois at Chicago, 842 West Taylor Street, Chicago, IL 60607, USA
{rrai5, jkc4, kreddy}@uic.edu
[2] Sequencing Core, Research Resources Center and Department of Biological Sciences, University of Illinois at Chicago, 835 S. Wolcott, Chicago, IL 60612, USA
greendna@uic.edu

Abstract. In the USA, municipal solid waste (MSW) landfills constitute one of the major anthropogenic sources of methane emissions. In the landfill cover soils employed at MSW landfills, aerobic methane-oxidizing bacteria (MOB) convert CH_4 to CO_2, thereby partially mitigating the CH_4 emissions to the atmosphere. In this study, culture-dependent and culture-independent techniques were employed to evaluate methane oxidation capacity and to characterize the microbial community in landfill cover soil. Microcosms with synthetic landfill gas headspace were used to measure potential methane oxidation rates in landfill cover soil and in methanotrophs-enriched microbial consortia. The results demonstrate that the enriched landfill cover soil supported the growth of a diverse group of methanotrophic and methylotrophic microorganisms, and were dominated by Type I methanotrophs showing positive correlation with CH_4 oxidation rates.

Keywords: Methanotrophs · CH_4 oxidation activity · Landfill cover soil
Microbial community

1 Introduction

In the USA, landfills are estimated to be the third largest anthropogenic source of CH_4 emissions making up 16.4% of the total CH_4 emissions in 2016 (USEPA 2018). Despite significant amount of methane emitted from landfills, it is estimated that between 10 and 90% is actively been consumed by the methane oxidizing bacteria (MOB) in landfill cover soils as reviewed by Semrau et al. (2010). MOB, also known as methanotrophs, are a subset of a larger microbial community called Methylotrophs. The methanotrophs utilize methane as a sole source of carbon and energy, whereas the methylotrophs use C1-compounds as their source of carbon and energy (Hanson and Hanson 1996). Methanotrophs within the phylum Proteobacteria are classified into three phylogenetically distinct groups: Type I, Type II and Type X methanotrophs, where Type I and

L. Zhan et al. (Eds.): ICEG 2018, ESE, pp. 308–316, 2019.
https://doi.org/10.1007/978-981-13-2224-2_38

Type X are grouped in Gamma-proteobacteria and Type II within Alpha-proteobacteria. Type I, Type II and Type X methanotrophs are ubiquitous in nature and are usually found in abundance where low to high concentration of methane prevails.

Landfill cover soils are typically dominated with either by Type I or Type II methanotrophs as reported in many studies (Yargicoglu and Reddy 2017; Cébron et al. 2007). Molecular ecology studies have utilized two distinct approaches for identification of methanotrophs from the environmental samples: cultivation-based enrichment and isolation approaches, and cultivation-independent molecular tools (Murrell et al. 1998) employing targeted amplification and sequencing of functional or structural genes or shotgun sequencing approaches. In some cases, both approaches are used concurrently, with molecular tools used to monitor and characterize enrichments and isolates. Targeted amplification protocols typically target structural genes such as the microbial small subunit ribosomal RNA genes (*i.e.*, 16S rRNA gene), as well as methane monooxygenase (MMO) genes and genes involved in C1 compound oxidation.

This study focuses on adopting PCR-based high-throughput amplicon sequencing technique to analyze microbial structure in landfill cover soil (LFCS) and in microcosms studies inoculated with LFCS. The specific objectives of this research were to: (1) characterize methanotrophic communities in enriched landfill cover soil using 16S rRNA gene analysis, (2) Conduct batch enrichments methanotrophic bacteria from LFCS, and follow enrichment patterns using 16S rRNA gene analysis, and (3) assess the relationship between methane oxidation rates and the relative abundance of methanotrophic community in LFCS. This study is a part of broader on-going study funded by the U.S. National Science Foundation with ultimate goal to evaluate system parameters that control microbial diversity and activity and design optimal and efficient biocover systems to mitigate CH_4 emissions at landfills.

2 Materials and Methods

2.1 Soil Enrichment

Soil was collected from Zion landfill site, located in Greater Chicago area, Illinois, USA. Soil samples were collected from an interim cover layer at a depth of ~1 to 2 feet and were shipped to the Geotechnical and Geoenvironmental Engineering Laboratory at the University of Illinois at Chicago (UIC) where it was stored at room temperature (23 °C). Soil samples were air dried (moisture content < 0.5%), pulverized and screened through 2 mm sieve prior to the inoculation of batch reactors. To obtain methanotroph-enriched consortia, approximately 5 g of sieved soil was mixed with 100 ml of modified NMS medium (Whittenbury et al. 1970) in a 500 ml serum vial and stoppered using long sleeved rubber septa. Approximately 80 ml of air from the headspace was replaced with equal volume of mix gas CH_4/CO_2 to achieve a headspace concentration of 7% CH_4 (v/v) and 7% CO_2 (v/v) balanced in air (86%) and were incubated for 20 days at 23 °C. To determine the activity of Methanotrophs and methane oxidation rates, gas samples were analyzed at regular intervals using Gas Chromatography (GC) and were monitored until the methane concentration dropped to less than 1%. To enrich the methanotrophic culture the mix gas (CH_4/CO_2) was replenished twice throughout the enrichment. The enriched soil was later stored in micro-centrifuge tubes and frozen at −20 °C for DNA extraction and molecular analysis. Similarly, the supernatant enriched with methanotrophic cells consortium were also pelletized in 2 ml micro-centrifuge tubes by centrifuging at 12,500 RPM for 15 min, decanting the supernatant and freezing at −20°C for DNA extraction and molecular analysis.

2.2 Culture Experiments

Prior to culture experiments, serum vials, rubber septa and pipettes were sterilized using a Napco Model 8000-DSE autoclave operated at >120 °C for a minimum of 60 min to ensure complete sterilization. The supernatant obtained from the enrichment mentioned above was used in this experiment. 1 ml of enriched culture was inoculated in 9 ml modified NMS medium (total of 10 ml), placed in 125 ml serum vials and sealed air tight using butyl rubber septa followed by crimp cap. Approximately 20 ml of air from the headspace was replaced with equal volume of synthetic landfill gas comprising of 50% (v/v) CH_4 and 50% (v/v) CO_2 to achieve a headspace concentration of 5% CH_4 (v/v) and 5% CO_2 (v/v) balanced in air (90%). To determine changes in the headspace concentration, gas samples were analyzed every alternate day using Gas Chromatography (GC) until the headspace concentration dropped to less than 1%. All the experiments were conducted in triplicates along with the controls (media-NMS). The rates of CH_4 oxidation were determined from linear regression analysis of methane concentration with respect to time based on zero-order kinetics observed during testing. pH of the culture along with controls (NMS only) were also measured at the beginning and end of the experiment to examine any changes in the pH due to microbial activity.

2.3 DNA Extraction, PCR Amplification and Next Generation Sequencing

To measure microbial diversity in enriched soils and in culture consortia, genomic DNA was extracted from the samples using DNeasy Power Soil Kit (Qiagen). Extractions were performed according to the manufacturer's instructions, with slight modifications. Briefly, samples were heated at 65 °C for 10 min before homogenization with FastPrep-24 5G bead-beating device (MP Biomedicals) at 6 m/s for 40 s. Genomic DNA was used as template for PCR amplification with primers 515F-modified and 926R (Walters et al. 2016), targeting the V4–V5 variable region of the microbial small subunit ribosomal RNA gene using a two-stage "targeted amplicon sequencing (TAS)" protocol (Green et al. 2015; Bybee et al. 2011). The primers contained 5′ common sequence tags (known as common sequence 1 and 2, CS1 and CS2) as described previously (Moonsamy et al. 2013). The CS1_515F and CS2_926R primer sequences were ACACTGACGACATGGTTCTACAGTGYCAGCMGCCGCGGTAA and TACGGTAGCAGAGACTTGGTCTCCGYCAATTYMTTTRAGTTT, respectively, with the underlined regions indicating the common sequence tags.

First stage PCR amplifications were performed in 10 μl reactions in 96-well plates, using the MyTaq HS 2X mastermix (Bioline, Taunton, MA). PCR conditions were 95 °C for 5 min, followed by 28 cycles of 95 °C for 30″, 50 °C for 60″ and 72 °C for 90″. Subsequently, a second PCR amplification was performed in 10 μl reactions in 96-well plates. Each well received a separate primer pair with a unique 10-base barcode, obtained from the Access Array Barcode Library for Illumina (Fluidigm, South San Francisco, CA; Item# 100-4876), as well as 1 μl of 1st stage PCR product. Cycling conditions were as follows: 95 °C for 5 min, followed by 8 cycles of 95 °C for 30″, 60 °C for 30″ and 72 °C for 30″. Libraries were loaded onto a MiSeq v3 flow cell and sequenced using an Illumina MiSeq sequencer. Raw sequence data were processed and merged using software package PEAR (Zhang et al. 2013), followed by quality checking (Q20), length trimming (>300 bp) and chimera checking using the UCHIME algorithm as compared with the Silva 119 16S.97 database (Edgar 2010). After chimera removal, the software package QIIME (Caporaso et al. 2010) was used to annotate sequences and generate annotation tables using a sub-OTU protocol. Briefly, all sequences were pooled, and unique sequences were de-replicated from the combined sequenced. Those sequences with counts greater than 10 were used as seed or master sequences for clustering. Low abundance sequences (fewer than 10) were queried against the master sequences using USEARCH to find the master sequence with a minimum percent identity of 98%; for matching sequences, the counts of the low abundance sequences were incorporated into the counts for the cluster. Taxonomic annotations were assigned to each seed and independent low abundance sequence using USEARCH and the Silva 119 reference database. Taxonomic and abundance data were merged into a single sequence table (seq table. biome) and summaries of absolute abundances of taxa were generated for all phyla, classes, orders, families, genera, and species present in the dataset. Library preparation, pooling and Illumina sequencing were performed at the University of Illinois at Chicago Sequencing Core (UICSQC), and basic bioinformatics processing of the data were performed at the UIC Research Informatics Core (RIC).

2.4 Gas Analysis

Gas samples were collected at regular time intervals and analyzed for CH_4, CO_2 and O_2 concentrations using an SRI 9300 Gas Chromatography (GC) equipped with thermal conductivity detector (TCD) and CTR-1 column that separates N_2 and O_2 for simultaneous analysis of CO_2, CH_4, O_2 and N_2. Gas samples were withdrawn using 1-ml syringe where 0.5 ml of gas sample was injected into the GC equipped with TCD. A calibration curve for a minimum of three points was established using high purity standard gas mixtures ranging from 1 to 50% CH_4 and CO_2.

3 Results and Discussion

3.1 Methane Oxidation

Figures 1(a) and (b) show a decrease in the headspace methane concentration with time in both enrichment and culture batch tests, respectively. An initial lag phase of four days was observed in enrichment batch tests, after which there was a rapid decrease in the methane headspace concentration, consistent with microbial methane oxidation. This is in agreement with prior studies showing similar lag phases (10–12 days) in microcosms inoculated with field samples without pre-incubation (Spokas and Bogner 2011). Culture batch tests also showed a lag phase of 24 h. Lag phases are generally due to the inoculation of microorganisms into the fresh media (NMS) and the time taken for responding or adjusting to the new environment. Methane oxidation rates were calculated using linear regression that followed zero-order kinetics. Maximum oxidation rates in the soil suspension and culture batch tests were calculated to be 123.3 μg CH_4/g dry-soil/day and 73.1 μg CH_4/ml/day, respectively. The pH of the culture sets was also measured at the beginning and end of the experiment to examine if changes in the pH had any impacts on methane oxidation rates. Results showed that the pH remained stable and was in the range of 6.7–7.1 throughout the experiment.

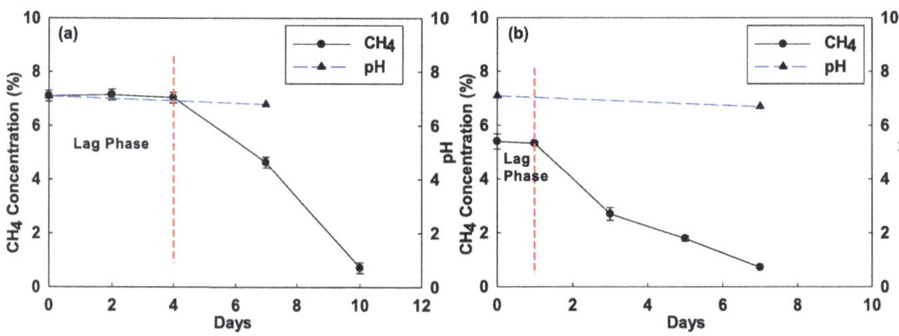

Fig. 1. Methane oxidation in: (a) Enriched soil, and (b) Culture consortium

3.2 Microbial Community Composition

The microbial community present in the enriched soil and culture consortium were analyzed using PCR amplification and high-throughput next generation sequencing of 16SrRNA genes. The results of the taxonomical classification for both samples are summarized below, with methanotrophic and methylotrophic taxa indicated.

Figure 2 shows the phylum-level composition of enriched soil and culture consortium microbial communities. The enriched soil sample was dominated by the phyla Proteobacteria (73%) and Bacteriodetes (9.1%), with lower levels of Acidobacteria (4.1%), Actinobacteria (2.8%), Verricomicrobia (2.31%), Firmicutes (1.1%), Planctomycetes (0.8%) and others. In the culture consortium, 79% of all sequences were annotated to the phylum Proteobacteria followed by Verrucomicrobia (5.7%), Bacteriodetes (2.9%) and others (<1%).

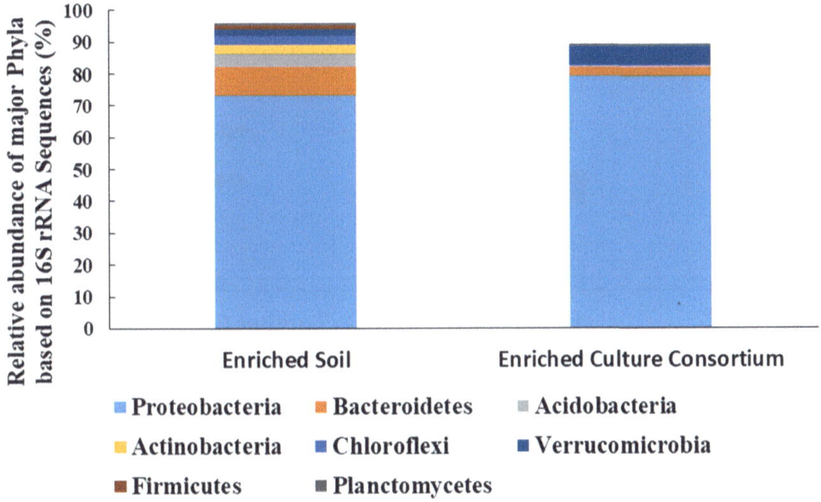

Fig. 2. Relative abundance (%) of sequences related to dominant phyla

Figure 3 shows the genus-level composition of enriched soil and culture consortium microbial communities. Sequences from different clades of methane-oxidizing bacteria, along with non-methane oxidizing methylotrophs, were identified in these samples. The most abundant methanotrophic taxa detected in both soil and cultures were *Methylobacter* accounting for 31% and 39% of the total 16SrRNA sequences identified, followed by *Methylovorus* (6.6% in soil and 4.7% in culture), and *Methylocystis* (2.6% in soil and 1.6% in culture). Some sequences recovered from these samples were not annotated at the taxonomic level of genus but could still be identified as being derived from putative methylotrophs. These sequences were derived from members of the families Methylophilaceae, Methylocystaceae, Methylococcacea, Crenotrichaceae, and Methylobacteriaceae. The majority (7.4%) of these sequences were derived from bacteria belonging to the family Methylophilaceae.

Kallistova et al. (2013) and Han et al. (2016) detected *Methylovorus* in landfill cover soil and were successful in cultivating them in laboratory, suggesting that their occurrence in landfill cover soils is not unusual and is consistent with the community composition found in the current study.

Methylobacter has been identified as one of the major genera of methanotrophic community present in abundance in most of the landfill cover soil (Yargicoglu and Reddy 2017; Cébron et al. 2007). *Methylobacter* are Type I methanotrophs that were found in abundance in the cover soil in the present study and could be responsible for oxidizing methane at faster rates, but this cannot be affirmed with our results alone as the 16SrRNA gene analysis does not determine the active members or function of the microbes/microbial community (Cébron et al. 2007). However, DNA profiling of microbial community structure can identify active microorganisms when sampling is performed longitudinally, and strong shifts are observed, or if analysis of RNA is performed.

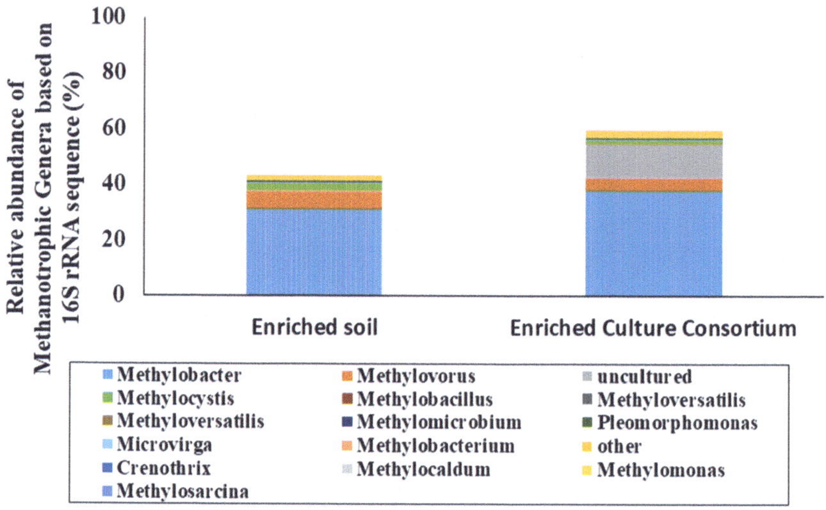

Fig. 3. Relative abundance (%) of sequences related to Methanotrophic genera

Overall, a number of methanotrophic genera (Type I and Type II) were detected in both soil and culture samples including *Methylobacter, Methylocystis, Methylomicrobium, Pleamorphomonas, Methylomonas, Methylobacterium, Methylosarcina, Methylocaldum, and Crenothrix*. More broadly, methylotrophic bacteria from the genera *Methylovorus, Methyloversatilis, Methylobacillus, Microvirga*, and from *OM43 clade* were detected. Sequences derived from methanotrophs were most frequently annotated as belonging to the genus *Methylobacter*, and second most frequently annotated as *Methylovorus glucosetrophus SIP3-4*. Due to the limited taxonomic resolution of the V4-V5 region of the microbial 16S rRNA gene, species-level annotation could not be obtained. Future studies in which shotgun metagenome sequencing (i.e., non-targeted, PCR-independent deep sequencing of microbial genomic DNA) will be performed, that will be used to identify taxa to the level of species. Shotgun metagenomic sequence data will also be used to identify taxon-specific DNA sequences of functional genes that will be used to develop quantitative assays for MMO messenger RNA molecules.

Of the total microbial community analyzed, the landfill cover soil that was enriched in laboratory with modified NMS showed a relative abundance of 39.5% of Type I methanotrophs, 1.5% of Type II methanotrophs, and 18.9% of methylotrophs in the culture consortium. Similarly, 33% of Type I methanotrophs, 2.8% of Type II methanotrophs, and 7.6% of methylotrophs were identified in enriched soil. From Fig. 4, it can be concluded that enrichment favored for cultivating majority of Type I methanotrophic bacteria and methylotrophic bacteria. The reason for not cultivating Type II methanotrophs in our cultures could possibly be the substrate (methane) limitation, as Type II methanotrophs usually dominate at high CH_4 concentration and low O_2 concentration or the nutrient rich NMS that usually promotes the growth of Type I methanotrophs inhibiting the growth of Type II methanotrophs (Wise et al. 1999). In addition, high methane oxidation potential from both enrichment and culture batch tests could possibly confirm the presence of abundant methanotrophic/methylotrophic community.

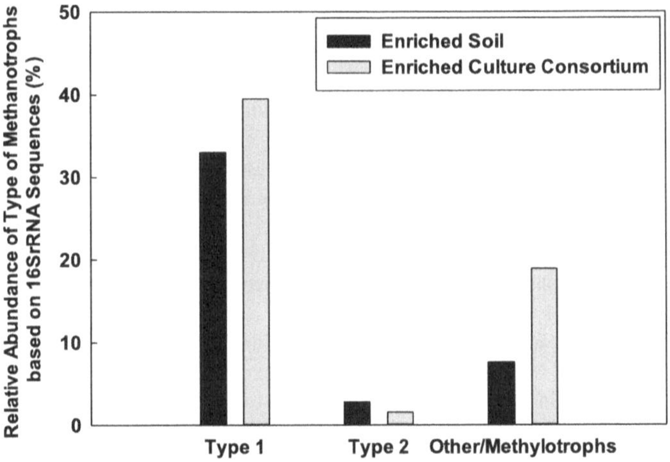

Fig. 4. Relative abundance (%) of sequences based on the types of methanotrophs

4 Conclusions

The following conclusions can be drawn from this study:

- Enrichment technique favored the maximum growth of Type I methanotrophs of the genera *Methylobacter* that were present in abundance constituting about 31% (enriched soil) and 39% (enriched culture) of the total 16SrRNA gene amplicon sequences identified. The growth of Type II methanotrophs were possibly inhibited due to lack of substrates needed for growth (1.5–2.6%). In future, shotgun metagenome sequencing shall be performed to identify taxa at the species level.
- High methane oxidation rates of 123.3 μg CH_4/g dry-soil/day (enriched soil) and 73.1 μg CH_4/ml/day (microbial culture) were observed in this study that correlate to the relative abundance of methanotrophic and methylotrophic community in both the enriched LFCS and the culture consortium.
- Overall, the enriched landfill cover soil was dominated by Type I methanotrophs (33%) followed by Methylotrophs (7.6%) and Type II methanotrophs (2.6%).

The present study provided initial insight into microbial diversity in the landfill cover soil under typical field condition. Additional DNA profiling and/or RNA analyses are being performed to identify active microorganisms. Series of microcosms are also being tested under variable moisture, pH, and temperature conditions to assess the resiliency of methanotrophs for methane oxidation. In addition, enhanced methane oxidation with an organic amendment such as biochar in the cover soil is also being investigated.

Acknowledgements. This project is funded by the U.S. National Science Foundation (Grant CMMI #1724773), which is gratefully acknowledged.

References

Bybee SM, Bracken-Grissom H, Haynes BD, Hermansen RA, Byers RL, Clement MJ et al (2011) Targeted amplicon sequencing (TAS): a scalable next-gen approach to multilocus, multitaxa phylogenetics. Genome Biol Evol 3:1312–1323

Caporaso JG, Kuczynski J, Stombaugh J, Bittinger K, Bushman FD, Costello EK et al (2010) QIIME allows analysis of high-throughput community sequencing data. Nat Methods 7 (5):335

Cébron A, Bodrossy L, Chen Y, Singer AC, Thompson IP, Prosser JI, Murrell JC (2007) Identity of active methanotrophs in landfill cover soil as revealed by DNA-stable isotope probing. FEMS Microbiol Ecol 62(1):12–23

Edgar RC (2010) Search and clustering orders of magnitude faster than BLAST. Bioinformatics 26(19):2460–2461

Green SJ, Venkatramanan R, Naqib A (2015) Deconstructing the polymerase chain reaction: understanding and correcting bias associated with primer degeneracies and primer-template mismatches. PLoS ONE 10(5):e0128122

Han JS, Mahanty B, Yoon SU, Kim CG (2016) Activity of a methanotrophic consortium isolated from landfill cover soil: response to temperature, pH, CO2, and porous adsorbent. Geomicrobiol J 33(10):878–885

Kallistova AY, Montonen L, Jurgens G, Münster U, Kevbrina MV, Nozhevnikova AN (2013) Culturable psychrotolerant methanotrophic bacteria in landfill cover soil. Microbiology 82 (6):847–855

Moonsamy PV, Williams T, Bonella P, Holcomb CL, Höglund BN, Hillman G, Simen BB (2013) High throughput HLA genotyping using 454 sequencing and the fluidigm access array™ system for simplified amplicon library preparation. HLA 81(3):141–149

Murrell JC, McDonald IR, Bourne DG (1998) Molecular methods for the study of methanotroph ecology. FEMS Microbiol Ecol 27(2):103–114

Semrau JD, DiSpirito AA, Yoon S (2010) Methanotrophs and copper. FEMS Microbiol Rev 34 (4):496–531

Spokas KA, Bogner JE (2011) Limits and dynamics of methane oxidation in landfill cover soils. Waste Manag 31(5):823–832

United States Environmental protection Agency (USEPA 2018) Inventory of greenhouse gas emissions and sinks (1990–2016)

Walters W, Hyde ER, Berg-Lyons D, Ackermann G, Humphrey G, Parada A, Apprill A (2016) Improved bacterial 16S rRNA gene (V4 and V4-5) and fungal internal transcribed spacer marker gene primers for microbial community surveys. Msystems 1(1):e00009–e00015

Wise MG, McArthur JV, Shimkets LJ (1999) Methanotroph diversity in landfill soil: isolation of novel type I and type II methanotrophs whose presence was suggested by culture-independent 16S ribosomal DNA analysis. Appl Environ Microbiol 65(11):4887–4897

Whittenbury R, Phillips KC, Wilkinson JF (1970) Enrichment, isolation and some properties of methane-utilizing bacteria. Microbiology 61(2):205–218

Yargicoglu EN, Reddy KR (2017) Microbial abundance and activity in biochar-amended landfill cover soils: evidence from large-scale column and field experiments. J Environ Eng 143 (9):04017058

Zhang J, Kobert K, Flouri T, Stamatakis A (2013) PEAR: a fast and accurate illumina paired-end reAd mergeR. Bioinformatics 30(5):614–620

Compressibility of Four Types of Granular Bentonite Materials

Guoliang Ma[1], Huyuan Zhang[1,2(✉)], and Zhenyan Su[1]

[1] School of Civil Engineering and Mechanics, Lanzhou University, Lanzhou, China
Zhanghuyuan@lzu.edu.cn

[2] Key Laboratory of Mechanics on Disaster and Environment in Western China (Lanzhou University), Ministry of Education, Lanzhou, China

Abstract. Granular bentonite materials are some types of buffer/backfill materials with good performance, which are constituted by bentonite pellets or by pellets and bentonite powder. The dry density is the key parameter which controls the engineering properties of buffer/backfill materials. The grain size distribution will influence compressibility and dry density of the compressed granular bentonite materials. In the present work, four types of granular bentonite materials and bentonite powder had been selected and the compressibility and the degree of compaction of these packing materials had been studied. The results showed that the granular bentonite material contains 30% (in weight) powder and 70% (in weight) pellets sized between 2 and 5 mm had the best cost performance, such as, higher dry density, higher compacted dry density, easy to produce.

Keywords: HLW disposal · Granular bentonite materials · Compressibility Compacted dry density

1 Introduction

Deep geology disposal method has been acknowledged as an effective way to dispose HLW. Beishan had been recognized as the preferred position for the depository which surrounded by granite. The pre-compacted blocks had been selected as the main buffer materials around canisters, while the tunnels and shafts, away from canisters, would be backfilled by materials compacted in situ [1]. Granular bentonite materials are one kind of these candidate backfilling materials.

The materials have several advantages, such as, high expansibility, low permeability, high radionuclide retardation properties, easy to compaction. The materials also can backfill irregular holes. Many laboratory tests and mock-up tests had been conducted to study the properties of granular bentonite materials in European countries. However, in China, in order to improve the performance of engineering barrier, much more work needs to be conducted on granular bentonite materials.

Hydro-mechanical properties of granular bentonite materials had been studied and the results showed that the performance of the materials is closely related to their dry

L. Zhan et al. (Eds.): ICEG 2018, ESE, pp. 317–325, 2019.
https://doi.org/10.1007/978-981-13-2224-2_39

densities [2–4]. The dry densities are higher, the performance of the materials is better. So, it is essential to compact the materials to a higher dry density.

Tang et al. [5] studied the compressibility of two types of granular bentonite materials and the results showed that the coarser one is easier to reach a higher dry density than the finer one. Hiroshi [6] studied the influence of grain size distribution and types of bentonite on the compaction properties, which reported that the plastic limit is the controllable parameter for the compressibility, while grain size distribution had little influence. Hoffman [2] studied the compressibility of granular bentonite materials have 1.3 g/cm^3 and 1.5 g/cm^3 in dry density and reported that the initial dry density has an obvious influence on compressibility.

In the present work, four types of granular bentonite materials and bentonite powder had been selected aimed to study the compressibility and degree of compaction of these materials.

2 Materials and Methods

2.1 Materials

The granular bentonite materials are constituted by bentonite pellets or by bentonite pellets and bentonite powder. Bentonite pellets are prepared from bentonite powder. The basic properties of the bentonite powder are given in Table 1. The compaction curve of the bentonite powder under 175 MPa is showed in Fig. 1. Processes to prepare pellets are showed in Fig. 2. Dry density of bentonite pellets is 1.79 g/cm^3 and water content of bentonite pellets is 3%. Some preheated bentonite powder had been maintained indoors and water content is 6.5%.

Table 1. Physical properties of bentonite powder

Particle density (g/cm^3)	Liquid limit (%)	Plastic limit (%)	Montmorillonite (%)	Percentage of particles with sizes < 75 mm
2.70	152	28	44.4	100

According to the size of samples and NHRI [7], the 5 mm (1/4 of the height of samples) was selected as the largest grain size. There were four types of granular bentonite materials had been selected. Their grain size distribution is showed in Fig. 3. When the same grain groups are used for granular bentonite materials, these grain size distributions would make the packing dry densities maximum [2, 8, 9].

The tested samples are prepared into cutting rings according to the test for minimum densities of sand [7]. The prepared samples are showed in Fig. 4. During the tests, bentonite powder had been served as a reference. The packing dry density of powder is 0.777 g/cm^3 and the initial water content is 6.5%.

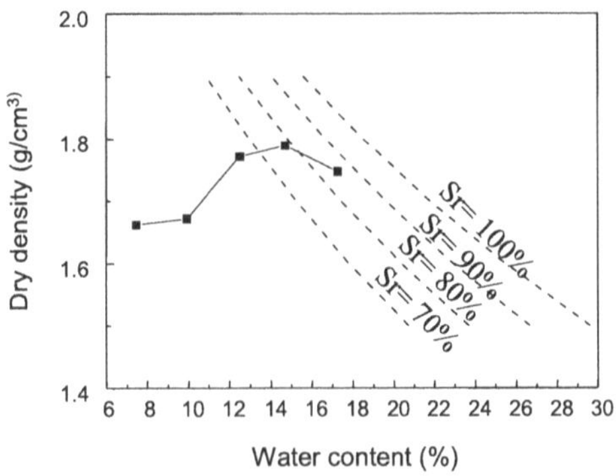

Fig. 1. Compaction curve of bentonite powder.

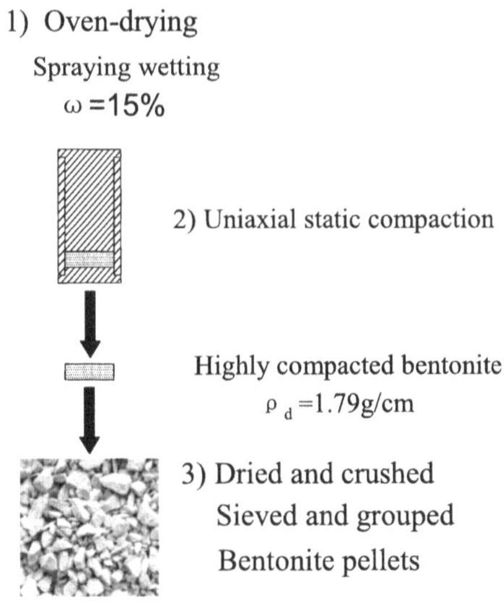

Fig. 2. Processes to prepare pellets

2.2 Method

The compressibility of the materials had been investigated by one-dimension oedometer. The loading path is showed in Fig. 5. From 1 to 5, the packing samples suffered loading (1–2), unloading (2–3), reloading (3–4) and unloading (4–5).

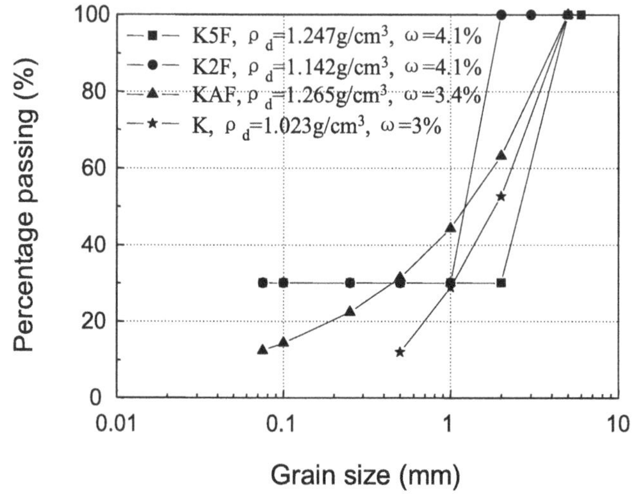

Fig. 3. Grain size distribution curves of granular bentonite

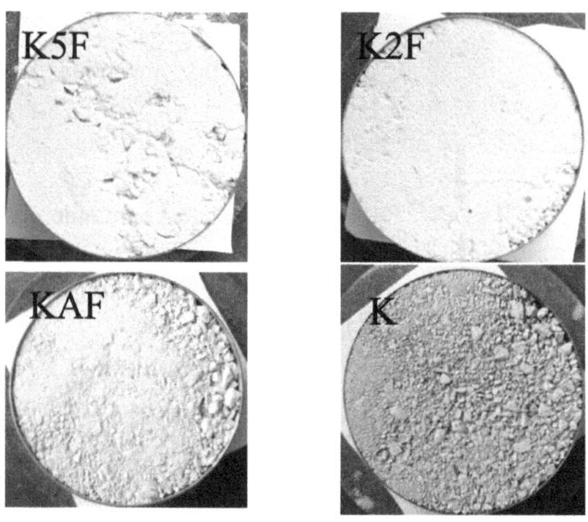

Fig. 4. Granular bentonite materials specimens for tests

3 Results and Discussion

3.1 Compression Curve and Degree of Compaction

Figure 6 shows the compression curve for granular bentonite materials. The curve for bentonite powder is also showed in Fig. 6. It can be seen that all of curves for granular bentonite materials are lower than that of powder. This means that the dry densities of granular bentonite materials after compression under each vertical stress are higher than that of powder. It shows that granular bentonite materials are easy to compaction.

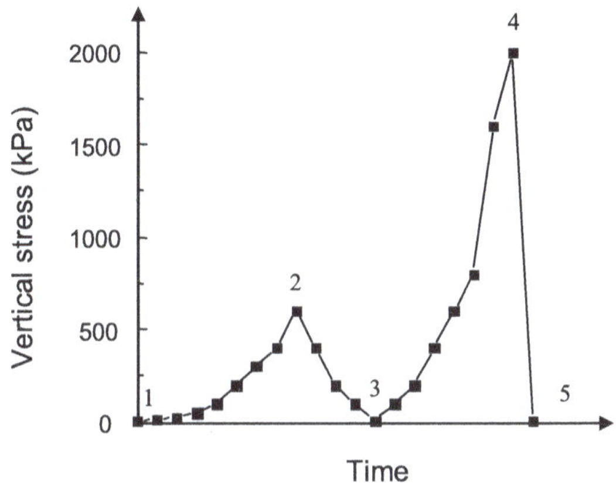

Fig. 5. Loading path for compressibility tests

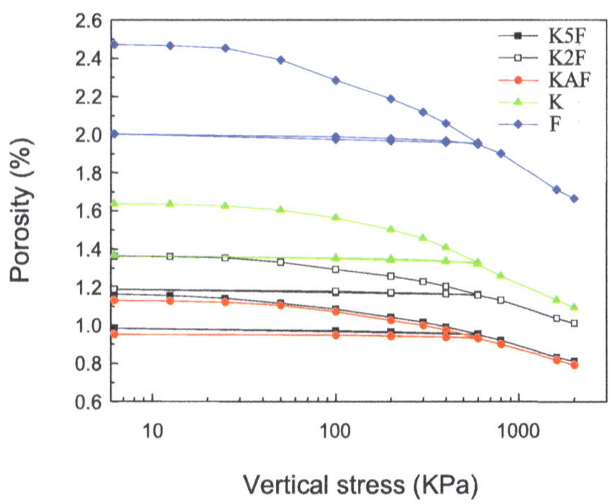

Fig. 6. Compression curves for granular bentonite materials

Among four granular bentonite materials, the curve for K is higher than those for other three. K is constituted by pellets whose sizes are between 0.5 and 5 mm, while the other three are constituted by pellets and powder. This means that compared to mixtures constituted by pellets solely, the mixtures with pellets and powder will obtain a higher dry density after compression.

Although there are 70% pellets (in weight) in two types of materials, the curve for K5F is lower than that for K2F. This may be caused by K5F contains pellets between 2 mm and 5 mm, while K2F contains pellets between 1 mm and 2 mm. This means that the coarser materials will be compacted to a higher dry density under the same vertical stress. This result is similar to that of Tang [5].

KAF has an ideal grain size distribution [8]. It can be seen from Fig. 6, with this gradation, KAF would reach the maximum packing dry density and degree of compaction, while K5F almost obtained the same dry densities after packing or compression.

As a result of too many grain groups in KAF materials, it is difficult to mix all the pellets homogeneously and it would cause the separation of coarser pellets and finer pellets. K5F materials are easier to mix, and the mixing of sand pellets and powder had been already studied by Zhang [10].

In the one hand, KAF had all of the pellets smaller than 5 mm, while K5F only contained the pellets between 2 and 5 mm, which was more accessible compared to other smaller pellets. In the other hand, there are 30% (in weight) powder in K5F, while 12% (in weight) in KAF, and powder need no more process, which can be used in their original form.

So, it can be said that the cost performance of K5F is higher than that of KAF, K2F and K.

3.2 Compression Index

Figure 7 shows the compression index of four granular bentonite materials and bentonite powder. It can be seen from Fig. 7 that compression index of F is larger than that of K5F, K2F, KAF and K.

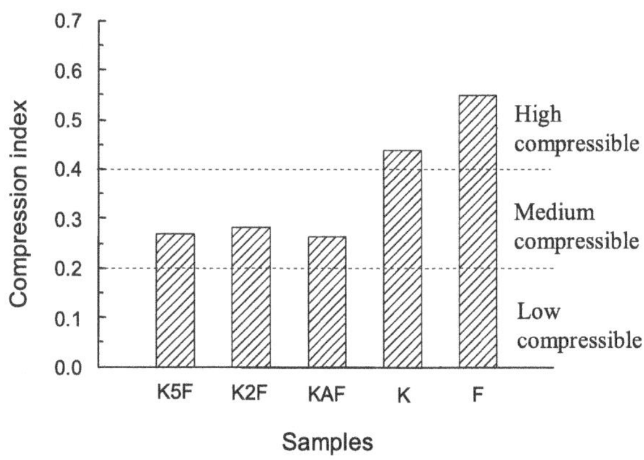

Fig. 7. Compression index of the tested materials

Compression index is in inverse proportion to initial packing dry density. The initial packing dry densities for KAF, K5F, K2F, K and F are 1.265 g/cm^3, 1.247 g/cm^3, 1.142 g/cm^3, 1.023 g/cm^3 and 0.777 g/cm^3, respectively. The compression indexes are 0.262, 0.267, 0.280, 0.436 and 0.546, respectively.

The powder F and granular bentonite material K is high-compressible, while granular bentonite materials K5F, K2F and KAF are medium-compressible.

3.3 Swelling Index

Figure 8 shows the swelling indexes for four granular bentonite materials and bentonite powder when materials suffered the maximum vertical stress of 600 kPa and 2000 kPa.

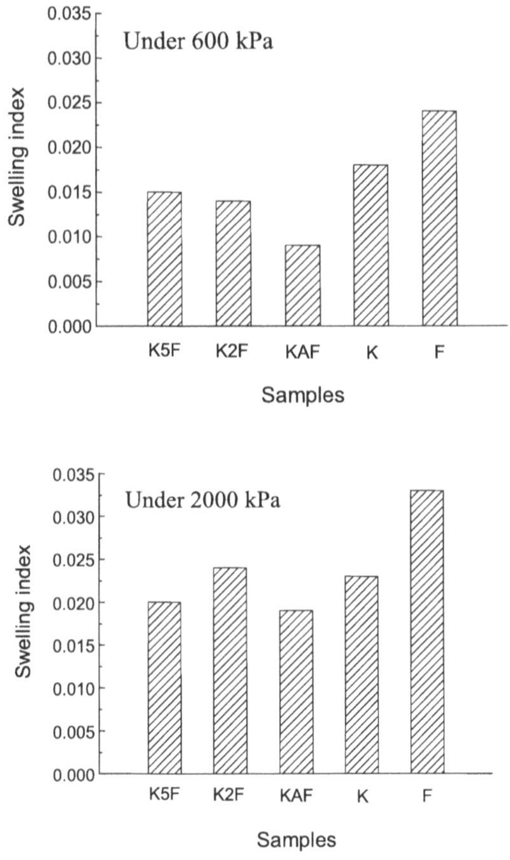

Fig. 8. Swelling index of the tested materials

From Fig. 8, it can be found that for the two cases, swelling indexes for all five materials are all smaller than 0.035. Powder F had the largest swelling index and the change of dry densities would be largest after removal of vertical stress. This means that the addition of pellets decreases the swelling behaviour of bentonite powder.

It also can be found that when the materials suffered a higher vertical stress, the swelling index would be larger. This means that as the vertical stress increases, the difference between dry densities before and after removal of vertical stress increases. Similar conclusions can be found from literature Stone [11].

4 Conclusions

In order to study the compressibility of four types of packing granular bentonite materials that would be candidates for HLW disposal tunnels. The one-dimension compression tests had been conducted on these four packing materials and packing bentonite powder. The results show that:

(1) After compressed under the same vertical stress, dry densities of granular bentonite materials are higher than those of bentonite powder.
(2) Among these four granular bentonite materials, the material contains 30% (in weight) powder and 70% (in weight) pellets sized between 2 and 5 mm had the best cost performance, such as, higher packing dry density, higher compacted dry density, easy to produce and so on.
(3) The compression index for powder is larger than those for granular bentonite materials, and the compression index is in inverse proportion to initial packing dry densities.
(4) The swelling indexes decrease as the pellets added into bentonite powder and the swelling indexes increase as the materials suffered a higher vertical stress.

Acknowledgement. The authors want to acknowledge the financial support provided by Project 41672261 supported by National Natural Science Foundation of China and the financial support provided by Project 2007-834 supported by the State Administration of Science, Technology and Industry for National Defence, PRC.

References

1. Zhang H, Cui S, Zhang M (2012) Swelling behaviors of GMZ bentonite-sand mixtures inundated in NaCl-Na_2SO_4 solutions. Nucl Eng Des 242:115–123
2. Hoffmann C, Alonso E, Romero E (2007) Hydro-mechanical behaviour of bentonite pellet mixtures. Phys Chem Earth 32:832–849
3. Pusch R, Bluemling P, Johnson L (2003) Performance of strongly compressed MX-80 pellets under repository-like conditions. Appl Clay Sci 23:239–244
4. Imbert C, Villar M (2006) Hydro-mechanical response of a bentonite pellets/powder mixture upon in filtration. Appl Clay Sci 32:197–209
5. Tang C, Cui Y, Tang A (2009) Compression behaviors of crushed COx argillite used as backfilling material in HLW repository. Chin J Rock Mech Eng 23(12):2459–2465 (in Chinese)
6. Ito H (2006) Compaction properties of granular bentonites. Appl Clay Sci 31:47–55
7. NHRI (1999) Specification of soil test. The ministry of water resources of the People's Republic of China, Beijing (in Chinese)
8. Mayor J, Garcia-Sineriz J, Alonso E (2005) Engineered barrier emplacement experiment in Opalinus Clay for the disposal of radioactive waste in underground repositories. Enresa Publications, Tecnicas
9. Zhang H, Wang X, Liu P (2016) Sealing and healing of compacted bentonite block joints in HLW disposal. Chin J Rock Mech Eng 35(Suppl 2):3605–3614 (in Chinese)

10. Zhang M, Zhang H, Jia L (2012) Method for sample preparation of buffer/backfilling materials in laboratory. J Build Mater 15(5):638–643 (in Chinese)
11. Stone J, Larson W (1980) Rebound of five one-dimensionally compressed unsaturated granular soils. Soil Sci Soc Am J 44(4):819–822

Effect of Powder Particle Size on the WRC of GMZ Bentonite

Bao Chen[1], Yiyi Huang[1(✉)], Kang Zhang[1], and Chilong Li[2]

[1] Key Laboratory of Geotechnical and Underground Engineering of Ministry of Education, Tongji University, Shanghai 200092, China
1410161@tongji.edu.cn
[2] Hunan Xiangjiang New Area Investment Group Co., Ltd., Changsha 410205, Hunan Province, China

Abstract. To unveil the effect of powder particle size on the water retention characteristic (WRC) of compacted GMZ bentonite under a state of constant volume, powder particle size and shape distributions of two GMZ bentonites (GMZ01 and GMZ06) were investigated by employing the dynamic digital imaging technology and WRC tests related to these two bentonites were conducted with both osmotic and air phase methods used for suction control. It turns out that under a state of constant volume, the water content increases gradually with the decreasing suction, and the effect of powder particle size on the WRC seems quite faint within a high suction range but relatively obvious within a low suction range.

Keywords: GMZ bentonite · Dynamic image analysis · Suction control
Particle shape and size · Water retention characteristic

1 Introduction

GMZ bentonite is a candidate buffer/backfill material for the intended HLW (high level radioactive waste) geological disposal repositories in China. Thus, due to the widely applied nuclear technology and the rapid development of nuclear power industry, it is estimated that huge amount of bentonite will be used in the near future, which may apply a relatively heavy economic burden on the national finance [1]. On other hand, for a given type of bentonite, the powder particle size will not only affect the performance of bentonite as buffer/backfill material, but also the manufacturing-handling cost. Generally, the smaller the powder particle size the more complex the manufacturing and handling process and thus the higher the cost. Hence, in an economic sense, it is better to use large powder particles. At the same time, in order to ensure the safety of the whole disposal system, it is of paramount importance to investigate the effect of powder particle size on the hydro-mechanical behavior of bentonite.

The water retention characteristic (WRC), one of the most important characteristics of the buffer materials, is defined by the relationship between suction and water content (gravimetric or volumetric). This relationship may affect significantly the hydro-mechanical behavior of these materials including the unsaturated hydraulic conductivity, the deformation characteristic and the shearing strength [2]. With the WRC, the

L. Zhan et al. (Eds.): ICEG 2018, ESE, pp. 326–334, 2019.
https://doi.org/10.1007/978-981-13-2224-2_40

seepage and strength behavior of bentonite can be predicted, the volume and microstructure changes can be analyzed, and the water transfer can be analyzed together with the hydraulic conductivity of materials [3–8]. Many works have been done focusing on the WRC of bentonite. However, to the authors' knowledge, the influence of powder particle size on the WRC has not been studied yet, which constitutes the main objective of this study.

As for the methods of particle size analysis, there are traditionally 3 methods, sieving analysis, elutriation analysis and laser diffraction method. With sieving analysis, a consecutive particle size distribution curve cannot be obtained for bentonite [9]; with elutriation analysis, the result can be inaccurate for the colloid aggregation caused by montmorillonite, the main ingredient in bentonite, that will be hydrated and swell in water, although well-defined particle size distribution can be obtained [10]. With laser diffraction method, the obtained particle size is actually the diameter of the equivalent sphere whose diffraction angle is exactly the same as that of the real particle which will bring about an inaccurate particle size that is larger than the real one when confronted with non-spherical particles [11, 12]. Dynamic digital imaging technology is a newly-born technology for analyzing the distributions of particle size and shape. It only requires simple operation procedures and short testing time, and allows dispersing the aggregative powder particles when the powder is dry. As a result, this technology has been widely applied in the powder-handling-and-testing industry [13–15].

In this paper, the powder particle size and shape distributions of two GMZ bentonites (GMZ01 and GMZ06) were analyzed using the dynamic digital imaging technology. Moreover, the WRC related to these two GMZ bentonites was also investigated under constant volume condition. The results obtained provide useful references for the selection and production of GMZ bentonite.

2 Methods

2.1 Experimental Material

The experimental material was a sodium bentonite taken from the GMZ bentonite mineral deposit in Inner Mongolia, whose main minerals are 74% montmorillonite, 9.0% cristobalite, 8.9% quartz and 5.8% feldspar. The raw bentonite was air-dried and then ground respectively into GMZ01 sample with a powder particle size corresponding to 200 mesh and GMZ06 sample with a powder particle size corresponding to 100 mesh. The main physical and chemical properties of this sodium GMZ bentonite are given in Table 1.

2.2 Specimen Preparation

40 compacted GMZ bentonite specimens were made using the aforementioned experimental material (including GMZ01 and GMZ06). Each specimen was marked with a code "Si" (i = 1–40) and confined in a special steel mold shown in Fig. 1. The main properties of these 40 specimens are displayed in Table 2.

Table 1. Physical and chemical parameters of the experimental material.

Parameter	Value
G_s	2.66
pH	8.68–9.86
Liquid limit (%)	276
Plastic limit (%)	37
Total specific surface area (m^2/g)	570
Main exchange ions (mmol/100 g)	Na^+ (43.36), Ca^{2+} (29.14) Mg^{2+} (12.33), K^+ (2.51)
Main minerals	Montmorillonite (74.0%) Christobalite (9.0%) Quartz (8.9%) Feldspar (5.8%)

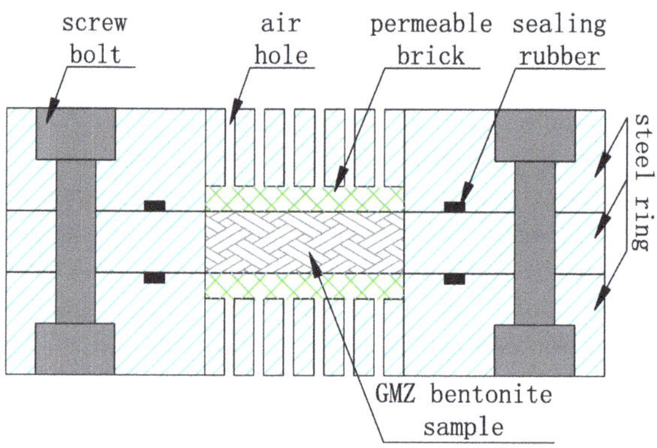

Fig. 1. Compacted GMZ bentonite sample confined in a special steel mold

Table 2. Properties of the specimens.

Specimen number	S1–S10	S11–S20	S21–S30	S31–S40
Bentonite type	GMZ01	GMZ01	GMZ06	GMZ06
Initial dry density (g/cm^3)	1.7	1.3	1.7	1.3
Diameter (cm)	2	2	2	2
Height (cm)	0.6	0.6	0.6	0.6
Residual water content (%)	7.69	7.69	9.72	9.72

2.3 Analysis of Particle Size and Shape

In this study, a particle analyzer Camsizer XT (Retsch Technology, Germany) was chosen to analyze the particle size and shape of the two GMZ bentonites, GMZ01 and GMZ06. The mass of bentonite powder to be tested was set as 5 g for both bentonites and the dispersion pressures for GMZ01 and GMZ06 were respectively 80 kPa and

40 kPa. The bentonite powder was dispersed and accelerated to a velocity of 50 m/s by compressed air. With a double-camera system, both particle size and particle shape were measured at a high accuracy.

2.4 WRC Experiment

For a high suction range (higher than 2 MPa), the vapor equilibrium method was employed to control suction. According to this method (see Fig. 2), a steel mold confining a GMZ bentonite specimen is put onto a dry dish, on the bottom of which a saturated saline solution is put to control the relative humidity [16, 17]. Over time, the water content of the specimen in the dry dish tends to a constant value if the dry dish is sealed well and the surrounding temperature remains unchanged. A definite water content value gives a corresponding definite suction value. Using this method, the suction value of a GMZ bentonite specimen can be controlled by simply changing the type of the saturated saline solution. In this study, the suction was controlled by 7 different saturated saline solutions. The types of saline solutions and the corresponding suction values are summarized in Table 3. These 7 saturated saline solutions respectively filled 7 dry dishes. In each dry dish, there were 4 GMZ bentonite specimens (S1; S11; S21; S31 in dish1, S2; S12; S22; S32 in dish2, and so forth). Each of the 28 specimens was weighed regularly until all the specimens stayed constant in weight, which served as the indication that all the specimens had reached the equilibrium in terms of relative humidity. The duration for all of the 28 specimens to reach equilibrium was approximately 20 days. Afterwards, the water content of each specimen was determined, allowing the determination of the water retention curve in the high suction range.

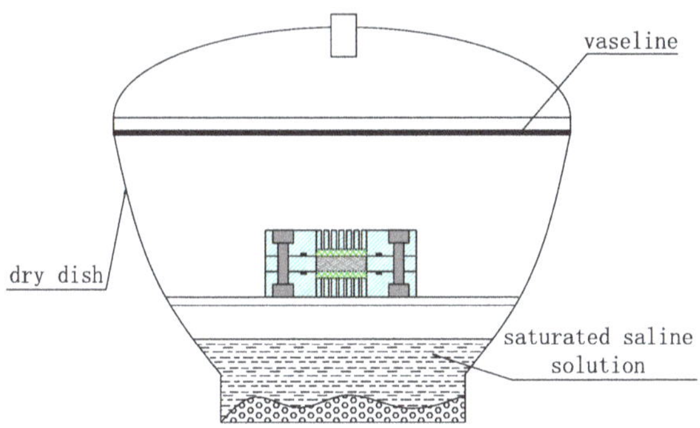

Fig. 2. Schematic diagram of the vapor equilibrium method

For a low suction range (lower than 2 MPa), the osmotic method was chosen for controlling suction. With this technique, water transfer is caused by the process of osmosis. As shown in Fig. 3, a GMZ bentonite specimen confined in a steel mold is put in a beaker full of PEG (polyethylene glycol whose molecular weight may vary from 1000 to 20000) solution and sealed by a polyethylene membrane. Moreover, the specimen is sandwiched between two semi-permeable membranes which are permeable

Table 3. Saturated saline solutions and the corresponding suctions.

Saturated saline solution	Suction (MPa)
K_2SO_4	4.2
KNO_3	9.0
KCl	21
NaCl	38
$Mg(NO_3)_2$	82
$Mg(Cl)_2$	150
LiCl	309

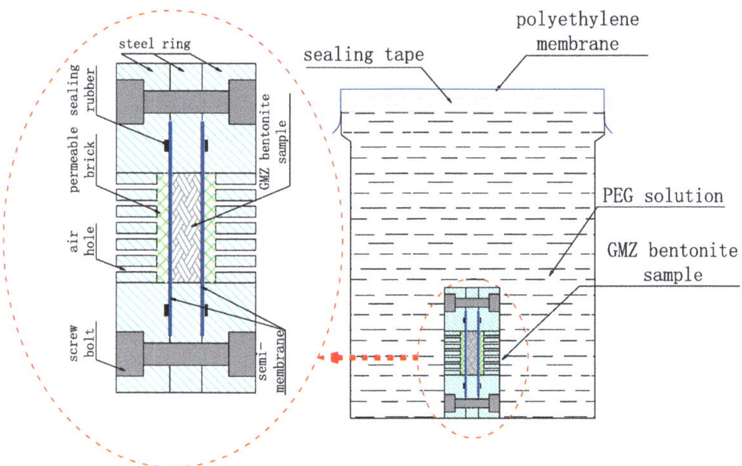

Fig. 3. Schematic diagram of the osmotic method

Table 4. Suction values and the corresponding B_r values, PEG and semi-membrane types.

Suction (MPa)	B_r (°)	PEG solution	Semi-membrane
1	20.8	PEG 6000	MWCO 3500
0.1	7.8	PEG 20000	MWCO 14000
0.01	2.6	PEG 20000	MWCO 14000

to the water molecules but impermeable to the large-sized PEG molecules, resulting in an osmotic suction applied to the specimen through the membranes. Finally, the suction in the sample reaches equilibrium with that in the PEG solution. Given that the quantity of the water transporting between the specimen and the PEG solution is quite limited with respect to the total volume of the PEG solution, the suction in the PEG solution can be regarded as constant during the whole test. Hence, in this osmotic system, the suction value depends solely on the concentration of the solution and the higher the concentration the higher the suction. The concentration of the PEG solution can be reflected by the Brix index B_r (°) and the higher the concentration the higher the B_r. The relationship between B_r (°) and suction s (MPa) was presented in [17]. With this relationship, the B_r values corresponding to the three chosen suction values (1 MPa, 0.1 MPa and 0.01 MPa) can be calculated and displayed in Table 4. An ATAGO hand-refractometer was employed to determine the B_r values. After preparation of these three PEG solutions, three beakers were used to contain them, respectively. The PEG solutions and the corresponding semi-membranes for suctions 1 MPa, 0.1 MPa and 0.01 MPa are also shown in Table 4. Each beaker contained four different specimens (S8; S18; S28; S38 in beaker1, S9; S19; S29; S39 in beaker2, S10; S20; S30; S40 in beaker3). The duration for all of the 12 specimens to reach equilibrium in suction was about 10 days. After the equilibrium, the water content of each specimen was determined, allowing the determination of the water retention curve for the low suction range.

3 Results and Discussions

3.1 Particle Size and Shape

The results for particle size and shape mainly include: (a) cumulative particle size distribution curves; (b) distributive particle size distribution curves; (c) particle shape distribution curves. Items (a) and (b) are respectively shown in Fig. 4(a) and (b), while item (c) is shown in Fig. 5. Some important characteristic parameters of particle size and shape are displayed in Table 5.

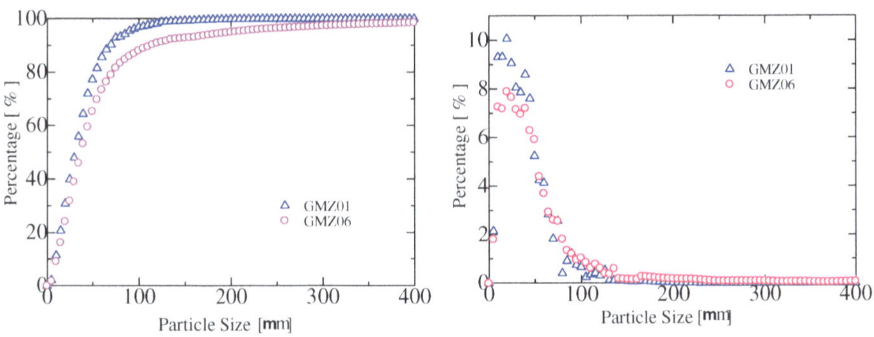

(a) Cumulative distributions (b) Distributive distributions

Fig. 4. Cumulative and distributive particle size distributions of GMZ01 and GMZ06.

From Fig. 4(a) and (b), it is found that the particle size at which the percentage reaches 100% is 250 μm for GMZ01 and 902 μm for GMZ06. That is to say, the breadth of the cumulative particle size distribution curve of GMZ06 is much larger than that of GMZ01, indicating that GMZ06 possesses a wider particle size range than GMZ01. From Fig. 4(b) it can be figured out that GMZ01 overdoes GMZ06 in the

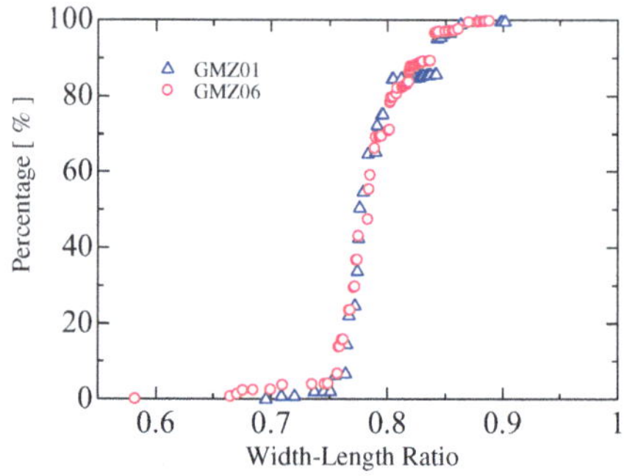

Fig. 5. Particle shape distributions of GMZ01 and GMZ06

Table 5. Characteristic parameters of particle size and shape.

Bentonite type	Largest particle size (μm)	D10; D50; D90 (μm)	Width-length ratio at 50%	Width-length ratio at 90%
GMZ01	250	9.24; 31.24; 69.05	0.776	0.842
GMZ06	902	10.63; 37.74; 112.88	0.783	0.837

percentage of particles less than 60 μm but surrenders to GMZ06 in the percentage of particles more than 60 μm, indicating that GMZ01 contains more fine grains than GMZ06. Both bentonites share scarcity in particles larger than 140 μm. Besides, the "D50" of GMZ01 is 31.24 μm while that of GMZ06 is 37.74 μm (Table 5), indicating that the average level of the particle size of GMZ06 is a little larger than that of GMZ01.

From Fig. 5, it appears obvious that the width-length ratios of GMZ01 and GMZ06 concentrate in a range from 0.7 to 0.9. This suggests that both bentonites possess non-spherical particle shape. Moreover, Table 5 shows that the width-length ratios corresponding to the percentages of 50% and 90% are respectively 0.776 and 0.842 for GMZ01, and respectively 0.783 and 0.837 for GMZ06. This shows that there is only a faint difference between GMZ01 and GMZ06 in terms of distribution of particle shape, and the grinding procedures have limited effect on the particle size distributions of both bentonites.

3.2 Water Retention Properties

With the results obtained in the WRC experiment, the water retention curves of "GMZ01; 1.3 g/cm^3", "GMZ01; 1.7 g/cm^3", "GMZ06; 1.3 g/cm^3" and "GMZ06; 1.7 g/cm^3" can be plotted in Fig. 6.

From Fig. 6, it appears obvious that in the high suction range (from 4.2 MPa to 309 MPa), neither the bentonite type nor the initial dry density affects the water retention curves. However, in the low suction range (from 0.01 MPa to 1.0 MPa), the effect of bentonite type is significant. Moreover, the lower the dry density the higher the

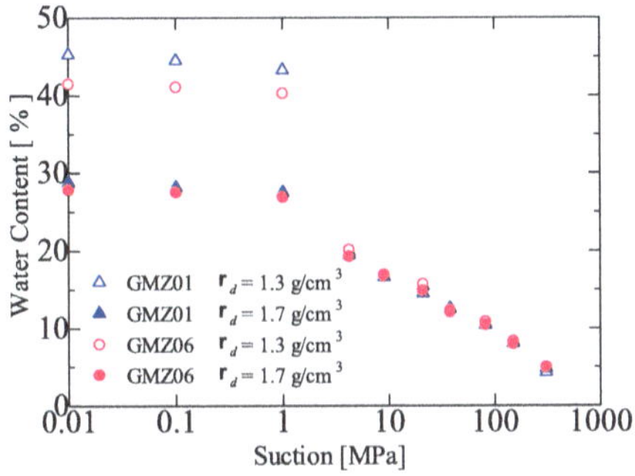

Fig. 6. Water retention curves of GMZ01 and GMZ06 at different initial dry densities

influence of the bentonite type: in the suction range from 0.01 MPa to 1.0 MPa, the curves of "GMZ01; 1.7 g/cm^3" and "GMZ06; 1.7 g/cm^3" almost overlap, while the curves of "GMZ01; 1.3 g/cm^3" and "GMZ06; 1.3 g/cm^3" keep a significant gap. This can be explained as follows: compacted soils are characterized by double structure with macro-pores between aggregates and micro-pore in the aggregates. As the high suctions are governed by intra-aggregate micro-pores, they are only dependent on the natural density of bentonite before grinding. Thus, the compacted sample density does not affect the water retention curve in the high suction range. By contrast, the low suctions are controlled by inter-aggregate macro-pores, and they are thus strongly dependent on the compacted sample density. As far as the difference between 1.3 g/cm^3 and 1.7 g/cm^3 densities are concerned, with a higher density as 1.7 g/cm^3 the size of macro-pores is decreased by mechanical compaction, thus the effect of density becomes less pronounced.

4 Conclusion

The particle size and shape of two different bentonites (GMZ01 and GMZ06) were determined and analyzed, and the water retention capacities of these two bentonites were investigated experimentally under constant volume condition. The obtained results allow the following conclusions to be drawn.

For the particle size distribution, GMZ06 possesses a wider range of particle size than GMZ01; within a same value of mass, GMZ01 contains more fine grains than GMZ06; GMZ06 overdoes GMZ01 in the average level of particle size. For the particle shape distribution, the width-length ratios of both bentonites concentrate in a range from 0.7 to 0.9 and both bentonites belong to bentonites with non-spherically irregular particle shape; the difference between the particle shape distributions of GMZ01 and GMZ06 is very little and the grinding procedures have limited effect on the particle shape distributions of the two bentonites.

The influence of particle size on the water retention curve is largely determined by the microstructure. For "low dry density and high suction" or "high dry density and low suction" or "high dry density and high suction", the influence of particle size on the water retention characteristic will be very faint. Only under a relatively low effective hydrostatic pressure (low dry density and low suction), the bentonite with more fine grains and a smaller average particle size (GMZ01) will have a higher water content than that with less fine grains and a larger average particle size (GMZ06).

Acknowledgements. This study was supported by the National Natural Science Foundation of China (41372270).

References

1. Liu YM, Xu GQ, Liu SF, Chen ZR (2001) Study on compactibility and swelling property of buffer/backfill material for HLW repository. Uranium Geol 17(1):44–47 (in Chinese)
2. Sun DA, Meng DL, Sun WJ, Liu YM (2011) Soil–water characteristic curves of two bentonites. Rock Soil Mech 32(4):973–979 (in Chinese)
3. Wiebe B, Graham J, Tang GX, Dixon D (1998) Influence of pressure, saturation, and temperature on the behaviour of unsaturated sand-bentonite. Can Geotech J 35:194–205
4. Blatz JA, Graham J, Chandler NA (2002) Influence of suction on the strength and stiffness of compacted sand-bentonite. Can Geotech J 39:1005–1015
5. Lloret A, Villar MV, Sanchez M, Gens A, Pintado X, Alonso EE (2003) Mechanical behaviour of heavily compacted bentonite under high suction changes. Géotechnique 53 (1):27–40
6. Cui YJ, Tang AM, Loiseau C, Delage P (2008) Determining the unsaturated hydraulic conductivity of a compacted sand-bentonite mixture under constant-volume and free-swell conditions. Phys Chem Earth 33:S462–S471
7. Ye WM, Cui YJ, Qian LX, Chen B (2009) An experimental study of the water transfer through confined compacted GMZ bentonite. Eng Geol 108:169–176
8. Ye WM, Zhang YW, Chen B, Zheng ZJ, Chen YG, Cui YJ (2012) Investigation on compression behaviour of highly compacted GMZ01 bentonite with suction and temperature control. Nucl Eng Des 252:11–18
9. Zhou ZX, Zheng SQ (2013) Application of laser particle size analyzer to the bentonite industry. China Non-metallic Min Ind Herald 4:12–13 (in Chinese)
10. Zou KS, Liu JC, Meng PF (2009) Optimum selection of particle size analysis of bentonite and quartz. J Earth Sci Environ 31(3):277–280 (in Chinese)
11. Jonasz M (1991) Size, shape, composition and structure of microparticles from light scattering. In: Principles, methods, and application of particle size analysis, pp 143–162
12. Cheng P, Cao S, Li XS (2001) Evaluation of a wide range laser particle size analyses and comparison with pipette and sieving methods. Acta Sedimentol Sin 19(3):449–455 (in Chinese)
13. Patchigolla K, Wilkinson D (2009) Crystal shape characterisation of dry samples using microscopic and dynamic image analysis. Part Part Syst Charact 26(4):171–178
14. Mohamed AB (2014) Micronized Ilmenite—a non-damaging non-sagging new weight material for drilling fluids. SPE Bergen One Day Seminar, Society of Petroleum Engineers
15. Strondl A, Lyckfeldt O, Brodin H, Ackelid U (2015) Characterization and control of powder properties for additive manufacturing. JOM 67(3):549–554
16. Schneider A (1960) Neue Diagramme zur Bestimmung der relativen Luftfeuchtigkeit über gesättigten wässerigen Salzlösungen und wässerigen Schwefelsäurelösungen bei verschiedenen Temperaturen. Holz als Rohund Werkstoff 18:269–272
17. Delage P, Howat MD, Cui YJ (1998) The relationship between suction and swelling properties in a heavily compacted unsaturated clay. Eng Geol 50(1):31–48

Applications of Geomembrane Cutoff Walls in Remediation of Contaminated Sites

Xuede Qian[1(✉)], Zhonghua Zheng[2], Zhi Guo[2], Changqing Qi[2], Liqi Liu[2], Yong Liu[2], Shengli Zhen[2], Shiwen Ding[2], Jing Jin[2], Yang Wang[2], and Yimin Ge[2]

[1] Michigan Department of Environmental Quality, 525 W Allegan Street, Lansing, MI 48909, USA
qianx@michigan.gov
[2] Beijing GeoEnviron Engineering & Technology, Inc., 9 Dijin Road, Beijing 100095, China

Abstract. A composite cutoff wall consisting of HDPE geomembrane combined with soil-bentonite backfill can be effectively applied to site remediation projects. This technique was firstly successfully applied by Beijing GeoEnviron Engineering & Technology, Inc. (BGE) in a copper-containing acidic liquid leakage project in China in 2011. Then, BGE successfully applied and promoted this technique in many other projects in recent 7 years. Some specific key sealing materials were developed to achieve the effective bonding between geomembrane and bedrock and clay aquifers. In addition, a special machine for installing geomembranes was also designed. The maximum geomembrane inserting depth on site reached to 32 m. Geomembrane composite cutoff wall is considered to be currently the safest and most effective vertical barrier technique to block horizontal pollution migration.

Keywords: HDPE geomembrane · Vertical barrier · Interlock Key-in

1 Overview

Geomembrane cutoff wall is a vertical barrier by using high-density polyethylene (HDPE) geomembrane as the main barrier material. The HDPE geomembrane is vertically inserted into the relatively impervious layer at a certain depth by vertical trenching or other methods. At the same time, the special grouting materials are grouted to seal the bottom of the geomembrane to bond between the geomembrane and the impervious layer. Both sides of the geomembrane need to be uniformly backfilled with low-permeability soil-bentonite mixture or plastic concrete mixtures to form a continuous, thorough and complete flexible vertical composite barrier to achieve the purpose of blocking the horizontal migration of pollutants (Fig. 1) (Koerner and Guglielmetti [1]; Zhen et al. [2]). Geosynthetic materials such as HDPE can improve the performance of a traditional slurry walls by (1) decreasing permeability of the barrier by as much as two orders of magnitude and (2) improving chemical resistance of the barrier. Interlocking HDPE has been developed to improve joint seals.

L. Zhan et al. (Eds.): ICEG 2018, ESE, pp. 335–342, 2019.
https://doi.org/10.1007/978-981-13-2224-2_41

Hydrophillic gaskets that expand when in contact with water can be used in the interlocks between the geomembrane panels (USEPA [3]).

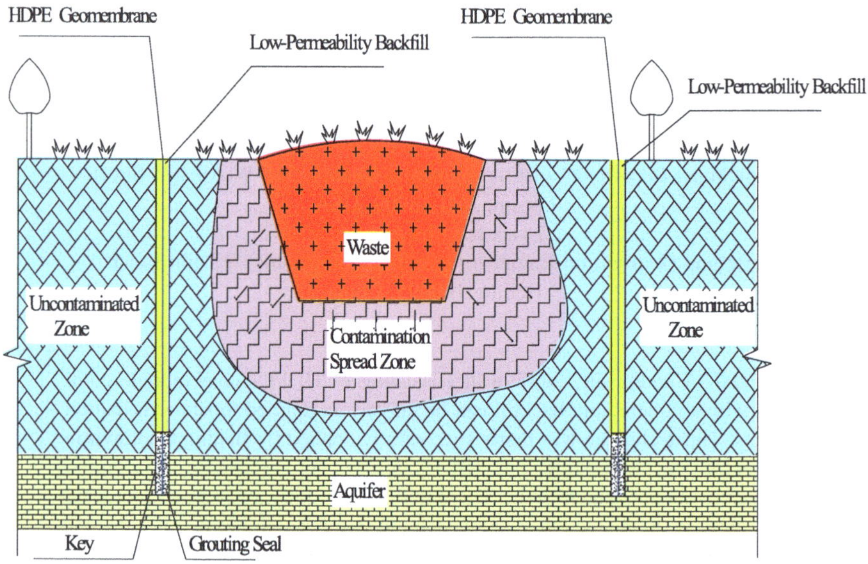

Fig. 1. Schematic diagram of the function of geomembrane composite cutoff wall

2 Design Considerations

Before designing and constructing a geomembrane cutoff wall, the depth and size of the contaminated site, the physical and chemical properties of the contaminated materials, and the depth of contamination should be fully ascertained. The grouting key sealing material should be able to withstand the long-term erosion and corrosion of the pollutants, and the permeability coefficient must be always less than 1.0×10^{-7} cm/s. The depth of vertical cutoff wall must be greater than the contaminated depth to block the pollutant migration.

The geotechnical and hydrogeological conditions of the site must be investigated in detail, including underground soil and rock strata structures, groundwater flow direction, aquifer nature and depth, topography, and site stability condition. The layout and depth of the cutoff wall should be determined according to the groundwater flow direction, the depth of the aquifer, the geological conditions above the aquifer, and the ground facilities around the contaminated site. The cutoff wall should be embedded in the aquifer with a permeability coefficient of not more than 1.0×10^{-7} cm/s, and the key-in depth should be not less than 1.0 m.

3 Installation Methods

There are several different installation methods currently available for the construction of geomembrane cutoff walls (Daniel and Koerner [4]; Bouazza et al. [5]). The slurry supported trench installation method is most widely used in China. This method begins by using a conventionally excavated and slurry supported trench and then the geomembrane panels are inserted into the slurry trench to the desired depth by utilizing a steel frame, mechanical roller or other methods.

3.1 Steel Frame Method

When a steel frame is used for installation of the geomembrane panels, a HDPE curtain panel is attached to the steel frame and inserted into the slurry supported trench using a crane (Fig. 2). When the geomembrane is lowered to the desired depth, the sheet is unhooked from the frame and hung to the steel support set across the trench (Fig. 3). Next, a second HDPE panel was attached to the frame and lifted into position, so the interlocks could be joined with the first panel. The second panel is then lowered once the panels are aligned. The frame is released from the HDPE, removed and reused as the panel is lowered to the bottom of the trench. This process was repeated to complete the installations of all geomembrane panels.

Fig. 2. A HDPE geomembrane panel fixed on a steel frame to be inserted into the slurry trench

3.2 Mechanical Roller Method

A special designed mechanical roller developed by Beijing GeoEnviron Engineering & Technology, Inc. (BGE) can also be used in geomembrane installation (Fig. 4). The use of the mechanical roller simplifies the placing and jointing of the HDPE panels, reduces the problems related to the handling of the panels in presence of wind and allows the installation of panels longer than 30 m. In order to install the HDPE panel, the mechanical roller with the panel is set at the edge of the trench. The HDPE panel can be continuously unrolled and lowered into the slurry supported trench. The interlock welded at the edge of the panel is sliding down in the inside of the interlock of the adjacent panel.

Fig. 3. A HDPE geomembrane panel hanged to steel support set across the trench

Fig. 4. A special designed mechanical roller used for geomembrane installation

Fig. 5. Combining use of frame and mechanical roller in geomembrane installation

By fixing the bottom of the geomembrane panel on the bottom beam of a steel frame and using the frame to apply downward force to the bottom edge of the panel, the geomembrane can be inserted into the bottom of the trench against the buoyancy of the slurry when a mechanical roller is used to install the geomembrane panels, especially installation in a deep trench (Fig. 5). This installation method was used in Huainan coal mining subsidence area project in Anhui and Binhai hazardous waste landfill project in Jiangsu, China by Beijing GeoEnviron Engineering & Technology, Inc. (BGE).

4 Interlocks and Panel Joint

The interlocking system of the geomembrane panels is one of the key elements governing the hydraulic performance of the geomembrane cutoff wall. The geomembrane sheets are manufactured into different widths of the panels with the interlocking joint pre-welded to the sheet to minimizing field welds (Fig. 6). The joints can be sealed with a hydrophilic rubber cord which can be fed into the joint during installation and that then swells upon contact with water (Fig. 7).

Fig. 6. An E-type interlock pre-welded at the edge of a 3-mm HDPE panel

Fig. 7. Installed panel joint with a hydrophilic rubber cord

5 Key Grouting Seal

The sealed connection between the bottom of the geomembrane and the subsurface aquifer (e.g., bedrock or clay layer) is another key issue to affect hydraulic performance of the geomembrane cutoff wall. The properties of the desired sealing material must meet the following basic requirements:

1. Adhesive property, that is, it can bond with HDPE panel under humid conditions, and can also bond well with the subsurface aquifer materials (i.e., rock or clay);
2. The sealing material has the property of resisting acid and alkaline corrosion;
3. The permeability coefficient of the sealing material must be not more than 1.0×10^{-7} cm/s; and

4. With sufficient engineering strength, the tensile strength of the sealing material is required to be not less than the strength of the bonded materials.

5.1 Bedrock Key Grouting Seal

After repeated selections and testing in the chemical laboratory, as well as the practical testing at the field, a special adhesive sealing material, a modified polymer material, was finally developed by Beijing GeoEnviron Engineering & Technology, Inc. (BGE) and National Chemical Building Materials Testing Center (NCBMTC) to achieve an effective bonding function between the HDPE geomembrane and the bedrock (Zheng et al. [6]). It can work to the harsh acidic environment and has good corrosion resistance under $2 \leq \text{pH} \leq 6$ conditions. It also has a long-term bonding effect and no any leakage.

On this basis, by optimizing the materials, adding high-strength cement and selected coarse sand, a type of modified polymer material concrete with high performance was developed. While maintaining low permeability, strong adhesion and corrosion resistance, the strength is increased. The compressive strength can reach 80.7 MPa. The cost is much reduced. it can be used in a wet condition without affecting performance.

The modified polymer material concrete has been used for the bedrock key sealing with geomembrane. It was successfully applied for the bedrock seal in the vertical geomembrane barrier project of Fujian Zijin Mining Company and for the concrete pipeline through the HDPE geomembrane in Xinjiang Urumqi project.

5.2 Clay Key Grouting Seal

The sealing materials for bonding geomembrane with adjacent clay in the key are required to achieve low permeability, proper fluidity, bondability, resistance to environmental corrosion, and clotting time adjustability.

After continuous testing and practicing, a kind of anti-corrosion sealant suitable for geomembrane in clay key, a modified plastic concrete, was developed by BGE. Different components and ratios can be formulated based on different project conditions. The effective bonding of HDPE geomembrane and clay layer can be achieved. Its slump is controlled to be greater than 250 mm to ensure to have an ability to use high pressure pump to conduct underwater pouring by using the tremie method. The material can withstand the contaminated groundwater and leachate in the actual condition and play a long-term sealing effect and no any leakage. This material has been used in many projects.

6 Applications

The geomembrane cutoff wall began to be used in Europe and U.S. at the end of the 20th century (Jefferies [7]). In 2011, this technique was firstly successfully applied by BGE in a copper-containing acidic liquid leakage event for Zijin Mining Company in Fujian, China. Then, BGE successfully applied and promoted the geomembrane cutoff walls in many other projects such as Qujing slag heap project in Yunnan, Jingjiang chemical contaminated site in Jiangsu, Leachate interception dam project of Urumqi landfill in Xinjiang, Huainan coal mining subsidence area project in Anhui, Zhongxiang landfill project in Hubei, Binhai hazardous waste landfill project in Jiangsu, Tiaoyuan non-standard landfill project in Hunan, Dafeng hazardous waste landfill project in Jiangsu, and Jiangyin pesticide contaminated site in Jiangsu. Moreover, a modified polymer material and a modified epoxy resin concrete were used to achieve the effective bonding between the HDPE geomembrane and bedrock aquifer in a strong acid condition. A specially formulated modified plastic concrete was used to achieve the effective sealing between the HDPE geomembrane and clay aquifer. In addition, a special machine for installing geomembranes has been developed. The maximum geomembrane inserting depth of 32 m has been successfully achieved and the maximum perimeter of the cutoff wall for a single project was up to 1500 m. Table 1 is a summary of the projects utilizing geomembrane as vertical barriers by BGE in China since 2011.

Table 1. Summary of projects utilizing geomembrane as vertical barriers in China

Location and year	Project site	GM (mm)	Installation method	Depth of wall (m)		Length of wall (m)	Type of interlock	Key sealing	Backfill
				Ave.	Max.				
Zijin 2011	Heap Leaching Field	2	Open excavation	30	30	350	Welding	MPM	Local soil
Qujing 2015	Mining Waste	3	Steel frame	18.2	30	740	E-type Interlock	MPC	SB
Jingjiang 2016	Chemical Waste	3	Mechanical roller	12.5	15	384	E-type Interlock	MPC	SB
Wulumuqi 2016	MSW Landfill	3	Mechanical roller	13.3	16.5	266	E-type Interlock	MPC, MPMC	Local soil
Huainan 2017	MSW Landfill	3	Mechanical roller and steel frame	8.3	15	1500	E-type Interlock	MPC	SB
Zhongxiang 2017	MSW Landfill	3	Mechanical roller	6.5	13	205	E-type Interlock	MPC	SB
Binhai 2018	HW Landfill	3	Mechanical roller and steel frame	18	20	1234	E-type Interlock	MPC	SB
Taoyuan 2018	MSW Landfill	3	Mechanical roller	6	7	50	E-type Interlock	MPC	SB
Dafeng 2018	HW Landfill	3	Steel frame	31	32	684	E-type Interlock	MPC	SB
Jiangyin 2018	Chemical and HW	3	Mechanical roller	13.5	13.5	1204	E-type Interlock	MPC	SB

HW: Hazardous Waste; MSW: Municipal Solid Waste; SB: Soil-Bentonite; MPM: Modified Polymer Material; MPMC: Modified Polymer Material Concrete; MPC: Modified Plastic Concrete.

7 Conclusions

The HDPE geomembrane has excellent barrier properties ($k \leq 1.0 \times 10^{-12}$ cm/s) and chemical resistance, as well as long-term service life. By using special key sealing materials, a composite cutoff wall consisting of HDPE geomembrane combined with soil-bentonite or plastic concrete can effectively be applied to the remediation of mining and heap leaching fields, petrochemical facilities, municipal and hazardous waste landfills, and contaminated sites. Geomembrane composite cutoff wall is considered to be currently the safest and most effective vertical barrier technique to block underground contaminated mass transport. Since 2011, this technique has been successfully and effectively applied in varies remediation projects in China.

References

1. Koerner RM, Guglielmetti JL (1995) Vertical barriers: geomembranes, assessment of barrier containment technologies, a comprehensive treatment for environmental remediation applications. In: Rumer RR, Mitchell JK (eds) International containment technology workshop, Baltimore, MD, USA, 29–31 August 1995, pp 95–118
2. Zhen S, Huo C, He Z, Zheng Z, Zheng F, Sun X, Liu L (2017) Application and comparative study of vertical barrier technology. Environ Sanit Eng 25(1):51–56
3. USEPA (1998) Evaluation of subsurface engineered barriers at waste sites. EPA 542-R-98-005, U.S. Environmental Protection Agency, Office of Solid Waste and Emergency Response, Washington, DC, USA
4. Daniel DE, Koerner RM (2000) On the use of geomembranes in vertical barriers. In: Advances in transportation and geoenvironmental systems using geosynthetics, proceedings of sessions of GeoDenver 2000, ASCE Geotechnical Special Publication No. 103, pp 81–93
5. Bouazza A, Zornberg JG, Adam D (2002) Geosynthetics in waste containment facilities: recent advances. In: Delmas P, Gourc JP, Girard H (eds) Proceedings of geosynthetics—7 ICG. Swets & Zeitlinger, Lisse, pp 445–507
6. Zheng Z, Zhu T, Luo B, Hong H, Ni X, Zhu H, Miao Z (2014) Applications of vertical ecological barrier system for prevention and control of mining groundwater pollutions. In: Proceedings of academic annual conference of Chinese society of environmental sciences. China Agricultural University Press, Beijing, pp 4385–4392
7. Jefferies SA (1997) The origins of the slurry trench cut-off and a review of cement-bentonite cut-off walls in the UK. Land Contam Reclam 5(3):239–245

Influence of Pore Distribution Characteristics on Relative Hydraulic Conductivity in Soil Covers—A Pore-Scale Numerical Investigation

Guangyao Li[1,2(✉)], Liangtong Zhan[1], and Sheng Dai[2]

[1] MOE Key Laboratory of Soft Soils and Environmental Engineering, Zhejiang University, Hangzhou 310058, China
guangyao@zju.edu.cn
[2] School of Civil and Environmental Engineering, Georgia Institute of Technology, Atlanta 30332, USA

Abstract. Relative hydraulic conductivity is an important input parameter for water balance models, which are commonly used to evaluate the performance of soil covers. In this paper, pore-network modeling was utilized to analyze the influence of pore distribution characteristics on relative hydraulic conductivity at various water saturations in soils. A drainage process of water slowly invaded by air was simulated to assign various saturations in pore networks. The pore networks have 30 × 30 × 30 pores, with log-normally distributed pore diameters in different mean values and standard deviations. Numerical results indicated that the increase in standard deviation and the decrease in the mean value of the pore diameters lead to a decrease in saturated hydraulic conductivity. Larger standard deviation or lower mean value of the pore diameters can result in more evident right-skewed pore diameter distribution. This provides more throats for water to flow at a given saturation, and thus a larger relative hydraulic conductivity in the pore network.

Keywords: Pore network modeling · Pore distribution characteristics
Drainage simulation · Relative hydraulic conductivity

1 Introduction

As one of the important components of waste containment facilities, the final cover systems are usually designed to minimize rainwater percolation into the underlying waste. In recent years, there has been an increasing interest in applying alternative covers to replace the conventional covers that employ resistive principles to prevent rainwater infiltration [1–4]. Alternative covers rely on water storage capacity to prevent rainwater percolation into the waste [1, 5, 6].

Water balance models are commonly used to simulate the hydrology of alternative covers. An input requirement for most water balance models is the relative hydraulic conductivity as soils used in alternative covers are generally unsaturated [7]. Many methods have been proposed to obtain the relative hydraulic conductivity, including steady and unsteady state methods [8]. Meerdink et al. [7] measured the relative hydraulic conductivity of two compacted barrier soils using the instantaneous profile

L. Zhan et al. (Eds.): ICEG 2018, ESE, pp. 343–350, 2019.
https://doi.org/10.1007/978-981-13-2224-2_42

method in both the laboratory and the field scales. Relevant results provide much valuable guidance for the design and construction of alternative covers. However, there has been limited data in published literature regarding the relationship between the pore distribution characteristics and the relative hydraulic conductivity. Fail to understand this relationship will not be able to provide insight into fundamental flow processes in soils.

This paper aims to investigate the influence of pore distribution characteristics on relative hydraulic conductivity in soils using pore-network modeling. Firstly, we assign the pore networks with different pore distribution characteristics, including different mean values and standard deviations of the pore diameters. Then we simulate a drainage process to assign various water saturations in the pore networks. Finally, relative hydraulic conductivity in these networks are computed and analyzed in the perspective of pore distribution characteristics and various saturations.

2 Pore Network Model

2.1 Methodology

The unsaturated hydraulic conductivity in soil covers associated with different pore characteristics is studied using pore network modeling, which is a widely used technique for simulating multiphase transport in porous media [9, 10].

In this study, the void spaces in network models were simplified as various spherical pores interconnected by cylindrical throats. The simple cubic packed pore structure was adopted in a 3D pore network, which meant one central pore is interconnected with six surrounding pores (i.e., pore connectivity = 6). Pore diameters were firstly specified to replicate realistic pore geometry of the soils. The throat diameters were then generated by assuming that the diameter of each throat is equal to the diameter of the smallest of its two adjacent pores. This kind of throat assignment scheme was chosen because it maintains physical consistency [10, 11].

2.2 Relative Hydraulic Conductivity

In saturated soils, all the pores and throats are occupied by water and water can flow through all the throats. Water flux in each throat, q_t, can be described based on Hagen-Poiseuille equation [12]:

$$q_t = \frac{\pi R_t^4 \Delta P_t}{8\mu \Delta L_t}, \tag{1}$$

where μ is the water viscosity. ΔP_t is the difference of water pressure between the two ends of the throat. ΔL_t and R_t are the length and radius of the throat respectively. At any pore site, there should be no net flow rate due to the mass conservation:

$$\Sigma q_i = 0, \tag{2}$$

Based on Eqs. (1) and (2), we can obtain the water pressure in all pores and the water flux through all throats. The saturated hydraulic conductivity in the soil, k_s, can be calculated by Eq. (3) according to Darcy's law:

$$k_s = \gamma \frac{Q}{\Delta P} \frac{L}{A}, \tag{3}$$

where γ is the unit weight of water. Q is the total water flux at the outlet of the network model. ΔP is the difference of water pressure between the inlet and outlet of the model. L and A are the length and cross-section area of the model respectively.

In unsaturated soils, part of the pores and throats are occupied by water and the others are occupied by gas. In this situation, the unsaturated hydraulic conductivity, k_u, could also be obtained using Eqs. (1)–(3) by assuming that water can only flow through the water-occupied pores and throats [11]. In other words, the water flux in the water-occupied throats can be computed by Eq. (1), and the water flux in the gas-occupied throats can be considered as nearly 0. Relative hydraulic conductivity is the ratio of unsaturated hydraulic conductivity over saturated hydraulic conductivity as described in Eq. (4). More details of the fundamentals of pore network modeling could be found in [9, 13].

$$k_r = \frac{k_u}{k_s} \tag{4}$$

2.3 Drainage Simulation

In order to obtain the relative hydraulic conductivity in soils at different water saturations, we simulated the drainage process of a wetting phase (water) slowly invaded by a non-wetting phase (air) in a 3D pore network [11]. The capillary pressure, P_c, is required for the air to penetrate the throat, and it can be estimated by the Young-Laplace equation [14]:

$$P_c = \frac{2\sigma \cos \theta}{R_t} \tag{5}$$

where σ is the surface tension of the water-air interface, θ is the contact angle between the two-phase contact line and the soil surface.

The following steps describe the algorithm for simulating drainage process in pore network modeling. First, the pore network is initially saturated (i.e., all pores and throats are filled with water). Then we assign the inlet boundary for the air to invade the water (the bottom side of the cubic model is chosen as the invasion inlet in this study). A low-capillary pressure is then selected to start the air invasion. At this given capillary pressure, all the throats in the model are scanned and marked as "open" if air can penetrate the throat according to Eq. (5). If a throat is invaded by air, the pores located at its two ends are marked as 'open' as well. Next, all 'open' throats and pores that are connected to the invasion inlet can form a cluster and the cluster is identified as invaded by air. All throats and pores that are not connected to the invasion inlet are returned to a 'closed' state. In this way, the invading front of the air only reaches pores that are both topologically accessible from the invasion inlet (i.e., through other invaded pores) and penetrable at the given capillary pressure. In addition, it is easy to calculate the water saturation at this moment by dividing the volume of 'closed' pores and throats over the volume of all pores and throats. By increasing the capillary pressure in small increments, the above step could be repeated and updated invasion state will appear until all pores and throats are penetrated by air. The drainage process is diagramed in Fig. 1, in which the surface tension is 0.072 N/m, the contact angle is 45° and the water viscosity is 8.93×10^{-4} Pa s.

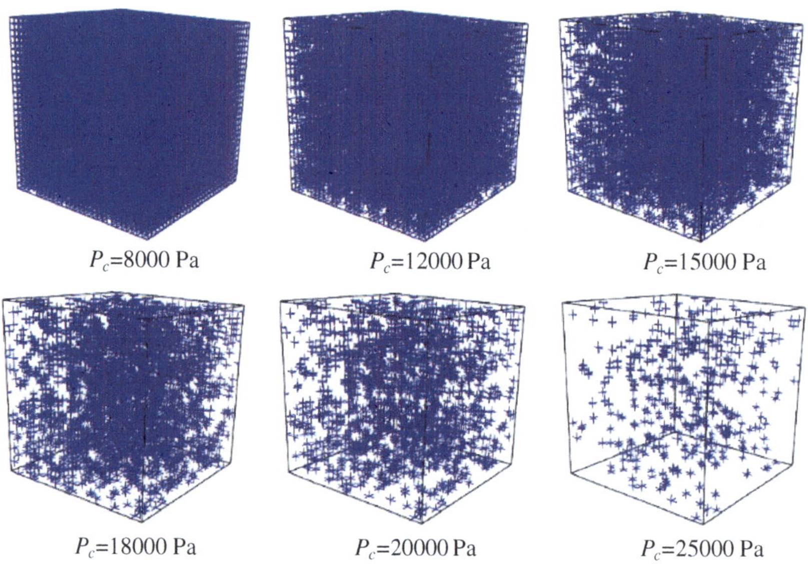

Fig. 1. Diagrams of drainage process in pore network modeling. Water is slowly invaded by air at different capillary pressures P_c. The water-occupied throats are shown in blue.

3 Numerical Results

Influence of pore characteristics on relative hydraulic conductivity in pore networks is investigated in this section. In order to obtain the relative hydraulic conductivity, we firstly calculated the saturated hydraulic conductivity in the networks. We generated the cubic pore networks with 30 × 30 × 30 pores. The distance of pore to pore center was 100 μm. Previous results indicated that natural sediments exhibit a log-normal distribution in pore size [15]. Herein, we assigned log-normally distributed pore diameters with different pore characteristics, including different mean values and standard deviations in the pore network. The water pressure of 0.6 kPa and 0.2 kPa were employed at the inlet and outlet boundary to compute the hydraulic conductivity. The fluid properties were the same as that used in last section.

Figure 2 shows the influence of standard deviation (σ_d) of pore diameters on saturated and relative hydraulic conductivity in the soils. The log-normally distributed pore diameters were generated with mean pore diameter μ_d = 10 μm and a maximum pore diameter of 100 μm. It can be seen that the saturated hydraulic conductivity decreases as the σ_d increases (similar results could be found in [13]). According to the pore diameter distribution curve shown in Fig. 2a, as the σ_d increases, the pore diameters become more non-uniform and are more evidently right-skewed, and this results in more relatively large pores in the soils. On the other hand, an increasing probability of large pores becoming surrounding by small throats appeared. That increases the probability of finding small throats along every potential flow path and so reduces the saturated hydraulic conductivity. For a given water saturation, the relative

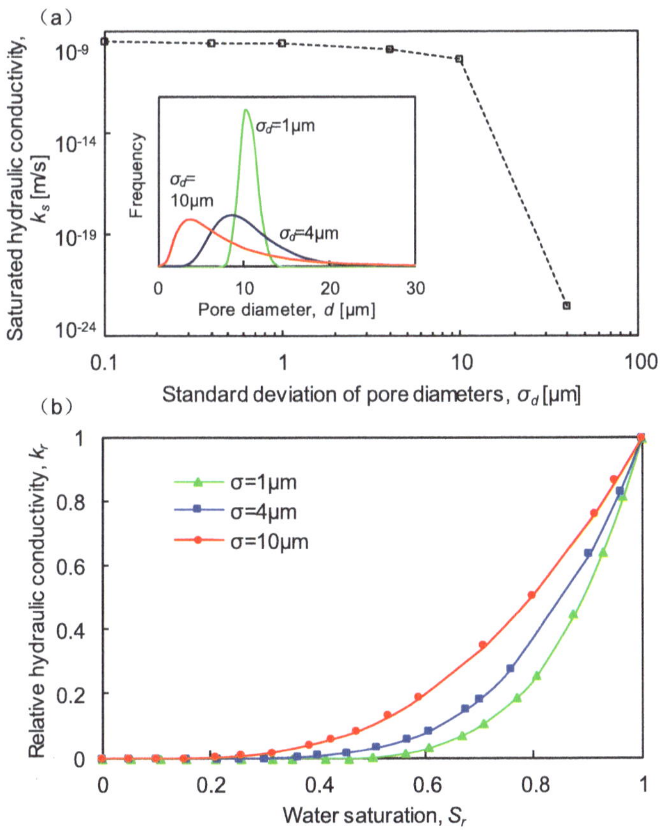

Fig. 2. Influence of the standard deviation of pore diameters on saturated and relative hydraulic conductivity in soils. All cases use a pore network with log-normally distributed pore diameters (mean pore diameter μ_d = 10 μm and connectivity cn = 6). (a) Pore diameter distribution curves and the variation of saturated hydraulic conductivity with standard deviation. (b) The variation of relative hydraulic conductivity with standard deviation.

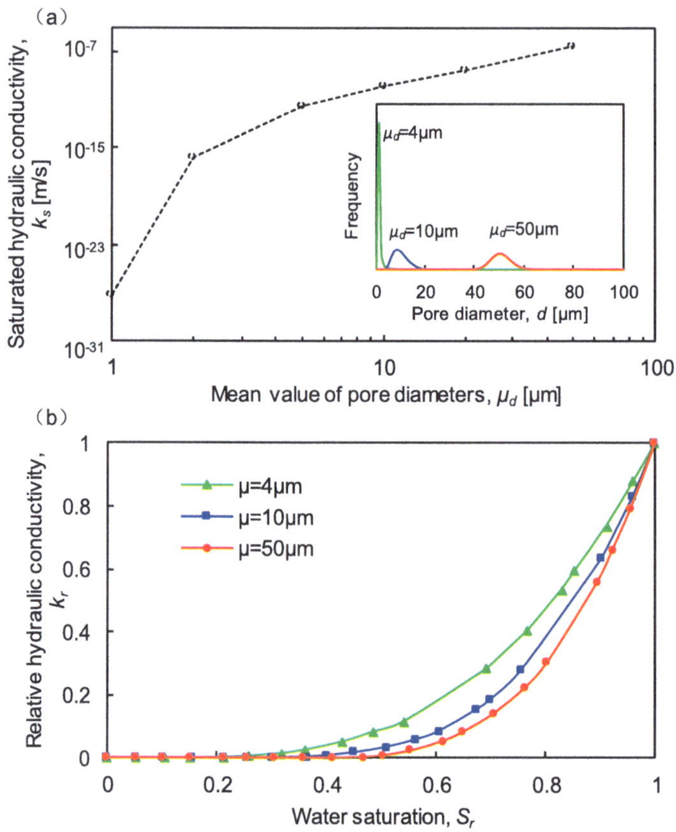

Fig. 3. Influence of the mean value of pore diameters on saturated and relative hydraulic conductivity in soils. All cases use a pore network with log-normally distributed pore diameters (standard deviation σ_d = 4 μm and connectivity cn = 6). (a) Pore diameter distribution curves and the variation of saturated hydraulic conductivity with the mean pore diameter. (b) The variation of relative hydraulic conductivity with the mean pore diameter.

hydraulic conductivity increases as the increase of standard deviation (see Fig. 2b). It can be found from the drainage process that the air will invade the large throats first due to their lower capillarity. In that way the number of water-occupied throats in the network with a larger σ_d is greater at a given saturation, and that provides more potential flow paths. Therefore, a larger relative hydraulic conductivity in the network with a larger σ_d is observed.

Figure 3 shows the influence of mean value of pore diameters on saturated and relative hydraulic conductivity in soils. The log-normally distributed pore diameters were generated with standard deviation σ_d = 4 μm and a maximum value of 100 μm. Figure 3a indicated that the saturated hydraulic conductivity increases as the mean pore diameters μ_d increases, which is consistent with the relation between the pore radius and saturation hydraulic conductivity described in Eqs. (1)–(3) (the saturation hydraulic conductivity is approximately proportional to R_t^4). As the decrease of μ_d, the pore size distribution is skewed to right even though the distribution range becomes more limited. It is because the right-skewed pore diameters that cause a larger relative hydraulic conductivity at the same saturation as shown in Fig. 3b, which is in agreement with the previously mentioned conclusion. That also demonstrated that the right-skewed distribution for the pore diameters is the dominated factor affecting the relative hydraulic conductivity rather than the pore diameters distribution range.

4 Conclusions

In this paper, pore network modeling was used to analyze the influence of pore distribution characteristics on relative hydraulic conductivity in soil covers. Log-normally distributed pore diameters with different mean values and standard deviations were assigned to the pore networks with $30 \times 30 \times 30$ pores. A drainage process was simulated to calculate the relative water conductivity in the pore networks with different water saturations. Based on the numerical results, the following conclusions can be drawn.

(1) An increase in the standard deviation of the pore diameters can increase the probability of large pores surrounded by small throats, and thus decrease the saturated hydraulic conductivity. An increase in the mean value of the pore diameters results in an increase in saturated hydraulic conductivity, which is in agreement with the description in Hagen-Poiseuille equation.
(2) Water drainage by air invasion is preferred in larger throats with lower capillarity. Thus, during drainage, the number of water-occupied throats at a given water saturation decreases in networks with more relatively large throats. More relatively larger throats occur in networks with larger standard deviation or lower mean value of the pore diameters. Such pore distribution characteristics can result in higher relative hydraulic conductivity.

Acknowledgements. The authors acknowledge financial support from the National Science Fund for Distinguished Young Scholars (No. 51625805) and program of China Scholarships Council (No. 201706320096).

References

1. Benson C, Abichou T, Albright W, Gee G, Roesler A (2001) Field evaluation of alternative earthen final covers. Int J Phytorem 3(1):105–127
2. Abichou T, Liu X, Tawfiq K (2006) Design considerations for lysimeters used to evaluate alternative earthen final covers. J Geotech Geoenviron Eng 132(12):1519–1525
3. Zhan LT, Li GY, Jiao WG, Wu T, Lan JW, Chen YM (2016) Field measurements of water storage capacity in a loess–gravel capillary barrier cover using rainfall simulation tests. Can Geotech J 54(11):1523–1536

4. Waugh WJ, Benson CH, Albright WH (2009) Sustainable covers for uranium mill tailings, USA: alternative design, performance, and renovation. In: ASME 2009 12th international conference on environmental remediation and radioactive waste management. American Society of Mechanical Engineers, Liverpool, pp 639–648
5. Khire MV, Benson CH, Bosscher PJ (1997) Water balance modeling of earthen final covers. J Geotech Geoenviron Eng 123(8):744–754
6. Khire MV, Benson CH, Bosscher PJ (2000) Capillary barriers: design variables and water balance. J Geotech Geoenviron Eng 126(8):695–708
7. Meerdink JS, Benson CH, Khire MV (1996) Unsaturated hydraulic conductivity of two compacted barrier soils. J Geotech Eng 122(7):565–576
8. Benson C, Gribb M (1997) Measuring unsaturated hydraulic conductivity in the laboratory and the field. Geotechnical Special Publication, Austin, pp 113–168
9. Fatt I (1956) The network model of porous media. Soc Petrol Eng 1956:207
10. Gostick J, Aghighi M, Hinebaugh J, Tranter T, Hoeh MA, Day H, Spellacy B, Sharqawy MH, Bazylak A, Burns A, Lehnert W (2016) OpenPNM: a pore network modeling package. Comput Sci Eng 18(4):60–74
11. Gostick JT, Ioannidis MA, Fowler MW, Pritzker MD (2007) Pore network modeling of fibrous gas diffusion layers for polymer electrolyte membrane fuel cells. J Power Sources 173(1):277–290
12. Sutera SP, Skalak R (1993) The history of Poiseuille's law. Annu Rev Fluid Mech 25(1):1–20
13. Jang J, Narsilio GA, Santamarina JC (2011) Hydraulic conductivity in spatially varying media—a pore-scale investigation. Geophys J Int 184(3):1167–1179
14. Adamson AW, Gast AP (1967) Physical chemistry of surfaces, 6th edn. Wiley, Hoboken
15. Phadnis HS, Santamarina JC (2011) Bacteria in sediments: pore size effects. Geotech Lett 1 (4):91–93

Prediction of Vegetation-Induced Soil Suction Using Numerical Modelling and AI

M. Indu Priya[1], Ankit Garg[2], S. Sreedeep[1], Ajit Sarmah[3], and Nik Norsyahariati Nik Daud[4(✉)]

[1] Department of Civil Engineering, Indian Institute of Technology Guwahati, Guwahati, India

[2] Department of Civil and Environmental Engineering, Shantou University, Shantou, China

[3] Department of Civil and Environmental Engineering, University of Auckland, Auckland, New Zealand

[4] Department of Civil Engineering, Universiti Putra Malaysia, Seri Kembangan, Malaysia

niknor@upm.edu.my

Abstract. Study of soil suction is important in design and implementation of slope stability and erosion control measures. In order to conduct a realistic analysis of performance of sustainable green infrastructure, it is essential to address the uncertainties in suction induced by vegetation due to variability in their leaf and root characteristics, evapotranspiration (ET) and initial conditions of the soil. The objective of this study is to investigate the combined influence leaf area index (LAI), root depth, ET rate and initial suction of soil on root water uptake-induced soil suction. A parametric numerical study was performed with 480 simulations using HYDRUS to carry out the finite element analysis. The study was done on completely decomposed granite (CDG) soil and vegetation species used was *Schefflera heptaphylla*. It was observed that although if independently considered, vegetation with higher LAI provided greater mechanical stability, when combined with higher ET rates or initial suction, the suction induced may lead to wilting of the vegetation. Artificial intelligence technique such as Artificial neural network (ANN) was used to predict matric suction at any given depth using the results obtained from the numerical simulations. Performance of the best model indicated that ANN was able to successfully predict the vegetation-induced matric suction.

Keywords: Soil matric suction · Numerical modelling · Artificial intelligence

1 Introduction

Matric suction of soil is an essential factor in determining the stability of slopes [1] and drying-wetting characteristics of soil surface [2, 3]. Shear stresses in soil are quantified using soil matric suction [4]. Implementation of bio-engineered slopes requires a good understanding of shear strength of soil as well as soil surface characteristics that influence the strength of soil [5, 6]. These studies clearly show that presence of vegetation has a considerable influence on soil strength. Moreover, shallow slope failures

L. Zhan et al. (Eds.): ICEG 2018, ESE, pp. 351–358, 2019.
https://doi.org/10.1007/978-981-13-2224-2_43

(i.e., failure plane less than 1 m deep) occur primarily due to heavy rainfall, especially during wet seasons [7, 8]. Presence of vegetation increases the slope stability by acting as fiber reinforcement and by increasing suction within the root zone compared to bare soil. Soil suction in vegetated soil is induced due to various factors such as physical characteristics of vegetation like leaf area index (LAI) and root depth [9], evapotranspiration (ET) rate [10] and initial state of the soil [11]. While these parameters have been independently analyzed in previous studies, their combined influence on suction has not been studied.

The objective of this study is to investigate the influence of these parameters on vegetation induced soil suction. To achieve this, a parametric numerical study was performed on 480 data sets to understand the effect of variability of the four parameters on suction. The results of these simulations were then utilized to predict the vegetation-induced soil suction at any given depth within the root zone. ANN was used to model the variability of suction with LAI, root depth, ET rate, initial suction and depth of observation as independent input parameters.

2 Soil and Vegetation Properties

The soil type selected for investigation is completely decomposed granite (CDG). The material properties and water retention curves of the bare soil in the domain was adopted from Leung et al. (2015) [12]. CDG is classified as a well-graded soil. Based on the measured particle size distribution and Atterberg limits, the CDG is classified as clayey sand, according to the Unified Soil Classification System, USCS. The vegetation type used for the investigation is *Schefflera heptaphylla*. This species is known for being drought resistant in warm climates [13, 14]. It is commonly found in tropical countries including India, Malaysia and Hong Kong and has been widely used for slope stabilization and green infrastructure projects.

3 Numerical Modelling

3.1 Finite Element Domain Geometry and Boundary Conditions

Numerical modelling of vegetated soil in the vadose zone is adopted through root-water uptake models. HYDRUS, a commercially available software, was used to model the root-water uptake induced suction [15]. The domain in which the study was done had a single tree on a flat ground. The finite element domain was axisymmetric and homogeneous, with a 5 m radius and 5 m depth (Fig. 1). The domain was partitioned into two zones as root zone (mesh size 0.001 m) and outside root zone (mesh size 0.065 m). The root zone was taken from the base of the vegetation to 0.45 m deep and 0.08 m lateral extent. Evapotranspiration was established as the atmospheric boundary condition. The allowable limit of suction at the atmospheric boundary was taken as 50000 kPa. No flux boundary conditions and free drainage were provided at the side boundary and bottom boundary respectively.

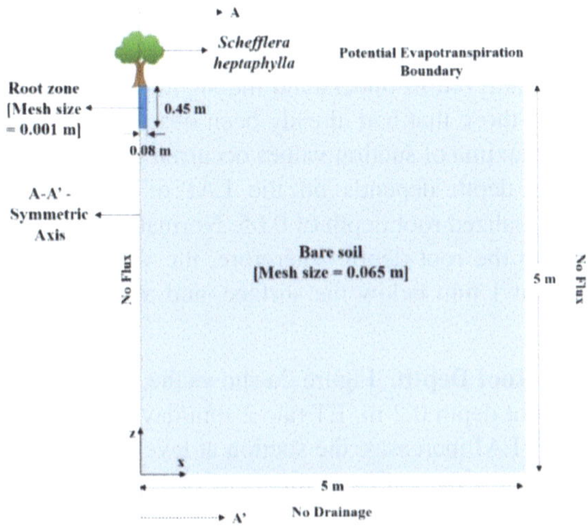

Fig. 1. Finite element model domain dimensions and boundary conditions

3.2 Input Parameters

The focus of this numerical study is to investigate the effects of LAI, root depth, ET rate and initial suction on matric suction induced in soil-root composite. The various LAI values considered were 1.0, 1.5, 2.0, 2.5, 3.0 and 3.5. These values were adopted Ref. [16]. Four root depths between 0.1 m and 0.4 m were considered. These values were typically found for matured *Schefflera heptaphylla* grown in CDG soil with a degree of compaction of 80% [12]. ET was varied from 1 to 5 mm/day. These values were adopted based on practical ET rates that could be found in tropical regions throughout a year [17]. Initial state of the soil was defined by varying the initial suction of the soil in the domain as 10, 25, 50 and 100 kPa. Initial suction of the soil could be affected by various environmental factors such as precipitation, irrigation etc. Field capacity of CDG is 25 kPa. T_{rf} depends on the LAI of a vegetation, it is also a variable input. Values of 'ψ_1', 'ψ_2' and 'ψ_4' are 0, 5 and 1500 kPa, whereas 'ψ_3' values corresponding to each LAI was obtained from Ref. [16].

3.3 Numerical Simulation Procedure

A comprehensive set of 480 numerical simulations were performed. Each of the 480 simulations was run for 10 h of drying. The maximum time step accuracy of the numerical analysis was taken as 0.01 h. At the end of 10 h, the suction profile at the vertical axis of the domain (A-A', Fig. 1) in the root zone was obtained.

3.4 Results and Discussion

Matric suction depends on depth of observation. At a near-surface depth, higher value of suction would cause loss of tensile strength of soil causing desiccation cracks which is undesirable [18]. In contrast, higher suction is desirable at larger depths since it increases the shear strength of soil, thereby increasing the stability of slopes [1]. In this study, vegetation induced suction profiles (induced suction vs. depth) obtained from 480 simulations were analyzed to understand the significance of each parameter. The profiles were similar to those that had already been observed in previous studies [19]. In every profile, local maxima of suction values occurred at the depth of maximum root area index (S_m). This depth depends on the LAI of the vegetation [16] and is approximately at a normalized root depth of 0.65. Normalized root depth is the ratio of depth of observation to the root depth. Therefore, the values of suction at the near-surface level (S_o), i.e. at 1 mm below the surface, and at sub-surface level (S_m) were analyzed.

Leaf Area Index and Root Depth. Figure 2a shows the variation of suction with LAI for a vegetation with root depth 0.2 m, ET rate 2 mm/day and initial suction 25 kPa. It can be observed that as LAI increases, the suction at level S_o decreases and at level S_m increases. This is because evaporation component of ET decreases and transpiration component of ET increases. Therefore, vegetation with higher LAI is more suitable in increasing the stability of the soil slopes. Variation of suction with respect to root depth for a vegetation with LAI 2.5, ET rate 2 mm/day and initial suction 25 kPa is shown in Fig. 2b. In the figure, it can be seen that at both levels, S_o and S_{m-}, induced suction decreases with increase in root depth. As root depth increases, the sensitivity of the induced suction also decreases. For an increase in root depth from 0.1 m to 0.2 m, suction decreased by 51.3% and 95.1% respectively at S_o and S_{m-}. Whereas, the decrease in suction for variation of root depth from 0.3 m to 0.4 m are comparatively lower at 7.5% and 17.1% respectively. Moreover, shallower roots (0.1 m) increased the suction from an initial value of 25 kPa to 1509 kPa at depth S_{m-}. However, deeper roots (0.4 m) just increased the suction to 39 kPa. This indicates that, while the effect of deeper roots leads to lower suction values at the surface, induced suction at the S_m level is not quite as effective for deeper roots compared to shallower roots.

Evapotranspiration and Initial Suction. Figure 2c illustrates the behavior of suction with respect to ET rate for a vegetation with LAI 2.5, root depth 0.2 m and initial suction 25 kPa. It can be seen that at both levels, S_o and S_{m-}, induced suction increases with increase in ET rate. This is consistent with the fact that as ET rate increases, more water loss occurs due to root water uptake and evaporation. Soil suction increases with the reduction in moisture content (as apparent in soil water retention curve (SWRC) of CDG, [12]). The rate of increase in suction for an increase in ET rate from 1 to 2 mm/day is 51% and 92.9% respectively at S_o and S_{m-}. However, this increase in more rapid for an increase in ET rate from 3 to 4 mm/day, at 220.6% and 576.2% respectively at S_o and S_{m-}. High ET rate leads to very high suction in the bulk of the soil.

Variation of vegetation-induced suction with respect to initial suction for vegetation with LAI 3.0, root depth 0.2 m and ET rate 2 mm/day is shown in Fig. 2d. As expected, as initial suction increases, induced suction also increases at both the levels. In case of initial suction 10 kPa, suction increased to 13 and 15 kPa respectively at S_o and S_m. However, at a higher initial suction of 100 kPa, the suction increased to 1030 and 1537 kPa respectively. Sensitivity of the system increased substantially for a higher initial suction.

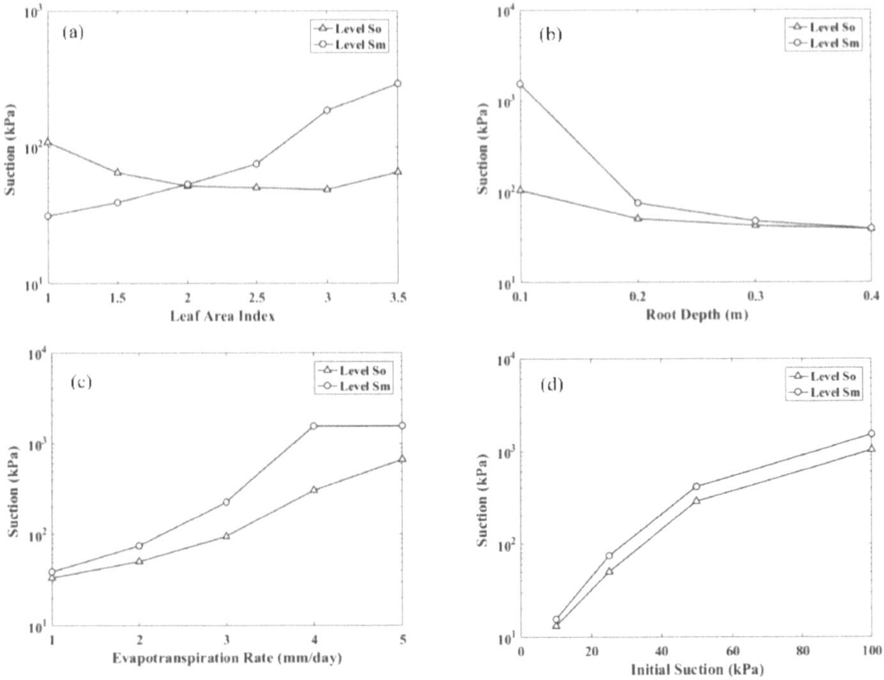

Fig. 2. Variation of suction with (a) LAI at root depth 0.2 m, ET 2 mm/day and initial suction 25 kPa (b) root depth at LAI 2.5, ET 2 mm/day and initial suction 25 kPa (c) ET at LAI 2.5, root depth 0.2 m and initial suction 25 kPa and (d) initial suction at LAI 2.5, root depth 0.2 m and ET 2 mm/day.

4 Modelling Using Artificial Neural Network (ANN)

In the current study, ANN is used to fit an approximate function for the data obtain through numerical simulation in the previous section. The objective is to predict the matric suction at any given depth of the vegetated soil provided the LAI, root depth, ET rate and initial suction. Input parameters for ANN modelling was LAI, root depth, ET rate, initial suction and normalized root depth. Thus, a total of 111,000 data points were obtained from the 480 simulations.

4.1 Implementation of ANN

ANN modelling was implemented using Neural Network Toolbox in MATLAB 2016b. The training parameters adopted are listed in Table 1. Training function used for all of the models is Levenberg-Marquardt algorithm ('trainlm' function). The training stops when any of the goals following are reached the earliest: (i) performance goal, (ii) maximum number of epochs or (iii) maximum number of validation failures.

Training algorithm uses 0.15 fractions of input data as testing fraction. In this case, the testing data amounts to 16650 data points. Transfer function used in the modelling was tan-sigmoid function.

Table 1. Training parameters used for ANN

Parameter type	Values
Training data size	94350
Testing data size	16650
Hidden layers	4
Neurons in each hidden layer	8–12
Training Function	Levenberg-Marquardt
Transfer function	Tan-sigmoid
Maximum epochs	2000
Minimum MSE	1
Minimum performance gradient	10^{-6}
Maximum validation fails	6

Since the data set is large, determining the number of hidden layers and number of neurons in each hidden layer is important to obtain a model with best performance. Performance is assessed with conventional metrics mean square errors (MSE), root mean square error (RMSE), correlation coefficient (R) and relative error (RE). Data set was shuffled before modelling to enhance the model's quality and predictive performance. Optimum number of layers was determined by running the model for several number of hidden layers and neurons in each hidden layers. Trial-and-error analysis indicated that with hidden layers of sizes [10, 11], further reduction of MSE was observed.

4.2 Performance Report of ANN

The best performing fit had the hidden layer sizes as [10, 11] with an RMSE of 6.886 and 10.267 kPa for the training and testing data respectively. Since the data set is large, RMSE is higher. A better indicator for good performance would be relative error. The average absolute RE for the test data set is 3.3% with a standard deviation of 3.38%. This indicates that 95% of the predicted data would have an absolute RE of less than 10%. Figure 3 shows the regression plots of target and predicted data for training and testing. Correlation coefficient (R) percentage of training and testing data are respectively 99.987% and 99.981% respectively. It can be observed that overfitting had not occurred in this case since R of testing data is very near to that of training data.

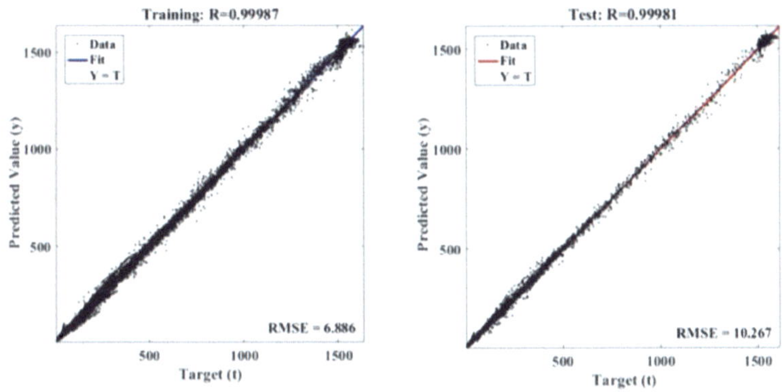

Fig. 3. Regression plot for training and testing data for the best performing ANN model

5 Conclusion and Future Work

To summarize, a parametric study was done on the independent variables LAI, root depth, ET and initial suction to obtain suction profiles in the root zone of vegetated soil. The results of this study indicates that the vegetation with higher LAIs is more suitable for slope stabilization compared to those with lower LAIs at any root depth, evapotranspiration rate and initial suction by increasing the suction within the root zone. Sensitivity of the system decreases with increasing root depth but increased substantially for a higher initial suction. This study gives us a good understanding of how the presence of *Schefflera heptaphylla* would affect the soil suction. Optimum range of all the parameters of *Schefflera heptaphylla* could be obtained based on the requirements of the bio-engineering project under consideration. The study can be reproduced for any species provided experimental results on each independent parameter are available. This knowledge can be used to design the bio-engineered slopes and embankments. The results of the numerical simulations were used in ANN to predict the suction induced at any given depth within the root zone. Performance indicators for the best model showed that ANN could be successfully utilized to predict vegetation-induced suction.

References

1. Zhang LL, Zheng YF, Zhang LM, Li X, Wang JH (2014) Probabilistic model calibration for soil slope under rainfall: effects of measurement duration and frequency in field monitoring. Géotechnique 64(5):365
2. Kodikara JK, Barbour SL, Fredlund DG (2000) Desiccation cracking of soil layers. Unsaturated soils Asia 90(5809):139
3. Tang CS, Cui YJ, Shi B, Tang AM, Liu C (2011) Desiccation and cracking behaviour of clay layer from slurry state under wetting–drying cycles. Geoderma 166(1):111–118
4. Fredlund DG, Morgenstern NR, Widger RA (1978) The shear strength of unsaturated soils. Can Geotech J 15(3):313–321

5. Pollen N (2007) Temporal and spatial variability in root reinforcement of streambanks: accounting for soil shear strength and moisture. Catena 69(3):197–205
6. Stokes A, Atger C, Bengough AG, Fourcaud T, Sidle RC (2009) Desirable plant root traits for protecting natural and engineered slopes against landslides. Plant Soil 324(1–2):1–30
7. Tsaparas I, Rahardjo H, Toll DG, Leong EC (2003) Infiltration characteristics of two instrumented residual soil slopes. Can Geotech J 40:1012–1032
8. Luino F (2005) Sequence of instability processes triggered by heavy rainfall in the northern Italy. Geomorphology 66(1):13–39
9. Ng CW, Garg A, Leung AK, Hau BCH (2016) Relationships between leaf and root area indices and soil suction induced during drying–wetting cycles. Ecol Eng 91:113–118
10. Wilson GW, Fredlund DG, Barbour SL (1997) The effect of soil suction on evaporative fluxes from soil surfaces. Can Geotech J 34(1):145–155
11. Ng CWW, Woon KX, Leung AK, Chu LM (2013) Experimental investigation of induced suction distribution in a grass-covered soil. Ecol Eng 52:219–223
12. Leung AK, Garg A, Ng CWW (2015) Effects of plant roots on soil-water retention and induced suction in vegetated soil. Eng Geol 193:183–197
13. Hau BC, Corlett RT (2003) Factors affecting the early survival and growth of native tree seedlings planted on a degraded hillside grassland in Hong Kong, China. Restor Ecol 11 (4):483–488
14. Hu L, Wang Z, Du H, Huang B (2010) Differential accumulation of dehydrins in response to water stress for hybrid and common bermudagrass genotypes differing in drought tolerance. J Plant Physiol 167(2):103–109
15. Kool JB, Van Genuchten MT (1991) Hydrus: One-dimensional Variably Saturated Flow and Transport Model, Including Hysteresis and Root Water Uptake; Version 3.3. US Salinity Laboratory
16. Garg A, Leung AK, Ng CWW (2015) Transpiration reduction and root distribution functions for a non-crop species Schefflera heptaphylla. Catena 135:78–82
17. Nag A, Adamala S, Raghuwanshi NS, Singh R, Bandyopadhyay A (2014) Estimation and ranking of reference evapotranspiration for different spatial scales in India. J Indian Water Resour Soc 34(3)
18. Tang CS, Shi B, Liu C, Suo WB, Gao L (2011) Experimental characterization of shrinkage and desiccation cracking in thin clay layer. Appl Clay Sci 52(1–2):69–77
19. Garg A, Ng CWW (2015) Investigation of soil density effect on suction induced due to root water uptake by Schefflera heptaphylla. J. Plant Nutr. Soil Sc 178(4):586–591

Laboratory Study of the Detection of Metal Contaminated Clay Layer Using Four-Electrode Resistivity Cone

Ya Chu, Songyu Liu(✉), Guojun Cai, Hanliang Bian, and Lei Xu

Institute of Geotechnical Engineering, Southeast University, Si-Pai-Lou 2, Nanjing, Jiangsu, China
136235507@qq.com, 101004004@seu.edu.cn

Abstract. Electrical resistivity measurements provide potentially powerful tool for detection of heavy metal contaminated soil, which is on the rise in the whole world with the boost of industrialization and urbanization. A four-electrode resistivity cone penetrometer (RCPT) test (non-destructive, continuous, reliable, and demonstrates strong correlation with subsurface information such as soil properties, stratigraphy, and the location of anomalies) was used to detect metal contaminated soil layers in this study. To investigate the relationship between electrical resistivity and unsaturated subsurface conditions with varying physical property and metal contamination, the polluted clays were synthesized with controlled metal ions concentration and moisture content of loading in a chamber. A series of tests of polluted clay was performed and the mechanism of these tests were analyzed by electrical conduction theory. For all tested soils, the results show that increasing the degree of saturation and metal ions concentration on the clay led to a reduction in the electrical resistivity of clay, while increasing the porosity led to a higher electrical resistivity for the clay. It indicates that altering the porosity and saturation of polluted clay sample changed the ionic movement within the clay's interlayer. In addition, data gotten by four-electrode soil resistivity box test was used to compare with the data measured from RCPT. The results showed that a linear approximation was sufficient in relating resistivity ρ_s of soil resistivity box to resistivity ρ_r of RCPT. In general, the RCPT was well used to examine the pollution of soil and the effects of remediation effectively in-suit.

Keywords: Resistivity cone penetrometer test · Resistivity
Metal contamination · Calibration chamber

1 Introduction

The invading of heavy metal ions in subsurface soils may led to the foundation failure and put negative impact on the environment (Du et al. 2011). As well known, the detection of contaminated soil layer is the first step to evaluate and remediate the contaminated sites (Yoon and Park 2001; Fukue et al. 1999). Whereas, the current chemical analysis, which was a widely used detecting technique, is very expansive and time consuming. Therefore, the present requirement is to find a potentially powerful

L. Zhan et al. (Eds.): ICEG 2018, ESE, pp. 359–366, 2019.
https://doi.org/10.1007/978-981-13-2224-2_44

tool, which is easy, quick and low cost technique, to detect the heavy metal contaminated soil layer in-situ.

Electrical resistivity is a non-invasive method that can provide useful information about moisture condition, structural characteristics, salinity and contamination of subsoil (Mitchell and Soga 2005; Yoon and Park 2001; Chu et al. 2017a). According to the Maxwell-Wagner theory, the electrical resistance (R_{mix}) of fine-grained soils should be expressed as the sum of the pore fluid resistance (R_w) and the soil particle resistance (R_s) as shown in Eq. 1 (Bussian 1983; Klein and Santamarina 2003; Glover 2010; Choo and Burns 2014).

$$R_{mix} = K_w R_w + K_s R_s \tag{1}$$

Where K_w is the resistivity factor of pore water, K_s is the resistivity factor of soil particle, K_w and K_s are all relate to the moisture condition, structural characteristics and salinity of soil.

The resistivity of soil is primarily dependent on the resistivity of pore water and soil particle, which was mainly depended on the concentration of metal ions in soil. Therefore, the impact of heavy metal ions on the electrical response of contaminated soils is very significant (Tang 2000; Su and Momayez 2017). Nowadays, the electrical resistivity method is widely used to distinguish the contaminated soil layer and evaluate the properties of heavy metal contaminated soils (Campanella and Weemees 1990; Chu et al. 2017a, b). However, the calibration of resistivity cone in laboratory are not well documented in the literature. Furthermore, due to the measuring mechanics of cone and box are different, the relationship between the resistivity of cone and box also should be discussed.

The goal of this study is to investigate the detection of heavy metal contaminated soil layers using resistivity cone and establish a relationship between the resistivity of cone and box. All the works can make the results of indoor research better applied on site. A self-developed calibration chamber was used to research the variation of resistivity of heavy metal contaminated soil on different condition. A series of resistivity tests of polluted clay was performed under different water content, density and ions concentration. In addition, data gotten by four-electrode soil resistivity box test was used to compare with the data measured from RCPT.

2 Materials and Instruments

2.1 Materials

The typical clay of Nanjing (Yangzi silt clay SC) was utilized in the tests. In addition, the soil samples from source were obtained from test pits of 2–3 m. This clay soil was widespread in eastern China. The main physical properties of these soils are represented in Table 1.

The heavy metal contaminant (HMC) of Zinc (2+) nitrate [$Zn(NO_3)_2 \cdot 6H_2O$, molecular weight (M:W) = 297.49] were chosen for this study. All chemicals (HMC) were purchased from Zhongdong Chemical Scientific and were used as received. The water

Table 1. The main physical properties of raw site soils.

Items	r/kN/m^3	Gs	LL/%	w_P/%	PI	Granularity content/%		
SC	19.20	2.72	37.73	23.07	14.66	20.39	70.49	9.12

used in all experimentation was deionized (Zhongdong Chemical Scientific E-pure, electrical conductivity ranging from 2.44 to 2.74 μs/cm at temperature 22.0 ± 0.5 °C).

2.2 Four-Electrode Resistivity Cone

Penetration test used a new four-electrode RCPT system developed by geotechnical engineering research Institute of Southeast University at the beginning of 2016, the structure chart of probe is shown in Fig. 1. The resistivity testing part of RCPT is mainly composed of four copper electrodes and internal circuit system. Through its internal circuit system with four electrode synchronization and continuously measuring the internal voltage change between the two electrodes, and the calculation program compiled according to the principle of Ohm's law to compute the resistivity of surrounding soils electrode size. The theoretical equation as follows:

$$\rho = \pi^2 \Delta V/(CI) \tag{2}$$

where ρ = electrical resistivity of soil; ΔV = voltage difference; $C = 1/(d + r_M) - 1/(d + r - r_M) - 1/(d + r_N) + 1/(d + r - r_N)$ (d, r_M, r_N and r are shown in Fig. 1); I = electrical current, unit A.

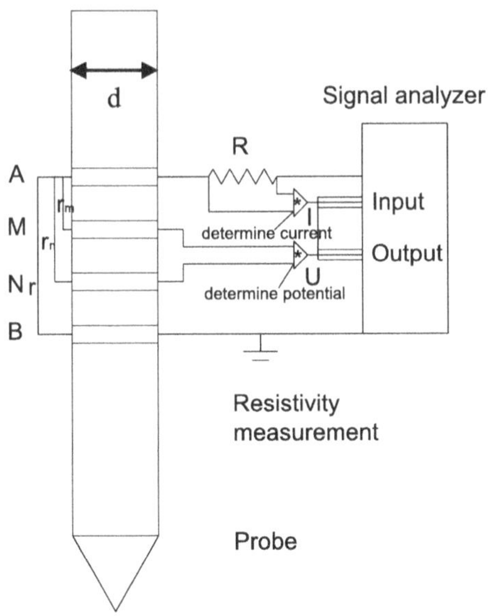

Fig. 1. The cone of electrical resistivity CPT

The resistivity experiments were also performed by a self-developed four electrodes soil resistivity testing cylinder in accordance with ASTM D G57-06. Electrical resistivity are obtained by Eq. 3:

$$\sigma = 1/\rho = L/(RA) = IL/(\Delta VA) \tag{3}$$

where R is resistance, σ is conductivity, ρ is electrical resistivity, ΔV is potential difference, A is a cross-sectional area, I is electric current, L is the distance between the electrodes.

2.3 Calibration Chamber Measurement

The calibration chamber system used in this research was self-development by writers, which includes penetration system, measurement system and sample system, as shown in Fig. 2.

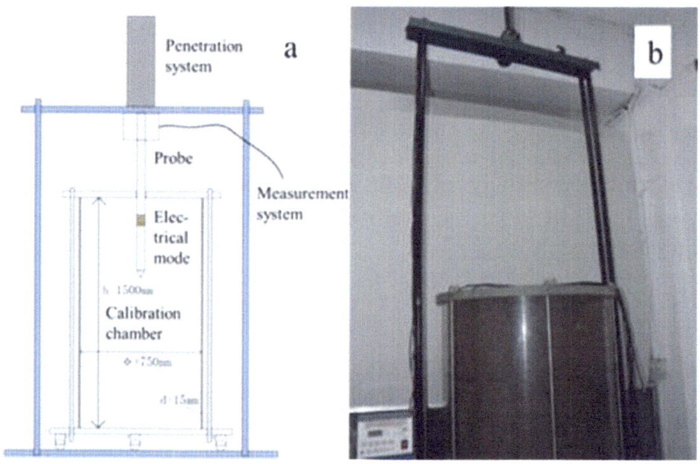

Fig. 2. Calibration chamber test system: (a) schematic of system; (b) picture of system

3 Results and Discussion

3.1 Results of RCPT

3.1.1 Impact of Moisture Content and Density

Figure 3 shows that the resistivity data of heavy metal contaminated soils with varying moisture content and density at controlled concentration. It can be illustrated that increasing moisture content cause a decrease in resistivity, whereas the resistivity decrease with the increase of density. These two important parameters (i.e. moisture content w and density) were the main impact factors of resistivity. The change of resistivity was caused by the change of structural properties (such as double diffuse layer and porosity) of soil sample. The red circle in Fig. 3 was the interface area of layers, in which the resistivity shows an irregular fluctuation. The above phenomenon allows the resistivity better to distinguish the soil layer.

It was shown in Fig. 3 that the resistivity of contaminated soil will decrease from 42 to 18 Ωm when moisture content increase from 13% to 19% at density 13.6 kN/m^3. Whereas when density increase from 13.6 to 14.9 kN/m^3, at moisture content 13%, the resistivity decease from 42 to 30 Ωm. With the addition of metal ions, the conductivity ability of particle and pore water will increase, therefore, value of resistivity decrease.

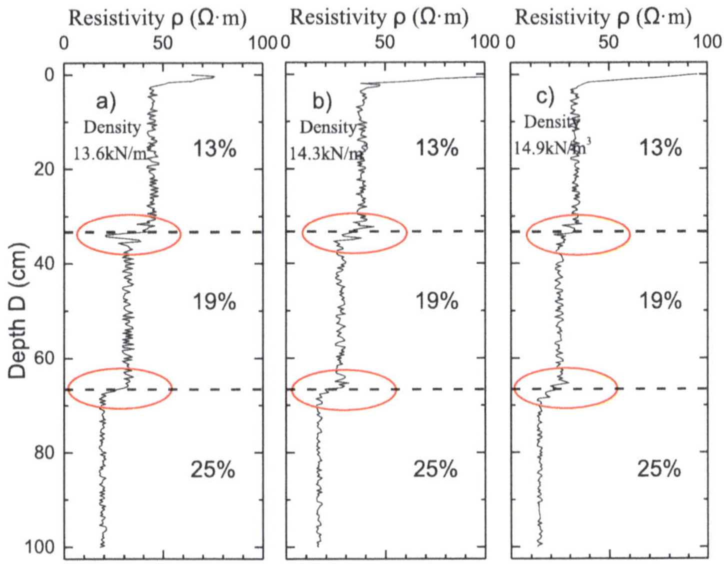

Fig. 3. Resistivity curve of heavy metal contaminated soil at varying moisture content and density with depth (ions concentration N = 500 mg/kg)

3.1.2 Impact of Metal Ions Concentration

Figure 4 shows that the resistivity of heavy metal contaminated soils with varying concentration and density at controlled moisture content. It can be illustrated that increasing concentration cause a decrease in resistivity, whereas the resistivity decrease with the increase of density. Therefore, the concentration was the anther main impact factor of resistivity. The change of resistivity was caused by the change of conductivity ability of soil sample with the addition of metal ions. The red circle in Fig. 4 also shows the interface of layers, in which the resistivity shows an irregular fluctuation. The above phenomenon allows the resistivity better to distinguish the soil layer at different pollution concentration.

It was shown in Fig. 4 that the resistivity of contaminated soil will decrease from 45 to 8 Ωm when concentration increase from 50 to 5000 mg/kg at density 13.6 kN/m^3. Whereas when density increase from 13.6 to 14.9 kN/m^3, at concentration 50 mg/kg, the resistivity decease from 45 to 31 Ωm. With the addition of density, the saturation increased and the conductive path of soil will increase. Therefore, value of resistivity decrease.

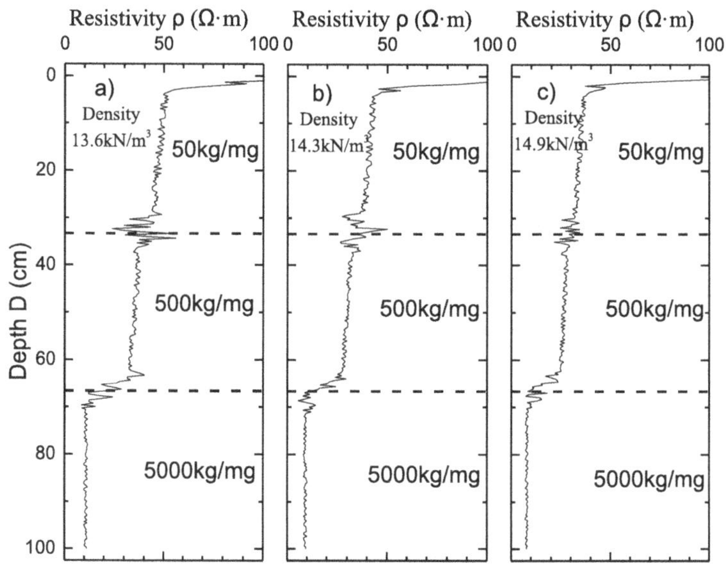

Fig. 4. Resistivity curve of heavy metal contaminated soil at varying concentration and density with depth (moisture content w = 19%)

3.2 Comparison of Resistivity Between Cone and Box

Comparison of resistivity measured by RCPT and soil box was shown in Fig. 5. Application of RCPT detecting the contaminated soil layer is judged by comprehensive comparing the data measured by soil box. Comparison of The correlation between RCPT and soil box is good. It shows that the relation of resistivity measured by the RCPT and soil box is basically 1:1. The fitting formula is as follows:

$$\rho_r = 0.996\rho_s \tag{4}$$

where ρ_r is resistivity measured by RCPT, Ωm; ρ_s is resistivity measured by soil box, Ωm.

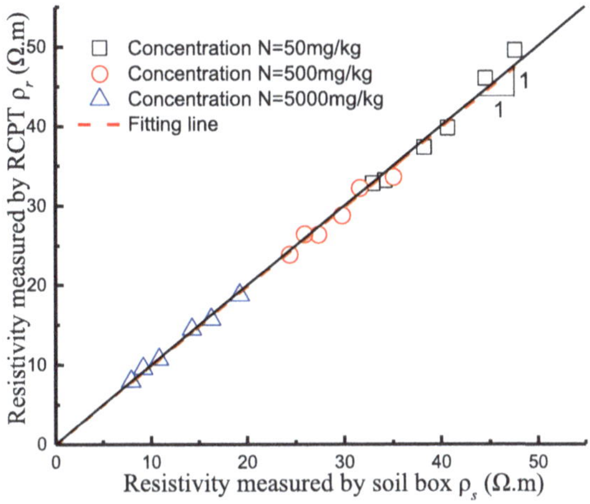

Fig. 5. Calibration chamber test system: (a) schematic of system; (b) picture of system

4 Conclusions

A four-electrode resistivity cone penetrometer (RCPT) test was performed in this study. These experimental findings of calibration chamber shows that resistivity can detect metal contaminated soil layers. The resistivity of heavy metal contaminated soils decreases with the increase of moisture content and metal ions concentration. Furthermore, the comparison between cone and box shows well, which means the powerful tool (RCPT) can be used to detect polluted soil layer.

Acknowledgment. Majority of the work presented in this paper was funded by the Key Project of Natural Science Foundation of China (Grant No. 41330641), the Research Innovation Project of Ordinary University Graduate Student of Jiangsu Province (Grant No. KYLX15_0140), and the postgraduate program of high-level university of national construction (201606090143).

References

Du YJ, Jin F, Liu SY et al (2011) Review of stabilization/solidification technique for remediation of heavy metals contaminated lands. Rock Soil Mech 32(1):116–124

Yoon GL, Park JB (2001) Sensitivity of leachate and fine contents on electrical resistivity variations of sandy soils. J Hazard Mater 84:147–161

Fukue M, Minato T, Horibe H, Taya N (1999) The micro-structures of clay given by resistivity measurements. Eng Geol 54(1):43–53

Mitchell JK, Soga K (2005) Fundamentals of soil behavior. Wiley, Hoboken

Chu Y, Liu S, Wang F, Cai G, Bian H (2017a) Estimation of heavy metal-contaminated soils' mechanical characteristics using electrical resistivity. Environ Sci Pollut Res Int 24 (15):13561–13575

Chu Y, Liu S, Bate B, Xu L (2017b) Evaluation on expansive performance of the expansive soil using electrical responses. J Appl Geophys 148:265–271

Tang LS (2000) Mechanical effect of chemical action of water on soil and analysis on its mechanism. J Sun yat-sen Univ (Nat Sci Ed) 39(4):104–109

Su O, Momayez M (2017) Indirect estimation of electrical resistivity by abrasion and physico-mechanical properties of rocks. J Appl Geophys 143:23–30

Campanella RG, Weemees I (1990) Development and use of an electrical resistivity cone for groundwater contamination studies. Can Geotech J 27(5):557–567

ASTM D G57-06 (2012) Standard test method for field measurement of soil resistivity using the Wenner four-electrode method. USA ASTM International

Study on Thermo-Hydro-Mechanical Coupling Behaviors of Buffer Material

Shengfei Cao[1,2(✉)], Yuemiao Liu[1,2], Jingli Xie[1,2], and Like Ma[1,2]

[1] Beijing Research Institute of Uranium Geology, Beijing 100029, China
csf831016@163.com
[2] CNNC Key Laboratory on Geological Disposal of High Level Radioactive Waste, Beijing 100029, China

Abstract. The concept model of high-level radioactive waste (HLW) repositories in deep geological media is based on a multi-barrier system. Buffer material is considered as an important artificial engineered barrier between the canister and the surrounding rock. Coupled thermo-hydro-mechanical (THM) phenomenon will occur in the buffer material under conditions of the heat generated by the radioactive decay and of the underground water and geo-stress supplied by the surrounding rock, et al. Therefore, it is considered of fundamental important for the evaluation of long-term behavior that the processes taking place in the near filed for the safety operation of the HLW repository. Following the need of understanding the coupled THM behaviors of bentonite, mock-up test is considered as one of the most effective approach. An experiment at middle scale whose internal diameter is 20 cm and inner length 45 cm is proposed as a complementary step for the China-Mock-up test. The middle one is carried out to obtain the bentonite parameters and to identify the processes by simulating the conditions of the China-Mock-up test. The aims intended are to know and understand the long-term behavior of the bentonite submitted to thermal and hydraulic gradients at the opposite direction. The characteristics of the bentonite related to swelling pressure, relative humidity and temperature are presented and interpreted. The results can provide design parameter and theoretical basis for HLW repository.

Keywords: High level radioactive waste geological disposal · Buffer material Bentonite · Thermo-hydro-mechanical coupling

1 Introduction

With the development of nuclear industry, safe disposal of the high level radioactive waste (HLW) has become more and more important, and it is a challenging task facing the scientific and technological world. The design of high-level radioactive waste repositories in deep geological media includes the construction of an engineered barrier around the waste containers constituted by a buffer material. Bentonite-based material has been chosen as buffer/backfill material in most disposal concepts because of its low permeability, heating conductivity, swelling capacity and retention properties, among other features. The behavior of HLW repository is determined, to a large extent, by the characteristics of the design and construction of the engineered barriers and especially

L. Zhan et al. (Eds.): ICEG 2018, ESE, pp. 367–375, 2019.
https://doi.org/10.1007/978-981-13-2224-2_45

by the changes that may occur in the mechanical, hydraulic, and geochemical properties as a result of the combined effects of heat generated by the radioactive decay and of the water and solutes supplied the surrounding rock. Therefore, it is considered of fundamental importance for the evaluation of long-term behavior that the processes taking place in the near field be understood and quantified (Villa and Gomez-Espina 2009).

In order to interpret the processes occurred in the bentonite under coupled thermo-hydro-mechanical (THM) conditions, mock-up test is considered as one of the most effective approach to observe the phenomenon. In recent years, a great number of experiments have been performed at underground research laboratories and room laboratories. Much attentions has been paid to the performance of tests at different scales, such as the large scale buffer material test FEBEX mock-up experiment at CIEMAT in Spain and FEBEX in-situ test at GTS in Switzerland (Martin and Barcala 2005; ENRESA 2000, 2006), a large-scale heating and hydration OPHELIE mock-up test and in-situ heater test PRACLAY experiments in Belgium (Li et al. 2006, 2010; Romero and Li 2006), the Mock-Up-CZ experiment at CTU in Czech Republic (Pacovsky et al. 2007; Svoboda and Vasicek 2010), and the engineering scale test KENTEX facility at KAERI in Korea (Lee et al. 2008; Cho et al. 2010). Thanks to these research works at different scales, the knowledge about the thermal, hydraulic and mechanical properties of the bentonite under the conditions of a deep geological repository has considerably improved.

At present, the preliminary concept of HLW repository in China is a shaft-tunnel structure, located in saturated zones in granites, as shown in Fig. 1. The Gaomiaozi (GMZ) bentonite has been selected as the candidate buffer/backfill material for China's HLW repository through a national screening and comparison (Liu and Wen 2003). Numerous experimental investigations on the basic laboratory research of GMZ bentonite have been performed. In order to further understand the performance of GMZ Na-bentonite under coupled THM conditions, a large-scale mock-up test, named China-Mock-up, has been constructed in our institute (Liu et al. 2013). However, the performance of large-scale test is complicated and time-consuming. Another mock-up test at middle scale is proposed as a complementary method for the China-Mock-up test. The middle one is carried out to obtain the parameters and to identify the processes in the bentonite which submitted to thermal and hydraulic gradients at the opposite direction.

2 THM Middle Mock-up Test

2.1 Material

The test has been performed with a GMZ Na-bentonite from the northern Chinese Inner Mongolia autonomous region, 300 km northwest from Beijing. The processing at the factory consisted on disaggregation and gently grinding, drying at natural condition and sieving by 200 meshes. The mineralogical composition of the GMZ Na-bentonite was analyzed by X-ray diffraction. the clay mineralogy is dominated by montmorillonite ($75 \pm 1\%$) which is the essential mineral needed to ensure the sealing properties.

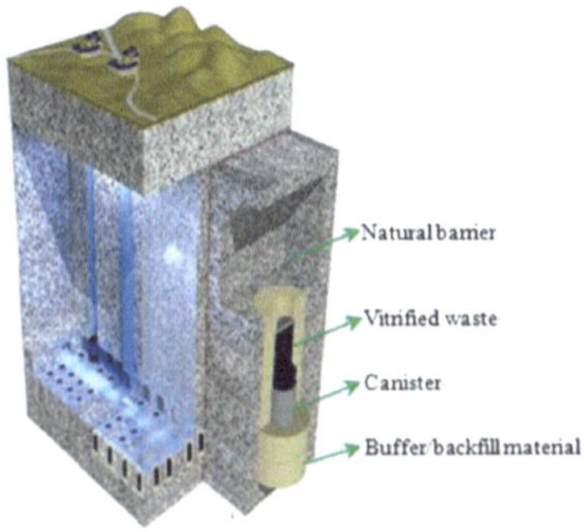

Fig. 1. Preliminary repository concept of HLW in China.

The GMZ bentonite also contains variable quantities of quartz (12 ± 1%), cristobalite (7 ± 1%), feldspar (4 ± 1%), calcite and kaolinite (1 ± 1%). The high content of montmorillonite results in a high cation exchange capacity (CEC = 77.30 meq/100 g), methylene blue exchange capacity of 102 mmol/100 g, alkali index of 1.14, a large plasticity index (I_p = 275), and large specific surface area (a_s = 570 m^2/g) (Liu and Wen 2003). The hygroscopic water content in equilibrium with the laboratory atmosphere is 8.7%, approximately.

A computer-controlled triaxial test machine in combination with specially designed stainless steel molds are used to compact the GMZ bentonite by mean of static uniaxial compaction of powder until reaching the desired density (1.8 g/cm^3) and height (50 mm) into compacted blocks with semicircular shape, as shown in Fig. 2 (180 mm diameter).

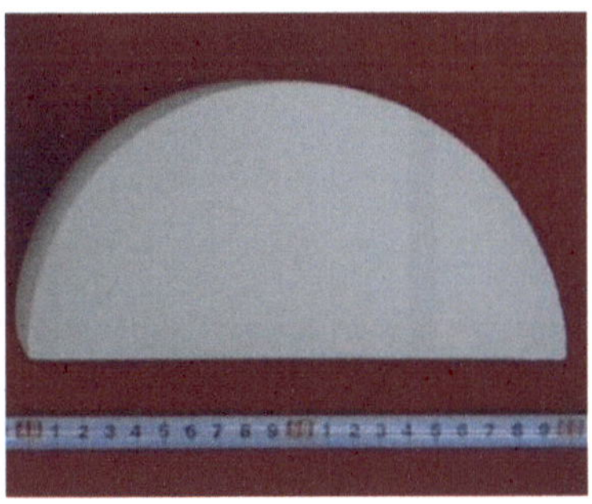

Fig. 2. Highly compacted GMZ Na-bentonite block

2.2 Description of the Experimental Infrastructure

This test was performed in a cylindrical cell which is a stainless steel container with internal diameter 20 cm, inner length 45 cm and thickness 1.5 cm. Two stainless steel pieces were covered on the top and bottom to avoid the deformation of the cell by bentonite swelling. The upper closing the cell was connected to hydration systems which simulate the water penetration. A stainless steel filter was placed on the top of the compacted bentonite block to avoid the erosion of water. The bottom part of the cell was a plane stainless steel heater which is automatically controlled by the electronic thermocouple to simulate the waste canister.

Nine levels of compacted bentonite blocks with its hygroscopic water content (around 8.7%) at an initial nominal dry density of 1.8 g/cm^3 were piled up in the cylindrical cell. Each level had two semicircular shape blocks. Small holes and slots were carved on or inside the bentonite blocks in order to place several kinds of sensors. The cell were instrumented with three relative humidity sensors (H1 ~ H3), seven temperature sensors (T1 ~ T7), five stress sensors (M1 ~ M5) and one water pressure sensor (M6). The data acquisition system can record all the data collected by the sensors automatically. The main components of the cell and the distribution of the blocks and the sensors are shown in Fig. 3.

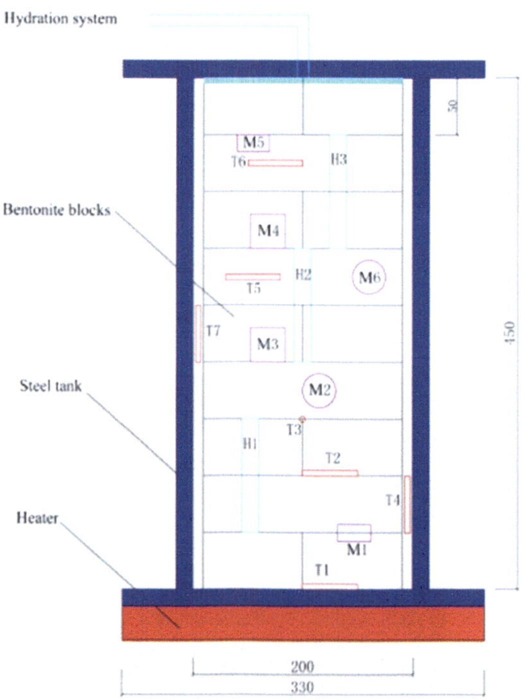

Fig. 3. Scheme of THM middle mock-up test

3 Analysis of Experimental Data and Discussion

The THM middle mock-up test was assembled completely in April 2013. After a pre-operational phase, the operational stage of the test was started on April 24th 2013, and the data indentified as the day 0 on the time scale. Once the sensors connected, the starting off sequence of the test was as follows: (1) Hydration with an artificial way started at day 15 and continued for 390 days; (2) Operated for 220 days at room temperature without heating; (3) The temperature of the heater was set from 30 °C to 90 °C at day 220, and maintained at 90 °C from day 232; (4) The heater was tuned off at day 405; (5) The test was dismantled when the temperature of the heater decreased to the room temperature. The process of water injection and heating is shown in Fig. 4.

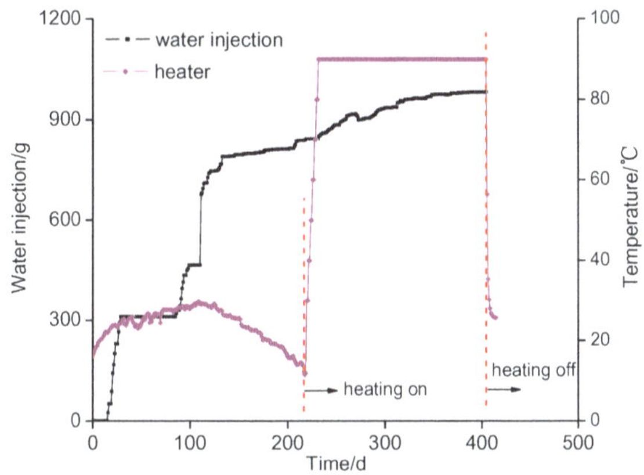

Fig. 4. The procedure of water injection and heating in the test

3.1 Temperature Evolution

Temperature inside the bentonite blocks was measured during the THM middle mock-up test. The temperature evolution at different positions is given in Fig. 5. According to the procedure described above, the test operated at room temperature at the beginning stage. The temperature in bentonite at different levels was mainly influenced by the seasonal and daily changes in the laboratory temperature. Unfortunately, there was a mistake for the temperature data acquisition during the heating stage. But when the heating system maintained at 90 °C, the temperature evolution shows that bentonoite has reached a relative balance state. It has to be mentioned here, the temperature increased sharply especially far from the heating part when the insulation material was installed at the outside of the cylindrical cell. It can be noticed that, the temperature has continuously increased with time from 300 days. There are two reasons to interpret this trend of the temperature. The temperature in the bentonite increased with the changes in the room temperature. It is obvious that the surrounding environment has a strong influence to the distribution of the temperature field. The temperature distribution is also influenced by a complex coupling mechanism. Considering the saturation will change thermal conductivity, the temperature distribution is also dependent on the saturation process in the compacted bentonite. Another phenomenon is that the temperature difference between the bentonite blocks becomes smaller. This means that the temperature in the bentonite is getting more and more stable easily in the future.

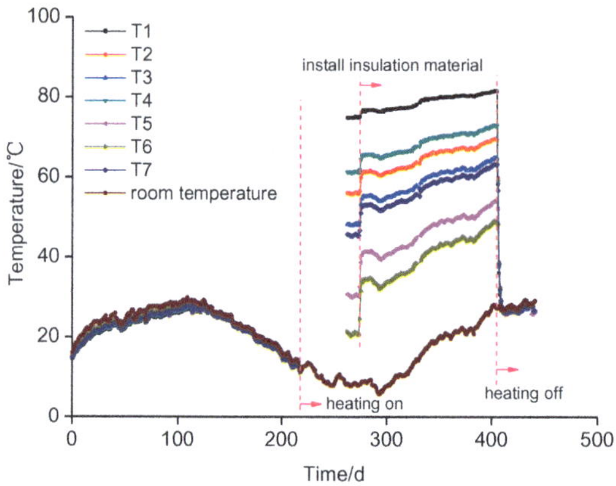

Fig. 5. Temperature distribution in bentonite blocks.

3.2 Relative Humidity Evolution

The evolution of the relative humidity at several distances from the heater in the compacted bentonite block is shown in Fig. 6. As illustrated, the compacted bentonite was progressively saturated from the beginning to the end. During this time, because of the extremely low permeability of the compacted bentonite, an increased in relative humidity was registered by the sensor H3 placed on the top of the cell and to a lesser extent, by the sensor H2 placed in the middle of the cell and H1 at the bottom of the cell. However, an unexpected phenomenon was observed for H1 in the beginning of hydration process. In the test, the gap between the inner steel container and the compacted bentonite blocks was filled with bentonite pellets. The water could penetrate into the bottom of the cell due to the gravity of the liquid water. So the sensor H1 was damaged by the liquid water at the beginning and later became normal because of the self-sealing performance of the bentonite. The relative humidity on the top of the cell (H3) increased because of the continuous hydration and reached almost saturated.

It also can be seen that the distance to the heater has a significant influence on the saturation processes. An initial difference in the relative humidity measured at different levels is observed after heating. A clear increase in relative humidity being registered by the two sensors H1 and H2 near the heater, which is probably related to the generation of the water vapor phase produced by the heating. This revel the quick migration of the water in the vapor phase from the bentonite near the heater zones to towards the cooler zones. However, as noticed, the relative humidity closest to the heater area is much more complex. With the increase of temperature, the decrease of H1 can be observed. The continuous heating produced the dry effect stronger than the

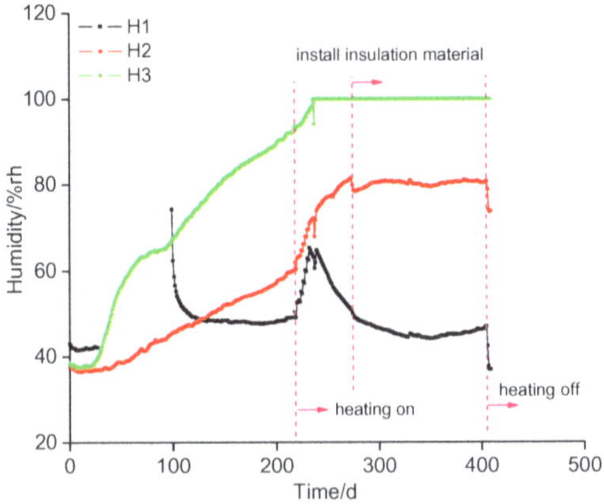

Fig. 6. Relative humidity evolution in bentonite blocks.

hydration effect that caused the decreasing relative humidity. After some time, with the increase of water injection, hydration overcomes the desiccation and the relative humidity increases.

3.3 Swelling Pressure Evolution

The mechanical response of the bentonite, which has important implications on the design and performance of the repository, is affected by hydration and temperature. The swelling pressure evolution in different levels is given in Fig. 7. As shown in the figure, the swelling pressure variation is rather limited or negligible in the beginning of the test because of the low permeability of the compacted bentonite. After, the swelling pressure M5 start to develop as soon as the water comes in contact with the bentonite closest to the hydration surface, as can be observed in Fig. 7. In other positions, no significant swelling pressure in compacted bentonite is recorded to present. However, the swelling pressures (M1 ~ M4) increase after the heating started. This evolution in the bentonite may be influenced by several mechanisms, including the thermal expansion induced by the high temperature and the swelling pressure generated by the bentonite saturation effect. Then, the swelling pressure decreases gradually after reaching a peak of 1.6 MPa. This could be attributed to the stress releasing induced by the initial gap between the bentonite blocks and pellets, and also the gaps between the sensors and the blocks. It can be seen that the swelling pressure decrease sharply when the heater turned off.

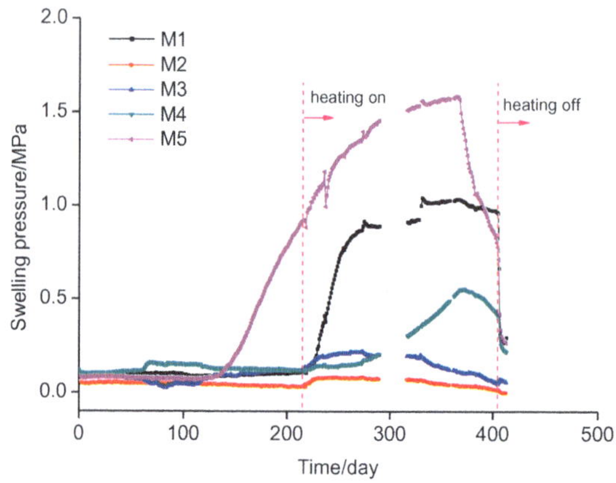

Fig. 7. Evolution of swelling pressure in bentonite blocks.

4 Conclusions

An experiment at middle scale is proposed as a complementary step for the China-Mock-up test. The aims intended are to know and understand the long-term behavior of the bentonite submitted to thermal and hydraulic gradients at the opposite direction. Based on the analysis and discussions of the recorded results, several preliminary conclusions can be drawn:

The temperature in the bentonite blocks has increased with time. The temperature distribution is affected by the distance from the heater and also influenced by the relationship between the thermal conductivity and the saturation procedure.

The relative humidity in the bentonite blocks is strongly influenced by the wetting effect by the hydration and the drying effect by the desiccation of the heater. And near the heating zones, the desiccation phenomenon is observed obviously.

The swelling pressure in the bentonite blocks is influenced by the saturation process and the thermal expansion. Because of the extremely low permeability and the inhomogeneous saturation process, the swelling pressure variation is relatively limited and inhomogeneous. And the stress releasing can also be seen due to the gaps between the bentonite blocks.

The information obtained can improve the knowledge on the behavior of compacted bentonite and will be helpful to interpret the China-mock-up test.

References

Villa MV, Gomez-Espina R (2009) Report on thermo-hydro-mechanical laboratory tests performed by CIEMAT on Febex Benonite 2004–2008

Martin PL, Barcala JM (2005) Large scale buffer material test: mock-up experiment at CIEMAT. Eng Geol 81:298–316

ENRESA (2000) FEBEX Project. Full-scale engineered barriers experiment for a deep geological repository for high level radioactive waste in crystalline host rock. Final Report. Publication Technical ENRESA 1/2000, Madrid

ENRESA (2006) FEBEX Full-scale Engineered Barriers Experiment, Updated Final Report 1994–2004. Publication Technical ENRESA 05-0/2006, Madrid

Li XL, Frédéric B, Johan B (2006) The Belgian HLW repository design and associated R&D on the THM behaviour of the host rock and EBS. Chin J Rock Mechan Eng 25(4):682–692

Li XL, Bastiaens P, VanMarcke P et al (2010) Design and development of large-scale in-situ PRACLAY heater experiment and horizontal high-level radioactive waste disposal gallery seal experiment in Belgian HADES. J Rock Mech Geotech Eng 2(2):103–110

Romero E, Li XL (2006) Thermo-hydro-mechanical characterization of OPHELIE backfill mixture. Chin J Rock Mechan Eng 25(4):733–740

Pacovsky J, Svoboda J, Zapletal L (2007) Saturation development in the bentonite barrier of the Mock-Up-CZ geotechnical experiment. Phys Chem Earth 32:767–779

Svoboda J, Vasicek R (2010) Preliminary geotechnical results from the Mock-Up-CZ experiment. Appl Clay Sci 47:139–146

Lee JO, Park JH, Cho WJ (2008) Engineering-scale test on the thermal-hydro-mechanical behaviors in the clay barrier of HLW repository. Ann Nucl Energy 35:1386–1396

Cho W-J, Lee J-O, Kwon S (2010) Analysis of thermo-hydro-mechanical process in the engineered barrier system of a high-level waste repository. Nucl Eng Des 240:1688–1698

Liu YM, Wen ZJ (2003) Study on clay-based materials for the repository of high level radioactive waste. J Mineral Petrol 23(4):42–45 (in Chinese)

Liu YM, Wang J, Cao SF et al (2013) A large-scale THMC experiment of buffer material for geological disposal of high level radioactive waste in China. Rock Soil Mech 34(10):2756–2789 (in Chinese)

Vane Shear Tests to Evaluate in Situ Stress State of a Soil-Bentonite Slurry Trench Wall

Jeffrey C. Evans[1(✉)], Yu-Ling Yang[1], and Daniel G. Ruffing[2]

[1] Department of Civil and Environmental Engineering, Bucknell University, Lewisburg, PA 17837, USA
evans@bucknell.edu
[2] Geo-Solutions, Inc., New Kensington, PA, USA

Abstract. Soil-bentonite slurry trench cutoff walls have been employed widely in the US as engineered barriers to control groundwater flow and contaminant migration in the subsurface. The hydraulic conductivity of the soil-bentonite backfill is stress dependent and to better understand the in-situ stress state of soil-bentonite slurry trench walls, a wall with a length of 194 m, depth of 7 m, and width of 0.9 m was constructed, instrumented, and investigated in-situ. As a part of the in-situ investigation, vane shear tests were conducted at different locations and depths to evaluate the shear strength distribution within the wall. The results indicate that the peak undrained shear strength value, S_u, exhibits a consistent trend with increasing depth. In general, two stages of S_u were observed: (1) at depth from 1 to 2.5 m, the S_u decreases as depth increases, presumably due to the effects of the cover soil and groundwater level; (2) at depth of 2.5 to 7 m, the S_u shows insignificant change with depth and falls into the range of 5 to 10 kPa. Using a constant shear strength to effective stress ratio, S_u/σ', the stresses predicted from the vane shear strength tests were in good agreement with those measured using earth pressure sensors embedded at the time of wall construction.

Keywords: Bentonite · Shear strength · Slurry wall · State of stress
Vane shear

1 Introduction

Soil-bentonite (SB) slurry trench cutoff wall backfill is commonly composed of a blend of in-situ excavated or borrow soil and slurry made from sodium bentonite and water. These cutoff walls have been employed widely in the US and Canada as engineered barriers to control the migration of groundwater as well as subsurface contaminants (D'Appolonia 1980; Evans 1993; Sharma and Reddy 2004). Construction steps include subsurface trench excavation under bentonite-water slurry, and then replacement of the slurry within the trench by SB backfill. As a seepage control structure, hydraulic conductivity of the wall is considered to be one of the major principle performance criterions, and its upper limit value is commonly specified as 1×10^{-9} m/s.

Hydraulic conductivity of the SB backfill is known to be stress dependent. Hydraulic conductivity decreases with increasing effective stress (Evans 1994;

L. Zhan et al. (Eds.): ICEG 2018, ESE, pp. 376–384, 2019.
https://doi.org/10.1007/978-981-13-2224-2_46

Shackelford 1994; Filz et al. 2001; Ruffing and Evans 2010; Evans and Huang 2016). Therefore, reliable estimation of the stress-state within the wall is of critical importance. However, there are a limited numbers of studies, particularly field studies, addressing the in-situ stress-state of SB walls (Ruffing et al. 2010, 2011, 2015; Ruffing and Evans 2010). Based on the above considerations, a SB wall was constructed, instrumented, and investigated at a site located 3 km east of the Bucknell University campus in Lewisburg, PA. Post installation, the wall has been subjected to a series of field and laboratory tests to investigate the barrier integrity, state-of-stress in the wall, and backfill properties. As a part of these investigations, vane shear tests were conducted in June/July 2017 at eight locations to evaluate the in-situ shear strength of the wall.

This paper presents results of the in-situ vane shear tests. Variations of the undrained shear strength (S_u) at different locations and depths are presented. A S_u to effective vertical stress ratio (S_u/σ') is suggested based on the stress measured by the earth pressure sensors.

2 Means and Methods

2.1 Cutoff Wall Site and Backfill Properties

The SB cutoff wall was set within a buffer zone between a permitted mining area of a commercial sand/gravel quarry, operated by Central Builders Supply, and a natural wetland known as the Montandon Marsh (Fig. 1). The subsurface of the project site consists of primarily alluvial materials within the footprint of historic flooding and river channels of the Susquehanna River. The wall, with a length of approximately 194 m, width of 0.9 m, and average depth of approximately 7 m, was constructed in July 2016 by Geo-Solutions, Inc. of New Kensington, PA, USA. The constructed wall was covered by a geotextile followed by 0.3 m of top soil over the geotextile.

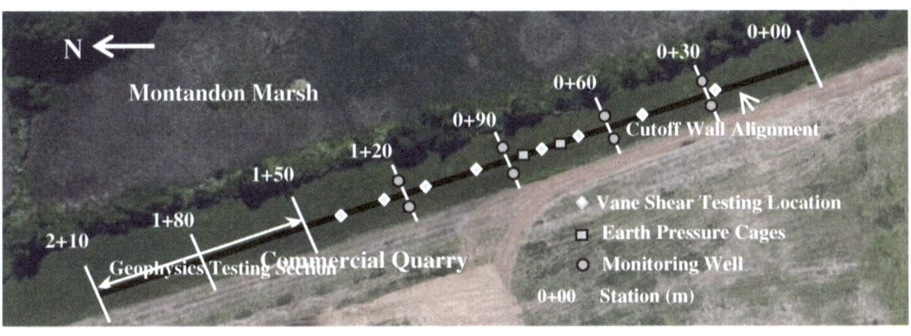

Fig. 1. Plan diagram of the cutoff wall with locations of the monitoring wells and the vane shear testing points.

The backfill within the wall consists of imported base soil from the mine site and enough of 5% bentonite-water slurry to create a workable material. The resulting total bentonite content of the backfill is approximately 1%, calculated by weight of bentonite to weight of soil. The fines content in the backfill ranges from 44% to 54%. Laboratory measurements of grab samples of the backfill indicate that the hydraulic conductivity ranges from 7.0×10^{-10} to 1.4×10^{-9} m/s as tested in flexible-wall triaxial tests using an effective stress of 21 kPa to 35 kPa. For more details, please see Evans and Ruffing (2017) and Malusis et al. (2017).

2.2 Vane Shear Tests

Vane shear tests were conducted at eight locations along the wall alignment, at an average distance interval of 16 m (Fig. 1). The vane shear tests were conducted in general accordance with ASTM D2573. Considering the softness of the backfill and to ensure acceptable torque resolution, a larger, custom-made field vane was selected. This vane has a tapered tip with a diameter of 0.089 m and vane height of 0.184 m consistent with the dimension ratios required by ASTM. The taper angles at the top and bottom of the vane are 47°. The vane shear test was performed at vertical increments of 0.5 m until the tip reached a depth of approximately 7.0 m, the wall bottom. At each 0.5 m increment of depth, torque readings were taken using a digital torque wrench. The maximum recorded torque was then used to calculate the undrained shear strength, S_u. As per ASTM, a slip coupling was used to determine rod friction. Average rotation rates of 0.1 degree per second were used with a variation of 0.05 to 0.2 degree per second, as permissible. Per the ASTM D2573, S_u was calculated using Eq. 1:

$$S_u = \frac{12 * T_{max}}{\pi D^2 * \left(\frac{D}{\cos(i_T)} + \frac{D}{\cos(i_B)} + 6H\right)} \tag{1}$$

where S_u is the undrained shear strength, T_{max} is the maximum torque value, D is the diameter of the vane, H is the height of the vane, i_T is the angle of taper at the top of the vane, and i_B is the angle of taper at the bottom of the vane.

2.3 Stress Measurement

In addition to the in-situ vane shear testing, the stress in the wall was directly measured using earth pressure sensors as shown in Fig. 2. The earth pressure sensors were mounted in an aluminum frame at transverse, longitudinal, and vertical directions to directly measure the total stress along with a pore pressure sensor to permit calculations of effective stress. In addition, pitch, roll and direction sensors were mounted on the frame to correct stress measurements for placement orientation. A total of four frames were installed, with three placed at 0 + 84 at depths of 2.4 m, 4.4 m, and 6.4 m, and one placed at 0 + 75 at depth of 6.2 m. For more information regarding the instrumentation, please see Evans and Ruffing (2017) and Malusis et al. (2017).

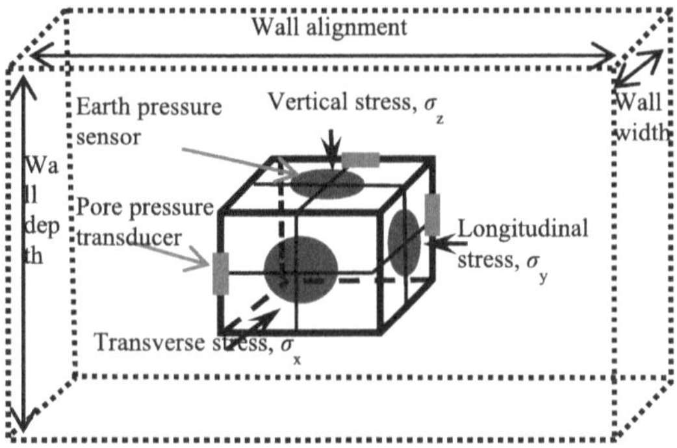

Fig. 2. Diagram of earth pressure cage installed within the trench wall.

3 Results and Discussion

3.1 Undrained Shear Strength

Based on the plasticity index of the backfill of 7 (Malusis et al. 2017) and ASTM D2573, a correction factor of 0.92 was used to correct the raw field S_u. The variations of the resultant S_u values with depth at eight different locations of the wall are shown in Fig. 3. In general, the S_u value at 1 to 7 m depth ranges from 3.9 to 14.8 kPa, which is consistent with the magnitude of shear strength results reported from other SB walls, e.g. Ruffing and Evans (2010, 2015) and Evans and Ryan (2005). Anomalous data points are marked with the "?" sign in Fig. 3. These anomalous points show S_u values that are substantially higher than the other data and are attributed to the presence of occasional oversize gravel and cobbles within the backfill. The groundwater levels presented in Fig. 3 were obtained from adjacent groundwater monitoring wells by linear interpolation.

As shown in Fig. 3, the S_u values at the eight different locations exhibit a consistent trend with respect to depth. Two distinct regions of S_u are shown: (1) the S_u decreases from about 13 kPa to 7 kPa as depth increases from 1 to 2.5 m; (2) at depths of 2.5 to 7 m, the S_u shows insignificant change with depth, and falls into the range of 5 to 10 kPa. The higher values of S_u near the top of the wall can be attributed to the localized effects of the weight of the cover soil and suction pressures from wet and dry cycles. Also, lateral deformation data from inclinometers installed along the sidewall alignment indicates that the maximum inward deformation occurred at the top of the wall with little to no movement at bottom of the wall (Evans et al. 2017). All of these factors contribute to higher stress, and relatedly strength, in the top 2.5 m of the wall. For depth of 2.5 to 7.0 m, less lateral deformation was observed, and expectedly the backfill within this depth exhibits relatively constant S_u values. At some testing locations, the trench bottom exhibits higher S_u values as compared to the backfill which is attributed to the vane penetrating into the native soil below the wall.

Figure 4 shows a composite data set of the average undrained shear strength value, $S_{u,ave}$, vs. depth. Note that the anomalous data points marked in Fig. 3 are not included

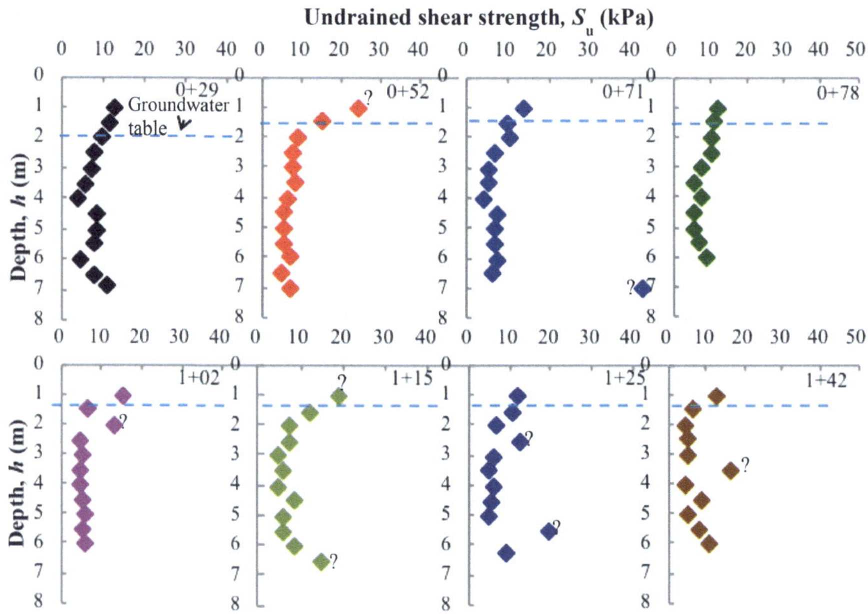

Fig. 3. Undrained shear strength at different test locations of the wall.

in the $S_{u,ave}$ value in Fig. 4. Figure 4 clearly shows the trend of decreasing S_u within the first 2.5 m depth and a slight increasing trend from 2.5 m to 7.0 m depth.

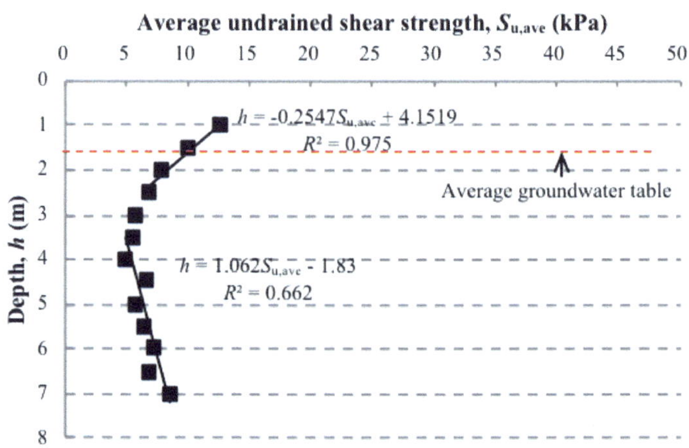

Fig. 4. Variation of average undrained shear strength with depth.

3.2 Relationship Between Shear Strength and Effective Stress

Soil mechanics tells us that there is a direct correlation between the effective confining stress and the shear strength (Lambe and Witman 1969; Holtz and Kovacs 1981). Moreover, shear strength has been related to the major principle stress, normally the vertical effective stress, σ'_v, for normally consolidated (NC) soils in the form of S_u/σ'_v by many researchers (Lambe and Witman 1969; Ruffing et al. 2015). In Ruffing et al. (2015) the authors conclude that the S_u/σ'_v value ranges from 0.1 to 0.3. In that study, the authors selected a S_u/σ' value of 0.22 and 0.30 to use undrained shear strength measurements to predict the horizontal effective stress in a SB wall.

The transverse, longitudinal, and vertical effective stresses of the study wall (denoted as σ'_x, σ'_y, and σ'_z, respectively) obtained from the pressure cells at 2.4 m, 4.4 m, and 6.4 m depths at location 0 + 80 and at 6.2 m depth of location 0 + 75 are plotted in Fig. 5. These data were collected one year after the wall construction. At both locations, both σ'_x and σ'_y increase, while σ'_z decreases as depth increases, with the horizontal effective stress (i.e., σ'_x and σ'_y) generally higher than the vertical one (i.e., σ'_z). The transverse effective stress, σ'_x, appears to be the major principle stress in the wall and is used to represent the horizontal effective stress (σ'_h) in the following analysis. This is consistent with the determination made in Ruffing et al. (2011) using Marchetti Dilatometer tests, and that assumed by Ruffing et al. (2015), who also indicated that the major principle stress is in the horizontal rather than the vertical direction within most SB walls.

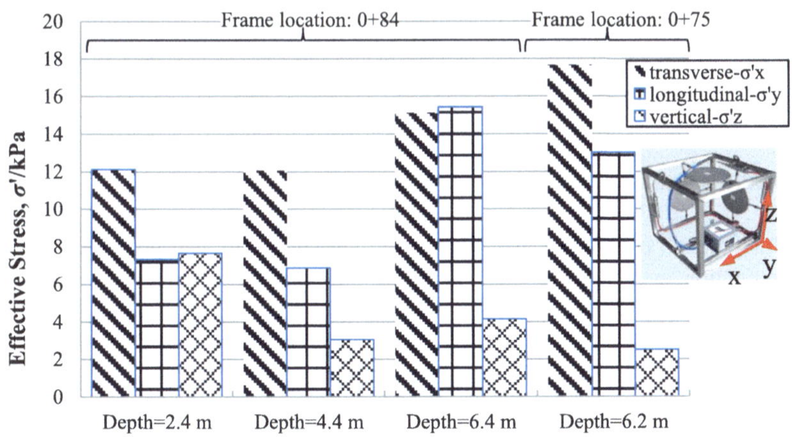

Fig. 5. Transverse, longitudinal, and vertical effective stresses at different depth in wall.

Figure 6 displays the variation of measured σ'_x with wall depth. In addition, the predicted horizontal effective stresses, σ'_h, using $S_{u,ave}/\sigma'$ of 0.2, 0.3, 0.4, 0.5, and 0.6 also plotted in Fig. 6 for comparison. The $S_{u,ave}/\sigma'$ values of 0.2 and 0.3 were chosen based upon the recommendations in Ruffing et al. (2015) which are also typical of normally consolidated soils deposited in a typical sedimentary geologic environment.

The $S_{u,ave}/\sigma'$ values were plotted after it was evident that the $S_{u,ave}/\sigma'$ for this wall is higher than 0.3. A review of Fig. 6 indicates that the best fit generally occurs at a $S_{u,ave}/\sigma'$ value of 0.5.

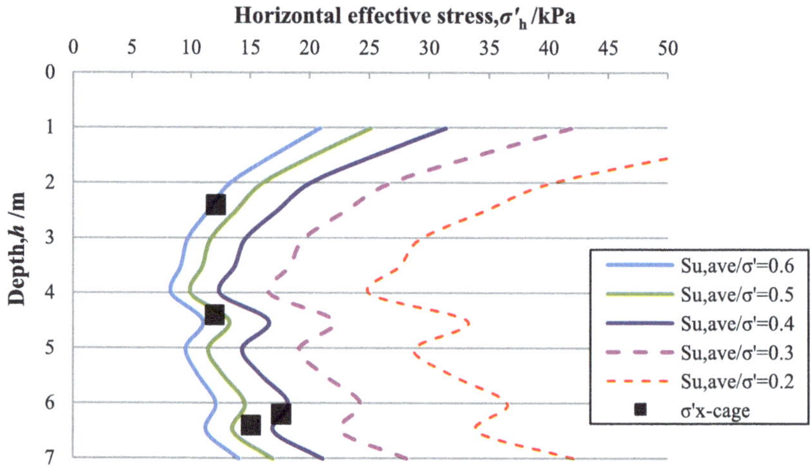

Fig. 6. Measured and predicted horizontal effective stress vs. depth.

This analysis shows that, for this project, the stresses are lower than would be predicted using S_u/σ' ratios of 0.3, as might be expected from the literature. Another way to think of this result is that the measured strengths are higher than those predicted using S_u/σ' of 0.3.

4 Conclusions

A SB slurry trench cutoff wall was constructed, instrumented, and investigated. One of the investigation methods, the vane shear test, was performed for direct measurement of the shear strength with the ultimate purpose of calculating the state-of-stress within the wall. The vane shear data show a decrease in strength for the shallow portion of the wall (from 1 to 2.5 m in depth) due to a combination of factors including decreasing effects of the soil cover weight, wetting and drying, unsaturated stresses, and greater inward lateral movement near the top of the wall. Below 2.5 m, the shear strength is essentially constant until the bottom of the wall at a 7 m depth, with shear strength values between 5 to 10 kPa. Based on the measured effective stress using earth pressure sensors, the transverse effective stress, σ'_x, is found to be the major principle stress in the wall. A $S_{u,ave}/\sigma'$ value of 0.5 correlates well with the data in this study.

Acknowledgements. The financial support of the National Science Foundation (Grant No. 1463198) is greatly appreciated. The authors would like to thank Central Builders Supply (Montandon, PA), CETCO (Hoffman Estates, IL) and Geo-Solutions, Inc. (New Kensington, PA) for their resources, bentonite supply and wall construction for this study. Special thanks goes to

James Gutelius, Director of Laboratories for the Department of Civil and Environmental Engineering, Bucknell University, as well as Landon Barlow, graduated student of Bucknell University for their great assistance and contribution on data collection. Finally, the authors appreciate the financial support through the generosity of Michael and Laureen Costa, which enabled Yuling Yang's postdoctoral research on this study.

References

D'Appolonia DJ (1980) Soil-bentonite slurry trench cutoffs. J Geotech Eng Div 106(4):399–417

Evans JC (1993) Vertical cutoff walls. In: Daniel DE (ed) Geotechnical practice for waste disposal. Chapman and Hall, London

Evans JC (1994) Hydraulic conductivity of vertical cutoff walls. In: Hydraulic conductivity and waste contaminant transport, ASTM STP 1142, pp 79–94 (1994)

Evans JC, Huang H (2016) Hydraulic conductivity of soil bentonite slurry walls. In: Proceedings of ASCE GeoChicago 2016 sustainability, energy, and the geoenvironment, ASCE GSP, Reston, VA, pp 548–577

Evans JC, Ruffing DG (2017) Design and construction of an experimental soil-bentonite cutoff wall. In: ASCE geotechnical frontiers 2017, ASCE GSP, Reston, VA, pp 164–174

Evans JC, Ryan CR (2005) Time-dependent strength behavior of soil-bentonite slurry wall backfill. In: Geo-frontiers congress 2005, waste containment and remediation, ASCE GSP, Reston, VA, pp 1–9

Evans J, Orobia M, Gutelius J, Ruffing D, Barlow L, Malusis M (2017) Soil-bentonite slurry trench cutoff wall lateral deformations, consolidation, stress transfer and hydraulic conductivity. In: Proceedings of the 2nd symposium on coupled phenomena in environmental geotechnics (CPEG2), Leeds, UK, pp 1–6

Filz et al (2001) Determining hydraulic conductivity of soil-bentonite using the API filter press. ASTM Geotech Test J 24(1):61–71

Holtz RD, Kovacs WD (1981) An introduction to geotechnical engineering. Prentice-Hall, Inc., Upper Saddle River, New Jersey

Lambe TW, Witman RV (1969) Soil mechanics, SI version. John Wiley and Sons, Inc., New York

Malusis MA., Evans JC, Gutelius J, Jacob RW, Ruffing DG, Barlow LC, Barchiori A (2017) Construction and monitoring of an instrumented soil-bentonite cutoff wall: field research case study. In: Proceedings of the 29th central Pennsylvania geotechnical conference proceedings, Hershey, PA, 25–27 January 2017

Ruffing DG, Evans JC (2010) In situ evaluation of a shallow soil bentonite slurry trench cutoff wall. In: Proceedings of the 6th international congress on environment geotechnics, New Delhi, India, Tata McGraw-Hill, pp 758–763. ISBN 13:9780070707108

Ruffing DG, Evans JC, Malusis MA (2010) Prediction of earth pressures in soil-bentonite cutoff walls. In: ASCE GeoFlorida 2010 advances in analysis, modeling and design GSP 199, ASCE GSP, Reston, VA, pp 2416–2425

Ruffing DG, Evans JC, Malusis MA (2011) Evaluations of lateral earth pressure in a soil-bentonite slurry trench cutoff wall. In: 14th Pan-American conference on soil mechanics and geotechnical engineering. CD-ROM. ISBN 978-0-920505-47-2

Ruffing DG, Evans JC, Ryan CR (2015) Strength and stress estimation in soil bentonite slurry trench cutoff walls using cone penetration test data. In: Proceedings of the international foundations congress and equipment expo, ASCE GSP, Reston, VA, pp 2567–2576

Ruffing DG, Evans JC (2015) Strength and stress estimation in soil bentonite slurry trench cutoff walls using cone penetration test data. In: The international foundations congress and equipment expo, San Antonio Texas ASCE GSP 256, pp 2567–2576
Shackelford CD (1994) Waste-soil interactions that alter hydraulic conductivity, ASTM STP, Reston, VA, pp 111–168
Sharma HD, Reddy KR (2004) Geoenvironmental engineering: site remediation, waste containment, and emerging waste management technologies. John Wiley & Sons Inc, Hoboken, New Jersey

Study on Soil-Water Characteristic Curves of Unsaturated Silty Clay of an Old Landfill Cover

Zhan-hong Qiu[1], Zan-cheng Zhu[1], Qiao Wang[2], Hai-jian Xie[2(✉)], and Chenxi Zhao[3]

[1] Institue of Coastal Environmental and Geotechnical Engineering, Taizhou University, Taizhou 318000, China

[2] MOE Key Laboratory of Soft Soils and Geoenvironmental Engineering, Zhejiang University, Hangzhou 310058, China
xiehaijian@zju.edu.cn

[3] Civil Engineering College, Shenyang Jianzhu University, Shenyang 110168, China

Abstract. In order to better understand the performance of silty clay under different conditions, an assessment of the water retention characteristics under different temperatures is required. Laboratory tests were carried out to measure the soil-water characteristic curves (SWCCs) of the silty clay at different temperatures by the filter paper method. The soil-water characteristic curves (SWCCs) in this study were shown in terms of moisture content, void ratio and degree of water saturation (S_r). Temperature has great influence on the SWCCs of the silty clay. Water retention capacity decreases with the increase of temperature, particularly at low suction range (0–500 kPa for matric suction). For a fixed suction value (e.g. 100 kPa), when the temperature increases from 20 to 60 °C, the water retention capacity of the soil decreases from 22.5% to 13.0%. It is also found that temperature has greater influence on matric suction for the case with $S_r \geq 0.2$ than that for the case with $S_r < 0.2$. The curve indicated that the correlation between silty clay void ratio and suction is approximately linear. It is noted that the values of soil matric suction with temperature of 40 °C appeared to be larger than the case with 20 °C and 60 °C for a given void ratio.

Keywords: Soil-water characteristic curve (SWCC) · Filter paper method Dry density · Temperature · Water content

1 Introduction

Landfill cover system commonly placing over landfills forms a barrier between the environment and the waste (EPA 1991; Benson et al. 2007). Compacted clay liner (CCL) is widely used as the part of the composite cover system for the municipal solid waste landfill (EPA 1991; Ng et al. 2015). Ng et al. (2015) proposed a three-layer capillary barrier landfill cover system consisting of a CCL, a gravelly sand layer and a silt layer. It is confirmed from both experimental and back-analysis results that no infiltrated water seeped through the proposed three-layer barrier system. It is also

L. Zhan et al. (Eds.): ICEG 2018, ESE, pp. 385–390, 2019.
https://doi.org/10.1007/978-981-13-2224-2_47

indicated that the permeability of CCL increases with time because of the desiccation and differential settlement (Suter et al. 1993; Ghazizade and Safari 2016).

The soil-water characteristic curve (SWCC) for a soil is the correlation between water content and suction for the soil. SWCCs play a vital role in both hydraulic and mechanical behavior of unsaturated soils. In order to get better understanding of the soil-related problems, it is necessary to take into account the relevant moisture holding capacity and stress of soils. Nowadays, there are many researches about the modelling of the SWCC of soils (Phoon et al. 2010; Pedroso and Williams 2010; Zhou et al. 2012). Phoon et al. (2010) demonstrates that a lognormal random vector is suitable to model the curve-fitting parameters of the SWCC. Factors influencing the soil's SWCCs have been widely investigated (Cai et al. 2010; Thyagaraj and Rao 2010; Gallage and Uchimura 2010; Roshani and Sedano 2016). Gallage and Uchimura (2010) investigated the effects of dry density and grain-size distribution on the SWCCs of four kinds of sandy soils and revealed that soils with a low density have lower air-entry value and residual suction than soils with a high dry density. Roshani and Sedano (2016) presented a rational approach to describe the effect of temperature on the SWCC under isothermal conditions of sandy soils.

CCLs are used as barrier systems and cover system in landfill design to minimize the escape of contaminant from landfills and prevent surface rainfall infiltrating into the waste. The SWCCs for the clay have been investigated (Tripathy et al. 2014; Li et al. 2017). Li et al. (2017) develop an experimental method to measure the SWCCs for cracked soils during drying-wetting cycles and provided necessary parameters for the prediction of SWCCs based on a theoretical model for cracked soils. There is not much studies about the influence of temperature on the SWCCs of CCL.

The objectives of this study are to investigate the effect of the temperature on the SWCCs of silty clay collected from Jiaojiang Shabei landfill of Taizhou, China. The relevance between soil matric suction and void ratio, water degree of saturation and moisture content were studied, respectively.

2 Materials and Methods

2.1 Sample Preparation

The soil samples used in this study were excavated from the first-stage project of Jiaojiang Shabei landfill cover system at the depth of 0.20–0.30 m in Taizhou, Zhejiang province of China. The soil samples are gray brown, slightly wet, at hard plastic state, fissure agenesis, poor viscosity, and with a small amount of gravels, hairs, etc. X diffraction analysis demonstrates that the soil mainly consists of quartz (QTZ), plagioclase (PL), chlorite (CHL), mica (MICA). The soil mainly consists of sand (70%) and silt (20%), with some clay (10%) fractions.

Other basic soil sample properties such as specific gravity, liquid limit had been measured by the GB/T 50123-1999 standards and are shown in Table 1. It can be seen from Table 1 that the liquid limit, plastic limit and plasticity index of soil samples are low, which belong to low liquid limit silty clay.

Table 1. Physical property index of silty clay

Moisture content ω_f/%	Specific density G_s	Liquid limit W_l/%	Plastic limit W_p/%	Plasticity indexI_p
23.2	2.737	31.35	18.64	12.71

2.2 Test Methods

The test was started with immersing the filter papers into the formalin solution until it is saturated, then the paper was dried in the 60 °C oven. The mentioned procedure above was to avoid mildewing of the papers and the soil samples.

The dry density of soil samples is 1.3 g/cm^3 which were obtained by applying different stress by jacks. Moisture content of the soil samples are 3–24% or 3–21% and which were pressed every 3%. The test temperature is 20 °C, 40 °C and 60 °C respectively. The experimental scheme is shown in Table 2.

Table 2. Experimental scheme of the soil samples

No.	Dry density ρ_d (g/cm^3)	Initial moisture content w_0 (%)	Temperature t/°C
1	1.3	3–24	20
2	1.3	3–24	40
3	1.3	3–21	60

Whatman No. 42 filter papers are adopted in this experiment. Zhu et al. (2016) covered these shortages in the rating curve equation. Therefore, the rating curve equation for the moisture content less than 20% can be expressed as follows (Zhu et al. 2016).

$$\log s = 2.56308 - 0.0957\omega_f \quad (2.89 \le \omega_f \le 21.61, T = 20\,°\text{C}) \tag{1}$$

$$\log s = 2.55437 - 0.11242\omega_f \quad \left(1.82 \le \omega_f \le 17.60, T = 20\,°\text{C}\right) \tag{2}$$

$$\log s = 2.57930 - 0.13743\omega_f \quad \left(1.37 \le \omega_f \le 13.89, T = 20\,°\text{C}\right) \tag{3}$$

where s is the total suction in MPa, ω_f is the moisture content of the filter paper in percentage (%) after the moisture balancing.

2.3 Results and Discussions

Figure 1 shows the soil SWCCs in terms of soil moisture content under different temperatures conditions for the initial dry density equals 1.3 g/cm^3. The water holding capacity of soil samples decreases with the increase of temperature. The smaller the suction, the more obvious the difference is. It is indicated that temperature has great influence on the SWCCs of the soil sample. The water retention capacity increased with the decrease of the temperature especially at the low suction range.

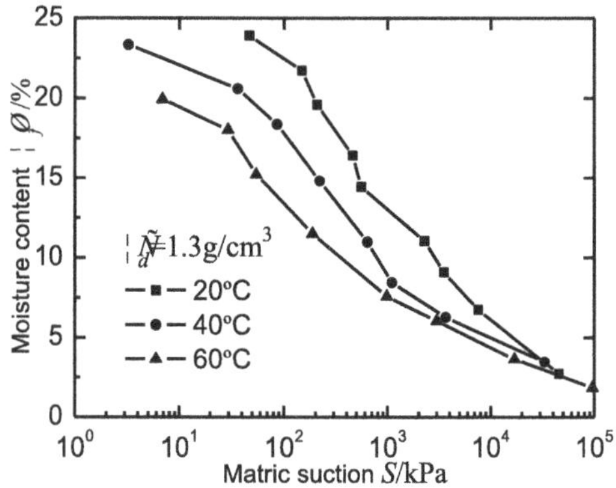

Fig. 1. Correlation between matric suction and moisture content

Figure 2 shows the SWCCs in terms of void ratio under different temperature conditions when initial dry density equals 1.3 g/cm^3. It is indicated that the void ratio of the soil samples increases with the decrease of temperature and the curve is almost overlapped. It is noted that the values of soil matric suction with temperature of 40 °C appeared to be larger than the case with 20 °C and 60 °C for a given void ratio. Comparison between the effect of suction change on void ratio and water content demonstrated that water content is more sensitive to the change of suction than the void ratio.

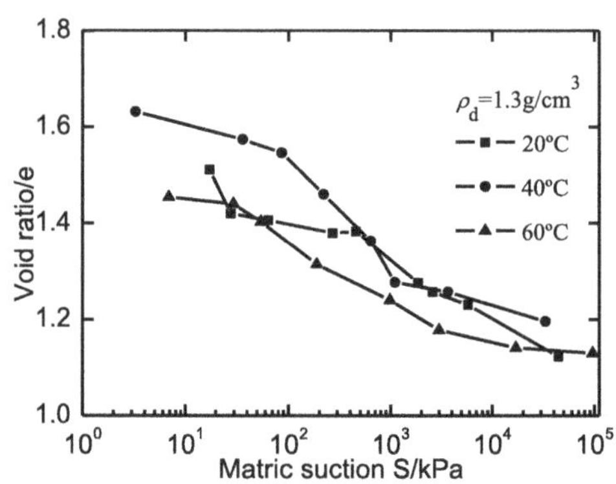

Fig. 2. Correlation between matric suction and void ratio

Figure 3 shows the SWCCs in terms of the degree of water saturation for the cases with different temperature when initial dry density equals 1.3 g/m^3. It is indicated that the temperature has greater influence on the suction of the soil samples for $S_r \geq 20\%$

(see Fig. 3). Or the saturation of soil sample is higher, the variation of suction is larger and the saturation is smaller, the influence of temperature difference on the suction is smaller. When the suction is constant, the saturation decreases with the increase of soil temperature, that is to say, the water holding capacity of the soil sample is reduced.

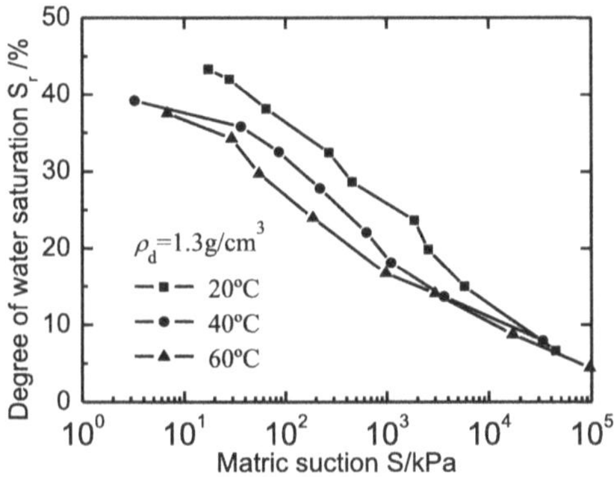

Fig. 3. Correlation between matric suction and saturation

3 Conclusions

The temperature has greater influence on the suction of the soil samples for $S_r \geq 20\%$. The water holding capacity of soil samples decreased with the increase of temperature and the smaller the suction, the more obvious the difference was. The values of soil matric suction with temperature of 40 °C appeared to be larger than the case with 20 ° C and 60 °C for a given void ratio. Comparison between the effect of suction change on void ratio and water content demonstrated that water content is more sensitive to the change of suction than the void ratio.

Acknowledgements. This work was financially supported by the National Natural Science Foundation of China under Grant No. 51578356 and No. 51478427.

References

Benson CH, Barlaz MA, Lane DT, Rawe JM (2007) Practice review of five bioreactor/recirculation landfills. Waste Manag 27(1):13–29

Cai GQ, Zhao CG, Liu Y (2010) Temperature effects on soil-water characteristic curve of unsaturated soils. Rock Soil Mech 31(4):1055–1060

EPA (1991) Design and construction of RCRA/CERCLA final covers. Environmental Protection Agency. Seminar Publication, EPA/625/4-91/025

Gallage CPK, Uchimura T (2010) Effects of dry density and grain size distribution on soil-water characteristic curves of sandy soils. Soils Found 50(1):161–172

Ghazizade MJ, Safari E (2016) Analysis of desiccation crack depth in three compacted clay liners exposed to annual cycle of atmospheric conditions with and without a geotextile cover. J Geotech Geoenviron Eng 143(3):1–6
Suits LD, Sheahan TC, Leong EC, He L, Rahardjo H (2002) Factors affecting the filter paper method for total and matric suction measurements. Geotech Test J 25(3):322–333
Li JH, Lu Z, Guo LB, Zhang LM (2017) Experimental study on soil-water characteristic curve for silty clay with desiccation cracks. Eng Geol 218:70–76
Ministry of Construction of China (MCC) (2007) Technical code for municipal solid waste sanitary landfill closure (CJJ 112-2007). China Architecture & Building Press, Beijing (in Chinese)
Ministry of Construction of China (MCC) (2004) Technical code for municipal solid waste sanitary landfill (CJJ 17-2004). China Architecture & Building Press, Beijing (in Chinese)
Ng CW, Liu J, Chen R, Xu J (2015) Physical and numerical modeling of an inclined three-layer (silt/gravelly sand/clay) capillary barrier cover system under extreme rainfall. Waste Manag 38:210–221
Pedroso DM, Williams DJ (2010) A novel approach for modelling soil–water characteristic curves with hysteresis. Comput Geotech 37(3):374–380
Phoon KK, Santoso A, Quek ST (2010) Probabilistic analysis of soil-water characteristic curves. J Geotech Geoenviron Eng 136(3):445–455
Roshani P, Sedano JÁI (2016) Incorporating temperature effects in soil-water characteristic curves. Indian Geotech J 46(3):309–318
Suter GW, Luxmoore RJ, Smith ED (1993) Compacted soil barriers at abandoned landfill sites are likely to fail in the long term. J Environ Qual 22(2):217–226
Thyagaraj T, Rao SM (2010) Influence of osmotic suction on the soil-water characteristic curves of compacted expansive clay. J Geotech Geoenviron Eng 136(12):1695–1702
Tripathy S, Tadza MYM, Thomas HR (2014) Soil-water characteristic curves of clays. Can Geotech J 51(8):869–883
Zhou AN, Sheng D, Carter JP (2012) Modelling the effect of initial density on soil-water characteristic curves. Géotechnique 62(8):669
Zhu Z, Sun DA, Zhou A, Qiu Z (2016) Calibration of two filter papers at different temperatures and its application to GMZ bentonite. Environ Earth Sci 75(6):1–11

Analysis of Consolidation Processing of Piezocone Penetration Test in Cutoff Wall

Xuepeng Li[1(✉)], Guojun Cai[1], Songyu Liu[1], and Yuchao Li[2,3]

[1] Institute of Geotechnical Engineering, Southeast University, Nanjing 210096, China
lixuepeng1984@126.com
[2] MOE Key Laboratory of Soft Soils and Geoenvironmental Engineering, Zhejiang University, Hangzhou 310058, China
[3] College of Civil Engineering and Architecture, Zhejiang University, Hongzhou 310058, China

Abstract. The piezocone penetration test (CPTU) is now widely used as an efficient and economical in-situ geotechnical investigation technology. The CPTU not only can provide the data of conventional cone penetration test (CPT), but also give a continuous profile of pore pressure. Furthermore, when the piezocone is halted in soil with a specific depth, the excess pore pressure produced by penetration will be dissipated until the value of zero. This paper investigates the consolidation processing after the piezocone is penetrated into soil-bentonite cutoff wall in a landfill site, China. A finite-element procedure is also used for simulating the consolidating processing of soil-bentonite backfills. The coefficient of consolidation and hydraulic conductivity of soil-bentonite backfills can be evaluated from the consolidation processing of excess pore pressure. The paper gives the necessity and importance of simulation of consolidation processing by finite-element procedure.

Keywords: Soil-bentonite · Consolidation processing · CPTU
Finite-element procedure

1 Introduction

The piezocone penetration test (CPTU), which gives measurement of tip resistance (q_t), sleeve friction (f_s), and pore pressure at the shoulder locations of cone (u_2), appears to be an extremely promising tool to evaluate the geotechnical parameters (Mayne and Campanella 2005; Cai et al. 2010; McQueen et al. 2015). Furthermore, the excess pore pressure, induced by penetration test, can be reduced or dissipated when the penetration device is halted at a specific depth of the soil (Teh and Houlsby 1991; Burns and Mayne 1998; Chai et al. 2015). The CPTU can record the dissipation processing of excess pore pressure. Numerous research studies have been published about this aspect of CPTU (Robertson and Wride 1998; Cai et al. 2011; Ha et al. 2014; Krage et al. 2015).

Most of the studies about the excess pore pressure dissipation are almost based on the field investigation and laboratory test. However, the above studies based on

L. Zhan et al. (Eds.): ICEG 2018, ESE, pp. 391–396, 2019.
https://doi.org/10.1007/978-981-13-2224-2_48

numerical simulation are very few. Chai et al. (2012) established an empirical method for interpreting the field value of the horizontal coefficient of consolidation (c_h) from "non-standard" dissipation curves by numerical simulation. Yu et al. (2000) used the finite-element procedure for the analysis of steady-state in clay.

This paper presents a novel finite-element approach for the analysis of consolidation processing of soil-bentonite cutoff wall. The main purpose of this present study is to test the spatial distribution of excess pore pressure at different times, which plays a very important role in evaluating the coefficient of consolidation of soil-bentonite backfills.

2 Materials and Methods

2.1 In-suit Soil-Bentonite Cutoff Wall

The soil-bentonite cutoff wall is located in Jingjiang city of Jiangsu province, China. The soil in the city can be divided into four layers. About 3 m sandy silt, 1 m silty clay, 3 m sandy silt and silty sandy from top to bottom soil layers are listed in the soil.

The cutoff wall is composed of excavated in-suit soil and bentonite. To maintain the trench stability, 5% bentonite and 95% water by weight are selected to produce a slump of 125 mm. The bentonite slurry is set aside at least 24 h. The soil-bentonite cutoff wall has the properties of 15 m length, 10 m depth and 0.6 m width. Ground water table depth at the test locations varies in the range 0.5–1.5 m.

The cutoff wall is composed of excavated in-suit soil and bentonite. To maintain the trench stability, 5% bentonite and 95% water by weight are selected to produce a slump of 125 mm. The bentonite slurry is set aside at least 24 h. The soil-bentonite cutoff wall has the properties of 15 m length, 10 m depth and 0.6 m width. Ground water table depth at the test locations varies in the range 0.5–1.5 m.

2.2 In-suit CPTU

The penetration device is imported from Hogentogler with 200 kN Digital Piezocone Penetrometer and data acquired using E4FCS software. The study uses a piezocone with the characteristic of 10 cm^2 cross sectional area, 150 cm^2 sleeve friction surface area and 35.6 mm cone bottom diameter. Six CPTU soundings and twelve pore pressure dissipation test are conducted in this study. The penetration devices are controlled by hydraulic piston to ensure piezocone penetration at a constant 2 cm/s speed.

2.3 Numerical Investigation

The process of penetration test is treated as a contact problem, where the cone and soil-bentonite cutoff wall are regarded initially as two separate bodies. Furthermore, the process of penetration is modeled as the continuous contact and deformation of those two bodies (Sheng et al. 2005; Walker and Yu 2006). The Modified Cam Clay (MCC) model was adopted to represent the stress-strain behavior of the soil-bentonite cutoff wall. The parameters adopted for the MCC model are listed in Table 1. Here, e_0 is the initial void ratio (the values of top or bottom surface of soil-bentonite backfills are 0.96 and 0.54). γ_t is the total unit weight; v *is the* Poisson's ratio; κ *and* λ are the rebound line and consolidation line in $e - \ln p'$ plot respectively; M is the stress ratio at failure; k_0 is the rest earth pressure coefficient; k is the hydraulic conductivity.

Table 1. Parameters adopted for the MCC model

e_0	γ_t (kN/m^3)	v	κ	λ	M	k_0	k (m/s)
0.96/0.54	17.7	0.33	0.01	0.1	1.2	0.56	4.1E−8

The initial stress state of the soil-bentonite cutoff wall is assumed to be geostatic. The depth of the model is 2 m and the width is 0.5 m. Appling a given force instantaneously to the top of piezocone and causes the piezocone to produce a slight vertical displacement, which induce the excess pore pressure of soil. And the next dissipation time is about 30 min.

3 Results and Discussion

The study gives the distribution of vertical stress, vertical displacement and the excess pore pressure of soil-bentonite cutoff wall by CPTU. The position of the piezocone is held fixed during the dissipation processing of excess pore pressure.

3.1 The Vertical Stress

Figure 1 gives the contours of the vertical stress of soil-bentonite cutoff wall. The vertical stress of the piezocone is largest. Furthermore, only considering the soil-bentonite cutoff wall, the distribution features are mainly around the cone shoulder. The largest value of vertical stress is about 496 kPa, which is listed near the cone shoulder. The vertical stress of soil-bentonite cutoff wall gradually decreases around the cone shoulder.

3.2 The Vertical Displacement

The penetration device is imported from Hogentogler with 200 kN Digital Piezocone Penetrometer and data acquired using E4FCS software. The study uses a piezocone with the characteristic of 10 cm^2 cross sectional area, 150 cm^2 sleeve friction surface area and 35.6 mm cone bottom diameter. Six CPTU soundings and twelve pore pressure dissipation test are conducted in this study. The penetration devices are controlled by hydraulic piston to ensure piezocone penetration at a constant 2 cm/s speed.

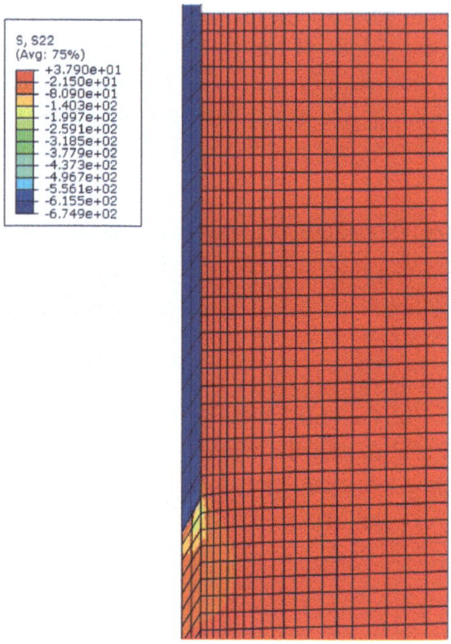

Fig. 1. Contours of the vertical stress.

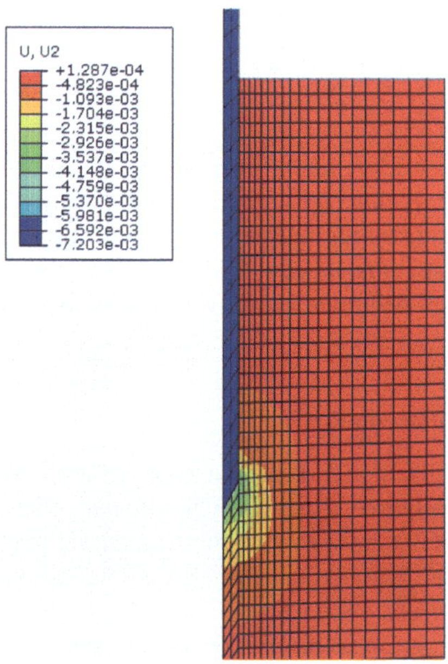

Fig. 2. Contours of the vertical displacement.

3.3 Excess Pore Pressure

Figure 3 gives the contours of excess pore pressure around the cone tip at the end of penetration test. It can be seen that the largest excess pore pressure value is 36 kPa, which is listed near the cone tip. And the excess pore pressure values at the periphery of the cone are smaller than values in the zone adjacent to it. The values of excess pore pressure around the cone tip are positive. However, those values above the cone shoulder are negative. The soil listed below the shoulder will move to the new location, and the new location is just above the shoulder. The possible mechanism can be explained as unloading effect (Teh and Houlsby 1991). Similarly, the distribution pattern of excess pore pressure at different dissipation time can also be obtained, and then the dissipation of excess pore pressure at the same location can be obtained, which is useful to the calculation of coefficient of consolidation.

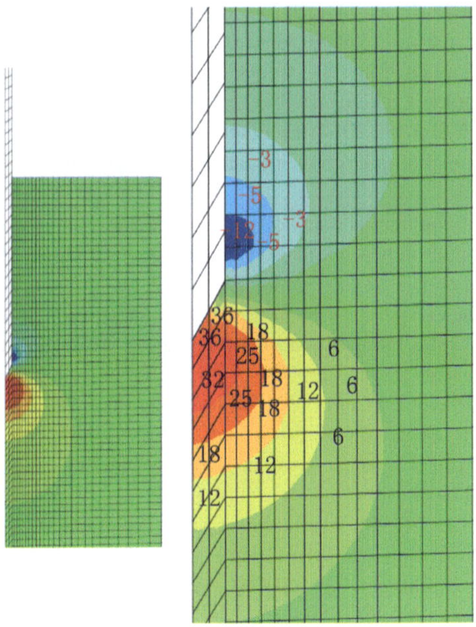

Fig. 3. Contours of the excess pore pressure.

4 Results and Discussion

A finite-element procedure is used for simulating the dissipation process of excess pore pressure of soil-bentonite backfills. The distribution of vertical stress, vertical displacement and the excess pore pressure of soil-bentonite are listed in this paper. The excess pore pressure values at the periphery of the cone are smaller than values in the zone adjacent to it. The dissipation of excess pore pressure is useful for the calculating the calculation of coefficient of consolidation.

Acknowledgements. Majority of the work presented in this paper was supported by the following supporting funds Organization: The National Natural Science Foundation of China (Grant No. 41672294) and The Foundation of Jiangsu Province Outstanding Youth (Grant No. BK20140027). We deeply appreciate your consideration of our manuscript, and we look forward to receiving comments from the reviewers.

References

Burns SE, Mayne PW (1998) Monotonic and dilatory pore-pressure decay during piezocone tests in clay. Can Geotech J 35(6):1063–1073

Cai GJ, Liu SY, Tong LY (2010) Field evaluation of deformation characteristics of a lacustrine clay deposit using seismic piezocone tests. Eng Geol 116(3–4):251–260

Cai GJ, Liu SY, Puppala AJ (2011) Comparison of CPT charts for soil classification using PCPT data: example from clay deposits in Jiangsu Province, China. Eng Geol 121(1–2):89–96

Chai JC, Sheng D, Carter JP et al (2012) Coefficient of consolidation from non-standard piezocone dissipation curves. Comput Geotech 41:13–22

Chai JC, Hossain MJ, Yuan DJ, Shen SL, Carter JP (2015) Pore pressures induced by piezocone penetration. Can Geotech J 53(3):540–550

Ha TG, Jang IS, Choo YS et al (2014) Evaluation of coefficient of consolidation for dilatory dissipation in piezocone test in overconsolidated cohesive soils. KSCE J Civ Eng 18(2):475–487

Krage CP, DeJong JT, Schnaid F (2015) Estimation of the coefficient of consolidation from incomplete cone penetration test dissipation tests. J Geotech Geoenviron Eng 141(2). Article number 06014016

Mayne PW, Campanella G (2005) Versatile site characterization by seismic piezocone. In: Proceedings of the 16th international conference on soil mechanics and geotechnical engineering, Osaka, Japan, 2, pp 721–724

McQueen W, Miller B, Mayne PW et al (2015) Piezocone dissipation tests at the Canadian Test Site No. 1, Gloucester, Ontario. Can Geotech J 53(5):884–888

Teh CI, Houlsby GT (1991) An analytical study of the cone penetration test in clay. Geotechnique 41(1):17–34

Robertson PK, Wride CE (1998) Evaluating cyclic liquefaction potential using the cone penetration test. Can Geotech J 35(3):442–459

Yu HS, Herrmann LR, Boulanger RW (2000) Analysis of steady cone penetration in clay. J Geotech Geoenviron Eng 126(7):594–605

Walker J, Yu HS (2006) Adaptive finite element analysis of cone penetration in clay. Acta Geotech 1(1):43–57

Sheng D, Eigenbrod KD, Wriggers P (2005) Finite element analysis of pile installation using large slip frictional contact. Comput Geotech 32:17–26

Testing of Leachate Levels at a Landfill with Multiple Intermediate Covering Layers

Haijie He(✉), Jiwu Lan, Yunmin Chen, and Pengcheng Ma

MOE Key Laboratory of Soft Soils and Geoenvironmental Engineering, Zhejiang University, Hangzhou 310058, China
{he_haijie, lanjiwu, chenyunmin, mapengcheng}@zju.edu.cn

Abstract. There is no leachate level testing method for a landfill with multiple intermediate cover layers. This paper provides an in-situ testing method for multi-layer perched leachate levels. In-situ tests were carried out at a landfill with multiple intermediate cover layers and the perched leachate level on each intermediate cover layer were obtained. The testing results showed that: (1) by drilling to a position 1–2 m away from the cover layer, the well casing was positioned and the in-well casing leachate level was reduced to the well casing bottom, and the leachate level recovery was observed so that the steady leachate level of this layer could be measured 24 h later; (2) the perched leachate level height on each intermediate cover layer were not consistent; (3) the low-permeability intermediate cover layers were the main course of the formation of the multi-layer perched leachate levels.

Keywords: Landfill · Leachate levels · Intermediate cover layer

1 Introduction

Landfill leachate is produced from rainfall, snowfall and landfill waste. Rain and melted snow can permeate the covering layers and then enter the landfill to form leachate [1]. Meanwhile, moisture of buried municipal solid waste (MSW) can also be transformed into leachate [2]. The main influential factors of leachate generation in North America are rainfall or snowfall, and the waste moisture content is basically less than 30% [3]. However, kitchen waste occupies a large proportion in MSW from developing countries, which has a high moisture content, and the moisture content of MSW from developing countries is pretty high. For example, the waste moisture content reaches 50–70% in China [4]. Kitchen waste contains intracellular water, which can be released during the MSW degradation and generate leachate continuous [5]. For the above reasons, leachate levels in developing countries landfills are generally higher than leachate levels in landfills of developed countries, such as America, so more complicated situations need to be confronted.

In China, the water level gauge, the time domain reflectometry (TDR) and the piezometer are mainly used for the practical testing of leachate levels. Zhang et al. [6] buried piezometers above and below the intermediate cover layer. Gravel and clay mud balls were refilled into the water-bearing layers and the damaged cover layers, respectively. In this way, the leachate levels on the upper and lower waste layers were

L. Zhan et al. (Eds.): ICEG 2018, ESE, pp. 397–403, 2019.
https://doi.org/10.1007/978-981-13-2224-2_49

isolated and pore pressures at different depths were measured. With the method, the leachate levels on and below one intermediate cover layer can be measured effectively. However, the operations are complicated and it is hardly applicable to a landfill with multiple intermediate cover layers. At presently, an effective and practical testing method for a landfill with multiple intermediate cover layers is absent.

The multi-layer water testing method in the traditional hydro-geological investigation refers to the water-bearing formation layers that are determined according to the existing materials or core records. Water is stopped by a water-stop plug on the intermediate cover layer. Then, the water level of each water-bearing layer is measured. Testing courses include a 73 mm or 89 mm well casing for a water stop, a filter pipe connected under the water-bearing layer, and 2 support trays welded 1–2 m higher than the top plate of the filter pipe, where the outer diameter is slightly smaller than the pore diameter of the water-stop well casing. The interval between the two support trays is smoothly wrapped layer by layer with kelp, and then fastened with cotton ropes to form a water-stop plug. After the water level measurement of this layer, the water-stop plug is pulled out, drilling is continued, and the next layer level is tested. Nevertheless, due to obvious waste cavity shrinkage, the hole wall of a waste borehole, which is drilled, may not be maintained, and placement of the filter pipe may become difficult. Cavity shrinkage may also take the place of waste during the leachate level recovery. Hence, the filter pipe is hardly pulled out, so testing of the next layer leachate level becomes difficult. For these reasons, the experiment makes the following improvements based on applying the well casing water stop method in a traditional geological survey: first, when the filter pipe cannot be placed, the drilling depth is 0.5–1 m higher than the well casing bottom face so a good hydraulic power guide area is formed in this section and the leachate can effectively permeate into the well casing. Secondly, a self-made water collection device is used to extract the remaining leachate inside the well casing so the influences brought by the leachate that remains in the well casing that are to be tested are avoided.

In order to investigate the leachate level distribution, especially the multi-layer perched leachate levels, five leachate level testing boreholes were drilled in a landfill located in Northwest China which was with multiple intermediate cover layers. An in-situ multi-layer leachate level testing method was proposed. The perched leachate level distribution in the landfill was obtained.

2 Site Description

The landfill where the multi-layer leachate level testing was conducted is located in Xi'an of Shanxi Province, China. As shown in Fig. 1, the area of the landfill is more than 0.73 km^2 and has a gross storage capacity of more than 49 million m^3. The current daily average landfilling capacity reaches 8000 t/d. There is a loess dam downstream from the landfill and the dam crest elevation is 509 m. The thickness of each MSW layer is approximately 10 m. The total waste thickness is approximately 70–80 m. There are 8 waste layers (the first layer to the eighth layer) from the bottom to the top. To prevent diffusion of the landfill gas and increased leachate caused by rainfall infiltration, intermediate cover layers were positioned on top of the filled waste (see Fig. 2). The landfill is located in a loess area surrounded by abundant loess resources. Hence, loess with a thickness of 10–30 cm was selected for the intermediate cover layers where one loess cover layer was positioned on top of each waste layer. The construction process of the intermediate cover layer was shown in Fig. 3.

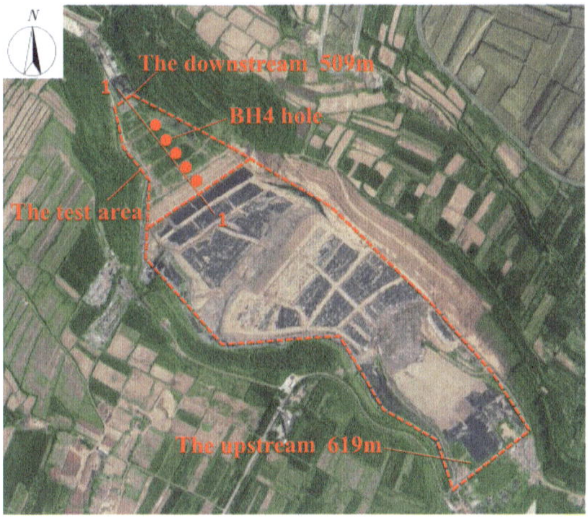

Fig. 1. Layout of landfill

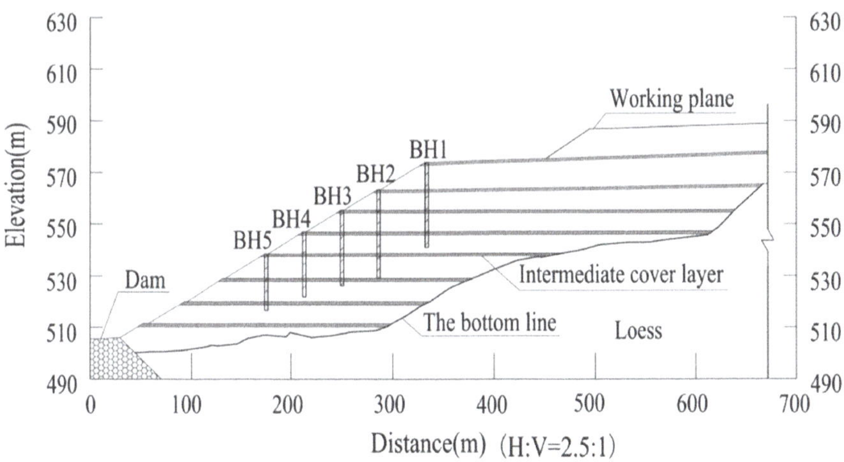

Fig. 2. 1-1 cross-section of landfill

Fig. 3. The construction process of the intermediate cover layer

3 Methods

As shown in Fig. 4, the steps of the leachate level test are as follows:

(a) Leachate level test of the first layer: drive the drill bit until near the first intermediate cover layer. Then advance the well casing to the bottom of the drill hole (see Fig. 5). After that, start leachate level recovering test and test the leachate level in the well casing every 2–4 h (see Fig. 6), then obtain the leachate level recovery curve of the first layer waste.

(b) Leachate level test of the second layer: the drill bit drills near the second intermediate cover layer and the well casing follows up to the bottom of the drill hole. Use a water extraction device to extract the leachate from the well casing until the leachate level drops to the bottom of the well casing. Test leachate level in the well casing every 2–4 h, and draw a leachate level recovery curve of second layer waste.

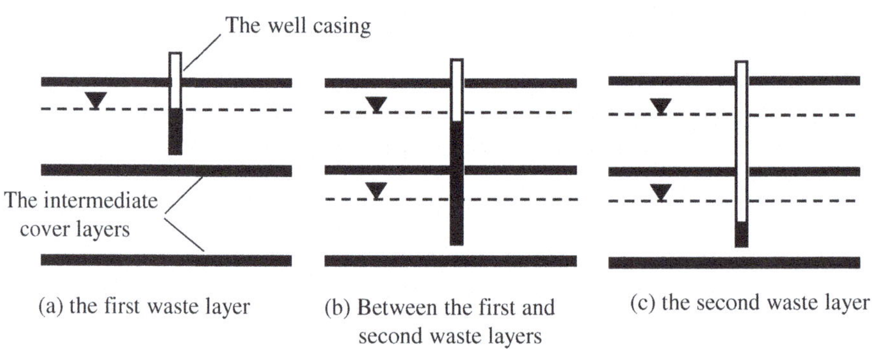

Fig. 4. Testing of multi-layer leachate levels.

(c) The following leachate level testing method are consistent with the second leachate level testing method.

Fig. 5. In-well casing leachate level reduction

Fig. 6. On-site testing for multi-layer leachate levels

4 Results and Discussion

As shown in Fig. 7, the first intermediate cover layer of the BH4 hole appears at 7.3–7.4 m. When testing the leachate level of this layer, the drilling depth was 7 m, the well casing depth was 7 m, and the leachate level in the well casing was reduced to 6.62 m. Figure 7 shows that during the recovery of the leachate level in this layer, the level rose from 6.62 m below the ground to 6.47 m below the ground in the recovery phase of 0–1 h, which was an increase of 0.15 m. After 20 h of standing, the leachate level rose to 4.22 m below the ground. After standing for 24 h, the leachate level was 4.08 m below the ground. The second intermediate cover layer appeared at 16.6–16.8 m. When testing the leachate level of this layer, the drilling depth was 14 m, the casing depth was 13.5 m, and the leachate level in the well casing was reduced to 13.15 m. During the recovery of the leachate level in this layer, the level rose from 13.15 m below the ground to 11.09 m below the ground during the recovery period of 0–4 h, an increase of 2.06 m. After standing for 24 h, the level was 7.82 m below the ground. When testing the third leachate level, the drilling depth was 25 m, the casing depth was 24 m, and the leachate level in the well casing was reduced to 24.36 m. During the recovery of the leachate level in this layer, the level rose from 24.36 m below the ground to 24.06 m below the ground in the recovery phase of 0–4 h, which was a 0.3 m increase. After standing for 24 h, the waste level was 23.65 m below the ground.

By collecting the leachate level test data for each test hole, the leachate distribution near the slope can be obtained. It is shown in Fig. 8 that the existence of the intermediate cover layers effectively stopped the vertical transport of the leachate inside the landfill where one layer of the perched leachate level appeared on each intermediate cover layer.

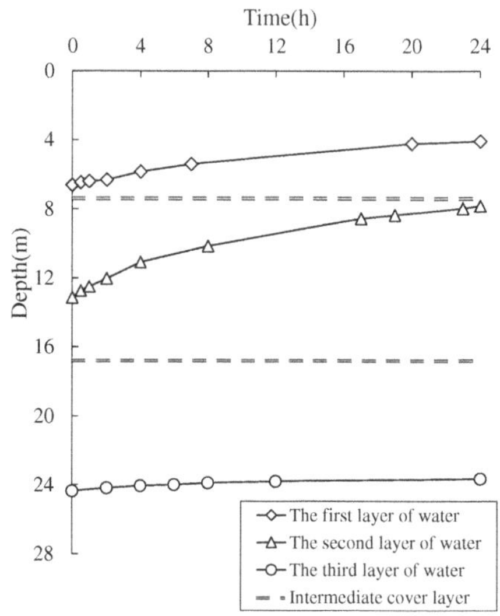

Fig. 7. Leachate level recovery curve of BH4

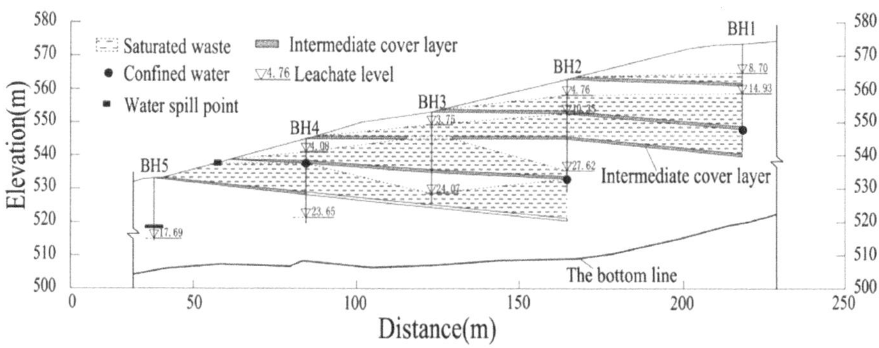

Fig. 8. Diagram of leachate distribution in testing area

5 Conclusions

The paper carried out investigation of on-site multi-layer leachate levels and comes up with the following conclusions based on on-site testing:

1. Drilling to the position 1–2 m away from the cover layer; the well casing is also advanced to this position; reduce the in-well casing leachate level to the well casing bottom; and a leachate level recovery curve is measured. On this basis, the steady leachate level on this cover layer is deemed as leachate level of this layer;
2. During leachate level recovery tests of each waste layer, recovery rates decreased as time went by, wherein standing of 24 h can satisfy testing requirements;

3. Existence of intermediate cover layers stopped vertical transport of leachate, one stagnant leachate level appeared on each intermediate cover layer, and stagnant level of each waste layer was not consistent.

Acknowledgements. The authors are very grateful for the financial support of the National Basic Research Program of China (973) via Grant No. 2012CB719800 and the National Natural Science Foundation of China via Grant No. 41502276.

References

1. Benson CH, Pliska RJ (1996) Final covers: HELP needs help form the field. Waste Age 27 (3):89–98
2. Feng SJ, Cao BY, Zhang X, Jiao Y (2015) Leachate recirculation in bioreactor landfills considering the stratification of MSW permeability. Environ Earth Sci 73(7):3349–3359
3. Qian XD, Koerner RM, Gray DH (2002) Geotechnical aspects of landfill design and construction. Prentice Hall, Upper Saddle River
4. Zhan LT, Zhan XJ, Lin WA (2014) Field and laboratory investigation on geotechnical properties of sewage sludge disposed in a pit at Changan landfill, Chengdu, China. Eng Geol 170(4):24–32
5. Zhan LT, Xu XB, Chen YM, Ma XF, Lan JW (2015) Dependence of gas collection efficiency on leachate level at wet municipal solid waste landfills and its improvement methods in China. J Geotech Geoenviron Eng 141(4):04015002
6. Zhang WJ, Zhang GG, Chen YM (2013) Analyses on a high leachate mound in a landfill of municipal solid waste in China. Environ Earth Sci 70(4):1747–1752

A Simplified Third-Type Inlet Boundary Condition Solution for Contaminate Transport Through Slurry Cut-off Walls

Guan-Nian Chen(✉), Yu-Chao Li, and Han Ke

MOE Key Laboratory of Soft Soils and Geoenvironmental Engineering, Zhejiang University, Hangzhou 310058, China

Abstract. Low-permeability slurry trench cut-off walls are commonly constructed as barriers for containment of subsurface point source pollution or as part of seepage control systems on contaminated sites. Scholars made achievements on design methods for first-type inlet boundary scenarios, but there are still lacking on third-type inlet method, as the choice of inlet boundary conditions is still controversial. A method to estimate wall thickness in slurry wall design in such boundary condition is proposed based on previous design method for first-type inlet. An error analysis is performed to investigate the impact of this assumption by comparing the results obtained by the proposed method to those from an analytical solution. For breakthrough criterion C^* of less than 0.1, which covers common practical situations, the relative error is not greater than 15%, demonstrating that the proposed method is suitable for design. An example is also presented to illustrate the procedure of implementing the proposed method for slurry wall design.

Keywords: Slurry wall · Wall thickness · Breakthrough time
Third-type inlet boundary

1 Introduction

Slurry trench cut-off walls are commonly constructed as barriers for containment of subsurface point source pollution or as part of seepage control systems on contaminated sites. In slurry wall design, the determination of wall thickness is typically made via a contaminant transport analysis rather than a seepage analysis, the transient contaminant transport through slurry walls can be regarded as a one-dimensional advective-dispersive process. Analytical solutions for different boundary conditions [1–4] are available to calculate the contaminant transport in slurry walls. However, the evaluation of these analytical solutions is nontrivial and generally requires the use of a computer [5]. A simplified design method for first-type inlet boundary condition was carried out previously [6], but there are discussions on selection of boundary conditions existing [7–12], so a new design method for third-type inlet boundary is also necessary.

A new method for determining the wall thickness of slurry walls considering third-type inlet boundary condition is proposed in this paper based on previous research. The error of the proposed method is investigated by comparing the results with those

L. Zhan et al. (Eds.): ICEG 2018, ESE, pp. 404–412, 2019.
https://doi.org/10.1007/978-981-13-2224-2_50

obtained by an analytical solution commonly used for slurry wall design. Finally, an example is presented to illustrate the procedure of implementing the proposed method for slurry wall design.

2 Methods and Error Analysis

As illustrated in Fig. 1, the slurry wall keys into the impervious soil layer. The backfill is assumed to be homogenous, fully saturated and non-deformable. The pore water flow in the backfill is assumed to be in a steady state condition. The contaminant migration in the slurry wall is regarded as a one-dimensional advective-dispersive process [13], the non-dimensionalized governing equation can be written as:

$$\frac{\partial C_r}{\partial T} = \frac{1}{P_L}\frac{\partial^2 C_r}{\partial X^2} - \frac{\partial C_r}{\partial X} \tag{1}$$

Where C_r is the dimensionless resident concentration, that is, resident concentration c_r divided by inlet reservoir concentration c_0, or c_r/c_0; X is the dimensionless distance (x/L), where x is the distance between from the upstream, L is the width of the wall; T is effective pore-volume flow(vt/LR), where t is time, R is the retardation factor and is assumed to be constant due to linear, instantaneous and reversible equilibrium adsorption of reactive contaminants, v is the seepage velocity and can be further written as kh/nL, n and k are the porosity and hydraulic conductivity of the backfill, respectively, h is the hydraulic head difference between the entrance and exit boundaries of the slurry wall and is assumed to be independent on L; and P_L is the Peclet number (vL/D_h, or written as kh/nD_h, where D_h is hydrodynamic dispersion coefficient), which represents the relative importance of advective migration to the dispersive migration [10], from the second expression, it can be found that the Peclet number is independent of wall thickness L if the water head difference h is assumed as a constant.

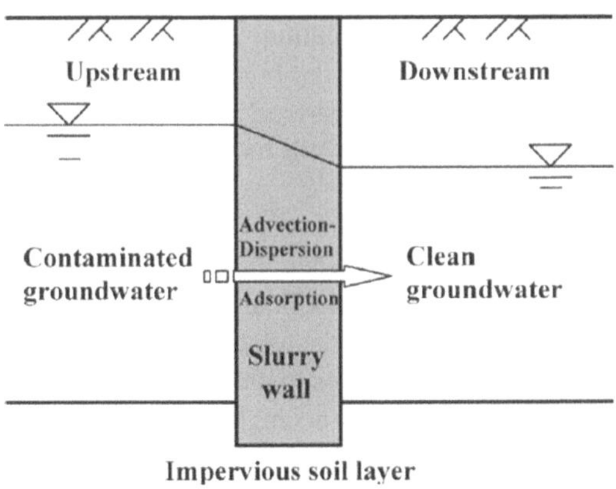

Fig. 1. Configuration of contaminant transport through a slurry trench cutoff wall

The backfill is assumed to be initially free of contaminant, which reasonably represents the field condition since slurry walls are commonly constructed using uncontaminated backfills. A constant flux at the entrance boundary (that is, $J = C_r(0,T) - \frac{1}{P_L}\frac{\partial C_r(0,T)}{\partial T} = 1$), is assumed to be the inlet boundary condition in this paper, following Prince [8] and van Genuchten [9], which is also called to be a third-type inlet boundary condition. The outlet boundary condition is assumed to be a semi-infinity boundary condition, correspondingly. The breakthrough time is commonly defined to be the time when the exit concentration reaches a specified value (c^*), which is often based on groundwater quality or other standards, or the dimensionless concentration reaches the corresponding breakthrough criterion C^*.

The analytical solution for previous conditions is [4]:

$$C_r(X,T) = \frac{1}{2}\mathrm{erfc}[\frac{1}{2}(X\sqrt{\frac{P_L}{T}} - \sqrt{P_L T})] + \sqrt{\frac{P_L T'}{\pi}}\mathrm{e}^{-\frac{P_L(X-T)^2}{4T}} - \frac{1}{2}(1 + P_L X + P_L T)\mathrm{e}^{XP_L}\mathrm{erfc}[\frac{1}{2}(X\sqrt{\frac{P_L}{T}} + \sqrt{P_L T})] \quad (2)$$

As Eq. 2 shows, the analytical solution is too complex to be used in engineering designs while there are also nonelementary functions existing in the solution. So, based on the previous first-type inlet simplify method [6], a first-step simplification for analytical solution Eq. 2 was firstly carried out (the full procedure is in Appendix 1), that is:

$$C* = \mathrm{erfc}[\frac{1}{2}(\sqrt{\frac{P_L}{T}} - \sqrt{P_L T})] - \mathrm{e}^{P_L}\mathrm{erfc}[\frac{1}{2}(\sqrt{\frac{P_L}{T}} + \sqrt{P_L T})] \quad (3)$$

The relative error of $\sqrt{P_L/T}$ (which is proportional to the calculated wall thickness L) obtained from the first-step simplification Eq. 3 with that by the analytical solution Eq. 2 is shown in the Fig. 2:

The relative error of the first-step simplification is rather similar to that of the decoupled method for first-type inlet solution, the maximum relative error for this simplification is no more than 5% when the breakthrough criterion is no higher than 0.1. However, alike the reason why the first-type analytical solution is hardly able to be used in engineering, this simplification is still too complex to be used, and need to be further simplified.

As Shackelford [10] demonstrated, the breakthrough curve is not sensitive to the choice of two inlet boundary conditions when the Peclet number is higher than 10, the decoupled design method for first-type inlet boundary condition still fits the third-type boundary in thus scenario. As calculated, the relative error of the first-type inlet method used in third-type boundary for scenarios with Peclet number higher than 10 and breakthrough criterion lower than 0.1 is no higher than 8%, which is an acceptable error for engineering.

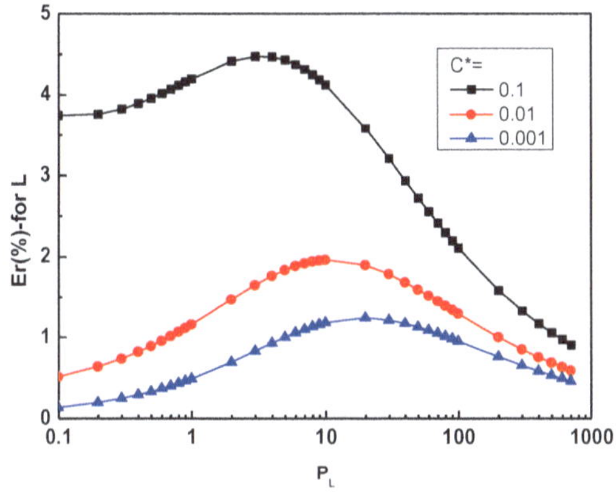

Fig. 2. Relative error of the first-step simplification

For dispersion controlled scenario, another simplification was carried out based on the first-step simplification, the derivation process is given in Appendix 2, that is:

$$\frac{C*}{P_{\mathrm{L}} e^{\frac{P_{\mathrm{L}}}{2}}} = 2\sqrt{\frac{T}{\pi P_{\mathrm{L}}}} e^{-\frac{P_{\mathrm{L}}}{4T}} - \mathrm{erfc}\left(\frac{1}{2}\sqrt{\frac{P_{\mathrm{L}}}{T}}\right) = f\left(\frac{P_{\mathrm{L}}}{T}\right) \tag{4}$$

Compared with analytical solution Eq. 4 and first-step simplification Eq. 3, Eq. 4 degenerates the function to a unary function, which is able to be further fitted. Alike the relative error calculation of the first-step simplification, the relative error of Eq. 4 is:

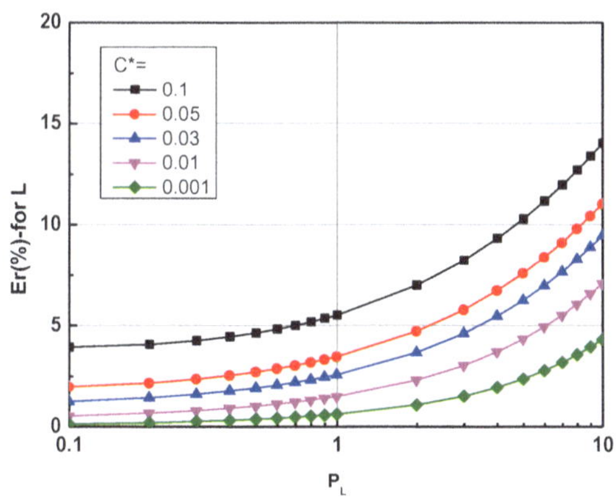

Fig. 3. Relative error of Eq. 4

The relative error of the equation under common scenarios is no higher than 15%, and it's increasing while the Peclet number increases due to the simplification is aimed on dispersion controlled scenarios. Comparing with the first-step simplification, the relative error of Eq. 4 is much higher but still acceptable.

Equation 4 provides a way to design dispersion controlled constructions, but it cannot be used in designing directly. Here a fitting formula was provided for most scenarios may be met in constructions, that breakthrough criterion C^* ranges from 0.001 to 0.1, Peclet number ranges from 0.1 to 10 (as Eq. 4 is carried out aiming at dispersion controlled scenarios), that is:

$$\sqrt{\frac{P_{\mathrm{L}}}{T}} = 12.55 - 12 \times \left(\frac{C*}{P_{\mathrm{L}} e^{\frac{P_{\mathrm{L}}}{2}}}\right)^{0.044} \tag{5}$$

The relative error of Eq. 5 comparing with Eq. 4 is no higher than 1% in common scenarios.

3 Example

Table 1. Value of parameters

	Parameter	Value
Hydrogeological condition	Porosity of the soil	0.4
	Hydraulic dispersion coefficient of the soil	1×10^{-9} m/s
	Water head difference	0.8 m
Property of contaminant (phenol)	Hydrodynamic dispersion coefficient	5×10^{-10} m^2/s
	Retardation factor	30
	Reservoir concentration	1.0 mg/L
	Breakthrough concentration	0.002 mg/L
Construction need	Service life	50 years

The procedure of implementing the proposed method to determine the wall thickness of slurry walls is illustrated in this section. The value of the parameters follows [6], which is listed in the Table 1.

The wall thickness of the slurry wall can be determined by the following steps using the proposed method:

Step 1: Calculate the ratio of the specified breakthrough exit concentration to the source concentration,

$$\frac{c^*}{c_0} = \frac{0.002}{1.0} = 0.002$$

Step 2: Calculate the value of P_L

$$P_L = \frac{kh}{nD_h} = \frac{1 \times 10^{-9} \times 0.8}{0.4 \times 5 \times 10^{-10}} = 4.0$$

As the Peclet number of the construction is less than 10 and the breakthrough criterion between value 0.001 and 0.1, Eq. 5 can be used in this design. For scenarios whose Peclet number is higher than 10, the designing may still use the simplified method for first-type inlet boundary, whose full procedure can be seen in [6].

Step 3: Calculate the effective breakthrough PVF via Eq. 5:

$$T = \frac{P_L}{[12.55 - 12 \times (\frac{C*}{P_L e^{\frac{P_L}{2}}})^{0.044}]^2} = 0.1823$$

Step 4: Calculate the wall thickness:

$$L = \frac{vt}{TR} = \frac{kht}{nTRL}, \ L = \sqrt{\frac{kht}{nTR}} = 0.7589\,\text{m}$$

Or step 3 and step 4 can be combined in a clearer form:

$$L == [12.55 - 12 \times (\frac{C*}{P_L e^{\frac{P_L}{2}}})^{0.044}]\sqrt{\frac{D_h t}{R}} = 0.7589\,\text{m}$$

The wall thickness fitting the construction need is 0.76 m.

4 Conclusion

A simplified method has been proposed to determine the thickness of slurry walls for third-type inlet scenarios. For dispersion controlled migrations (low Peclet number), a new simplification was carried out based on the first-type design method; for advection controlled scenarios (high Peclet number), the first-type design method is still suitable for designs. An error analysis is performed to investigate the impact of this assumption. The relative error in terms of wall thickness is not greater than 15% when the breakthrough criterion of c^*/c_0 is less than 0.1, which covers common practical situations in slurry wall design and is recommended as the limitation of the simplified method. An example was laid out to show the procedure of using the proposed method.

Acknowledgements. The financial supports received from the National Natural Science Foundation of China (NSFC) by grant No. 41672284 and the Science Technology Department of Zhejiang Province by grant No. 2015C03021 are gratefully acknowledged.

Appendix 1

It is noted that the dimensionless flux concentration of the constant flux inlet boundary is the same with the dimensionless resident concentration of a constant concentration inlet boundary [1, 2, 4]:

$$C_f(X,T) = \frac{1}{2}\mathrm{erfc}[\frac{1}{2}(X\sqrt{\frac{P_L}{T}} - \sqrt{P_L T})] + \frac{1}{2}e^{XP_L}\mathrm{erfc}[\frac{1}{2}(X\sqrt{\frac{P_L}{T}} + \sqrt{P_L T})] \tag{A1}$$

Where C_f is volume average flux divided by c_0, according to the Law of conservation of mass, the dimensionalized flux concentration and the dimensionalized resident concentration suits the following equation:

$$-\frac{\partial C_f}{\partial X} = \frac{\partial C_r}{\partial T} \tag{A2}$$

As described in previous research [6], there is an alternative solution for first-type inlet scenario solution which will lead to a rather low relative error, to be written in equation form, that is:

$$\begin{aligned} &\frac{1}{2}\mathrm{erfc}[\frac{1}{2}(X\sqrt{\frac{P_L}{T}} - \sqrt{P_L T})] + \frac{1}{2}e^{XP}\mathrm{erfc}[\frac{1}{2}(X\sqrt{\frac{P_L}{T}} + \sqrt{P_L T})] \\ &\cong \mathrm{erfc}[\frac{1}{2}(X\sqrt{\frac{P_L}{T}} - \sqrt{P_L T})] \end{aligned} \tag{A3}$$

And Eq. A3 can also be used in third-type inlet boundary condition as an alternative solution for flux concentration, as Eq. A1 describes. The alternative solution can be further translated into resident concentration form using the translation formula Eq. A2. To reduce the amount of calculation, Eq. A4 only calculates the concentration of X = 1, that is, the downstream of the wall:

$$\begin{aligned} C_r &= \int_0^T -\frac{\partial \mathrm{erfc}[\frac{1}{2}(X\sqrt{\frac{P_L}{\tau}} - \sqrt{P_L \tau})]}{\partial X}\Big|_{X=1} d\tau \\ &= \int_0^T \sqrt{\frac{P_L}{\pi\tau}} e^{-\frac{P_L(1-\tau)^2}{4\tau}} d\tau \\ &= \mathrm{erfc}[\frac{1}{2}(\sqrt{\frac{P_L}{T}} - \sqrt{P_L T})] - e^{P_L}\mathrm{erfc}[\frac{1}{2}(\sqrt{\frac{P_L}{T}} + \sqrt{P_L T})] \end{aligned} \tag{A4}$$

For wall designs, C_r in Eq. A4 is replaced with breakthrough criterion C^*, thus the corresponding T is the breakthrough PVF. The converted formula is the first-step simplification (Eq. 3).

Appendix 2

The second step of simplification is based on the integration formula in Eq. A4,

$$C_r = \int_0^T \sqrt{\frac{P_L}{\pi\tau}} e^{-\frac{P_L(1-\tau)^2}{4\tau}} d\tau = \int_0^T \sqrt{\frac{P_L}{\pi\tau}} e^{-\frac{P_L}{4\tau}+\frac{P_L}{2}-\frac{P_L\tau}{4}} d\tau \quad (A5)$$

As described previously, the second simplification aims at low Peclet number scenarios, besides, considering the$\tau$is an infinitesimal, an attempt to simplify the formula was made as follow:.

$$\begin{aligned} C_r &= \int_0^T \sqrt{\frac{P_L}{\pi\tau}} e^{-\frac{P_L}{4\tau}+\frac{P_L}{2}-\frac{P_L\tau}{4}} d\tau \cong \int_0^T \sqrt{\frac{P_L}{\pi\tau}} e^{-\frac{P_L}{4\tau}+\frac{P_L}{2}} d\tau \\ &= P_L e^{\frac{P_L}{2}} [2\sqrt{\frac{T}{\pi P_L}} e^{-\frac{P_L}{4T}} - \operatorname{erfc}(\frac{1}{2}\sqrt{\frac{P_L}{T}})] \end{aligned} \quad (A6)$$

Same with Appendix 1, C_r in Eq. A6 is changed into C^* in designs, that is, Eq. 4.

References

1. Lapidus L, Amundson NR (1952) Mathematics of adsorption in beds. The effect of longitudinal diffusion in ion exchange and chromatographic columns. J Phys Chem-Us 56 (8):984–988
2. Ogata A, Banks RB (1961) A solution of the differential equation of longitudinal dispersion in porous media. U. S. geological survey professional paper 411-A, pp. A1–A9
3. Brenner H (1962) The diffusion model of longitudinal mixing in beds of finite length - numerical values. Chem Eng Sci 17(4):229–243
4. Lindstrom FT, Haque R, Freed VH, Boersma L (1967) Theory on the movement of some herbicides in soils: linear diffusion and convection of chemicals in soils. Environ Sci Technol 1(7):561–565
5. Rowe RK, Quigley RM, Brachman RWI, Booker JR (2004) Barrier systems for waste disposal facilities. E & FN Spon, Abingdon, Oxon
6. Chen G-N, Cleall PJ, Li Y-C, Yu Z-X, Ke H, Chen Y-M (2018) A decoupled advection-dispersion method for determining wall thickness of slurry trench cut-off walls. Int J Geomech ASCE 18(5):06018007
7. Rabideau A, Khandelwal A (1998) Boundary conditions for modeling transport in vertical barriers. J Environ Eng-ASCE 124(11):1135–1139
8. Prince MJ, Maneval JE, Evans JC (2000) Analysis of boundary conditions for contaminant transport through adsorptive, low-permeability slurry trench cutoff walls. In: Proceedings of environmental geotechnics, ASCE, pp. 58–72
9. van Genuchten MT, Parker JC (1984) Boundary conditions for displacement experiments through short laboratory soil columns. Soil Sci Soc Am J 48(4):703–708
10. Shackelford CD (1994) Critical concepts for column testing. J Geotech Eng ASCE 120 (10):1804–1828

11. Shackelford CD (1995) Analytical models for cumulative mass column testing. In: Proceedings of ASCE specialty conference. geoenvironment 2000, pp. 355–372. ASCE, New York
12. Shackelford CD (1995) Cumulative mass approach for column testing. J Geotech Eng ASCE 121(10):696–703
13. Freeze RA, Cherry JA (1979) Groundwater. Prentice-Hall Inc., Englewood Cliffs

Accumulated Vertical Strain Without Confining Pressure for Compacted Bentonite Due to Hydration Effort

Tomoyoshi Nishimura[1(✉)] and Junichi Koseki[2]

[1] Ashikaga University, Ashikaga, Japan
tomo@ashitech.ac.jp
[2] The University of Tokyo, Tokyo, Japan

Abstract. To use high dense compacted bentonite or bentonite-sand mixture material are accepted as engineered barriers for high level nuclear waste disposal waste. The bentonite as engineered barrier is initially in unsaturated condition before pore water coming from natural host rock. The bentonite subjected to thermo-hydro-mechanical actions by natural geological situation and the canisters of high-level radioactive disposal waste. This study conducted out the creep test for compacted bentonite, and the developed relative humidity circulation triaxial apparatus. Two different initial water contents were prepared to make the specimens at constant dry density. As parameter for this creep test, the stress ratio was defined using specified unconfined compressive strength. All specimens were applied without confining pressure. The vapor transfers due to changing of relative humidity induced some hydration efforts that were controlled using vapor pressure technique. Performing the creep test in this study is of couple hydro-mechanical and is applied to identification for two efforts (i.e. strain due to hydration and stain by deviator compression stress). Investigation of creep behavior that accumulated strains associated to couple thermo-mechanical actions are measured at constant deviator compression stress.

Keywords: Bentonite · Creep test · Suction

1 Introduction

1.1 Back Ground

To use high dense compacted bentonite or bentonite-sand mixture material are accepted as engineered barriers for high level nuclear waste disposal waste. The bentonite as engineered barrier is initially in unsaturated condition before pore water coming from natural host rock. The bentonite subjected to thermo-hydro-mechanical actions by natural geological situation and the canisters of high-level radioactive disposal waste. The coupled thermo-hydro-mechanical actions is essential to the safe design of the barrier system during a considerable long time (Delage 2013). Also, the bentonite buffers in radioactive waste disposal experience cyclic hydration such as drying and wetting paths in suction during various construction stages of the design life (Alonso et al. 2005). The cyclic suction change is gained by natural environmental fluctuations,

L. Zhan et al. (Eds.): ICEG 2018, ESE, pp. 413–420, 2019.
https://doi.org/10.1007/978-981-13-2224-2_51

and it is important features on similar with experience stress of design of buffers. The bentonite or expansive clay subjected to the hydration develop swelling and deformation, and accumulation of deformations is concerned. The hydration closely related to soil suction defined by Sillers and Fredlund (2001), which can explain as water potential in the soil-water system. The vapor pressure technique is useful to control the pore fluid in vapor diffusion for high plastic clay. The hydro-mechanical of compacted bentonite using vapor pressure technique was studied in previous works. In these reports, oedometer test was a common method, and it was find out that bentonite mechanical properties were influenced by soil suction from many experimental results. Also, hydration induced by changing of suction closely related to swelling of bentonite. As far as, heavy compacted bentonites have high suction, vapor pressure technique is effectively one of experimental method to interpret hydro-mechanical behavior.

1.2 Purpose of This Study

This study conducted out the creep test for heavy compacted bentonite, and the developed relative humidity circulation triaxial apparatus was used to observe the accumulated vertical strain as creep deformation. Two difference initial water content were prepared to make the specimens at constant dry density that water contents of 8.0% and 17.2%, and dry density was 1.600 g/cm^3. The applied creep stress (i.e. deviator compression stresses) were determined from unconfined compressive strength at initial water content. As parameter for this creep test, the stress ratio was defined between specified unconfined compressive strength and creep stress. All specimens were applied without confining pressure in modified triaxial cell. The vapor transferred due to changing of relative humidity induced some hydration efforts that were controlled using vapor pressure technique.

Performing the creep test in this study is of couple hydro-mechanical and is applied to identification for two efforts (i.e. strain due to hydration and stain by deviator compression stress). Investigation of creep behavior that accumulated strains associated to couple thermo-mechanical actions are measured at constant deviator compression stress. Consequently, this study evidenced that some of heavy compacted bentonites described the creep failure with accumulation of strain.

2 Soil Material and Test Procedure

2.1 Soil Sample

This testing program used Sodium bentonite, which was produced as Kunigeru V1, and it was one of most investigated bentonites such as the swelling properties in the Japan. As mineral components feature, SiO_2 occupied 62% to all components. Other characters such as grain size distribution were mentioned in the detail sheet published product company. The bentonite included so much fine particles that was more than 80%. The water retention activity is important feature which described as the soil-water characteristic curve (i.e. SWCC). The SWCC was one of key parameters for establishing of mathematical models, and basic physical parameter (Sillers and Fredlund

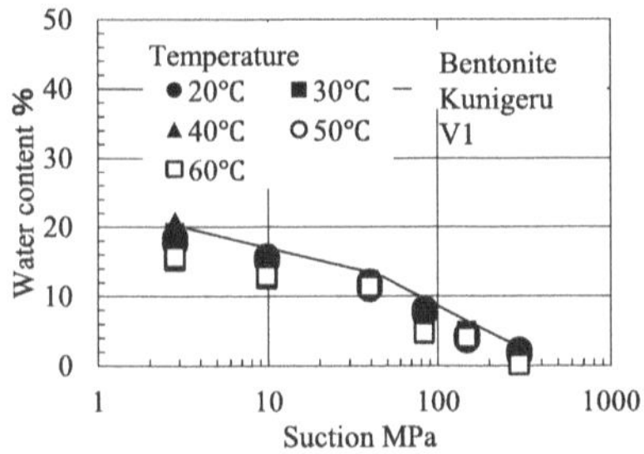

Fig. 1. SWCC of bentonite used in this test.

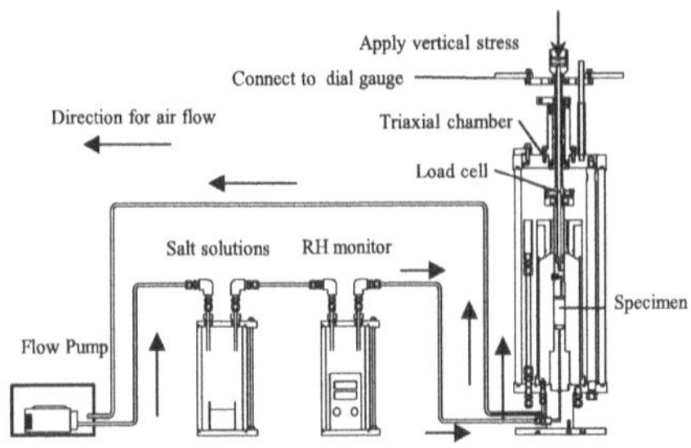

Fig. 2. Modified creep test apparatus.

2001; Sillers et al. 2001). The soil-water characteristic curve of bentonite used in this testing measured and was indicated in Fig. 1. The determined SWCCs had a wide range from 2.8 MPa to 296 MPa which was obtained using the vapor pressure technique. Using a vapor pressure technique for high suction measurement is no longer a special suction application. Also, the SWCC was controlled with various temperature, which had a range from 20°to 60°.

Water content decreased with suction in drying portion that it was possible to define one curve having an inflection point. So, unique opinion associated with increasing of temperature was not confirmed from these data sets. Nishimura and Koseki (2014) reported the hydration effort induced by RH 98% that the compacted unsaturated bentonite described the increment of both water content and degree of saturation.

This study conducted to unconfined compression test and creep test for compacted bentonite. The unconfined compression tests were performed with and without effort of hydration, and the unconfined compressive strengths were used for determination of creep stresses. The unconfined compression tests were of relevance to consider creep behavior of bentonite. Moreover, compression tests were conducted out for the bentonite subjected to hydration under RH 98%. It was certain that the influence of hydration on reduction of shear resistance or development of void structure.

The creep test in this testing program meant that only vertical stress was applied without lateral confining pressure, and axial strain was measured. The deformation in lateral direct was directly observed. The specimens were statically compacted with two initial water contents (i.e. 8.0% and 17.2%) under constant dry density, and it was 1.600 g/cm^3. Its value generally focused on investigation of mechanical properties for radioactive waste barrier system. Statically compaction was applied through one layer in the stiffness steel mold for avoiding the influence of non-uniformity. Finally, the specimens had a diameter of 50 mm, and a height of 100 mm.

2.2 Creep Test Apparatus with Controlling Suction

The modified creep test apparatus was shown in Fig. 2. The apparatus had a conventional cyclic relative humidity control system (Nishimura and Koseki 2013), which applied a required RH using salt solutions. The vapor pressure technique has been accepted as conventional suction control method as well as pressure plate technique using ceramic disk plate. For expansive materials as bentonites, vapor pressure technique seems to be adequately suitable.

The creep test apparatus is similar to conventional triaxial compression apparatus which some components are installed, and it consists of triaxial chamber, air cyclic flow pump, the chamber including salt solution, conventional RH display, displacement sensor and loading cell. Air cyclic flow pump makes stability air flow circulation that steady air flow had at least 10 kPa in pressure. Salt solutions were possible to control RH in circulation air, it was conventional in testing procedure.

The specimens in the chamber were applied creep stress and hydration effort. The hydration effort induced by changing of RH in the chamber. As above mentioned, this test no applied confining pressure. While the creep test was maintained, vertical displacements were measured. The creep stresses were defined using unconfined compressive strength.

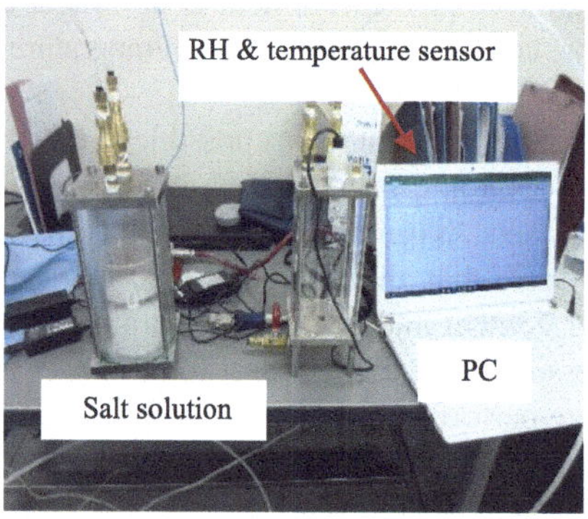

Fig. 3. Measurement of RH.

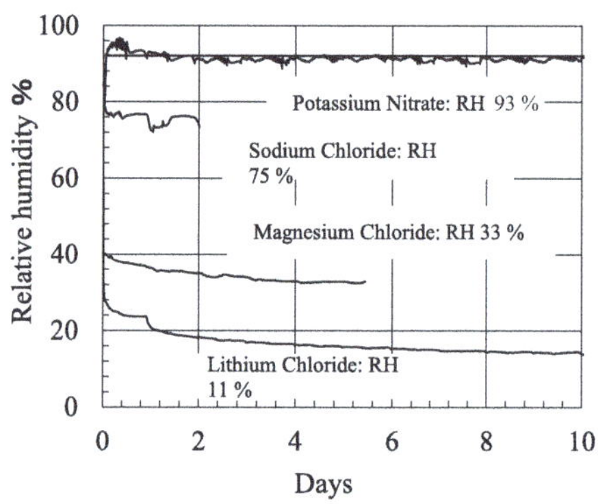

Fig. 4. Measured RH with various salt solutions

2.3 Measurement of RH for Air in Circulation Flow

The vapor pressure technique had significant in this testing that high suctions were applied to specimens and revealed the influence of hydration effort on bentonite properties. Many experimental investigations were reported or accepted into widely geotechnical engineering. It is so important to make assurance to RH in circulation system as above mentioned in Fig. 2. Some papers reported RH values created by salt solutions under some temperatures. Previously, RH was measured using an equipment shown in Fig. 3 under maintain of air circulations. The system was consisted of two chambers, conventional pump and PC. The installed sensor having white color measured both RH and temperature. Among two chambers and conventional pump were connected flexible tube, and the system was completely closed (i.e. closure). Four different salt solutions were used which were Potassium Nitrate, Sodium Chloride, Magnesium Chloride and Lithium Chloride, respectively (Fig. 4). Each of RH values corresponding to salt solutions were described in the figure which defined in Test method for water retentivity of soils (JGS 0151-2009). To use vapor pressure technique in the modified creep test apparatus is certain for applying identified high suctions.

Also, it was important matter in maintain good quality that how to establish determination equilibrium condition under RH controlling for specimens. This study performed to measure weight of specimen during air circulations as shown in Fig. 5. The specimen was put into chamber on electrical weigh balance. The weight balance was connected to PC, and measured weight was saved into PC under an interval time. As one case, water content was described with elapsed time which had a dimeter of 3.8 m and height of 7.6 cm. Dry density and water content at beginning of test were 1.600 g/cm^3 and 17.2%, respectively. RH of 98% at 20° induced by Potassium Sulfate was maintained in the chamber. Water content smoothly increased with time, and it approached to be stabilization at ten days as shown in Fig. 6. When test performance continued to one month, water content indicated 18.5% and it was possible to evaluate that twenty days reached to be equilibrium to high suctions.

Fig. 5. Measurement of weight of specimen.

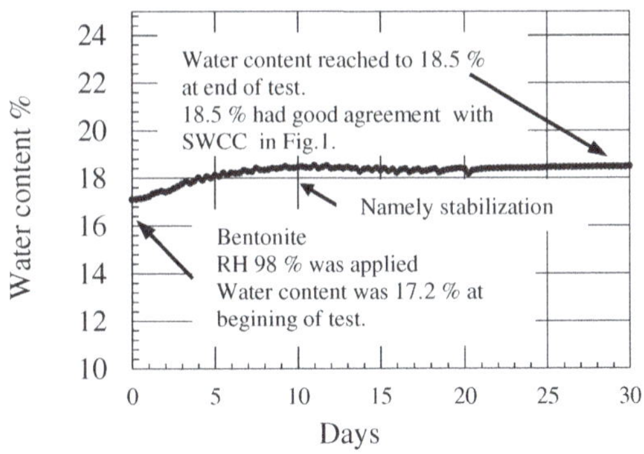

Fig. 6. Water content approached to equilibrium.

2.4 Unconfined Compressive Strength for Determination of Creep Stress

The unsaturated compacted bentonites both with and without hydration effort were performed the unconfined compression test as shown in Figs. 7 and 8. It was obviously that the strength of bentonite depends on initial water content. The influence of hydration on unconfined compressive strength was investigated that the bentonite subjected to hydration induced by RH 98% before applied compression processing. The obtained stress-strain curve described that bentonites without hydration effort produced large increment of deviator stress at beginning shear, and it seemed linear increment to strain.

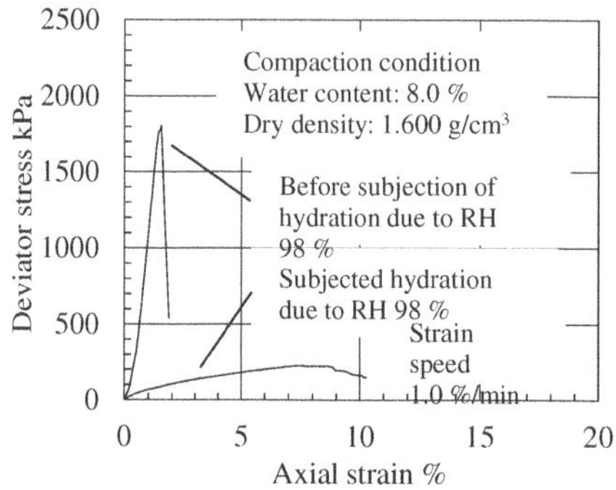

Fig. 7. Stress-strain curve for water content of 8.0%.

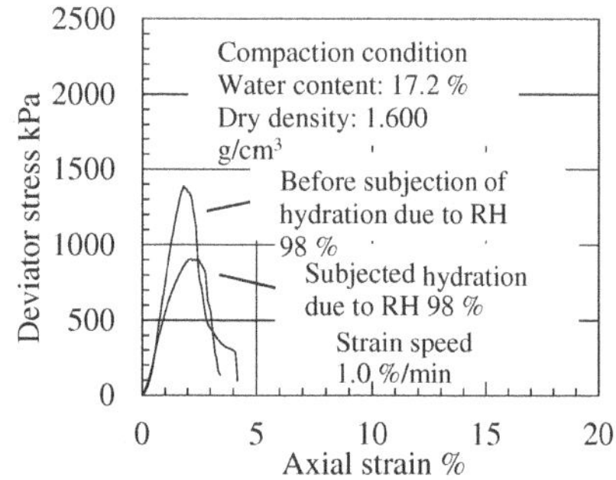

Fig. 8. Stress-strain curve for water content of 17.2%.

Other hands, the bentonite subjected to RH 98% indicated rough damages that the reduction compared initial compressive strengths.

Here, constant deviator stress as creep stress was selected using unconfined compressive strength in creep test. Each creep stresses and stress ratios were summarized in Table 1. The stress ratio was defined that the rate was creep stress in unconfined compressive strength.

Table 1. Summaries of creep test including hydration effort conditions.

Water content at compaction %	Creep stress kPa	Ratio (i.e. rate of deviator stress to unconfined compressive strength)	Hydration effort due to RH	Salt solutions
8.0	549.3	0.3	98%	Potassium Sulfate
17.2	416.5	0.3	Increment from 93% to 98%	Ammonium Dihydrogen and Potassium Sul

3 Creep Behavior of Bentonite Subjected to Hydration Effort

3.1 Compacted Water Content of 8.0%

The creep tests were conducted for bentonite with compacted water content of 8.0% when circulation air remained RH 98% during creep stress of 549.3 kPa (i.e. corresponding to ratio of 0.30). At beginning of test, both creep stress and hydration effort were applied to specimens in same time, and the shrinkage strain occurred at once. The shrinkage vertical strain described short period, and vertical strain transfer into expansion way. The expansion behavior caused by hydration effort due to applying RH 98% that increment of soil moisture mentioned as shown in Fig. 6 attributed the expansions. Then development of expansion deformation continued so long term that it

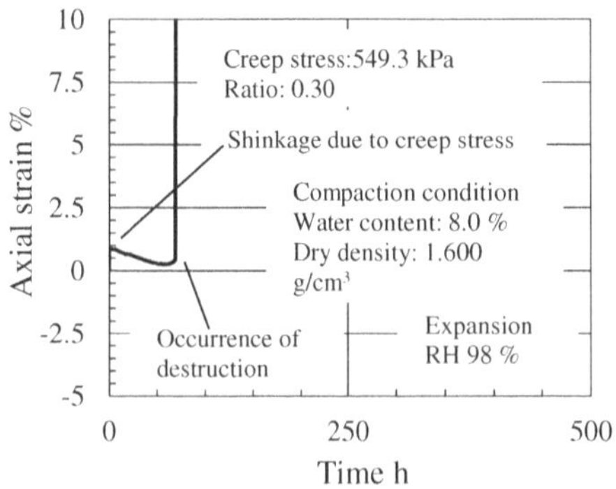

Fig. 9. Creep behavior for water content of 8.0%.

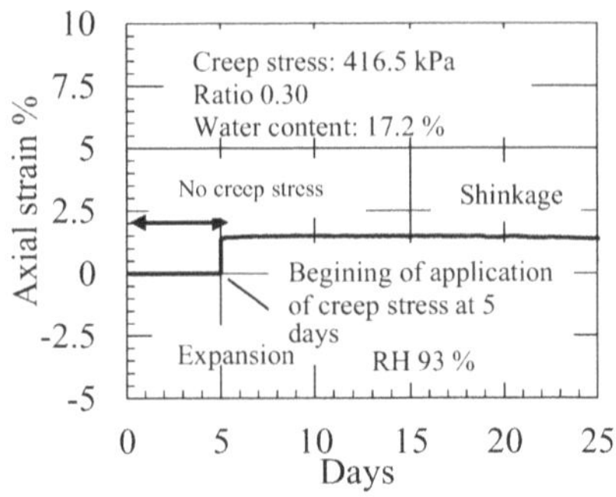

Fig. 10. Creep behavior for water content of 17.2%.

was indicated linear growth in strain. Because of weak void structure due to hydration effort, specimen indicated shrinkage strain against expansive strain that was up to failure or destruction. The failure was observed at 65 days in testing time. Thus, hydration effort induced the creep break to the compacted bentonite.

3.2 Compacted Water Content of 17.2%

The specimen with compacted water content of 17.2% was applied one creep stress and two hydration efforts (i.e. RH 93% and 98%). A test processing was as followings. At beginning of test, the specimen was placed in RH 93% without creep stress period of five days. Subsequently, 416.5 kPa of creep stress was loaded under constant RH 93%, which was coincident stress ratio of 0.30. When elapsed days was 25 days, RH increased to 98% that suction decreased from 9.8 MPa to 2.8 MPa, and measurement of vertical deformation maintained up to destruction. The creep behavior was shown in Figs. 10, 11 and 12. The deformation under RH 93% was observed without creep stress for five days which was ignored as shown in Fig. 10. Then a creep stress of 416.5 kPa was applied at five days while RH was 93%. At once vertical strain was produced, the strain approached to 1.7% though one day. Both creep stress and hydration effort as two external affects were 416.5 kPa and RH 93% were controlled in constant as shown in Fig. 11. The vertical strain had a small changing at long period, and it seemed to be small expansion. At elapsed time was 25 days, RH increased from 93% to 98% that suction decreased from 9.8 MPa to 2.8 MPa. The creep behavior had an expansion from 25 days to 35 days, and subsequently the strain accordingly transferred to shrinkage. Then test was over 65 days, shrinkage produce so large, much incline continued till 96 days. Finally, creep failure was observed as shown in Fig. 12.

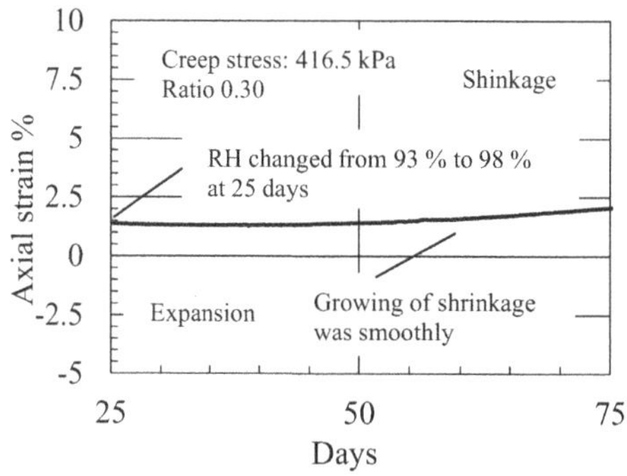

Fig. 11. Creep behavior for water content of 17.2%.

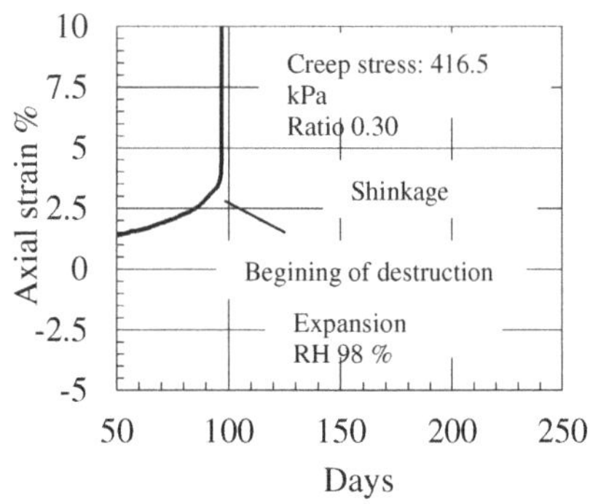

Fig. 12. Creep behavior for water content of 17.2%.

4 Conclusions

This study conducted out creep test using modified creep test apparatus which was possible to apply hydration to specimen using vapor pressure technique. Controlled suction was over 2.8 MPa at least that corresponded to RH 98%. The required creep stresses were determined form result of unconfined compressive strength. Two different water contents at compaction were used for bentonite. The compacted bentonite described up to destruction due to creep stress and hydration.

References

Alonso EE, Romero E, Hoffmann C, Garcıa-Escudero E (2005) Expansive bentonite–sand mixtures in cyclic controlled-suction drying and wetting. Eng Geol

Delage P (2013) On the thermal impact on the excavation damaged zone around deep radioactive waste disposal. J Rock Mech Geotech Eng 5:179–190

Sillers WS, Fredlund DG (2001) Statistical assessment of soil-water characteristic curve models for geotechnical engineering. Can Geotech J 38:1297–1313

Sillers WS, Fredlund DG, Zakerzadeh N (2001) Mathematical attributes of some soil-water characteristic curve models. Geotech Geolog Eng 19:243–283

Nishimura T, Koseki J (2013) Influence of shear speed on direct shear strength for compacted bentonite with different soil suctions. In: Third international conference on geotechnique, construction materials and environment, Nagoya, Geo-Mate 2013, pp 598–603 (2013)

Nishimura T, Koseki J (2014) Deformation of compacted bentonite due to changes in relative humidity. In: Proceedings of the sixth international conference on unsaturated soils, UNSAT2014, pp 923–928 (2014)

Chemical Compatibility of CMC-Treated Bentonite Under Heavy Metal Contaminants and Landfill Leachate

R.-D. Fan[1,2], S.-Y. Liu[1,2], Y.-J. Du[1,2(✉)], Krishna R. Reddy[3], and Y.-L. Yang[1,2]

[1] Institute of Geotechnical Engineering, Southeast University, Nanjing, China
duyanjun@seu.edu.cn
[2] Jiangsu Key Laboratory of Urban Underground Engineering and Environmental Safety, Southeast University, Nanjing, China
[3] Department of Civil and Materials Engineering, University of Illinois at Chicago, Chicago, IL 60607, USA

Abstract. Carboxymethyle cellulose (CMC) is considered to enhance the chemical compatibility of bentonite in terms of hydraulic conductivity. In this study, a series of modified fluid loss (MFL) tests is conducted to investigate the hydraulic conductivity (*k*) of CMC-treated bentonite when exposed to heavy metal contaminant and landfill leachate. CMC content in bentonite is set at 0% (i.e., untreated bentontie) and 10% (i.e., CMC-treated). Lead-zinc nitrate mixture is selected as representative heavy metal contaminants; and calcium chloride is also used for comparison. Total metal concentrations of the chemical liquid ranged from 0 to 20 mmol/L. The result indicated that minimal change in *k* of CMC-bentonite is found from the MFL test; whereas the *k* value of untreated bentonite significantly increased with increased Pb-Zn and Ca concentration. The *k* value of untreated bentonite is approximately 20 times higher than that of CMC-treated bentonite for a given void ratio. In addition, it is found that the impact of landfill leachate on the hydraulic conductivity is insignificant for CMC-treated bentonite. The *k* of CMC-treated bentonite considerably lower than that of untreated bentonite when exposed to the leachate.

Keywords: Bentonite · Carboxymethyle cellulose · Heavy metal Landfill leachate · Hydraulic conductivity

1 Introduction

Engineered soil barriers are extensively used for contaminated groundwater and waste containment (Sharma and Reddy 2004). High quality sodium bentonite (Na-bentonite) is one of the most common materials that used in the engineered barriers. However, Natural Na-bentonites may not available at some sites such as China and India, while Ca-bentonites and sodium modified Ca-bentonites are abundant as alternative to make up engineered barriers (Fan et al. 2014; Du et al. 2015a, b), which show relatively lower swell potential when compared with natural Na-bentonite due to their different proportion of exchangeable cations. Moreover, inorganic salt solutions with medium to

L. Zhan et al. (Eds.): ICEG 2018, ESE, pp. 421–429, 2019.
https://doi.org/10.1007/978-981-13-2224-2_52

high metal concentrations (e.g., 100 mmol/L) lead to a considerable increase in hydraulic conductivity of bentonites, GCLs, and sand/Na-bentonite backfills (Yong et al. 2009; Malusis and McKeehan 2013; Shackelford and Sample-Lord 2014; Du et al. 2015a) due to the compression of the diffuse double layer (DDL) around bentonite particles.

Creative efforts are therefore being conducted to enhance the chemical compatibility of natural Na-bentonites using polymeric materials, e.g., bentonite-polymer composite (BPC), multiswellable bentonite (MSB), and carboxymethyl cellulose-treated bentonite (HYPER clay) (Bohnhoff and Shackelford 2014; Di Emidio et al. 2015; Yang et al. 2017). Results indicate that the hydraulic conductivity of these treated bentonites could decrease by approximately ten to ten thousand-times when compared with their corresponding untreated ones when permeated with inorganic solutions, including KCl, NaCl, $CaCl_2$, NaOH, and HNO_3 (Katsumi et al. 2008; Scalia and Benson 2011; Bohnhoff and Shackelford 2014; Di Emidio et al. 2015). However, study on the impact of heavy metal contaminants and landfill leachate on the hydraulic conductivity of polymer-treated bentonite is limited.

In this study, a series of swell index and modified modified fluid loss (MFL) tests is conducted to investigate the hydraulic conductivity of carboxymethyle cellulose (CMC)-treated Ca-bentonite when exposed to heavy metal contaminant and landfill leachate. CMC is used to treat the Ca-bentonite in this study because: (1) CMC shows high potential to improve the hydraulic conductivity of Na-bentonite (i.e., Hyper clay) when exposed to KCl and $CaCl_2$ solutions (Di Emidio et al. 2015); (2) Standardized production of CMC-treated bentonite is practicable as CMC has been used as a standard thickening agent for bentonite-water slurry in drilling engineering.

2 Materials and Methods

The untreated bentonite, classified as Ca-bentonite, is provided by MUFENF mineral processing plant in Zhenjiang City, China. The resulting swell index and liquid limit is 16.5 mL/2 g and 269%, respectively. CMC used in this study is provided by ACROS Organics™. The average molecular weight of the CMC is 700000; and the degree of substitution is 0.9.

The method for CMC-treated Ca-bentonite preparation is based on Di Emidio et al. (2015). Predetermined mass of CMC is mixed with predetermined volume of deionized water (Type IV DIW as per ASTM D1193) with temperature of 60 °C for 30 min using a paddle mixer. After that, predetermined mass of Ca-bentonite (denoted as CMC0) is mixed with the CMC-DIW water mixture for 3 h using a paddle mixer. A water bath is used to control the temperature of 60 °C during the mixture. The CMC/Ca-bentonite slurry is then dried in an oven at 105 °C, in which mass ratio of CMC and Ca-bentonite is 10% by dry dry weight. Finally, the dried CMC-treated bentonite (denoted as CMC10) is cracked and sieved through the No. 200 sieve.

Lead and zincnitrate ($Pb(NO_3)_2$-$Zn(NO_3)_2$) mixture is selected as target contaminants in this study to simulate common heavy metal contaminants found in groundwater. The Pb-Zn mixture is prepared at different concentrations using DIW; and the ratio of molar concentration of Pb and that of Zn in the mixture was 1.0. The landfill leachate is collected from Zion Landfill, Illinois, United States. Measured pH and EC of the leachate are 7.27 and 1048 mS/cm, respectively.

The properties of CMC-treated bentonite, including liquid limit (w_L), plastic limit (w_P), swell index under DIW and chemical liquid (*SI*), organic matter (*OM*), and pH are measured as per ASTM standards. It should be noticed that ASTM D2974 provides two methods for determining organic matter. In this study, the *OM* value is determined by igniting oven dried CMC-treated bentonite from moisture content determination in a furnace at 440 °C according to Test Method C, from ASTM D2974. In addition, Fourier transform infrared (FTIR) spectroscopy is conducted to characterize the chemical structure of inorganic minerals from clay minerals and clay-organic compounds. The FTIR analysis is conducted in Modern Analysis Center, Nanjing University.

In this study, modified fluid loss (MFL) test is used to evaluate the hydraulic conductivity of the bentonites. The MFL test is considered as a quick method for investigating chemical compatibility of bentonite compared with conventional permeability test. This method is also applied to evaluate the impact of landfill leachate, acid conditions, and heavy metal solutions on the hydraulic conductivity of GCLs and bentonties (Rosin-Paumier et al. 2011; Liu et al. 2013, 2016). The testing procedure is in accordance with that proposed by Chung and Daniel (2008). Predetermined mass of bentonite is mixed with 350 mL chemical solution (i.e., Pb-Zn mixture and leachate) to form a bentonite-chemical solution slurry for 5 min using a paddle mixer in 500-mL plastic bottle. The bentonite content in the slurry, defined as ratio of dry mass of bentonite to mass of slurry, is controlled to 10% and 6% for CMC0 and CMC 10 samples, respectively. The slurry is then vibrated and hydrated for 24–48 h using a shaker. The overall air pressure (P_0) is set at 400 kPa during the MFL test.

3 Results and Discussion

Table 1 presents the properties of CMC-treated Ca-bentonite as well as HYPER clays reported in previous studies (Di Emidio 2010; Malusis and Di Emidio 2014). The results indicate that the organic matter, liquid limit, plastic limit, and swell index all considerably increase due to CMC treatment. The liquid limit and swell index of CMC10 increase by 110% and 164% compared with untreated Ca-bentonite (CMC0). CMC-treated Ca-bentonite possesses slightly lower liquid limit and swell index than those of HYPER clays in the literature.

Figure 1 illustrates the result of FTIR spectra from pure CMC, untreated Ca-bentontie, and CMC-treated Ca-bentonites. The presence of the carboxymethyl ether group in the CMC used in this study can be identified from bands at 2918, 1605, 1420, 1327, 1269, and 718 cm^{-1}. The strong band at 2918 cm^{-1} is characteristic of CH_2 stretching vibration [$\nu(CH_2)$]. Stretching vibration of COO- [ν(COO-)] from carboxyl group (-COOH) occurs at both 1605 and 1420 cm^{-1}, in which band at 1420 cm^{-1} can also attributed to the bending vibration of CH_2 connected with COOH [$\nu(CH_2)_{COOH}$]. The bands at 1327 and 718 cm^{-1} result from the bending vibrations of HCO [δ(HCO)] and COO- [δ(COO-)]; while the stretching vibration of C-O [ν(C-O)] gives a band at 1056 cm^{-1} (Wiley and Atalla 1987). In addition, carboxylic acid (R-COOH) leads to a C-OH stretching vibration [ν(C-OH)], as illustrated by the band at 1269 cm^{-1}.

Table 1. Properties of clean CMC-treated bentonite in this study and CMC-treated Na-bentonite reported in previous study

Sample ID	Organic matter, *OM* (%)	Liquid limit, w_L (%)	Plastic limit, w_P (%)	Specific gravity, G_s	pH	Swell index in DIW (mL/2 g)	References
CMC0	1.18	269	34	2.66	10.33	16.5	This study
CMC10	10.39	565	55	2.61	9.26	43.5	This study
HC2	N.A.	673	73	N.A.	N.A.	38	Malusis and Di Emidio (2014)
HC8	N.A.	742	61	N.A.	N.A.	48	Malusis and Di Emidio (2014)
HYPER clay	N.A.	650.5	N.A.	2.53	N.A.	37	Di Emidio (2010)

N.A.: not available

For untreated Ca-bentonite (i.e., CMC0), the band at 3622 cm^{-1} is characteristic of the stretching vibration of -OH [ν(OH)] bonded with Al^{3+} cations. Absorbed water from the Ca-bentonite results in both stretching vibration (3440 cm^{-1}) and bending vibration (1645 cm^{-1}). Strong band at 1036 cm^{-1} is considered as the Si-O stretching from the silica sheet of montmorillonite (Madejová 2003). The Si-O-Si stretching vibrations [ν(Si-O-Si)] at 797, 778, and 694 cm^{-1} indicates the presence of quartz.

The result of CMC-treatment can therefore be estimated by comparing the spectra of CMC-treated Ca-bentonites with that of CMC0 and pure CMC. The bands due to $\nu(CH_2)$ and δ(HCO) occurs at 2923 cm^{-1} and 1317 to 1320 cm^{-1}. Characteristic bands at 1645, 1430, and 1056 cm^{-1} in the spectra pure CMC noticeably move to the right with an increase in CMC content in bentonite, indicating the composite of ν(COO-) and $\delta(H_2O)$, the composite of ν(COO-), $\nu(CH_2)_{COOH}$, and $\nu(CO_3^{2-})$ as well as the composite of ν(C-O) and ν(Si-O). Thus, it can be concluded that CMC chains existed in clay sheet lead to a more open interlayer between bentonite particles, i.e., more hydrophilic (Qiu and Yu 2008; Di Emidio 2015), leading to an increase in liquid limit and swell index under DIW of CMC-treated Ca-bentonite compared to untreated one.

Figure 2 presents variation in swell index and swell index ratio (*SIR*) with metal concentration under Pb-Zn solution. The swell index ratio is defined as the ratio of swell index of Pb-Zn contaminated sample to that of clean sample, representing the chemical compatibility of bentonite in terms of swell index. The result shows that the swell index of CMC-treated Ca-bentonites are higher than that of untreated Ca-bentonite for a given metal concentration. The swell index considerably decreases with increased metal concentration for both bentonites. It is evident that the resulting *SIR* decreases with increased metal concentration. However, CMC-treated Ca-bentonite do not show a relatively higher *SIR* value compared with their corresponding untreated bentonite. Slight decrease in *SIR* of CMC10 is found compared with that of CMC0 under Pb-Zn solution with metal concentration ranging from 50 to 500 mmol/L. Similar result is reported by Di Emidio (2010), in which the *SIR* value of HYPER clay

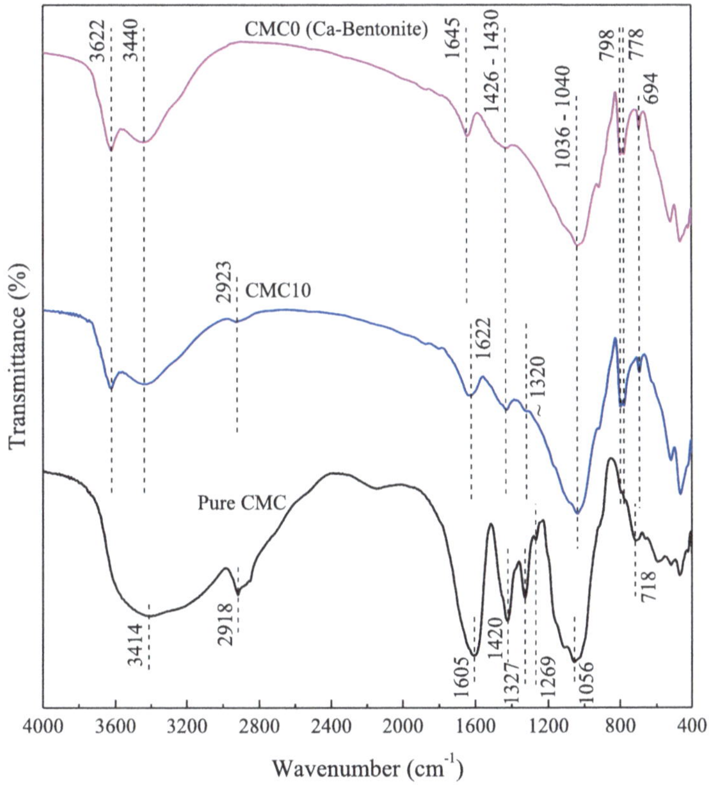

Fig. 1. FTIR spectra of pure CMC, untreated Ca-bentonite, and CMC-treated Ca-bentonite

10% to 40% lower than that of untreated Na-bentonite, which is more noticeable under a higher metal concentration.

Figure 3 presents the relationship between metal concentration and the hydraulic conductivity obtained from the MFL test. To simplify, the hydraulic conductivity that obtained from chemical solutions and DIW water is hereafter denoted as k_C and k_W, respectively. It is found that the CMC treatment results in a considerable decrease in k when compared with untreated Ca-bentonite for a given metal concentration. This result is more noticeable with an increase in metal concentration. For Pb-Zn contamination, the k_C of CMC-treated Ca-bentonites keeps consistent with the corresponding k_W value when metal concentration is lower than 10 mmol/L; and it tends to increase when metal concentration increases to 20 mmol/L. However, the k_C of untreated Ca-bentonite starts to increase from metal concentration of 1 mmol/L due to Pb-Zn contamination.

It has been understood that the increase in k for untreated bentonites is due to the compression of the DDL around bentonite particles. The reasons for the beneficial effect of CMC treatment is due to the chemical interaction between carboxylate (COO-) groups from CMC chain and cation (i.e., K^+, Ca^{2+}, Pb^{2+}, Zn^{2+}) in the form of either unidentate, bidentate, or bridging (Nara et al. 1996). As a result, DDL around bentonite

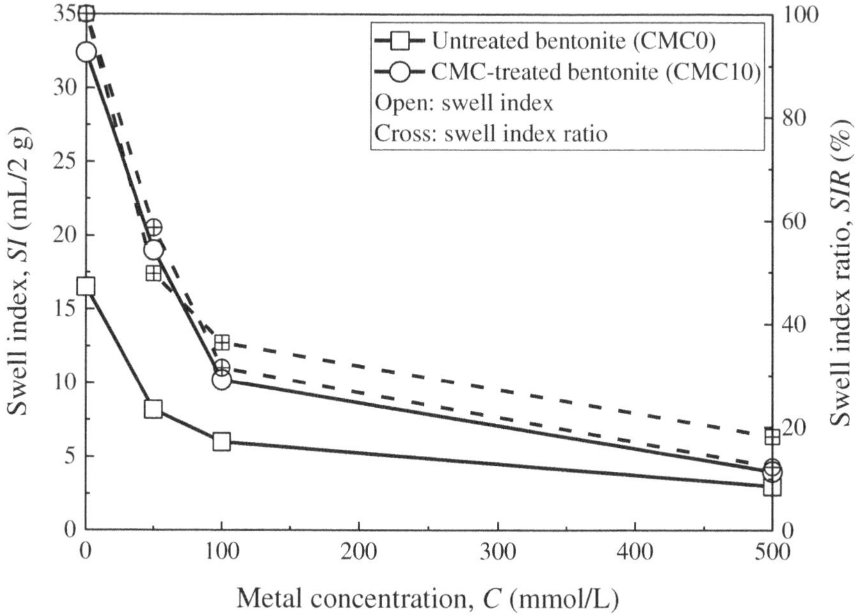

Fig. 2. Variation in swell index and swell index ratio (ratio of swell index of contaminated sample to that of clean sample) with metal concentration under Pb-Zn solution

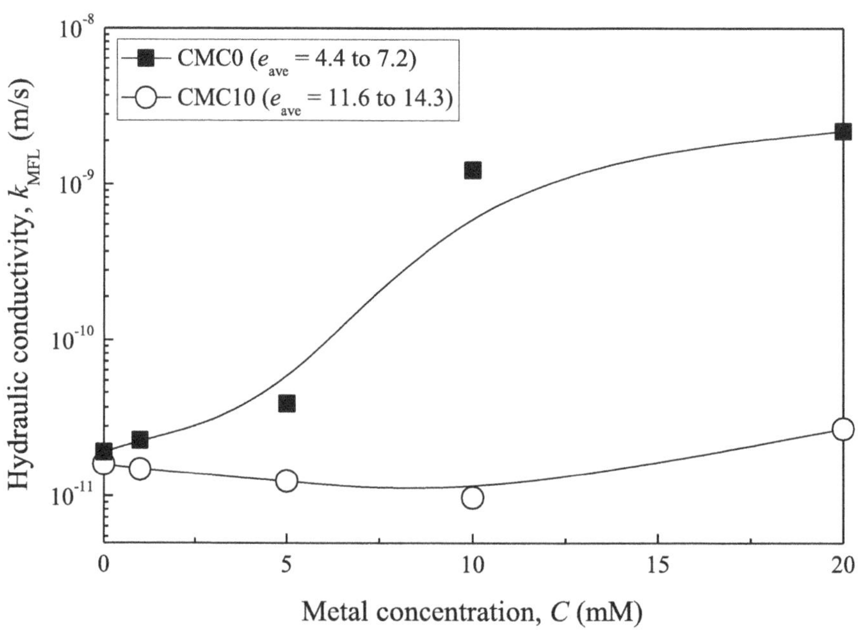

Fig. 3. Relationship between metal concentration and hydraulic conductivity

particles is protected to be compressed when cation from solution is available to CMC chains. However, the exhaustion of CMC chains implies an increase in k and swell potential, e.g., different degrees of increase in k and decrease in SI of CMC-treated bentonites when exposed to ionic solution with metal concentration >10 mmol/L.

Figure 4 presents the impact of landfill leachate on the hydraulic conductivity of CMC-treated Ca-bentonite as well as untreated Ca-bentonite. In addition, the average void ratio of bentonite cake and the pH of the bentonite-leachate slurry are also presented in Fig. 4. The result shows that the k value of CMC10 decreases 48 times when compared with that of untreated Ca-bentonite (CMC0). The relatively higher k of CMC10 under DIW compared with that of CMC10 under leachate is due to its corresponding higher average void ratio of bentonite. Thus, it can be concluded that CMC treatment significantly contributes to the decrease in k of Ca-bentonite when exposed to landfill leachate.

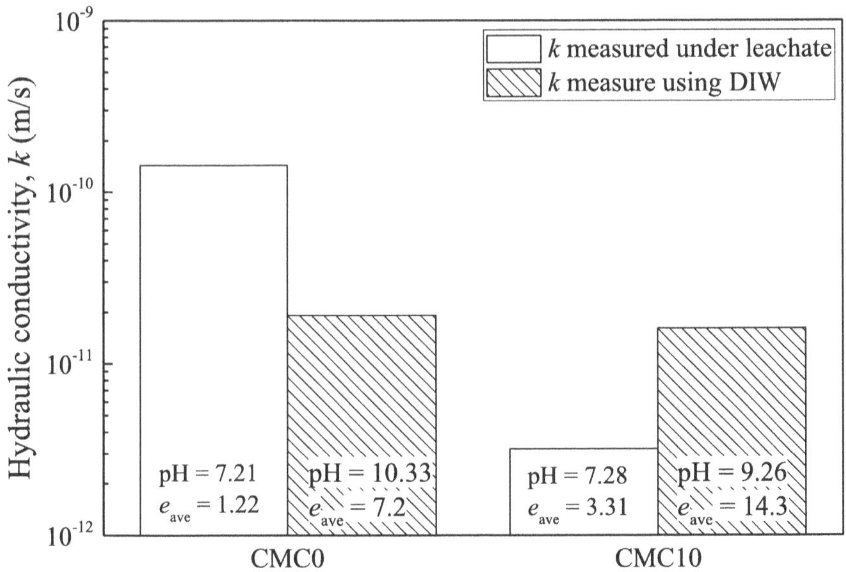

Fig. 4. Variation in hydraulic conductivity and pH with CMC content in bentonite

4 Conclusions

This study investigates the properties and chemical compatibility of CMC-treated Ca-bentonite under Pb-Zn solution and landfill leachate. The following conclusions can be drawn:

(1) The liquid limit, plastic limit and swell index of Ca-bentonite increase after CMC treatment, which is attributed to the intercalation of CMC chains into clay sheet. The CMC-treated Ca-bentonite with 10% CMC possesses a comparable swell index with that of HYPER clay.

(2) The CMC-treatment is well illustrated by FTIR analysis as characteristic bands representing carboxymethyl groups are found from CMC-treated Ca-bentonite. A more significant displacement of bands mainly due to CH_2 and COO- groups vibration indicate a more consumption of CMC chains by Ca-bentonite.
(3) The swell index of CMC-treated Ca-bentonites under Pb-Zn solution is considerably higher than that of untreated Ca-bentonite. However, CMC-treatment is unlikely to promote swell potential when exposed to Pb-Zn solution with high metal concentration (e.g., 100 and 500 mmol/L).
(4) The results from modified fluid loss test indicate that CMC-treated Ca-bentonites possess a lower hydraulic conductivity under Pb-Zn solution and landfill leachate compared with untreated one. The beneficial effect of CMC-treatment is due to the sorption of metal on CMC chains, which protects the double diffusion layer around bentonite particles to be compressed.

Acknowledgements. The authors are grateful for the financial support of the National Natural Science Foundation of China (Grant No. 51278100, 41330641 and 41472258), Key Program of Natural Science Foundation of Jiangsu Province (Grant No. BE2017715).

References

Bohnhoff GL, Shackelford CD (2014) Hydraulic conductivity of polymerized bentonite-amended backfills. J Geotech Geoenviron Eng 140(3):04013028

Chung J, Daniel DE (2008) Modified fluid loss test as an improved measure of hydraulic conductivity for bentonite. Geotech Test J ASTM 31(3):243–251

Di Emidio G (2010) Hydraulic and chemico-osmotic performance of polymer treated clays. Ghent University, Ghent

Di Emidio G, Mazzieri F, Verastegui-Flores RD, Van Impe W, Bezuijen A (2015) Polymer-treated bentonite clay for chemical-resistant geosynthetic clay liners. Geosynth Int 22(1):125–137

Du YJ, Fan RD, Reddy KR, Liu SY, Yang YL (2015a) Impacts of presence of lead contamination in clayey soil–calcium bentonite cutoff wall backfills. Appl Clay Sci 108:111–122

Du YJ, Fan RD, Liu SY, Reddy KR, Jin F (2015b) Workability, compressibility and hydraulic conductivity of zeolite-amended clayey soil/calcium-bentonite backfills for slurry-trench cutoff walls. Eng Geol 195:258–268

Fan RD, Du YJ, Reddy KR, Liu SY, Yang YL (2014) Compressibility and hydraulic conductivity of clayey soil mixed with calcium bentonite for slurry wall backfill: Initial assessment. Appl Clay Sci 101:119–127

Katsumi T, Ishimori H, Onikata M, Fukagawa R (2008) Long-term barrier performance of modified bentonite materials against sodium and calcium permeant solutions. Geotext Geomembr 26(1):14–30

Liu SY, Fan RD, Du YJ, Yang YL (2016) Modified fluid loss test to measure the hydraulic conductivity of heavy metal-contaminated bentonite filter cakes. Paper presented at the Geo-Chicago 2016: sustainable geoenvironmental systems, Chicago, IL

Liu Y, Gates WP, Bouazza A, Rowe RK (2013) Fluid loss as a quick method to evaluate hydraulic conductivity of geosynthetic clay liners under acidic conditions. Can Geotech J 51 (2):158–163

Madejová J (2003) FTIR techniques in clay mineral studies. Vib Spectrosc 31(1):1–10

Malusis MA, McKeehan MD (2013) Chemical compatibility of model soil-bentonite backfill containing multiswellable bentonite. J Geotech Geoenviron Eng 139(2):189–198
Qiu H, Yu J (2008) Polyacrylate/(carboxymethylcellulose modified montmorillonite) superabsorbent nanocomposite: preparation and water absorbency. J Appl Polym Sci 107(1):118–123
Under Heavy MetalUnder Heavy MetalRosin-Paumier S, Touze-Foltz N, Pantet A (2011) Impact of a synthetic leachate on permittivity of GCLs measured by filter press and oedopermeameter tests. Geotext Geomembr 29(3):211–221
Scalia J IV, Benson CH (2011) Hydraulic conductivity of geosynthetic clay liners exhumed from landfill final covers with composite barriers. J Geotech Geoenviron Eng 137(1):1–13
Shackelford CD, Sample-Lord KM (eds) (2014) Hydraulic conductivity and compatibility of bentonite for hydraulic containment barriers, vol GSP 233. ASCE, Reston
Sharma HD, Reddy KR (2004) Geoenvironmental engineering: site remediation, waste containment, and emerging waste management technologies. Wiley, New York
Yang YL, Du YJ, Reddy KR, Fan RD (2017) Phosphate-amended sand/Ca-bentonite mixtures as slurry trench wall backfills: assessment of workability, compressibility and hydraulic conductivity. Appl Clay Sci 142:120–127
Yong RN, Ouhadi VR, Goodarzi AR (2009) Effect of Cu^{2+} ions and buffering capacity on smectite microstructure and performance. J Geotech Geoenviron Eng 135(12):1981–1985

Durability of Alkali-Activated Slag - Bentonite Cutoff Wall Exposed to Sodium Sulfate and Pb-Zn Solution

Haoliang Wu, Jing Ni, Luo Zeng, Mengyu Huang, and Yanjun Du(✉)

Jiangsu Key Laboratory of Urban Underground Engineering & Environmental Safety, Institute of Geotechnical Engineering, Southeast University, Nanjing 210096, China
duyanjun@seu.edu.cn

Abstract. This paper addresses the reactive magnesia (MgO) activated ground granulated blast furnace slag (GGBS) blends with bentonite (GMB)as the cutoff wall mixture, compared with Portland Cement with bentonite (CB) for enhancing durability performance. A range of tests have been conducted to investigate the durability of the two batches cutoff wall (GMB and CB) as subjected to sodium sulfated and Pb-Zn solution, including unconfined compressive strength (UCS), hydraulic conductivity, and microstructure analyses with Fourier transform infrared spectroscopy (FTIR), X-ray diffraction (XRD) and scanning electron microscopy (SEM). The influence of curing age, GMB and CB content, and soaking solution type on the durability are also addressed. The UCS of GMB mixture is lower 73.6% and 11.2% than the CB reference after curing 28-days and 90-days. The hydraulic conductivity of GMB is much lower than the cutoff wall commonly accepted limits of 10^{-6}–10^{-7} cm/s, regardless of soaking in sodium sulfated and Pb-Zn solution. The microstructural analyses show the hydrotalcite-like phases (Ht) could be formed during the hydration of GMB, and C-S-H could be founded in GMB and CB mixture. The characteristic H-O-H bond stretching is noticed around 3460 and 3620 cm^{-1}, followed by shallow wide transmittances, which indicates the presence of Portlandite ($Ca(OH)_2$) and brucite ($Mg(OH)_2$). The brucite and Ht are confirmed in the SEM images of GMB mixture.

Keywords: Cutoff wall · MgO · GGBS · Durability · Sodium Sulfated

1 Introduction

The low-permeability and cost-effective vertical cutoff walls have been widely employed extensively to control seepage and contaminant migration in groundwater around the worldwide [1–5]. It is increasingly important to construct a downstream cutoff wall for preventing groundwater pollution in China due to the development of modern cities [6]. There are many types of cutoff wall in different countries. In the USA, the technique of chose is the soil-bentonite slurry wall [2, 3, 13]. The cement-bentonite and cement-bentonite-soil cutoff wall are preferred in European countries [4, 7]. The current trend is to consider using granulated ground blast furnace slag (GGBS) as a replacement for 80%

L. Zhan et al. (Eds.): ICEG 2018, ESE, pp. 430–439, 2019.
https://doi.org/10.1007/978-981-13-2224-2_53

Portland cement in the United Kingdom [4, 7]. The most commonly used methods to construct low permeability cutoff wall is the trench technique. In general, the trenched vertical barriers can reach 60 m in depths and 0.6 to 1.2 m in thickness [8, 9]. The EPA [1] has defined slurry cutoff walls as a vertical trench excavated along the perimeter of a site, using bentonite slurry for support, and then backfilled with a mixture of low-permeability (1×10^{-8} m/s or less) material. Ryan [8] pointed out that the primary design parameters of the slurry wall were permeability, strength, compressibility, and durability. Although the PC is most commonly used binder in cement-bentonite (CB) or soil-cement-bentonite (SCB) cutoff wall, GGBS is an alternative replacement material used in geotechnical and geoenvironmental engineering [5, 10, 11]. Moreover, the reactive magnesia (MgO) is a suggested activator to stimulate the dissolution of GGBS [5, 10–12]. The GGBS-MgO (GM) forms C-S-H gel and Ht as a principle hydration product during hydration process [5, 10, 12]. However, few studies have investigated the change in mechanic and durability performance on the GMB cutoff wall.

This paper aims to investigate the recycling the industrial byproduct of GGBS active with MgO, to produce low-permeability material for applying into trenched vertical cutoff walls. The mechanic characteristic of the cutoff wall is performed by measuring UCS, hydraulic conductivity (k) and strength loss. The reasons for the changes in these macro-properties are analyzed based on XRD, FTIR, and SEM. The results are useful for promoting the applying of recycling materials into design and construction vertical cutoff wall.

2 Materials and Methods

2.1 Materials

The mixtures are fabricated in the laboratory with Nanjing local sandy clay with bentonite and binders. The physical and mechanical properties of the soil mixtures are presented in Table 1. The water content is tested by using ASTM D 698 test method (ASTM, 2012). The Atterberg limits are performed using ASTM D4318 test method (ASTM, 2003). The specific gravity of the soil which is determined using ASTM D854 (ASTM, 2014), and the pH is performed according to the standard method, ASTM D4972 (ASTM, 2001). Light-burned magnesia (MgO) is provided by Shanghai Future Trading Company, with an activity coefficient of 90–100 s and 87.95% purity. Bentonite physical properties specifications require standard suspension properties such as viscometer dial reading to be a minimum of 30 at 600 rpm and yield point/plastic viscosity ratio to be a maximum of 3. The dominant mineral of bentonite was montmorillonite according to X-ray diffraction analysis. The basal spacing of the montmorillonite was identified as 15.48 Å, indicating that the bentonite used in this study is Ca-bentonite. Tap water is used in all mixtures with a pH of 8.0 and total dissolved salt of 176 ppm. Different water percentages are added to each mix to maintain a standard slump ranging between 100 and 200 mm as recommended by Ryan and Day [8]. Table 2 shows the physicochemical properties and mineralogical compositions of the sandy-clay, GGBS, MgO, and PC used for this study. Based on the X-ray diffraction analysis, the dominant minerals those materials are listed.

Table 1. Physical and chemical properties of sandy-clay and bentonite

Index	Sandy-clay		Bentonite
Moisture, (%)	4.81	0.66	11.2
pH	7.32	7.3	8.6
Specific gravity, G_s	2.66	2.64	2.66
Plastic limit, w_P (%)	20.6	–	54
Liquid limit, w_L (%)	45.3	–	697
Grain size distribution (%)			
Clay (<0.002 mm)	5.6	–	99
Silt (0.002–0.02 mm)	14.2	–	1
Sand (0.02–2 mm)	80.2	100	–

Table 2. Chemical compositions of sandy-clay, GGBS, MgO and PC used

Oxide Chemistry[a]	Sandy-clay (%)	GGBS (%)	MgO (%)	PC (%)
CaO	0.41	34.0	0.23	54.97
Al_2O_3	33.76	17.9	0.28	14.4
MgO	0.06	6.02	87.95	3.46
K_2O	0.15	0.64	0.01	0.07
SiO_2	42.73	34.3	0.28	19.19
Fe_2O_3	5.13	1.02	–	2.13
O_3	0.07	1.64	0.45	2.54
MnO	0.11	0.28	0.01	0.01

[a]Mineral composition is analyzed by X-ray fluorescence method using ARL9800XP + XRF spectrometry

2.2 Preparation of Mixture

The first step of the mixtures fabrication in the laboratory consists of mixing dry materials (sandy-clay, bentonite, GGBS, MgO and PC). The dry materials are weighted according to mix proportions presented in Table 3. The Mixtures are thoroughly mixed with a high-shear mixer with a mixing speed of 4500 rpm for 10-min period. The backfill materials for trenched cutoff wall should have a slump ranging between 100 mm and 150 mm to achieve the optimum workability and smooth consistency accordance with other researchers. In this research, water is added to achieve a standard slump within the range of 100 to 200 mm as recommended by Ryan and Day [8] to provide practical workability for in-situ placement. For each group mixture, the specimens are cast into a cylindrical PVC mold (50 mm in diameter and 100 mm in height) in five equal height layers. Entrapped air bubbles are carefully eliminated by vibrating with hand during the sample preparation process. The specimens are wrapped with vinyl bags and then cured under the standard condition (20 ± 2 °C, relative humidity 95%) in the laboratory prior to testing.

Table 3. Proportion of mixture in laboratory testing program

Id. of test specimen		GMB5	GMB10	CB5	CB10
Weight ratio (by sand-clay)	Sandy-clay	100	100	100	100
	Bentonite	5	10	5	10
	GGBS	4.5	9	0	0
	MgO	0.5	1	0	0
	PC	0	0	5	10

2.3 Experiment Testing Program

Unconfined compression tests are performed on a standard low-pressure capacity machine conforming to the ASTM D 4219 (ASTM, 2008). Samples are 50 mm in diameter and 100 mm in height. All samples are tested at a constant rate of displacement of 1.0 mm/min with a constant reading record for every 15 s. Hydraulic conductivity test is conducted per as ASTM D 5084 (ASTM, 2016). The samples are saturated with a backpressure of 300 kPa for 23 ± 0.5 h. The immersion test is performed in according to the ASTM C267 (ASTM, 2012). After 28 days' curing, the specimens are subjected to a soaking test under 4000 mg/L sodium sulfate as suggested by Garvin and Hayles [14]. Both lead and zinc are considered by the quality standard for groundwater as fourth drinking water contaminates with recommended limiting (maximum) concentration of 0.1 mg/L and 5 mg/L, respectively [15]. The UCS and weight loss of the specimens were monitored at 7, 28, 60, 90 and 120 days after immersion.

3 Results and Discussion

3.1 Unconfined Compressive Strength

Figure 1 plots the variation of UCS (q_u) with GMB and CB with the changes of curing age and the soaking solution. There is a clear trend that the q_u of CB decreases with the immersion time, regardless of soaking solution. However, the q_u of GMB increases with increasing curing time as subjected to Na_2SO_4, while a decrease in the Pb-Zn exposure. As compared the CB and GMB in the Na_2SO_4, the q_u of CB5 is higher 2.1 times than the q_u of GMB5 at 90-days. Furthermore, the 120-days' q_u of the CB10 decreases by 24.39% and 7.98% as compared to its 0-d, 28-days, respectively. Whereas the q_u of GMB10 increases 17.32% and 10.26% as the immersion time increased from 0-d and 28-days to 120-days. This trend could be mainly attributed to that CB mixture might be corroded by sulfate ion [16], such as the hydration products of C-S-H and Ca $(OH)_2$ easily reactive with freedom SO_4^{2-} ion. However, the GGBS could be activated by sulfate and then results in high durability [17]. For the Pb-Zn solution, the CB and GMB specimens decrease with the curing time, which is attributed to that the heavy metal restrains the hydration produce of C-S-H and Ht.

5

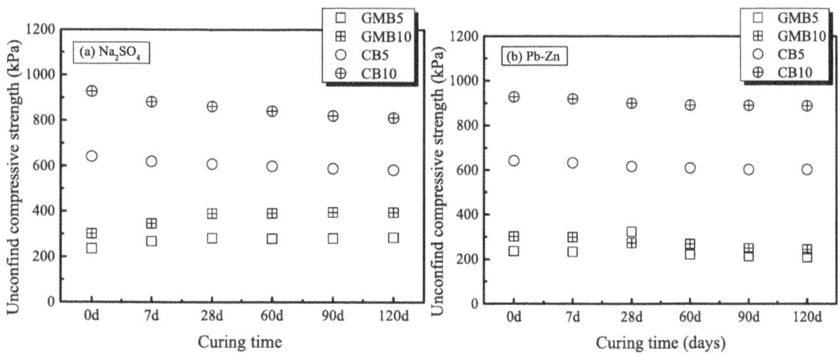

Fig. 1. Variation of unconfined compressive strength with curing time and soaking solution

3.2 Strength Loss

Strength augments of GMB specimens and strength loss of CB specimens with curing time are shown in Fig. 2. For the sulfate acid solution, the experimental results distinguish show that the augment of GMB5 and GMB10 after 120 days curing are 32.3% and 18.9% than the specimen before soaking, respectively. While the strength loss of CB5 and CB10 is 8.91% and 13.86%, respectively, at the same curing condition. The durability could be improved by the increased GMB and CB content for the Na_2SO_4 soaking solution as shown in Fig. 2(a). The strength loss of GMB and CB are both increased with immersion time, while the strength loss of GMB is tinily than CB at the curing time as shown in Fig. 2(b). These phenomena mean that the GMB pose higher durability than CB as exposed to contaminated environmental [18].

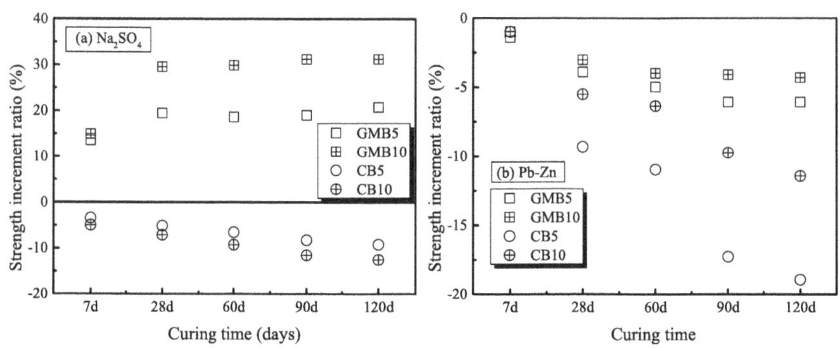

Fig. 2. Strength increment ratio of GMB and CB specimens with immersion time

3.3 Hydraulic Conductivity

Figure 3 shows the variations of hydraulic conductivity of GMB and CB as subjected to Na_2SO_4 and Pb-Zn solution. The hydraulic conductivities of specimens significantly reduce as increased GMB and CB content, regardless of soaking condition. This indicates that higher GMB and CB content have better antifouling properties as proposed by [19]. The higher content GMB and CB improve the density and decrease the porosity for the specimens during the hydration progress on the standard condition. As compared to CB, the GMB show a relative lower hydraulic conductivity on all cases, which is attributed by the especially hydration products (Ht) of GMB are buffer than C-S-H formed in CB. For this study, the hydraulic conductivity of GM5 and GM10 satisfied the commonly accepted limits (1×10^{-7} cm/s) after soaking in Na_2SO_4 and Pb-Zn solution.

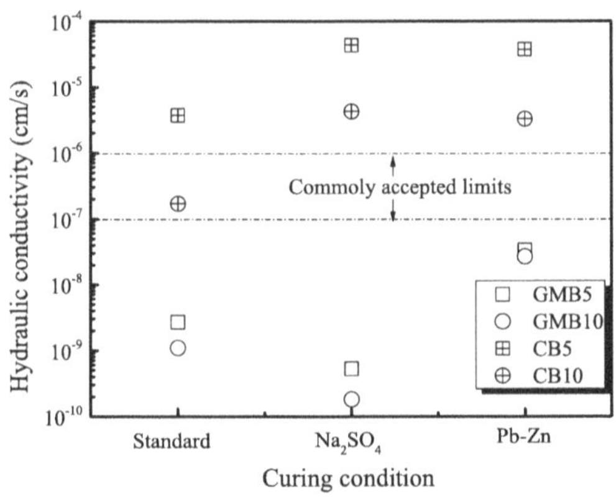

Fig. 3. Hydraulic conductivity results after soaking in sodium sulfate and Pb-Zn solution

3.4 FTIR Test Results

Figure 4 gives the infrared spectra as the GMB10 and CB10 subjected to Na_2SO_4 and Pb-Zn solution after 120 days. The similar infrared bands are observed on the GMB10 and CB10, regardless of soaking solution. The water component at the characteristic H-O-H bond stretching is noted around 3620 cm^{-1} followed by shallow wide transmittances, which indicated the presence of Portlandite: $Ca(OH)_2$ and brucite ($Mg(OH)_2$) [20]. The indication of ettringite (AFt) presence is confirmed by the sulfate ($-SO_4^{2-}$) stretching vibrations around 1116 and 1118 cm^{-1} as shown in Fig. 4(a). The vibrational modes occur at approximately 1430 cm^{-1} in this study is attributed by the $CaCO_3$, which formed from the reaction of atmospheric CO_2 with the $Ca(OH)_2$ along with hydration as shown in Fig. 4(b) [21]. The deep banding around 980 cm^{-1} indicates the Si-O bonding for possible C-S-H presence. Stretching bending occurs at 530 cm^{-1} that could be an indication of magnesium incorporation in the C-S-H.

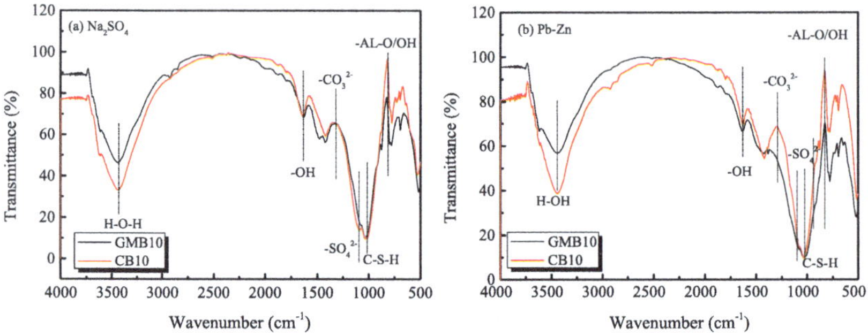

Fig. 4. FTIR analysis on the GMB and CB mixture

3.5 XRD Results

Figure 5 shows the XRD patterns of the mixture mix with GMB10 and CB10 as subjected to Na_2SO_4 and Pb-Zn solution after120 days. In the XRD patterns, the characteristic peaks of quartz (SiO_2), $Mg(OH)_2$, $CaCO_3$, $Ca(OH)_2$, C-S-H and bentonite are observed. There is an obvious drop in the level of the $CaCO_3$ phase in the Na_2SO_4 as comparing to the Pb-Zn solution. It is possible that the $Ca(OH)_2$ consume less CO_2 in the Na_2SO_4 solution. The C_3S and C_2S phases are reduced as compared the GMB to CB. It can also be observed that the intensity of bentonite peak decrease as the GMB and CB exposed to Pb-Zn solution [22]. This reduction in the peak intensity is attributed to the irregular stacking (or random orientation) of the clay platelets derived from the exfoliated clusters [23].

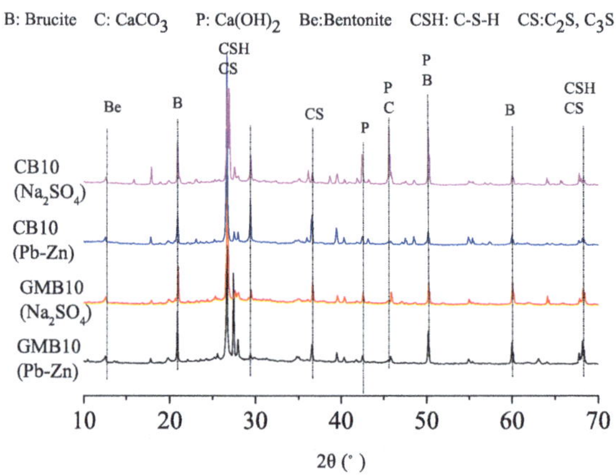

Fig. 5. X-ray diffractograms results of major hydration products

3.6 SEM Results

Figure 6 shows the 1000 × magnification and 5000 × magnification SEM images of the GMB10 and CB10 after 90-days standard curing. It can be seen from Fig. 6 that hydration products are taking part in the reaction include $Mg(OH)_2$ (i.e., the hydration product of MgO). It is also found that C-S-H in the CB. The CB10 exhibits more pore voids distributed between particles and therefore higher hydraulic conductivity as shown in Fig. 3. The 5000 × magnification image illustrates that the flake-like Mg $(OH)_2$ $CaCO_3$, C-S-H and Pb-Zn hydration product are formed during the hydration as shown in Fig. 6(b) and (d). These phenomena are consistent with those present in Figs. 4 and 5.

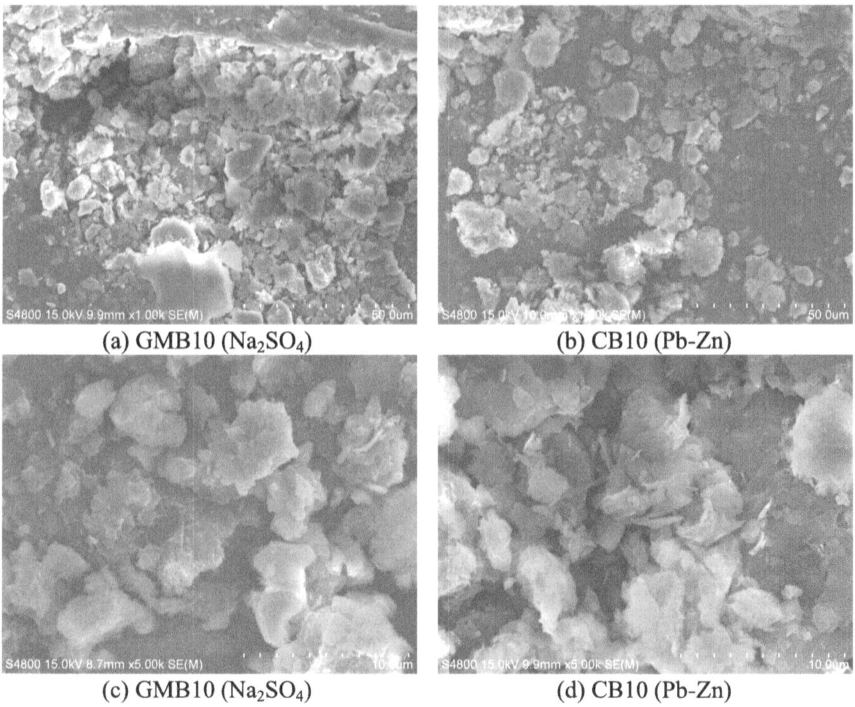

Fig. 6. SEM images major hydration products of GMB10 and CB10 after soaking.

4 Conclusions

This study investigated the durability of MgO active GGBS blends with bentonite for the cutoff wall mixture, via a series of soaking tests. The study confirmed that GMB could effectively replace the CB to achieve optimal durability performance, regarding low permeability and sufficient strength. The following are specific conclusions:

(1) The strength of GMB mixture increases while CB mixture decreases with curing time in Na_2SO_4 soaking condition. For the heavy metal solution, the CB and GMB mixture both decrease with the immersion time. The durability could be improved by the increased GMB and CB content for the Na_2SO_4 and Pb-Zn solution.
(2) The hydraulic conductivities of specimens significantly reduce as increased GMB and CB content, regardless of soaking condition. For this study, the hydraulic conductivity of GM5 and GM10 satisfied the commonly accepted limits (1×10^{-7} cm/s) after soaking in Na_2SO_4 and Pb-Zn solution.
(3) For the microanalysis, the hydration product such as $Ca(OH)_2$, $Mg(OH)_2$, C-S-H, SiO_2, and $CaCO_3$ can be confirmed by FTIR and XRD pattern. For the SEM images, the $Mg(OH)_2$ and C-S-H and Pb/Zn hydration product can be found.

Acknowledgments. This research is financially supported by Financial support for this research is partially obtained from the Environmental Protection Scientific Research Project of Jiangsu Province (Grant No. 2016031), National Natural Science Foundation of China (Grant No. 41472258), Natural Science Foundation of Jiangsu Province (Grant No. BK2012022), Colleges and Universities in Jiangsu Province Plans to Graduate Research and Innovation (KYLX16_0242), and the Scientific Research Foundation of Graduate School of Southeast University (Grant No. YBJJ1735).

References

1. EPA (1998) Evaluation of subsurface engineered barriers at waste sites. In: United States environmental protection agency. EPA 542-R-98-005 EPA Office of Research and Development's (ORD)
2. Evans JC, Shackelford CD, Yeo SS, Henning J (2008) Membrane behavior of soil-bentonite slurry-trench cutoff walls. Soil Sediment Contam 17(4):316–322
3. Malusis MA, Evans JC, McLane MH, Woodward NR (2008) A miniature cone for measuring the slump of soil-bentonite cutoff wall backfill. Geotech Test J 31(5):373–380
4. Soga K, Joshi K (2015) Cement bentonite cutoff walls for polluted sites. In: Multilevel modeling of secure systems in QoP-ML, p 149
5. Du YJ, Fan RD, Liu SY, Reddy KR, Jin F (2015) Workability, compressibility and hydraulic conductivity of zeolite-amended clayey soil/calcium-bentonite backfills for slurry-trench cutoff walls. Eng Geol 195:258–268
6. China Ministry of Environmental Protection (2010) The First National Pollution Source Census Bulletin. China Ministry of Environmental Protection, Beijing, pp 3–15
7. Opdyke SM, Evans JC (2005) Slag-cement-bentonite slurry walls. J Geotech Geoenvironmental Eng 131(6):673–681
8. Ryan CR, Day SR (2002) Soil-cement-bentonite slurry walls. In: Deep foundations: an international perspective on theory, design, construction, and performance, pp 713–727
9. Koch D (2002) Bentonites as a basic material for technical base liners and site encapsulation cut-off walls. Appl Clay Sci 21(1–2):1–11
10. Yi Y, Liska M, Al-Tabbaa A (2013) Properties of two model soils stabilized with different blends and contents of GGBS, MgO, lime, and PC. J Mater Civil Eng 26:267–274
11. Jin F, Gu K, Al-Tabbaa A (2015) Strength and hydration properties of reactive MgO-activated ground granulated blastfurnace slag paste. Cem Concr Compos 57:8–16

12. Wu HL, Jin F, Bo YL, Du YJ, Zheng JX (2018) Leaching and microstructural properties of lead contaminated kaolin stabilized by GGBS-MgO in semi-dynamic leaching tests. Constr Build Mater 172:626–634
13. D'Appolonia DJ (1980) Soil-bentonite slurry trench cutoffs. J Geotech Eng Div 106(4):399–417
14. Garvin SL, Hayles CS (1999) The chemical compatibility of cement–bentonite cut-off wall material. Constr Build Mater 13(6):329–341
15. China Ministry Environmental Protection Agency (2015) Environmental Quality Standard for Drink Water
16. Neville A (2004) The confused world of sulfate attack on concrete. Cem Concr Res 34 (8):1275–1296
17. Matschei T, Bellmann F, Stark J (2005) Hydration behavior of sulphate-activated slag cements. Adv Cem Res 17(4):167–178
18. Du YJ, Bo YL, Jin F, Liu CY (2016) Durability of reactive magnesia-activated slag-stabilized low plasticity clay subjected to drying–wetting cycle. Eur J Environ Civ Eng 20 (2):215–230
19. Xu H, Zhu W, Qian X, Wang S, Fan X (2016) Studies on hydraulic conductivity and compressibility of backfills for soil-bentonite cutoff walls. Appl Clay Sci 132:326–335
20. Angulo-Ramírez DE, de Gutiérrez RM, Puertas F (2017) Alkali-activated portland blast-furnace slag cement: mechanical properties and hydration. Constr Build Mater 140:119–128
21. Liu J, Yu Q, Zuo Z, Yang F, Duan W, Qin Q (2017) Blast furnace slag obtained from dry granulation method as a component in slag cement. Constr Build Mater 131:381–387
22. Yang YL, Reddy K, Du YJ, Fan R (2017) SHMP amended calcium bentonite for slurry trench cutoff walls: workability and microstructure characteristics. Can Geotech J (ja)
23. Sinha Ray S, Okamoto K, Okamoto M (2003) Structure − property relationship in biodegradable poly (butylene succinate)/layered silicate nanocomposites. Macromolecules 36(7):2355–2367

Experimental Study on Swelling Property and Gas Flow in a Compacted Bentonite-Sand Sample

Hong-Yang Ni[1,2], Jiang-Feng Liu[1,2(✉)], Jing-Na Guo[1], Xu-Lou Cao[1], and Shuai-Bing Song[1]

[1] The State Key Laboratory for GeoMechanics and Deep Underground Engineering, and School of Mechanics and Civil Engineering, China University of Mining and Technology, Xuzhou 221116, China
jeafliu@hotmail.com
[2] The State Key Laboratory of Oil and Gas Reservoir Geology and Exploitation, Southwest Petroleum University, Chengdu 610500, Sichuan, China

Abstract. This paper focuses on the experimental research on swelling properties gas migration in a compacted bentonite-sand plug. The aim is to investigate the swelling behaviour, gas breakthrough pressure as well as its sealing efficiency. Results show that bentonite-sand swelling is incomplete from 4 MPa gas pressure, so that gas breakthrough occurs at very low pressures (on the order of 1 MPa). Gas permeability of the bentonite-sand plug is affected by the variation in the confining pressure, so that tightness to gas can be achieved at a certain coupled hydro-mechanical condition. Specially, when the confining pressure is equal to 7 MPa and the water saturation degree is 86–91%, gas permeability can decrease to 10^{-20} m^2.

Keywords: High-level radioactive waste · Bentonite · Gas permeability

1 Introduction

Nuclear waste disposal is a worldwide problem for many countries. Currently, many countries are looking to place high-level radioactive waste in long-term storage in deep rock formations, such as the Callovo-Oxfordian argillite in France and granite in China. In this context, different kinds of clay are placed around the disposal tank as a buffer material, or as a sealing material to seal the underground chamber, in order to form a low-permeability area around the disposal tank [2–4, 10, 11]. After a lot of research and comparison, scientists discovered that bentonite or bentonite based materials are good candidate materials. Bentonite has low permeability, good expansion and adsorption capacity, and self-healing ability. In spite of this, several situations should be investigated to evaluate the robustness of this solution. Firstly, in situ, if bentonite-sand does not display proper sealing efficiency, the space between the buffer (or sealing) material and the gallery wall may allow the passage of ground water. Second, after the bentonite-sand mixture absorbs groundwater, the seal will be obtained due to its swelling property [6]. Meanwhile, hydrogen gas generation is expected. This gas

L. Zhan et al. (Eds.): ICEG 2018, ESE, pp. 440–445, 2019.
https://doi.org/10.1007/978-981-13-2224-2_54

generation is due to humid corrosion, organic matter degradation and water radiolysis. The gas pressure induces by this process-may hinder the swelling and sealing properties of bentonite. Lastly, for the bentonite engineering barrier, due to the intrusion of groundwater from the surrounding, a water saturation gradient will form from the periphery of the barrier to the core [5]. Therefore, the bentonite having a higher water content in the peripheral portion will come into contact with the inner wall of the surrounding rock due to expansion (e.g. argillite and granite). Due to the interaction of the forces and the limitations of the surrounding rock, the bentonite in the outer part will squeeze (or apply stress to) the internal bentonite blocks with a lower water content. Therefore, the problem is how to evaluate the gas tightness of the unsaturated bentonite in the middle part of the barrier under stress.

In more practical terms, an extensive experimental program has been performed at our laboratory in order to investigate the swelling and gas tightness of a bentonite-sand mixture in presence of both water and gas.

2 Materials and Experimental Methodology

2.1 Materials and Specimen Preparation

Base on the previous research performed by Andra [1], the materials consisting a mix of 30% sand and 70% WH2 bentonite provided by CEA (France). In order to obtain the target moisture content, the bentonite-sand powder is first placed in a humidity controlled desiccator (RH = 85%). After mass balance, the powders were placed in a cylindrical steel sleeve and subjected to a 12 MPa axial force for compaction. The prepared sample had a dry density of 1.77 ± 0.1 g/cm^3 and a water content of 12%. For gas permeability tests, the sample is of 12.5 or 25 mm height and 37 mm diameter; and for swelling test and gas breakthrough test, the relevant dimension is of 25 mm height and 42.5 mm diameter.

2.2 Experimental Scheme

The design of this set-up is to reproduce the in-situ situation where the compacted bentonite sample is rigidly constrained by the surrounding rock (E.g. argillite tube or Plexiglas-aluminum^{TM} tube), see Fig. 1. Then, the tube is put in the triaxial cell and wrapped in a Viton^{TM} jacket. Confining oil is utilized to impose a hydrostatic stress P_c of 12 MPa. This is approximately the *in situ* lithostatic pressure at about 500 m depth, given that argillite density is evaluated at 2.4 g/cm^3. Water pressure is applied by a Gilson^{TM}–type pump from the bottom side with 4 MPa. In our research, three series of tests are performed, see Table 1.

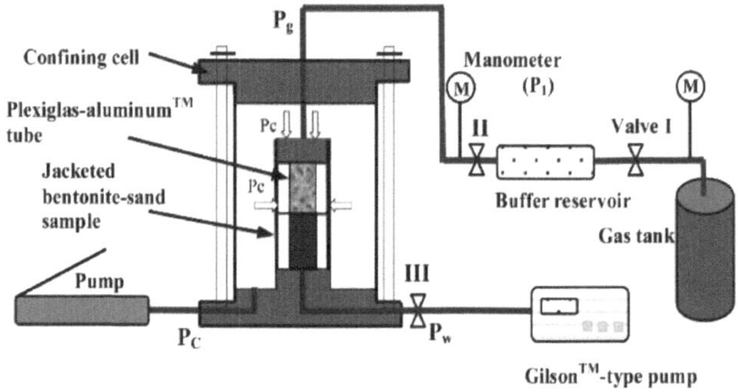

Fig. 1. Schematic diagram of swelling test: gas is inject form top side and water is injected form the downside

Table 1. Definition of tests preformed

Number	Boundary conditions		Note
	P_g (MPa)	P_w (MPa)	
B3	4	4	1. Direct contact with a completely water saturated sample; 2. Sample was put into a Plexiglas-aluminum™ tube
A7	0	4	1. Direct contact with water (4 MPa); 2. Sample was put into an argillite tube
S3-9–S3-13	0	0	Gas permeability tests

3 Experimental Results and Analysis

3.1 Swelling and Gas Breakthrough Test of B3 and A7

The evolution law of the swelling pressure of the B3 sample is shown in Fig. 2, as a function of time. It is can be see that there are two stable swelling pressures during the whole test. The first equilibrium swelling pressure, 10.72 MPa, named *total swelling pressure*, is obtained after 60 days swelling, which includes the contact pressure between the bentonite skeleton and the inner wall of the tube, pore water pressure and gas pressure. The second equilibrium swelling pressure, 6.93 MPa, called *effective swelling pressure*, which is caused by the action of bentonite solid matrix on the inner wall of the tube. In addition, for the swelling of the plug with an argillite tube, see Fig. 3, it is clear that the sample becomes fully saturated after 12–14 days swelling, which can be judged by the evolution of the water permeability. Now a problem arises, namely, the sealing efficiency of the two different situations. Therefore, gas breakthrough tests are performed to answer this question.

As shown in Fig. 4, two different gas breakthrough pressures are measured, namely, *discontinuous gas breakthrough pressure* and *continuous gas breakthrough*

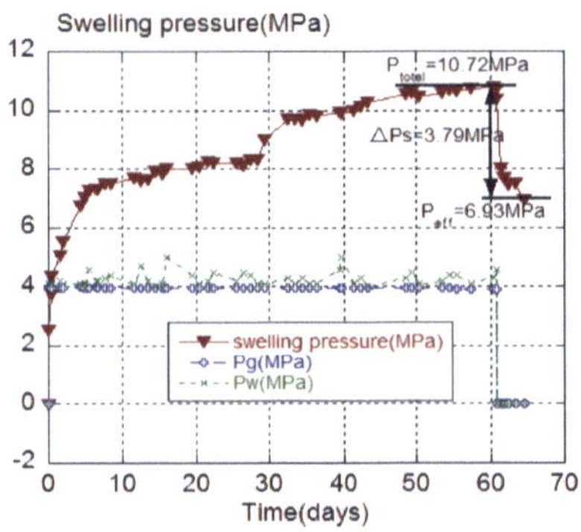

Fig. 2. Swelling test of sample B3

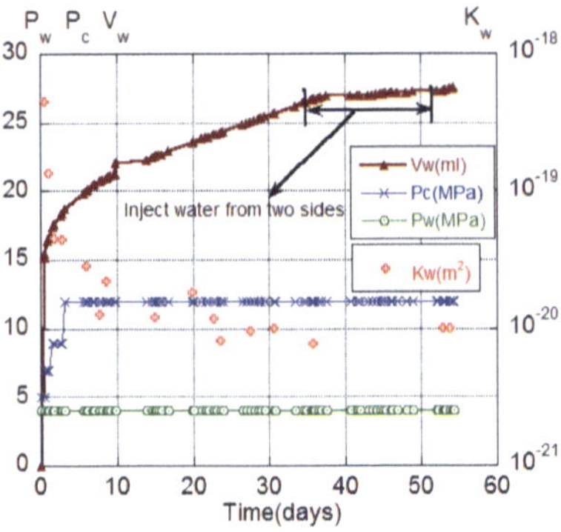

Fig. 3. Swelling test of sample A7

pressure. Regarding the gas migration mechanisms, it is either the capillary snap-off (discontinuous) or permeation (continuous) [8, 9]. A pressure sensor (±1–10 mbar) is used to record the pressure evolution at the outlet of triaxial cell, and a gas detector is used to periodically detect gas spillage. If the gas can be detected in the first few seconds, and then the gas disappears, we define this phenomenon as a discontinuous gas breakthrough. However, if gas passage phenomena can still be detected after 20 min, 1 h or longer time, we deem this phenomenon as gas passage continuous. As reportedly by Liu et al. [7], continuous gas breakthrough pressure is close to or higher than the effective swelling pressure of a fully saturated bentonite-sand plug, namely, about 7 MPa. This means that the sample A7, swelling with an argillite tube and without gas pressure during the swelling process, is fully saturated; while the sample B3, swelling with both water pressure and gas pressure, is only partially saturated. Therefore, *in situ*, gas pressure has to be taken into account when considering the sealing efficiency of the disposal pit.

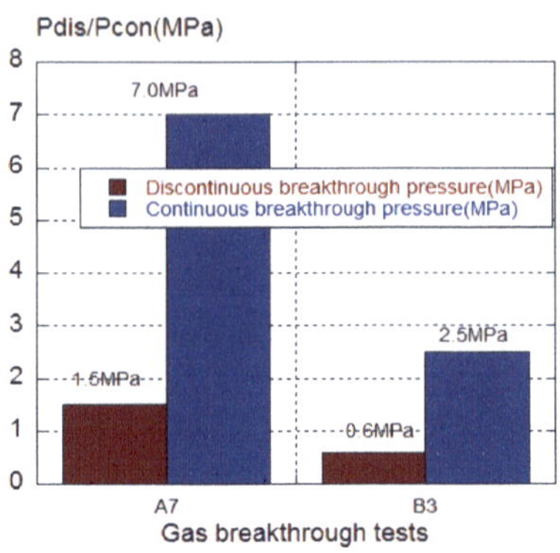

Fig. 4. Gas breakthrough test of sample A7 and B3

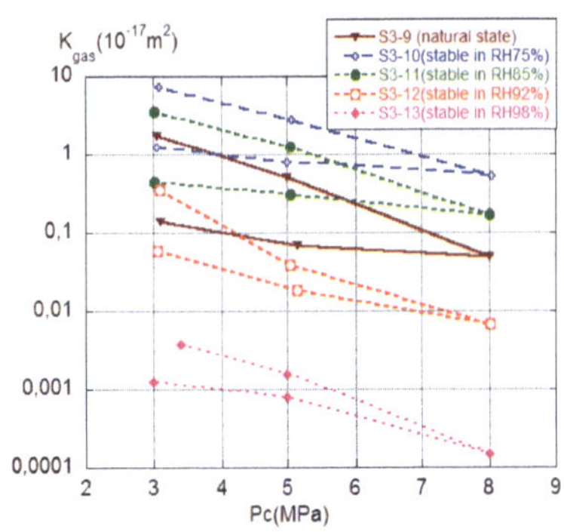

Fig. 5. Gas permeability of sample S3 series at different P_c and S_w

3.2 Gas Permeability Test of Series S3

Field tests have shown that the bentonite barrier has a significant water saturation gradient from the periphery to the core [12]. Thus, it is necessary to research the sealing ability of the bentonite-sand mixture under different water saturation level. As presented in Fig. 4, we found that the gas permeability of the the partially saturated samples (S_w = 72.8% and 91.7%) under 5 MPa is very low. When considering the difference in their initial gas permeability, the gas permeability of these samples follows the same evolutionary law under confining pressure. This means that tightness to gas will be achieved under confinement even through the bentonite/sand plug is not fully saturated.

4 Conclusions

This study aimed at researching water/gas migration processes through bentonite-sand mixture. Additionally, we investigate the effect of gas pressure on the sealing effect of the bentonite-sand mixture and the gas permeability of the samples with different saturation and confinement level. Our results show that the sample is only partially saturated when 4 MPa gas pressure is performed during the swelling process. For claystone-bentonite plug, continuous gas passage is found when gas injection pressure is more than 7 MPa, namely, the effective swelling pressure of a fully saturated bentonite/sand plug. In addition, for samples with an initial saturation of 86% to 91%, gas tightness can be obtained under a certain pressure.

Acknowledgments. The authors are grateful to the financial support from the Fundamental Research Funds for the Central Universities (Grant No. 2018ZZCX04).

References

1. Bosgiraud JM (2004) Scope of work. Design, fabrication, assembly, handling and packaging of buffer rings. Technical report C CC ASTE 04-0520/B, ANDRA
2. Cui SL, Zhang HY, Zhang M (2012) Swelling characteristics of compacted GMZ bentonite–sand mixtures as a buffer/backfill material in China. Eng Geol 141–142:65–73
3. Davy CA, Skoczylas F, Lebon P, Dubois T (2009) Gas migration properties through a bentonite/argillite interface. Appl Clay Sci 42(3):639–648
4. Liu JF, Song SB, Ni HY et al (2018) Research on gas migration properties in a saturated bentonite/sand mixture under flexible boundary conditions. Soils Found 58(1):97–109
5. Liu JF, Skoczylas F, Talandier J et al (2016) Dismantling of the EB experiment: experimental research on the retrieved GBM and bentonite blocks. Nucl Eng Des 300:297–307
6. Liu JF, Skoczylas F, Talandier J (2015) Gas permeability of a compacted bentonite–sand mixture: coupled effects of water content, dry density, and confining pressure. Can Geotech J 52(99):1–9
7. Liu JF, Song Y, Skoczylas F, Liu J (2016) Gas migration through water-saturated bentonite-sand mixtures, COx argillite and their interfaces. Can Geotech J 53(1):60–71

8. Marschall P, Horseman S, Gimmi T (2005) Characterisation of gas transport properties of the Opalinus Clay, a potential host rock formation for radioactive waste disposal. Oil Gas Sci Technol 60(1):121–139
9. Rossen WR (2000) Snap-off in constricted tubes and porous media. Colloids Surf A Physicochem Eng Asp 166:101–107
10. Sun DA, Zhang L, Li J, Zhang B (2015) Evaluation and prediction of the swelling pressures of GMZ bentonites saturated with saline solution. Appl Clay Sci 105:207–216
11. Ye WM, Chen YG, Chen B et al (2010) Advances on the knowledge of the buffer/backfill properties of heavily-compacted GMZ bentonite. Eng Geol 116(1):12–20
12. Villar MV, Iglesias RJ, Gutiérrez-Álvarez C et al (2018) Hydraulic and mechanical properties of compacted bentonite after 18 years in barrier conditions. Appl Clay Sci. https://doi.org/10.1016/j.clay.2017.12.045

Fluid Indicator Test (FIT) for Screening the Hydraulic Conductivity of Enhanced Bentonites to Inorganic Aqueous Solutions

A. Norris(✉), J. Scalia, and C. Shackelford

Department of Civil and Environmental Engineering, Colorado State University, Fort Collins, CO 80521, USA
anna.norris@colostate.edu

Abstract. Over the past 30 years, geosynthetic clay liners (GCLs) have been demonstrated to be a valuable barrier technology in numerous environmental containment applications. However, past research has also shown that the hydraulic conductivity (*k*) of GCLs is sensitive to the electrolyte concentration of the permeant liquid. At low effective stresses, a sufficiently high electrolyte concentration has the potential to result in a multiple order of magnitude higher *k* than the same GCL exhibits to water. To expand the boundaries of GCL functionality, recent research has focused on the use of enhanced (e.g., polymer modified) bentonites (EBs) for improving the resilience of bentonites, and GCLs, subjected to high concentration electrolyte solutions. These EBs have frequently incorporated anionic polymers (e.g. sodium polyacrylate) into the bentonite using a plethora of methods, (e.g. dry mixing, wet mixing, in-situ polymerization.) However, existing standard indicator tests (swell index, fluid loss, modified Atterberg limits) for GCL *k* may not always be effective for EB-GCLs. A rapid fluid indicator test (FIT) for screening high *k* versus low *k* of EB-GCLs and unmodified sodium bentonites, is proposed. Preliminary tests were conducted on granular sodium bentonite and an EB (bentonite polymer composite). Results of the FIT were compared with the swell index and *k* for these materials over a range of $CaCl_2$ solutions. The results support that a passing FIT generally correlates to low *k* for the same fluid under low stress conditions.

Keywords: Geosynthetic clay liners · Enhanced bentonite · Fluid indicator test

1 Introduction

Geosynthetic clay liners (GCLs) are used in environmental containment systems due to the low hydraulic conductivity (*k*) of sodium bentonite (NaB) to water [1, 2]. Unfortunately, NaB may not maintain low *k* when exposed to inorganic waste leachates with high concentrations of dissolved ions, particularly at low effective stresses [1, 2]. Enhanced bentonites (EBs), a composite material composed of NaB blended with polymers or high molecular weight organic molecules, are a potential solution to the instability of natural bentonite in many containment applications [3–5]. Enhanced bentonites have shown improved performance (lower *k*, lower rates of diffusion, greater membrane efficiency) relative to unamended NaB when exposed to high concentration

L. Zhan et al. (Eds.): ICEG 2018, ESE, pp. 446–453, 2019.
https://doi.org/10.1007/978-981-13-2224-2_55

inorganic aqueous solutions, but the mechanisms underlying these improvements are largely unknown [5].

The *k* of GCLs containing NaB is controlled primarily by the swelling of NaB component [1, 2, 5]. The test used to measure the swelling capability of a NaB in a specific solution is the swell index (SI) defined by ASTM 5890 [6]. The SI is used to screen the *k* of GCLs when both tests are conducted with the same solution [2, 7]. However, swelling (and Atterberg limits or fluid loss) do not universally correlate to the *k* of EB-GCLs [5]. For example, the relationship between *k* and SI is shown for a NaB and an EB-GCL (bentonite polymer composite, BPC) in Fig. 1. For NaB, a low SI correlates with increased *k* ($>1 \times 10^{-10}$ m/s). However, the opposite trend was observed for BPC, where the lowest SI correlates to lower k ($\ll 1 \times 10^{-11}$ m/s). These data provide an illustration of how the SI test may not be an effective indicator test for EB-GCLs.

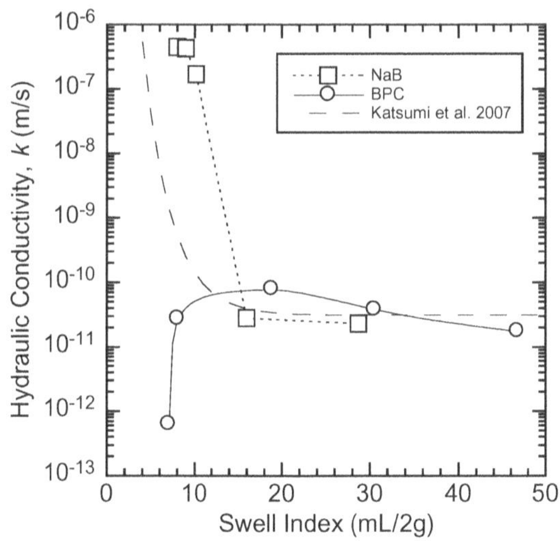

Fig. 1. Swell index as a function of $CaCl_2$ concentration for NaB and BPC, reproduced from [5], and the logarithmic trend fitted by [7].

The purpose of this study was to develop a rapid and simple indicator test for GCL *k* in various aqueous inorganic solutions that is effective for both GCLs containing unmodified bentonites and EBs. A new method is described and compared to *k* and SI in 5, 20, 50, 200, and 500 mM calcium chloride ($CaCl_2$) solutions.

2 Materials

The NaB used in this study, Volclay CG-50, was obtained from the Colloid Environmental Technologies Company (CETCO) of Hoffman Estates, IL, USA, and is the same NaB used in Bentomat GCLs as well as other studies [5]. Bentonite-polymer composite is prepared by free radical initiated additional polymerization acrylic acid within a bentonite slurry to create a covalent crosslinked polymer network. Properties of NaB and BPC (cation exchange capacity, mineralogical composition, polymer content) are reported in [5].

Solutions of 5, 20, 50, 200, and 500 mM $CaCl_2$ were used in this study. The electrical conductivities of these solutions were 1.22 mS/m (5 mM $CaCl_2$), 4.50 mS/m (20 mM $CaCl_2$), 10.4 mS/m (50 mM $CaCl_2$), 38.5 mS/m (200 mM $CaCl_2$), and 86.5 mS/m (500 mM $CaCl_2$). The pH of these solutions were near neutral (5.7–6.9). Solutions were prepared by dissolving calcium chloride di-hydrate ($CaCl_2 + 2H_2O$) in deionized water, and were stored in collapsible carboys with no headspace.

3 Methods

The fluid indicator test (FIT) apparatus is a pseudo-rigid wall permeameter as shown in Fig. 2. A hollow acrylic cylinder (outside diameter, D_o = 63.5 mm) contains sandwiched layers of mesh aluminum screens, filter paper, and NaB or EB forming a pseudo-GCL (with filter paper in leu of geotextiles; specimen diameter, D = 38.1 mm). The cylinder base is fitted with 6.35-mm outside diameter tubing that is used to supply or receive water for hydration and flow. A second hollow acrylic cylinder (outside diameter, D = 38.1 mm; inside diameter, D_i = 25.4) sits above the pseudo GCL and within the outer acrylic cylinder. The 38.1-mm diameter cylinder is filled with the selected liquid in both the hydration and testing phases. Weights are added to the top of the inner acrylic cylinder to apply a total normal stress of 6 kPa to the pseudo GCL. A flexible membrane and O-rings surround the connection between the two cylinders.

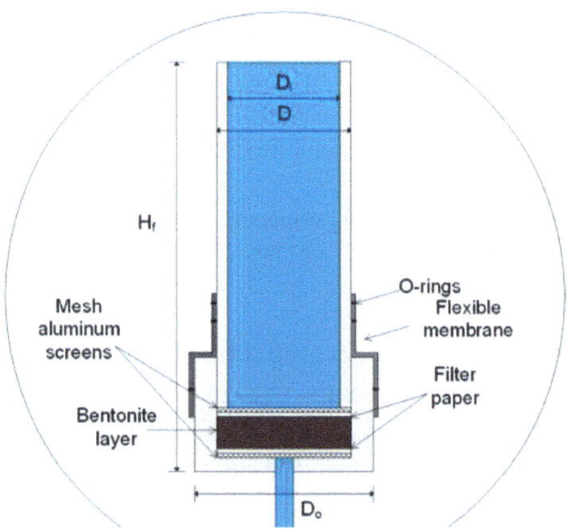

Fig. 2. Fluid indicator test (FIT) apparatus set up.

The FIT is run in two stages, a hydration stage (Fig. 3a), and a flow stage (Fig. 3b). The FIT apparatus (Fig. 2) is placed on a stand with height, H_S, of 225 mm, fitted with a hole for the 6.35-mm outside diameter tubing. The tubing is connected to a vertical burette at a height, H_B of 400 mm above the lab floor. To control the flow of solution, a valve is included near the base of the burette. After the pseudo-GCL is assembled within the FIT apparatus, and weight(s) are placed atop the apparatus yielding a applied normal stress of 6 kPa, a small amount (<10 mL) of solution is added to burette to removal of air bubbles from the tubing. During this step, the bentonite should not be allowed to contact the solution (i.e., the bentonite should not be allowed to hydrate). Once all air is removed from the tubing, solution is added to the burette, hydrating the bentonite from the bottom up until the system reaches equilibrium at a solution height, H_w of 1.0 m above the lab floor. The system is then left to hydrate for 24 h while the bentonite hydrates under the 6 kPa total normal stress.

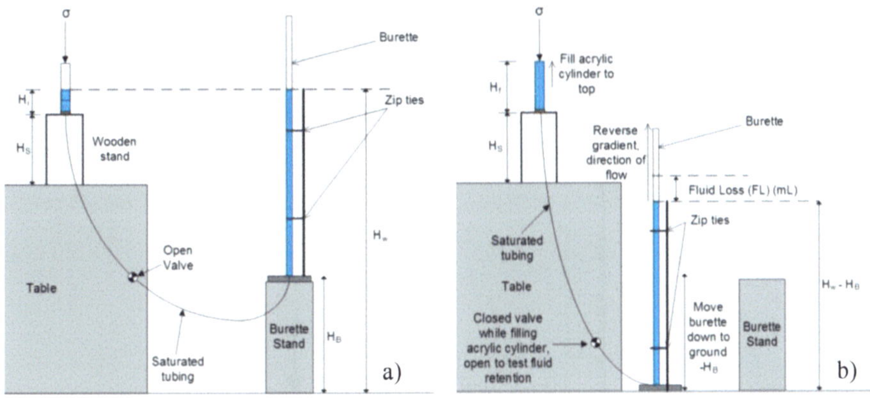

Fig. 3. Fluid indicator test (FIT) (a) setup (left) and (b) measurement (right).

After 24 h hydration period, the valve is closed, and the burette is moved to the lab floor ($-H_B$). The acrylic cylinder above the pseudo-GCL is filled with solution through a hole in the weights ($\sim$100 mL total volume, H_f = 18 cm). To initiate flow, the valve is opened and the volumetric flow through the pseudo-GCL is measured for a period of 5 min. The fluid retention (FR) is calculated by subtracting the flow through the pseudo-GCL from 100 mL. If FR = 100 mL, the specimen passes the FIT and is anticipated to potentially have a low k to the testing solution under low effective stress conditions. If the FR < 100 mL the specimen fails the FIT, and is antipated to potentially have a high k to the testing solution under low effective strsess conditions.

4 Results

The results of the FIT test for the NaB and BPC are shown in Fig. 4 and summarized in Table 1. These results are compared with the k and SI for the same material in the same solution [5]. The FIT test generally captures hydraulic compatibility (low or high k to a given solution) of both NaB and EB. Two outliers are evident in Fig. 4. The NaB tested with 50 mM $CaCl_2$ solution produced a passing result for the FIT but when hydrated and permeated with the same solution produced a high k. The reason for this outlier is unknown and is under investigation. The BPC with 500 mM $CaCl_2$ solution also produced an anomaly with a failing FIT but a low k. This outlier is explained by the polymer (sodium polyacrylic acid) behavior in the 500 mM $CaCl_2$ solution. Polymer in BPC acts to clog the pores of the GCL [5] via a tacky polymer hydrogel within the bentonite matrix. When hydrated with 500 mM $CaCl_2$, the network of polymer within BPC interconnects and tightens resulting in shrinkage of the specimen away from the sidewalls, producing sidewall leakage upon execution of the FIT test. Sidewall leakage was proven for this experimental combination by dying the specimen during the FIT as shown in Fig. 5. Preliminary tests of BPC with 500 mM $CaCl_2$ under a higher total stress (20 kPa) produced a passing FIT, correlating to the low measured k of this material-permeant solution combination. The higher load enabled maintenance BPC edge seal preventing sidewall leakage.

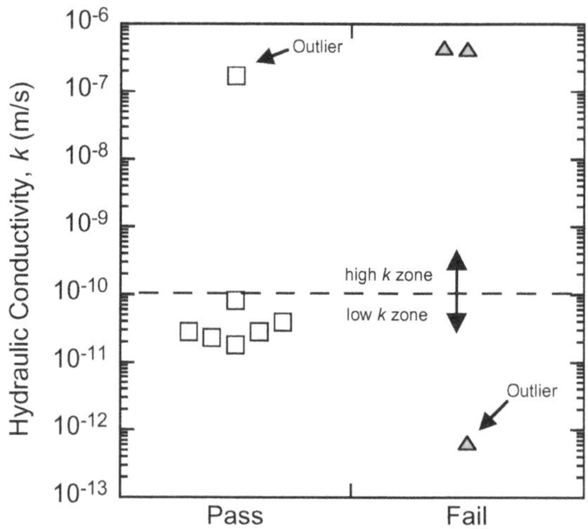

Fig. 4. Hydraulic conductivity (k) as a function of fluid indicator test (FIT) pass or fail for NaB and BPC. FIT passes when fluid retention = 100 mL, FIT fails when fluid retention <100 mL).

5 Discussion

The FIT results shown in Fig. 4 illustrate the potential use of this method as an indicator for high or low k of NaBs and EBs with potential permeant solutions. For comparison SI as an indicator test, an empirical equation for k based on SI from the literature [7] was used calculate the expected a predicted $k(k_c)$ from SI

$$\log(k_c/c) = \exp(\mathrm{a}(\mathrm{SI} - \mathrm{b})) \quad (1)$$

where k_c is in cm/s, and a, b, and c are fitting parameters equal to −0.31, 8.69 mL/2 g solid, and 3.09×10^{-9} cm/s, respectively, and SI is the measured swell index in mL/2 g.

Table 1. Swell index (SI), hydraulic conductivity (k), and fluid retention (FR) of Na-bentonite (NaB) and bentonite-polymer composite (BPC) in $CaCl_2$ solution. SI and k data taken from [5].

Conc. $CaCl_2$ (mM)	NaB			BPC		
	SI (mL/2 g)	k (m/s)	FR (mL)	SI (mL/2 g)	k (m/s)	FR (mL)
5	28.7	2.3×10^{-11}	100	46.7	1.8×10^{-11}	100
20	16.0	2.8×10^{-11}	100	30.5	3.9×10^{-11}	100
50	10.2	1.7×10^{-7}	100	18.8	8.1×10^{-11}	100
200	9.0	4.3×10^{-7}	97	8.0	2.8×10^{-11}	100
500	8.0	4.5×10^{-7}	94	7.0	6.5×10^{-11}	0

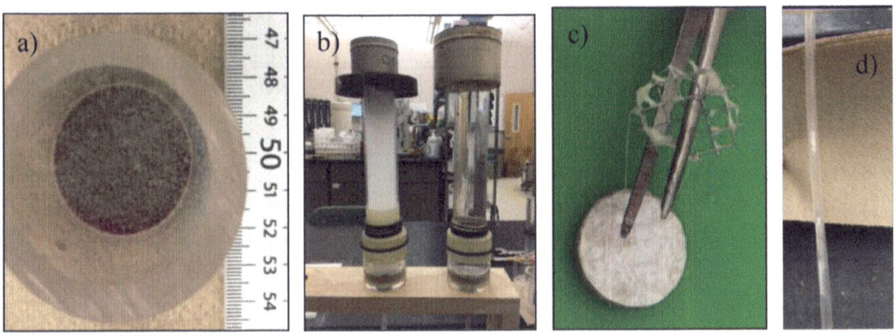

Fig. 5. Photographs of polymer effects on FIT: (a) preferential flow of BPC with 500 mM $CaCl_2$; (b) visible dispersion of polymer in inflow column of BPC tested with 50 mM $CaCl_2$ (left) versus no visible dispersion of BPC tested in 200 mM $CaCl_2$ (right); (c) crosslinked polymer on inflow side of BPC tested with 200 mM $CaCl_2$; (d) polymer eluted into outflow line of BPC tested in 200 mM $CaCl_2$.

Measured k, k_m, are compared to those calculated using Eq. 1, k_c, in Fig. 6. The data plotted shows agreement between k_c and k_m when SI is above 14 mL/2 g. This swell index is close to the lower limit (15 mL/2 g) that [7] noted was related to low k. A SI of 14 mL/2 g is also noted by [8] as a minimum SI associated with low k at low effective stress. The poor correlation zone shown in Fig. 6 illustrates that Eq. 1 predicts a k that is > two orders of magnitude lower than k_m for NaB specimens with a SI < 14 mL/2 g. Equation 1 also predicts a k_c > one order of magnitude higher for BPC specimens with a SI < 14 mL/2 g.

Unlike the SI test, the FIT generally matched low and high k of the materials tested. The FIT test was also qualitatively useful in enabling observations of BPC polymer behavior. As shown in Fig. 5b–d, polymer elution was visible in the testing and breakdown of the FIT with $CaCl_2$ solutions. In relatively dilute solutions (≤ 50 mM $CaCl_2$), the polymer was dispersed into solution upon filling the acrylic inflow cylinder (Fig. 5b). In 200 mM $CaCl_2$, the creation of solid polymer precipitates was visible (Fig. 5c). As shown previously in Fig. 5, the BPC FIT tested in 500 mM $CaCl_2$ caused catastrophically cross-linked polymer chains within the pore matrix such that no

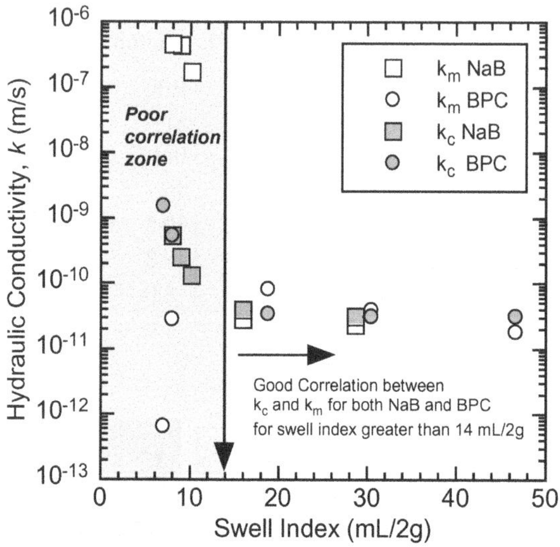

Fig. 6. Hydraulic conductivity (k) as a function of swell index (SI) for sodium bentonite (NaB) and bentonite polymer composite (BPC). Measured k, (k_m) from [5]; calculated k, (k_c) calculated using Equation A reported in [7] with SI from [5] used to calculate k_c.

dispersion or elution occurred. The photographs shown in Fig. 5 illustrate the potential utility of visual observations of EB hydration in different aqueous solutions for an indicator of polymer migration, and behavior. Efforts to standardize quantification of mobilized polymer are ongoing.

6 Conclusions and Recommendations

A fluid indicator test (FIT) was developed as a rapid indicator test to predict the hydraulic conductivity (k) of geosynthetic clay liners (GCLs), and GCLs containing enhanced bentonites (EBs). Results of the FIT illustrate the pass/fail criterion developed provided a better indication of high and low k for bentonite-polymer composite (BPC) than the swell index (SI) test. The FIT also allowed for qualitative observations of polymer behavior (e.g. elution and crosslinking) that were not observed upon inspection of the SI test. The FIT requires further testing to prove if the pass/fail criterion can accurately predict the high or low measured k of other EBs produced with different polymers and with other material preparation techniques. Additional study is also needed to with more complex aqueous inorganic solutions that may be more representative of field solutions.

References

1. Ruhl JL, Daniel DE (1997) Geosynthetic clay liners permeated with chemical solutions and leachates. J Geotech Geoenviron Eng ASCE 123(4):369–381
2. Shackelford C, Benson C, Katsumi T, Edil T, Lin L (1000) Evaluating the hydraulic conductivity of GCLs permeated with non-standard liquids. Geotext Geomembr 18(2–3):133–161
3. Onikata M, Kondo M, Kamon M (1996) Development and characterization of a multiswellable bentonite. In: Kamon M (ed) Environmental geotechnics. A A Balkema, Rotterdam, pp 587–590
4. Di Emidio G (2010) Hydraulic and chemico-osmotic performance of polymer treated clays. Ghent University, Faculty of Engineering, Ghent
5. Scalia J, Benson C, Bohnhoff G, Edil T, Shackelford C (2014) Long-term hydraulic conductivity of a bentonite-polymer composite permeated with aggressive inorganic solutions. J Geotech Geoenviron Eng ASCE 140(3):04013025-1–04013025-13
6. ASTM D5890 (2018) Standard test method for swell index of clay mineral component of geosynthetic clay liner. ASTM International, West Conshohocken
7. Katsumi T, Ishimori H, Ogawa A, Yoshikawa K, Hanamoto K, Fukagawa R (2007) Hydraulic conductivity of nonprehydrated geosynthetic clay liners permeated with inorganic solutions and waste leachates. Soils Found 47(1):79–96
8. Scalia J, Benson C, Finnegan M (2018) Alternate procedures for swell index testing of granular bentonite from GCLs. Geotech Test J Rev (2018)

Gas Breakthrough of Compacted Bentonite/Sand-Argillite Interfaces

Shuai-Bing Song[1,2], Jiang-Feng Liu[1,2(✉)], Xu-Lou Cao[1,2], Yong-Gui Chen[1], Hong-Yang Ni[2], Jia-Rui Zhang[2], and Yan Ma[2]

[1] Key Laboratory of Geotechnical and Underground Engineering of Ministry of Education, Tongji University, Shanghai, China
jeafliu@hotmail.com

[2] The State Key Laboratory for GeoMechanics and Deep Underground Engineering, and School of Mechanics and Civil Engineering, China University of Mining and Technology, Xuzhou 221116, China

Abstract. This paper focuses on gas breakthrough tests through bentonite-argillite interface. The main contribution of our study is to provide insights into the swelling kinetics, gas breakthrough pressure of these materials. During the tests, bentonite/sand plugs and bentonite-argillite plugs are used to perform swelling tests and gas breakthrough tests. Experimental results show that continuous gas passage through fully saturated bentonite/sand is not obtained until 10.5 MPa gas pressure. When a bentonite/sand plug swells inside a smooth metal tube, gas passage occurs at 7–8 MPa, which is similar to the effective swelling pressure of bentonite, as measured by independent swelling experiments. For argillite alone, saturated with water until sealing (water permeability on the order of 10^{-20}–10^{-21} m^2), gas breakthrough may occur at as low a pressure as 0.2 MPa and up to 6 MPa for undisturbed argillite. Finally, also after full water saturation, continuous gas breakthrough through the bentonite-argillite interface is detected at between 7 to 8 MPa gas pressure. It is interpreted that the argillite rock or the bentonite-argillite interface are both preferential pathways for gas migration when all materials become fully saturated.

Keywords: Swelling clay · Bentonite-sand plug · Argillite
Bentonite/argillite interface · Gas breakthrough

1 Introduction

The long-term safety of nuclear waste repositories is an important issue in all industrialized countries with a significant nuclear programme. The design of long-term nuclear waste repositories includes the building of an engineered barrier around the waste containers, which aims to create a "low permeable zone" around them (Alonso et al. 2006; Cui et al. 2012; Sun et al. 2015; Xu et al. 2003). Bentonite/sand mixtures are good candidate materials, due to their low permeability (when fully water-saturated), swelling and self-healing ability, together with high expandability and fair ductility. They have been chosen by several industrialized countries as a buffer and backfill material to separate radioactive waste from the surrounding host rock. The host

L. Zhan et al. (Eds.): ICEG 2018, ESE, pp. 454–460, 2019.
https://doi.org/10.1007/978-981-13-2224-2_56

rock is either made of granite (Sweden, Canada and China), or claystone (France, Switzerland), or clay (Belgium).

After filling of the disposal pits, sealing will be obtained using an engineered barrier constituted of compacted blocks of swelling clay (namely, bentonite), and concrete. Underground water seepage should ensure proper bentonite/sand plug swelling and hence, disposal pit sealing. Nevertheless, on the long term, humid corrosion may occur and induce hydrogen gas production and pressurization inside the repository tunnel (Xu et al. 2017; Liu et al. 2014; Liu et al. 2016; Push and Forsberg 1983; Ye et al. 2017).

After water saturation and swelling of bentonite, a weaker zone is expected as regards gas flow: it could be either the host rock, or the bentonite buffer, or the contact zone between the rock host and bentonite. This issue is clearly present in the storage waste project initiated by Andra (the French National Agency) at Bure (East of France). At the laboratory scale, the questions we propose to answer is the following: if gas is injected through a water-saturated mixed plug (Fig. 1) constituted of a swollen bentonite/sand plug placed within an argillite cylinder (in order to reproduce the tunnel seal), at what pressure and where will this gas flow through (argillite, bentonite/sand mixture or through their interface)? This question is extremely important for the evaluation of the sealing efficiency of the whole engineered barrier in situ. In order to answer these questions, as required by Andra, a series of tests have been performed to investigate the swelling and water-saturation of bentonite/sand plug, of the bentonite-argillite interface and of argillite (in other studies), as well as their gas breakthrough pressure (both discontinuous and continuous).

2 Materials and Experimental Methods

2.1 Materials and Sample Preparation

Based on preliminary studies performed by Andra (Bosgiraud 2004; Gatabin 2005), the materials used in our experiments consist of a mix of (30% TH1000 silica sand + 70% GELCLAY WH2 bentonite) from a single powder mix provided by CEA (France), batch reference RE08015. The grain-size distribution of TH1000 silica sand is similar to that of bentonite powder, which aims to make the mixture more homogeneous, see Table 1.

In order to obtain the adequate water content, before the compaction operation, the bentonite-sand mixture is let in an atmosphere at controlled relative humidity (RH = 85%). After mass stabilization of the powder mix at 85% relative humidity, it is compacted up to 12 MPa axial pressure in a dedicated steel mould in order to achieve $\rho_d = 1.77 \pm 0.1$ g/cm^3 apparent dry density and $w = 15.2\% \pm 1\%$ water content (Liu et al. 2014). For each manufactured series of bentonite/sand plugs, one first sample is made to assess ρ_d and w: if these are in the adequate range, several plugs are made subsequently; otherwise, powder is let to stabilize longer.

For the argillite tube, after coring, both the top and bottom surfaces are ground to ensure proper flatness see Fig. 1.

Fig. 1. Photograph of an argillite tube (left) and a bentonite/sand plug (right)

2.2 Experimental Method

The design of all sample mock-ups aims to reproduce the in-situ situation where the compacted bentonite/sand plug is rigidly constrained by the host rock, either inside a argillite tube or inside a Plexiglas-aluminium^{TM} tube, see Table 1. E1 test corresponds to the swelling of a bentonite/sand plug inside a rough (grooved) Plexiglas-aluminum^{TM} tube. This aims at preventing gas from migrating through this interface. The second test type F2 is performed on a bentonite/sand plug swelling inside a tube of argillite. For all tests, swelling test is carried out first, until full water saturation, followed by gas breakthrough test (GBT).

The sample (tube + bentonite/sand plug) is placed inside a triaxial cell and wrapped in a Vitton^{TM} jacket. Confining oil is used to impose a hydrostatic stress P_c of 12 MPa, see Fig. 2a and b. This is approximately the in situ lithostatic pressure at about 500 m depth, given that argillite density is evaluated at 2.4 g/cm^3. For sample water saturation, fluid is injected by a Gilson^{TM}–type pump from the upstream side at the *in situ* average pressure of 4 MPa, see Fig. 2a.

Table 1. Definition of all test types performed. Imposed water pressure is 4 MPa. for samples F2, the inner tube surface is smooth, while for sample E1, the inner tube surface is grooved, i.e. rough, which makes potential gas passage more difficult.

Number	Sample type	Test conditions
E1	Bentonite/sand plug inside a grooved Plexiglass^{TM}-aluminium tube	1. Direct contact with water
		2. Presence of an initial clearance between the tube/plug
F2	Bentonite/sand plug inside a smooth argillite tube	1. Direct contact with water
		2. Presence of an initial clearance between the tube/plug

In our experiment, the static method is chosen, although it lasts longer, as it is possible to determine more precisely the *in situ* case by increasing gas pressure gradually until the occurrence of the final gas breakthrough. Besides, this method also allows to observe different phases of gas flow during the process: expulsion of water, intermittent gas flow and continuous gas flow, while the observation of these phases is impossible with other methods. When performing gas breakthrough tests, gas is injected from the upstream side and gas detection is conducted in a downstream chamber by both a manometer (±1–10 mbar relative) and a dedicated argon gas detector ($\pm 10^{-4}$ml/s). The gas pressure is increased regularly (by 1 MPa steps every one to three days) until discontinuous and continuous gas flows are detected on the other side, see Fig. 2b. More details can be found in Liu et al. (2018).

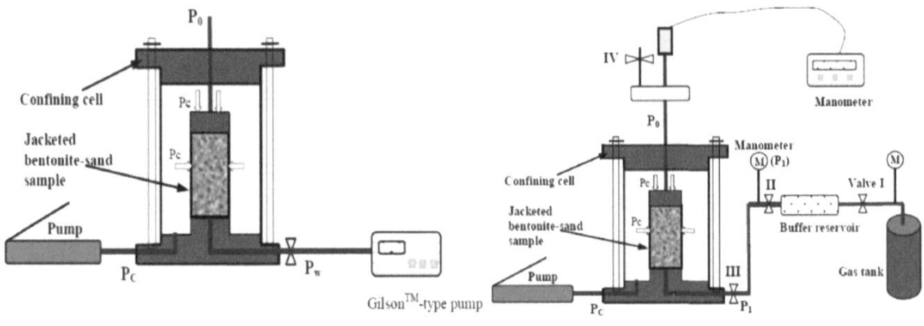

(a) (b)

Fig. 2. (a) Schematic diagram of the swelling test; (b) schematic diagram of the gas breakthrough test (GBT)

3 Results and Analysis

3.1 Water-Injection Test

Figure 3 shows water permeability versus time for sample E1 and sample F2. The saturation kinetics of samples E1 and F2 are similar: water permeability becomes stabilized after 5–10 days water injection. The initial water permeability of sample E1 is 8.72×10^{-16} m^2, which is higher when compared with the initial water permeability of sample F2 (1.92×10^{-18} m^2). This is as expected, because there exists an initial clearance between the plug and the tube. Therefore, the measured $K_{ini\text{-}w}$ (initial water permeability) is greater than the actual value for a fully swollen plug. Several minutes after the start of water injection, it is stopped to let the plug swell and wait for the occurrence of hydraulic cut off, whereby the bentonite/sand plug comes in contact with the tube (no more clearance). Water is injected again 5 h later, and water permeability decreases down to 3.09×10^{-19} m^2, which is attributed to the increase in saturation of the plug, and to the seal of the interface due to swelling of bentonite. Then, water permeability continues decreasing until reaching a stable value of 0.872–2.92×10^{-20} m^2. Water permeability of sample F2 is measured after the occurrence of hydraulic cut-off. Similarly to sample E1, water permeability of sample F2 becomes stable after 5–10 days saturation and the stable value is also close to 1×10^{-20} m^2.

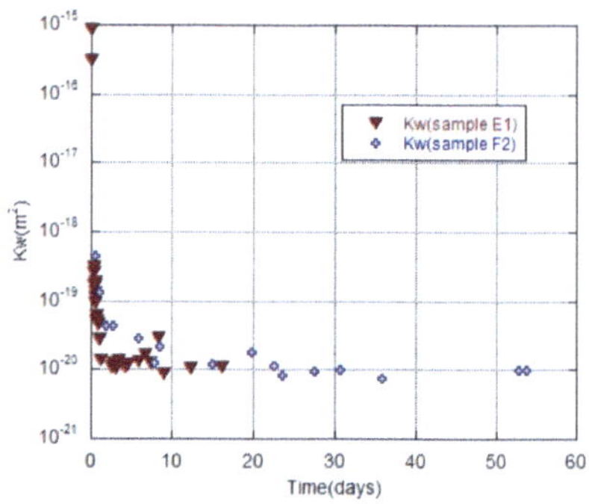

Fig. 3. Evolution of water permeability versus time for samples E1 and F2

The swelling process of argillite is presented in detail in (M'Jahad 2012). The same trends as for bentonite/sand plugs are observed. The initial water permeability of argillite is of $5\text{–}6\times10^{-20}$ m^2, which is smaller than for samples E1 and F2. However, the fully water-saturated permeability of argillite is similar to the bentonite/sand plugs and the bentonite/sand plug-argillite assembly, with values of about 1.13×10^{-20} m^2 (varying between $0.32\text{–}1.89 \times 10^{-20}$ m^2). This means that the saturated water permeability of argillite, the bentonite/sand plug and the bentonite/sand plug-argillite assembly are of the same order of magnitude ($10^{-20}\text{–}10^{-21}$ m^2). From water permeability alone, it is then expected that tightness to gas can be obtained for bentonite/sand plug, argillite and bentonite/sand plug-argillite assembly when they become completely saturated.

3.2 Gas Breakthrough Test

Gas breakthrough tests are performed after sample full water saturation. Prior to GBT, water injection is stopped for two or three days to let the water pressure become zero, and then gas is injected. Gas pressure is always smaller than confining pressure, i.e. below 12 MPa, to avoid oil leakage through to the sample. As shown in Fig. 4, for sample E1 (bentonite/sand plug swollen inside a grooved tube), discontinuous gas breakthrough pressure is of 5 MPa, while continuous gas flow is not detected up to 10.5 MPa gas pressure. This means that the fully water-saturated bentonite/sand plug is impermeable to gas, which is favorable to the sealing efficiency of the disposal pit. Discontinuous gas breakthrough pressure of sample F2 (bentonite-sand plug/argillite assembly) is observed at 3.6 MPa, while its continuous gas breakthrough pressure is 7 MPa.

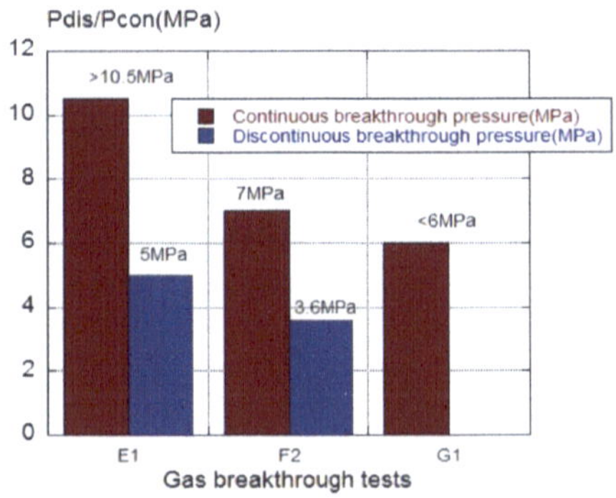

Fig. 4. Results of gas breakthrough tests: samples E1, F2 and G1

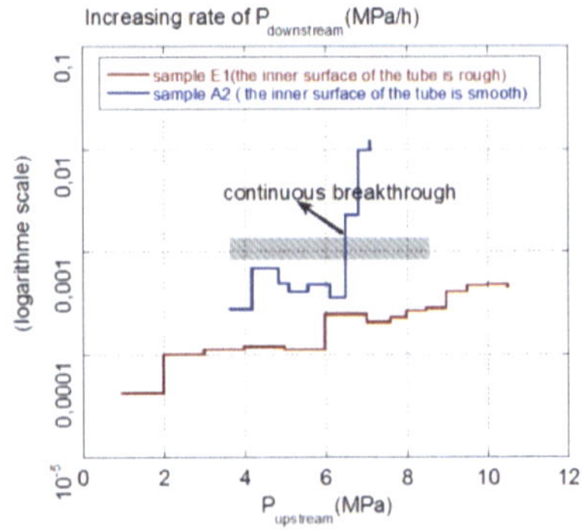

Fig. 5. Relationship between the upstream gas pressure and the increasing rate of downstream gas pressure: sample A2 (the inner surface of the Plexiglas-Aluminium tube is smooth) and sample E1 (the inner surface of the Plexiglas-Aluminium tube is rough) (Liu et al. 2014)

Breakthrough pressures of sample F2 (both discontinuous and continuous) are similar to those of bentonite/sand plugs swollen into a smooth Plexiglas-Aluminium™ tube, see Liu et al. (2012). Besides, for argillite alone, continuous gas breakthrough pressure is measured at values ranging from 0.2 MPa and up to 6 MPa for undisturbed argillite (Davy et al. 2012). As proposed in Liu et al. (2013), gas transfers through the interface between the bentonite/sand plug and the Plexiglas-Aluminium™ tube, when gas pressure is similar or higher than the effective swelling pressure (between 7 to 8 MPa), see Fig. 5. For gas pathways through the assembly of bentonite/sand plug-argillite, this means that there exists two possible cases: either through the interface or through argillite, which will be checked by experiments with a grooved argillite tube.

4 Conclusions

This study aimed at investigating water and gas transport processes in bentonite/and plugs, and bentonite/sand plug-argillite interfaces. Our results show that the swelling kinetics of a bentonite/sand plug and of an argillite plug are similar: the fully water saturated state is achieved after about 10 days water injection, which is assessed by their water permeability value on the order of 10^{-20}–10^{-21} m^2. Gas breakthrough tests show that continuous gas breakthrough pressure of bentonite-sand plugs is higher than 10.5 MPa. Due to the limits of our experimental device, we did not obtain continuous gas breakthrough for bentonite-sand plugs when swollen inside a grooved tube. For argillite, this value is lower, with values consistently smaller than 6 MPa. As proposed in Liu et al (2014), gas transfers through the interface between the bentonite/sand plug and the Plexiglas-Aluminium tube, when gas pressure is similar to or higher than the effective swelling pressure (between 7 to 8 MPa). This means that the interface or the argillite host rock are two possible pathways for gas migration, when all clayey materials become fully water saturated.

Acknowledgments. The authors are grateful for the support provided by the Open Fund (KLE-TJGE-B1703) of Key Laboratory of Geotechnical and Underground Engineering of Ministry of Education (Tongji University), the Natural Science Foundation of Jiangsu Province of China (BK20160249) and the Fundamental Research Funds for the Central Universities (China University of Mining and Technology) (2017QNA29).

References

Alonso EE, Olivella S, Arnedo D (2006) Mechanisms of gas transport in clay barriers. J Iber Geol 32(2):175–196

Cui SL, Zhang HY, Zhang M (2012) Swelling characteristics of compacted GMZ bentonite–sand mixtures as a buffer/backfill material in China. Eng Geol 141–142:65–73

Bosgiraud J (2004) Scope of work design, fabrication, assembly, handling and packaging of buffer rings. Technical report C CC ASTE 04-0520/B, ANDRA

Gatabin C (2005) Selection and THM characterization of the buffer material. Technical report RT DPC/SCCME 05-704-B, ANDRA

Liu JF, Davy CA, Talandier J, Skoczylas F (2014) Effect of gas pressure on the sealing efficiency of compacted bentonite-sand plugs. J Contam Hydrol 170:10–27

Liu JF, Song Y, Skoczylas F, Liu J (2016) Gas migration through water-saturated bentonite-sand mixtures, COx argillite and their interfaces. Can Geotech J 53(1):60–71

Liu JF, Song SB, Ni HY, Cao XL, Pu H, Mao XB, Skoczylas F (2018) Research on the gas migration properties in a saturated bentonite/sand mixture under flexible boundary condition. Soils Found 58(1):97–109

M'Jahad S (2012) Impact de la fissuration sur les propriétés de rétention d`eau et de transport des géométraux. Application au stockage profond des dechets radioactifs. PhD thesis, Ecole Centrale de Lille, France (in French)

Pusch R, Forsberg, T (1983) Gas migration through bentonite clay. Technical report TR. 83-71. SKBF-KBS, SKBF-KBS technical report

Sun DA, Zhang L, Li J, Zhang B (2015) Evaluation and prediction of the swelling pressures of GMZ bentonites saturated with saline solution. Appl Clay Sci 105:207–216

Xu L, Ye WM, Ye B (2017) Gas breakthrough in saturated compacted GMZ bentonite under rigid boundary conditions. Can Geotech J 54(8):1139–1149

Ye WM, Su W, Chen YG et al (2017) Membrane behavior of compacted GMZ bentonite and its granite mixture. Environ Earth Sci 76(20):683

Hydraulic Conductivity and Sorption Capacity of Special Barrier Materials in Inorganic Solutions

Evelina Fratalocchi(✉), Jonathan Domizi, and Francesco Mazzieri

Università Politecnica delle Marche, Ancona, Italy
e.fratalocchi@univpm.it

Abstract. The hydraulic performance and migration parameters of cementitious mixtures are currently under investigation by an experimental study. The main aim of the research is to investigate the diffusion coefficient and sorption capacity of the mixtures and their interaction with inorganic sulphate solutions at different concentrations, as a function of their composition and curing time. The paper shows and discusses the results related to the hydraulic performance and the sorption capacity as a function of the mixtures composition. Simulations of solute migration through cut-offs made by the different mixtures investigated show the importance of taking into account their sorption capacity to optimize the design.

Keywords: Hydraulic conductivity · Sorption · Cut-off

1 Introduction

Cut-off walls to isolate polluted soils need to be designed with materials able to limit pollutant migration as much as possible and as long as possible. In particular, they are required to act as barriers against advective and diffusive fluxes and to have sorption capacity. Contaminants should be adsorbed without any deleterious interaction with the solid phase of the cut-off wall material.

The literature is rich as far as the hydraulic conductivity of traditional mixtures is concerned (cement-bentonite and soil-bentonite mixtures, plastic concrete), whereas few data and research are available in the literature concerning diffusion, sorption, chemical interaction and durability. Values of the effective diffusion coefficient, D*, of the order of 10^{-10} m^2/s and no sorption capacity are typically suggested for different cement-bentonite mixtures related to some inorganic compound.

A research is currently going on to investigate the hydraulic performance and migration parameters of different cement-bentonite mixtures, able to give very low hydraulic conductivity at brief curing ($k < 10^{-10}$ m/s after 1 month). With such low k values, the diffusive flux is predominant; therefore, the main aim of the research is to investigate the diffusion coefficient and sorption capacity of these mixtures as well as their interaction with inorganic neutral or acidic sulphate solutions at different concentrations, as a function of their composition and curing time. The paper shows the

L. Zhan et al. (Eds.): ICEG 2018, ESE, pp. 461–469, 2019.
https://doi.org/10.1007/978-981-13-2224-2_57

results related to the hydraulic performance and the sorption capacity with curing time of cementitious mixtures for saline sulphate solutions.

2 Materials and Testing Methods

Two special cementitious mixtures were investigated together with a conventional cement-bentonite (CB) mixture typically used for plastic cut-off walls. The special mixtures are registered as Solidur® by Buzzi Unicem S.p.A (Italy) and here distinguished as Solidur.A and Solidur.B. Their composition includes blast furnace slag, cement and bentonite in different proportion; in both cases, the solid/water ratio is 0.33. The Solidur mixtures are prepared mixing the ready-made powder with water by means of a high turbulence mixer for 5 min. The CB mixture consists of activated sodium bentonite, bentonite/water ratio = 0.045, slag cement, cement/water = 0.22. The mixing procedure are those recommended by ETC8 (1993).

Soon after mixing, the main properties at the fluid state of all the mixtures were measured (Table 1). The mixtures were then poured in plastic moulds properly prepared to limit sample disturbance during extraction (Fratalocchi et al. 2013). The moulds were stored under water until the curing time for sample testing.

Table 1. Composition and main properties of the investigated mixtures at the fluid state.

Mixture	Solid/water	Marsh viscosity (s)	Bleeding (%)	Unit weight (kN/m^3)
CB	0.27	49	1.0	11.5
Solidur.A	0.33	43	1.5	11.8
Solidur.B	0.33	49	0.5	11.8

Solution of K_2SO_4 at different concentrations (1, 2, 5, 15 and 30 g/l) were used to evaluate the sorption capacity of the CB and of the Solidur mixtures with reference to K^+ and SO_4^{2-}. Batch tests were performed to the purpose according the procedures for fine grained soils (US EPA, 1992) on samples of the mixtures crumbled to pass the 2 mm sieve. The specific procedures are described in details in Fratalocchi et al. (2013).

On the basis of the results in the literature of cement pastes and CB mixtures, the batch tests were performed with a contact time of 72 h, that was demonstrated to be sufficient for chemical equilibrium.

Some of the sulphate solutions were also used as permeant for hydraulic conductivity measurements, with the procedures described in Fratalocchi et al. (2013).

3 Results and Discussion

3.1 Hydraulic Conductivity

Cementitious mixtures are attacked by sulphate solutions due to the reaction between SO_4^{2-} and dissociated calcium hydroxide and, partially, CSH, with the consequent precipitation of gypsum and release of OH^-. The subsequent damaging reactions take place between the sulphate and the hydrated calcium aluminate or calcium aluminate monosulphate hydrate (Bensted 1995; Gollop and Taylor 1992, 1995) to produce ettringite. The resulting volume expansion and fissuring give rise to preferential paths and to an increase in k.

Figure 1 shows the k values with curing time measured on the Solidur samples permeated with water and with the K_2SO_4 solution at concentration of 15 g/l. The excellent performance of both mixtures in water is maintained with the aggressive solution. The very low hydraulic conductivity of both Solidur mixtures at brief curing (10^{-9}–10^{-10} m/s at 20–30 days) is able to strongly reduce the pore volumes of flow, PV, under a given hydraulic gradient. Consequently, after several months of continuous permeation, notwithstanding a high hydraulic gradient (ranging from 60 to 120 in the tests), only about 0.1–0.2 PV passed through the samples (the K_2SO_4 concentration at the effluent was always zero). The same permeation tests performed on a more permeable mixture at brief curing (Fig. 2) show a rapid increase of k when the sample is permeated with the salt solution, after about 2 months of curing, corresponding to about 2.5 PV (Fratalocchi et al. 2013). The number of PV corresponding to the increase of k was indeed identified as a reference parameter to evaluate the durability of typical cement-bentonite mixture.

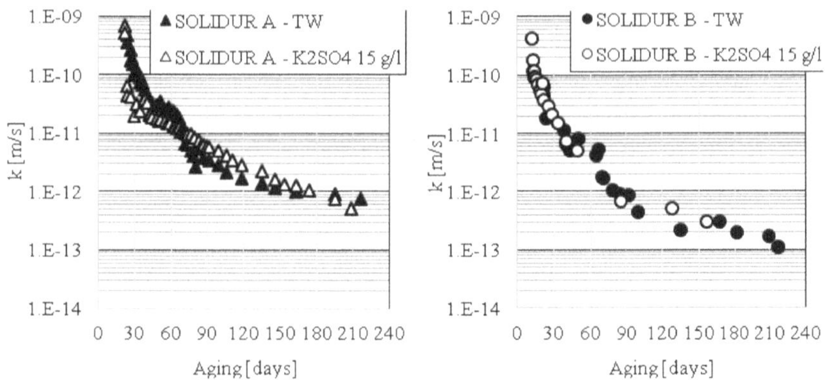

Fig. 1. Hydraulic conductivity vs curing time of the Solidur mixtures permeated with tap water (TW) and with the K_2SO_4 solution at the concentration of 15 g/l.

3.2 Sorption

Sorption is here intended as the overall mechanisms removing contaminant from solution, e.g. cation exchange, precipitation, adsorption and any binding between solutes and the solid matrix.

It is important to point out that, for CB mixtures, both sorption isotherms of K^+ and SO_4^{2-} from batch tests were found to be in good agreement with those from column tests (Fratalocchi et al. 2013) which reflect the actual in situ migration conditions.

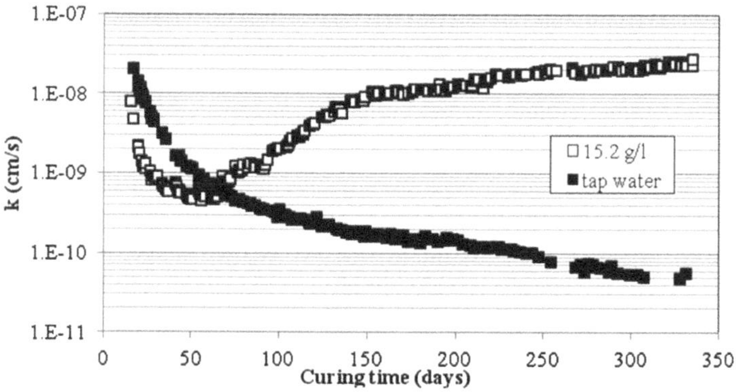

Fig. 2. Hydraulic conductivity vs curing time of the CB mixture permeated with tap water and with the K_2SO_4 solution at the concentration of 15 g/l (Fratalocchi et al. 2013).

Figure 3 shows the results of the batch tests with reference to sorption of K^+ of the CB mixture, whereas Figs. 4 and 5 show those related to the Solidur.A and Solidur.B, respectively. All the results demonstrate that, within the wide range of concentration investigated, the potassium sorption is always well modelled by a linear isotherm:

$$S = k_d C \tag{1}$$

where S = absorbed mass of solute per unit mass of solid [–], C = equilibrium concentration of solute $[ML^{-3}]$, k_d = distribution coefficient $[L^3M^{-1}]$. The same type of isotherm is documented in the literature for sorption of K^+ on CB mixtures and on cement pastes in contact with KOH solution. In particular, batch tests performed on calcium hydro-silicate (Hong and Glasser 1999) gave k_d values ranging from about 0.5 to 6 ml/g depending on Ca/Si ratio. The k_d values of the tested mixtures are close to the highest values of this range and correspond to low Ca/Si ratios, typical of blended cements, such as those used in the CB and Solidur mixtures investigated. The main sorption mechanism of K^+ on the mixtures can be considered as the same of cement pastes: the binding of alkali in hardened cement pastes containing blast-furnace slag cement or Portland cement is restricted to sorption at the CSH or CASH phases.

Sorption of K^+ by bentonite is negligible since its cation exchange capacity (<100 meq/100 g) and its mass (about 10–20% of the total solid mass in the different mixtures) give rise to a negligible mass of absorbed K^+ in comparison with the overall K^+ mass absorbed by the mixtures.

Figure 6 shows a comparison among the k_d values of the three mixtures as a function of the curing time. For the CB and the Solidur mixtures, the sorption capacity appreciably increases with curing time. It is also evident the higher sorption capacity of the Solidur mixtures in comparison with the CB one and that the greater the solid content in the mixture, the higher the increase of the k_d value with curing. The increasing k_d values with curing time are consistent with the sorption mechanisms related to the CSH and CASH phases.

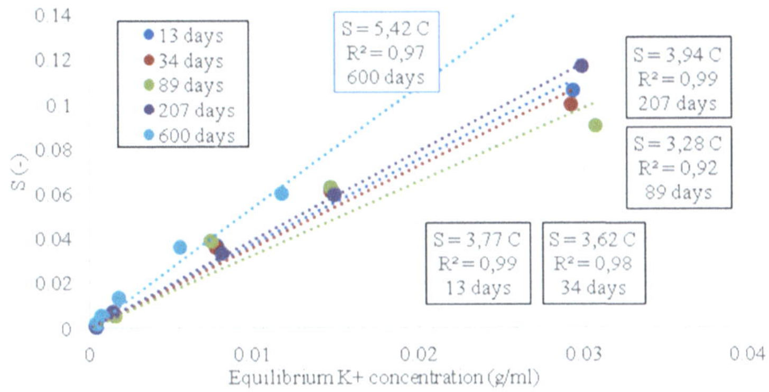

Fig. 3. Potassium linear isotherms at different curing time of the CB mixture (modified from Fratalocchi et al. 2013).

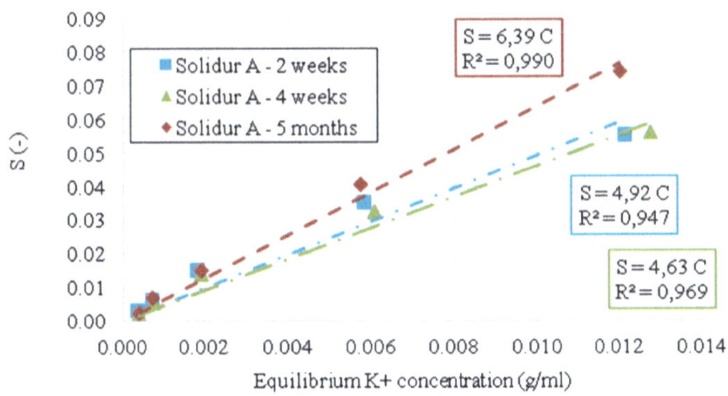

Fig. 4. Potassium linear isotherms at different curing time of the Solidur.A mixture

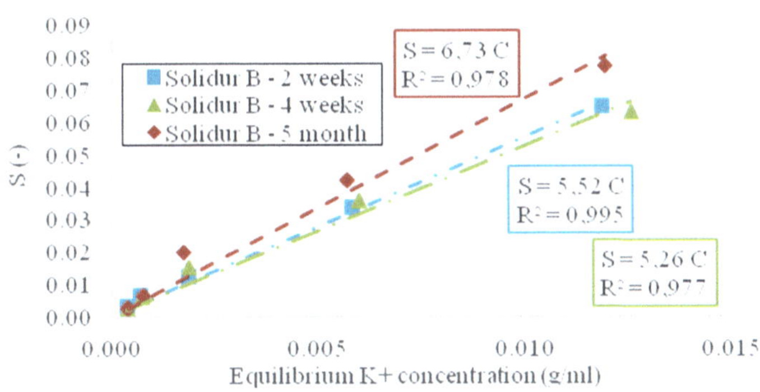

Fig. 5. Potassium linear isotherms at different curing time of the Solidur.B mixture

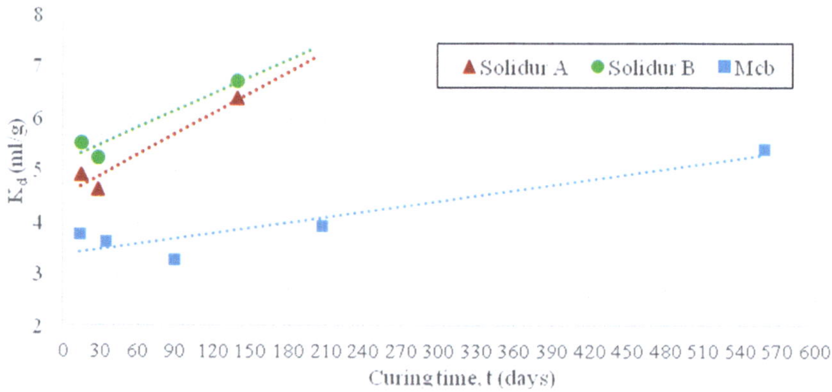

Fig. 6. Distribution coefficient of K^+ vs. curing time for the mixtures investigated.

The increasing k_d with curing time demonstrates that both for Solidur and CB mixture sorption capacity can be determined at brief curing to safely analyze the migration of potassium through a cut-off. The k_d values of potassium also demonstrate the importance of sorption when analyzing its migration through cut-offs. To this regard, Fig. 7 shows the breakthrough curves of potassium through hypothetical cut-off walls made by the different mixtures investigated. All the cut-offs have been supposed 0.6 m thick and of porosity, n = 0.6. Migration due to pure diffusion has been considered, assuming for potassium $D^* = 2 \times 10^{-10}$ m^2/s. The assumption of pure diffusion is reasonable considering the very low hydraulic conductivity for all the mixtures just after one month of curing (Figs. 1 and 2). The simulations were performed by the code Pollute.v7 assuming: semi-infinite thickness, initial absence of concentration in the cut-off and constant concentration at the boundary. The concentration at the distance = 0.6 m from the boundary was assumed to be the downstream concentration of the actual barrier. k_d values were assumed at 1 month of curing, equal to 3.6 ml/g for the CB cut-off, 4.6 and 5.2 ml/g for the Solidur.A and B, respectively (Fig. 6).

Figure 7 clearly shows the importance, when designing a cut-off wall, of taking into account the sorption capacity of the mixture and of selecting the best mixture considering its sorption capacity. For example, the simulation shows that, supposing a maximum allowable relative concentration of 0.4 downstream of the cut-off wall, if sorption is neglected, the wall would be efficient for about 25 years whereas it would be efficient much longer due to its sorption capacity (about 5 times). Using the Solidur. B mixture the time of efficiency would be about 40% longer than that of the CB mixture.

With reference to sorption of SO_4^{2-}, the Freundlich isotherm gave the best fitting:

$$S = k_f C^{\varepsilon} \tag{2}$$

where k_f $[L^3M^{-1}]$ and ε [–] are constant. Figures 8, 9 and 10 show the batch test results at different curing times fitted by Eq. (2). For the CB mixture, sorption of SO_4^{2-} appreciably decreases with curing time, due to the significant decrease in porosity and

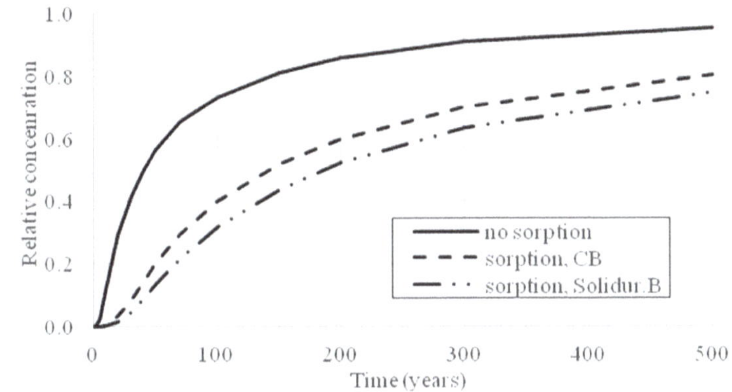

Fig. 7. Effect of sorption on the breakthrough curves due to diffusion of K^+ through the same cut-off wall made by the mixtures investigated.

in the mean void size of the mixtures (sorption sites tend to be less accessible, Fratalocchi et al. 2013). For the same reason, sulphate sorption of both Solidur mixtures, which have a solid/water ratio higher than that of the CB mixture, results to be significantly lower than that of the CB mixture, and independent of the curing time (Figs. 9 and 10). The very low sulphate sorption of the Solidur mixtures, which means minor interaction between sulphate and cement hydration products, is consistent with the hydraulic conductivity results of Fig. 2. All the results related to the sulphate interaction are consistent with those by Soga and Joshi (2010) who found that CB mixtures can better resist sulphate attack after hydration owing to lower mass flux through the mixture.

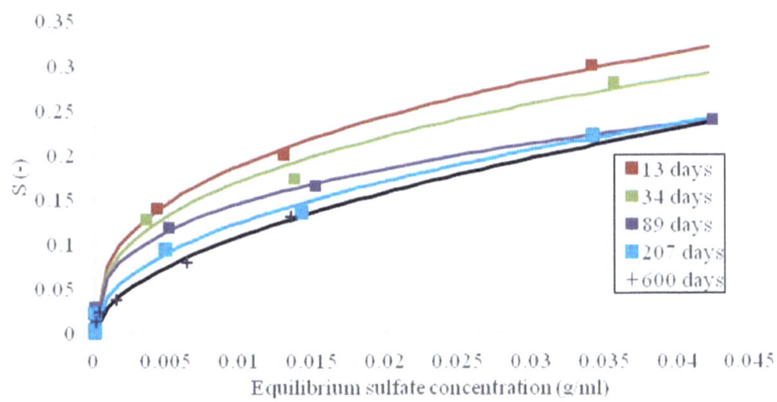

Fig. 8. Sulphate sorption isotherms at different curing (in the legend) of the CB mixture.

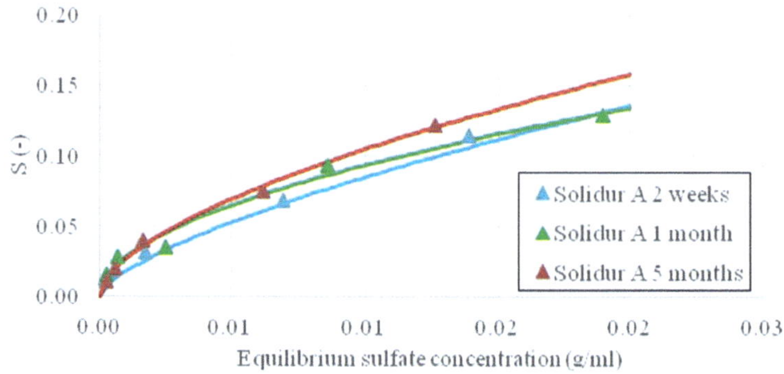

Fig. 9. Sulphate sorption isotherms at different curing time of the Solidur.A mixture

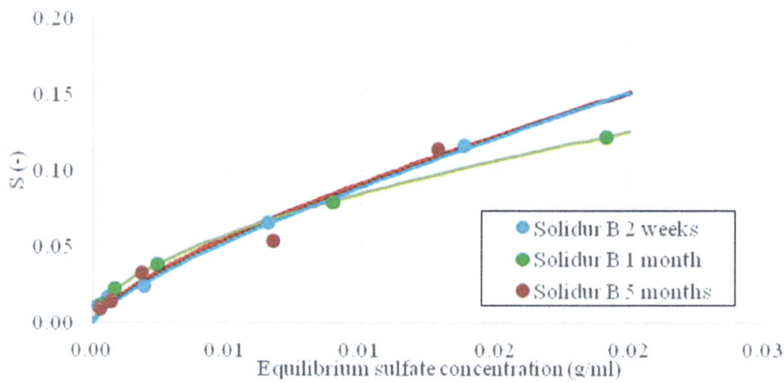

Fig. 10. Sulphate sorption isotherms at different curing time of the Solidur.B mixture

4 Conclusions

For the mixtures investigated, the sorption of K^+ resulted to be noticeable and well described by a linear isotherm; it increases with curing time and solid content

The results showed that choosing a mixture having a very low hydraulic conductivity at brief curing is fundamental to reduce adverse effects on k due to interaction with SO_4^{2-}. In particular, the two special mixtures investigated showed minor interaction with sulphate, resulting in very low k values even after several months of continuous permeation with a solution of K_2SO_4 at the constant concentration of 15 g/l.

The research also showed that the mixture to be used for cut-off walls should be selected considering the performance not only in terms of advection, but also taking into account its sorption capacity.

References

Bensted J (1995) The standardisation of sulphate-resisting cements. Part I + part II. World Cem 7:47–50

ETC8 (1993). Technical recommendation 'GLC'. Geotechnics of landfills and contaminated land. German Geotechnical Society Ed. for the ISSMFE

Fratalocchi E, Brianzoni V, Di Sante M, Mazzieri F, Pasqualini E (2013) Migration of aggressive solutions through cement-bentonite slurry walls. In: Coupled phenomena in environmental geotechnics, CRC Press, pp 243–252

Gollop RS, Taylor HFW (1992) Microstructural and microanalytical studies of sulphate attack. I. Ordinary Portland cement paste. Cem Concr Res 22(6):1027–1038

Gollop RS, Taylor HFW (1995) Micro-structural and microanalytical studies of sulphate attack. III. Sulphate-resisting Portland cement: reactions with sodium and magnesium sulphate solutions. Cem Concr Res 25(7):1581–1590

Hong SY, Glasser FP (1999) Alkali binding in cement pastes. Part I. The C-S-H phase. Cem Concr Res 29:1893–1903

Soga K, Joshi K (2010) Long-term engineering performance of cement-bentonite cut-off walls: a case study. In: Proceedings of VI ICEG, New Delhi, vol 1, pp 151–164

Hydraulic Conductivity of Modified Bentonites After Wet and Dry Cycles

M. De Camillis(✉), G. Di Emidio, A. Bezuijen, and R. D. Verastegui-Flores

Laboratory of Geotechnics, Ghent University, Technologiepark, 905, Zwijnaarde, 9052 Ghent, Belgium
micheladecamillis@gmail.com

Abstract. Geosynthetic clay liners (GCLs) are often used as barriers for landfill liners and cover systems due to the high swelling ability of the bentonite in water that results in low hydraulic conductivity. However, contact with electrolyte solutions and leachate can impede the osmotic swelling of the bentonite and cause an increase in both liquid and contaminant flux through the GCL.

In this framework, different modified bentonites (e.g. multiswellable bentonite, dense pre-hydrated GCL, HYPER clay) have been developed to improve their chemical resistance in aggressive environments.

The purpose of this paper is to provide an overview of the hydraulic performance of the recently developed chemically modified bentonites subjected to wet and dry cycles.

Hydraulic conductivity results of modified bentonites are promising. However, the enhanced performance may still be impaired with aggressive solutions after wet and dry cycles.

Keywords: Bentonite · Hydraulic conductivity · Modified bentonites
Wet and dry cycles

1 Introduction

GCLs are often used as barriers for landfill liners and cover systems due to their high swelling ability that results in low hydraulic conductivity. However, several investigations showed that electrolyte solutions and leachate can impede the osmotic swelling of the bentonite and cause an increase in hydraulic conductivity (Shackelford et al. 2000; Jo et al. 2001; Katsumi et al. 2008; Bradshaw et al. 2016; Tian et al. 2016).

A number of laboratory studies focused on the hydraulic conductivity of bentonites in contact with various permeant liquids (Shackelford et al. 2000; Jo et al. 2005; Meer and Benson 2007; Katsumi et al. 2008; Hosney and Rowe 2014; Tian et al. 2016). To determine the effect of contact of the bentonite with an electrolyte solution, Meer and Benson (Meer and Benson 2007) performed hydraulic conductivity tests on exhumed conventional GCLs demonstrating the detrimental effect of permeation with aggressive liquids on the hydraulic performance. Laboratory studies conducted by Tian et al. (2016) confirmed the harmful effect of leachate on the performance of conventional

L. Zhan et al. (Eds.): ICEG 2018, ESE, pp. 470–480, 2019.
https://doi.org/10.1007/978-981-13-2224-2_58

GCLs and reported an increase of hydraulic conductivity by approximately a factor of 10 during the initial 40 pore volume of flows.

In addition, seasonal temperature changes and rainfalls may further deteriorate the barrier performance of the liners subjected to wet and dry cycles. The consequences of this process are an increase of hydraulic conductivity and loss of self-healing capacity, due to the combination of compression of the diffuse double layer combined with cracks formed during desiccation. For instance, Lin and Benson (2000) assessed the impact of wet and dry cycles on a conventional GCL. The authors reported an increase of the permeability from 3.8×10^{-11} m/s up to 2.8×10^{-9} m/s during the fifth cycle and 7.6×10^{-8} m/s during the sixth cycle with 0.0125 M $CaCl_2$.

In this framework, different types of modified bentonites have been developed in order to improve their chemical resistance in various applications and in presence of aggressive environments.

The purpose of this paper is to provide an overview of the hydraulic performance of the recently developed chemically modified bentonites subjected to wet and dry cycles.

2 Modified Bentonites

2.1 Multi Swellable Bentonite (MSB)

Low hydraulic conductivity of the bentonites contained in GCLs is attributed to osmotic swelling. Osmotic swelling is the attraction of a significant amount of water promoted by a gradient in concentration between the interparticle and the interlayer pore fluid (Likos et al. 2010). The osmotic swelling might be inhibited by high concentrated solutions (Petrov and Rowe 1997; Shackelford et al. 2000; Kolstad et al. 2004; Katsumi et al. 2007). In fact, the exchangeable cations in the interlayers of clays play an important role in the swelling ability of bentonite. If the exchange sites are occupied by multivalent cations, i.e. Ca-montmorillonite, water intercalation is limited and the interlayers decrease due to the attraction of the multivalent cations to the crystal interlayers (McBride 1994).

In this framework, propylene carbonate (PC) can be utilized as a swelling activation material to natural bentonite. Multi-swellable bentonite (MSB) is an amended bentonite produced by Hojun Corporation, Gunma, Japan. The bentonite treatment was developed by Kondo (1996) and the preparation procedure is outlined in Onikata et al. (1999).

The principle is to modify sodium bentonite by adding Propylene Carbonate (PC), an organic compound, in order to activate the osmotic swelling capacity of the clay. Propylene Carbonate is an aprotic polar organic solvent, which means it has a hydrogen atom bound to an oxygen. This feature allows PC to donate protons (H^+) to reagents displaying hydrogen bonding.

MSB is prepared by dissolving PC in water and then adding sodium bentonite. The slurry is then dried and pulverized to obtain a powder. Through this treatment the propylene carbonate is placed in the interlayer of the smectite and attracts numerous water molecules. This enhances swelling ability even in presence of permeant containing polyvalent cations or a high concentration of monovalent cations.

Figure 1 compares the results of an untreated sodium bentonite with MSB containing 45% PC mixed with varying concentration of NaCl (from 0.1 to 2 M NaCl). The treatment improved the basal distance, d(001)-values, up to 1 M NaCl. In particular, the spacing of MSB with 0.75 M NaCl was 45 Å, whereas the spacing of untreated bentonite decreased to 20 Å already with 0.3 M NaCl. Therefore, the range of osmotic swelling of MSB in electrolyte solution can be as high as 0.75 M, which is much higher than the concentration of natural seawater (SW).

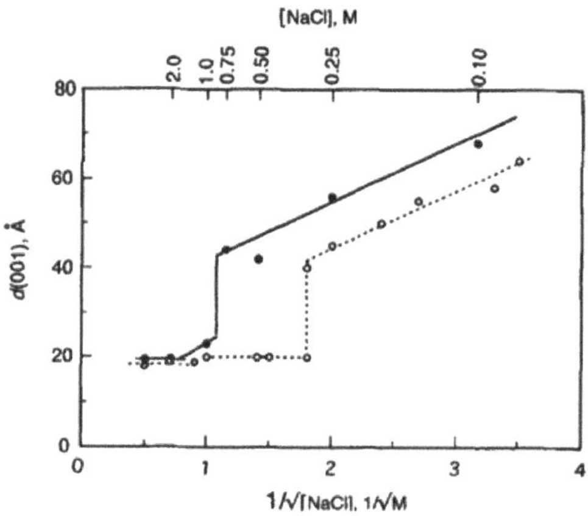

Fig. 1. The d(001)-values of Na-rich montmorillonite in different concentrations of NaCl aqueous solutions with (closed points) and without (open points) PC (Onikata et al. 1999)

Onikata et al. (1999) suggested that the PC molecules are bound to the interlayer cations through water molecules. These molecules are strongly bound to the interlayer cations so that heating at 105 °C cannot remove the water molecules.

2.2 Dense-Prehydrated GCL (DPH GCL)

DPH-GCLs have been developed in the mid-90 s and they are currently available in the market (Flynn and Carter 1998). They combine the beneficial aspects of prehydration and densification on the hydraulic conductivity of sodium bentonite.

The compression of the interlayer is prevented in the short term by prehydrating bentonite with a dilute solution (Shackelford et al. 2000). In this way, water molecules hydrate the clay surface during the crystalline swelling and then they are attracted to the interlayer and osmotic swelling occurs. As a consequence, most of the interlayer water is bound to the clay and it does not take part to hydraulic flow (McBride 1994). Therefore, the hydraulic conductivity of the bentonite is reduced.

DPH-GCL is a factory prehydrated-consolidated sodium bentonite sandwiched between two geotextiles. First, a dilute polymeric solution is prepared by mixing carboxymethyl cellulose (CMC), sodium polyacrylate (Na-PAAS) and methanol. Then, sodium bentonite is added to the liquid in a high shear mixer. The obtained paste is calendered under vacuum into bentonite sheet with low void ratio (e = 1.5). At the end, the bentonite sheet is placed between a high strength woven polypropylene geotextile and a perforated polyester scrim geotextile.

As described in Flynn and Carter (1998), the modified bentonite is made more workable thanks to the addition of methanol. Na-PAAS is used to avoid ion exchange as it replaces the sodium cations coated to the clay, whereas CMC works as anti-fungicidal, lubricant and improves the flexibility of the material. The barrier efficiency is then improved against electrolyte solution. The main feature of DPH GCL is the vacuum calendering (or extrusion) step. During extrusion a pressure is applied to increase the density of the product by eliminating voids. This, together with the vacuum step that has removed air, has the effect of compressing the material to a high density. The molecules are moved slightly closer together promoting the formation of a complex from the intercalated polyacrylate (Flynn and Carter 1998).

2.3 HYPER Clay

HYPER clay technology was developed by Di Emidio (2010). The principle of HYPER clay is to combine powdered bentonite with an anionic polymer (sodium carboxymethyl cellulose, Na-CMC). The treatment consists of dissolving the polymer in water and then adding the base clay. The material is mixed with a mechanical stirrer for 30 min to increase the surface area available for polymer adsorption. This slurry is then oven dried at 105 °C for 16 h to irreversibly adsorb the polymer. The HYPER clay is then ground first manually, using a mortar and pestle, and then mechanically using a Retsch Mortar Grinder RM 200.

Di Emidio (2010) demonstrated through X-ray diffraction that the polymer has intercalated in the interlayer region between platelets and that the interplatelet spacing increases with increasing polymer dosage (Fig. 2). These results suggest the permanence of the polymer in the interlayer region after addition. Moreover, long-term hydraulic conductivity tests showed the enhanced performance of HYPER clay compared to untreated bentonite (Di Emidio et al. 2015).

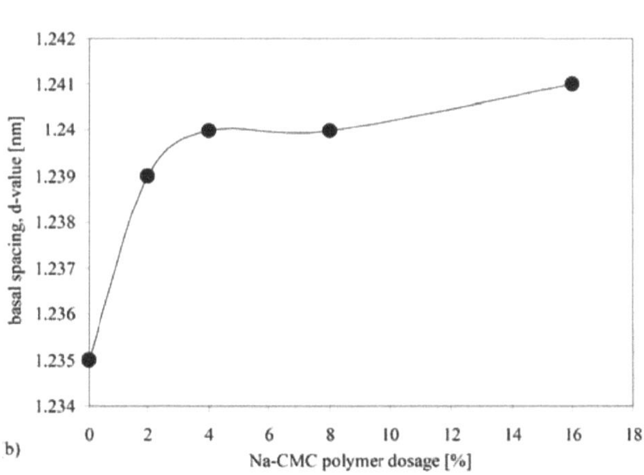

Fig. 2. Increase of the basal spacing of HYPER clay with increasing polymer dosage (Di Emidio 2010)

3 Wet and Dry Cycles on Modified Bentonites: Testing Procedures

Mazzieri and Pasqualini (2008) tested a DPH GCL through four wet and dry cycles with 0.0125 $CaCl_2$ (Ionic Strength, IS = 0.075 M) dried at 35 °C, to represent site conditions, until constant weight. Material characteristics are listed in Table 1. A vertical load of 6 kPa was applied during the cycles to represent an overburden pressure. The samples were rewetted in oedometer cells and the hydraulic conductivity was then measured during each cycle under an effective stress of 12.5 kPa.

Mazzieri et al. (2016) also subjected a DPH GCL and a MSB to five wet and dry cycles with seawater (IS = 0.71 M) prior to test their hydraulic conductivity. An overview of the material characteristics is provided in Table 1. A vertical load of 6 kPa was also applied during the cycles. The wet cycles were performed in oedometer tests until constant swelling was achieved. Test specimens were dried at 35 °C, under 6 kPa, until constant weight. At the end of four wet and dry cycles, the hydraulic conductivity was measured with 12.5 kPa effective stress during permeation.

Table 1. Material characteristics of DPH GCL, MSB and HYPER clay (De Camillis 2018; Di Emidio 2010; Mazzieri et al. 2016)

	DPH GCL	MSB	HYPER clay
Mass per unit area (kg/m^2)	5.43	4.7	4.5
Void ratio CEC [meq/100gr]	1.5 51.8	1.7 52.6	2.45–3.66 90.07
Exchangeable cations			
Ca^{2+}	NA	NA	53.46
K^+	NA	NA	2.89
Mg^{2+}	NA	NA	16.04
Na^+	NA	NA	88.38
Chemical composition (%)			
Al_2O_3	NA	NA	15.76
CaO	NA	NA	5.10
Fe_2O_3	NA	NA	4.04
K_2O	NA	NA	0.74
MgO	NA	NA	2.93
MnO	NA	NA	0.06
Na_2O	NA	NA	3.40
P_2O_5	NA	NA	0.16
SiO_2	NA	NA	47.67
TiO_2	NA	NA	0.71
LOI	NA	NA	19.25

**NA* = not available

The hydraulic conductivity to seawater (IS = 0.77 M) of GCLs containing HYPER clay (GCL_HC) treated with 8% of CMC was measured by De Camillis et al. (2018) following the Standard FprCEN/TS 14417. Material characteristics of HYPER clay are presented in Table 1. Although the Standard recommended a drying temperature of 105 °C, other temperatures were also used to be more representative of the reality and to investigate the impact of drying temperature on the hydraulic performance of GCL prototypes. Specimens 200 by 200 mm were first subjected to four wet and dry cycles and then a sample of 100 mm diameter was cut and tested for its hydraulic conductivity. The cycles consisted of hydrating the GCLs for 48 h and dry them at 40 °C, 60 °C or 105 °C under an overburden pressure of 4 kPa until the water content was lower than 25%. The hydration was either with prehydration with deionized water (DW + 3SW) or with direct contact with the solution (4SW). The hydraulic conductivity to seawater was then measured using an effective stress of 27.5 kPa.

4 Hydraulic Conductivity Tests

At the end of the wet and dry cycles, the specimens were inspected and transferred to flexible wall permeameter for permeability testing. This permeameter is connected to Pressure Control Panels by means of a bladder accumulator. The bladder accumulator also allows to sample the outlet for chemical analysis.

The following formula is used to determine the hydraulic conductivity based on the ASTM D5084:

$$k = \frac{aL}{2A\Delta t} \ln\left(\frac{h_1}{h_2}\right)$$

where:

K = hydraulic conductivity of the material [m/s]
a = area of the reservoirs containing the inflow and outflow liquid [m^2]
L = Length of the specimen [m]
A = cross-sectional area of the specimen [m^2]
h1 = Head loss across the specimen at t1 [m]
h2 = Head loss across the specimen at t2 [m]

5 Wet and Dry Cycles on Modified Bentonites: Overview of Literature Results

Figure 3 shows an overview of literature results of modified bentonites subjected to wet and dry cycles (Mazzieri and Pasqualini 2008; Mazzieri et al. 2016; De Camillis et al. 2018). As described in more detail later on in this section, test results include different test conditions.

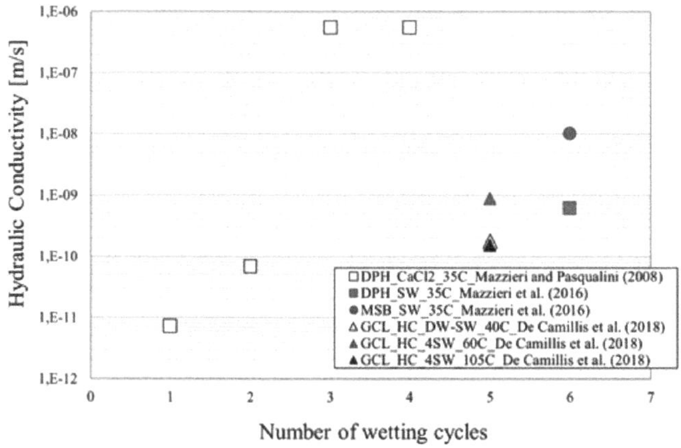

Fig. 3. Hydraulic conductivity of a DPH GCL throughout wet and dry cycles with 0.0125 $CaCl_2$ (data from (Mazzieri and Pasqualini 2008). Hydraulic conductivity of a DPH GCL and a MSB to seawater after five wet and dry cycles (data from (Mazzieri et al. 2016). Hydraulic conductivity of a GCL containing HYPER clay to seawater after four wet and dry cycles (data from (De Camillis et al. 2018)

As shown in Fig. 3, the hydraulic conductivity of DPH_$CaCl_2$ with the calcium chloride solution remained low after the second wetting cycle. Thereafter, the permeability increased strongly to a final value of 5.5×10^{-8} m/s. A possible explanation is the occurrence of cation exchange that converted the bentonite into a calcium exchanged form. This was demonstrated by testing a small portion of dry bentonite through swell index test (Mazzieri and Pasqualini 2008). The authors reported a swell index of 9 ml/2 g that is a typical value of a calcium bentonite.

Additionally, small desiccation cracks that occurred during drying might not be healed upon rehydration. Although preferential flow paths were not found using an anionic tracer in the permeant solution, microcracks invisible to the naked eye were probably present, that were responsible for the high permeability. The obtained hydraulic conductivities are very similar to the values obtained by Lin and Benson (2000) for a conventional GCL.

A companion paper was published by Mazzieri (2011) to simulate milder desiccation states. Six wet and dry cycles were performed on DPH GCL samples with 0.0125 M $CaCl_2$ and visible cracks appeared already at dehydration water content of 65%. It was noticed that the size of the cracks increased with decreasing water content, as described also by Thiel et al. (2006). The hydraulic performance of the DPH GCL subjected to partial desiccation with 0.0125 M $CaCl_2$ was affected when the water content at the end of the drying cycles was < 35%, whereas for water contents > 45% the material appeared able to recover the hydraulic performance.

The effect of wet and dry ageing in seawater of a DPH GCL and of a MSB was investigated by Mazzieri et al. (2016). Permeation of the DPH GCL with seawater led to an increase of the hydraulic conductivity of about three orders of magnitude compared to the unaged DPH GCL permeated with SW ($2 \times 10^{-10-1} \times 10^{-9}$ m/s versus

$1\text{–}2 \times 10^{-12}$ m/s, respectively). The difference in permeability was linked to the presence of cracks found during a post-test visual inspection.

The hydraulic conductivity of a MSB was measured at the end of the five wet and dry cycles and it resulted on the order of 10^{-8} m/s (Fig. 3). Considering for simplicity seawater as a NaCl solution, the Na^+ concentration would be 0.501 M which is lower than the limit for which MSB is known to exhibit osmotic swelling (0.75 M). However, the sample was affected by crack formation and the reduction of the swelling ability was likely linked to the elution of the amendment, to the presence of cracks and to the increase in electrolyte concentration beyond 0.75 M NaCl. Therefore, bonding of the amending agents to the clay appeared to be non-permanent.

De Camillis et al. (2018) described the behavior of HYPER clay (Figs. 3 and 4) after four wet and dry cycles in seawater at different drying temperatures (40 °C, 60 °C and 105 °C). However, high drying temperature are unlike in the field unless of extreme conditions, such us for a GCL placed under a black geomembrane (Rowe 2014).

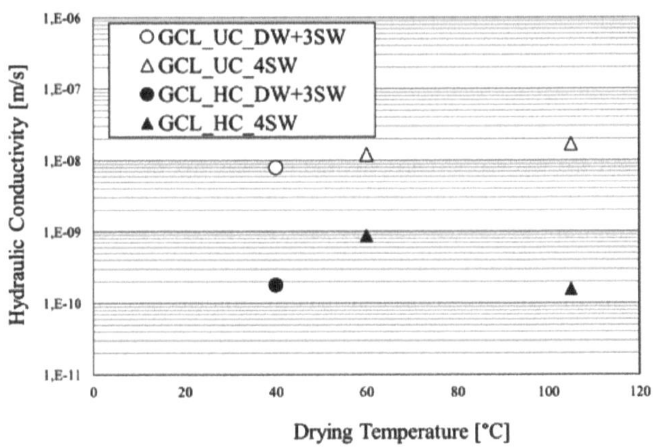

Fig. 4. Hydraulic conductivity of GCL prototypes containing untreated bentonite (GCL_UC) and HYPER clay 8% (GCL_HC) after four wet and dry cycles with prehydration (DW + 3SW) and direct exposure to seawater (4SW) at 40 °C, 60 °C and 105 °C (data from (De Camillis et al. 2018)

The hydraulic conductivity of HYPER clay was almost constant at each drying condition (as shown in Fig. 4) and it resulted to be 1.79×10^{-10} m/s, 8.83×10^{-10} m/s and 1.6×10^{-10} m/s at 40 °C, 60 °C and 105 °C respectively. The slightly lower permeability at 40 °C compared to 60 °C might be related more to the prehydration than to temperature impact.

Contrary to expectation, a clear trend of increasing permeability with increasing drying temperature was not found. On the contrary, the impact of different drying temperature was associated to the self-healing timing rather than to the hydraulic conductivity (as shown in (De Camillis et al. 2018).

Figure 3 shows that DPH GCL and GCL_HC presented hydraulic conductivities to seawater after wet and dry cycles lower than the limit value of 10^{-9} m/s at given boundary conditions.

Nevertheless, the GCLs tested by De Camillis et al. (2018) were: (1) subjected to an effective stress of 27.5 kPa, versus 12.5 kPa of the DPH GCLs and MSB, (2) hydrated with the most aggressive solution (IS = 0.77 M (De Camillis et al. 2016) versus IS = 0.075 M (Mazzieri and Pasqualini 2008) and IS = 0.71 M (Mazzieri et al. 2016)), (3) subjected to the highest drying temperatures.

In addition, the needle-punched GCL used with HYPER clay might produce more cracks compared to the MSB poured in the oedometer as powder bentonite and the DPH-GCL (not needle-punched) where the bentonite and geotextiles are united by simply pressing them together. However, all specimens tested presented cracks after being subjected to wet and dry cycles irrespective of solution concentration and dry temperature.

6 Conclusions

This paper summarized the results of research conducted to evaluate the hydraulic performance after wet and dry cycles of modified bentonites (DPH GCL, MSB and HYPER clay) used for hydraulic and chemical containment applications.

At given boundary conditions, DPH GCL and GCL containing HYPER clay still presented a permeability to seawater lower than the limit value of 10^{-9} m/s. However, all specimens tested presented cracks after being subjected to wet and dry cycles irrespective of solution concentration and drying temperature.

With respect to previous research results on modified bentonites, wet and dry cycles have been shown to influence the hydraulic performance of the materials investigated. However, the extent of this behavior varies depending on the boundary conditions: effective stress, type and concentration of salt solutions, number of cycles, test procedure. Contrary to expectation, the drying temperature did not show to have an impact on the hydraulic conductivity, but it influenced the healing time. Care is recommended on the testing conditions to be used to better represent the reality: e.g. standard wet and dry testing should be performed at actual temperatures expected in the site.

Despite modified bentonites showed promising performances, further work needs to be done to establish the influence of wet and dry cycles on the hydraulic behavior of these innovative materials.

References

Bradshaw SL, Asce AM, Benson CH, Asce F, Rauen TL (2016) Hydraulic conductivity of geosynthetic clay liners to recirculated municipal solid waste leachates. J Geotech Geoenviron Eng 142(2):04015074

De Camillis M (2018) Experimental and numerical study of wet and dry cycles on an innovative polymer treated clay for geosynthetic clay liners. PhD thesis, Ghent University

De Camillis M, Di Emidio G, Bezuijen A, Verástegui-Flores RD (2018) Impact of wet-dry cycles on the swelling ability and hydraulic conductivity of a polymer modified bentonite. In: 11th international conference on geosynthetics, Seoul

Di Emidio G (2010) Hydraulic and chemico-osmotic performance of polymer treated clays. PhD thesis, Ghent University

Di Emidio G, Mazzieri F, Verastegui-Flores RD, Van Impe W, Bezuijen A (2015) Polymer-treated bentonite clay for chemical-resistant geosynthetic clay liners. Geosynth Int 22(1):1–13

Flynn BN, Carter GC (1998) Waterproofing material and method of fabrication therefor

Hosney MS, Rowe RK (2014) Performance of GCL after 10 years in service in the arctic. J Geotech Geoenviron Eng 140(10):04014056

Jo HY, Benson CH, Shackelford CD, Lee J, Edil TB (2005) Long-term hydraulic conductivity of a geosynthetic clay liner permeated with inorganic salt solutions. J Geotech Geoenviron Eng 131(4):405–417

Jo HY, Katsumi T, Benson CH, Edil TB (2001) Hudraulic conductivity and swelling of nonprehydrated GCLs permeated with single-species salt solutions. J Geotech Geoenviron Eng 127(6):557–567

Katsumi T, Ishimori H, Ogawa A, Yoshikawa K, Hanamoto K, Fukagawa R (2007) Hydraulic conductivity of nonprehydrated geosynthetic clay liners permeated with inorganic solutions and waste leachates. Soil Found Jpn Geotech Soc 47(1):79–96

Katsumi T, Ishimori H, Onikata M, Fukagawa R (2008) Long-term barrier performance of modified bentonite materials against sodium and calcium permeant solutions. Geotext Geomembr 26(1):14–30

Kolstad DC, Benson CH, Edil TB (2004) Hydraulic conductivity and swell of nonprehydrated geosynthetic clay liners permeated with multispecies inorganic solutions. J Geotech Geoenviron Eng 130(12):1236–1249

Kondo M (1996) Method of activation of clay and activated clay

Likos WJ, Bowders JJ, Gates WP (2010) Mineralogy and engineering properties of Bentonite. In: Geosynthetic clay liners for waste containment facilities, pp 31–53

Lin LC, Benson CH (2000) Effect of wet-dry cycling on swelling and hydraulic conductivity of GCLs. J Geotech Geoenviron Eng 40(1):40–49

Mazzieri F (2011) Impact of desiccation and cation exchange on the hydraulic conductivity of factory–prehydrated GCLs. In: Geo-frontiers 2011: advances in geotechnical engineering, ASCE, Reston, VA, pp 976–985

Mazzieri F, Di Emidio G, Pasqualini E (2016) Effect of wet-and-dry ageing in seawater on the swelling properties and hydraulic conductivity of two amended bentonites. Appl Clay Sci 142 (6):40–51

Mazzieri F, Pasqualini E (2008) Effect of dry/wet cycles and cation exchange on the permeability of a dense prehydrated GCL. In: Fourth international conference on geosynthetic edited by N. Dixon

McBride MB (1994) Environmental chemistry of soils. Oxford University Press, New York

Meer SR, Benson CH (2007) Hydraulic conductivity of geosynthetic clay liners exhumed from landfill final covers. J Geotech Geoenviron Eng 133(5):550–563

Onikata M, Kondo M, Hayashi N, Yamanaka S (1999) Complex formation of cation-exchanged montmorillonites with propylene carbonate: osmotic swelling in aqueous electrolyte solutions. Clays Clay Miner 47(5):672–677

Petrov RJ, Rowe RK (1997) Geosynthetic clay liner (GCL) - chemical compatibility by hydraulic conductivity testing and factors impacting its performance. Can Geotech J 34(6):863–885

Rowe RK (2014) Performance of GCLS in liners for landfill and mining applications. Environ Geotech 1(1):3–21

Shackelford CD, Benson CH, Katsumi T, Edil TB, Lin L (2000) Evaluating the hydraulic conductivity of GCLs permeated with non-standard liquids. Geotext Geomembr 18(2–4):133–161

Tian K, Benson CH, Likos WJ (2016) Hydraulic conductivity of geosynthetic clay liners to low-level radioactive waste leachate. J Geotech Geoenviron Eng 142(8):1–12

Inducing Hydrophobicity to Improve Long Term Engineering Performance of Kaolinite Clay

Aisha M. S. Haquie(✉) and Megan L. Hart

University of Missouri Kansas City, Kansas City, MO 64110, USA
amh8x4@mail.umkc.edu, hartme@umkc.edu

Abstract. Engineered compacted earthen liners are significantly important for the disposal of municipal and industrial solid wastes. Compacted clay liners strain during seasonal wet and dry fluctuations, providing pathways for leachate to contaminate surrounding soil and groundwater. Although geosynthetic clay liners are utilized, GCLs may be infeasible in areas that undergo cyclic wet and dry seasonal or groundwater fluctuations. This study focuses on improving the properties of kaolinite silty-clay by minimizing permeability through induced hydrophobicity. Siloxane, a water repellent material, was mixed with kaolinite soil at concentrations between 0.5% to 20% by weight to determine optimal induced hydrophobicity for retention of mass during cyclic wetting and drying. Hydrophobicity was measured using sessile drop methodology and water drop penetration time. Fine particles are preferentially influenced, resulting in soil gradation differences on the portion passing No. 200 sieve. Resistance to water penetration was improved, resulting in greater adherence between soil particles without inducing any crystalline structural changes. Volume losses and moisture absorption from wetting and drying cycles were reduced in optimally treated siloxanated clay molds, as compared to untreated samples. These experimental results indicate siloxane treated kaolinite clay possess improved engineering characteristics, and have potential to improve the useful lifetime of clay liners.

Keywords: Kaolinite · Siloxane · Hydrophobic · Contact angle
Grain size · Wetting and drying cycle · SEM

1 Introduction

Clay soils complicate engineering design for civil infrastructure due to their potential for shrinking and swelling, low permeability, or alteration of composition caused by external impetus such as deicing fluids, construction, spills, etc. Buildings and other structures built on clay soils are potentially prone to differential settlement, which may instigate significant structural damage. However, the low permeability of clays has been exploited for use in lagoons or landfills applications as barrier liner systems designed to prevent the migration of unwanted chemicals or materials leaking into the greater environment. Municipal solid waste disposal landfills have traditionally relied upon the use of compacted earthen liners or barriers for isolation of the waste material. Solid waste disposal landfill sites have greatly relied upon the use of compacted clay

L. Zhan et al. (Eds.): ICEG 2018, ESE, pp. 481–488, 2019.
https://doi.org/10.1007/978-981-13-2224-2_59

barriers for safe isolation of the waste materials (Suter et al. 1993). Due to small particle size, high specific surface area, low permeability, and complex porous structure, naturally occurring clay minerals show the lowest permeability of traditionally used geomaterials (Choi et al. 2016; Cuevas et al. 2009). Traditional low permeability clay soils are used in landfill barriers in order to comply with the Code of Federal Regulations requirements of preventing the migration of hazardous materials, and possess a hydraulic conductivity of no more than 1×10^{-7} cm/s (40 CFR 264.301).

With active monitoring and maintenance, designed clay liners may function effectively throughout their lifecycle, but natural physical and biological processes can be expected to cause the barrier to fail in the long term (Suter et al. 1993). Seepage of pollutants and contaminants through the clay lining barriers, and also diffusion of heavy metals, has been reported in a number of studies (Goodall and Quigley 1997; Johnson et al. 1989; Zhang et al. 2004; Chalermyanont et al. 2009). Potential for expansion and shrinkage during wetting and drying cycles under load can induce strain and fractures, providing pathways for leachate to contaminate soil and pollute surrounding soil and groundwater. High desiccation rates can cause the development of strain energy that gives rise to simultaneous dense cracking (Costa et al. 2013). In order to prevent leakage, strain or desiccation cracks must be prevented. One method of preventing shrinkage is to modify the clay to decrease permeability by the addition of hydrophobic materials.

Hydrophobic treatments have been shown to reduce permeability in expansive clays, as well as reducing the potential to swell and shrink (Daniels and Hourani 2009; Daniels et al. 2009). Kaolinite is typically considered a non-expansive clay, but is found abundantly throughout the continental United States and Canada and is used extensively for liner systems due to its relative abundance (Tong 2000; King 2005; Hamilton 1980, Suter et al. 1993). For this study, naturally occurring soil was directly mixed with increasing concentrations of siloxane by weight in order to determine if engineering properties were enhanced.

2 Materials and Methods

This study focuses on improving the engineering performance and properties of the regionally extensively Greenton silty clay by inducing hydrophobicity, thus minimizing permeability to water. The Greenton silty clay consists of very deep to deep, poorly drained, impermeable soils formed in thick loess layers and the underlying residuum from clay shales and thinly bedded limestone (NRCS). Greenton silty clay was formed on upland slopes which range from 2 to 25% (NRCS). While the taxonomic designation indicates that the soil is smectite in derivation, only kaolinite was present in the soil used in this study.

2.1 Materials

Greenton silty clay for the study was obtained locally from an active road cut located at 39°09′20.7″N 94°30′23.6″W in North Kansas City, Missouri. Soil samples were visually classified as brownish grey in color with fine-grained texture and classified as a SC, silty clay (ASTM D2488). Greenton soil drape is deep and consists of multiple thin stratum throughout the 65 foot road cut exposure. A representative, disturbed sample was collected over the face of the exposure, requiring several iterative sieve analyses to be conducted in order to represent an "average" for the soil. Specific gravity of the soil was averaged and measured to be 2.64 g/cm^3 (ASTM D854). The finer portions of the soil were classified via Atterberg limits as medium plastic materials with a liquid limit of 50 and a plasticity index of 15 (ASTM D422). The in situ moisture content of the collected aggregate sample was approximately 15% (ASTM D2216.) Siloxane used in this study was commercially available, which is an organic based silane, having an appearance of cloudy white liquid, odorless, with a specific gravity of 0.96. Type I deionized water (ASTM D1193), having 18.0 MΩ-cm resistivity, was used in all soil mixtures.

2.2 Sample Preparation

Collected soil was first thoroughly air dried in shallow, large pans, followed by crushing in a jaw rock crusher in order to disaggregate the rock like clumped fragments formed during drying of plastic soils. Further refinement of the soil was conducted using mortar and pestle to insure that the soil was thoroughly powdered. Dried soil was thoroughly mixed with siloxane, by stirring micropipetted siloxane at 0.5%, 2.5%, 5%, 7.5%, 10%, 15% and 20% by weight to dried soil. Deionized water is then added to the soil-siloxane mixture until a fluid consistency was reached. The mixture was then stirred for 2 min with a Hobart mechanical stirrer at medium speed, to insure complete dispersion and thorough incorporation of siloxane with the soil particles. The samples were then oven dried at 105 °C for 24 h prior to being re-ground with mortar and pestle prior to further experimental testing.

2.3 Methods

To investigate the engineering effects of water repellency and the viability of use of siloxane, various experiments were carried out on the following characteristics of soil: specific gravity, particle size distribution, soil-water contact angle, and the wetting and drying cycle. All the experiments were carried out using ASTM standards.

Scanning Electron Microscopy. The structures of the original clay soil and the 5 wt% and 10 wt% siloxanated clay were analyzed using scanning electron microscopy (SEM) under 50 microns magnification. Energy Dispersive X-ray Spectroscopy (EDS) was also carried out on the sample with the same resolution to provide an estimate of qualitative and quantitative compositions of various elements present.

X-ray diffraction (XRD) was used to ascertain the mineralogy of the aggregated soil. The soil consisted primarily of kaolinite and quartz bearing minerals, with minor amounts of potassium feldspars. X-Ray diffraction showed no crystallographic alteration of the clay particles when coated with 5% by weight siloxane, which indicates that the siloxane coats the external basal and edge surfaces of the particles but does not penetrate the actual minerals. An idea of the characteristics depth of field can be found from the high resolution SEM images, for the control, and 5% and 10% siloxanated clay are shown in Figs. 1, 2 and 3. It can be seen from the SEM images that the

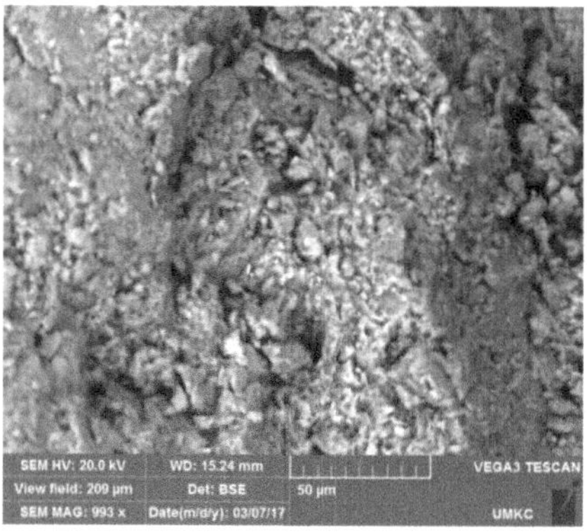

Fig. 1. Natural kaolinite clay

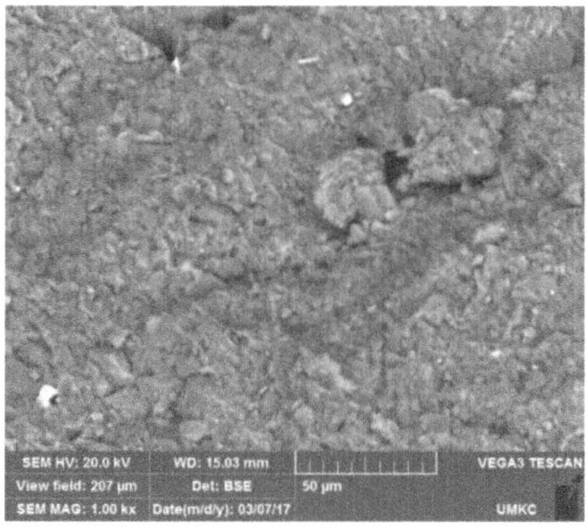

Fig. 2. Kaolinite mixed with 5% siloxane by mass

siloxane causes formation of smoother clay particles with greater face surfaces and lesser edge surfaces, thus reducing the permeability through the soil. At 5% siloxane concentration, the clay particles seem closer together and form smooth surfaces. At 10% concentration of siloxane, the scanned surface showed much more smooth showing better coverage of siloxane over the clay particles, thus enhancing the water repellency property.

Specific Gravity and Particle Size Distribution. A 25 ml water pycnometer was used to find the specific gravity of soil samples (ASTM D854). Approximately 5 grams of each, namely, the unsiloxanated clay, 5 wt% siloxanated clay and 10 wt% siloxanated clay were used in the experiment. Hydrometer analysis was used to find the distribution of the fines (particles passing sieve no. 200 mesh) (ASTM D422).

Soil Water Repellency. Soil water repellency was measured using the soil-water contact angle. The contact angle was determined by the sessile drop technique (SDT), which measures the contact angle between water droplet and the soil surface, as shown in Fig. 4. 15 grams of each sample were first crushed thoroughly using mortar and pestle, and kept on a glass plate within a circular area of 20 cm^2 to ensure consistency in thickness of all the samples. Using a 5 ml syringe, one drop of water was placed on the sample, and a high resolution camera was used to capture the side view image, as soon as the drop falls on the soil surface. Contact angle was then measured by processing the image with image analysis software called ImageJ (Rasband 1997).

Wetting and Drying Cycle. The standard wetting and drying tests for compacted soil samples was carried out (ASTM D559). Four cylindrical molds were prepared, each with a diameter of 2 inches and length of 4 inches approximately. They were compacted at optimum moisture content which was experimentally measured. Two samples were made with the untreated clay and another two samples was made with 7.5 wt% siloxanated clay. The optimum amount of siloxane, 7.5 wt%, as found from the contact

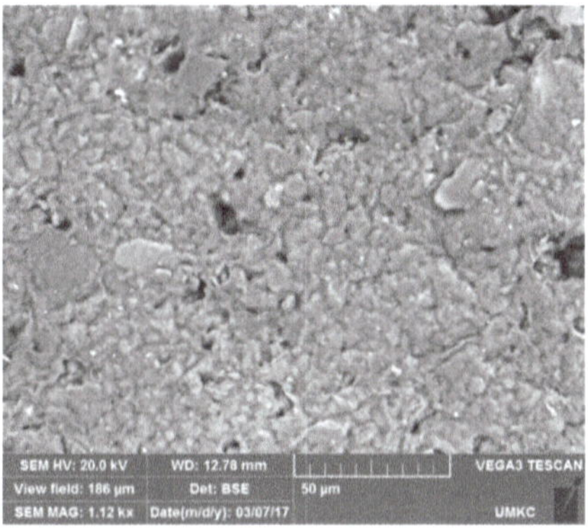

Fig. 3. Kaolinite mixed with 10% siloxane

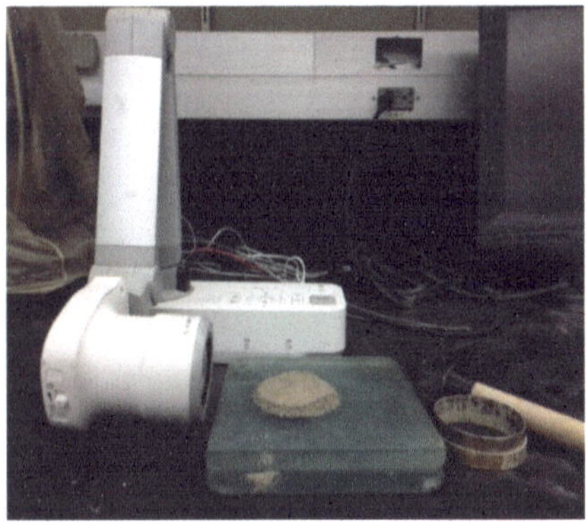

Fig. 4. Experimental set up for contact angle measurement

angle measurement, was thus used to prepare the experimental specimens. The top and bottom of the samples were capped with porous stones with filter papers in between. The samples were then wrapped with plastic film to avoid disintegration during wetting cycles.

After measuring the initial mass, the samples were submerged in water for 8 h which is the wetting cycle at controlled temperature of 25 °C, followed by drying cycle, which is keeping the samples in oven at 70 °C for 48 h. The cycles were carried out until sample failure, where the eventually sample breaks from increasing strain. The temperature is enough to slowly disintegrate the sample and ultimately break it into two or three pieces, but not aggressive enough to change the structure of the soil. The mass and volume of the sample molds at each cycle were recorded.

3 Results and Discussion

3.1 Specific Gravity and Particle Size Distribution

Preliminary work indicates that there were no changes in the coarse portions of the soil materials, which is consistent with literature (de Jonge et al. 1999). However, the fine particles passing Sieve no. 200, which is about 40% of the clay soil, showed significant changes with increasing siloxane concentrations. From Fig. 5, it can be seen that at 0.01 mm grain size, the percentage passing for untreated clay, 5 wt% and 10 wt% of siloxanated clay, is 38%, 34% and 15% respectively. Results from Pearson correlation analysis showed high correlations between water repellency indices and smaller grain sizes (de Jonge et al. 1999; Wijewardana et al. 2015).

It was also found that increasing siloxane reduced the specific gravity of the clay. The kaolinite had a specific gravity of 2.64, whereas the 5% and 10% siloxane clay had a value of 1.82 and 1.51, respectively. The reason for the decrease in the specific gravity can be because the siloxane makes the clay particles water repellent, thus making them more buoyant. The reduction in specific gravity indicates that the siloxane

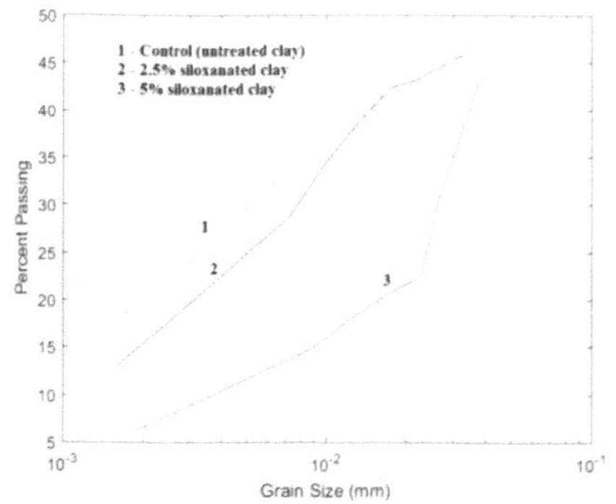

Fig. 5. Hydrometer analysis on fine particles passing sieve no. 200

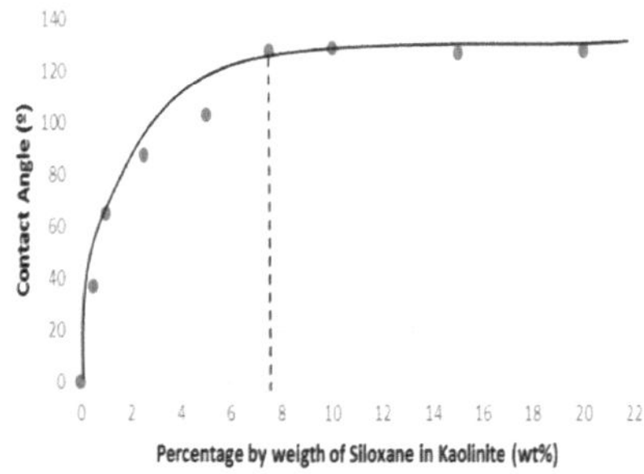

Fig. 6. Plot showing contact angles at various concentrations of siloxane

has a significant effect on the buoyancy of the material as well as the potential for water repellency which may be problematic for long term performance of siloxane modified clays.

3.2 Soil Water Repellency

Figure 6 illustrates the contact angle measured at various concentration of siloxane in kaolinite clay. It can be seen from the plot that, the contact angle increases up to a maximum of 128º and then remains constant. Therefore, 7.5 wt% of siloxane in kaolinite gives the optimum level of hydrophobicity since it demonstrates the maximum contact angle. The most accurate measurement for soil water repellency can be made from the measurement of contact angle as soon as the water drop falls on the material (TAPPI 1997). The siloxane layer on the kaolinite contributes to a low free surface energy, due to which the contact angles increases. The angle also depends on the roughness of the surface and the chemical heterogeneity of the solid involved (Lourenço et al. 2015).

3.3 Wetting and Drying Cycle

Data indicated that the mass and volume loss of the sample during six cycles was mitigated by application of siloxane. No apparent swelling was visible in the specimens during the wetting cycle, but there was a gain in weight due to the absorption of water. Shrinkage was seen in all the specimens during drying cycles. In the first cycle, there was about 6.3% gain in mass in the control specimens, compared to 4.3% in the specimens treated with siloxane. After the second wetting cycle, the control had a 13.5% weight gain, and the siloxanated specimens had a 11.5% weight gain. The weight gain reduced from the third cycle, being at 7.3% for the control specimen, compared to 5% for the siloxanated specimen which remained consistent in the fourth and fifth cycle. It is seen that the percentage weight gain in the control specimens is always higher and is also fairly constant. The specimens treated with siloxane showed a much reduced gain in moisture, due to their water repellency property.

During the drying cycles cracks were visible, both vertically and horizontally along the specimens. Loss of mass and volume was clearly visible since there is considerable amount of soil materials crumbling off the edges and surfaces of the control molds, and very little amount of the soils falling off the edges from the siloxanated specimens. After the fifth drying cycle, the control specimens failed, whereas the siloxanated specimens lost 26% weight by mass. Although the samples did not show much swelling during the wetting cycle, there was progressively more shrinkage seen in the samples as the drying cycles increased.

It was also found out that during the wetting and drying cycles, the permeability of the experimental samples increased, showing reduction of the strength of hydrophobicity. This was carried out using water drop penetration test, where the time taken for water droplet to completely penetrate the sample was measured. The average time reduced from 105 s at the beginning of the cycle, to 85 s at the fifth cycle.

4 Conclusions

The optimum hydrophobicity that can be achieved using siloxane in kaolinite was evaluated in the study through measurements of contact angles at various concentrations of siloxane. The following conclusions can be drawn from the experimental results:

- Mixing siloxane in kaolinite results in physical adsorption of siloxane onto the surface of kaolinite, without changing the crystalline structure of the clay, as seen in the XRD results.
- An optimum amount of 7.5 wt% of siloxane in kaolinite was found which presented a maximum contact angle of 128°.
- With the increasing concentration of siloxane, the grain size of fines is seen to increase, and the specific gravity decreases.
- The wetting cycle showed an average moisture gain of 7% by mass for the control specimens compared to 3.7% mass gain for the siloxane treated specimens.
- The control specimens failed after the fifth drying cycle while the siloxane treated specimens had a mass loss of 26% after the fifth drying cycle.

References

ASTM D1193-06 (2011) Standard Specification for Reagent Water, ASTM International, West Conshohocken, PA. www.astm.org

ASTM D2216-10 (2010) Standard Test Methods for Laboratory Determination of Water (Moisture) Content of Soil and Rock by Mass, ASTM International, West Conshohocken, PA. www.astm.org

ASTM D2488-17 (2017) Standard Practice for Description and Identification of Soils (Visual-Manual Procedures), ASTM International, West Conshohocken, PA. www.astm.org
ASTM D422-63(2007)e2 (2007) Standard Test Method for Particle-Size Analysis of Soils (Withdrawn 2016), ASTM International, West Conshohocken, PA. www.astm.org
ASTM D559/ D559 M-15 (2015) Standard Test Methods for Wetting and Drying Compacted Soil-Cement Mixtures, ASTM International, West Conshohocken, PA. www.astm.org
ASTM D854-14 (2014) Standard Test Methods for Specific Gravity of Soil Solids by Water Pycnometer, ASTM International, West Conshohocken, PA. www.astm.org
Choi Y, Choo H, Yun TS, Lee C, Lee W (2016) Engineering characteristics of chemically treated water-repellent kaolin. Materials 9(12):978
Cuevas J, Leguey S, Garralón A, Rastrero MR, Procopio JR, Sevilla MT, SánchezJiménez N, Abad RR, Garrido A (2009) Behavior of kaolinite and illite-based clays as landfill barriers. Appl Clay Sci 42(3):497–509
Daniels JL, Hourani MS (2009) Soil improvement with organo-silane. In: Advances in ground improvement: research to practice in the United States and China, pp 217–224
Daniels JL, Hourani MS, Harper LS (2009) Organo-silane chemistry: a water repellent technology for coal ash and soils. In: Proceedings, pp 4–7
de Jonge LW, Jacobsen OH, Moldrup P (1999) Soil water repellency: effects of water content, temperature, and particle size. Soil Sci Soc Am J 63(3):437–442
Hamilton JJ (1980) Behavior of expansive soils in western Canada. In: Expansive Soils, pp 815–833. ASCE
King MK (2005) Expansive soil and expansive clay. https://geology.com/articles/expansive-soil.shtml. Accessed 26 Mar 2018
Lourenço SDN, Woche SK, Bachmann J, Saulick Y (2015) Wettability of crushed air-dried minerals. Géotechnique Letters 5(3):173–177
NRCS - Natural Resources Conservation Service, United States Department of Agriculture. https://soilseries.sc.egov.usda.gov/OSD_Docs/G/GREENTON.html. Accessed on 26 Mar 2018
Rasband WS (1997) ImageJ, U. S. National Institutes of Health, Bethesda, Maryland, USA. http://imagej.nih.gov/ij/, 1997–2018
Suter GW, Luxmoore RJ, Smith ED (1993) Compacted soil barriers at abandoned landfill sites are likely to fail in the long term. J Environ Qual 22(2):217–226
TAPPI (1997) T 558 om-97 Surface wettability and absorbency of sheeted materials using an automated contact angle tester
Tong KW (2000) Introduction to clay minerals and soils. http://www.oakton.edu/user/4/billtong/eas100/clays.htm. Accessed 26 Mar 2018
Wijewardana NS, Kawamoto K, Moldrup P, Komatsu T, Kurukulasuriya LC, Priyankara NH (2015) Characterization of water repellency for hydrophobized grains with different geometries and sizes. Environ Earth Sci 74(7):5525–5539
The State Duma (2009) Technical regulations on the safety of buildings and structures, The State Duma, Moscow, Russia

Influence Factors of Swelling Characteristics of GMZ Bentonite

Yufeng Gao[1,2(✉)], Yuemiao Liu[1,2], Jingli Xie[1,2], Shengfei Cao[1,2], Like Ma[1,2], and Qiang Tong[1,2]

[1] CNNC Key Laboratory on Geological Disposal of High-Level Radioactive Waste, Beijing, China
hdyygaoyf@163.com
[2] Beijing Research Institute of Uranium Geology, Beijing 100029, China

Abstract. The swelling characteristics of bentonite is important to evaluate the long-term performance of deep geological repository of nuclear waste. A series of wetting tests on GMZ sodium and calcium bentonite were performed to investigate the swelling characteristics under different conditions. The influences of initial dry density and particle size were studied at different temperature. The test results show that the swelling deformation due to wetting is mainly affected by the dry density and little affected by particle size. The swelling ratio of GMZ bentonite increases with the increase of initial dry density. However, only the swelling ratio of sodium bentonite increases with the rise of temperature. The process of swelling deformation is a single or multi-stage mode depending on the initial dry density. Finally, the sample of the bentonite is subjected to a rapid swelling with the increase of the initial dry density. The swelling property of sodium bentonite is much higher than that of calcium bentonite. All results are very important for the study of swelling properties of GMZ bentonite.

Keywords: GMZ bentonite · Swelling characteristics
Temperature · Initial dry density · Particle size

1 Introduction

Swelling characteristics is the key performance of buffer/backfill material in deep disposal of high-level radioactive waste [1]. GMZ bentonite has been chosen as the first choice for using as buffer/backfill materials for HLW repository in China [2, 3]. The basic functions of the materials used in the waste repositories request among others a sufficient swelling property and low hydraulic conductivity in order to provide long-term stability to the barrier system under environmental pressure and behavior of the waste loads [4]. Therefore, it is of great practical significance to study the influence factors of swelling characteristics of GMZ bentonite.

In recent years, many scholars have carried out some researches on the swelling characteristics of bentonite under different conditions. Some foreign scholars studied the influence of temperature on the swelling pressure [5–7] and swelling deformation [8, 9] of compacted FEBEX bentonite and MX-80 bentonite respectively. Some domestic scholars carried out research on the swelling property of GMZ bentonite at

L. Zhan et al. (Eds.): ICEG 2018, ESE, pp. 489–496, 2019.
https://doi.org/10.1007/978-981-13-2224-2_60

different temperature [10, 11]. Temperature changes affect important swelling pressure, swelling and collapse behavior [12–14].

In this paper, the free swelling tests of bentonite with different initial dry density were carried out at different temperatures to study the effects of temperature and initial dry density on swelling properties of GMZ bentonite.

2 Swelling Test

2.1 Materials

The tests have been performed with the GMZ bentonite, which was screened through 200 mesh (sodium bentonite (200) and calcium bentonite) or 80 mesh (sodium bentonite (80)). The main mineral composition of bentonite used for testing is montmorillonite, and the associated minerals are quartz, feldspar and muscovite. The cation exchange capacity (CEC) of sodium bentonite varies from 59 mmol/100 g, and the major exchangeable cations are: Ca (18 mmol/100 g), Mg (11 mmol/100 g), Na (29 mmol/100 g) and K (1 mmol/100 g). The cation exchange capacity (CEC) of calcium bentonite varies from 58 mmol/100 g, and the major exchangeable cations are: Ca (29 mmol/100 g), Mg (20 mmol/100 g), Na (8 mmol/100 g) and K (1 mmol/100 g).

2.2 Free Swelling Test

Free swelling test of compacted bentonite (1.0–1.9 Mg/m^3) were carried out at different temperature (25 °C, 40 °C, 50 °C, 60 °C) to study the swelling behavior of GMZ bentonite. The compacted samples are placed in the metal device with diameter of 50 mm. Two pairs of non-woven fabric and porous stones are covered top and bottom surfaces of the specimens and the supply deionized water drained through the bottom porous stone. The swelling deformation is measured by dial gauge which minimum value is 0.01 mm to understand the swelling change of GMZ bentonite under no loading condition. Firstly, it remains free to swell to stable at the room temperature (25 °C), and then changes the deionized water temperature to test the swelling deformation.

3 Results and Analysis

3.1 Temperature

The relationship between swelling ratio of sodium bentonite and elapsed time for various temperature are shown in Fig. 1. In this figure, the free swelling test is divided into two stages, namely free swelling at room temperature (25 °C) from the beginning to the 123rd day and free swelling under variable temperature later (Fig. 1b). Figure 1b shows the deionized water temperature setting process is 25 °C, 40 °C, 25 °C, 40 °C, 25 °C, 50 °C, 25 °C, 50 °C, 60 °C, 25 °C, 60 °C, 25 °C, and the corresponding time node is 0, 123rd day, 132nd day, 141st day, 145th day, 146th day, 153rd day, 161st day, 163rd day, 166th day, 167th day, 168th day. At initial stage of swelling deformation test swelling rate of sample was high, then the swelling rate was basically unchanged from the 35th day to the 123rd day. The swelling ratio increased significantly when the deionized water temperature was set to 40 °C at the 123rd day. However, the swelling rate remained constant when the temperature is down to 25 °C. The swelling ratio increased to varying degrees when the temperature rose to 40 °C, 50 °C and 60 °C.

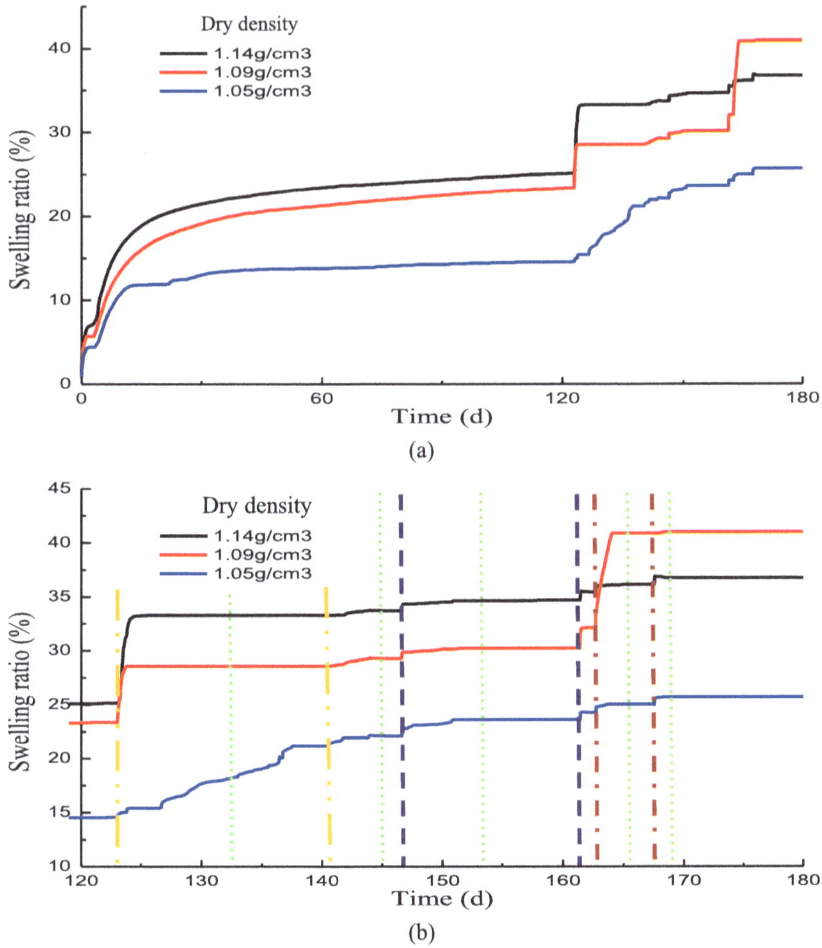

Fig. 1. The curves of swelling ratio of sodium bentonite (80) at different temperature conditions.

Figure 2 shows the swelling ratio of calcium bentonite of initial dry density of (1.0–1.9) Mg/m^3 at different temperature conditions. The free swelling test of calcium bentonite is only 48 days less than that of sodium bentonite in the stage of free swelling at the room temperature (25 °C), what the rest are the same. At initial stage of swelling deformation test swelling rate of sample was high, then the swelling rate was basically the same when the temperature increased to 40 °C, 50 °C and 60 °C. In Fig. 2, it is difficult to find the effect of temperature on the swelling property of calcium bentonite.

This study finds the temperature has greater influence on the swelling property of sodium bentonite. The swelling ratio of sodium bentonite increases in varying degrees with the increase of the temperature. Zhang et al. [10] have found that the swelling strain decreases with the increase of temperature; the swelling duration decreases with the increase of temperature, while the influence of temperature on the swelling rate is different under vertical loads. However, Ye et al. [11] have confirmed the swelling pressure of GMZ01 bentonite decreases as the pH value of NaOH solutions increases, while the decreasing rate significantly depends on temperature.

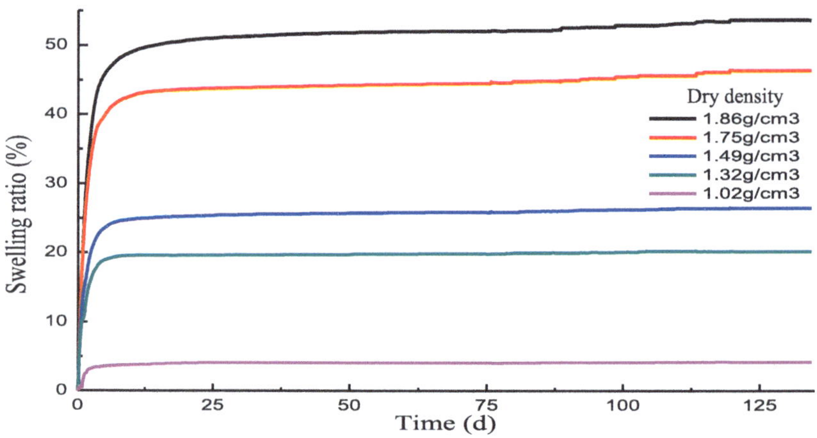

Fig. 2. The curves of swelling ratio of calcium bentonite with different initial dry density at different temperature conditions.

3.2 Dry Density

The experimental results of the relationship between swelling ratio and dry density of sodium bentonite (80) are shown in Fig. 3. In Fig. 3, it is clear that higher dry density provides higher swelling ratio. As can be seen in Fig. 3a, the process of increasing the swelling ratio is different at initial stage of free swelling deformation test. The sample of the bentonite with dry density greater than 1.66 Mg/m^3 is subjected to a rapid swelling. However, a slow stage will appear when the dry density is less than 1.66 Mg/m^3. Figure 3b shows the sample will swell in the "three stage" model: (I) The first stage is initial swelling stage, and the sample swells rapidly. (II) The second stage is the stagnation stage. The swelling rate decreases sharply when the swelling ratio reaches a certain value, and a "platform" appears in the time curve. At this stage, no swelling or swelling is very slow. (III) The third stage is the re-swelling stage. The specimen will swell again until the stability is faster. Figure 4 shows the relationship between swelling ratio and initial dry density of (1.05–1.86) Mg/m^3 at the room temperature (25 °C). The swelling properties of sodium bentonite in Fig. 4 is the same as that in Fig. 3. The samples with similar initial dry density are similar in swelling ratio.

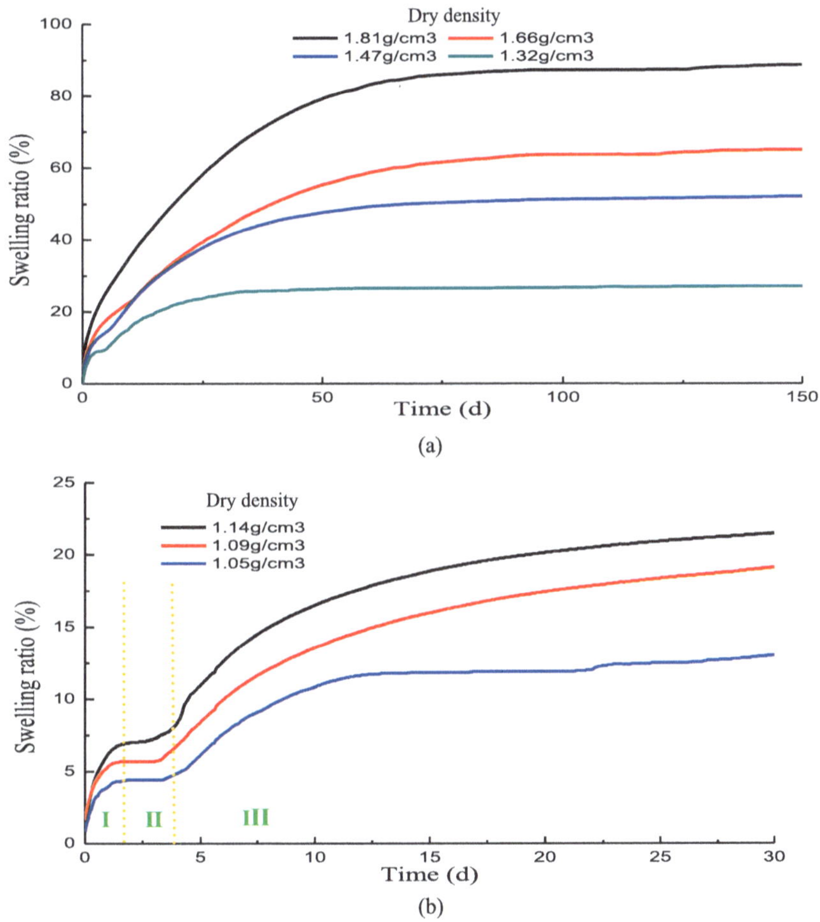

Fig. 3. The curves of swelling ratio of sodium bentonite (80) with different initial dry density.

Figure 5 shows the relationship between swelling ratio and initial dry density of calcium bentonite at the room temperature (25 °C). It is obvious that the swelling ratio of calcium bentonite increases with the increase of initial dry density. The variation law of calcium bentonite with dry density greater than 1.32 Mg/m^3 in Fig. 5 is the same as that of sodium bentonite in Fig. 3. However, the swelling ratio of sodium bentonite is higher than that of calcium bentonite with similar initial dry density. What's more, the variation in the swelling rate of calcium bentonite with dry density less than 1.04 Mg/m^3 is different from the sodium bentonite at initial stage of free swelling test. The swelling ratio of calcium bentonite will go through "four stage", which reduced first before the re-swelling stage, and the swelling ratio distribution shows similar to a "double-peak" structure (Fig. 5b).

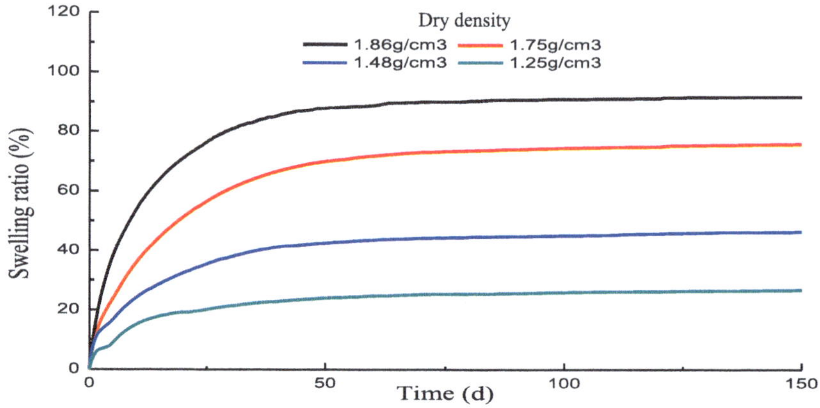

Fig. 4. The curves of swelling ratio of sodium bentonite (200) with different initial dry density.

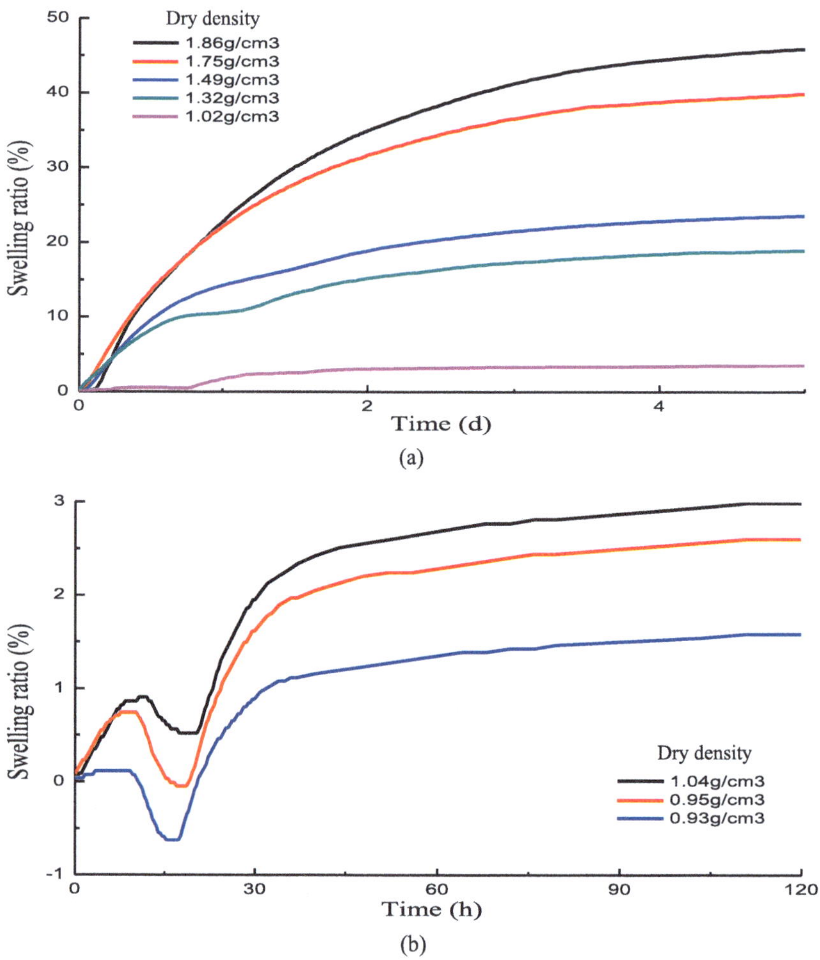

Fig. 5. The curves of swelling ratio of calcium bentonite with different initial dry density.

This study shows the swelling ratio increases with the increase of initial dry density. The process of increasing the swelling ratio is a single or multi-stage mode depending on the initial dry density. When the initial dry density is relatively small, the swelling of sample takes on a multi-stage mode. Finally, the sample of the bentonite is subjected to a rapid swelling with the increase of the initial dry density. The maximum swelling pressure of the mixture is exponential increase with initial dry density [15].

3.3 Component

Montmorillonite is the main component of bentonite, and it is the main factor affecting swelling performance. Research shows that different component of bentonite has obvious difference in swelling property. The contrast of Figs. 3 and 5 show that the swelling ratio of sodium bentonite is higher than that of calcium bentonite with similar initial dry density. Next, the influence of different montmorillonite content on swelling property of GMZ bentonite will be studied.

3.4 Particle Size

This study shows the change of particle size may have little effect on the swelling property due to the particle size of bentonite was very small. In Figs. 3 and 4, It is clearly found that the swelling ratio of the sodium bentonite with the same initial dry density is similar, whether it was screened through 200 mesh or 80 mesh. The swelling property of sodium bentonite is higher than that of calcium bentonite with similar initial dry density when those were screened through 200 mesh (Figs. 4 and 5a).

4 Conclusions

Based on the results of the free swelling tests of the GMZ bentonite with different initial dry density at different temperature conditions, the following conclusions can be drawn.

Temperature and initial dry density of GMZ bentonite are the important factor to influence the swelling behavior. The temperature has greater influence on the swelling property of sodium bentonite. However, it has no obvious effect on the swelling property of calcium bentonite. The swelling ratio increases with the increase of initial dry density. The process of increasing the swelling ratio is a single or multi-stage mode depending on the initial dry density. Finally, the sample of the bentonite is subjected to a rapid swelling with the increase of the initial dry density. The swelling deformation due to wetting is mainly affected by the initial dry density and little affected by particle size.

References

1. Liu YM, Wen ZJ (2003) Study on clay-based materials for the repository of high level radioactive waste. Mineral Petrol 23(4):42–45
2. Liu YM, Chen ZR (2001) Bentonite from Gaomiaozi, inner Mongolia as an ideal buffer/backfilling material in handling highly radioactive wastes—a feasibility study. Acta Mineral Sin 21(3):541–543
3. Wang J, Su R, Chen WM, Guo YH et al (2006) Deep geological disposal of high-level radioactive wastes in china. Chin J Rock Mech Eng 25(4):649–658
4. Liu YM, Cai MF, Wang J (2007) Compressibility of buffer material for HLW disposal in China. Uranium Geol 23(2):91–95
5. Pusch R, Karlnland O, Hokmark H (1990) GMM-a general microstructural model for qualitative and quantitative studies of smectite clays. SKB, Stockholm, Sweden, pp 90–94
6. Lloret A, Romero E, Villar MV (2004) FEBEX II project: final report on thermos-hydro-mechanical laboratory tests. Enresa, Madrid, Spain, pp 62–72
7. Villar MV, Lloret A (2008) Influence of dry density and water content on the swelling of a compacted bentonite. Appl Clay Sci 39(1):38–49
8. Arifin YF (2008) Thermo-hydro-mechanical behavior of compacted bentonite-sand mixture: an experimental study. Bauhaus University, Weimar, Germany
9. Villar MV, Lloret A (2004) Influence of temperature on the hydro-mechanical behaviour of a compacted bentonite. Appl Clay Sci 26:337–350
10. Zhang FZ, Fang ZD, Qin B et al (2017) Study on bentonite's swelling strain behaviour at different temperatures. J Logist Eng Univ 33(2):23–27
11. Ye WM, Zheng ZJ, Chen B et al (2014) Effects of pH and temperature on the swelling pressure and hydraulic conductivity of compacted GMZ01 bentonite. Environ Earth Sci 68 (1):192–198
12. Sultan N, Delage P, Cui YJ (2002) Temperature effects on the volume change behavior of Boom clay. Eng Geol 64:135–145
13. Shirazi SM, Kazama H, Chim-oye W (2006) Temperature and density effect on swelling characteristics and permeability of bentonite. Aust Geomech 41(2):89–100
14. Onal M (2007) Swelling and cation exchange capacity relationship for the samples obtained from a bentonite by acid activations and heat treatments. Appl Clay Sci 37:74–80
15. Zhang HY, Cui SL, Liu JS et al (2010) Experimental study of swelling pressure of compacted bentonite-sand mixture. Rock Soil Mech 31(10):3087–3095

Laboratory Investigation of Anisotropic Shrinkage of HLW Buffer Block: An Attempt Using Digital Image Correlation Method

Huyuan Zhang[1,2(✉)], Yu Tan[1], Dongjin He[1], and Gang Luo[1]

[1] School of Civil Engineering and Mechanics, Lanzhou University, Rd 222 South Tianshui, Lanzhou 730000, Gansu, China
zhanghuyuan@lzu.edu.cn

[2] Key Laboratory of Mechanics on Disaster and Environment in Western China (Lanzhou University), Ministry of Education, Lanzhou 730000, China

Abstract. Anisotropic shrinkage of compacted bentonite blocks threatens the security of a buffer barrier in a geological disposal project for high-level radioactive wastes (HLW). In this investigation, fan-shaped buffer blocks on an industrial scale were compacted and exposed to an indoor environment to dry naturally. Digital image correlation (DIC) was used in an attempt to monitor the blocks' shrinkage deformation in a vertical (z) direction, and dial indicators were used to measure horizontal (r and θ) shrinkage for comparison. Before and after the blocks desiccation, pore size distributions were investigated by mercury intrusion porosimetry (MIP) to analyze the shrinkage behavior in the microstructure, and the micro-fissuring was investigated by X-ray computed tomography (CT) in terms of internal visual structure. The initial evaporation stage on the industrial scale and over compaction of the blocks was not detected. In horizontal directions, isotropic shrinkage strains were observed, which were lesser than 2%. Linear shrinkage strains decreased with the increase of the initial lengths of the measuring lines, indicating that block shrinkage is mainly concentrated on the external part, while the internal part shows poorer shrinkage ability. CT images showed that no obviously micro-fissuring was found in a block before drying it and MIP results demonstrated that both macrostructure and microstructure at a 5 cm depth slightly changed after block desiccation. Vertical shrinkage, monitored by DIC, exhibited a poorer ability compared to the horizontal direction, which had almost equal linear strains though its initial measuring lines lengths were shorter. In future research, more attention should be paid to the lateral shrinkage of buffer blocks.

Keywords: High-level radioactive waste (HLW) · Bentonite block Desiccation · Anisotropy shrinkage · Digital image correlation (DIC)

1 Introduction

Deep geological disposal is widely accepted as a promising way to dispose of high-level radioactive wastes (HLW) in line with the development of nuclear science. In Chinese disposal project, compacted GMZ bentonite block can be used to build an

L. Zhan et al. (Eds.): ICEG 2018, ESE, pp. 497–505, 2019.
https://doi.org/10.1007/978-981-13-2224-2_61

engineering barrier and its quality guarantees the security of the disposal project. Other countries, such as Spain [1], Czech Republic [2], and Sweden [3], have attached great significance to the buffer block research. If bentonite blocks were exposed to the atmosphere during it storage and transportation, they would dry and shrink due to the relatively higher compacted moisture content [4]. Therefore, in the construction of buffer barrier, blocks joints will widen in these shrunken blocks and the function of the barrier, such as hyposmosis, will be seriously damaged.

Research studies on the shrinkage of buffer blocks on an in industrial scale are limited. SKB [4] reported the shrinkage of compacted bentonite specimens in a laboratory that were exposed to 35% relative ambient humidity. Different shrinkages between blocks and specimens may be found because of scale effects. Peron [5] focused on the desiccation of a paste-like, fine-grained soil, but such work conducted on paste-like specimens cannot replace the research of compacted blocks. Moreover, anisotropic shrinkage is enlarged by the blocks industrial scale. There is an urgency to guarantee the quality of buffer blocks in terms of shrinkage to participate with a plan for building a Chinese underground laboratory for HLW disposal in 2020.

Considering large-scale buffer blocks, there is a lack of accuracy in measuring their shrinkage strain by traditional equipment, such as a digital caliper. Since digital image correlation (DIC) was put forward [6, 7] to precisely measure the full-field strain, many researchers [8, 9] have made contributions to this technique being applied to engineering to measure concrete strains and cracks.

The purpose of this work is to investigate the anisotropic shrinkage of buffer blocks comprised bentonite and sand. These were compacted on an industrial scale and exposed in an indoor condition to monitor the desiccation shrinkage and moisture content. In a vertical direction, block shrinkage strain was measured by DIC, and the horizontal shrinkage strain was measured by dial indicators for comparison. Mercury intrusion porosimetry (MIP) and X-ray computed tomography (CT) were used to investigate the pore size distributions and visual fissures, respectively. Base on the test results, the three-direction (3D) shrinkage characteristics of buffer blocks are noted and now the priorities for further research on buffer block can be suggested.

2 Materials and Methods

2.1 Material Characterization

The GMZ bentonite used in this work was mined in Inner Mongolia, China. As reported, quartz sand was added to bentonite at a 7:3 mass ratio, which improves the buffer materials' properties, such as reinforcing thermal conductivity [10] without undermining the buffer's permeability [11] and swell property [12]. The quartz sands were obtained from Yongdeng, China. Material properties are summarized in Table 1.

2.2 Buffer Block Preparations

Until now, the construction of an engineering buffer has not been determined for China's HLW disposal. An idealized conception of a buffer barrier (Fig. 1), with patent

Table 1. Basic properties of bentonite and quart sand used.

GMZ bentonite					Sand		
Particle diameter	Air dry moisture contents	Liquid limit	Plastic limit	Specific gravity	Particle diameter	Silica contents	Specific gravity
Main < 2 (μm)	9 (%)	160.7 (%)	30.65 (%)	2.70 (g/cm^3)	0.2–2.0 (mm)	>99 (%)	2.65 (g/cm^3)

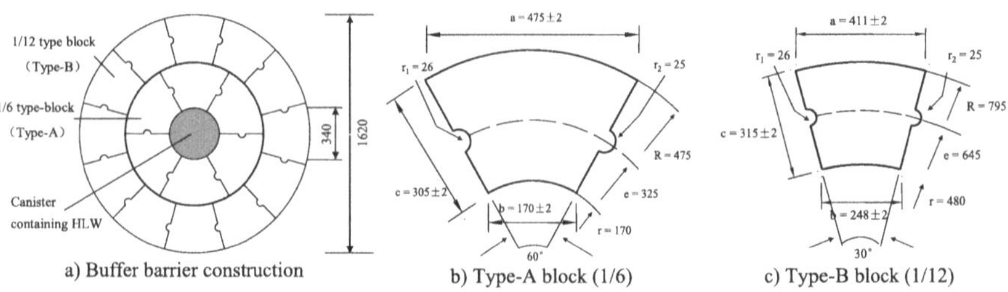

Fig. 1. The conception of constructing the compacted blocks together (unit: mm)

authorization, was put forward following other in-country projects [1, 2, 13]. Bentonite-sand mixtures were compacted to prepare Type-B blocks at about 10 cm height by using a uniaxial compressor in this work. Before the compaction, bentonite powders and quartz sands with a 7:3 mass ratio were uniformly mixed to obtain 18% moisture contents after spraying water and were then sealed for preservation in plastic bags for 48 h to reach a uniformly humidifying. After compaction, blocks were sealed in plastic wraps for 1 month to uniformly humidify and stabilize the shape.

2.3 Desiccated Condition

As reported, the Beishan area in Gansu Province is a promising site for disposing of HLW in China [14]. In this setting, the local climate is a key factor for block shrinkage during transportation and storage. In this work, the blocks were desiccated at the relative humidity of the atmosphere of Lanzhou, where is the capital of Gansu Province. Test results have contributed to the quality control for these buffer blocks.

2.4 Shrinkage Strain Measurements

After correlating certain subsets of pixels in the reference image with their homologous ones in the deformation image [15, 16], DIC measures strain by catching special pixels created by speckles. In this work, white paint was spotted on a block vertical surface and then black points were added on these white spots to create speckles. By this method, not field strain but linear strain on a block was measured vertically by DIC. A quarter of the block was used to measure the strain (Fig. 2a). Another quarter of the block at the same ambient was used to record moisture content by interval weighing,

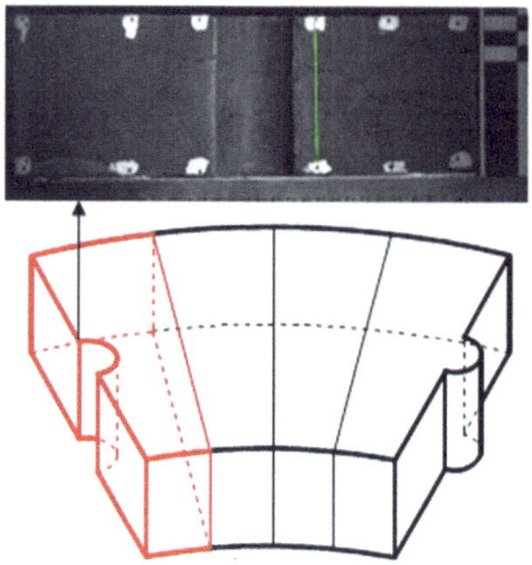

Fig. 2a. Block shrinkage monitor in vertical direction by DIC

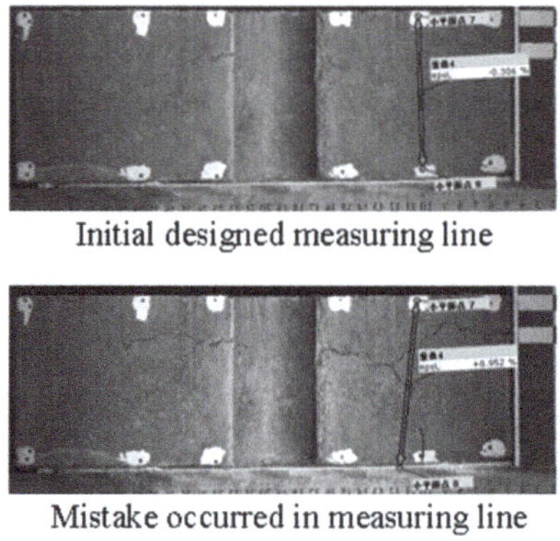

Fig. 2b. An example of a DIC mistake

measure the vertical height by a digital caliper (0.01 mm precision) for comparison. The height of the quarter block was the average value of six measuring lines to minimize a measuring error. In this work, the horizontal shrinkage strain of a buffer block (unpublished) is also presented as a comparison.

2.5 X-Ray CT Tests

X-ray CT is a non-invasive technique that can be used to visualize the interior of objects based on the attenuation of an electromagnetic wave. This technology has been widely used in geotechnique and soil science [17–19]. To observe the internal of the block apparently, the spatial resolution of the CT device is 24 LP/cm (i.e., 0.208 mm), the block is sawed into a cuboid segment of about 10 × 10 × 20 cm in size.

3 Experimental Results

3.1 Evaporation Process

In this test, a quarter of a block and a whole block were dried at indoor condition at Lanzhou and their moisture contents were recorded by interval weighs as showed in Fig. 3. These two curves follow a similar variation trend, which decrement rates of moisture content gradually decrease as shown by the curves tending to horizontal. In general, the evaporation curve can be divided into an initial stage, a falling stage and a residual stage [20]. At the initial stage, the evaporation rate almost stays constant which is accompanied by a linear decrease of moisture content as time lapsed. At the falling stage, the water evaporation rate sharply decreases and then gradually becomes constant by turning into the residual stage. However, the initial stage was not observed in block evaporation as shown in Fig. 4. As reported, the falling evaporation rate is the result of the evaporating interface transforming from the material surface to inside. Hierarchical evaporation occurs in buffer blocks due to their industrial scale, such as a block's external part evaporating initially and its internal part evaporating later. The internal part, a large portion of an industrial scale block, exhibits falling evaporation for a long period in comparison to the external part's evaporation. This phenomenon leads to trouble in detecting the short duration of the initial evaporation stage. Besides, moisture in an overly compaction block lacks connectivity, so its internal moisture is hard to transform in a timely way to the block surface to keep up a constant evaporate rate. This feature also contributes to the undetectability of the initial evaporation stage.

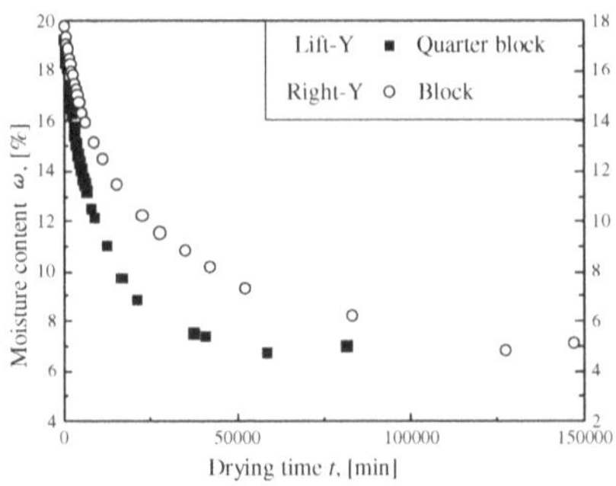

Fig. 3. Evolution of water content with time

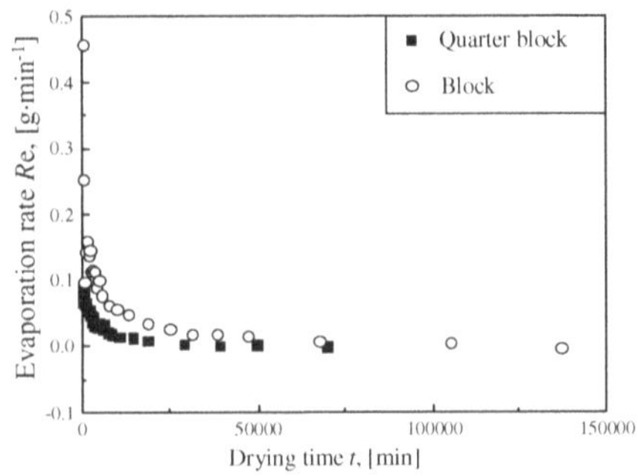

Fig. 4. Evolution of evaporation rate with time

3.2 Shrinkage Strain

According to preliminary research, isotropy horizontal shrinkages in r and θ directions were measured by a specially designed device equipped with 20 fixed dial indicators. Linear shrinkage strain in each direction takes the average value of 5 measuring lines by 10 fixed dial indicators, and the average initial length of measuring lines in r and θ direction is 315 mm and 336 mm, respectively.

In a vertical direction, speckles were painted to create 6 measuring lines to monitor block vertical shrinkage by DIC. However, these speckles were not always precisely recognized by DIC as shown in Fig. 2b. After carefully inspecting the analysis process, we determined only 1 measuring line combined 2 speckles in a vertical direction, and this was adopted for the analysis that showed no obvious mistake.

Vertical linear shrinkage strains as measured by both DIC and digital caliper were presented in Fig. 5, as well as strains in r and θ direction. Parallel blocks, used to measure vertical strain and horizontal strain, respectively, were slightly different in the initial moisture due to imprecise block preparation. For a better comparison, top-X axis was used to make strain curves start at the same point. In Fig. 5, vertical strain measured by DIC presents almost the same variations with a horizontal direction at initial shrinkage. And the result of vertical shrinkage strain measured by digital caliper is similar to DIC results, but it is indeterminate because the whole shrinkage in a vertical direction is only about 2 mm. The final horizontal strains are less than 2%.

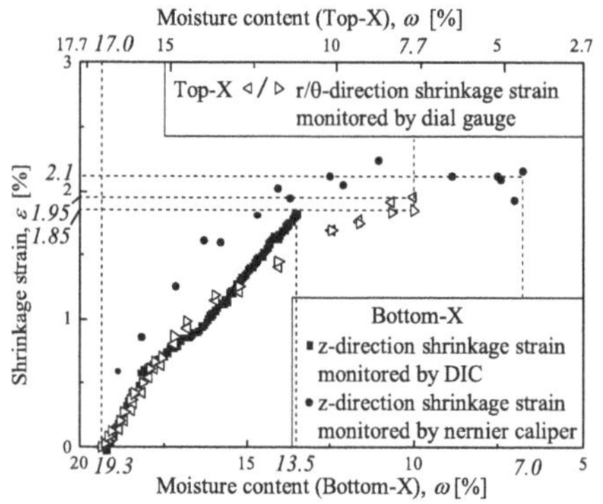

Fig. 5. Shrinkage strain curves in r/θ/z direction

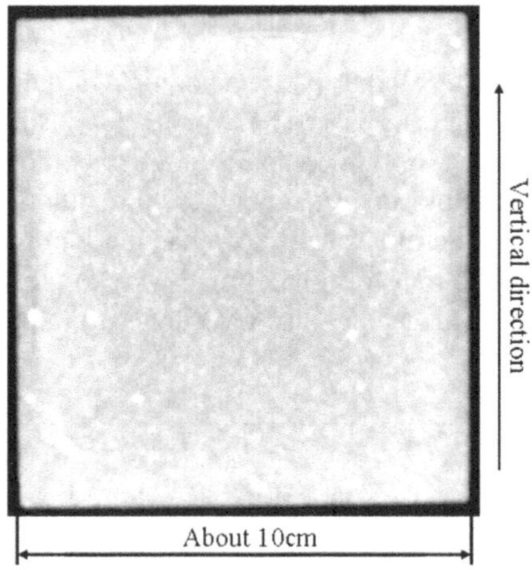

Fig. 6. An example of CT image for a part of block with 18% moisture content

4 Discussions

In the horizontal shrinkage of a block, linear strains decreased with an increase of the initial length of measuring lines. While shrinkage variations of each measuring lines with a different initial length were nearly identical. It is conjectured that block shrinkage strain is mainly concentrated on the external part and the internal part shows poor shrinkage ability.

In order to test this conjecture, pore distributions of block internal parts before and after block desiccation were investigated by MIP tests. A dried specimen was taken at 5 cm depths under a dried block top face and side face, and an un-dried parallel block was used to obtain an un-dry specimen. Specimens should be freeze-dried to keep their porous characteristics without any seriously shrinkage during moisture desorption. Comparing the cumulative pore volume obtained from the un-dried and dried block, Fig. 7a shows that the intruded mercury volume of a block after drying is remarkable higher than for a block before drying. However, the mine difference of intruded mercury volume cumulated at the initial of MIP tests, and the peaks of pore size density curves of blocks before and after drying were similar (as showed in Fig. 7b) compared to Hashmei [21] results. According to Lloret [22], pore size around 0.06 μm and 6 μm can be recognized as a microstructure and a macrostructure respectively, and pore sizes larger than 200 μm (Fig. 7b) are considered as visual structures. It can be distinguished that the block's macrostructure is basically unchanged and its microstructure has slightly accumulated after desiccation, while the visual structure remarkably increased. Visual structure detected by a CT scan uses a cuboid segment taken from an undried block. No visual fissures were detected by CT images (Fig. 6) as the MIP tests have shown. The visual structure in un-dried block tested by MIP is considered as the result

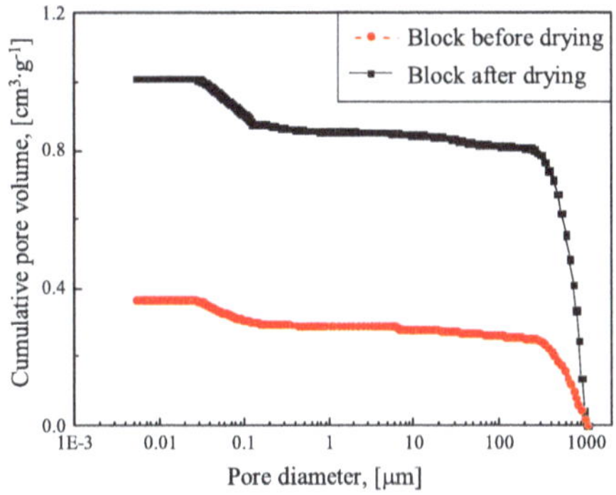

Fig. 7a. Cumulative pore volume curves

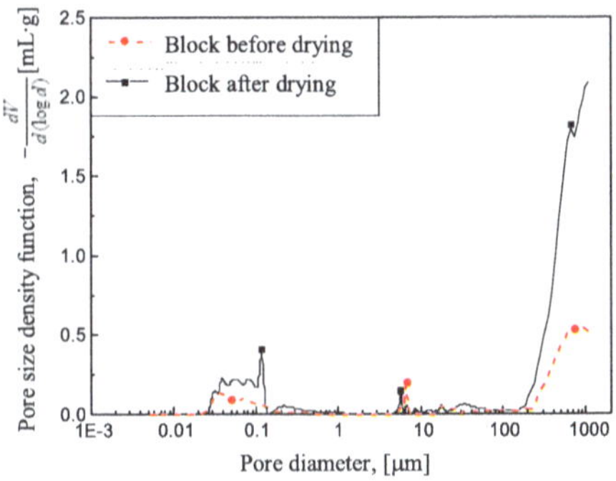

Fig. 7b. Differential pore volume curves

of a disturbance during sample preparation. The visual structure of dried blocks cannot change seriously (Fig. 7b) given the circumstances that the macrostructure and the microstructure were not changed remarkably after drying. So a conclusion can be drawn that shrinkage is apparently not detected by MIP and CT tests at a 5 cm depth of a block, which corresponds to our conjecture based on the results of the horizontal strain.

Figure 5 shows that linear shrinkage strain in a vertical direction corresponds to the strain in a horizontal direction. However, the initial length of the vertical measuring line, around 10.5 cm, is shorter than the average horizontal lengths. Even take an un-vaporized bottom into consideration, measuring line initial length in a vertical direction is still shorter than half a length of horizontal measuring lines. As remarkable shrinkage was not detected below the 5 cm depth of the block, shrinkage strain should increase with the decrease of measuring line's initial length when the shrinkage ability keeps constant. The correspondence of vertical and horizontal shrinkage means that with a shorter initial length, the block has poorer shrinkage ability. As a result, poorer shrinkage ability in a vertical direction corresponds to the situation that compaction was exerted in a vertical direction, and uniaxial compaction leading to well orientation of smectite platelets [23] may restrain the vertical shrinkage of blocks.

5 Conclusions

(1) The initial stage was not detected in block evaporation and so the constant evaporation ratio did not keep periods in comparison to the longer duration of the falling and residual stages. Block's distinguishing features, industrial scale, and over-compaction contributes to this result of bringing about hierarchical evaporation from the block's external part to its internal part and restricting the connectivity of the block moisture.

(2) 5 cm depth of blocks did not remarkably shrink according to similar pore distributions before and after desiccation as tested by MIP under the circumstance that no obvious visual fissures were detected by CT scan in an un-dried block.

(3) Vertical shrinkage ability presented by DIC was poorer than in a horizontal direction because vertical strain was similar to horizontal direction and the initial lengths of the measuring lines were obviously shorter. Vertical compaction is partly responsible for the poorer shrinkage ability. The final shrinkage strain of buffer is about 2%, and much attention should be paid to lateral shrinkage of buffer block.

Acknowledgments. The authors are grateful to the National Nature Science Foundation of China (41672261) and Fundamental Research Funds for the Central Universities (lzujbky-2017-ct02) for their financial support.

References

1. Martin PL, Barcala JM (2005) Large scale buffer material test: mock-up experiment at CIEMAT. Eng Geol 81(3):298–316
2. Pacovský J, Svoboda J, Zapletal L (2007) Saturation development in the bentonite barrier of the Mock-Up-CZ geotechnical experiment. Phys Chem Earth 32(8–14):767–779
3. Svensk K (2010) Design, production and initial state of the buffer. TR-10-15, SKB, Swedish
4. Torbjorn S, ULF Nlisson, L Andersson (2016) Investigation of parameters influencing bentonite block quality. P-16-06, SKB, Swedish
5. Peron H, Hueckel T, Laloui L et al (2009) Fundamentals of desiccation cracking of fine-grained soils: experimental characterisation and mechanisms identification. Can Geotech J 46(10):1177–1201
6. Yamaguchi I (1981) A laser-speckle strain gauge. J Phys E Sci Instrum 14(11):1270–1273
7. Peters WH, Ranson WF (1982) Digital imaging techniques in experimental stress analysis. Opt Eng 21(3):213427
8. Corr D, Accardi M, Graham-Brady L et al (2007) Digital image correlation analysis of interfacial debonding properties and fracture behavior in concrete. Eng Fract Mech 74(1–2):109–121
9. Helm JD (2008) Digital image correlation for specimens with multiple growing cracks. Exp Mech 48(6):753–762
10. Chen H, Zhang HY, Guo YQ et al (2014) Measurement and prediction of thermal propreties of bentonite-sand mixture as buffer backfilling materials for high level radioactive waste. Chin J Rock Mech Eng S2:4312–4320
11. Zhou L, Zhang HY, Yan M et al (2013) Laboratory determination of migration of Eu(III) in compacted bentonite–sand mixtures as buffer/backfill material for high-level waste disposal. Appl Radiat Isot 82:139–144
12. Cui SL, Zhang HY, Zhang M (2012) Swelling characteristics of compacted GMZ bentonite–sand mixtures as a buffer/backfill material in China. Eng Geol 141:65–73
13. Villar MV, Iglesias RJ, Gutierrez AC et al (2018) Hydraulic and mechanical properties of compacted bentonite after 18 years in barrier conditions. Appl Clay Sci 160:49–57
14. Zhao XG, Wang J, Chen F et al (2016) Experimental investigations on the thermal conductivity characteristics of Beishan granitic rocks for China's HLW disposal. Tectonophysics 683:124–137
15. Wang LL, Zhang GQ, Hallais S et al (2017) Swelling of shales: a multiscale experimental investigation. Energy Fuel 31(10):10442–10451

16. Wang LL, Tang CS, Shi B et al (2018) Nucleation and propagation mechanisms of soil desiccation cracks. Eng Geol 238:27–35
17. Raynaud S, Ngan-Tillard D, Desrues J et al (2008) Brittle-to-ductile transition in Beaucaire marl from triaxial tests under the CT-scanner. Int J Rock Mech Min Sci 45(5):653–671
18. Helliwell JR, Sturrock CJ, Grayling KM et al (2013) Applications of X-ray computed tomography for examining biophysical interactions and structural development in soil systems: a review. Eur J Soil Sci 64(3):279–297
19. Soto-Gómez D, Pérez-Rodríguez P, Vázquez-Juiz L et al (2018) Linking pore network characteristics extracted from CT images to the transport of solute and colloid tracers in soils under different tillage managements. Soil Tillage Res 177:145–154
20. Zhang Y, Ye WM, Chen B et al (2016) Desiccation of NaCl-contaminated soil of earthen heritages in the Site of Yar City, northwest China. Appl Clay Sci 124:1–10
21. Hashemi MA, Massart TJ, Salager S et al (2015) Pore scale characterization of lime-treated sand–bentonite mixtures. Appl Clay Sci 111:50–60
22. Lloret MA, Villar MV, Sanchez M et al (2003) Mechanical behaviour of heavily compacted bentonite under high suction changes. Geotechnique 53(1):27–40
23. Lee JO, Choi H, Lee JY (2016) Thermal conductivity of compacted bentonite as a buffer material for a high-level radioactive waste repository. Ann Nucl Energy 94:848–855

Modeling Solute Transport Through Geosynthetic Clay Liners Permeated with Inorganic Solutions

Francesco Mazzieri(✉) and Evelina Fratalocchi

Università Politecnica delle Marche, Via Brecce Bianche 12, Ancona, Italy
f.mazzieri@staff.univpm.it

Abstract. GCLs are employed in containment applications, where both advection and diffusion need to be evaluated as possible migration mechanisms of contaminants. Permeation column tests were carried out on a GCL using two synthetic multispecies inorganic solutions of different ionic strength and containing equimolar concentration of heavy metals (Pb, Zn, Cu). The hydraulic conductivity (k) of the GCL to inorganic solutions increased by one order of magnitude relative to permeation with water. The solute breakthrough curves were interpreted using the software Pollute®, which allows modeling solute transport in case of variable transport parameters and nonlinear equilibrium sorption. The experimental data were best fitted by assuming effective diffusion coefficients of solute species increasing with time.

Keywords: Geosynthetic Clay Liners · Permeability · Migration Modeling

1 Introduction

Geosynthetic Clay Liners (GCLs) consisting of a thin layer of bentonite clay incorporated between two geotextiles or glued to a geomembrane have been widely used as hydraulic and pollutant barriers. The content of sodium montmorillonite is the key factor for the characteristically low hydraulic conductivity of GCLs. Sodium bentonite can exhibit considerable adsorption of cations, including heavy metals, owing to the large cation exchange capacity (CEC 80–100 meq/100 g) and specific surface area (up to 800 m^2/g). Heavy metals are commonly found in landfills leachates and in water treatment systems of the mining industry (Lange et al. 2007). Various investigations have addressed the permeability and retention capacity of GCLs in the presence of metal-rich liquids. Lake et al. (2007) performed permeability and diffusion tests using a needle-punched GCL and aluminum-rich solutions. The GCL was prehydrated with water and preserved a hydraulic conductivity, k, lower than 5×10^{-11} m/s upon permeation with the aluminum solution. Diffusion testing established a diffusion coefficient of aluminum of 1.5×10^{-10} m^2/s and an equivalent distribution coefficient, K_d, of 30 mg/L. Lange et al. (2007) investigated permeability and metals retention in a needle-punched GCL permeated with two synthetic mining leachates. The k increased by up to one order of magnitude upon of permeation with the synthetic leachates.

L. Zhan et al. (Eds.): ICEG 2018, ESE, pp. 506–514, 2019.
https://doi.org/10.1007/978-981-13-2224-2_62

On the other hand, the GCL showed strong attenuation of metals and metalloids present in the leachates, attributed to cation exchange, precipitation or co-precipitation. Lange et al. (2009) investigated diffusion and sorption of metals in a needle-punched GCL in the presence of various heavy metals leachates. The average diffusion coefficients for Cu, Cd, Zn, Fe and Ni covered a narrow range from 0.67×10^{-10} m^2/s to 0.89×10^{-10} m^2/s.

A few studies have investigated the relationship between the changes in the microfabric of bentonites and the presence of heavy metals. Bouchelaghem and Jozja (2009) showed that saturating a Mg-bentonite with Pb^{2+} ions leads to an increase in its permeability. Scanning electron microscopy and X-ray diffraction analysis revealed that the increase in the permeability value was associated to fabric changes either at the nanometric scale (increase of particle size) or at the micrometric scale (microfissuration). Souli et al. (2008) showed that increasing the zinc concentration in a bentonite leads to an increase in the particle size due to the compression of the double layer and to a decrease in the repulsive forces between particles. The X-ray diffraction results showed that bentonite saturated with the zinc solution at 0.01 M concentration presents a flocculated (isotropic) fabric, whereas, at 1 M zinc concentration, the fabric becomes more anisotropic, leaving larger pores between particles, while interparticular pores become smaller. The result of these mechanisms is shorter flow paths and, as a consequence, higher permeability.

Based on the results reported above, migration of leachates containing heavy metals into water-hydrated bentonite may lead to particle fabric rearrangement and to an increase in k. This phenomenon can be, in part, counterbalanced if metal solids precipitate and clog the pore space (Lake et al. 2007). In the presence of an hydraulic gradient, an increase in k will result in an increase in the advective solute flux through the GCL. In principle, the solute transport parameters related to clay fabric (e.g., the effective diffusion coefficient, D^*) will also be affected. On the other hand, adsorption of metal solutes to the surface and/or precipitation will retard the overall mass transport. Due to many interacting phenomena, analyzing and describing the transport of metals through bentonite or GCLs is a complex task. This paper presents the results of permeation column tests performed on a needle-punched GCL using two synthetic solutions containing metals (Pb, Zn, Cu) at two different concentration levels. The purposes of the study were: (1) to observe the effect of the solutions on the hydraulic performance of the GCLs; (2) to determine effluent pore water chemistry; (3) to simulate the experimental breakthrough curves with a theoretical migration model.

2 Materials and Methods

The GCL used in this study is needle-punched. It consists of an upper nonwoven polypropylene geotextile (300 g/m^2) and a carrier woven geotextile (200 g/m^2), encapsulating a nominal 4200 g/m^2 of powdered natural sodium bentonite. The upper geotextile is impregnated with powdered bentonite (800 g/m^2). The bentonite has cation exchange capacity, CEC of 98 meq/100 g, with sodium (77 meq/100 g) and calcium (18 meq/100 g) and magnesium (8 meq/100 g) as the main adsorbed cations. Further details on the GCL used in the study are given in Mazzieri et al. (2013).

GCLs specimens were first hydrated and permeated with distilled water (DW) and subsequently permeated with two synthetic solutions (SL = Synthetic Leachate and DSL = Dilute Synthetic Leachate) of different ionic strength, *I*, containing metal cations. Distilled water (DW) had Electrical Conductance, EC, of 0.014 mS/cm and pH = 6.2. The SL was prepared by dissolving copper, zinc and lead nitrate salts (Merck, p.a.) in DW in order to obtain for each metal a target concentration equal to 25 mM, and adjusting with HNO_3 to a target pH = 2.0. The DSL solution was prepared by diluting (1:10 ratio) the SL with DW. The SL had EC = 15.85 mS/cm, ionic strength, *I* = 0.235, pH = 2.0 (average values). The DSL had EC = 2.10 mS/cm, I = 0.0242, pH = 2.8 (average values). The concentration of metals in either SL and DSL solutions are higher than the maximum concentration reported in typical waste leachates or mining treatment waters (Lange et al. 2007). However, relatively high concentrations were used in order to simulate severe conditions for metal loading and acidity and to accelerate chemical interaction.

Permeability tests were carried out in 100 mm diameter flexible-wall permeameters (FWP). The GCL specimen was placed in the test cell, hydrated with DW without applied gradient for 7 days and then permeated with DW (1st stage) to achieve saturation and to determine the reference *k*. Permeation was obtained by applying a cell pressure of 103.4 kPa and pore pressures of 89.6 kPa and 68.9 kPa (back-pressure) at the inlet and outlet end, respectively. The permeant liquid was then switched to the synthetic solution (2nd stage). The paper reports on two tests, performed with the SL and DSL as permeant liquids in the second stage. Termination criteria outlined in ASTM D-6676 were generally followed. Bladder accumulators where used both in the inflow and outflow lines; the effluent solutions were accumulated in the bladder and sampled over regular intervals for pH, EC and concentrations of chemicals species. The measured concentrations thus represent average incremental concentrations (Shackleford and Redmond 1995). The chemical analyses were carried out by a Dionex ICS 1000 Ionic Chromatograph.

3 Results and Discussion

3.1 Macroscopic Effects of Permeation: Hydraulic Conductivity and Volume Changes

The hydraulic conductivity, *k*, is plotted versus the net number of pore volumes of flow (Shackleford and Redmond 1995) *NPV*, in Fig. 1a and b for SL and DSL, respectively (curves in the top part of the graph). The *k* values during permeation with DW were in the range 1–2.5 × 10^{-11} m/s. Upon switching to the synthetic solutions the *k* remained constant or slightly decreased (up to *NPV* ~ 1.0 and to *NPV* ~ 7.0 for the SL and DSL, respectively). Gas bubbles were observed in the drainage lines during this stage, likely as a result of neutralization reactions between HCO_3 anions present in the alkaline pore water of bentonite and H_3O^+ cations introduced with the acidic permeant solutions, with consequent release of gaseous CO_2. Thereafter *k* increased gradually up to *NPV* ~ 7.0 and up to *NPV* ~ 50 for the SL and DSL, respectively. The *k* value remained thereafter relatively constant until the termination of the test (*NPV* ~ 45 and *NPV* ~ 75 for SL and DSL, respectively). The *k* values at test termination were 7.9 × 10^{-11} m/s and 2.0 × 10^{-10} m/s for SL and DSL, respectively.

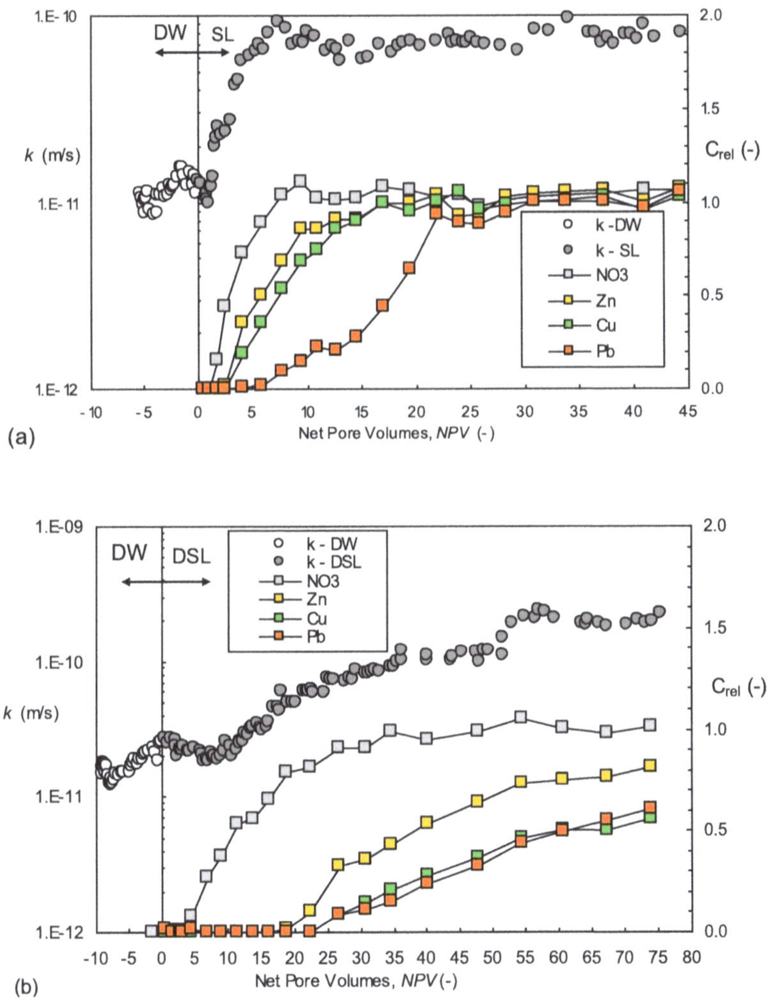

Fig. 1. Hydraulic conductivity (left axis) and solute breakthrough curves (right axis) versus *NPV*: (a) test with SL as permeant solution; (b) test with DSL as permeant solution.

The current height of the GCL specimen, *L*, was calculated from the initial height and from the measurement of the vertical displacements of the top cap of the permeameter by means of a rigid rod connected to a dial indicator. Permeation with the synthetic solutions was associated with a decrease in *L* with respect to the first stage of permeation with DW (L_{DW}). The corresponding maximum vertical strains (i.e., $\Delta L/L_{DW}$) were equal to 4% and 5.2% for permeation with SL and DSL, respectively. The volume of pores and *NPV* were calculated with reference to L_{DW} and to the corresponding total porosity (n_{DW}).

The increase in *k* and the reduction in volume relative to DW is consistent with the migration of divalent metal cations into the GCL. The slower rate of increase in *k* with *NPV* observed during permeation with DSL relative to SL is related to the lower entry rate of metal cations into the GCL, due to the lower input concentration. Divalent cations substitute monovalent Na^+ cations on bentonite surface exchange sites. Bentonite retains less water when surface cations are multivalent, therefore exchange of monovalent cations (Na^+) for divalent (e.g., Pb^{2+}) on the surface sites determines a volume reduction at macroscopic level, as observed in this study. Moreover, the fraction of immobile ("bound") water is less, the flow paths around the particles become less tortuous, resulting in higher *k* (Jo et al. 2004). Protons can also replace Na^+ in exchange site, reduce the swell volume and increase the hydraulic conductivity of bentonite. However, in both solutions the concentration of H_3O^+ was about 7.5 times lower than that of divalent cations, therefore the exchange for metals cations was likely predominant.

3.2 Chemical Composition of Effluent Solutions

The solute breakthrough curves (plots of relative concentration $C_{rel} = C/C_0$ versus *NPV*, where C = measured effluent concentration over a sampling interval, C_0 = influent concentration) of the chemical species are shown in Fig. 1a and b (colored symbols in the lower part of the graphs). C_0 was assumed as the average concentration of individual species measured in the inlet bladder. The attenuation order for metals observed in column test, inferred from the delay of breakthrough curves, was $Pb^{2+} > Cu^{2+} > Zn^{2+}$). Desorption of Na^+ and Ca^{2+} in the effluent solutions indicated that metal cations were actively retained in the GCL by cation exchange. For all metals, attenuation was higher (i.e., breakthrough was delayed) with DSL relative to SL permeant solution. This is likely a result of the different seepage velocities during the test as well as to nonlinear sorption. Batch tests performed on bentonite sampled from the GCL showed for all metals strongly nonlinear sorption isotherms (Mazzieri 2012). The batch tests were performed using the SL solution at various dilution ratios (including 1:10, e.g., DSL) rather than monospecies solutions, in order to account for competition between cations for exchange sites on bentonite surface.

The delay in the breakthrough curve of nitrate in Fig. 2a and b suggests retardation that can, at least in part, be attributed to the dilution of the effluent solution by DW present in the bladder accumulator from the first stage of the tests (Mazzieri et al. 2015). One-point sorption tests performed with the bentonite and each permeant solution (SL and DSL) confirmed the retention of nitrate, presumably resulting from anion adsorption to protonated edge sites of montmorillonite particles under acidic conditions (Mazzieri and Pasqualini 2016). The pH of the effluent solution was alkaline ($7 < pH < 9$) during permeation of the GCL with DW, whereas upon permeation with the SL (inflow, $pH_{in} = 2.0$) a gradual decline occurred to pH $\sim$ 2.5 when the test was terminated. Upon permeation with the DSL ($pH_{in} = 2.8$), the effluent solution pH remained practically constant ($\sim$7.0) for $0 < NPV < 15$ and then gradually decreased to pH $\sim$ 5.2 at test termination.

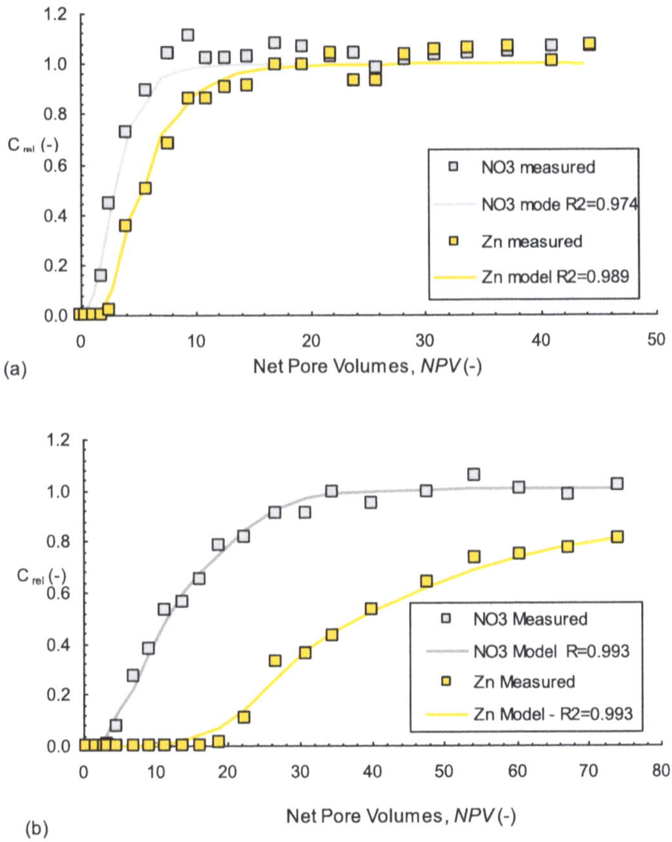

Fig. 2. Comparison of experimental and modeled solute breakthrough curves of nitrate and zinc: (a) test with SL as permeant solution; (b) test with DSL as permeant solution.

3.3 Solute Migration Modeling

This study aimed at modeling the solute transport through the GCL and assessing the transport parameters of individual ion species. The variable seepage velocity observed during permeation and nonlinear sorption relationships prevented the use of closed-form solutions to the solute transport equation (ADRE). The software POLLUTEv7 (Rowe and Booker 2005), which allows taking into account variable seepage velocity and nonlinear sorption was then used.

The observed trend in Darcy velocity versus time was discretized in intervals in which the velocity was assumed to be constant or to vary linearly with time. The effective seepage velocity was calculated from the total porosity n_{DW} (assumed constant). The entrance boundary condition was of constant concentration (C_0). To model the measured effluent concentrations, a fictitious "aquifer" was introduced (representing the porous disk placed underneath the specimen in the testing apparatus) using the "fixed flow" boundary condition available in the software. The "aquifer" was subjected to a horizontal liquid flux equal to the measured Darcy flux through the specimen (continuity condition). The effluent concentration over a given sampling interval was calculated as the ratio of the total mass exiting the "aquifer" and the liquid volume accumulated in the bladder, taking into account dilution induced by the residual (i.e., not drainable) volume of the bladder, V_{res}. ($\sim$22 mL), at each sampling (Mazzieri et al. 2015). The effective diffusion coefficient, D^*, was assigned for each time interval. Mechanical dispersion was assumed be to negligible.

As different combinations of sorption and diffusion parameters can result in similar breakthrough curves, modeling was assisted by separate batch sorption tests (Mazzieri 2012), which indicated that solute-bentonite sorption behaviour was well represented by Langmuir isotherms:

$$S = \frac{S_m b C_e}{1 + b C_e} \tag{1}$$

where S is the solute mass retained per unit mass of soil, S_m is the maximum sorption capacity, b is a parameter, and C_e is the equilibrium concentration. S_m was estimated from the mass balances of column tests, whereas b was treated as a fitting parameter.

The results of modeling reported herein are those of nitrate and zinc, which displayed the fastest transport through the GCL. Mass balances between influent and effluent solutions in column tests allowed estimating the nitrate masses retained per unit mass of soil, equal to 7.4 mg/g (SL test) and 10.4 mg/g (DLS test), respectively. These values were assigned to the S_m parameters of the isotherms. In the case of zinc, the mass balance for the SL test gave a retained mass of 6.2 mg/g (assigned to S_m). The same value of S_m was used model sorption of zinc in the DSL test.

The fitting of measured relative concentrations with simulated relative concentrations is shown in Fig. 2a and b for the SL and DSL tests, respectively. The relative concentration in a given sampling interval was plotted at the center of corresponding *NPV* interval. The goodness-of-fit was evaluated by the coefficient of determination R^2 (also reported in the figures). The experimental data could not be effectively fitted using a constant diffusion coefficient D^*. In the case of SL, the best fit for nitrate was obtained with b = 5.2 L/g and D^* increasing from 0.4×10^{-10} to 3.1×10^{-10} m^2/s whereas for zinc, the best fit was obtained with b = 31.1 L/g and D^* increasing from 0.1×10^{-10} to 1×10^{-10} m^2/s. In the case of DSL, the best fit for nitrate was obtained with b = 5.2 L/g and D^* increasing from 0.9×10^{-10} to 7.6×10^{-10} m^2/s, whereas for zinc, the best fit was obtained with b = 31.1 L/g and D^* increasing from 0.6×10^{-10} to 6.1×10^{-10} m^2/s. The S_m (6.2 mg/g) and b (31.1 L/g) values used to model sorption of zinc in column tests are close to S_m (7.6 mg/g) and b (15.9 L/g) values determined from batch tests.

The range of D^* values obtained for zinc is higher than that reported by Lange et al. (2009). Possible reasons for the differences are the GCL type, the testing method (diffusion vs. column testing) and the interpretation approach (constant diffusion coefficients and linear sorption vs. variable diffusion coefficients and nonlinear sorption in this study). Similar to *k*, the increase in D^* obtained in the present study can be interpreted as due to a change in clay microfabric as a result of interaction with metals, leading to less tortuous travel paths for solutes around clay particles.

4 Conclusions

Permeability column tests were performed using a needle-punched GCL and two synthetic solutions (denoted as SL and DSL) of different ionic strengths containing heavy metals (Pb, Zn, Cu). The GCL was first permeated with water. Subsequent permeation with either solution increased *k* by a factor 8–9. The rate of increase in *k* was lower for DSL than for SL, due to the lower entry rate of divalent cations into the GCL. The breakthrough curves of nitrate and zinc were modeled using the software POLLUTEv7. Sorption was represented by the Langmuir model. Under the adopted assumptions, fitting of the experimental data required D^* values increasing with time by a factor ranging from 7.5 to 10. Similar to *k*, the increase in D^* can be interpreted as due to a change in clay microfabric due to interaction with metals, leading to less tortuous travel paths for solutes around clay particles. In practical applications where prediction of breakthrough times of solutes is critical, the possible variability of transport properties with time as a result of soil-solute interactions should be carefully considered.

References

Bouchelaghem F, Jozja M (2009) Multi-scale study of permeability evolution of a bentonite clay owing to pollutant transport: part II. Application to a MG-bentonite. Eng Geol 108:286–294

Jo HY, Benson CH, Edil TB (2004) Hydraulic conductivity and cation exchange in non-prehydrated and prehydrated bentonite permeated with weak inorganic salt solutions. Clays Clay Min 52(6):661–679

Lake CB, Cardenas G, Goreham V, Gagnon GA (2007) Aluminum migration through a geosynthetic clay liner. Geosynth Int 14(4):201–209

Lange K, Rowe RK, Jamieson H (2007) Metal retention in geosynthetic clay liners following permeation by different mining solutions. Geosynth Int 14(3):178–187

Lange K, Rowe RK, Jamiesion H (2009) Diffusion of metals in geosynthetic clay liners. Geosynth Int 16(1):11–27

Mazzieri F (2012) Assessment of heavy metals retention in GCLs by column and batch tests. GSP 225, GeoCongress 2012, ASCE, Reston/VA, pp 3447–3456

Mazzieri F, Di Sante M, Fratalocchi E (2015) Dilution effect of the apparatus on contaminant transport parameters assessed by column testing. J Geotech Geoenv Engrg 141(9)

Mazzieri F, Di Emidio G, Fratalocchi E, Di Sante M, Pasqualini E (2013) Permeation of two GCLs with an acidic metal-rich synthetic leachate. Geotext Geomembr 40(5):1–11

Mazzieri F, Pasqualini E (2016) Hydraulic conductivity and solute transport in two GCLs permeated with a synthetic solution. GSP 273, Geochicago, pp 156–167

Rowe KR, Booker JR (2005) POLLUTE v7. Distributed by GAEA Environmental Engineering Ltd., Whithby

Shackleford CD, Redmond PL (1995) Solute breakthrough curves for processed kaolin at low flow rates. J Geotech Eng 121(1):17–32

Souli H, Fleureau J-M, Trabelsi AM, Besnard M (2008) Physicochemical analysis of permeability changes in the presence of zinc. Geoderma 145:1–7

Permeability Control of Compacted Landfill Liners and Covers

Belkacem Moussai(✉)

University of Science and Technology Houari Boumedienne, Algiers, Algeria
bmoussai@yahoo.ca

Abstract. The most common soil barriers used to contain waste units consists of fine grained clayey soils. For these soil barriers, the permeability is the parameter of primary concern. This parameter can be determined either directly by performing in situ tests or indirectly through the check of the dry unit weight and water content frequently measured in situ. In fact, the measurement of soil permeability in situ is not an easy task; the permeability test takes long time to perform, and makes difficult any intervention to improve the quality of the soil barrier. Therefore, the simple way to check the permeability of compacted soils in situ is through measuring the dry unit weight and water content. This paper presents a methodology for the study and indirect control of the permeability of soils compacted in situ. According to the quality sought for the soil barrier, this methodology allows laboratory definition of an acceptable zone in terms of the dry unit weight and water content. This zone must be complied with during construction and serves as a criterion for in situ checking of permeability.

Keywords: Landfill · Compacted soil · Permeability
Control of the permeability

1 Introduction

For soil barriers used to contain waste units, the best approach to control their permeability in situ is to define in laboratory a range of unit weights and water contents where we will be sure to obtain the required quality of the structure. This range should be complied with during construction.

Mundell and Bailley (1985) and Daniel and Benson (1990) suggested to determine an area on the diagram $\gamma_d - w$ on which the permeability is less than or equal to 10^{-9} m/s (case of landfill liners and covers). So, it is necessary to make permeability measurements to define the limits of this area.

Daniel and Kwang (1993) have established a zone of compaction function of γ_d and w to meet both permeability, shrinkage potential and shear strength objectives. However, the established area of compaction does not take into account the phenomena of overcompaction.

L. Zhan et al. (Eds.): ICEG 2018, ESE, pp. 515–520, 2019.
https://doi.org/10.1007/978-981-13-2224-2_63

2 Methodology

The proposed methodology consists of determining a range of dry unit weights and water contents necessary to define the acceptable zone of permeability. This zone may be modified to account for other factors which may affect the engineering behaviour of the earth structure.

The recommended procedure is as follow:

- Compaction of soil with at least two different energies which span the range of energies expected in the field, using the appropriate method of compaction.
- Measurement of the permeability of several samples (three samples at least along each compaction curve: at the optimum and at dry and wet side of the optimum) for each compactive effort.
- Define the suitable zone of the permeability (based only on the permeability considerations). This can be done by plotting permeability contour for the required value of the permeability (kc) in the diagram $\gamma_d - \mathrm{w}$.
- Defining the acceptable zone of permeability by adjusting the suitable zone to account for other factors which may affect the engineering behaviour of the earth structure such as shear strength, cracks due to shrinkage or differential settlements, etc.

After the definition of the acceptable zone of the permeability in laboratory, this one will be used as guidance for construction. The compaction parameters (w and γ_d) determined in situ should fall within the zone defined by the relation between the parameters k, w and γ_d.

This methodology presents the following advantages:

- Simple control: from the parameters w and γ_d frequently measured in situ, we can make judgement about the engineering performance of the earth structure.
- This procedure ensures a second control. It allows deciding for a given soil whether it ensures the required performance of the earth structure.

In general, the soils which are more sensitive to water are difficult to be compacted at a relative compaction more than 95% when its water content is close to saturation. With this procedure we can compact easily this type of soil and obtain the required quality of the earth structure (γ_{dmin} defined in the acceptable zone can easily exceeded).

3 Definition of the Acceptable Zone of the Permeability

The material used for this study is silty clay of low plasticity having the following characteristics:

% finer than 0.2 mm = 98
% finer than 0.080 mm = 93
% finer than 0.002 mm = 21
Liquid limit = 27%,
Plastic index = 6%
Specific surface = 42 m^2/g

The permeability is carried out on samples compacted on one lift at different water contents and compactive efforts directly on the compaction mold (90 mm diameter and 40 mm height). The soil is compacted using three compactive efforts: 15, 25 and 45 blows of standard Proctor hammer.

The purpose of using a compaction mold with dimensions less than the Proctor mold is to allow the saturation of the soil sample and therefore to complete the permeability test within a reasonable time (about 10 days).

The permeability tests were performed in a rigid wall permeameter with constant head. The percolated fluid is water and the coefficient of permeability is measured after the establishment of a permanent regime. Figure 1 gives the results of the compaction and permeability obtained under these conditions.

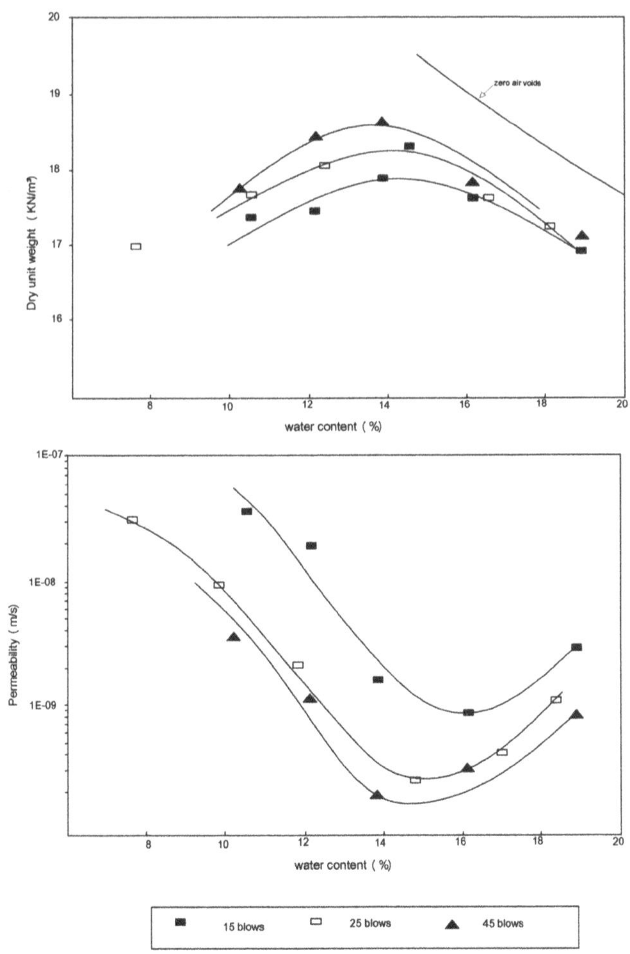

Fig. 1. Relationship between permeability, unit weight and water content

Let's take for example a permeability of control $k_c = 10^{-9}$ m/s (case of landfill liners and covers), the suitable zone of permeability may be defined by plotting permeability contur for $k_c = 10^{-9}$ m/s in the diagram $\gamma_d - w$ (see Fig. 2). However, the acceptable zone should be defined taking into account other factors that may affect the engineering behavior of the earth structure.

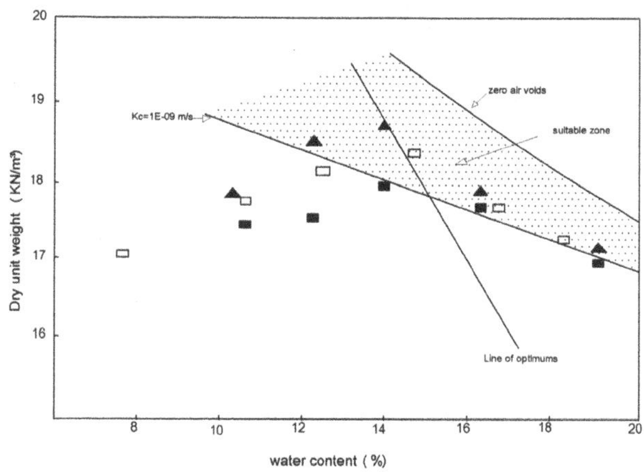

Fig. 2. Definition of the suitable zone of permeability

For this case the following conditions should be taken into account (Fig. 3):

- The soil should be compacted in the wet side of the optimum in order to avoid the cracks due to differential settlement after an eventual wetting of the soil. Furthermore, the clods in the wet side can be well compacted and the void between clods will be reduced.

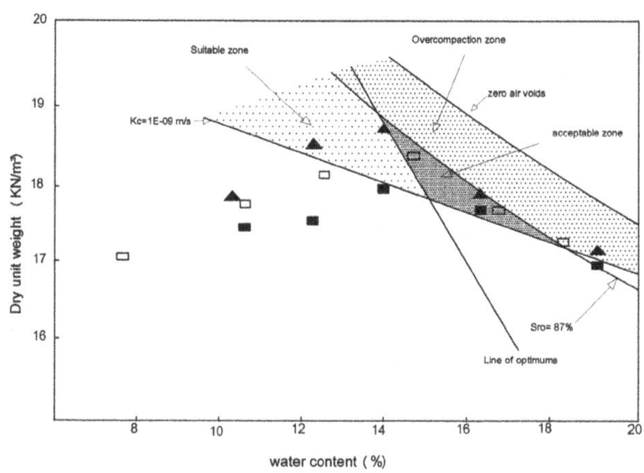

Fig. 3. Definition of the acceptable zone of permeability

- The degree of saturation of the soil should not exceed a certain amount during compaction ($S_{ro} < 100\%$) in order to avoid the overcompaction phenomena (the value of S_{ro} depends on the permeability of soil and the compactive effort).

A test fill program was conducted on this soil in the Road Experiment Center – France. The compaction was performed using a pneumatic roller having 25 kN weight per tire and a tire pressure of 600 kPa. During compaction we noticed that at about 18% of water content which corresponds to 87% of degree of saturation the over-compaction phenomena started to appear. So for this case to avoid overcompaction phenomena of the soil a degree of saturation of 87% should not be exceeded.

Note that the permeability contours established from laboratory tests do not correspond necessarily to that obtained in situ. The research carried out by several authors such as Olson and Daniel (1981) and Daniel (1984) show that, more often the permeability of soils compacted in situ is greater than that measured in laboratory. So, it is necessary to take into account the difference between the permeability results obtained in laboratory and in situ during the definition of the acceptable zone of permeability.

4 Conclusions

In the case of impervious barriers, the proposed methodology to define a compaction zone in the plan $\gamma_d - w$, where we are sure to obtain the required coefficient of permeability has important practical consequences. This methodology leads directly to field specifications, whether for work execution or control criteria.

This procedure is very interesting especially for the clayey soils where any modification of natural water content is a difficult task. With this procedure we can compact the soil at its natural water content within the limits of the acceptable zone of the permeability defined in advance by laboratory tests.

The methodology elaborated here, if it is confirmed by further tests in laboratory as well as in situ will have without doubt important consequences on the conduction of field work.

References

Daniel DE, Benson CH (1990) Water content – density criteria for compacted soil liners. J Geotech Eng 116(12):1811–1830

Mundell JA, Bailley B (1985) The design and testing of a compacted clay barrier layer to limit percolation through landfill covers. In: Hydraulic barriers in soil and rock, vol 874. ASTM Special Technical Publication, Philadelphia, pp 246–262

Daniel DE, Kwang KW (1993) Compacted clay liners and covers for arid sites. J Geotech Eng 119(2):223–237

Daniel DE (1984) Predicting hydraulic conductivity of clay liners. J Geotech Eng ASCE 110 (2):285–300

Olson RE, Daniel DE (1981) Measurement of the hydraulic conductivity of fine grained soils. In: Zimmie TF, Riggs CO (eds) Permeability and groundwater contaminant transport. ASTM Special Technical Publication, vol 874. ASTM, Philadelphia, pp 18–64

Boutwell GP, Hedges C (1989) Evaluation of waste retention liners by multivariate statistics. In: Proceedings of the 12th international conference on soil mechanics and foundation engineering, Rio De Janeiro, Brazil, pp 815–818

Research on the Sealing Ability of Saturated Compacted Bentonite/Sand Mixture

Xu-Lou Cao[1,2], Jiang-Feng Liu[1,2(✉)], Wen-Bo Bao[1], Si-Qi Ye[1], Jing-Na Guo[1], and Ji Huang[1]

[1] The State Key Laboratory for GeoMechanics and Deep Underground Engineering, and School of Mechanics and Civil Engineering, China University of Mining and Technology, Xuzhou 221116, China
[2] Key Laboratory of Geotechnical and Underground Engineering of Ministry of Education, Tongji University, Shanghai, China
jeafliu@hotmail.com

Abstract. This paper focuses on the experimental research of water and gas transport in a compacted bentonite/sand mixture under flexible condition. The main contribution of our study is to provide insights into water and gas transport properties in the compacted bentonite/sand mixture. Experimental results indicate that water tightness can be achieved after full saturation, and the water permeability of the sample is at the magnitude of 10^{-20} m^2. Gas breakthrough tests show that gas migration properties depend on both external stress state and internal fluid pressure. In addition, change of the external stress also has a hysteresis effect on the gas transport properties. At certain stage, the increasing rate of downstream gas pressure continue decreasing even though the gas pressure was increased while the confining pressure was kept constant. During the whole process, gas migration properties were determined by the coupling effects of confining pressure and gas pressure.

Keywords: Bentonite/sand mixture · Water tightness · Gas migration Coupling effects

1 Introduction

Long-term radioactive waste management usually considers final disposal in a deep geological repository. This includes an engineered barrier system working in conjunction with the surrounding host rocks to minimize the migration of radioactivity. In the design of radioactive waste repositories at great depths, bentonite or bentonite-based materials (a bentonite/sand mixture used in this study) are generally chosen as buffer materials to separate the radioactive waste canisters from the host rock or as backfill material to seal the access tunnels of the disposal pits (Cui et al. 2008; Liu et al. 2016; Ye et al. 2012). Bentonite or bentonite-based materials are usually compacted at an intermediate water content, and they are progressively wetted by water coming from the host rock formation (King et al. 2001). With time, the clay plug will swell due to the absorption of water, closing up any construction clearance and sealing the repository. The clay plug and the host rock will form a low-permeability continuous medium that will significantly limit the migration of radionuclide that are released from a canister.

L. Zhan et al. (Eds.): ICEG 2018, ESE, pp. 521–528, 2019.
https://doi.org/10.1007/978-981-13-2224-2_64

As the repository system evolves, gases may be produced, such as hydrogen from the corrosion of metals and from the radiolysis of water, and radon from the radioactive decay of some of the waste. If present, biodegradable wastes can also produce carbon dioxide and methane. In low-permeability clay formations with a distinct matrix porosity, gas transport through the matrix is controlled not only by the hydraulic and mechanical properties of the sample (i.e. water saturation, porosity, strength), but also the hydro-mechanical state (i.e. pore water pressure, stress state).

Previous researches have indicated that gas migration property is closely related to experimental devices (smooth or grooved inner surface of the tube/cell) and the confinement conditions (e.g., rigidly confined/semi-rigid/flexible cell or tube) (Davy et al. 2009; Gallé 2000; Harrington et al. 2012a, b; Liu et al. 2016). Unlike our previous tests, which were performed with semi-rigid tubes (e.g., aluminium-plexiglas/aluminium/COx argillite tube), compacted bentonite/sand sample was put in the flexible Viton™ membrane directly to perform water-injection test and gas breakthrough test. The main purpose of this study is to investigate that gas migration property in saturated bentonite/sand mixture under flexible condition and its comparison with our previous tests with semi-rigid tubes.

2 Materials and Experimental Methods

2.1 Materials and Sample Preparation

Based on the preliminary studies performed by Andra (The French National Radioactive Waste Management Agency), the materials used in our experiments consisted of a mix (30% TH1000 silica sand + 70% MX80 bentonite) from a single-powder mix provided by CEA (France), batch reference RE08015. The MX80 bentonite is a natural sodium bentonite, which contained 75%–90% of montmorillonite and has a specific gravity of 2.74 g/cm^3. The siliceous sand used in the mixture was made of 99.0% quartz, with a specific gravity of 2.65 g/cm^3. In order to obtain the adequate water content, before the compaction operation, the bentonite/sand mixture is let in an atmosphere at a controlled relative humidity (RH). After mass equilibrium state of the powder mix at a given RH of 85%, it is compacted up to 12 MPa of axial pressure in a dedicated steel mould in order to achieve d = 1.77 ± 0.1 g/cm^3 (Liu et al. 2016) apparent dry density and w = 15.2% ± 1% water content. The compacted samples have a diameter of 37 ± 0.5 mm and a height of 25 ± 1 mm.

2.2 Experimental Method

The whole experimental process contained (1) water-injection test and (2) gas breakthrough test (GBT) (Fig. 1). Firstly, water was injected from the bottom of the triaxial cell with 1.5 MPa of water pressure. Previous studies indicated that gas migration in compacted bentonite/sand mixture is very sensitive to the confining pressure (Liu et al. 2015; Villa et al. 2012). In order to avoid affecting the pore structure of the sample,

3 MPa of confining pressure was chosen at this stage. The water used in this study was specially prepared to reproduce the COx pore water from Andra's Underground Research Laboratory (URL) at Bure, France. Water-injection time usually lasts 400–500 h to ensure that the sample is fully saturated. When the sample is fully saturated, water permeability (K_w) can be calculated with Darcy's law.

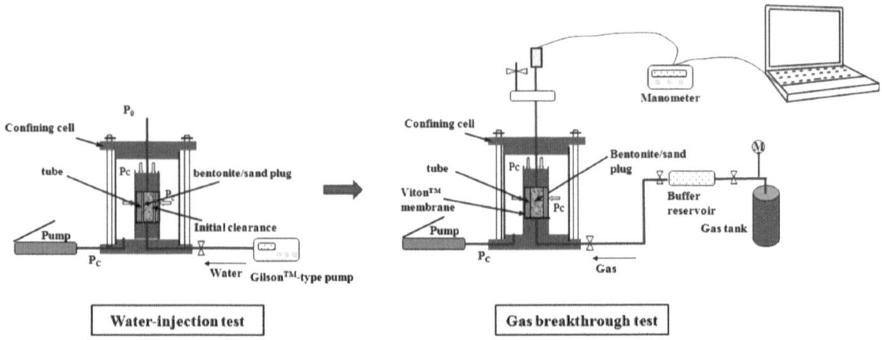

Fig. 1. Schematic diagram of water injection and GBTs.

As soon as the sample was fully saturated, gas breakthrough test would be performed. Gas breakthrough test is a better method to understand the gas tightness of the material, which can help us to judge the gas migration property in the material. First of all, water-injection test would be stopped for 24 h to allow the internal water pressure to equilibrate at atmospheric pressure. Then the inlet of the triaxial cell, previously connected with water pump, was connected to an inert argon gas source followed by a buffer reservoir. Gas pressure was increased from 1 MPa, and by 1 MPa step every one to three days until continuous gas flow was detected at the downstream side. More details can be found in Table 1.

Table 1. Summary of water-injection test

No.	Boundary conditions		Note
	Water-injection test (MPa)	Gas breakthrough test (MPa)	Sample was wrapped by flexible Viton™ membrane directly before put into the triaxial cell
WI-1	$P_g = 0$,	(1) $P_c = 7 (P_g < 6)$	
WI-2	$P_w = 1.5$, $P_c = 3$	(2) P_c exceeds P_g by 1–2 MPa ($P_g > 6$)	

Note: WI means without tube.

3 Experimental Results and Analysis

3.1 Water-Injection Test

Water-injection test lasts more than 200 h. Currently, there is a difficulty to judge whether full saturation is achieved after longtime water injection. Since the sample was put in the triaxial cell and could not be removed immediately because of the following gas breakthrough test. The only method is to calculate the water flow rate. We assume that full saturation is obtained when the value of water flow rate is stable.

As soon as full water saturation is obtained, water permeability is calculated. As presented in Fig. 2, the values of K_w were in the order of magnitude of 10^{-20} m^2, which hints that water tightness can be achieved after longtime water injection (i.e. 10^{-20} m^2–10^{-21} m^2). This is favorable for the bentonite/sand mixture to prevent the migration of nuclides in situ. The values of K_w measured in this study are larger than those measured in previous study (Liu et al. 2016), which is also related to the hydro-mechanical state of the test. As indicated by Liu et al. (2014), the swelling capacity of compacted bentonite/sand mixture under free swelling condition is very large when contacting with water. It means that larger swelling occur under flexible condition than that of semi-rigid condition, which lead to the increase of the porosity and further affect its water permeability.

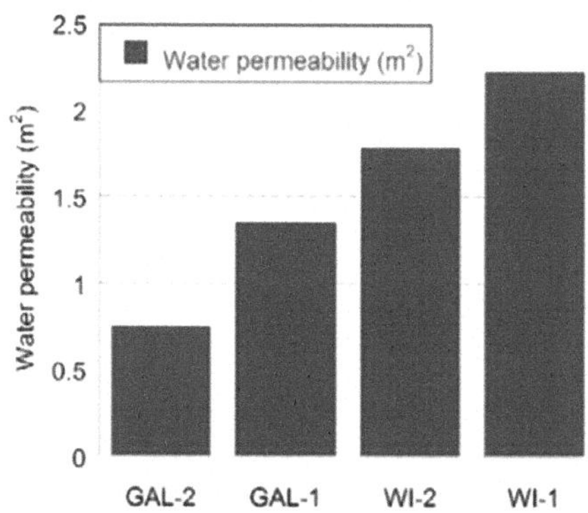

Fig. 2. Water permeability of bentonite-sand samples (WI-1 and WI-2), bentonite-sand sample + grooved aluminum tube (GAL means grooved aluminum tube).

3.2 Gas Breakthrough Test

3.2.1 Effect of Confining Pressure and Gas Pressure on the Gas Migration Property

Figure 3(a)–(d) present the results of gas breakthrough tests of sample WI-1, and the whole process can be divided into four phases. During the first and second phases that confining pressure was kept at 7.8 MPa, the injected gas pressure was increased gradually until 6.5 MPa. At first, we can find that the increasing rate (Q_g) of downstream gas pressure does not increase with the increase in upstream gas pressure until

6.5 MPa (Figs. 3(a) and (b)). This phenomenon may be related to the applied confining pressure and gas pressure. An increase in confining pressure will lead to a decrease in the pore volume, which contribute to the decrease in Q_g. Meanwhile, an increase in gas pressure will increase the pore volume, which contributes to the increase in Q_g. In addition, the creep effect caused by the two effects also play an important role in the gas migration process.

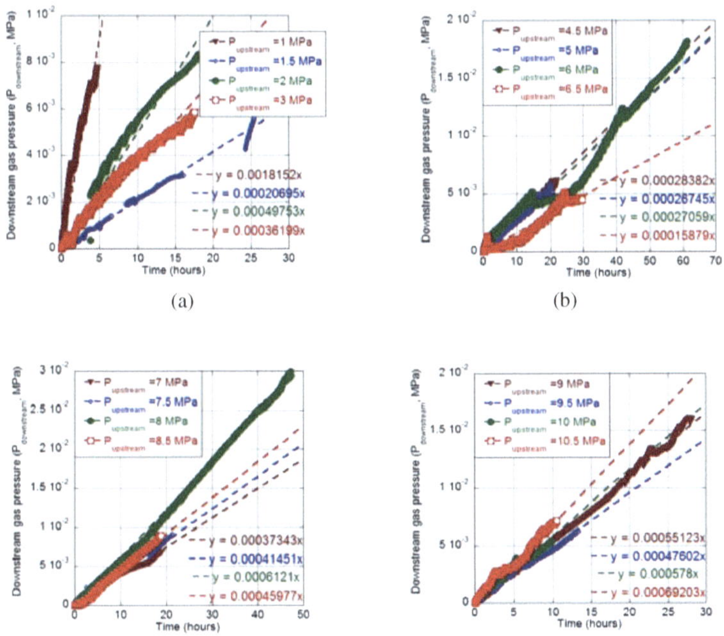

(c) (d)

Fig. 3. Evolution of gas pressure at downstream side with time, while (a) $P_{upstream}$ = 1–6 MPa; (b) $P_{upstream}$ = 7–10.7 MPa (for WI2).

From $P_{upstream}$ = 1 MPa to 1.5 MPa, Q_g decreased while the confining pressure was maintained at 7.8 MPa. It means that the creep effect caused by confining pressure plays a predominant role in the gas migration process. In fact, the deformation of the pore structure caused by the confining pressure usually needs a quite long period to achieve the stabilization. From 1.5 MPa to 2.0 MPa, Q_g increased while the confining pressure was maintained at 7.8 MPa. This means gas pressure plays a predominant role in the gas migration process.

From $P_{upstream}$ = 7 MPa to 10.5 MPa, confining pressure was always higher than confining pressure by 1 MPa to avoid oil leakage. We can find that Q_g increases and decreases at different stage, which hints that confining pressure, gas pressure and the creep effects caused by the two factors are predominant factors at different stage. During the whole process no gas flow was detected at downstream side until

$P_{upstream}$ = 9.0 MPa; however, the detected gas flow was discontinuous and unstable. No continuous gas flow was found during the whole gas injection test.

3.2.2 Comparison of Gas Migration Property Under Flexible and Semi-rigid Conditions

When comparing with our previous tests with semi-rigid tube, it is found that increasing rate of downstream gas pressure of current study is higher than those of all previous tests when the upstream gas pressure was lower than 2 MPa (Fig. 4). This phenomenon has a close relationship with their states after water-injection tests, since different confining pressures (12 MPa vs. 3 MPa) and different confinement conditions (semi-rigid tube vs. Viton™ membrane) lead to different pore structures of the samples and thus cause the different gas migration properties. After that, gas migration property of this study is close to the case with grooved aluminum tube. However, the Q_g of this study is always higher than that of the case with grooved aluminum tube, but lower than the case with smooth aluminum tube and smooth/grooved COx argillite tube.

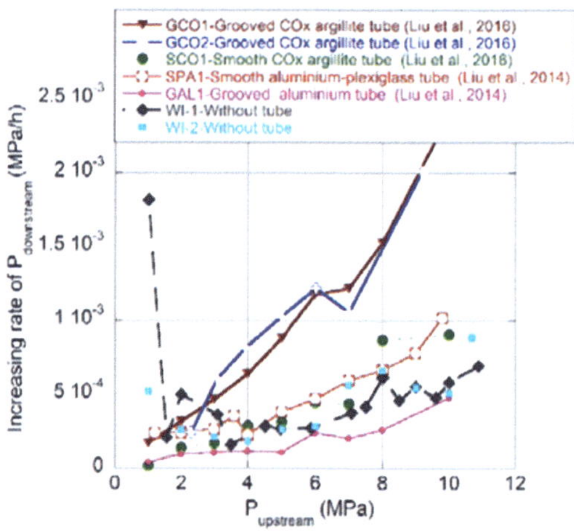

Fig. 4. Evolution of gas flow rate at downstream side with $P_{upstream}$ for (a) WI2 and (b) comparison with other tests (Liu et al. 2018) (Note: GCO means grooved CO_x argillite tube; SCO means smooth COx argillite tube; SPA means smooth aluminium-plexiglass tube and GAL means grooved aluminium tube).

As indicated by Liu et al. (2015), the interface and COx argillite are more likely to be gas migration pathways when comparing with saturated bentonite/sand mixture. This is why the increasing rate of downstream gas pressure of this study is lower than the cases with smooth aluminum tube and smooth/grooved COx argillite tube. When comparing with the case of grooved aluminum tube, their differences are related to the different stress conditions around the sample and different pore structures formed during the previous water-injection tests.

4 Conclusions

This study aims to investigate the hydraulic and gas migration properties in a fully saturated bentonite–sand mixture under flexible condition. After full water saturation, the water permeability of the bentonite/sand mixture is at the magnitude of 10^{-20} m^2, hinting that the water-tightness of bentonite/sand mixture is achieved. Further gas breakthrough test indicates that gas transport properties in fully saturated bentonite/sand mixture are closely related to the confinement conditions, external stress state, internal fluid pressure and the creep effect. At some stages the increasing rate of downstream gas pressure continue to decrease even through the confining pressure was kept constant and the gas pressure was increased gradually. This means that confining pressure has a hysteresis effect on the gas transport property. Also at some other stages, the effect of gas pressure has a larger influence effect on the gas transport property than that of confining pressure. When comparing with previous studies with semi-rigid tubes, it is found that gas migration property is more sensitive to the change of confining pressure and gas pressure due to flexible confinement condition.

Acknowledgments. The authors are grateful for the support provided by the Natural Science Foundation of Jiangsu Province of China (BK20160249), the Open Fund (KLE-TJGE-B1703) of Key Laboratory of Geotechnical and Underground Engineering of Ministry of Education (Tongji University) and the Fundamental Research Funds for the Central Universities (China University of Mining and Technology) (2017QNA29).

References

Cui YJ, Tang AM, Loiseau C, Delage P (2008) Determining the unsaturated hydraulic conductivity of a compacted sand–bentonite mixture under constant-volume and free-swell conditions. Phys Chem Earth Parts A/B/C 33:S462–S471

Davy CA, Skoczylas F, Lebon P, Dubois T (2009) Gas migration properties through a bentonite/argillite interface. Appl Clay Sci 42:639–648

Harrington JF, Milodowski AE, Graham CC, Rushton JC, Cuss RJ (2012a) Evidence for gas-induced pathways in clay using a nanoparticle injection technique. Mineral Mag 76:3327–3336

Harrington JF, De la Vaissiére R, Noy DJ, Cuss RJ, Talandier J (2012b) Gas flow in Callovo-Oxfordian claystone (COx): results from laboratory and field-scale measurements. Mineral Mag 76:3303–3318

King F, Ahonen L, Taxén C, Vuorinen U, Werme L (2001) Copper corrosion under expected conditions in a deep geologic repository, vol 1. Svensk Kärnbränslehantering AB, Stockholm

Liu JF, Skoczylas F, Liu J (2014) Experimental research on water retention and gas permeability of compacted bentonite/sand mixtures. Soils Found 54:1027–1038

Liu JF, Skoczylas F, Talandier J (2015) Gas permeability of a compacted bentonite-sand mixture: coupled effects of water content, dry density and confining pressure. Can Geotech J 52(99):1–9

Liu JF, Song Y, Skoczylas F, Liu J (2016) Gas migration through water-saturated bentonite-sand mixtures. COx argillite and their interfaces. Can Geotech J 53(1):60–71

Liu JF, Song SB, Ni HY, Cao XL, Pu H, Mao XB, Skoczylas F (2018) Research on the gas migration properties in a saturated bentonite/sand mixture under flexible boundary condition. Soils Found 58(1):97–109

Ye WM, Wan M, Chen B, Chen YG, Cui YJ, Wang J (2012) Temperature effects on the unsaturated permeability of the densely compacted GMZ01 bentonite under confined conditions. Eng Geol 126:1–7

Salinity and Clay Mineralogy Effects on Consolidation Behavior of Pure Clays

Tongwei Zhang[1(✉)], Yongfeng Deng[2], Haocheng Xue[2], Xingjun Zhang[3], and Shijun Wang[4]

[1] Key Laboratory of Mechanics on Disaster and Environment in Western China, College of Civil Engineering and Mechanics, Lanzhou University, Lanzhou 730000, China
ztw@lzu.edu.cn
[2] Institute of Geotechnical Engineering, Transportation College, Southeast University, Nanjing, China
[3] Gansu Henlu Transportation Survey and Design Co., Ltd., Lanzhou, China
[4] Economy and Technology Research Institute, State Grid Gansu Electric Power Company, Lanzhou, China

Abstract. Pure clays or sand-clay mixtures with different clay mineral are widely used in liners or vertical barriers of waste landfill, and the pore water was characterized by its high salinity in coastal area. Thus, it is essential to understand the salinity effects on their consolidation behavior. In this study, oedemeter and SEM tests were conducted on four pure clays with different mineralogy. Several aspects of soil behavior, such as compression index, the end of primary consolidation stage and micro-structure, were investigated in view of pore water salinity effect. The results show that higher ionic concentration significantly decreases the compression index C_c and liquid limits (*LL*) for clays containing montmorillonite. While the period of primary consolidation was shorten under the impact of water salinity. For kaolinite with or without sodium chloride, a similar C_c, *LL* and the ending time of primary consolidation stage t_p were observed. In view of flocculation, the salinity effect on consolidation behavior was interpreted. Increasing in salinity allows a flocculation and particle readjustment at constant water content for montmorillonitic clay. Then, larger inter-particle spaces and effective void ratio increased the permeability and decreased the drainage period, and larger internal frictions decreased the compression index C_c and *LL*.

Keywords: Pure clays · Water salinity · Clay minerals
Consolidation behaviour · Micro-structure

1 Introduction

Previous studies revealed that changing of pore water chemistry (such as cation concentration, cation valence and dielectric constant etc.) has significant influences on physical and mechanical behaviour of clays, especially on swelling clays (Yukselen-Aksoy et al. 2008; Abdullah et al. 1997; Bolt 1956; Di Maio et al. 2004; Kaya and Fang 2000; Oren and Kaya 2003). Bolt (1956) firstly stated that the compressibility of

L. Zhan et al. (Eds.): ICEG 2018, ESE, pp. 529–536, 2019.
https://doi.org/10.1007/978-981-13-2224-2_65

pure clay suspensions could be accounted well by considering the interaction between double electric layers around clay particles. Di Maio et al. (2004) presented that an increase in pore solution concentration may cause a reduction in compressibility of compacted bentonite. Gajo and Maines (2007) investigated the mechanical behaviour of active clays affected by pore fluid acidity and alkalinity. Zhang et al. (2016) and Ye et al. (2017) systematically investigated the influences of infiltrating solutions on volume change behaviour and hydraulic conductivity of highly compacted GMZ bentonite. However, the salinity and clay mineral effect on their consolidation behaviour of pure clays is still few concerned.

This paper focuses on the pore water salinity and clay mineral effect on the consolidation behaviour and micro-structure of four soils at high initial water contents (0.6–1.5 times liquid limits). To simulate this variety of mineral composition and porewater salinity, a mixture of commercial kaolin, bentonite and NaCl solution was employed in this investigation. The salinity effect on the compression index C_c and the ending time of primary consolidation were examined. With the help of micro-structure observations, a reasonable mechanism of salinity effect on macro consolidation behaviour was suggested.

2 Experimental Programs

2.1 Materials

To prepare specimens in laboratory, commercial kaolin and bentonite from Jiangsu province in China, were mixed with different proportions based on the dry weight ratio. It should be noted that the composition of mixtures was selected based on the typical marine clay deposited in Lianyungang area in Jiangsu Province. The basic physical properties of materials are shown in Table 1. Liquid limit (*LL*) and plastic limit (*PL*) was determined using Casagrande method according to British standard BS1377-2 (1990). Besides, the specific surface area (SSA) was determined using the Ethylene Glycol Monoethyl Ether (EGME) method (Cerato and Lutenegger 2002; Fan et al. 2014). By the X-ray diffraction analysis, the mineral compositions were presented in Table 2.

Table 1. Physical properties of the considered materials

Soil types	Specific gravity G_s	*LL* (%)	*PL* (%)	*PI* (%)	SSA (m^2/g)
Kaolinite	2.66	33	16	17	46
Bentonite	2.73	301	95	206	378.5

Table 2. Mineral composition of the kaolinite and bentonite

Materials	Mineral composition (%)								
	Primary mineral (%)						Clay mineral (%)		
	Quartz	K-feldspar	Plagioclase	Calcite	Dolomite	Siderite	Illite	Kaolinite	Smectite
Kaolinite	4.0	5.1	17.1	3.5	2.1	0		68.2	
Bentonite	1.3	0.3	0	0.1	0	2.0	3.9	1.9	90.5

2.2 Specimen Preparation

The pure kaolin or mixed clays were added with distilled water or sodium chloride solutions at a mass concentration of 1%, 3%, 5% and 10% (equal to 0.17 mol/L, 0.51 mol/L, 0.86 mol/L and 1.7 mol/L). For the artificial mixed clays, three mass ratios of kaolin to bentonite at 19/1, 9/1 and 4/1 were designed. A preset mass of dry powder and solutions were thoroughly stirred for at least 10 min using a paddle mixer. Thereafter, the slurries were cured in a sealed plastic chamber for a minimum of 24 h to homogenize the distribution of the pore liquid. The indexes of mixtures were summarized in Table 3. The selected water contents were set as 0.6 to 1.2 times of *LL*. Hereafter, for K5%B95% as an example, 'B' and 'K' in sample identification is shorted for 'bentonite' and 'kaolin' clay, and the value represents the percentage of bentonite and kaolin clay in the mixture.

Table 3. Initial state of samples

Sample ID	Bentonite content (%)	G_s	w_0 (%)	*LL* (%)	e_0	w_0/LL
K	0	2.66	41 ± 1	33	1.08 ± 0.01	1.2
B5%K95%	5	2.664	42 ± 1	35	1.10 ± 0.02	1.2
B10%K90%	10	2.667	43 ± 1	42	1.13 ± 0.02	1.0
B20%K80%	20	2.674	33 ± 1	52	0.92 ± 0.01	0.6
B20%K80%Micro	20	2.674	63 ± 1	52	1.78 ± 0.02	1.2
KMicro	0	2.66	43 ± 1%	33	1.24	1.3

2.3 Oedometer Test

A total of 24 specimens were prepared by mixing kaolin and bentonite at different proportions. These mixtures were divided into 6 groups: K, B5%K95%, B10%K90%, B20%K80%, B20%K80%Micro and KMicro. The first 4 groups were used to conduct oedometer tests, and each specimen was mixed with NaCl solution at concentrations 0%, 1%, 3%, 5% and 10%. The mixtures were carefully installed to the oedometer ring. Then, samples were first saturated by soaking with the equivalent concentrations of sodium chloride for 48 h. For the slurry samples ($w_0 > LL$), vertical loads 1 kPa, 1.8 kPa, 3.2 kPa and 5.8 kPa were supplied step by step in order to avoid squeezing the soil from the ring (Hong et al. 2010). For sample B20%K80% ($w_0 < LL$), an initial load 12.5 kPa was applied avoiding the swelling influence. Each load was kept for at least 24 h until the displacement satisfied the standard requirement (ASTM-D2435 2011). Additionally, the organic plastic oedometer cell was applied in order to keep the concentration in the cell and prevent the cell erosion.

2.4 Micro-structural Investigation

As mentioned above, the last 2 groups (B20%K80%Micro and KMicro) were used to perform SEM (Scanning electronic microscopy) tests, aiming to clarify the mechanism of salinity effect on the intrinsic compression behavior of soft clay. Samples were consolidated by stepwise applying the vertical load from 1 kPa to 100 kPa, and then removed from oedometer cells. Thereafter, all samples were carefully trimmed to appropriate sizes and then immersed in liquid nitrogen for instant freezing (Zhang et al. 2014). Finally, they were lyophilized and vacuum coated with a layer of gold to conduct SEM tests.

3 Results and Discussions

3.1 Salinity Effects on Compression Index C_c

As a common knowledge, the compression index C_c was defined by the slope of straight portion of compression curve. The Evolutions of compression curves, *LL* and C_c changing with concentrations of NaCl solution were presented in Fig. 1(a)–(b). It shows that the *LL* and C_c of B5%K95%, B10%K90% and B20%K80% all decrease with pore water salinity, whilst those of kaolinic clay are relatively stable. Note that the variation trend of *LL* shown in Fig. 1(b) is consistent with the results in previous publications (Abdullah et al. 1997; Di Maio et al. 2004; Kaya and Fang 2000; Oren and Kaya 2003; Sridharan and Rao 1975; Yukselen-Aksoy et al. 2008). Among them, Sridharan and Rao (1975) proposed a mechanism on the *LL* of the clay, i.e., for montmonrillonite, *LL* is predominantly controlled by the thickness of double diffusion layer. Thickness suppression of the double diffusion layer caused by increasing the salt concentration, leads to *LL*'s reduction. The similar variation trends of Fig. 1(a) and (b) indicate that the mechanism accounting for pore water salinity effect on compression behavior of clays is probably related to the suppression of double diffusion layer.

3.2 Salinity Effects on Consolidation Behavior

The primary consolidation stage is defined in terms of the change in effective vertical stress. In the absence of pore water pressure measurements, the end of primary consolidation stage is defined using settlement observations. The elapsed time t_p represents the time required for the completion of primary consolidation. The typical relationships between void ratio e and elapse time t of four clays are shown in Fig. 2(a)–(d). As illustrated in Fig. 2(a), the primary consolidation ended in 10 min for kaolinite mixed with distilled water and 10% NaCl solutions, and the ending point of primary consolidation t_p were not significantly affected by pore water salinity and applied stresses. In contrary, the results as shown in Fig. 2(b)–(d) indicate that the period of primary

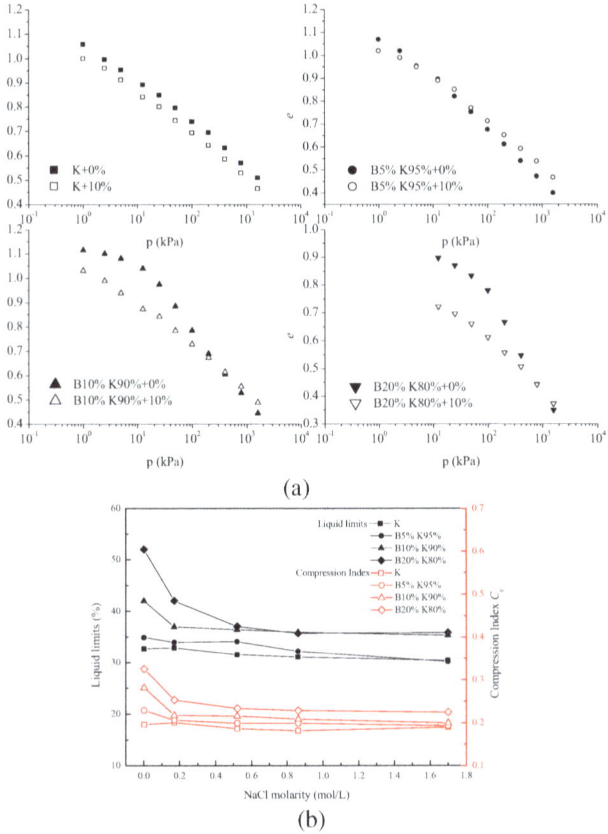

Fig. 1. Liquid limits *LL* and compression index C_c *vs.* NaCl molarity: (a) Compression curves; (b) C_c *vs.* NaCl molarity and *LL vs.* NaCl molarity

consolidation behavior of clays containing montmorillonite are apparently influenced by water salinity. The t_p of B5%K95%, B10%K90% and B20%K80% decreased by nearly an order of magnitude when the saline concentration increased to 10%. According to Terzaghi's one-dimensional consolidation theory, the decreasing of t_p represents a shorter drainage time and a higher consolidation coefficient. The permeability coefficient *k* is directly proportional to consolidation coefficient C_v, and it can be concluded that the permeability of pure clays containing montmorillonite increased with pore water salinity increasing. The mechanism attributed to the above findings was further discussed by micro-structure observations.

3.3 Microstructure and Mechanism

The salinity effect on microstructures was examined by binary SEM images (magnified 5000 times) as shown in Fig. 3. The fabric element of specimen B20%K80% Micro-01 hydrated with distilled water (Fig. 3a) exists a shape of thin platelet; whilst that of

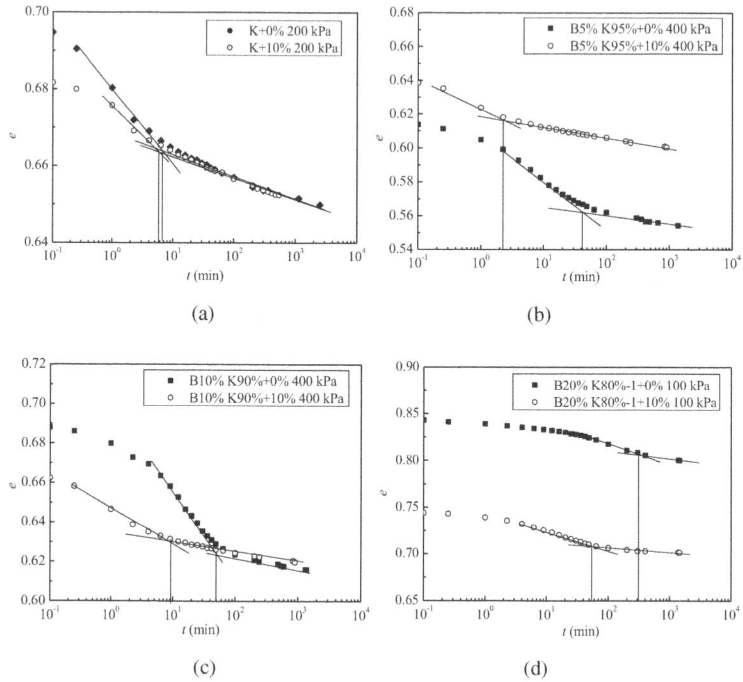

Fig. 2. The *e*-lg*t* curves of specimens: (a) K with 0% and 10% under 200 kPa; (b) B5%K95% with 0% and 10% under 400 kPa; (c) B10%K90% with 0% and 10% under 400 kPa; (d) B20% K80% with 0% and 10% under 100 kPa

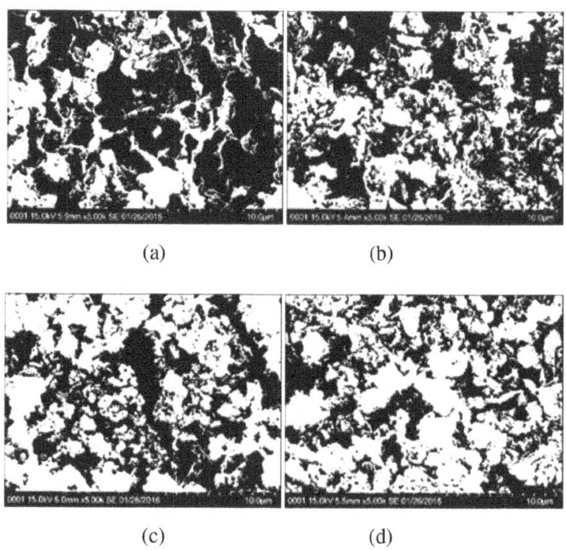

Fig. 3. SEM photos (magnified 5000 times): (a) B20%K80%Micro-01; (b) B20%K80%Micro-02; (c) KMicro-01; (b) KMicro-02;

specimen B20%K80%Micro-02 salinized with 0.86 mol/L NaCl solution (Fig. 3b) presents a shape of bulky pellet. On the other hand, for specimens only composed of kaolin (Fig. 3c and d), the similar arrangement and shape of bulky pellet are identified whatever with or without porewater salinity. Hence flocculation mechanism could be introduced to interpret the salinity effect on normalized compression behavior for reconstituted clay, especially for that with rich montmorillonite.

4 Conclusions

To better understand the salinity and clay mineral effects on the consolidation behaviour of clays, oedometer tests and SEM tests were conducted on four artificial clays. The main conclusions can be drawn as belows:

(1) It shows that the *LL* and C_c of pure clays containing bentonite all decreased with pore water salinity, whilst those of kaolinic clay are relatively stable.
(2) The ending points of primary consolidation of kaolinite were not significantly affected by pore water salinity. In contrary, the period of primary consolidation for clays containing montmorillonite decreased with pore water salinity.
(3) The SEM observation shows that the fabric element of B20%K80% with a shape of thin platelet tend to be flocculated when the water salinity increased, whilst the micro-structure of kaolinite were not influenced by the water salinity.

Acknowledgements. This study is supported by Fundamental Research Funds for the Central Universities (Grant lzujbky-2018-8), and the National Natural Science Foundation of China (General Program, Grant Nos. 51378117; 41572280, 41790443, 41525010) and Postgraduate Research & Practice Innovation Program of Jiangsu Province (Grant No. KYCX17-0131).

References

Abdullah WS, Alzoubi MS, Alshibli KA (1997) On the physicochemical aspects of compacted clay compressibility. Can Geotech J 34(4):551–559

Bolt GH (1956) Physico-chemical analysis of the compressibility of pure clays. Géotechnique 2 (6):86–93

Cerato AB, Lutenegger AJ (2002) Determination of surface area of fine-grained soils by the ethylene (EGME) method. Geotech Test J 25(3):315–321

Di Maio C, Santoli L, Schiavone P (2004) Volume change behaviour of clays: the influence of mineral composition, pore fluid composition and stress state. Mech Mater 36(5–6):435–451

Fan RD, Du YJ, Reddy KR, Liu SY, Yang YL (2014) Compressibility and hydraulic conductivity of clayey soil mixed with calcium bentonite for slurry wall backfill: Initial assessment. Appl Clay Sci 101:119–127

Gajo A, Maines M (2007) Mechanical effects of aqueous solutions of inorganic acids and bases on a natural active clay. Géotechnique 57(8):687–699

Kaya A, Fang HY (2000) The effects of organic fluids on physicochemical parameters of fine-grained soils. Can Geotech J 37(5):943–950

Oren AH, Kaya A (2003) Some engineering aspects of homoionized mixed clay minerals. Environ Monit Assess 84(1–2):85–98

Sridharan A, Rao GV (1975) Mechanisms controlling the liquid limit of clays. In: Proceeding of the Istanbul conference on soil mechanics and foundation engineering, pp 75–84

Ye W-M, Zhang F, Chen Y-G, Chen B, Cui Y-J (2017) Influences of salt solutions and salinization-desalinization processes on the volume change of compacted GMZ01 bentonite. Eng Geol 222:140–145

Yukselen-Aksoy Y, Kaya A, Oren AH (2008) Seawater effect on consistency limits and compressibility characteristics of clays. Eng Geol 102(1–2):54–61

Zhang F, Ye W-M, Chen Y-G, Chen B, Jun Cui, Yu (2016) Influences of salt solution concentration and vertical stress during saturation on the volume change behavior of compacted GMZ01 bentonite. Eng Geol 207:48–55

Zhang T, Yue X, Deng Y, Zhang D, Liu S (2014) Mechanical behaviour and micro-structure of cement-stabilised marine clay with a metakaolin agent. Constr Build Mater 73:51–57

Sorption of Lead to Slurry Trench Cutoff Wall Backfills Comprised of SHMP-Amended Ca-Bentonite

Yu-Ling Yang[1], Krishna R. Reddy[2], Yan-Jun Du[1(✉)], and Ri-Dong Fan[1]

[1] Jiangsu Key Laboratory of Urban Underground Engineering and Environmental Safety, Institute of Geotechnical Engineering, Southeast University, Nanjing 210096, China
duyanjun@seu.edu.cn
[2] Department of Civil and Materials Engineering, University of Illinois at Chicago, Chicago, IL 60607, USA

Abstract. This study investigated effects of phosphate dispersant on lead (typical heavy metal) sorption capacity of the sand/Ca-bentonite backfill for slurry trench cutoff walls. Several series of batch sorption tests were conducted using backfills consisting of sand and 20% Ca-bentonite with/without dispersant, as well as amended/un-amended Ca-bentonite alone. First, amended/un-amended backfills and amended Ca-bentonite were prepared. Next, aqueous solutions with 0 to 40,000 mg/L Pb concentration were prepared. Finally, the backfills and the Ca-bentonites were mixed with the solution in liquids to solid ratio of 10 and shaken for 24 h. The mixture was then centrifuged, and the supernatant was separated and analyzed for Pb concentration. The results show that the relationship between the amount of sorption and equilibrium concentration of Pb was well quantified by Langmuir and Freundlich isotherm models. Phosphate dispersant significantly increased Pb sorption capacity of both the sand/Ca-bentonite backfill and Ca-bentonite. The sorption process of the Pb onto the backfills or Ca-bentonites was found to be favorable. Overall, phosphate dispersant is shown to be a promising additive to sand/Ca-bentontie backfill in slurry trench cutoff wall construction for effective containment of Pb-laden groundwater.

Keywords: Calcium bentonite · Slurry walls · Sorption · Lead

1 Introduction

Soil-bentonite (SB) slurry trench cutoff walls composed of in-situ excavated soil or appropriate borrowed soil and high quality sodium bentonite (Na-bentonite) are commonly used in the U.S. and Canada as contaminated groundwater containment systems [1, 2]. The low hydraulic conductivity and the contaminant sorption capacity are two key mechanisms controlling containment performance of the walls. Recently, greater attention is being paid on improving sorption capacity of the SB walls to extend

L. Zhan et al. (Eds.): ICEG 2018, ESE, pp. 537–543, 2019.
https://doi.org/10.1007/978-981-13-2224-2_66

the breakthrough time for contaminant migrating within the walls, particularly when relatively high hydraulic gradient is applied horizontally across the wall [3–9].

In country such as China, where high quality Na-bentonite is scarce and contaminant containment is urgently needed, commonly available calcium bentonite (Ca-bentonite) is considered as an alternate material for use. However, Ca-bentonite exhibits lower swelling and sorption capacities than those of Na-bentonite, leading to unacceptable higher hydraulic conductivity and lower contaminant retention capacity of sand/Ca-bentonite backfills, thus proper modification is needed for the Ca-bentonite before its use.

In previous studies, the authors have shown that sodium hexametaphosphate (SHMP), one of phosphate dispersants, has positive effects on improving workability of Ca-bentonite/water slurry and decreasing hydraulic conductivity of the sand/Ca-bentonite backfills [10–12]. According to these studies, the optimum SHMP dosage for the amended Ca-bentonite is recommended to be 2% (dry weight basis), and the optimum amended Ca-bentonite content for the sand/Ca-bentonite backfill is recommended to be 20% (dry weight basis). However, by date, sorption of industrial contaminants to the SHMP amended sand/Ca-bentonite backfills for slurry trench cutoff walls has not been investigated.

Based on the above mentioned considerations, this study is performed to address the following issues: (1) assessing sorption behavior of Pb to amended/un-amended Ca-bentonites and sand/Ca-bentonite backfills by conducting batch sorption tests; (2) analyzing the sorption test data by means of Langmuir and Freundlich isotherm models; and (3) comparing Pb sorption capacity of the amended backfill and bentonite with that of the un-amended backfill and bentonite.

2 Materials and Methods

2.1 Constituent Soils and Sample Preparation

A powdered un-amended Ca-bentonite (CaB) and SHMP-amended Ca-bentonite (SHMP-CaB), a clean sand, and granular SHMP were used as the solid materials in this study. The CaB with trade name of Panther Creek® 200 was provided by the Colloid Environmental Technologies Company (CETCO, Hoffman Estates, IL, USA). The SHMP-CaB was manufactured in laboratory by mixing the CaB into a SHMP-distilled water solution with SHMP concentration of 10 g/L. The soil-to-water ratio of the resulting slurry was approximately 1:2 (g/mL), and the optimum SHMP-to-bentonite ratio was 2% (dry weight basis). The slurry was then subjected to curing, oven dried, milled and screened (using 0.075 mm standard sieve) following procedures described in previous publications by the authors [11, 12]. The sand known as Three River Sand was obtained from a quarry in Beloit, WI, USA. The used SHMP was obtained from Humboldt Manufacturing Co., Elgin, IL, USA. The bentonite contents of the amended/un-amended sand/Ca-bentonite mixtures were controlled at 20% and 100% (bentonite alone) by dry weight.

A stock solution with Pb concentration of 40000 mg/L was created by dissolving predetermined amount of lead nitrate, $Pb(NO_3)_2$, into deionized water. The sorbate solutions were then prepared by diluting the stock solution to initial Pb concentration (C_0) ranging from 5 to 40,000 mg/L. The pH and electrical conductivity (*EC*) of the sorbate solutions are plotted as a function of C_0 as presented in Fig. 1.

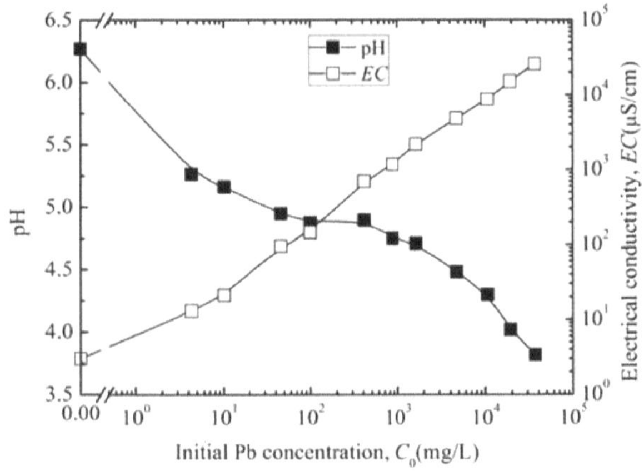

Fig. 1. pH and electrical conductivity (*EC*) of the sorbate solutions versus initial Pb concentration

2.2 Batch Sorption Tests

The batch sorption tests were performed as per ASTM D4646, except that a solid-to-liquid ratio of 1:10 was used in this study as per literature [13]. 10 g of sorbent and 100 mL of sorbate solution were placed in a 200 mL plastic bottle and mixed at approximately 29 rpm for 24 h using a mechanical tumbler at room temperature of 20 °C. The supernatant in the plastic bottle was then extracted and subjected to centrifuge process. After that, the centrifuged supernatant was filtered using 0.45 μm filter paper, and acidized to pH lower than 2 using concentrated nitric acid. The Pb concentration of the resultant supernatant was analyzed using an atomic absorption spectrophotometer.

3 Results and Analysis

3.1 Sorption Results

The amount of Pb sorbed to unit dry mass of the bentonites or backfills (q_e, with unit of mg/kg) was calculated based on the mass of the dry backfill or bentonite (M_s, which is 0.01 kg here), volume of added Pb solution (V, with value of 0.1 L in this study), initial Pb concentration (C_0, with unit of mg/L), and sorbate concentration at equilibrium (C_e, with unit of mg/L) using the following equation:

$$q_e = \frac{(C_0 - C_e)M_s}{V_L} \tag{1}$$

Figure 2 depicts the variation in amount of sorption, q_e, with equilibrium sorbate concentration, C_e, for amended bentonite and backfill (SHMP-CaB and SHMP-20CaB) as well as un-amended bentonite and backfill (CaB and 20CaB). It can be seen from Fig. 2 that q_e-C_e shows significant nonlinear relationship regardless of sorbent types. As C_e value increases, there is a first sharp and then gradual increasing trend in the q_e value. The q_e-C_e curves of the SHMP-CaB and SHMP-20CaB are found to surpass that of their un-amended counterpart, i.e., CaB and 20CaB. It is worthy to note that although only 4 visible data points are presented in Fig. 1 for each sorbents, there are additional 8 data points contained in the range of C_e values lower than 30 mg/L.

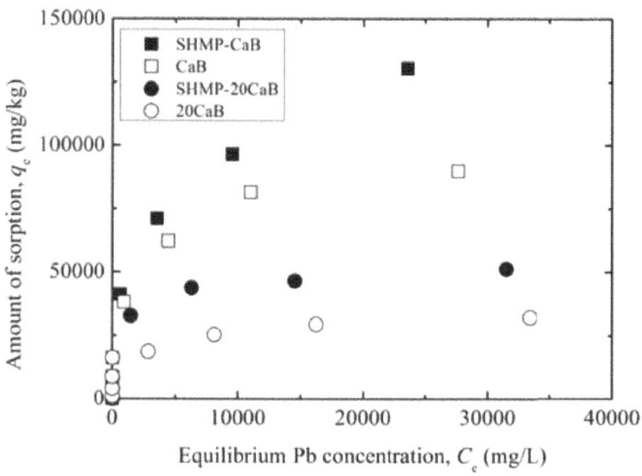

Fig. 2. Measured q_e-C_e data for Pb sorbed to amended/un-amended Ca-bentonites or backfills.

3.2 Sorption Isotherm Models

Three sorption isotherm models, specifically Langmuir and Freundlich models, were investigated to represent the q_e-C_e relationship. The Langmuir model is given by the following equation [14]:

$$q_e = \frac{K_L q_{m,L} C_e}{1 + K_L C_e} \tag{2}$$

where $q_{m,L}$ is the sorption capacity (unit of mg/kg); K_L is the Langmuir isotherm constant (unit of L/mg). The Freundlich model is given as follows [15]:

$$q_e = K_F C_e^{1/n_F} \tag{3}$$

where the K_F is the Freundlich constant (unit of L/kg); n_F is the sorption intensity which is dimensionless parameter. An n_F value between 1 to 10 is associated with favorable sorption.

Figures 3 and 4 show q_e-C_e relationships fitted via a least-square-root method by Langmuir and Freundlich models, respectively. The measured data are well fitted by the two models with high coefficient of determination (r^2) values. For example, a $0.912 \leq r^2 \leq 0.965$ for Langmuir model and a $0.913 \leq r^2 \leq 0.990$ for Freundlich model, respectively. The fitted parameters of Langmuir and Freundlich models are tabulated in Table 1.

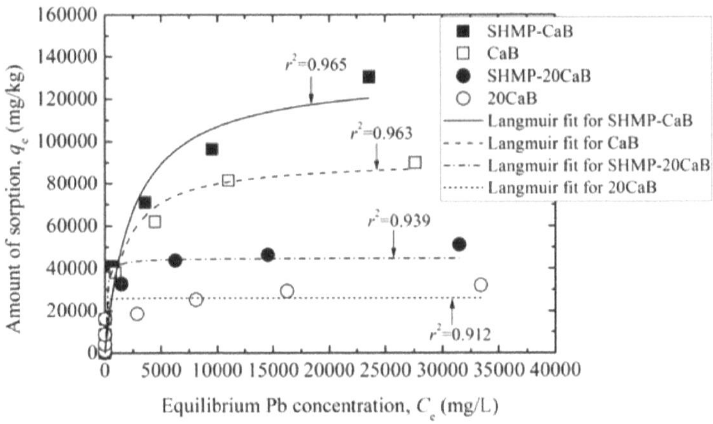

Fig. 3. Langmuir fitting results for amended/un-amended Ca-bentonites or backfills.

Based on the Langmuir model, sorption capacities of the amended/un-amended bentonite are estimated as 132,592 mg/kg versus 91,467 mg/kg, and the sorption capacities of the amended/un-amended backfill are 44,824 mg/kg versus 26,044 mg/kg. This indicates an increase in sorption capacity of the bentonite and backfill by factors of 45% and 72%, respectively, after the SHMP amendment. The increase in Pb sorption capacity of the amended sorbents is attributed to: (1) increased surface area of the CaB by the SHMP. The grain size and surface area of the CaB is significantly increased after amended by the SHMP [16], thus provided more sites for Pb sorption. The CaB possesses lower surface area, therefore corresponds to lower sorption capacity; (2) increased surface charge density of the CaB by SHMP. The SHMP-CaB exhibits more negative zeta potential as compared to the CaB [16], resulting in increased negative surface charge density on the amended bentonite particles and thus is beneficial to the sorption of Pb; (3) interaction between the SHMP and Pb might also be a reason for greater sorption capacity of the SHMP-CaB. The sorption intensity (n_F) for all of the sorbents (ranging from 2.81 to 5.31) is higher than 1.0, implying a favorable sorption.

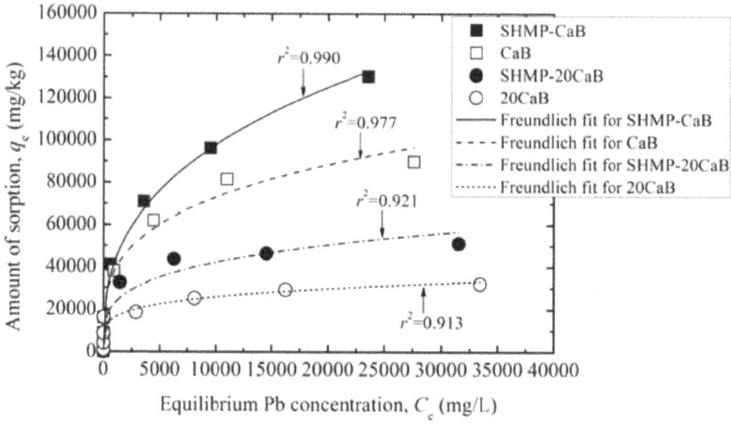

Fig. 4. Freundlich fitting results for amended/un-amended Ca-bentonites or backfills.

Table 1. Summary of best-fit Langmuir and Freundlich, model parameters for Pb sorbed to un-amended and SHMP-amended bentonites and backfill mixtures.

Fitting parameters	CaB	SHMP-CaB	20CaB	SHMP-20CaB
Langmuir model				
$q_{m,L}$ (mg/kg)	91,467	132,592	26,044	44,824
K_L (L/mg)	0.000687	0.000416	0.146	0.0103
r^2	0.963	0.965	0.912	0.939
Freundlich model				
K_F (L/kg)	6,014	3,700	4,602	4,070
n_F	3.69	2.81	5.31	3.94
r^2	0.977	0.990	0.913	0.921

4 Conclusions

This study investigated the sorption behavior of Pb to SHMP amended sand/Ca-bentonite backfills and Ca-bentonite alone via batch sorption tests. The results indicated that all of the tested materials displayed nonlinear sorption to Pb, which were well fitted by Langmuir and Freundlich models. The SHMP amendment increased Pb sorption capacity relative to that of un-amended bentonite and backfill by a factor of 45% and 72%, respectively. The enhanced sorption was attributed to the increased surface area and charge density of the bentonite after SHMP amendment, as well as reaction between the SHMP and Pb.

Acknowledgements. Financial support for this project was provided by Primary Research & Development Plan of Jiangsu Province (Grant No. BE2017715) and National Natural Science Foundation of China (Grant Nos. 41330641, 51278100 and 41472258). The authors thank the Colloid Environmental Technologies Co. (CETCO) for providing the bentonite used in this study, and the China Scholarship Council for making it possible to undertake this research at the University of Illinois at Chicago.

References

1. D'Appolonia DJ (1980) Soil-bentonite slurry trench cutoffs. J Geotech Eng Div 106(4):399–417
2. Sharma HD, Reddy KR (2004) Geoenvironmental engineering: site remediation, waste containment, and emerging waste management technologies. Wiley, Hoboken
3. Evans JC, Adams TL, Prince MJ (1997) Metals attenuation in minerally enhanced slurry walls. In: 1997 International containment technology conference, National Technical Information Service, pp 679–687. Springfield, VA
4. Du YJ, Hayashi S, Xu YF (2004) Some factors controlling the adsorption of potassium ions on clayey soils. Appl Clay Sci 27(3):209–213
5. Du YJ, Hayashi S (2006) A study on sorption properties of Cd^{2+} on Ariake clay for evaluating its potential use as a landfill barrier material. Appl Clay Sci 32(1):14–24
6. Malusis MA, Maneva JE, Barben EJ, Shackelford CD, Daniels ER (2010) Influence of adsorption on phenol transport through soil–bentonite vertical barriers amended with activated carbon. J Contam Hydrol 116(1):58–72
7. Fan RD, Du YJ, Liu SY, Yang YL (2014) Sorption of Pb(II) from aqueous solution to clayey soil/calcium-bentonite backfills for slurry-trench walls. In: 7th International congress on environmental geotechnics, pp. 1566–1573. Engineers Australia, Melbourne
8. Hong CS, Shackelford CD, Malusis MA (2016) Adsorptive behavior of zeolite-amended backfills for enhanced metals containment. J Geotech Geoenviron Eng 142(7):04016021
9. Yang YL (2017) Contaminant containment performances of soil-phosphate amended calcium bentonite vertical cutoff walls. Ph.D. thesis, Southeast University, Nanjing, China
10. Du YJ, Yang YL, Fan RD, Wang F (2016) Effects of phosphate dispersants on the liquid limit, sediment volume and apparent viscosity of clayey soil/calcium-bentonite slurry wall backfills. KSCE J Civ Eng 20(2):670–678
11. Yang YL, Du YJ, Reddy KR, Fan RD (2017) Phosphate-amended sand/Ca-bentonite mixtures as slurry trench wall backfills: assessment of workability, compressibility and hydraulic conductivity. Appl Clay Sci 142:120–127
12. Yang YL, Reddy K, Du YJ, Fan RD (2018) Short-term hydraulic conductivity and consolidation properties of soil-bentonite backfills exposed to CCR-impacted groundwater. J Geotech Geoenviron Eng 144(6):04018025
13. Reddy KR, Xie T, Dastgheibi S (2014) Removal of heavy metals from urban stormwater runoff using different filter materials. J Environ Chem Eng 2(1):282–292
14. Langmuir I (1916) The constitution and fundamental properties of solids and liquids. Part I. Solids. J Am Chem Soc 38(11):2221–2295
15. Freundlich HMF (1906) Over the adsorption in solution. J Chem Phys 57:385–471
16. Yang YL, Reddy K, Du YJ, Fan RD (2018) SHMP amended calcium bentonite for slurry trench cutoff walls: workability and microstructure characteristics. Can Geotech J 55(4):528–537

Unsaturated Hydraulic Conductivity of Compacted Bentonite: Revisit of Microstructure Effects

Tian Chen[1], Majid Sedighi[1(✉)], Andrey Jivkov[1], and Suresh C. Seetharam[2]

[1] School of Mechanical, Aerospace and Civil Engineering, The University of Manchester, Manchester M13 9PL, UK
majid.sedighi@manchester.ac.uk
[2] Engineered and Geosystems Analysis Unit, Belgian Nuclear Research Centre, SCK-CEN, 2400 Mol, Belgium

Abstract. Developing a sound theoretical description for the evolution of unsaturated hydraulic conductivity of compacted bentonite during re-saturation has remained a challenge because of the complex microstructure effects influenced by physico-chemical phenomena. This paper proposes a new relationship for the unsaturated hydraulic conductivity of compacted bentonite based on Kozeny-Carman relationship. A new relationship is also proposed for the relative permeability of compacted bentonite that incorporates the effects of microstructure swelling. The application of the new relationship for the case of compacted GMZ bentonite provides well correlated results with the experimental data reported in the literature.

Keywords: Compacted bentonite · Clay microstructure · Unsaturated Permeability

1 Introduction

In the context of geological disposal concepts for high level nuclear waste in crystalline rock, partially saturated compacted bentonite is to be emplaced in the depositional holes as buffer material. The clay buffer will undergo a re-saturation phase due to moisture intrusion from the host rock. Experimental observations have shown that the moisture flow and chemical transport in smectite rich clays are strongly influenced by swelling/shrinkage of the clay microstructure (Pusch and Yong 2006). The application of conventional unsaturated hydraulic conductivity relationships in predicting the moisture flow and chemical transport processes in compacted bentonite might overestimate the rate of re-saturation or chloride transport against the experimental observations (e.g. Sedighi et al. 2012; 2018). Limited experimental and theoretical approaches exist that can provide a sound description of the evolution of unsaturated hydraulic conductivity of compacted bentonite.

In this paper, various phenomenological aspects of microstructure of compacted bentonite and its evolution during wetting process are re-visited to develop a theoretical

L. Zhan et al. (Eds.): ICEG 2018, ESE, pp. 544–550, 2019.
https://doi.org/10.1007/978-981-13-2224-2_67

description of pore system. The established Kozeny-Carman relationship (Carman 1939; Kozeny 1927) for permeability of porous media is adopted and modified to describe the intrinsic permeability of compacted bentonite under confined conditions. A relationship for the relative permeability of compacted bentonite is proposed by incorporating the effects of swelling/shrinkage of the clay microstructure. The relationship proposed for the hydraulic conductivity of compacted bentonite is tested against a series of experimental results from the literature.

2 Microstructure Evolution in Compacted Bentonite

The pore system of compacted bentonite can be described by two scales of porosity: (i) micro porosity that includes the spaces between the unit layers of smectite (microstructure) and (ii) macro porosity that represents the pore spaces between the particles and aggregates of particles (Fig. 1). Macro/meso pores are the only major water conduits. Water molecules are exchanged between the micro and macro/meso pores. The water in micro pores has different properties compared with free water and is widely considered as an immobile phase (Sedighi and Thomas 2014). The interlayer hydration or dehydration process can only alter the distance between the clay microstructure units (interlayer distance) by adding or removing discrete layers of water in micro-porosity. During wetting/re-saturation, the size of micro porosity increases as the results of adsorption of water molecules into the interlayer space. Consequently, under confined conditions (i.e. swelling is constrained), the macro porosity can shrink, thus reducing the space for water to flow. By increasing the water content in smectite, the number of unit layers in a particle consistently decreases and the larger aggregates are divided into smaller particles (Saiyouri et al. 2004) that affects the tortuosity of the porous system. Figure 1 presents a schematic diagram of the evolution of pore system in compacted bentonite during re-saturation at constant total volume.

The volume of interlayer water can be described as:

$$V_{IL} = d_{IL} N A_s \tag{1}$$

where, d_{IL} is the thickness of interlayer water (i.e. interlayer distance), N stands for the total number of unit layer of smectite (TOT) in the sample and A_s is the surface area of a TOT.

The thickness of interlayer water varies with water content due to adsorption of one to three layers of water in the interlayer space. The variations of interlayer water with the total volumetric water content follow a hydrodynamic adsorption behaviour that can be simplified by a quasi-dynamic equation (Azizian 2004), given as:

$$d_{IL} = d - d_0 = 9[1 - exp(b\theta_l)] \tag{2}$$

where, d and d_0 is the basal spacing and thickness of single TOT sheet, respectively. b is a constant that is related to the water adsorption isotherm of smectite ($b = -2.63$

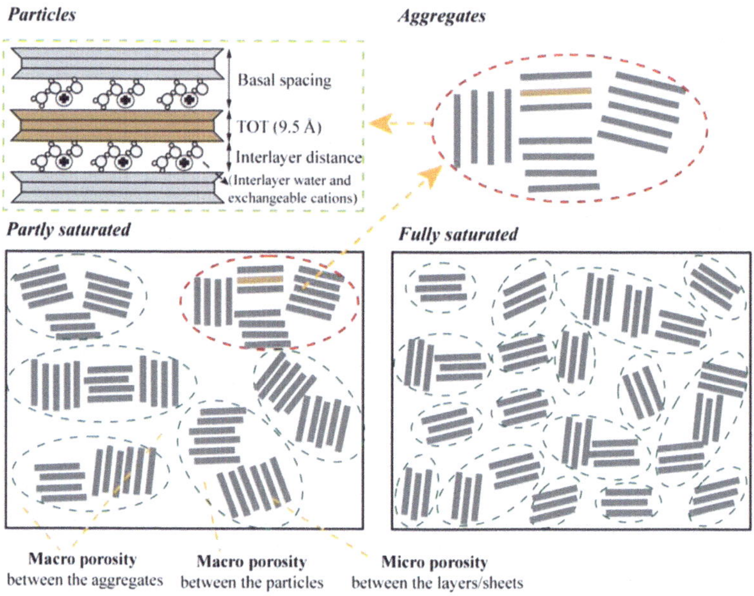

Fig. 1. Schematic diagram of pore system evolution of compacted bentonite during wetting.

provides a good agreement with water/vapour adsorption data of smectite (Holmboe et al. 2012; Villar et al. 2012). θ_l is the total volumetric water content.

The total number of TOT layers in a sample (N) can be given as:

$$N = \frac{m_{smectite}}{m_s} n_c \tag{3}$$

where, $m_{smectite}$ is the total mass of smectite in the soil, m_s is the mass of single particle containing the number of n_c TOTs and n_c is the number of stacked TOT layers per particle that varies with water content (Saiyouri et al. 2004).

The surface area of a TOT (A_s) can be calculated from the total specific surface area (A_{total}) and mass of smectite ($m_{smectite}$) in the soil:

$$A_s = \frac{m_{smectite} A_{total}}{2N} \tag{4}$$

where, A_{total} is the total specific surface area.

The micro porosity (ϕ_{micro}) and macro porosity (ϕ_{macro}) can therefore be calculated as:

$$\phi_{micro} = 0.5 d_{IL} A_{total} \rho_d^s \tag{5}$$

$$\phi_{macro} = \phi_t - \phi_{micro} \tag{6}$$

where, ρ_d^s is the dry density of smectite in the soil and ϕ_t is the total porosity

3 Saturated Hydraulic Conductivity of Compacted Bentonite

Kozeny-Carman relationship (KC) is an established theoretical approach that provides a relationship between porosity and permeability in porous media (Carman 1939; Kozeny 1927). However, it has been shown that the KC for compacted clays does not provide a good correlation with experimental observations (Chapuis and Aubertin 2003). The hypothesis presented here is that since part of the pore water in smectite is adsorbed in the microstructure and it is regarded as an immobile phase, the system of compacted bentonite should then be conceptualised by a dual porosity system, in which only the macro porosity contributes to the flow. Accordingly, the total porosity in KC relationship should be replaced by the macro porosity to provide a realistic pore system of compacted bentonite.

A modified Kozeny-Carman relationship for saturated hydraulic conductivity of compacted bentonite is therefore proposed that reflects the effects of microstructure on flow behaviour through the use of macro porosity in the relationship:

$$k_{sat} = \left(\frac{\gamma_w}{\mu_w}\right)\left(\frac{C_s}{\rho_d^2\tau^2A_{ext}^2}\right)\frac{\phi_{macro}^3}{(1-\phi_{macro})^2} \tag{7}$$

where, γ_w and μ_w are the unit weight and viscosity of water, respectively. C_s is a shape factor. ρ_d is the dry density of soil. τ is the tortuosity and A_{ext} is the external specific surface area (i.e. $A_{ext} = A_{total}/n_c$).

4 Unsaturated Hydraulic Conductivity of Compacted Bentonite

The unsaturated hydraulic conductivity of compacted bentonite can be presented as a function of its saturated hydraulic conductivity (Mualem 1976), given as:

$$k_{unsat} = k_r k_{sat} \tag{8}$$

where, k_r is the relative permeability that is commonly described as a function of water content (Mualem 1976):

$$k_r = \left(\frac{\theta_l - \theta_r}{\theta_{sat} - \theta_r}\right)^{\alpha} \tag{9}$$

where, θ_{sat} and θ_r are the saturated and residual volumetric water contents, respectively. α is a constant parameter. θ_r is calculated from the relationships between water content and suction based on experimental water/vapour adsorption data.

The relative permeability relationships are mainly based on observations and measurements in coarse grained soils where the entire porosity is available for water flow. However, in the case of compacted bentonite a considerable proportion of water is located in micro porosity, which is almost immobile. To account for this, a revised relationship for relative permeability is proposed as follows:

$$k_r = \left(\frac{\theta_{macro} - \theta_{macro,r}}{\phi_{macro} - \theta_{macro,r}} \right)^{\alpha} \tag{10}$$

where, θ_{macro} is the volumetric water content in macro porosity and $\theta_{macro,r}$ is the residual macro water content. The value of θ_{macro} is calculated from ϕ_{micro} (Eq. 5) and total volumetric water content (θ_l) at a given soil water potential or suction ($\theta_{macro} = \theta_l - \phi_{micro}$).

5 Application to Compacted GMZ Bentonite

The relationship proposed for hydraulic conductivity is applied for the case of GMZ bentonite, compacted at a dry density of 1.7 Mg/m^3. GMZ bentonite contains 75.4% Na-smectite (Ye et al. 2009). Figure 2 presents the variations of micro porosity and macro porosity with suction for GMZ bentonite calculated from Eqs. 5 and 6. Figure 2 presents a comparison of micro porosity with those for compacted MX-80 bentonite calculated by Sedighi and Thomas (2014) based on a geochemical modelling approach. The results are compatible with those presented for compacted MX-80. Both MX-80 and GMZ are sodium smectite-rich clays although, the smectite content of the GMZ bentonite is lower than that of the MX-80 bentonite. It can be observed that at saturation, ϕ_{micro} is approximately 2.6 times larger than that of ϕ_{macro} that accounts only for 27.9% of the total porosity. Substituting all variables and constants, the saturated hydraulic conductivity of compacted GMZ is found to be 1.01×10^{-13} (m/s). The value obtained is very close to the value reported based on experimental tests (e.g. 1.18×10^{-13} by Ye et al. (2009)). In this study we use van Genuchten's relationship between water content and suction (Van Genuchten 1980):

$$\theta_l = \frac{\theta_{sat} - \theta_r}{[1 + (\alpha s)^n]^{(1-1/n)}} \tag{11}$$

where, s is suction. α and n are fitting parameters.

Figure 3 presents the variations of unsaturated hydraulic conductivity of compacted GMZ bentonite with total volumetric water content. The relative permeability is calculated using Eqs. 9 and 10 for comparison with experimental data presented by Ye et al. (2009). For both approaches the value of α is assumed to be 3 that is acceptable value across in the literature. The results obtained by using the proposed relationship for the relative permeability correlate well with the experimental data. In comparison with the classic approach that considers total volumetric water content/porosity, the new relationship provides a better correlation with the experimental results.

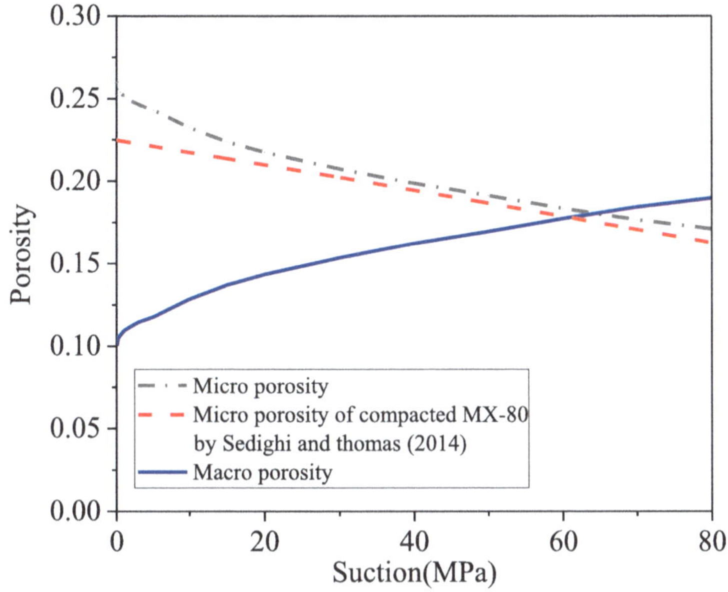

Fig. 2. Variations of micro and macro porosity of compacted bentonite with suction.

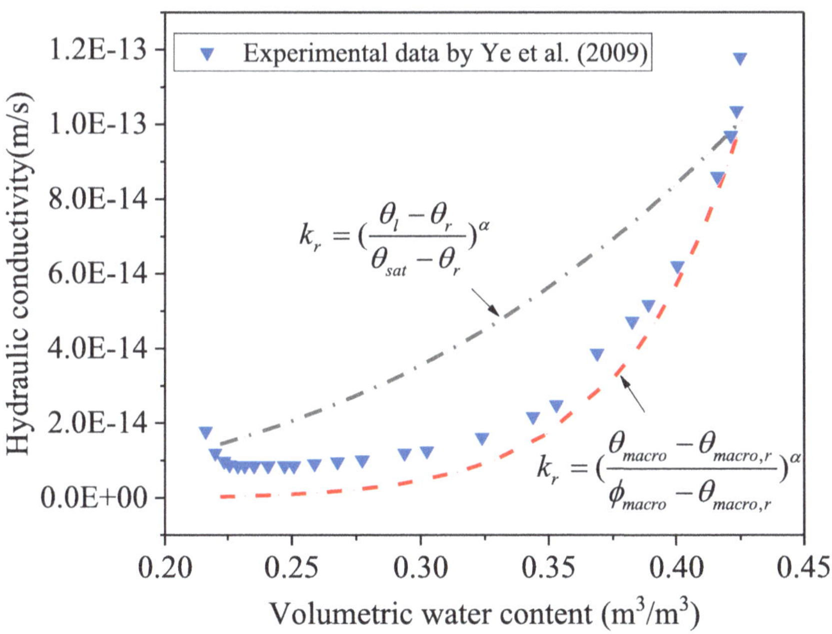

Fig. 3. Variation of unsaturated hydraulic conductivity of compacted GMZ with water content.

6 Conclusions

A new approach for describing the effects of microstructure on hydraulic conductivity of compacted bentonite was presented. The Kozeny-Carman relationship was revisited to include the dual porosity system of compacted bentonite. It was shown that by using the revised relative permeability relationship proposed in this paper, an improved correlation with experimental data for the case of compacted GMZ bentonite is achieved. The new developments have important applications in predicting the flow behaviour of smectite rich clays in which a more accurate description of the pore system evolution is included.

Acknowledgements. Financial support through the University of Manchester's President Doctoral Scholarship Award for Tian Chen (1st author) is gratefully acknowledged.

References

Azizian S (2004) Kinetic models of sorption: a theoretical analysis. J Colloid Interface Sci 276 (1):47–52

Carman PC (1939) Permeability of saturated sands, soils and clays. J Agric Sci 29(2):262–273

Chapuis RP, Aubertin M (2003) On the use of the Kozeny-Carman equation to predict the hydraulic conductivity of soils. Can Geotech J 40(3):616–628

Holmboe M, Wold S, Jonsson M (2012) Porosity investigation of compacted bentonite using XRD profile modeling. J Contam Hydrol 128(1–4):19–32

Kozeny J (1927) Uber kapillare leitung der wasser in boden. R Acad Sci Vienna Proc Cl I 136:271–306

Mualem Y (1976) A new model for predicting the hydraulic conductivity of unsaturated porous media. Water Resour Res 12(3):513–522

Pusch R, Yong RN (2006) Microstructure of smectite clays and engineering performance. Taylor and Francis, New York

Saiyouri N, Tessier D, Hicher PY (2004) Experimental study of swelling in unsaturated compacted clays. Clay Miner 39(4):469–479

Sedighi M, Thomas HR, Vardon PJ (2012) On the reactive transport of chemicals in unsaturated expansive soils. In: The 4th international conference on problematic soils, Wuhan, China, pp 211–218

Sedighi M, Thomas HR (2014) Micro porosity evolution in compacted swelling clays — A chemical approach. Appl Clay Sci 101:608–618

Sedighi M, Thomas HR, Vardon PJ (2018) Reactive transport of chemical in compacted bentonite under nonisothermal water infirltration. J Geotech Geoenviornmental Eng ASCE (Accepted)

Van Genuchten MT (1980) A closed-form equation for predicting the hydraulic conductivity of unsaturated soils. Soil Sci Soc Am J 44(5):892–898

Villar M, Gómez-Espina R, Gutiérrez-Nebot L (2012) Basal spacings of smectite in compacted bentonite. Appl Clay Sci 65:95–105

Ye WM, Cui YJ, Qian LX, Chen B (2009) An experimental study of the water transfer through confined compacted GMZ bentonite. Eng Geol 108(3–4):169–176

A Prediction Method for Swelling Deformation of Bentonite and Its Sand-Mixtures in Salt Solution

G. S. Xiang(✉) and Y. Wang

Civil Engineering Department, Anhui University of Technology, Maanshan 243002, China
xianggsh2011@163.com

Abstract. The swelling deformation tests of Tsukinuno bentonite and its sand-mixtures were performed in distilled water and NaCl solution with different concentration. The results showed the swelling strain decreased with the increase in the vertical pressure, the salt concentration, and the sand content in the mixtures. The swelling properties of the bentonite can be characterized by a unique fractal e_m-p_e relation, where the effective stress p_e incorporating with osmotic suction was applied to take the effect of salinity into consideration. And the coefficient *K*, as the physical constant of bentonite can be estimated by diffuse double layer (DDL) model in the case of distilled water. The applicability of the fractal e_m-p_e relation has been confirmed by comparing the predicted results with swelling deformation tests on Tsukinuno bentonites in various salt solution concentrations.

Keywords: Bentonite-sand mixture · Swelling · Fractal · Effective stress NaCl solution

1 Introduction

Compacted bentonite-sand mixtures have been proposed as suitable buffer/backfill materials for construction of the artificial barrier in HLW (Ye et al. 2015), because of the low hydraulic conductivity, high swelling property, and good self-sealing capacities of bentonite.

Many studies about the swelling behaviors of bentonite and its sand mixtures in salt solution are available in literature (Zhang et al. 2012; Chen et al. 2016). The swelling properties are smaller in higher salt concentration. But few methods can accurately predict the swelling characteristics of bentonite in a salt solution. The DDL model was found to work excellently for osmotic swelling in compacted bentonites in a dilute salt solution (Tripathy et al. 2004; Xiang et al. 2014), whereas has great limitations in estimating crystalline swelling and is not suitable for specimens in strong solution (Liu 2013). Later, Rao and Thyagaraj (2007) pointed out that an increase in salt concentration of pore fluid leads to an increase in osmotic suction, which acts as an additive effective stress component leading a decline in swelling potential. The effective stress incorporated with osmotic suction was furtherly expressed by Xu et al. (2014), and the

L. Zhan et al. (Eds.): ICEG 2018, ESE, pp. 551–557, 2019.
https://doi.org/10.1007/978-981-13-2224-2_68

swelling properties of bentonite in several of salt solution concentrations can be expressed by a unique curve of e_m-p_e, where e_m is the void ratio of montmorillonite, p_e is the effective stress. It offers a new method for estimating the swelling properties of bentonite in salt solution.

However, the sand content also has great influence on the swelling properties of the bentonite-sand mixture. How to estimate the swelling properties of the bentonite-sand mixture is a question. In the present paper, experimental tests of swelling characteristics were conducted on Tsukinuno bentonite and its sand mixtures in NaCl solution. The swelling of mixtures in different salt solution were predicted by the fractal swelling model. In this model, the swelling coefficient K is estimated by DDL model in the case of distilled water. There is a well agreement between the estimations and experimental data of swelling deformation for the tested bentonite-sand mixture.

2 Materials

The commercial bentonite used in this paper is the Tsukinuno bentonite, a commercial bentonite produced at the Tsukinuno mine in Japan, it contains about 48% Na-montmorillonite. Typical profile of Tsukinuno bentonite was shown in Table 1 (Xu et al. 2003). Standard sand used in this study were produced in the city of Xiamen in Fujian province of China, the physical indexes with the sand were as follows: specific gravity $G_s = 2.65$, granularity $D = 0.5 - 1.0$ mm, SiO_2 content $C_s = 99.45\%$. After bentonite and sand under the dry condition were mixed at a given ratio, the bentonite contents α of the bentonite-sand mixtures were 60%, 80% and 100%, respectively. The distilled water was added to the mixtures until the desired water content ($16 \pm 0.5\%$) was reached. Bentonite and its sand mixtures were statically compacted to cylindrical specimens of diameter 60 mm, height 10 mm, and dry density 1.75 g/cm^3. Liquids used in this study included distilled water and solutions with different concentration (0.5, 1.0 mol/L) by dissolving sodium chloride (NaCl), they were abbreviated as DW, 0.5 M and 1.0 M in following figures and tables, respectively.

Table 1. Physical properties of Tsukinuno bentonite

Physical properties	Value
Specific gravity	2.79 g/cm^3
Liquid limit	473.9%
Plastic limit	26.6%
Plasticity index	447.3
Content of smectite	48%
Cation exchange capacity	0.73 mEq/g

The specimens were put into oedometers, and then distilled water or NaCl solution was injected into the oedometer. A precast stress (12.5, 25, 50, 100, 200 400 or 800 kPa) was applied to the compacted specimen. The displacement was measured by a dial gauge. The change in height of the specimen with duration was recorded every 12 h until the change in reading of the dial gauge in 36 h did not exceed 0.001 mm.

3 Results of Swelling Deformation Tests

Figure 1 illustrates the ε_{max}-σ relation for bentonite–sand mixtures at different bentonite/sand ratios. The figures revealed that an increase in salt concentration of the pore fluid or in vertical pressure favors the reduction in swell potential of mixture specimens; contrarily, the increase in bentonite content increases the swell strain.

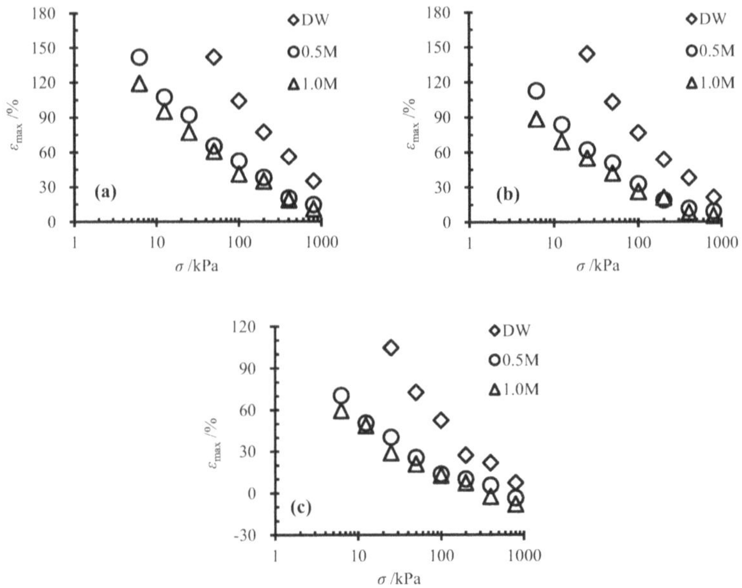

Fig. 1. ε_{max}-σ relation for (a) α = 100% (pure bentonite), (b) α = 80%, and (c) α = 60% bentonite-sand mixtures in different solution.

According to the result of mineral composition test (Xu et al. 2003), montmorillonite is the only mineral responsible for the water uptake in Tsukinuno bentonite and its sand mixture. The swelling deformation of the mixture can also be characterized by the void ratio of montmorillonite e_m, that is defined as the ratio of water volume at full saturation V_w to the volume of montmorillonite V_m. Sun et al. (2009):

$$e_m = \frac{V_w}{V_m} = e\frac{\rho_m}{\rho_s \alpha \beta} \tag{1}$$

where e is the void ratio at full saturation, ρ_m is the grain density of montmorillonite, ρ_s is the grain density of the bentonite-sand mixture, α is the bentonite content in the mixture (in mass), and β is the montmorillonite content in the bentonite (in mass).

Figure 2 reflects the relationship between the void ratio of montmorillonite e_m and the vertical pressure σ. Similar as the swelling strain, e_m decreases with the increase in salt concentration of the pore fluid and in vertical pressure. But unlike the swelling strain, e_m of pure bentonite is basically the same as that of the bentonite-sand mixtures when they are under the same pressure and in the same solution.

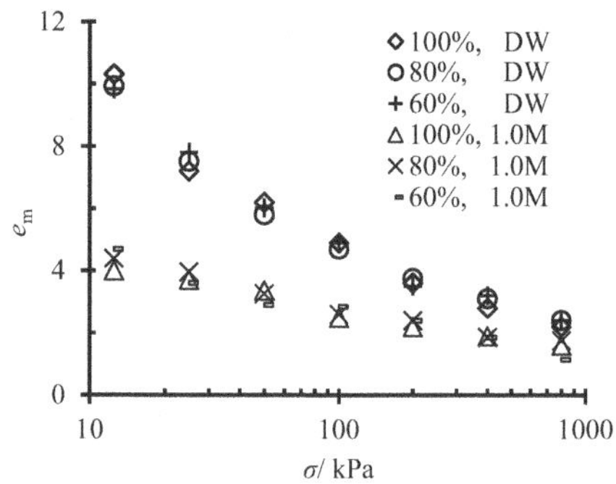

Fig. 2. e_m-σ relation for bentonite-sand mixtures.

4 Correlation of e_m to p_e

When compacted specimens are inundated with distilled water, the effective stress (p_e) is balanced with the vertical pressure (σ) applied in swelling tests, that is

$$p_e = \sigma \tag{2}$$

But in salt solution, osmotic suction arises with an increase in ion concentration, and acts as an additional effective stress component that reduces the swelling potential of compacted clay specimens. The effective stress can be represented as (Xu et al. 2014)

$$p_e = \sigma + \pi\left(\frac{p}{\pi}\right)^{D_s-2} \tag{3}$$

where D_s is the surface fractal dimension, π is osmotic suction of pore water and can be expressed as (Rao and Thyagaraj 2007)

$$\pi = iMRT \tag{4}$$

where M is the molar concentration of the salt solution, R is the universal gas constant, T is absolute temperature in Kelvin, i is the Van't Hoff factor and is expressed as, for NaCl, $i = 2$.

The correlation of the void ratio e_m to the effective stress p_e is given by Xu et al. (2014)

$$e_{\mathrm{m}} = Kp_{\mathrm{e}}^{D_s-3} \tag{5}$$

The surface fractal dimension of bentonite in salt solution can be determined by using nitrogen adsorption. Using this method, the effect of concentration of NaCl solution on the surface fractal dimension of Tsukinuno bentonite was investigated (Xiang et al. 2014). The change value of the surface fractal dimension was just about 0.1. Therefore, the fractal dimension (D_s) of Tsukinuno bentonite is determined as 2.65. Moreover, the unique curve of e_m-p_e has been validated by the experimental data four bentonites in salt solution with different concentration (Xu et al. 2014). Thus, the swelling coefficient of montmorillonite K also has no concern with salinity of pore water, it is an inherent parameter of clay.

The DDL model performs excellently in osmotic swelling for clays in diluted solutions (Liu 2013). The coefficient K in the fractal model (Eq. 5) can be estimated by solving the simultaneous equations of the fractal model and the DDL model in the case of osmotic swelling in distilled water. Osmotic swelling is generally assumed to follow the completion crystalline swelling, and DDLs begin to develop within the interlayer space. Thus, the average distance d distance between the unit montmorillonite layers should be larger when crystalline swelling has fully taken place (Xiang et al. 2014)

$$d > \frac{t + R_{\mathrm{ion}}}{t} d_{\mathrm{m}} + R_{\mathrm{ion}} \tag{6}$$

where R_{ion} is the radius of exchangeable cations, and t is the thickness of a montmorillonite layer, d_m is the fixed distance between the unit layers. For Ca-montmorillonite and Na-montmorillonite d_m is approximately 0.9 nm and 1.2 nm, respectively (Liu 2013).

At the micro-adsorption of water into montmorillonite layers, the montmorillonite void ratio e_m can be expressed by Komine and Ogata (1996) as

$$e_{\mathrm{m}} = \frac{d - R_{\mathrm{ion}}}{t + R_{\mathrm{ion}}} \tag{7}$$

When the DDLs develop well and the distance between two-clay platelet is large enough, the van der Waals attractive forces can be ignored, then the interaction force p between two platelet can be simplified (Israelachvili 1985):

$$p = 64nkT \tanh^2(\mathrm{Z}/4)\mathrm{e}^{-2\kappa d} \times 10^{-3}\ (\mathrm{kPa}) \tag{8}$$

where n is the ion concentration of porewater, k is the Boltzmann constant, T is the absolute temperature, κ is reciprocal Debye length, z is the scaled surface potential of the isolated particle.

By this method, the swelling coefficient K of the tested bentonite is estimated as 24 (Xiang et al. 2014). Therefore, substitution the $K = 24$ and $D_s = 2.65$ into Eq. 5, a general e_m-p_e relation in the swelling process can be expressed for the Tsukinuno bentonite as

$$e_m = 24p_e^{-0.35} \tag{9}$$

Figure 3 illustrated predicted results by Eq. (10) with the swelling experimental data of pure bentonite. Published data of pure Tsukinuno bentonite in distilled water from Xu et al. (2003) were quoted. The e_m-p_e relation of the Tsukinuno bentonite predicted by the fractal method fitted well with the experimental data. Figure 3 showed that the e_m-p_e relation also fitted well with the experimental data for bentonite-sand mixtures. Compare these figures, it can be seen the amount of water absorbed by a per unit volume of montmorillonite is a constant value under a given effective stress.

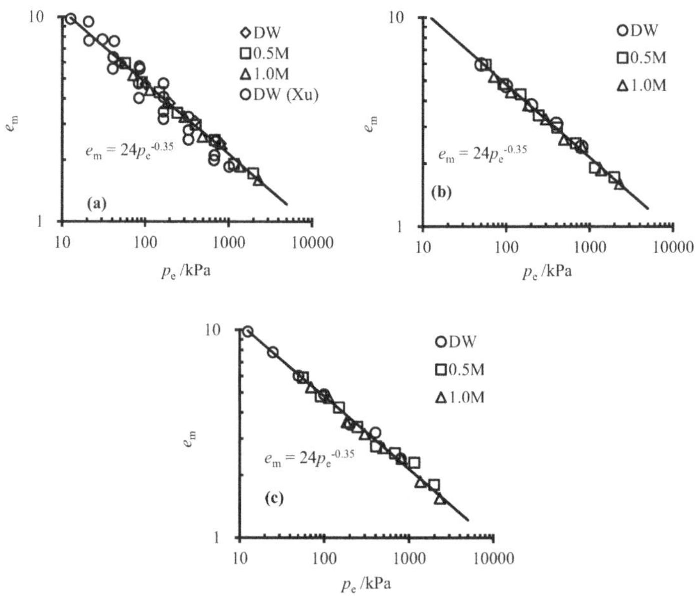

Fig. 3. Comparison between estimation e_m-p_e relation from Eq. (8) and tested results for (a) $\alpha = 100\%$, (b) $\alpha = 80\%$, and (c) $\alpha = 60\%$ bentonite-sand mixtures.

5 Conclusion

(1) The sand or bentonite content has great influence on the swelling characteristics of the bentonite-sand mixture. The swelling strain always increases with the increase in bentonite content. But the bentonite and its sand mixtures have the same e_m value when they under the same pressure and in the same solution.

(2) The swelling fractal model was introduced to explain the swelling characteristics of bentonite-sand mixture in salt solution. In this model, the effective stress incorporating osmotic suction was applied to take the effect of salinity on swelling into consideration. The e_m-p_e relation of the mixtures with predicted by the fractal method fitted well with the experimental data. It offers a new method to predict the sealing property of buffer/backfill materials affected by salt solution.

Acknowledgments. The National Nature Science Foundation of China (Grant No. 41630633 and No. 41702311) and the Nature Science Foundation of Anhui province of China (Grant No. 1708085QE99) are acknowledged for their financial support.

References

Chen YG, Zhu CM, Ye WM, Cui YJ et al (2016) Effects of solution concentration and vertical stress on the swelling behavior of compacted GMZ01 bentonite. Appl Clay Sci 124:11–20

Israelachvili JN (1985) Intermolecular and surface forces. Aeademic Press, London

Liu LC (2013) Prediction of swelling pressures of different types of bentonite in dilute solutions. Colloid Surf A 434(19):303–318

Rao SM, Thyagaraj T (2007) Swell-compression behaviour of compacted clays under chemical gradients. Can Geotech J 44(5):520–532

Sun DA, Cui HB, Sun WJ (2009) Swelling of compacted sand–bentonite mixtures. Appl Clay Sci 43:485–492

Tripathy S, Sridharan A, Schanz T (2004) Swelling pressures of compacted bentonites from diffuse double layer theory. Can Geotech J 41(3):437–450

Xiang GS, Xu YF, Jiang H (2014) Surface fractal dimension of bentonite and its application in calculation of swelling deformation. Surf Rev Lett 21(5):1450074

Xu YF, Matsuoka H, Sun DA (2003) Swelling characteristics of fractal-textured bentonite and its mixtures. Appl Clay Sci 22:197–209

Xu YF, Xiang GS, Jiang H et al (2014) Role of osmotic suction in volume change of clays in salt solution. Appl Clay Sci 101:354–361

Ye WM, Zhu CM, Chen YG et al (2015) Influence of salt solutions on the swelling behavior of the compacted GMZ01 bentonite. Environ Earth Sci 74(1):793–802

Zhang HY, Cui SL, Zhang M et al (2012) Swelling behaviors of GMZ bentonite-sand mixtures inundated in NaCl–Na_2SO_4 solutions. Nucl Eng Des 242:115–123

Hydraulic Conductivity of Soil-Bentonite Cutoff Walls Constructed at the Ground Containing Cement Stabilized Soil

Atsushi Takai(✉), Kazuki Yamaguchi, Toru Inui, and Takeshi Katsumi

Kyoto University, Yoshida-honmachi, Sakyo-ku, Kyoto 6068501, Japan
takai.atsushi.2s@kyoto-u.ac.jp

Abstract. Because of high deformability even after construction, soil-bentonite (SB) cutoff walls, which are constructed by blending slurry and powder bentonite with in situ soil, are occasionally conjunct with soil-cement (SC) walls or cement stabilized soil having relatively high strength in the case that both strength and barrier performance is partly required. In this study, barrier performance of SBs containing cement at different proportions was experimentally evaluated using a flexible-wall permeameter. Specifically, SCs made with cement proportions of 100, 300, 500 kg/m^3 were crushed after certain periods of curing time and used as a parent material of SBs. Experimental results demonstrate that hydraulic conductivity of SB containing cement can sufficiently decrease with time when bentonite is blended three hours after cement addition, while barrier performance of SB is insufficiently high when bentonite is added after 7-day and 28-day curing. Thus, a bentonite powder content of 100 kg/m^3, which is a conventional proportion for construction of SB walls, is insufficient when cement stabilized soil exists in the ground.

Keywords: Bentonite · Containment · Cement · Hydraulic conductivity

1 Introduction

Soil-bentonite (SB), which is a mixture of slurry and powder bentonite and in situ soil, is one of the most widely used barrier materials for cutoff walls because of its extremely low hydraulic conductivity (*k*) as well as sufficient deformability and high vertical homogeneity (Takai et al. 2016). Because of this softness, SB cutoff walls are frequently used in conjunction with soil-cement (SC) walls or cement stabilized soil when strength is partly required. Hydraulic barrier performance of SB is attributed to the swelling of bentonite, which is known to be restricted by inorganic solutions and nonpolar liquids (Norrish and Quirk 1954, Mesri and Olson 1971). Numerous researches on the bentonitic clay barriers have indicated that inorganic solutions containing polyvalent cations can lead to increase in *k* value with relatively lower concentration ($\sim$0.1 mol/L) (e.g. Petrov et al. 1997, Katsumi et al. 2008 and Takai et al. 2013). This is because the existence of these polyvalent cations results in the development of thinner diffuse double layers and the reduced osmotic swelling.

L. Zhan et al. (Eds.): ICEG 2018, ESE, pp. 558–566, 2019.
https://doi.org/10.1007/978-981-13-2224-2_69

When SB cutoff walls are constructed at the ground containing cement stabilized soil, hydration of bentonite in SB should be accordingly inhibited due to chemicals released from cement. Although barrier performance of soil-cement-bentonite walls have been studied in these decades (e.g. Ryan and Day 2002 and Opdyke and Evans 2005), compatibility of SB cutoff walls with soil stabilization has not been studied yet. In this study, barrier performance of SB containing cement in its parent material was evaluated by the hydraulic conductivity test. In particular, SCs made with cement proportions of 100, 300, 500 kg/m^3 were crushed after curing and used as a parent material of SBs.

2 Methodologies

2.1 Materials and Specimen Preparation

As a parent material of SCs, we used decomposed granite soil sieved through a 4.75-mm opening screen. Table 1 shows the basic properties of the soil. Blast-furnace slag cement (Type B, Sumitomo Osaka Cement Co., Ltd.) with a particle density of 3.04 g/cm^3 and sodium bentonite (Kunimine Industries Co., Ltd.) with a particle density of 2.79 g/cm^3, a montmorillonite fraction of 48.0% and a cation exchange capacity of 0.732 meq/g were used for preparing SCs and SBs.

Table 1. Basic properties of granite soil used

Particle density (g/cm^3)		2.65
Particle size distribution (%)	Gravel [>2 mm]	29.8
	Sand [0.075–2 mm]	61.1
	Fines [<0.075 mm]	9.0
Average particle size (mm)		1.11
Coefficient of uniformity		17.1

First, a water content of the granite soil was adjusted to 23%, which corresponds to saturated sandy ground with a void ratio of 0.60. Then, for preparation of SC, cement slurry made with equal mass of distilled water and the cement (a water/cement ratio is 1) was added to the wet soil and sufficiently mixed using a soil mixer. Contents of cement powder (C_{CP}) of 0 kg/m^3, 100 kg/m^3, 300 kg/m^3, and 500 kg/m^3 were applied in this study, considering typical proportions of soil stabilization method. The SC mixture was carefully poured into a plastic mold with a diameter of 5.0 cm and a height of 10.0 cm to avoid the presence of air bubbles and covered with a plastic film, according to JGS 0821 (2009). The SC mixture in the mold was cured for 3 h, 7 days, and 28 days, under a constant temperature of 20 °C. The cured SC was crushed using a 2.5-kg rammer so that all fraction could pass through a 4.75-mm opening screen. The crushed fractions were sieved through the screen at each 5 blows to avoid overcrushing of the material. Note that SCs with 3-hour curing could not be crushed because of insufficient solidification. It was confirmed that particle size distribution curves of the all crushed SCs were almost the same with that of the granite soil used. Bentonite slurry of 10% concentration by weight basis was added and blended with the parent material —the granite soil or the crushed SC—using the soil mixer, in a 10 kg/m^3 proportion, to achieve 150 mm flow value of parent material-slurry mixture according to JIS R5201 (1997). Then, bentonite powder was added to the mixture with a proportion of 100 kg/m^3 and sufficiently re-mixed using the mixer. Proportions of each material in SBs can be summarized on weight basis, as shown in Table 2, where values are expressed as ratios to 1.00 dry soil.

Table 2. Proportions of each material in SBs on weight basis

C_{CP} in parent material (kg/m^3)	Soil	Water[a]	Cement	Bentonite
0 (pure SB)	1.00	0.29	0	0.08
100	1.00	0.37	0.07	0.09
300	1.00	0.52	0.21	0.11
500	1.00	0.67	0.34	0.13

[a] Summation of pore water and water fractions in cement and bentonite slurries.

2.2 Hydraulic Conductivity Test

In this study, flexible-wall permeameters were used for all cases according to ASTM D 5084. First, the prepared SB was carefully poured into a consolidation cell with a diameter of 60 mm and a height of 20 mm. After the specimen in the consolidation cell was saturated using a vacuum deaerator for at least 2 days, the specimen was pre-consolidated in a consolidation test apparatus with consolidation pressures of 19.6 kPa for 24 h as first loading and 39.2 kPa for additional 24 h as second loading. After the pre-consolidated SB specimen was removed from the cell without

Table 3. Experimental conditions of hydraulic conductivity test

Case	C_{CP} (kg/m^3)	Curing period	Void ratio before permeation
No cement	0	0	0.71
500-3H (1)	500	3 h	0.89
500-3H (2)			0.81
300-3H (1)	300		0.74
300-3H (2)			0.74
100-3H (1)	100		0.75
100-3H (2)			0.71
500-7D (1)	500	7 days	1.18
500-7D (2)			1.20
300-7D (1)	300		1.07
300-7D (2)			1.03
100-7D (1)	100		0.83
100-7D (2)			0.91

(continued)

Table 3. (*continued*)

Case	C_{CP} (kg/m^3)	Curing period	Void ratio before permeation
500-28D (1)	500	28 days	1.24
500-28D (2)			1.26
300-28D (1)	300		1.11
300-28D (2)			1.09
100-28D (1)	100		0.98
100-28D (2)			0.98

disturbance, the cylindrical specimen was placed to a flexible-wall permeameter. Confining pressure of 30 kPa was applied to the specimen in all cases. Distilled water was used as permeant. The hydraulic gradient during permeation was approximately 30–50.

Table 3 summarizes all conditions of the hydraulic conductivity test. In this table, void ratio after the pre-consolidation before permeation, calculated from particle densities of granite soil and cement and bentonite powders, is shown. In the no-cement-addition case, SB was poured into the consolidation cell with a wet density of 1.80 g/cm^3, which was pre-determined according to field data. In other cases, wet densities filled in the cell was determined so that dry density of granite soil can be uniform in all cases. In all cases except the no-cement-addition case, two duplicate specimens were subjected the test.

During permeation, effluent volume was measured at regular intervals to calculate k values. In this study, k of 1.0×10^{-9} m/s was used as the performance-based criteria. This value was determined by considering travel time of contaminants with same groundwater level difference inside/outside the cutoff wall are equivalent with the conditions of $k = 1.0 \times 10^{-7}$ m/s with L (thickness) = 5.0 m and $k = 1.0 \times 10^{-9}$ m/s with $L = 0.5$ m, which is a standard thickness of SB walls.

2.3 Chemical Analysis

For evaluation of the magnitude of chemical impacts on k of SB, the concentration of cations leached from the crushed SCs was quantified. The concentrations of aluminum, calcium, potassium, and sodium, which are the main component of the cement used, were measured. A conventional batch leaching test method was applied in this study. In particular, the crushed SC sample of 20.0 g, sieved through a 2.00-mm opening screen, was added to 200-mL distilled water in a plastic bottle to achieve a liquid/solid ratio of 10. The solid-liquid mixture was, then, shaken at room temperature at 200 rpm for 6 h in the horizontal direction. After shaking, the sample was centrifuged under 3000 rpm for 60 min. After centrifugation, supernatant was filtered with a 0.45-μm membrane filter and subjected to chemical analysis using ICP-OES (Agilent Technologies, 700 Series). Based on the concentrations of measured cations, ionic strength was also calculated. The ionic strength is the sum of the product of the molar concentration and the charge of each ion.

3 Results and Discussion

3.1 Results of Hydraulic Conductivity Test

Figure 1 shows changes in *k* values of SBs made with SCs crushed after 3-h curing as well as of SB without cement addition. Pore volumes of flow (PVF) in the left figure is the cumulative flow volume normalized by the pore volume of each specimen. SB without cement addition showed a nearly constant *k* of approximately 4.5×10^{-11} m/s during permeation. It is obvious that *k* values of SBs made with SCs crushed after 3-h curing decreased over time regardless of C_{CP}, although being rather high at the beginning of permeation, compared with SB without cement addition (= 4.5×10^{-11} m/s). This decrease in *k* values is attributed to progression of cement hydration. The *k* values of SBs with C_{CP} = 100 and 300 kg/m^3 decreased from an order of magnitude of 10^{-9} m/s to ones of 10^{-11} m/s eventually. Another important finding is that the decrease in *k* values is governed not by PVF but by elapsed time. Although large differences between two duplicate specimens seem to be observed in the left figure of Fig. 1, those differences were negligible in the right figure. These results support that SBs made with SCs with C_{CP} = 100 and 300 kg/m^3, crushed after 3-h curing, can perform a sufficient *k* value smaller than 1.0×10^{-9} m/s, which is the performance-based criterion in this study, within 2 elapsed days.

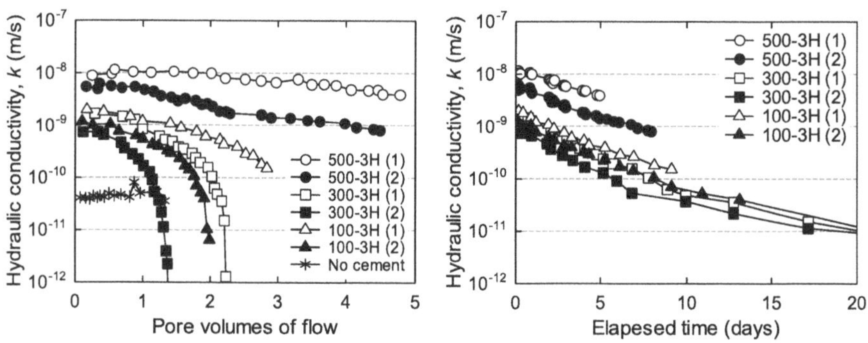

Fig. 1. Changes in *k* values of SBs without cement addition and of SBs made with SCs cured for 3 h; (left) *k* vs pore volumes of flow, (right) *k* vs elapsed time.

k values of SBs made with SCs crushed after 7-day curing are shown in Fig. 2. SBs made with SCs crushed after 7-day curing showed obviously little change in *k* values over time, different from the tendencies of 3-h curing. Same trends were observed in the cases of 28-day curing. The *k* values, regardless of C_{CP}, were in the range from 1.0×10^{-7} m/s to 1.0×10^{-8} m/s approximately, which are considerably higher than the performance-based criteria of 1.0×10^{-9} m/s. These results verify that the barrier performance of SB cutoff walls constructed at the ground containing cement stabilized soil can significantly worsened when constructed after a sufficient time has elapsed since cement addition.

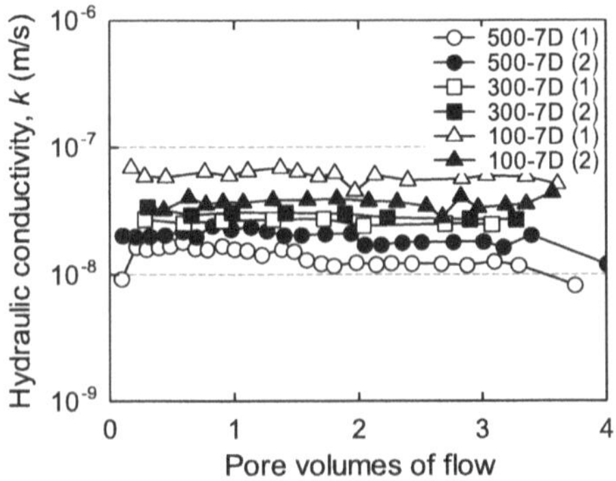

Fig. 2. Changes in *k* values of SBs made with SCs cured for 7 days.

Figure 3 shows the relationship between *k* values and void ratio before permeation. The *k* value for SBs made with SCs crushed after 7-day and 28-day curing are of the average value during permeation, while that value after 3-h curing are of the initial value of permeation. As shown in Table 2, larger C_{CP} lead to larger void ratio at any curing periods, because a relative amount of water in a soil matrix increase with an increase in the amount of cement hydrate. Accordingly, *k* of SB decreased with an increase in the C_{CP} value for the cases of 7-day and 28-day curing because cement hydrate could fill void space in SBs. In contrast, *k* of SB increased with an increase in C_{CP} for the case of 3-h curing. This result implies that the cement addition negatively influences hydraulic barrier performance of SBs. It can be concluded that cement addition with a large proportion has a positive effect on k of SB after a sufficient time has elapsed since the beginning of cement hydration, while having a negative effect when bentonite is blended almost at the same time.

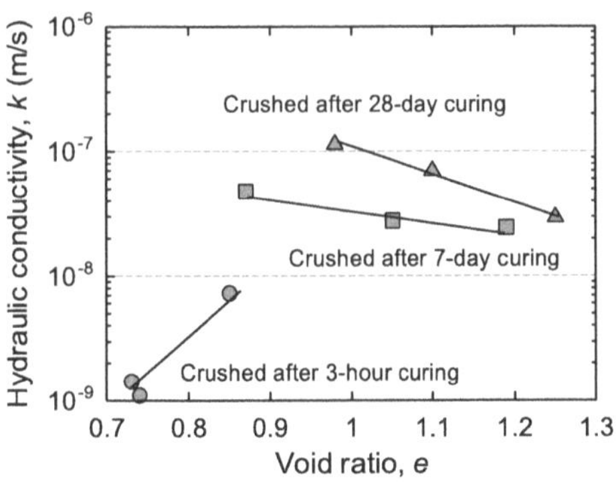

Fig. 3. Relationship between *k* and void ratio

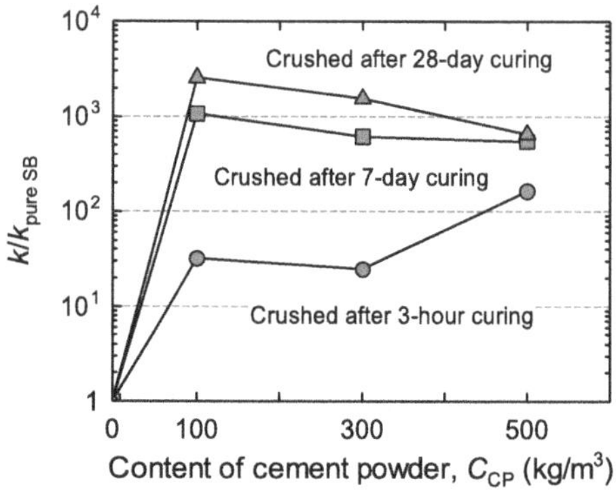

Fig. 4. Relationship between $k/k_{pure\ SB}$ and content of cement powder, C_{CP}

Figure 4 shows the relationship between k of SB normalized by that of pure SB (without cement) and C_{CP}. In this figure, a larger $k/k_{pure\ SB}$ value means worse hydraulic barrier performance of SB compared to pure SB. As can be seen from this figure, by the presence of cement stabilized soil in the ground, k of SB increased up to 2600 times when bentonite was mixed 7 days and 28 days after cement addition. Since cement hydration is progressed over time, the effect of cement addition on k values of SBs was relatively small in the cases of 3-h curing.

3.2 Results of Chemical Analysis

The relationship between $k/k_{pure\ SB}$ and ionic strength of solution calculated by chemical analysis is shown in Fig. 5. It is obvious that longer curing period and higher C_{CP} lead to lower ionic strength probably because calcium is consumed during the progression of cement hydration. In the case of 3-h curing, a linear relationship between $k/k_{pure\ SB}$ and ionic strength was not observed, same as other results. Since the hydraulic conductivity largely increased even when the ionic strength is approximately 0.01 mol/L, it is also obvious that the bentonite swelling is greatly inhibited by the presence of cement stabilized soil.

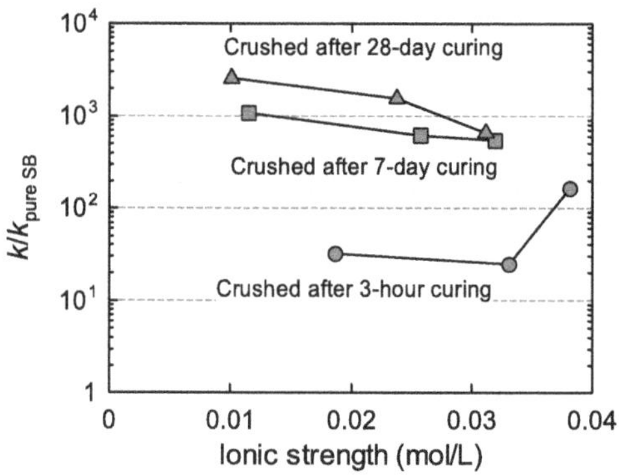

Fig. 5. Relationship between $k/k_{pure\ SB}$ and ionic strength of crushed SCs

From these observations, a bentonite powder content of 100 kg/m^3, which is a typical proportion for construction of SB walls in the general ground, is insufficient when cement stabilized soil exists in the ground where SB cutoff walls are installed and when SB cutoff walls are constructed in conjunction with SC walls. As Takai et al. (2013) noted, in such cases, increase in the bentonite powder content such as 150 kg/m^3 needs to be considered.

4 Conclusions

In this study, the authors studied hydraulic conductivity of soil-bentonite (SB) made using crushed soil-cement (SC) with different cement proportions and curing periods, and found the following items:

(1) Hydraulic conductivity (k) value of SBs made with SCs after 3-h curing decreased with time regardless of cement content, although being rather high at the beginning of permeation. SBs made with SCs crushed after 7-day and 28-day curing showed almost constant k values over time.
(2) The k values, regardless of cement content, were in the order of 1.0×10^{-8} m/s approximately. Thus, the barrier performance of SB cutoff walls constructed at the ground containing cement stabilized soil is significantly worsened when it is constructed after the elapse of sufficient time from cement addition.
(3) A content of bentonite powder of 100 kg/m^3, which is a conventional proportion, is insufficient when cement stabilized soil exists in the ground where SB cutoff walls are installed.

Acknowledgement. The authors acknowledge to Mr. G. Araki, Mr. N. Ukaji and Mr. K. Ikeda (Raito Kogyo Co., Ltd.) for their great support to this research work.

References

Katsumi T, Ishimori H, Onikata M, Fukagawa R (2008) Long-term barrier performance of modified bentonite materials against sodium and calcium permeant solutions. Geotext Geomembr 26(1):14–30
Mesri G, Olson RE (1971) Mechanisms controlling the permeability of clays. Clays Clay Miner 19:151–158
Norrish K, Quirk J (1954) Crystalline swelling of montmorillonite: Use of electrolytes to control swelling. Nature 173:255–257
Opdyke SM, Evans JC (2005) Slag-cement-bentonite slurry walls. J. Geotech Geoenviron Eng 131(6):673–681
Petrov RJ, Rowe RK, Quigley RM (1997) Selected factors influencing GCL hydraulic conductivity. J Geotech Geoenviron Eng 123:683–695

Ryan CR, Day SR (2002) Soil-cement-bentonite slurry walls. In: International deep foundations congress 2002, ASCE, pp 713–737

Takai A, Inui T, Katsumi T, Kamon M, Araki S (2013) Hydraulic barrier performance of soil bentonite mixture cutoff wall. In: Manassero M, Dominijanni A, Foti S, Musso G (eds) Coupled phenomena in environmental geotechnics – from theoretical and experimental research to practical applications, pp 707–714. CRC Press

Takai A, Inui T, Katsumi T (2016) Evaluating the hydraulic barrier performance of soil-bentonite cutoff walls using the piezocone penetration test. Soils Found 56(2):277–290

Impact of Heavy Metals on Consolidation Properties of Bentonite

Saswati Ray(✉), Bismoy Roy Chowdhury, Anil Kumar Mishra, and Ajay S. Kalamdhad

Department of Civil Engineering, Indian Institute of Technology, Guwahati, India
r.saswati@iitg.ernet.in

Abstract. Bentonite is frequently used as a liner material at the waste disposal site because of its high swelling ability, high contaminant adsorption capacity and low hydraulic conductivity. It acts as a barrier material by preventing and minimising the ingress of pollutant to contaminate the ground water resource. Metals present in the landfill leachates may shrink the thickness of the diffuse double layer (DDL) of bentonite, which in turn reduce its effectiveness as a liner material. Therefore, it is very essential to investigate the behaviour of bentonite in the existence of the different pollutant present in the leachate. Consolidation is one of the important behavior of bentonite, which is required to be investigated for settlement analysis of landfill liner. This study was performed to investigate the influence of copper (Cu^{2+}) and zinc (Zn^{2+}) solutions of various concentrations, on the consolidation behaviour of bentonite. Various consolidation parameters, such as compression index (C_c), coefficient of consolidation (c_v), volume change (m_v), time requires for the completion of 90% of consolidation (t_{90}), were studied in the presence of heavy metals of concentration of 0, 500 and 1000 mg/L. The results indicated that with the increase in concentration of heavy metals the C_c, m_v, and t_{90} of the bentonite decreases; whereas, c_v increases. The study concluded that the presence of the heavy metals in the lechate increases the rate of consolidation of the compacted bentonite. The outcomes of this investigation may provide a general suggestion for evaluating the liner performance in the presence of different kinds of pollutants present in the lechate.

Keywords: Bentonite · Consolidation characteristics · Heavy metal Landfill liner · Leachate

1 Introduction

In the recent years, heavy metal contamination has become a severe environmental and worldwide public health concern. Mining, foundries and smelters, and other metal-based industrial activities are the point source zones where the pollution is exceptionally prominent [1–3]. The exponential increment of the usage of various domestic, industrial, agricultural and technological applications are causing the rise in human exposure drastically [3].

L. Zhan et al. (Eds.): ICEG 2018, ESE, pp. 567–574, 2019.
https://doi.org/10.1007/978-981-13-2224-2_70

Presence of heavy metals may cause an adverse effect on the human health such as disturbing the metabolism, cellular and tissue damage and various diseases if their consumption exceeds the permissible limits. It has been reported that heavy metal ion can cause carcinogenesis or apoptosis and cell cycle modulation by interacting with cell components that may lead to the damaging of DNA [4–6]. Generally, leachate generated from the waste discarded in the waste disposal sites is the potential source of heavy metal. In landfills, clay liner is used as a barrier material to prevent the contamination of groundwater and surrounding environment by inhibiting the migration of leachate. Bentonite is commonly used as a barrier material in landfills due to its low hydraulic conductivity, high sorption and high swelling capacity [7]. Various researchers in the past have studied the effect of different metal ions on the change in the behaviour of bentonite. Ouhadi et al. [8] studied the consolidation behaviour of bentonite in the presence of heavy metal at different pH and observed that at low pH and high metal concentration the microstructural change in the soil occurs. They also concluded that heavy metals have the different onset of precipitation and the rheological performance of soil is controlled by the osmotic phenomenon. Nakano et al. [9] conducted various tests to determine hydraulic conductivity and lead adsorption behaviour of three Japanese and one U.S. bentonite and concluded that carbonate precipitate as $PbCO_3$ on the bentonites and it also shows a leading role in the low lead solution. They also found that, due to the presence of high montmorillonite content and swelling capacity, the U.S. bentonite exhibits a lower hydraulic conductivity in comparison the Japanese bentonites. Du et al. [10] investigated the influence of various levels of lead concentration on calcium–bentonite backfills/bentonite clay and concluded that with the increase in the concentration of lead, the pH, liquid limit and compression index of the soil decreases. Since the consolidation properties of the bentonite depends on the type of heavy metals and concentrations; therefore, it is essential to study the change in the behaviour of bentonite soil in the presence of various concentration of the metal solution. The present work focuses on the impact of heavy metals on the consolidation properties of bentonite. Bentonite was exposed to various concentrations of zinc and copper solution to investigate different consolidation parameters, such as coefficient of consolidation (c_v), coefficient of volume change (m_v), time to complete 90% of the consolidation (t_{90}) and compression index (C_c) and their change when they are exposed to various heavy metal concentration.

2 Materials and Methods

2.1 Bentonite

The bentonite used in this investigation was procured from Rajasthan state of India. The mineral composition of the bentonite is listed in Table 1.

The grain size distribution of the bentonite was determined by hydrometer analysis as per as ASTM D422 [11]. Ethylene Glycol Monoethyl Ether (EGME) method was followed for determining the specific surface area (SSA) of the bentonite [12]. Attergbergs limit and free swelling capacity were determined according to ASTM D4318 [13] and ASTM D5890 [14]. Cation exchange capacity (CEC) and type of exchangeable

Table 1. Mineral composition of bentonite

Property	Bentonite
Liquid limit (%)	305.5
Plasticity index (%)	269.5
Free swelling (ml/2 g)	20
Clay content (%)	61
Cation exchange capacity (CEC) (meq/100 g)	36.2
Specific surface area (SSA) (m^2/g)	340.4
Optimum moisture content (OMC) (%)	33.5
Maximum dry density (MDD) (gm/cc)	1.30

cations of the bentonite was examined by following the procedure explained by Chapman & Pratt [15, 16]. The compaction parameter of the bentonite was determined as per ASTM D698 [17].

2.2 Permeant Liquid

The metal solutions used in the study were comprised of Cu^{2+} and Zn^{2+}. These metal solutions were selected to examine the influence of various concentrations on the consolidation behaviour of bentonite. By dissolving the powdered salts in deionized (DI) water different concentration of copper and zinc i.e. 500 and 1000 mg/L was prepared. As a reference solution, DI water was used. From the literature, it has been observed that maximum concentration of heavy metal in the leachate is 1000 mg/L [18]. Therefore, in the present study the maximum concentration of 1000 mg/L was considered.

2.3 Determination of Consolidation Characteristics

The consolidation test was conducted on bentonite as per as ASTM D2435 (ASTM 1996) [19]. Bentonite was compacted at its OMC. DI water was added to the bentonite to bring the initial water content correspond to its OMC and then kept in the moisture-controlled desiccator for 24 h to attain equilibrium. The bentonite sample was then compacted statically to its MDD inside the oedometer ring to prepare a sample of 60 mm diameter and 15 mm thickness and with an initial void ratio of 1.06. The setup was then placed in the loading frame inside the consolidation cell and then immersed in deionized (DI) water, Cu^{2+} and Zn^{2+}solution of different concentrations and was permitted to swell. After the end of the swelling the void ratio of the sample increased to 1.519, 1.43, 1.39 for DI water, 500 and 1000 mg/l of copper solution and 1.48 and 1.38 for 500 and 1000 mg/l of zinc solution, respectively.

After the end of the swelling, the sample was consolidated with load increment ratio one started from 4.9 kPa up to a maximum vertical effective consolidating pressure 784.5 kPa. By mean of the dial gauge, the thickness of the sample was measured for every load increment. The change in the void ratio triggered with the load increment was computed as

$$\Delta e = \frac{\Delta H(1 + e_0)}{H} \tag{1}$$

Where ΔH is the change in the thickness of soil sample; H is the initial thickness of the soil sample due to; e_0 is the initial void ratio; Δe is the change in void ratio.

By applying Taylor square-root-of-time method [20] at every load increment a time-settlement plot was plotted and then the coefficient of consolidation (c_v) was calculated.

$$c_v = \frac{D^2 T_v}{t_{90}} \tag{2}$$

Where t_{90} is the time for 90% degree of consolidation; T_v is the time factor (For 90% of consolidation = 0.848).

The coefficient of volume change (m_v) was determined for every increment in pressure as in Eq. (3)

$$m_v = -\frac{\Delta e}{\Delta \sigma (1 + e_0)} \tag{3}$$

Where $\Delta\sigma$ is the change in pressure.

The compressibility behaviour of the soil can be indicated by compression index (C_c) of a soil and was computed as the slope of the straight line portion of the void ratio–pressure (e-log P) curve as

$$C_c = \frac{e_i - e_j}{\log \frac{p_i}{p_j}} \tag{4}$$

Where e_i and e_j are the void ratio corresponding to the consolidation pressure of p_i and p_j at ith and jth steps of loading respectively.

3 Results and Discussion

3.1 Coefficient of Volume Change (Mv)

Figure 1 indicates the plot between the coefficient of volume change (m_v) and the consolidation pressures at various concentration of Cu^{2+} and Zn^{2+} for the bentonite clay. The plot indicates that with the increase in consolidation pressure, the m_v primarily increases and then decreases. This is because initially, the void ratio was high at a lesser consolidation pressure. Further, with an increase in pressure higher decrease in void ratio took place causing in a higher value of m_v. However, the value of m_v decreased further with the increase in pressure once it attains the peak value. The plot also illustrates that with an increase in heavy metal concentration the m_v decreased.

This is because in case of DI water, the thickness of diffuse double layer (DDL) is large and the bentonite clay experiences a higher compression caused by the application of load subsequently in a higher m_v. However, the thickness of DDL gets reduced in the presence of various concentration of heavy metal.

Coefficient of Consolidation (c_v)
The plot in Fig. 2 illustrates that with an increase in consolidation pressure the c_v values decreased. This signifies that due to the presence of the metal ion in the pore fluid the degree of consolidation increases with the application of pressure. Interaction of the heavy metal with the leachate causes a rapid settlement of the barrier material in the landfill disposal sites [21]. Similar kind of trend was identified by Robinson and Allam [22] for montmorillonite soil; however, Mishra et al. [23] observed the opposite trend for sand-bentonite mixtures. The process that controls the compressibility behaviour of the soil samples is responsible for this opposite trend. The mechanical aspects mainly control the compressibility behaviour of sand–bentonite mixture [21] rises with the rise in consolidation pressure. However, a range of attractive and repulsive forces is produced because of physicochemical factors, which control the compressibility behaviour of bentonite clay [22].

3.2 Coefficient of Consolidation (CV)

Figure 2 also demonstrate that with the increase in metal concentration the c_v increases which signifies higher rate of consolidation when the soil sample is permeated with various metal solutions. With the increase in metal concentration, the thickness of DDL reduces causing the decrease in repulsive forces between the clay particles, subsequently soil sample consolidated more rapidly which results in a increase of the c_v value. Figure 2 shows that, with the rise in consolidation pressure from 19.6 kPa to 784.5 kPa for bentonite clay permeated DI water, the c_v decreases from 1.9×10^{-4} to

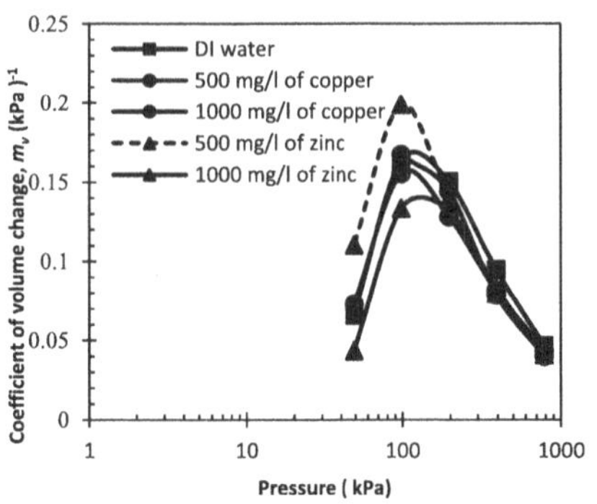

Fig. 1. Coefficient of volume change and consolidation pressures of Bentonite at various heavy-metal concentration

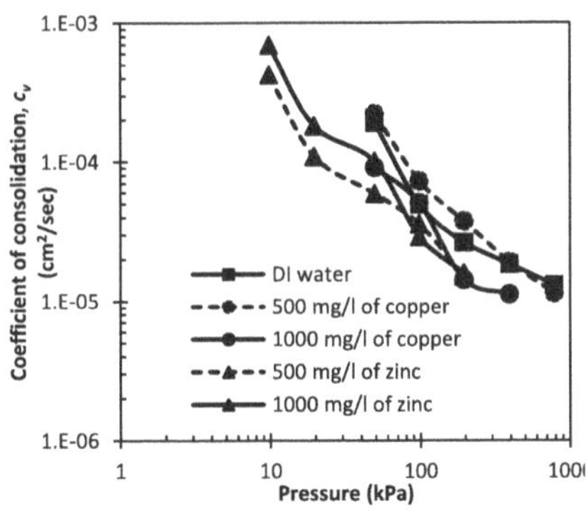

Fig. 2. Coefficient of consolidation and consolidation pressures of Bentonite at various heavy-metal concentration

1.31×10^{-5} cm^2/s (14.5 times). However, the c_v declined from 1×10^{-4} to $1.12\ 10^{-5}$ cm^2/s (8.93 times) when the soil sample was permeated with 1000 mg/L of Cu^{2+} solutions.

3.3 Time to Complete 90% of Consolidation (t90)

The plot between pressure and time for 90% of settlement (t_{90}) in Fig. 3 shows that by the increase in the concentration of Cu^{2+} and Zn^{2+}, t_{90} for the bentonite sample decreases. Since the c_v increases with an increase in the metal concentration, the time required to achieve a desired degree of consolidation decreases resulting in a reduction in t_{90}.

On the contrary, t_{90} increase with the increase in consolidation pressure. With an increase in the consolidation pressure, the clay plates move to a closer distance because of which the repulsion between clay plates increases, which prevents further movement of the plates and results in a decrease in the c_v and increase in t_{90}. The t_{90} value increased from 63.2 min to 564.0 min when the bentonite sample is permeated with DI water and exposed to a consolidation pressure from 9.8 kPa to 784.5 kPa. However, with 500 mg/L concentration of copper it increased from 49.03 min to 620 min. Similarly, the t_{90} increased from 106 min to 552.2 min with further increase in concentration to 1000 mg/L.

3.4 Compression Index (Cc)

The plot in the Fig. 4 depicts the relationship between C_c and various metal concentrations. The plot illustrates that with the increase in the concentration of Cu^{2+} and Zn^{2+} from 0 (DI water) to 1000 mg/L the C_c decreased from 0.64 to 0.58 (10.38%) and 0.64 to 0.54 (15.62%) respectively. This is due to increase in the metal concentration the clay particles come closer to each other and become aggregated. As a result, the thickness of DDL decreases, thus resisting settlement and causing the decrease in C_c value.

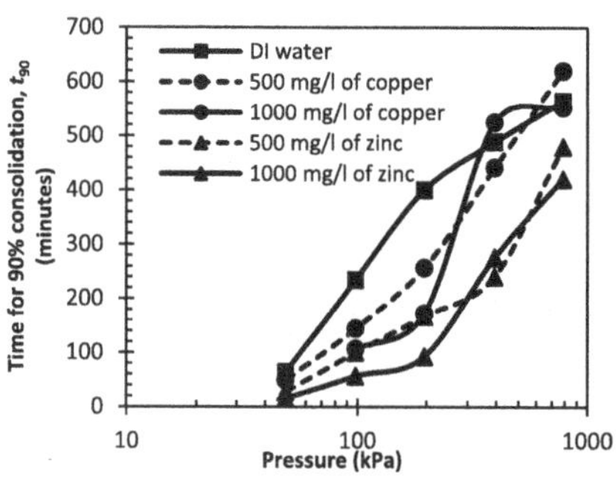

Fig. 3. Time for 90% of consolidation and pressure of bentonite at various heavy-metal concentrations

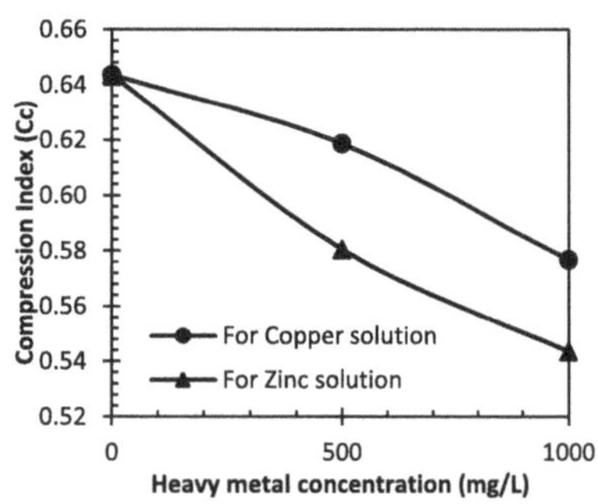

Fig. 4. Compression index of bentonite at different concentrations of heavy metal solutions

4 Conclusions

The present study aimed to investigate the behaviour of bentonite clay with various concentration of Cu^{2+} and Zn^{2+}. The investigation revealed that many factors such as type and concentration of metal and composition of bentonite could influence the consolidation parameters (C_c, m_v, c_v, and t_{90}) of bentonite. From the results, it can be depicted that with the increase in the concentration of heavy metal contaminant the void ratio decreases. Likewise, it was indicated from the graph that with an increase in heavy metal concentration C_c decreased. The plot suggests that with an increase in the concentration of Cu^{2+} and Zn^{2+}, m_v decreased. The reason is due to the implementation of consolidation pressure in the presence of various metal concentration, results in the decline in compressibility, that in turn influence the m_v. The study showed that with an increase in heavy metal concentration the c_v increased. It was also revealed from the investigation that increases in consolidation pressure lead to increase in the value of t_{90}.

References

1. Fergusson JE (1990) The heavy elements: chemistry, environmental impact and health effects. Pergamon Press, Oxford
2. Bradl H (2002) Heavy metals in the environment: origin, interaction and remediation, vol 6. Academic Press, London
3. He ZL, Yang XE, Stoffella PJ (2005) Trace elements in agroecosystems and impacts on the environment. J Trace Elem Med Biol 19(2–3):125–140 [PubMed: 16325528]
4. Chang LW, Magos L, Suzuki T (1996) Toxicology of metals. CRC Press, Boca Raton
5. Wang S, Shi X (2001) Molecular mechanisms of metal toxicity and carcinogenesis. Mol Cell Biochem 222:3–9 [PubMed: 11678608]
6. Beyersmann D, Hartwig A (2008) Carcinogenic metal compounds: recent insight into molecular and cellular mechanisms. Arch Toxicol 82(8):493–512 [PubMed: 18496671]
7. Dutta J, Mishra AK (2016) Consolidation behaviour of bentonite in the presence of salt solution. Appl Clay Sci 120:61–69
8. Ouhadi VR, Yong RN, Sedighi M (2006) Influence of heavy metal contaminants at variable pH regimes on rheological behaviour of bentonite. Appl Clay Sci 32:217–231
9. Nakano A, Li LY, Ohtsubo M, Mishra AK (2008) Lead retention mechanisms and hydraulic conductivity studies of various bentonites for geoenvironment applications. Environ Technol 29:505–514
10. Du YJ, Fan RD, Reddy KR, Liu SY, Yang YL (2015) Impacts of presence of lead contamination in clayey soil-calcium bentonite cutoff wall backfills. Appl Clay Sci 108:111–122
11. ASTM Standard test method for particle-size analysis of soils (2002) D 422-63. American Society for Testing and Materials, Philadelphia
12. Cerato AB, Lutenegger AJ (2002) Determination of surface area of fine-grained soils by the ethylene glycol monoethyl ether (EGME) method. Geotech Test J 25:1–7
13. ASTM Standard test methods for liquid limit, plastic limit, and plasticity index of soils (2000) D 4318. American Society for Testing and Materials, Philadelphia
14. ASTM Standard test method for swell index of clay mineral component of geosynthetic clay liners (2001) D 5890. American Society for Testing and Materials, Philadelphia

15. Chapman HD (1965) Cation exchange capacity. Methods of soil analysis, part 2 chemical and microbiological properties, 2nd edn. Soil Science Society of America, Madison, pp 891–895
16. Pratt PF (1965) Sodium, methods of soil analysis, part 2 chemical and microbiological properties, 2nd edn. Soil Science Society of America, Madison, pp 1031–1034
17. ASTM D698 Standard Test Methods for Laboratory Compaction Characteristics of Soil Using Standard Effort (2012) American Society for Testing and Materials, Philadelphia
18. Prudent P, Domeizel M, Massiani C (1996) Chemical sequential extraction as a decision-making tool: application to municipal solid waste and its individual constituents. Sci Total Environ 178:55–61
19. ASTM Standard test method for one-dimensional consolidation properties of soils (1996) D 2435. American Society for Testing and Materials, Philadelphia
20. Taylor DW (1948) Fundamentals of soil mechanics. Wiley, New York
21. Ouhadi VR, Sedighi M (2003) Variation of experimental results of oedometer testing due to the changes of pore fluid. In: Deformation characteristics of geomaterials, pp 299–304. Taylor & Francis, New York
22. Robinson RG, Allam MM (1998) Effect of clay mineralogy on coefficient of consolidation. Clays Clay Miner 46(5):596–600
23. Mishra AK, Ohtsubo M, Li LY, Higashi T (2010) Influence of the bentonite on the consolidation behaviour of soil bentonite mixtures. Carbonates Evaporites 25(1):43–49

Influences of Effective Confining Stresses on the Chemical Compatibility of Backfills for Soil-Bentonite Cut-off Walls

Haoqing Xu(✉), Wei Zhu, Shengwei Wang, Shi Shu, and Xihui Fan

School of Architecture and Civil Engineering, Jiangsu University of Science and Technology, Zhenjiang 212003, China
hankinxu@163.com

Abstract. The chemical compatibility of backfills for soil-bentonite vertical cut-off walls is examined in this paper. The use of soil-bentonite cut-off walls as hydraulic barriers for waste containment facilities is on the rise. Their low hydraulic conductivity to water is attributed to the bentonite in the soil-bentonite backfills, which is composed primarily of smectite mineral. However, there have been concerns as to the compatibility of soil-bentonite backfills, since bentonite is very sensitive to chemical effects and this can lead to an increase in hydraulic conductivity. Thus, Fujian standard sand was employed for simulating a sandy soil layer, and Hebei Bentonite and Jiangning Clay were served as additives for studying the chemical compatibility of sand-bentonite backfills. The present study focuses on the influences of the stress level on the hydraulic conductivity, bound water content and effective porosity of soil-bentonite backfills under the situation of calcium chloride solutions with various concentrations.

Keywords: Bound water · Chemical compatibility
Effective confining stresses · Effective pore void · Soil-bentonite backfills

1 Introduction

Landfill-induced subsurface soil and water pollutions represent a significant environmental issue, which is in urgent need of effective measures for containment. Vertical cut-off walls have been used for pollution control since 1970, and the technology is accepted and regarded as an effective way of isolating hazardous waste and preventing the migration of contaminants. Soil-bentonite vertical cut-off walls can be used extensively for Chinese landfill remediation projects due to its low permeability and construction cost (Du et al. 2015; Li et al. 2015; Hong and Shackelford 2017).

A number of studies conducted primarily to assess factors that potentially influence chemical compatibility of clays used as barriers have provided some insight into expected behavior (Katsumi et al. 2008; Mishra et al. 2009; Malusis and Mckeehan

Project: National Science & Technology Major Project of China (2013ZX07113001); Natural Science Foundation of Jiangsu Province (BK20170501).

L. Zhan et al. (Eds.): ICEG 2018, ESE, pp. 575–582, 2019.
https://doi.org/10.1007/978-981-13-2224-2_71

2013). The results indicate that qualitative predictions of chemical incompatibility can be made based mainly on three considerations: (1) the type and properties of the fluid, (2) the type and properties of the soil and (3) the physical conditions imposed on the barrier materials.

First, the higher the concentration and/or the charge of cations in inorganic solutions, the greater the possibility for chemical incompatibility. Decreasing pH tends to increase the potential for chemical incompatibility. High concentrations of hydrophilic organic compounds and pure-phase organic fluids can cause incompatibility due to clay shrinkage when the dielectric constant of the fluid is significantly lower than that for water. In addition, the potential for incompatibility usually increases as the activity, defined as the plasticity index divided by the percent of clay-size particles, of the soil increases. Lastly, three factors relating to the physical conditions are particularly important: ① the stresses imposed on the barrier soil, ② the initial level or degree of hydration of the barrier material and ③ the hydraulic gradient of the fluid across the barrier (Lee and Benson 2000; Mishra et al. 2009; Hudak 2016). In short, the higher the effective confining stress in the soil, the less susceptible the soil is to adverse consequences of chemical attack. However, the influence of stress level on chemical compatibility of soil-bentonite cut-off wall backfills has not been evaluated extensively.

2 Experimental Studies

2.1 Materials

Domestically produced Fujian standard sand (Sand-F) was employed for simulating the sandy soil layer at the site. The basic physical properties of Sand-F are given in Table 1.

Table 1. Physical properties of Sand-F

Specific gravity G_s	Uniformity coefficient	Maximum dry density (g/cm^3)	Minimum dry density (g/cm^3)	Maximum void ratio	Minimum void ratio
2.64	5.99	1.74	1.43	0.85	0.52

Two types of clays were chosen in the study, viz., Hebei Bentonite produced in Hebei Province, China (Bent-H) and Jiangning Clay produced in Jiangning District of Nanjing, China (Clay-J). X-ray diffraction analysis of Bent-H and Clay-J was performed by a Rigaku D/max-rC rotating anode X-ray powder diffractometer. Air-dried powdered specimens (particle size <75 μm) of Bent-H and Clay-J were used. The X-ray source was a Cu anode operating at 40 kV and 100 mA using Cu $K\alpha$ radiation (λ = 1.5406 Å). The run speed was 3°/min. Data were collected between 5° and 40° in 2θ increments.

Mineralogical analysis of X-ray diffraction pattern of Bent-H and Clay-J was performed by comparing with the X-ray powder diffraction standard files. X-ray diffraction analysis shows that the dominant minerals of the Bent-H and Clay-J are montmorillonite (48 wt%) and illite (58 wt%), respectively. The basic physical properties of Bent-H and Clay-J are listed in Table 2. All soils were oven-dried at 105 °C for 24 to 48 h as pretreatment before adopted in tests.

Table 2. Basic physical properties of Bent-H and Clay-J (Xu et al. 2016)

Type	G_s	w_L (%)	w_p (%)	Exchangeable cations (cmol/kg)				Swell index (mL/2 g)
				Na^+	K^+	Ca^{2+}	Mg^{2+}	
Bent-H	2.75	181	53	78.93	1.32	32.82	0.59	17.0
Clay-J	2.72	48	24	4.96	2.11	17.15	0.11	1.6

The grain-size distribution curves for the soils used in this study are manifested in Fig. 1. Sand-F was tested by using sieve analysis. The grain-size distributions of Bent-H and Clay-J were determined by a laser grain-size analyser Mastersizer 2000. On the basis of Unified Soil Classification System (ASTM 2011b), sand-F is a poorly graded sand (SP) and Bent-H and Clay-J are classified as high-plasticity clay (CH) and low-plasticity clay (CL), respectively. The hydraulic conductivities of Bent-H and Clay-J for deionised water are 6.3×10^{-10} cm/s and 2.3×10^{-8} cm/s under an effective confining stress of 50 kPa.

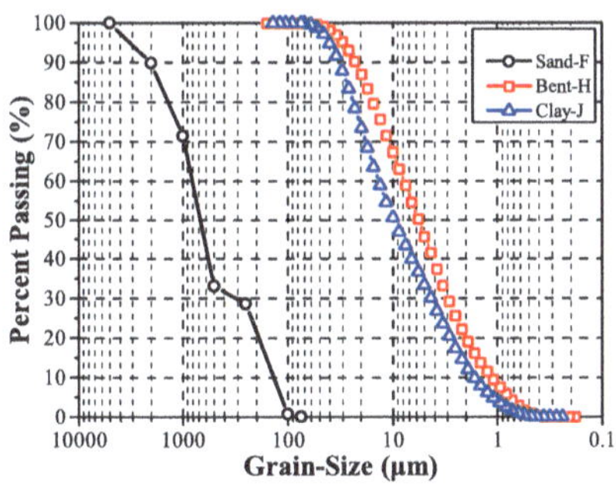

Fig. 1. Grain-size distribution curves

2.2 Specimen Preparation

Sand-F was uniformly mixed with a certain amount of Bent-H or Clay-J and then stirred well with 5% concentration slurry to form as similar to concrete pouring mortar-like specimen. In order to simulate the backfills in the practical projects, the slump and density of the test specimens were controlled at a range from 75–120 mm and from 1.45–1.80 g/cm^3, respectively (Malusis and Mckeehan 2013). Note that the content of bentonite/clay means the dry weight percentage of bentonite/clay to sand-bentonite/clay mixture. The bentonite/clay content of the specimens included both the dry bentonite added to the backfills and the bentonite that was contributed from mixing with the 5% bentonite slurry.

2.3 Improved Flexible-Wall Permeability Test

Due to the poor self-supporting characteristics of specimens, a flexible-wall permeameter was improved, namely a cutting ring with a lot of opened pores having a diameter of 3 mm was added at the periphery of the specimens (Malusis et al. 2009; Hong et al. 2012; Zhu et al. 2014; Xu et al. 2016). The specimens can be made not only well self-reliance, but also applied effective confining stress around the specimens. Compared to the rigid-wall permeameter, the improved flexible-wall permeameter could effectively prevent leakage of the side walls and in turn be able to simulate the stress conditions of the cut-off wall in the actual projects. The diameter and height of the specimens were 70 mm and 40 mm. The deionised water or calcium chloride solutions was served respectively as the permeant fluids in the tests. The applied hydraulic gradient was 70–80 during the tests. The flexible-wall permeability tests were referred to ASTM D5084.

2.4 Measurement of Moisture Content

To understand the bound forms of the pore water, a moisture measurement test was performed by using a high speed centrifuge separation method (Zhu et al. 2007; Wang et al. 2018). After the permeability test, the specimens were cut with a centrifuge cutting ring (h = 50.9 mm, D = 49.9 mm) and placed in a high-speed refrigerated centrifuge. Then, a high-speed centrifuge separation method was adopted to determine the water characteristic curve. The bound water content of specimens means the weight percentage of bound water to pore water in soils.

3 Results and Discussion

3.1 Hydraulic Conductivity of Sand-Bentonite/Clay Mixtures Versus Effective Confining Stress for $CaCl_2$ Solutions

To investigate the relationship between the confining pressure and the hydraulic conductivity of the consolidated sand-bentonite (5% Bent-H) and sand-clay (30% Clay-J) mixtures permeated with the calcium chloride solutions of different concentrations ($c(Ca^{2+})$ = 0, 2500, 5000, 10000 mg/L), the consolidation of the soils and the permeation of the solutions were conducted in improved flexible-wall permeameters. Figure 2 demonstrates the variations of the hydraulic conductivity of sand-bentonite/clay mixtures for electrolytic chemical solutions with the effective confining stress.

With the increase of effective confining stress, the hydraulic conductivities of sand-bentonite and sand-clay mixtures for $CaCl_2$ solutions are reduced to different extents. When the effective stress was higher than 400 kPa, the hydraulic conductivity still decline but the decreasing tendency had decelerated remarkably, which meant that the subsequent application of confining pressure only played an extremely limited influence on minimising the hydraulic conductivity.

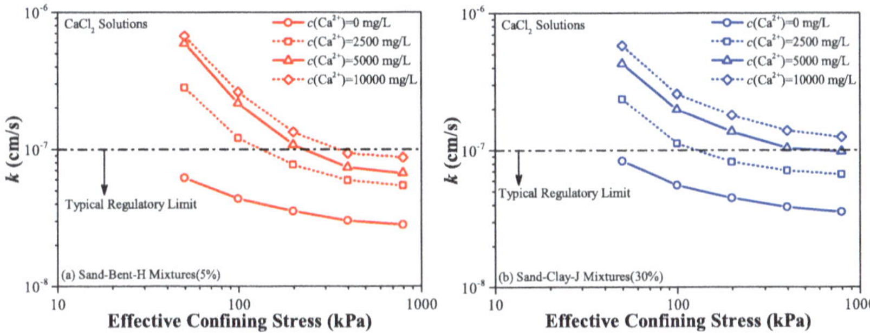

Fig. 2. Hydraulic conductivity *k* of soils vs. effective stress for $CaCl_2$ solutions

As manifested in Fig. 2, at a low imposed effective confining pressure at the time of permeation, there would be a major raising in hydraulic conductivity of sand-Bent-H or sand-Clay-J mixture as the permeant liquid was changed from deionised water to calcium chloride solution with increasing calcium chloride concentration. Consolidation during permeation could partly mitigate the impacts of clay-fluid interaction on hydraulic conductivity. The influence was far less obvious at larger confining pressure and the hydraulic conductivities for sand-bentonite/clay mixtures to calcium chloride solutions were still relatively low. Moreover, the influence of consolidation on compatibility for Bent-H is greater than that for Clay-J.

3.2 Bound Water Content of Sand-Bentonite/Clay Mixtures Versus Effective Confining Stress for $CaCl_2$ Solutions

The water characteristic curves of sand-bentonite (5% Bent-H) and sand-clay (30% Clay-J) mixtures were obtained from the centrifuge tests. The *pF* values are employed for indicating the size of the soil water potential, which is the logarithm value of suction water head. Lebedev (1936) suggested to use $pF = 3.8$ as the critical point between the bound water and free water. Gardner (1937) proposed that the high speed centrifugation method could be adopted to determine the bound water content of clay soils.

Accordingly, the relationships between bound water content and effective confining stress of sand-Bent-H and sand Clay-J mixtures, which were obtained from the moisture measurement tests, are demonstrated in Fig. 3. When the concentration of cation contained in the permeant solution was unvarying, the bound water content of soil mixture decreased constantly with the increase of the effective confining stress. However, as effective stress was added to 400 kPa, the bound water content of soil mixture presented a relatively slight downward trend.

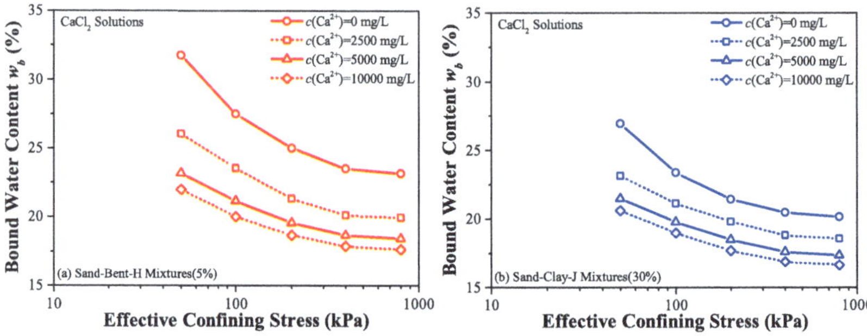

Fig. 3. Bound water content w_b of soils vs. effective stress for $CaCl_2$ solutions

3.3 Effective Porosity of Sand-Bentonite/Clay Mixtures Versus Effective Confining Stress for $CaCl_2$ Solutions

Numerous investigators (Hueckel 1997; Petrov et al. 1997) indicated the consolidation of soil to lower the pore space available for the liquid to flow. The hydraulic performance is intimately related to the pore space, thus the effective pore can be calculated as the total pore subtracting the volume occupied by the bound water.

Based on the analysis above, the computing formula for the effective porosity n_{eff} of the saturated soil might be simplified as follows:

$$n_{eff} = \frac{V_w - V_b}{V} = \frac{V_w}{V} - \frac{V_w}{V}\frac{V_b}{V_w} \approx \frac{V_w}{V} - \frac{V_w}{V}\frac{m_b}{m_w} \approx n(1 - w_b) \tag{1}$$

where V, V_w and V_b represent the soil volume, pore volume and bound water volume respectively; m_w and m_b represent the pore water weight and bound water weight respectively; n is the porosity and w_b is the bound water content in soil pores.

As a result, the changes of the effective porosity of sand-bentonite/clay mixtures with stress are manifested in Fig. 4. Though the $CaCl_2$ solutions with high ionic concentration increased the hydraulic conductivity of soil mixture, larger stress decreased the effective porosity of mixture even for the electrolytic chemical solution so that the confined soil can exhibit a relatively good hydraulic performance of less than 1.0×10^{-7} cm/s even to the calcium chloride solution. The consolidation of mixture could accelerated the solid phase and immobile water phase relative to the mobile water phase, which means the effective porosity fell off, so that the consolidated soil could obstruct the flow even to $CaCl_2$ solution.

In the low concentration level, the clay could be sufficiently swelling so that the bound water content is large and the hydraulic conductivity is low. The calcium chloride solution with the high concentration, nonetheless, reduces the swelling capacity of the clay. Accordingly, the sand-bentonite/clay mixture permeated with the chemical solution having the large concentration indicates the low bound water content and the high hydraulic conductivity at a low confining stress. As the confining stress is improved from 50 kPa to 400 kPa, the hydraulic conductivity is reduced with the effective porosity. The degree of this variation is especially significant in the high concentration.

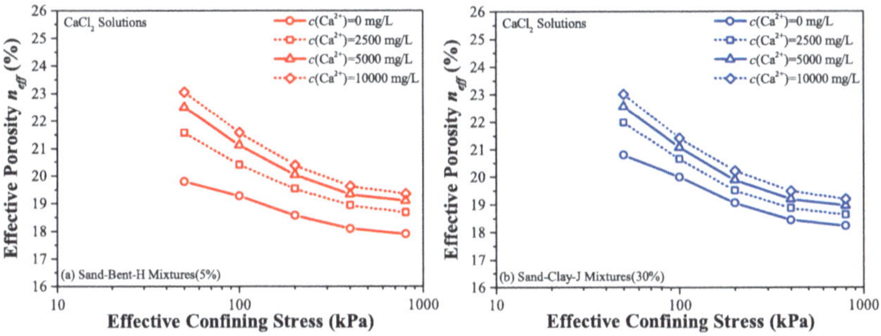

Fig. 4. Effective porosity n_{eff} of soils vs. effective stress for $CaCl_2$ solutions

4 Conclusions

The consolidation pressure affects the hydraulic conductivity in two opposed effects, dependent of the applied stress. In a low confining stress, the hydraulic conductivity is strongly dependent on the ionic concentration of the calcium chloride solution. The permeant solution with the strong ionic strength seriously increases the hydraulic conductivity due to insufficient swelling of the clay. In contrast, the hydraulic conductivity of sand-bentonite/clay mixture which was confined at the high effective pressure of 400 kPa was about $k \approx 1.0 \times 10^{-7}$ cm/s regardless of the concentration of calcium chloride solution. It is because the solid phase and immobile water phase are relatively improved by the consolidation. Thus, the effective porosity which influences the free mobility of the electrolytic chemical solution is narrowed.

References

Du YJ, Fan RD, Liu SY, Reddy KR, Jin F (2015) Workability, compressibility and hydraulic conductivity of zeolite-amended clayey soil/calcium-bentonite backfills for slurry-trench cutoff walls. Eng Geol 195:258–268

Gardner R (1937) A method of measuring the capillary tension of soil moisture over a wide moisture range. Soil Sci 43(4):277–284

Hong CS, Shackelford CD (2017) Mixed metals migration through zeolite-amended backfills for vertical cut-off walls. Environ Geotech 1–17

Hong CS, Shackelford CD, Malusis MA (2012) Consolidation and hydraulic conductivity of zeolite-amended soil-bentonite backfills. J Geotech Geoenvironmental Eng 138(1):15–25

Hudak PF (2016) Evaluation of non-pumped wells with slurry cutoff walls for containing and removing contaminated groundwater. Environ Earth Sci 75(1):1–5

Hueckel T (1997) Chemo-plasticity of clays subjected to stress and flow of a single contaminant. Int J Numer Anal Methods Geomech 21(1):43–72

Katsumi T, Ishimori H, Onikata M, Fukagawa R (2008) Long-term barrier performance of modified bentonite materials against sodium and calcium permeant solutions. Geotext Geomembr 26(1):14–30

Lebedev AF (1936) Soil and groundwaters. The Academy of Sciences of the USSR, Moskva

Lee T, Benson CH (2000) Flow past bench-scale vertical ground-water cutoff walls. J Geotech Geoenvironmental Eng 126(6):511–520

Li YC, Cleall PJ, Wen YD, Chen YM, Pan Q (2015) Stresses in soil–bentonite slurry trench cut-off walls. Géotechnique 65(10):1–8

Malusis MA, Barben EJ, Evans JC (2009) Hydraulic conductivity and compressibility of soil-bentonite backfill amended with activated carbon. J Geotech Geoenvironmental Eng 135 (5):664–672

Malusis MA, Mckeehan MD (2013) Chemical compatibility of model soil-bentonite backfill containing multiswellable bentonite. J Geotech Geoenvironmental Eng 139(2):189–198

Mishra AK, Ohtsubo M, Li LY, Higashi T, Junboum P (2009) Effect of salt of various concentrations on liquid limit, and hydraulic conductivity of different soil-bentonite mixtures. Environ Geol 57(5):1145–1153

Petrov RJ, Rowe LRK, Quigley RM, Rowe RK (1997) Comparison of laboratory-measured GCL hydraulic conductivity based on three permeameter types. Geotech Test J 20(1):49–62

Wang S, Zhu W, Fei K, Xu C, Zhang N (2018) Study on non-darcian flow sand-clay mixtures. Appl Clay Sci 151:102–108

Xu H, Zhu W, Qian X, Wang S, Fan X (2016) Studies on hydraulic conductivity and compressibility of backfills for soil-bentonite cutoff walls. Appl Clay Sci 132–133:326–335

Zhu W, Wang R, Zuo J, Lin C, Min F (2014) Improved isotropically consolidated undrained triaxial test method for non-self-supporting materials. Geotech Test J 37(4):20130114

Zhu W, Zhang CL, Chiu ACF (2007) Soil-water transfer mechanism for solidified dredged materials. J Geotech Geoenvironmental Eng 133(5):588–598

Effects of Soil Fabric on the Thermal Expansion Coefficient of Clay with Reference to the Crack Formation in Landfill Barriers

Q. Y. Mu(✉), C. W. W. Ng, C. Zhou, and H. J. Liao

Xi'an Jiaotong University, Xi'an, People's Republic of China
qingyimu@mail.xjtu.edu.cn

Abstract. Thermal expansion coefficient (TEC) of clay is an important parameter for evaluating the crack formation in landfill barriers during heating and cooling. To investigate the influences of soil fabric on the TEC of clay, saturated reconstituted, intact and recompacted loess were tested using a thermo-controlled invar oedometer. The linear TEC of each specimen was measured through thermal volume changes during cooling. Results showed that the linear TEC of the reconstituted specimen was almost three times larger than those of the intact and recompacted specimens. On the other hand, the void ratio has an insignificant effect on the linear TEC of reconstituted specimen. These observed differences in the linear TECs of reconstituted, intact and recompacted specimens are likely attributed to different distributions of clay particles in soil specimens (with 28% clay content). In the intact and recompacted specimens, most of clay particles formed silt-size aggregates which were less sensitive to temperature changes than clay particles founded in the reconstituted specimen.

Keywords: Thermal expansion coefficient · Loess · Oedometer test

1 Introduction

Landfill barriers (i.e., covers and liners), which are subjected to cyclic heating and cooling, tend to crack (Southen and Rowe 2005; Sadek et al. 2007). Due to cracking, preferential flow paths for both water and gas are formulated and consequently lowering the barrier efficiency (Ng et al. 2015; Zhan et al. 2017). Thermal expansion coefficient (TEC) of soil medium is an important parameter for the crack analysis of landfill barriers during heating and cooling. Previous studies considered the TEC of soil is a constant once soil grains are in mineral to mineral contact (Campanella and Mitchell 1968; Khalili et al. 2010). The effects of soil fabric, which defines as the arrangement of soil particles (Mitchell and Soga 2005), on the TEC of soil have not been fully understood.

The principal objective of this study is to investigate the effects of soil fabric on the linear TEC of soil. To meet this objective, a low-plasticity clay was tested at three different states (i.e., reconstituted, intact and recompacted). Through temperature-controlled oedometer tests, thermal volume changes of these three types of specimens were measured and the linear TECs were determined through the volume changes

L. Zhan et al. (Eds.): ICEG 2018, ESE, pp. 583–588, 2019.
https://doi.org/10.1007/978-981-13-2224-2_72

during cooling. The results were interpreted and compared to reveal the effects of soil fabric on the TEC of soil.

2 Test Apparatus

Figure 1 shows the schematic diagram of a temperature-controlled oedometer, which was modified from a conventional oedometer by adding a heating-cooling system. Furthermore, the volumetric TEC of invar ($5 \times 10^{-7}\,°C^{-1}$) is almost negligible compared to that of a soil skeleton (Di Donna and Laloui 2015; Vega and McCartney 2015). Therefore, the invar oedometer ring was used to minimise the thermal strain of the oedometer ring. More details of the test apparatus were given by Ng et al. (2017).

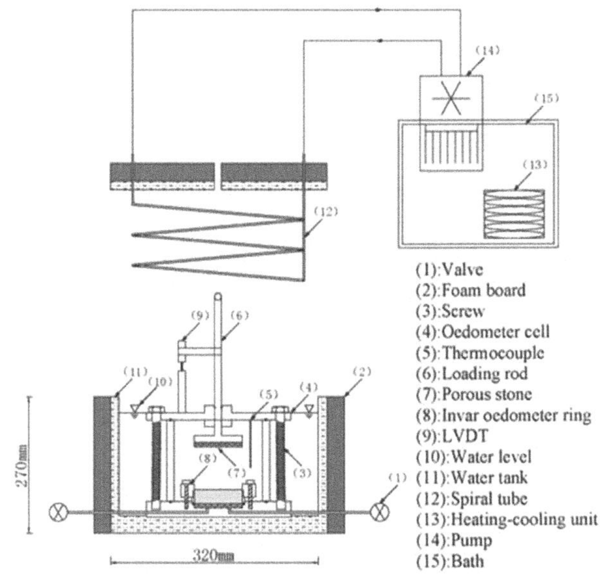

Fig. 1. Schematic diagram of the temperature-controlled oedometer (Ng et al. 2017)

3 Test Soil and Test Program

A loess taken from Xi'an in China was used in this study. The liquid limit and plastic limit were 36% and 19%, respectively. The soil consists of 28% clay and 72% silt particles. According to the Unified Soil Classification System (ASTM 2006), the soil is classified as clay of low plasticity (CL). More physical properties of the test soil were described in detail by Ng et al. (2017).

In this study, two series of cyclic heating and cooling tests were carried out on saturated normally consolidated loess. For the first series, volume changes of reconstituted, intact and recompacted specimens under cyclic thermal loads were measured. The linear TEC of each specimen was measured through the volume changes during cooling. It should be noted that apart from soil fabric, the reconstituted, intact and recompacted loess also had different void ratios. Therefore, the influence of void ratio on the linear TEC of reconstituted specimen was studied in the second test series. Details of the test program were shown in Table 1.

Table 1. Details of the test program

Test series	Test ID	Specimen preparation method	Consolidation stress (kPa)	e_0 (void ratio, initial state)	e_1 (void ratio, after mechanical consolidation)
I	R	Reconstituted	50	1.442	0.812
	I	Intact		1.278	1.095
	RC	Recompacted		1.271	0.907
II	R	Reconstituted	50	1.442	0.812
			100	1.442	0.734
			200	1.442	0.658

4 Specimen Preparation

Three different specimens were prepared: reconstituted, intact and recompacted. To prepare the reconstituted specimen, the air-dried soil was thoroughly mixed with de-aired water with a gravimetric water content of 1.5 times the liquid limit (water content: 54%). The reconstituted specimen was obtained by consolidating the slurry with an effective vertical stress of 50 kPa in a one-dimensional consolidometer.

To prepare the intact specimen, the undisturbed block sample was collected from a depth of 3.5 m in situ. Then an oedometer ring (diameter = 70 mm; height = 25 mm) was pushed into the block sample. The top and bottom of the specimen were trimmed with a wire saw. The initial water content of the intact specimen is 9.2%.

To prepare the recompacted specimen, the air-dried soil was mixed with de-aired water with a gravimetric water content of 9.2%, which was the same as that of the intact specimen. The mixture was kept in an air-tight plastic bag for moisture equalisation. The mixtures were statically compacted into the oedometer ring. The target dry density was 1.18 g/cm^3, which was similar to that of the intact specimen. The soil states of each specimen after preparation are shown in Table 1.

5 Test Procedures

After specimen preparation, each specimen was set up individually in the oedometer cell. De-aired water was added to the cell to submerge and saturate the soil specimen. After 24 h, the effective vertical stress of each specimen was increased to the predefined value (i.e., 50, 100 and 200 kPa). The drainage valve was kept open for the dissipation of excess pore water pressure. The consolidation was considered to be finished when the volumetric strain rate was less than 0.025% within 24 h (Romero et al. 2003).

After consolidation, each specimen was subjected to cyclic thermal loads at a constant effective vertical stress. In the heating phase, the soil temperature was increased from room temperature (25 °C) to 70 °C at 2 °C per hour. In the cooling phase, the temperature was decreased from 70 °C to 15 °C at 5 °C per hour. A cyclic thermal test was stopped when the incremental plastic vertical strain induced by a thermal cycle fell below 0.025% (close to the accuracy of the LVDT measurements, ±0.02%).

6 Interpretations of Experimental Results

6.1 Effects of Soil Fabric on Thermal Expansion Coefficient

Figure 2 shows the thermal axial strains of reconstituted, intact and recompacted specimens during cooling. The linear TEC of reconstituted specimen was calculated with volume changes of six cooling paths. For the intact and recompacted specimens, similar approach was used to determine the linear TEC but with volume changes of five cooling paths. The error bars of the volume changes were also shown in the figure. It can be seen that the linear TECs for the reconstituted, intact and recompacted specimens are 6.4×10^{-6} °C^{-1}, 2.7×10^{-6} °C^{-1} and 2.0×10^{-6} °C^{-1}, respectively. It is clear that the linear TEC of the reconstituted specimen is almost three times larger than those of the intact and recompacted specimens. The difference in the linear TECs between these three specimens possibly results from the different distributions of clay particles in soil (with 28% clay content). As revealed in the SEM observations by Ng et al. (2018), clay particles form the soil skeleton of the reconstituted specimen. On the contrary, in the intact and recompacted specimens, most of the clay particles are gathered in a small number of silt-size aggregates. Silt particles play a significant role in the soil skeleton formation of intact and recompacted specimens. It is well known that the soil skeleton formed by clay particles is much more sensitive to temperature changes than that formed by silt particles. It should be noted that some theoretical and experimental studies reported in the literature assumed that the TEC of soil depends on its mineral only (Campanella and Mitchell 1968; Khalili et al. 2010). Obviously, these theoretical studies should be improved by considering the influences of soil fabric on the TEC.

6.2 Effects of Void Ratio on the Thermal Expansion Coefficient

Figure 3 shows the relationship between linear thermal expansion coefficient of reconstituted specimen and void ratio. The determination of linear TEC was similar to the method described in the Sect. 6.1. The linear TECs of reconstituted specimen at void ratios of 0.812, 0.734 and 0.658 are 6.4×10^{-5} °C^{-1}, 9.1×10^{-5} °C^{-1} and 7.2×10^{-5} °C^{-1}, respectively. The result implies that the void ratio has an insignificant effect on the linear TECs of the reconstituted specimen. This conclusion is consistent with previous studies on the homogeneous porous materials (Khalili et al. 2010). The TEC of the soil skeletal is independent of the void ratio when soil grains are in mineral to mineral contact (Campanella and Mitchell 1968; Khalili et al. 2010).

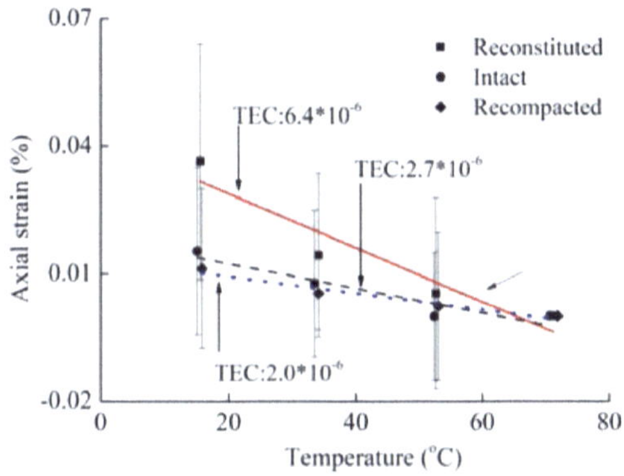

Fig. 2. Variations of thermal axial strains of the reconstituted, intact and recompacted specimens during cooling

Furthermore, the experimental results likely could explain the similar linear TECs of the saturated recompacted and intact specimens (see Fig. 2). The soil skeleton of both the intact and recompacted specimens is mainly formed by silt particles. Therefore, the linear TECs of intact and recompacted specimens are expected to be similar once the silt particles are in mineral-to-mineral contact.

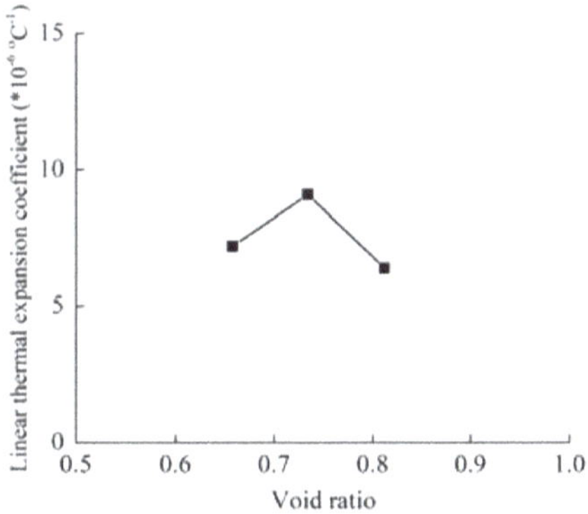

Fig. 3. Relationship between linear thermal expansion coefficient of the reconstituted specimen and void ratio

7 Conclusions

To evaluate the crack formation in landfill covers and liners during heating and cooling, the TEC of clay with different fabrics were investigated. The linear TEC of the reconstituted specimen was almost three times larger than that of the intact and recompacted specimens. On the other hand, the void ratio has insignificant effects on the linear TEC. These observed differences in linear TECs of reconstituted, intact and recompacted specimens are likely attributed to the different soil fabrics among the three types of specimens. The intact and recompacted specimens are composed of many silt-size clay aggregates but not in reconstituted one. It is well-known that clay particles are much more sensitive to temperature changes than silt particles. In addition, the similar linear TECs of the recompacted and intact specimens are because that the soil skeleton of both recompacted and intact specimens is mainly formed by silt particles.

Acknowledgements. The authors would like to acknowledge the financial support provided by the Research Grants Council of the Hong Kong Special Administrative Region (HKSAR) through the research grants 16216116 and the National Science Foundation of China (NSFC) through the research grant 51778166, 51509041 and 41630639.

References

ASTM (2006) Standard practice for classification of soils for engineering purposes (unified soil classification system). American Society for Testing and Materials, West Conshohocken

Campanella RG, Mitchell JK (1968) Influence of temperature variations on soil behaviour. J Soil Mech Found Eng Am Soc Civ Eng 94:709–734

Di Donna A, Laloui L (2015) Numerical analysis of the geotechnical behaviour of energy piles. Int J Numer Anal Methods Geomech 39(8):861–888

Khalili N, Uchaipichat A, Javadi AA (2010) Skeletal thermal expansion coefficient and thermo-hydro-mechanical constitutive relations for saturated homogeneous porous media. Mech Mater 42(6):593–598

Sadek S, Ghanimeh S, El-Fadel M (2007) Predicted performance of clay-barrier landfill covers in arid and semi-arid environments. Waste Manag 27(4):572–583

Southen JM, Rowe RK (2005) Laboratory investigation of geosynthetic clay liner desiccation in a composite liner subjected to thermal gradients. J Geotech Geoenvironmental 131(7):925–935

Mitchell JK, Soga K (2005) Fundamentals of soil behavior, vol 3. Wiley, New York

Ng CWW, Chen ZK, Coo JL, Chen R, Zhou C (2015) Gas breakthrough and emission through unsaturated compacted clay in landfill final cover. Waste Manag 44:155–163

Ng CWW, Mu QY, Zhou C (2017) Effects of boundary conditions on cyclic thermal strains of clay and sand. Géotechnique Lett 7(1):73–78

Ng CWW, Mu QY, Zhou C (2018) Effects of specimen preparation method on the volume change of clay under cyclic thermal loads. Géotechnique. https://doi.org/10.1680/jgeot.16.P.293

Romero E, Gens A, Lloret A (2003) Suction effects on a compacted clay under non-isothermal conditions. Géotechnique 53(1):65–81

Vega A, McCartney JS (2015) Cyclic heating effects on thermal volume change of silt. Environ Geotech 2(5):257–268

Zhan LT, Li GY, Jiao WG, Wu T, Lan JW, Chen YM (2017) Field measurements of water storage capacity in a loess–gravel capillary barrier cover using rainfall simulation tests. Can Geotech J 54(11):1523–1536

Effect of Initial Compaction State on Erosion Potential for Cover Liner

Himanshu Kumar[1], Sanandam Bordoloi[1], S. Sreedeep[1], and Ankit Garg[2(✉)]

[1] Indian Institute of Technology, Guwahati, India
[2] Shantou University, Guangdong, China
ankit@stu.edu.cn

Abstract. Multi-layered cover system (MLCS) is constructed upon hazardous landfills that has reached its full potential. Soil erosion is a primary stressor of surface soils used in MLCS, which ultimately results in exposure of lower layers. The effect of compaction state is crucial to assess erosion potential of compacted surface soil used MLCS. The erosion potential needs to be studied as a function of initial compaction state, soil type and flow rate, under high intensity rainfall conditions. For this purpose, a well-established pin hole test was used to evaluate the erosion potential of soil. A total of 324 pin-hole tests on four loamy soils were conducted and the individual effect of initial moisture content, soil density and flow rate was discussed. It is observed that compaction process in top surface of MLCS can be done at wet of optimum moisture content (OMC) with highest maximum dry density (MDD) to minimize risk of erosion related failure. The compaction state of (OMC+5, 1.05MDD) is suitable for top surface design as it showcases significantly lower erosion rates than the permissible erosion rate for cover systems.

Keywords: Soil erosion · Pin hole test · Multi layered cover system Compaction state

1 Introduction

Industrialization has led to the increase in hazardous wastes that requires to be disposed safely (EIA 2014). In this regard, safe and sustainable waste containment systems help to prevent the ill effects posed by unregulated waste disposal in open lands (USEPA 1989). Multi-layered cover systems (MLCS; Fig. 1a), inhibits contaminant interaction from atmosphere and is used for such waste disposal (Sadasivam and Reddy 2015). There are multitude of studies which deal with design considerations of bottom clay liner system (Stark 2015; Müller and Wöhlecke 2017). Only limited studies have focused in considering performance and designs methods for top cover systems (Albright et al. 2004; Benson et al. 2007). Surface layer of MLCS play a major role in the serviceability of the whole landfill system as it can lead to uncontrolled seepage and contamination, if not properly designed. The requirement for assessment of the performance of top surface layer in MLCS has been recently felt, due to the attainment of full capacity of landfills (Moraci et al. 2017; Rubinos and Spagnoli 2018). Soil erosion

L. Zhan et al. (Eds.): ICEG 2018, ESE, pp. 589–597, 2019.
https://doi.org/10.1007/978-981-13-2224-2_73

is a primary stressor of landfill cover failures, especially in regions with high intensity of precipitation and low vegetation growth (Anderson and Stormont 1997). Progressive surface erosion due to rainfall runoff is reported to damage lower barrier layers and ultimately induce failure of MLCS (Zekkos et al. 2014; Lavigne et al. 2014). Most of these forensic studies reported that the failure of cover system was due to lack of resistance by top surface against erosion and desiccation (Koerner and Daniel 1997). The United States Department of energy (DOE 2000) had proposed on alternative landfill cover system designs by introducing a vegetation layer for minimizing soil-erosion at top surface (Fig. 1a). However, early plant establishment period (where grass layer engulfs the top surface) takes nearly three months (Bordoloi et al. 2018a), during which the top surface is highly susceptible for erosion. Thus, an erosion assessment of top soil is of importance to predict the maximum potential erosion during extreme rainfall events (Fig. 1b).

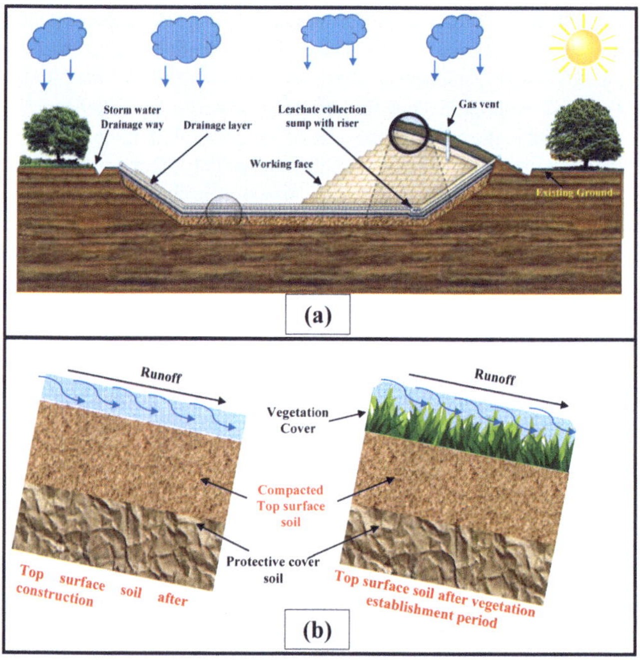

Fig. 1. Diagram of MLCS and erosion susceptibility after construction

Surface erosion is one of the major problems in geotechnical and geo-environmental infrastructures (Al-Kaisi 2000). There are many long term field studies, that attempted to measure soil erosion in sloped surface (Fang et al. 2015; Lei et al. 2006; Morgan 2009). However, parameters like compaction homogeneity and controlled runoff to study erosion mechanism; becomes difficult in field studies. Pin hole test is one of the most commonly adopted index tests to measure the erosion rate of soil (Sherard et al. 1976; Reddi et al. 2000; Wan and Fell 2004). Pinhole test is cheap, less time-consuming and easy to use which facilitates assessment of high amount of soil samples with controlled compaction conditions for the soil.

The main objective of this study is to explore the effect of initial compaction state on erosion rate of surface soil for application in cover surface of MLCS. Four loamy soils were investigated in the current study at nine different compaction states and three different runoff flow rates corresponding to three forms of rainfall, which include heavy, excessive and cloud. A total of 324 erosion tests were performed in the study using an in-house developed pin-hole test setup.

2 Material and Methodology

2.1 Soil Properties

In this study, nine-soil sample is collected from the different part the Assam, India. The designation of the soils is mentioned in the Table 1. The range of specific gravity, plastic limit and liquid limit for the soils are 2.640–2.776, 16.10%–25.01% and 22.91%–49.10%. The engineering properties - the range of optimum moisture content (OMC) and maximum dry density (MDD) are 10.65%–22.10% and 1.583%–1.801%. The total silt and clay content (grain size less than 75 micron) range from nearly (16.44%–48.60%). This percentage of soil composition is designated as PF75, which was recently identified as an important deciding parameter to predict soil infiltration (Bordoloi et al. 2018b). All the soil belongs to similar class "SC" according to unified soil classification system (USCS) as shown in Table 1. This study emphasis the top layer of the multi layered cover system (MLCS). Therefore, pure clay and sand are not suitable for top layer. Clay and sand also have very low and high infiltration that leads to high and low runoff, respectively.

2.2 Testing Procedure and Sample Preparation

The pin hole test (ASTM D4647) for measuring soil erosion rate was adopted in the current study. The flow cell in the test setup consists of compacted cylindrical soil sample with a pre-inserted hole in the middle for allowing runoff water to pass through it. The flow through the cell originates from a deionized water reservoir and is regulated for its discharge capacity. The flow is measured by flow meter before the runoff passes through the sample. The effluent after passing through the soil sample is measured by a collecting unit. The pin hole test gives the quantitative measure of the soil erosion rate from the flow cell. When the shear stress due to runoff acting on particles exceeds the critical shear strength, erosion is initiated by the mechanism of detachment, transportation and deposition (Nearing et al. 1989). The shear strength of soil is majorly dependent on its initial density, moisture content and soil type (Bordoloi et al. 2018c). The flow cell was designed in-house so that the compaction conditions (i.e. initial density and moisture content) of the soil sample will be controlled. The experimental matrix is shown in Table 2. The compaction states are selected across (OMC, MDD)

Table 1. Basic characterization of selected soils

Soil type			Shillong Brown Soil	Maligaon Brown Soil	Guwahati Hill Soil	IIT Hill Soil
Soil designation			SBS	MBS	GHS	IITHS
Basic index properties		Specific gravity	2.640	2.650	2.728	2.776
	Percentage Composition (%)	Gravel (>4.75 mm)	27.94	7.15	1.60	3.20
		Coarse sand (2.00 mm–4.75 mm)	2.30	1.80	4.64	3.90
		Medium sand (0.425 mm–2.00 mm)	22.70	23.90	29.00	17.00
		Fine sand (0.075 mm–0.425 mm)	28.10	30.50	48.32	27.30
		Silt (0.002 mm–0.075 mm)	17.30	34.10	16.44	38.90
		Clay (<0.002 mm)	1.70	2.50	0.00	9.70
		Total Silt & Clay Content (%)	19.00	36.60	16.44	48.60
	Atterberg Limits	Liquid limit (LL)	28.87	22.91	36.71	49.10
		Plastic limit (PL)	18.74	16.10	18.80	25.01
		Shrinkage limit (SL)	7.22	12.03	12.96	10.57
		Plasticity Index (PI) %	10.13	6.81	17.91	24.09
Engineering properties	USCS Classification		SC	SC	SC	SC
	Optimum Moisture content (%)		10.65	12.30	13.50	22.10
	Maximum Dry Density (g/cc)		1.830	1.934	1.801	1.583

Table 2. Experimental matrix for the selected soils

SI. No.	Soil Type	Compaction conditions		Flow Rate (ml/min)
		Water Content (%)	Density (g/cc)	
1	SBS	−5% OMC, OMC, +5% MDD	0.95MDD, MDD, 1.05MDD	115, 165, 235
2	MBS	−5% OMC, OMC, +5% MDD	0.95MDD, MDD, 1.05MDD	115, 165, 235
3	GHS	−5% OMC, OMC, +5% MDD	0.95MDD, MDD, 1.05MDD	115, 165, 235
4	IITHS	−5% OMC, OMC, +3% MDD	0.9MDD, 0.95MDD, MDD	115, 165, 235

condition taking in both dry and wet side of OMC (Yamshi 2017) so that it does not pass the zero air void line. Rainfall corresponding to heavy, excessive and cloud burst which are prone to significant erosion events (Boardman 2006) were selected for this work. As per the rainfall events, corresponding pin flow rates of 110 ml/min, 165 ml/min, and 230 ml/min were selected.

Figure 2(a–f) depicts the sequential steps for sample preparation for conducting the pinhole erosion test. Oven dried soil is uniformly mixed with calculated volume of water and kept inside the desiccator for 24 h. This ensures the uniformity of moisture throughout the soil mixture. The mixed soil is compacted by filling requisite weight the mixture inside the flow cell as density and volume of the flow cell is previously known. A tamper attached on the top of the flow cell helps to compact the mix by static compaction method (Reddy 1993). The hole within the cylindrical soil sample is achieved by drilling it with an auger using mechanical pressure. The final soil sample has dimensions of diameter 2.5 cm, 5 cm length and having a 7 mm hole through its center.

Figure 2(g) shows the pin-hole test setup. The water flows through the flow cell and the flow is continued for 8 min. The effluent coming out from the outlet passes through a Whitman filter paper of grade 12 which filters the soil particles eroded from the compacted soil sample. Dry eroded soil mass is obtained by drying the filter with eroded soil (refer Fig. 2i). The eroded mass being measured; final radius is calculated at

Fig. 2. Systematic (a–f) preparation of sample for pin hole test and (g–j) experimental procedure

the end of the reference time from eroded mass-radius relationship as given in Eq. 1 (Haghighi et al. 2012). Thus, erosion rate in $gcm^{-2}s^{-1}$ is obtained by Eq. 2.

$$R(t) = \sqrt{\frac{M}{\pi \times \rho \times h} + R(0)^2} \tag{1}$$

$$\dot{\varepsilon} = \frac{M}{A \times t} \tag{2}$$

Where,

$\dot{\varepsilon}$ = Erosion rate per unit area per unit time $\left(\frac{g}{cm^2 s}\right)$
M = Eroded Mass (g)
A = Surface Area of hole (cm^2)
t = reference time (sec)
R(t) = Average radius of hole after test (cm)
R(0) = Initial radius of hole (cm)
ρ = dry density of soil $\left(\frac{g}{cm^3}\right)$
g = gravimetric acceleration $(\frac{cm}{s^2})$

3 Results and Discussion

This study is carried out for four sandy clay type soil (classification as per UCSC classification in Table 1). All four soils are low plastic soils with the plastic limit range of 16.10% to 25.01%. For each soil, 81-soil sample is tested for 115 ml/min, 165 ml/min and 230 ml/min flowrate at 9-compaction state. The effects of compaction state, flow rate and particle size distribution are observed and plotted in Fig. 3.

Effect of Compaction and Flow Rate on Erosion of Soil
The effects of compaction state and flowrate on soil erosion can be observed in two ways - (1) constant density and (2) constant moisture content. Figure 3 shows that for a constant density, as initial moisture content increases, the erosion rate gradually reduces. The effect of moisture content on erosion rate follows an inverse relationship. It means for a particular flow rate, at low water content, erosion rate is high. In higher moisture content, cohesion force is developed in the along the thin film of water around the soil particles (Chepil 1956; Benahmed and Bonelli 2012); On the other hand, due to particle orientation, dry soil has relatively high flocculated soil structure than the wet soil structure which is dispersed (Lambe 1958). Therefore, there is edge to face interaction in drier soils that causes high shear force on the soil particles as runoff occurs through the soil matrix. High shear force can detach the particles and start relatively high erosion. Furthermore, disperse clay particles have orientation parallel to the water that so they experience less shear force at same flow rate that causes low erosion rate (Grissinger 1966). Figure 3 shows that for a constant moisture content, as

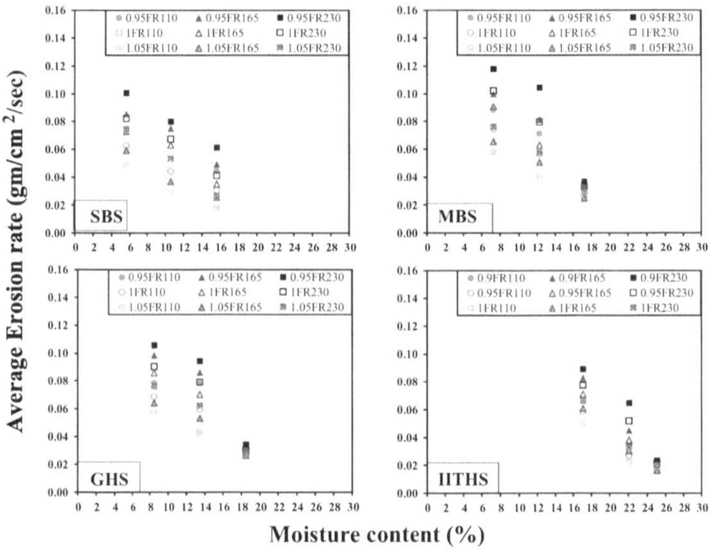

Fig. 3. Average erosion rate at various compaction conditions and flow rates for the selected soils.

density is increased, the erosion rate decreases for all soils. At low density, particles interlocking forces are relatively less than the particles at high density. To dislodge the surface particles in highly dense soil, shear stress required is relatively high. The effect of flow rate can be seen from the Fig. 3. It is seen that with higher flow rate the erosion rate is increases. High flow rate develops the high shear stress at the surface particles that causes high erosion rate. In all 4 soils, it is seen that erosion rate is highest for the compaction condition −5% OMC and 0.9% MDD and for the SBS soil shown by dotted circle in Fig. 3. Soil with high clay content (IITHS) experienced relatively low erosion rate than the low clay content soils (SBS, MBS, GHS). This is due to cohesive forces are prominent at finer fractions due to higher percentage of clay fraction.

4 Concluding Remarks

This work investigates the effect of compaction state on erosion rate of compacted soil for application in cover surface of MLCS. A total of 324 tests were conducted on 4 soils by varying the initial moisture content, soil density and flow rates. Based on the results and discussion, the following conclusions are put forward:

1. The erosion rate decreases significantly with increase in the compacted moisture content due to cohesive force of adsorbed water films and particle orientation in to dispersed structure.
2. The increase in soil density results in a relative decrease in soil erosion rate and is attributed to the higher interlocking of particles.

3. The compaction state of (OMC+5, 1.05MDD) is suitable for top surface design as it showcases significantly lower erosion rates than the permissible erosion rate for cover systems.

References

Albright WH, Benson CH, Gee GW, Roesler AC, Abichou T, Apiwantragoon P, Lyles BF, Rock SA (2004) Field water balance of landfill final covers. J Environ Qual 33(6):2317–2332

Al-Kaisi M (2000) Soil erosion: an agricultural production challenge

Anderson CE, Stormont JC (1997) Prediction of long-term erosion from landfill covers in the southwest (No. CONF-970208–Proc.). USDOE, Washington, DC

ASTM D4647/D4647M-13 (2013) Standard test methods for identification and classification of dispersive clay soils by the pinhole test. ASTM International, West Conshohocken. www.astm.org

Benahmed N, Bonelli S (2012) Internal erosion of cohesive soils: laboratory parametric study. In: 6th International conference on scour and erosion (8-p). SHF

Bennabi A, Karoui T, Benamar A, Wang HQ (2012) Some elements of comparison between two laboratory devices for soil erosion testing. In: International conference on scour and erosion, ICSE6, pp 1089–1096

Benson CH, Sawangsuriya A, Trzebiatowski B, Albright WH (2007) Post construction changes in the hydraulic properties of water balance cover soils. J Geotech Geoenviron Eng 133 (4):349–359

Boardman J (2006) Soil erosion science: reflections on the limitations of current approaches. Catena 68(2–3):73–86

Bordoloi S, Hussain R, Gadi VK, Bora H, Sahoo L, Karangat R, Garg A, Sreedeep S (2018a) Monitoring soil cracking and plant parameters for a mixed grass species. Géotechnique Lett 8(1):49–55

Bordoloi S, Kashyap V, Garg A, Sreedeep S, Wei L, Andriyas S (2018c) Measurement of mechanical characteristics of fiber from a novel invasive weed: a comprehensive comparison with fibers from agricultural crops. Measurement 113:62–70

Bordoloi S, Yamsani SK, Garg A, Sreedeep S, Borah S (2015) Study on the efficacy of harmful weed species Eicchornia crassipes for soil reinforcement. Ecol Eng 85:218–222

Bordoloi S, Garg A, Yamsani SK, Sekharan S, (2018b) Critical assessment of infiltration measurements for soils with varying fine content using a mini disk infiltrometer. J Test Eval 47(2)

Chepil WS (1956) Influence of moisture on erodibility of soil by wind. Soil Sci Soc Am J 20(2):288–292

Choi J, Choi YH, Lim KJ, Shin YC (2005) Soil erosion measurement and control techniques. Food Fertil Technol Cent

DOE (2000) U.S. Environmental Protection Agency Innovation technology summary report Alternative Landfill Cover, Sandia National Laboratories Report, DOE/EM-0558

EIA U (2014) Monthly energy review. Energy Information Administration, US

Fang H, Sun L, Tang Z (2015) Effects of rainfall and slope on runoff, soil erosion and rill development: an experimental study using two loess soils. Hydrol Process 29(11):2649–2658

Grissinger EH (1966) Resistance of selected clay systems to erosion by water. Water Resour Res 2(1):131–138

Haghighi I, Chevalier C, Duc M, Guédon S, Reiffsteck P (2012) Improvement of hole erosion test and results on reference soils. J Geotech Geoenviron Eng 139(2):330–339

Koerner RM, Daniel DE (1997) Final covers for solid waste landfills and abandoned dumps. Thomas Telford

Kumar YS, Kumar S, Sekharan S, Ranjan RR (2017) Determination of soil erosion index for surface soils of landfill covers. Environ Geotech 1–8

Lambe TW (1958) The engineering behavior of compacted clay. J Soil Mech Found Div 84 (2):1–35

Lavigne F, Wassmer P, Gomez C, Davies TA, Hadmoko DS, Iskandarsyah TYW, Gaillard JC, Fort M, Texier P, Heng MB, Pratomo I (2014) The 21 February 2005, catastrophic waste avalanche at Leuwigajah dumpsite, Bandung, Indonesia. Geoenviron Disasters 1(1):10

Lei T, Pan Y, Liu H, Zhan W, Yuan J (2006) A run off-on-ponding method and models for the transient infiltration capability process of sloped soil surface under rainfall and erosion impacts. J Hydrol 319(1–4):216–226

Lo IM, Mak RK, Lee SC (1997) Modified clays for waste containment and pollutant attenuation. J Environ Eng 123(1):25–32

Moraci N, Busana S, Cortellazzo G, Favaretti M, Mandaglio MC, Schepis M (2017) Design and construction of a compacted clay liner in the cover system of a MSW landfill using not standard procedures. Can Geotech J, (ja)

Morgan RPC (2009) Soil erosion and conservation. John Wiley & Sons

Müller WW, Wöhlecke A (2017) Zero leakage? Landfill liner and capping systems in Germany. Environ Geotech, 1–9

Nearing MA, Foster GR, Lane LJ, Finkner SC (1989) A process-based soil erosion model for USDA-Water Erosion Prediction Project technology. Trans ASAE 32(5):1587–1593

Reddy KR, Yargicoglu EN, Yue D, Yaghoubi P (2014) Enhanced microbial methane oxidation in landfill cover soil amended with biochar. J Geotech Geoenviron Eng 140(9):04014047

Rubinos DA, Spagnoli G (2018) Utilization of waste products as alternative landfill liner and cover materials–a critical review. Crit. Rev Environ Sci Technol 1–63

Sadasivam BY, Reddy KR (2015) Adsorption and transport of methane in landfill cover soil amended with waste-wood biochars. J Environ Manag 158:11–23

Sherard JL, Steele EF, Decker RS, Dunnigan LP (1976) Pinhole test for identifying dispersive soils. J Geotech Eng Div 102(1):69–85

Stark TD (2015) Evaluation of a four-component composite landfill liner system. Environ Geotech 4(4):257–273

USEPA-1989, United States Environmental Protection Agency (1989) Final covers on hazardous waste landfills and surface impoundments. Technical Guidance Document

Venkatarama-Reddy BV, Jagadish KS (1993) The static compaction of soils. Geotechnique 43(2)

Wan CF, Fell R (2004) Investigation of rate of erosion of soils in embankment dams. J Geotech Geoenviron Eng 130(4):373–380

Zekkos D, Vlachakis VS, Athanasopoulos GA (2014) The 2010 Xerolakka landfill slope instability. Environ Geotech 1(1):56–65

Application of Geosynthetics in Geo-Environmental Engineering

(Edited by Sifa Xu and Weilie Zou)

Effect of Consolidation on VOC Transport Through a GM/GCL Composite Liner System

Hefu Pu[1(✉)], Charles D. Shackelford[2], and Patrick J. Fox[3]

[1] Huazhong University of Science and Technology, Wuhan 430074, Hubei, China
puh@hust.edu.cn
[2] Colorado State University, Fort Collins, CO 80523, USA
[3] Pennsylvania State University, University Park, PA 16802, USA

Abstract. This paper presents the results of a numerical investigation of the transport of a volatile organic compound (VOC) through a composite liner system comprising a geomembrane (GM) overlying and in intimate contact with a geosynthetic clay liner (GCL), which is commonly used as an engineered barrier for containment of solid waste. The simulations were performed using the established CST2 model (i.e., Consolidation and Solute Transport 2). The results are presented in the form of VOC mass flux, cumulative VOC mass outflow, and distribution of VOC concentration within the GCL. Consolidation of the 10-mm-thick GCL is shown to result in a decrease in the thickness, void ratio, and effective diffusion coefficient of 41%, 49%, and 72%, respectively, such that the steady-state TCE mass flux and cumulative TCE mass outflow at the bottom of the GCL decreased by 47% and 43%, respectively, relative to results for traditional diffusive simulations that ignore the effect of consolidation. Thus, despite the thinness of the GCL, the effect of consolidation on the transport properties of the GCL can result in a significant decrease in VOC transport through a composite GM/GCL liner system.

Keywords: Consolidation · Contaminant transport · Geomembrane Geosynthetic clay liner · Numerical modelling

1 Introduction

To separate waste materials and the associated landfill leachate from the surrounding geology, composite liner systems comprising a geomembrane (GM) overlain by and in intimate contact with a geosynthetic clay liner (GCL) are commonly used as an engineered barrier. Traditionally, performance evaluations of contaminant transport for such liner systems are conducted using advective-diffusive models that assume a rigid GCL and ignore transient advection and concurrent changes in material properties caused by GCL consolidation (Foose 2002). In reality, however, placement of waste materials applies a vertical stress to bottom barrier systems that causes the GCL to consolidate, and, therefore, impact the contaminant transport process. In particular, excess pore-water pressures generated by the applied stress increase advective transport, whereas the consolidation process reduces the porosity of the clay which, in turn, reduces the

L. Zhan et al. (Eds.): ICEG 2018, ESE, pp. 601–607, 2019.
https://doi.org/10.1007/978-981-13-2224-2_74

hydraulic conductivity and effective diffusion coefficient of the bentonite component of the GCL. For example, Petrov and Rowe (1997) and Kang and Shackelford (2010) reported that consolidation caused the hydraulic conductivity of the GCLs to decrease by more than one order of magnitude. However, the effects of liner consolidation on contaminant transport through landfill bottom liner systems have not received much attention until recently (e.g., Fox 2007b; Lewis et al. 2009; Xie et al. 2016).

A previous study by Pu et al. (2016a) presented a numerical investigation of contaminant transport through a composite GM/CCL bottom liner system considering the effects of CCL consolidation. This paper presents a similar study for a GM/GCL composite liner. A more thorough evaluation of VOC transport through a composite GM/GCL liner is provided by Pu et al. (2016b).

2 Numerical Model

CST2 (Consolidation and Solute Transport 2) is a numerical model for simulating coupled large-strain consolidation and contaminant transport in saturated soil. The configuration for CST2 is shown in Fig. 1. A homogeneous, saturated soil layer of initial height H_o is treated as an idealized two-phase material. The solid phase is represented as a column of elements. The vertical coordinate and solid element coordinate are defined as positive upward (against gravity) from a fixed datum at the bottom of the soil layer. The pore liquid (e.g., pore water) is also represented as a column of elements. Each liquid element has an initial solute concentration and initial dissolved solute mass. The initial sorbed (solid-phase) concentration of the solute for each solid element is assumed to be in equilibrium with the local fluid concentration.

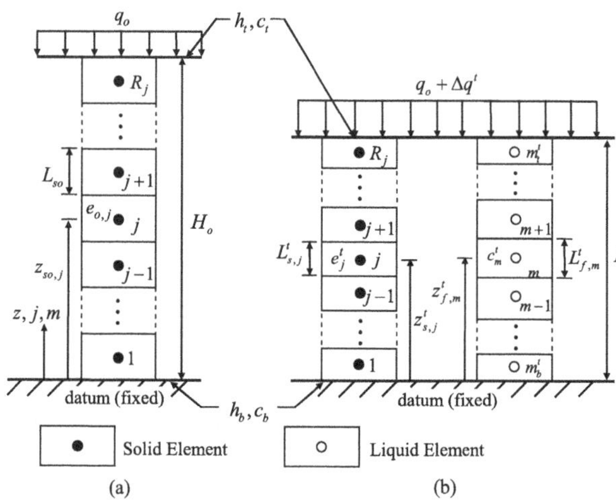

Fig. 1. Configuration for CST2: (a) initial configuration; (b) configuration during loading (Fox and Lee 2008)

Developed on the basis of the CS2 method (Fox and Berles 1997), the consolidation module for CST2 is 1D and considers vertical strain, soil self-weight, general constitutive relationships, relative velocity of the liquid and solid phases, varying compressibility and hydraulic conductivity during consolidation, time-dependent loading, unloading/reloading effects, and an externally-applied hydraulic gradient. The contaminant transport module considers advection, diffusion, mechanical dispersion, linear or nonlinear sorption, equilibrium or nonequilibrium (i.e., kinetic) sorption, and a soil porosity-dependent effective diffusion coefficient. The top and bottom boundary conditions for solute transport can be specified as prescribed concentration (Type I), concentration gradient (Type II), or solute mass flux (Type III). Further details of CST2 are provided by Fox (2007a, b) and Fox and Lee (2008).

3 Numerical Simulations

CST2 was used to investigate the impact of GCL consolidation on contaminant transport through a GM/GCL composite liner system. The initial geometry of the liner system is shown in Fig. 2. From top to bottom, the liner system consists of a leachate collection system (LCS), a 1.5-mm-thick intact high density polyethylene (HDPE) GM, and a 10-mm-thick GCL, and subgrade layer. The subgrade can represent an underlying leachate detection layer typical of double-composite liner systems for hazardous waste containment with the GM/GCL composite liner representing the primary (top) composite liner, or a natural soil subgrade in the case where the GM/GCL is the secondary (bottom) liner of a double-composite liner system or the only liner system. The top boundary of the GCL is undrained due to presence of the intact GM which is assumed to be without defects (e.g., puncture holes or defective seams between adjacent GM panels). The bottom boundary of the GCL is drained and at constant atmospheric pressure. The GM holds a constant head of leachate of 0.3 m in the overlying LCS. The leachate contains a volatile organic compound, such as trichloroethylene (TCE), at a constant concentration, c_o, of 100 mg/L that is assumed to be sufficiently dilute so as to not change the GCL properties. The GCL is initially uncontaminated and has a constant, zero concentration at the bottom. This bottom concentration condition ($c = 0$) allows for diffusion across the interface that produces the highest estimate of contaminant mass flux and, therefore, is conservative. TCE undergoes diffusive transport through the GM prior to reaching the GCL, and then undergoes advective-dispersive transport through the underlying GCL with sorption onto the bentonite and geotextiles components of the GCL. TCE transport parameters for the GM are taken from experimental data reported by Sangam and Rowe (2001), with a GM diffusion coefficient, D_{GM}, of 4×10^{-13} m^2/s and partition coefficient, K_{GM}, of 85. The TCE transport parameters for the GCL are taken from experimental data reported by Kim et al. (2001) and Lake and Rowe (2004).

GCL compressibility and hydraulic conductivity constitutive relationships are taken from experimental data reported by Kang and Shackelford (2010). The bentonite component has liquid and plastic limits of 478 and 39, respectively, and was classified as a high plasticity clay (CH) based on the Unified Soil Classification System. The compressibility relationship can be expressed as $e = 4.7 - 1.44 \log(\sigma'/34.5)$, where e is the

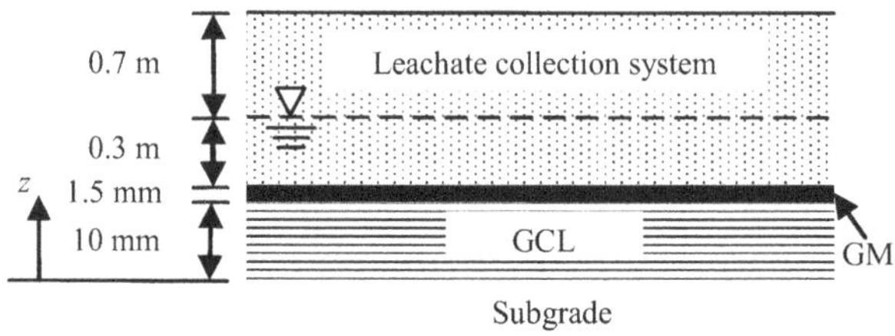

Fig. 2. Initial geometry for GM/CCL liner system

void ratio, and σ' is the vertical effective stress (kPa). The hydraulic conductivity relationship can be expressed as $e = 25.12 + 1.97\ \log(k)$, where k is the vertical hydraulic conductivity (m/s). The effective diffusion coefficient, D^*, for TCE in the GCL changes with solid element porosity as $D^* = D_o\ (n)^M$, where D_o is the free-solution (aqueous-phase) diffusion coefficient, n is the GCL bulk porosity, and M is an exponent (Lerman 1978). Based on values reported by Lake and Rowe (2004), $D_o = 8.6 \times 10^{-10}$ m^2/s and $M = 8.67$ (Pu et al. 2016b). Mechanical dispersion is neglected because the GCL is thin with low values of seepage velocity, v_s, resulting from the low hydraulic conductivity (e.g., Sleep et al. 2006). TCE sorption is characterized as linear and in equilibrium, with an equivalent distribution coefficient, K_d, to both the bentonite and the geotextile components of the GCL of 7.94 mL/g (Pu et al. 2016b). Decay of TCE (e.g., due to biological processes) is conservatively assumed to be negligible.

Initial stress conditions are calculated with the assumption that the GCL is saturated and initially in hydraulic equilibrium (i.e., no flow), with a total hydraulic head of zero ($h = 0$) taken with respect to the elevation, z, of 0 prior to waste placement (see Fig. 2). The initial effective stress at the top of the GCL, σ'_o, is calculated to be 28.2 kPa (Pu et al. 2016b). Starting at time $t = 0$, waste is placed continuously at an assumed rate of 100 kPa/yr for a total loading period of 10 years, giving a final applied stress, Δq, of 1000 kPa. The total elapsed time for each simulation is 40 years. These conditions correspond to a municipal solid waste landfill with a final waste height of approximately 70 to 90 m, a 10-year filling period, and a 30-year post-closure period. Although transport parameters are constant for each simulation, spatial and temporal changes in n (or e, where $e = n/(1 - n)$) and v_s cause variations of hydrodynamic dispersion, retardation factor, and Peclet number during consolidation (Pu et al. 2016b).

4 Simulation Results

Plots of applied stress (Δq), GCL compression (ΔH), and bottom boundary fluid outflow rate (v) for the composite GM/GCL bottom liner system are presented in Fig. 3. The Δq increased linearly to a final value of 1000 kPa at 10 years and remained constant thereafter. Consolidation was finished immediately after the end of loading period (t = 10 yr) because the GCL is thin, and the resulting final GCL compression was 4.1 mm corresponding to a final strain of 41%. Due to consolidation, the e of the GCL decreased from 5.09 to 2.58 and, as a result, D^* decreased by 72% (from

1.8×10^{-10} to 0.5×10^{-10} m^2/s). The liquid outflow rate, v, at the bottom boundary of the GCL was relatively high at the beginning of loading, gradually decreased as excess pore pressures dissipated, and then decreased rapidly after the completion of waste placement (t = 10 yr). A sudden decrease in v occurs at t = 10 yr, because the transition of v from a non-zero value at the end of the loading period (t = 10 yr) to a steady-state value of zero (v = 0) requires less than one day because of the thinness of the GCL.

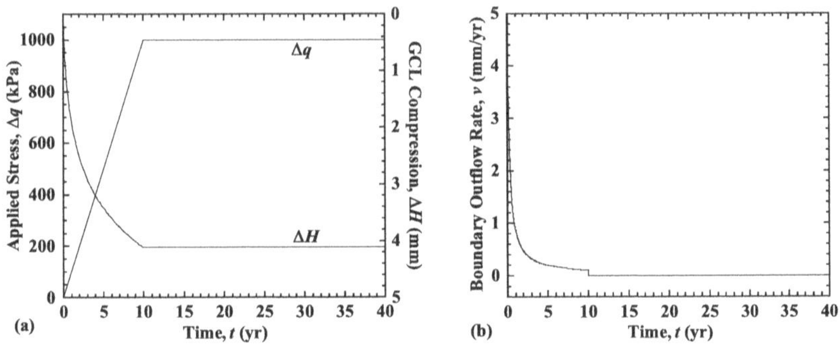

Fig. 3. Consolidation response: (a) applied stress and GCL compression; (b) bottom boundary fluid outflow rate (from Pu et al. 2016b)

Simulations for TCE transport through the composite GM/GCL bottom liner system were conducted for two cases. The case "C" simulations corresponded to consolidation-induced transport followed by diffusive transport after completion of consolidation. The case "NC" simulations neglected consolidation effects and corresponded to diffusive transport using initial conditions for the liner system (i.e., constant initial GCL thickness and constant profiles of initial e and initial D^*) over the entire 40-yr simulation period. Results for both cases C and NC were obtained using the CST2 model with appropriate input parameters (Pu et al. 2016b).

Simulation results for TCE mass flux, F, at the base of the GCL are presented in Fig. 4(a). TCE breakthrough for the case NC simulations occurred almost immediately and reached steady state within just a few days. TCE mass flux for the case C simulations was the same as that for the case NC simulation in the very early stages ($t < 0.04$ yr), and then decreased with elapsed time due to consolidation-induced reduction in D^*, and then ultimately reached steady state after the completion of consolidation (t = 10 yr). Steady-state mass flux for the case NC simulations is 87% higher than that for the case C simulations because D^* for the case NC simulations is much higher.

The advective mass flux, diffusive mass flux, and total mass flux for the case C simulations during the 10-year consolidation period are shown in Fig. 4(b). Total mass flux is the sum of advective mass flux and diffusive mass flux, and is equal to the mass flux shown in Fig. 4(a). As shown in Fig. 4(b), the advective mass flux was always less than 0.001 mg/m^2-yr, whereas the diffusive mass flux is at least five orders of magnitude higher than the advective mass flux. The consolidation-induced advective

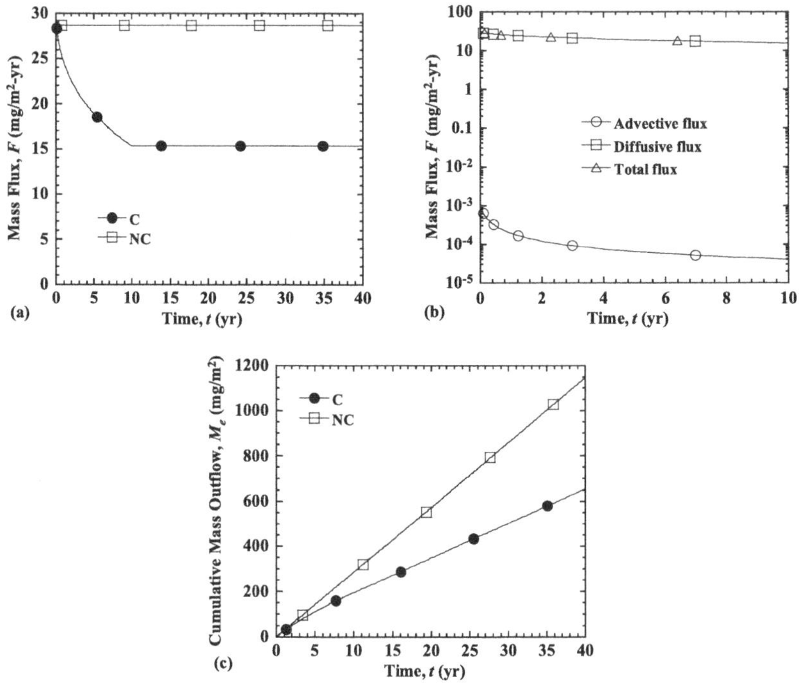

Fig. 4. Simulation results for the base of the GCL: (a) TCE mass flux; (b) TCE advective mass flux vs. diffusive mass flux for case C; (c) cumulative TCE mass outflow

transport is negligible because the GCL is very thin and has low hydraulic conductivity and, thus, the consolidation-induced liquid flow rate is negligible. As a result, the total mass flux is almost entirely attributed to diffusion.

The results for cumulative TCE mass outflow M_e are presented in Fig. 4(c). As the values of F are proportional to the slopes of the M_e curves, the relative differences between the results of the two simulation cases shown in Fig. 4(c) were consistent with those shown in Fig. 4(a). Thus, M_e for the case C simulations were initially the same as those for the case NC simulations, but then gradually decreased relative to the results for the case NC simulations due to consolidation-induced reduction in D^*.

5 Conclusions

GCL consolidation can have a significant impact on TCE transport through a composite GM/GCL liner system both during consolidation and long after consolidation has stopped. Although consolidation-induced advection is found to be negligible, consolidation can still significantly affect TCE transport through the GM/GCL system via alteration in the transport properties (i.e., thickness, porosity, and effective diffusion coefficient) of the GCL. Because of consolidation, the GCL thickness, void ratio, and effective diffusion coefficient decreased by 41%, 49%, and 72%, respectively. As a result, the final, steady-state TCE mass flux and cumulative TCE mass outflow at the bottom of the GCL decreased by 47% and 43% relative to results for traditional diffusive simulations without considering the consolidation effects.

Acknowledgements. Financial support for this research was provided by the National Natural Science Foundation of China (Grant No. 51678268) and the U.S. National Science Foundation (Grant Nos. CMMI-1001023, CMMI-0969346, and CMMI-1622781). This support is gratefully acknowledged.

References

Foose GJ (2002) Transit-time design for diffusion through composite liners. J Geotech Geoenvironmental Eng 128(7):590–601

Fox PJ (2007a) Coupled large strain consolidation and solute transport. I: Model development. J Geotech Geoenvironmental Eng 133(1):3–15

Fox PJ (2007b) Coupled large strain consolidation and solute transport. II: Model verification and simulation results. J Geotech Geoenvironmental Eng 133(1):16–29

Fox PJ, Berles JD (1997) CS2: a piecewise-linear model for large strain consolidation. Int J Numer Anal Methods Geomech 21(7):453–475

Fox PJ, Lee J (2008) Model for consolidation-induced solute transport with nonlinear and nonequilibrium sorption. Int J Geomech 8(3):188–198

Kang JB, Shackelford CD (2010) Consolidation of a geosynthetic clay liner under isotropic states of stress. J Geotech Geoenvironmental Eng 136(1):253–259

Kim JY, Edil TB, Park JK (2001) Volatile organic compound (VOC) transport through compacted clay. J Geotech Geoenvironmental Eng 127(2):126–134

Lake CB, Rowe RK (2004) Volatile organic compound diffusion and sorption coefficients for a needle-punched GCL. Geosynth Int 11(4):257–272

Lerman A (1978) Chemical exchange across sediment-water interface. Annu Rev Earth Planet Sci 6:281–303

Lewis TW, Pivonka P, Smith DW (2009) Theoretical investigation of the effects of consolidation on contaminant transport through clay barriers. Int J Numer Anal Methods Geomech 33 (1):95–116

Petrov RJ, Rowe RK (1997) Geosynthetic clay liner (GCL) – chemical compatibility by hydraulic conductivity testing and factors impacting its performance. Can Geotech J 34 (6):863–885

Pu H, Fox PJ, Shackelford CD (2016a) Assessment of consolidation-induced contaminant transport for compacted clay liner systems. J Geotech Geoenvironmental Eng 142 (3):04015091

Pu H, Shackelford CD, Fox PJ (2016b) Assessment of consolidation-induced VOC transport for a GML/GCL composite liner system. J Geotech Geoenviron Eng 142(11):04016053

Sangam HP, Rowe RK (2001) Migration of dilute aqueous organic pollutants through HDPE geomembranes. Geotext Geomembr 19(6):329–357

Sleep BE, Shackelford CD, Parker JC (2006) Modeling of fluid transport through barriers (Chapter 2). In: Chien CC, Inyang HI, Everett LG (eds) Barrier systems for environmental contaminant containment and treatment. CRC Press, Boca Raton, pp 71–141

Xie H, Yan H, Feng S, Wang Q, Chen P (2016) An analytical model for contaminant transport in landfill composite liners considering coupled effect of consolidation, diffusion, and degradation. Environ Sci Pollut Res 23(19):1–14

Effect of GCL on Moisture and Temperature Fields of Highway Subgrade Through Numerical Analysis

Feng Liu[1], Zhibin Liu[1(✉)], Feng Yin[2], Shujian Zhang[3], and Yi Wang[1]

[1] School of Transportation, Southeast University, Nanjing 210096, China
seulzb@seu.edu.cn

[2] Inner Mongolia Highway Construction and Development Co. Ltd., Hohhot 010051, China

[3] Jilin Provincial High Class Highway Construction Bureau, Changchun 130062, China

Abstract. The performance of highway subgrade is affected not only by soil type but also by other parameters such as water and temperature. Distribution and variation of temperature in subgrade affect the stability of highway subgrade in seasonal frozen area. In seasonal frozen areas, water accumulation in highway subgrade can cause frost heave and thaw settlement, which will finally affect the function of road structures. Due to its waterproof performance, geosynthetic clay liner (GCL) is mainly used in engineering practice as an effective material for sealing, isolation and leakage prevention. So it may be introduced into highway subgrade to provide possible moisture and temperature adjustment. Through water and heat coupling analysis of two kinds of subgrade models, following conclusions can be drawn. GCL can effectively prevent rainfall infiltration and adjust moisture field of subgrade. GCL in the subgrade can significantly decrease the maximum freezing depth and thereby reduce the frost damage risk of the highway subgrade in Seasonal frozen soil area.

Keywords: GCL · Moisture and temperature fields · Numerical simulation

1 Introduction

Frost damage of road subgrade has been a frequently happened problem that impairs the road construction and maintenance in northern and western China. As water accumulation in subgrade soil is one of the key reasons leading to above damages in cold regions, it's important to prevent water from migrate into the road subgrade [1]. On the other hand, frozen subgrade soil is also very sensitive to thermal disturbance. According to the study of Sheng et al., when the saturated soil of fine particle is below 0 °C, some pore water will solidify into pore ice [2].

The water migration and temperature change of frozen soil are a coupled problem of water and heat. Many scholars have studied this problem so far. Based on the theory of water flow and heat transfer in unsaturated soils, Tai et al. established a moisture-thermal coupling differential equation in frozen soils [3]. Taylor et al. established a coupled numerical model of water and heat based on heat transfer theory and unsaturated soil seepage theory [4].

L. Zhan et al. (Eds.): ICEG 2018, ESE, pp. 608–615, 2019.
https://doi.org/10.1007/978-981-13-2224-2_75

Traditional treatment technologies such as replacement method cannot completely eliminate and avoid the water migration and frost heave of the subgrade soil [5]. Previous research work indicates that the geosynthetic clay liner (GCL) is an effective material of sealing, isolation and leakage prevention. Based on the meteorological conditions and soil properties along a highway in Inner Mongolia, this paper uses GeoStudio software to numerically study the moisture and temperature fields of highway subgrade with and without GCL respectively. Since there are few studies about application of GCL in subgrade, the research results here can provide theoretical reference for related engineering practice.

2 Numerical Model and Material Parameters

Due to geometrical symmetry of the model, half of the subgrade is used for simulation. The width of the subgrade surface is 7 m; the height of the subgrade from top surface to the ground is 4.3 m. The gradient of the slope is 1:1.5, and the width of the natural surface extends to 7 m outside the foot of the slope. The model is meshed into quadrilateral and triangles and the size of the grid is 0.5 m * 0.5 m.

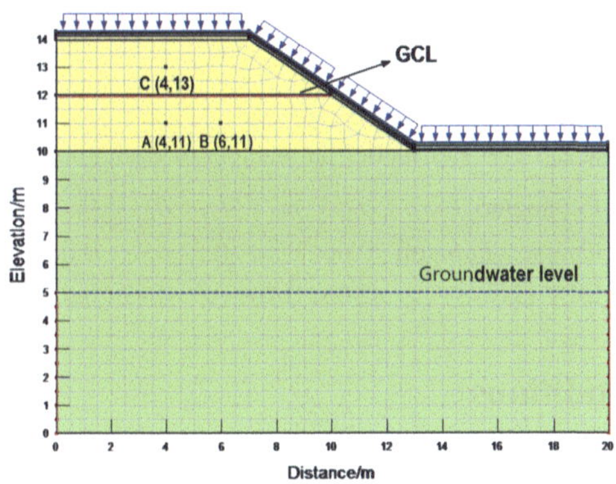

Fig. 1. The numerical model of subgrade with GCL

Table 1. The parameters of each soil layer.

Soil layers	Position [m]	Density [kg/m^3]	Heat capacity [J/(kg · °C)]	Thermal conductivity [W/(m · °C)]	Permeability coefficient [cm/s]	Water content [%]
Coarse grain soil	10–14.3	2060	860	1.41	1e−4	6
Silty clay	0–10	1540	1466	1.13	1e−7	5

As shown in Fig. 1, GCL is placed in the position of 2 m below the subgrade surface. Rowe et al. studied the basic properties of GCL, and found that it has high strength and good corrosion resistance. Furthermore, its permeability coefficient can reach 10^{-9} cm/s [6]. According to the study of the thermal conductivity of sodium bentonite by Liu et al., the thermal conductivity of GCL is about 0.5 W/(m · K) [7].

The model consists of two soil layers including a layer of natural silty clay from 0 to 10 m, and a second layer of coarse grain soil from 10 to 14.3 m. Parameters of each soil layer are shown in Table 1.

Xu et al. found that the Van Genuchten model (VG model) was suitable to fit the soil water characteristic curve for coarse soil [8]. Therefore, the VG model is used here to define the soil water characteristic curve of each soil layer. The specific expression of the VG model is [9]:

$$S_e = \frac{\theta - \theta_r}{\theta_s - \theta_r} = [1 + |\alpha h|^n]^{-m} \quad h<0 \tag{1}$$

$$S_e = 1 \quad h \geqslant 0$$

Where S_e - saturation; θ - volume water content (L^3L^{-3}); θ_r and θ_s - residual water content and saturated water content (L^3L^{-3}); h - pressure head (L); α (L^{-1}) and n - curve shape parameters; $m = 1 - 1/n$. The VG model parameters of each soil layer are shown in Table 2.

Table 2. The VG model parameters of each soil layer.

Parameters	a_0	m	l	θ_r	θ_s
Coarse grain soil	0.66	0.14	0.5	0.4	0.04
Silty clay	2.65	0.26	0.5	0.42	0.05

3 Boundary Conditions

This simulation analyzes the change of roadbed moisture and temperature in 1 year, from July of first year to June of next year. The initial temperature is 16 °C. The upper boundary of the temperature field is based on temperature variation data of Inner Mongolia [10], and the formula used here is:

$$T = 6.04 + 18.96 * \sin(2\pi t/365 + 1.41) \tag{2}$$

Where T - temperature (°C); t - Time (d)

The left and right boundaries of the temperature field are defined as adiabatic boundaries, and the boundary of temperature field at the bottom is set as constant temperature of 9 °C. The top of the model is defined as rainfall boundary, which is set according to the meteorological data in Ulanhot, Inner Mongolia. The rainfall intensity is 50 mm/d from May to September, and 10 mm/d from October to April of next year. The capillary rise of groundwater is taken into account and the depth of groundwater level is 5 m. The left and right boundary conditions of moisture field are defined as constant head. The initial moisture field is the water content of each soil layer.

4 Results of Simulation

4.1 Analysis of Moisture Field

The distribution of initial pore water pressure of the two subgrade models is shown in Fig. 2. Figure 3(a) and (b) show the distribution at 90 days (October) when the temperature is high and the rainfall reached maximum value. As shown in Fig. 3(a), after 3 months of rainfall, the pore water pressure in the subgrade model without GCL obviously becomes higher, which may result in the decrease of both matrix suction and the strength of subgrade soil. On the contrary, the pore water pressure in the subgrade with GCL is almost unchanged. In addition, the pore water pressure above the GCL slightly increases, which indicates that there may be a short term water accumulation.

Figure 4(a) and (b) show the distribution of pore water pressure at 210 days (February) of two subgrade models. At this time, the temperature and the rainfall are significantly decreased. It can be seen from Fig. 4(a) that the pore water pressure of the subgrade model without GCL is smaller than that at 90 days due to the decrease of rainfall. Figure 4(b) shows that the pore water pressure at 90 days is basically unchanged, and is still much lower than the pore water pressure of subgrade model without GCL, which further illustrates the remarkable water isolation ability of GCL.

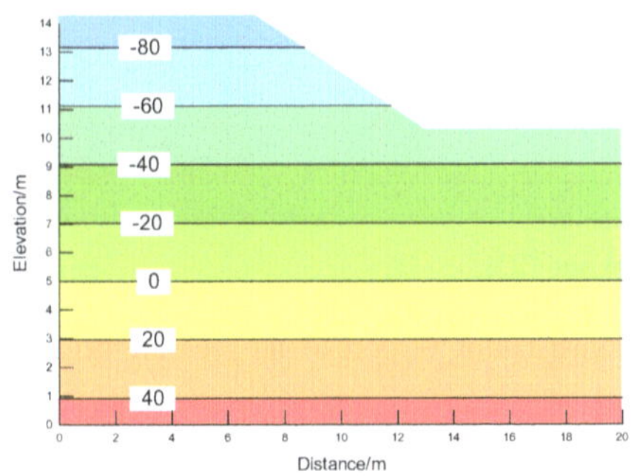

Fig. 2. The initial state of pore water pressure (kPa)

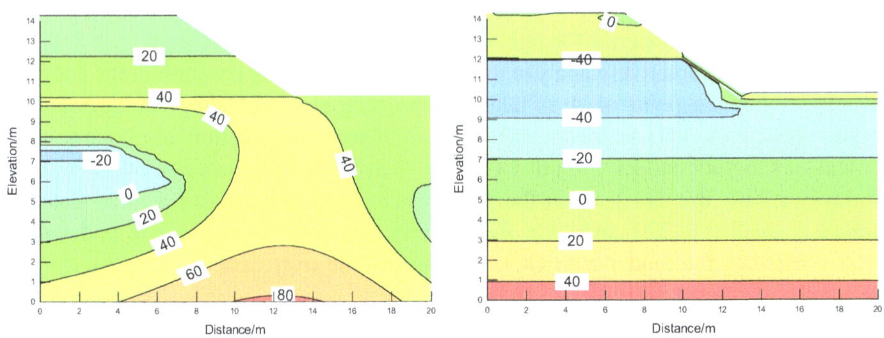

(a) Without GCL (b) With GCL

Fig. 3. The distribution of pore water pressure (kPa) at 90 days

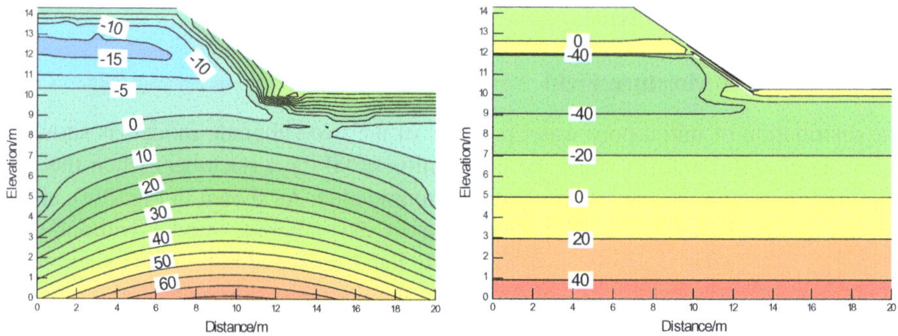

(a) Without GCL (b) With GCL

Fig. 4. The distribution of pore water pressure (kPa) at 210 days.

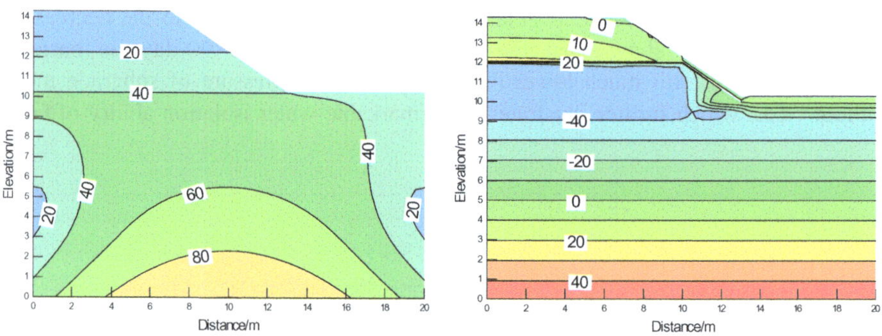

(a) Without GCL (b) With GCL

Fig. 5. The distribution of pore water pressure (kPa) at 300 days.

When it is 300 days (May of next year), the rainfall and the temperature increase again. It can be seen from Fig. 5 that pore water pressure of the subgrade model without GCL increases obviously, which may lead to the decrease of the matrix suction and the strength of the subgrade. However, the pore water pressure of the subgrade model with GCL is basically the same, which indicates that GCL can work well as an effective waterproof material for rather long time.

In order to quantitatively analyze the change of pore water pressure in the subgrade model, a reference point of A is taken at the bottom of the subgrade soil, and its coordinates are (4, 11). As shown in Fig. 6, the pore water pressure and volume water content of subgrade model without GCL rise sharply when there's heavy rain (e.g. 0–90 days or 270–360 days). Then the pore water pressure decreases a little when the rainfall is small (90–270 days). On the contrary, the pore water pressure and volume water content of the model with GCL increase a little over time.

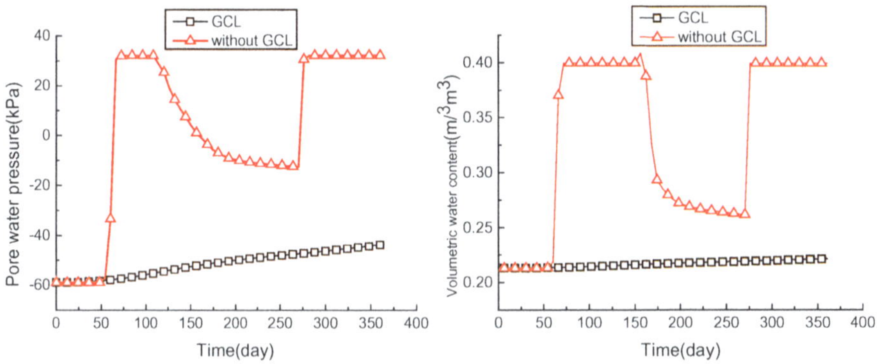

Fig. 6. Curves of pore water pressure and volume water content at (4, 11).

The pore water pressure and volume water content above GCL are analyzed too. Figure 7 shows the curve of pore water pressure and volume water content at the point of B, whose coordinates are (4, 13). During the rainfall, the pore water pressure and volume water content above GCL are slightly higher than that without GCL, which indicates that GCL may lead to water accumulation above the material, but as the rainfall decreases, the pore water pressure and volume water content gradually fall down. Therefore, a layer of drainage material is suggested to be added above the GCL so that water can be discharged effectively after rainfall.

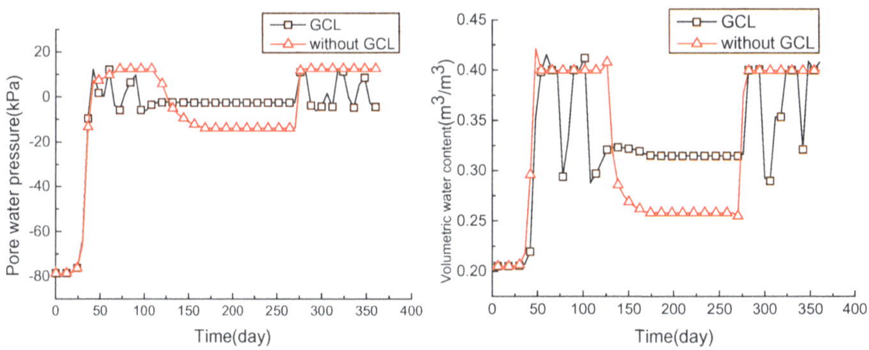

Fig. 7. Curves of pore water pressure and volume water content at the position (4, 13).

4.2 Analysis of Temperature Field

Figure 8 shows the distribution of temperature field in two subgrade models at 240 days (February). The equal potential line at 0 °C shows the maximum freezing depth of the subgrade. It can be seen from Fig. 8(a) that if without GCL in the subgrade, the maximum freezing depth of the subgrade center reaches 2.7 m, which is also consistent with the maximum freezing depth in Ulanhot, and the maximum freezing depth below the subgrade shoulder is about 3 m, higher than that in the subgrade center. Figure 8(b) shows the distribution of the temperature field in the subgrade model with GCL. The maximum freezing depth of the subgrade center is reduced to 2.3 m when GCL is embedded. The maximum freezing depth of the subgrade model is reduced by 40 cm, and the maximum freezing depth below the shoulder is also reduced from 3 m to 2.7 m, which is reduced by 30 cm.

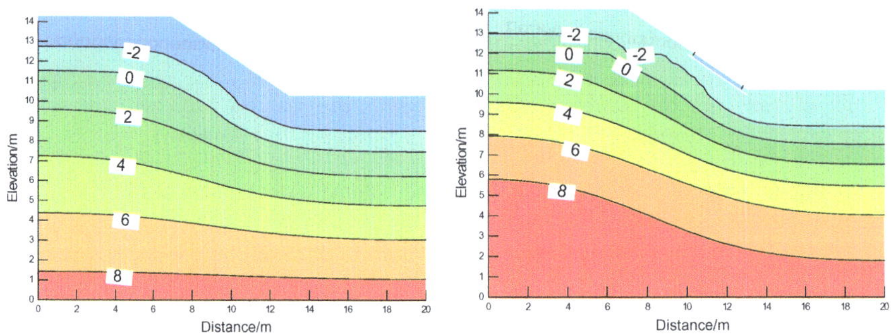

(a) Without GCL (b) With GCL

Fig. 8. The distribution of temperature (°C) at 250 days.

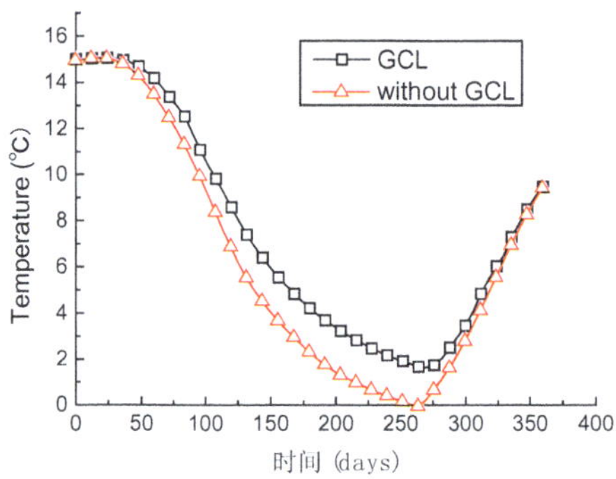

Fig. 9. Curve of temperature at (6, 11).

The point C whose coordinates are (6, 11) is taken to quantitatively analyze the change of temperature over time. As shown in Fig. 9, the temperature of the subgrade model without GCL is lower than that with GCL. At 250 days, the temperature of the point C without GCL reached 0 °C, which indicates that the maximum freezing depth of the model has reached 3 m. At the same time the temperature of the point C in the model with GCL is still above 0 °C, indicating that the maximum freezing depth is less than 3 m. The simulation results strongly show that GCL can effectively regulate the temperature field and reduce the maximum frozen depth of the subgrade.

5 Conclusions

In order to study the influence of GCL material on subgrade moisture and temperature fields, the water and heat coupling analysis of two kinds of subgrade models are made in this research. Following conclusions can be drawn:

(1) GCL can effectively reduce the pore water pressure and volume water content in the subgrade under the condition of rainfall; GCL can effectively prevent rainfall infiltration and adjust the moisture field of subgrade.
(2) GCL can significantly reduce the maximum freezing depth of the subgrade, decreasing the frost damage of the highway subgrade in seasonal frozen soil area.
(3) Small amount of water may be accumulated in the area above the GCL although it will decrease with time. Thus, it is suggested that a drainage material be added above GCL to discharge water effectively.

Acknowledgement. This research was sponsored by the National Natural Science Foundation of China (No. 41672280, 41330641). The authors are also grateful for the financial supports from Traffic Science and technology projects of Inner Mongolia and Jilin Province.

References

1. Rainwater NR (1999) Comprehensive monitoring systems for measuring subgrade moisture conditions. J Transp Eng 125(5):439–448
2. Sheng DC, Zhang S, He Z (2014) Assessing frost susceptibility of soils. Chin J Rock Mech Eng 33(2):594–605 (in Chinese)
3. Tai B et al (2017) Numerical modelling of anti-frost heave measures of high-speed railway subgrade in cold regions. Cold Reg Sci Technol 141:28–35
4. Harlan RL (1973) Analysis of coupled heat-fluid transport in partially frozen soil. Water Resour Res 9(5):1314–1323
5. Xu J, Niu FJ, Li AM et al (2010) Analysis of the prevention effect of thermal-insulation method on frost heave of railway subgrade in seasonal frozen regions. J China Railw Soc 32 (6):124–131 (in Chinese)
6. Xu SH, Chen DL (2002) Suitability of models describing soil water retention curve. Acta Pedologica Sinica 39(4):498–504 (in Chinese)
7. van Genuchten MT (1980) A closed-form equation for predicting the hydraulic conductivity of un-saturated soil. Soil Sci Soc Am J 44(5):892–898
8. Rowe RK, Rentz AK, Brachman RWI et al (2016) Effect of GCL type on downslope erosion in an exposed composite liner. J Geotech Geoenvironmental Eng 142(12):04016074
9. Liu Y, Cai M, Ju W (2007) Thermal properties of buffer material for high-level radioactive waste disposal. Chin J Rock Mech Eng 26:3891–3896 (in Chinese)
10. Jia H (2015) Analysis and calculation of groundwater resources in the Southern Valley of Inner Mongolia Autonomous Region Ulanhot. Master thesis, China University of Geosciences (in Chinese)

A Review of Experimental and Prediction Methods for Assessing the Freezing Characteristic Curve of GCLs

G. G. Carnero-Guzman[1(✉)], A. Bouazza[1], W. P. Gates[2], and R. K. Rowe[3]

[1] Department of Civil Engineering, Monash University, Clayton, Australia
gonzalo.guzman@monash.edu
[2] Institute for Frontier Materials, Deakin University, Geelong, Australia
[3] Geoengineering Centre at Queen's-RMC, Department of Civil Engineering, Queen's University, Kingston, Canada

Abstract. Geosynthetic clay liners (GCLs) are an important part of composite hydraulic barriers in environmental projects, with recent applications in harsh conditions such as in Antarctica. To assure an adequate hydraulic performance of the GCL, the bentonite of the GCL needs to be well-hydrated. However, the sub-zero temperatures attained in Antarctica freeze the water inside the bentonite, and as a result, the hydration process stops with potential consequences on the hydraulic performance of the GCL. To predict the impact of freezing on GCL performance, it is essential to obtain the unsaturated freeze property functions (UFPFs) for the bentonite. The freezing characteristic curve (FCC), which relates the unfrozen water content with freezing temperatures, is the first of the UFPFs, and can be obtained experimentally or by prediction methods. This paper reviews the concepts leading to the determination of the FCC, its relationship with the other UFPFs and its importance for predicting GCL performance in cold regions.

Keywords: Geosynthetic clay liners · Unsaturated behavior
Unsaturated freezing property functions · Cold regions engineering

1 Introduction

Geosynthetic clay liners (GCLs) are formed of a thin layer of bentonite (5–10 mm) enclosed between two geotextiles. As hydraulic barriers, GCLs have become an important element in environmental protection projects [1]. Usually, GCLs are installed in the field at low gravimetric water content (≈10%), and hydrate by absorbing water from the subsoil. As highlighted in the literature [e.g. 2] a well-hydrated GCL (highly saturated) becomes an effective barrier against gas or liquid flow. However, the GCL hydration process is influenced by external factors such as environmental and weather conditions. Specifically, in extremely cold regions such as the Arctic and Antarctic regions, temperatures < 0 °C and daily freeze-thaw (F-T) cycles during summer seasons can influence the water uptake of GCLs [3]. At negative temperatures, ice lenses can form in the hydrated bentonite, which subsequently melts during F-T cycles and

L. Zhan et al. (Eds.): ICEG 2018, ESE, pp. 616–623, 2019.
https://doi.org/10.1007/978-981-13-2224-2_76

can result in cracks within the bentonite which have been shown to adversely affect the sealing properties of the barrier system [3, 4]. Thus, when GCLs are used in circumpolar regions, design considerations should include the freezing and thawing phenomenon, similarly to the considerations adopted in the study of permafrost over the last few decades, for which predictive models have been created to assess the water-heat transport phenomena at freezing temperatures [5].

Harlan [6] proposed the first one-dimensional water-heat transport model, later modified and experimentally validated [e.g. 7, 8] to incorporate specific applications (e.g., conduction-advection, phase change, solute transport, etc.). Recently, more sophisticated models, such as multi-dimensional, multi-phase and regional-scale surface models, have been applied in investigations of climate change effects on permafrost [e.g. 9, 10]. In these (sometimes complex) models, the unsaturated freeze property functions (UFPFs) of geomaterials are used to simulate water-heat transport in frozen porous media [e.g. 11, 12]. UFPFs are similar to the unsaturated soil property functions applied in unsaturated soil mechanics problems, except that, by using UFPFs, thermodynamic equilibrium conditions during phase change can be simulated. UFPFs describe the relationship between freezing temperatures and the following parameters: (a) unfrozen water content, known as the freezing characteristic curve (FCC), (b) hydraulic conductivity, and (c) thermal conductivity. This paper focuses on the FCC, the definition of which is the first step towards a more accurate prediction of GCL performance in cold regions.

2 The Freezing Characteristic Curve (FCC)

FCC describes the coexistence of unfrozen water and ice at freezing temperatures and serves to identify the ice formation stages in porous media. Evaluation of the FCC is important since the unfrozen water content is known to affect thermal, hydraulic and stress properties of soils [5]. The freezing point (FP) is the lowest temperature at which ice is absent (around 0 °C). Lower temperatures than the FP will lead to proportional increases in the amount of ice or a proportional reduction in the unfrozen water fraction as temperatures decreases [13]. At any given freezing temperature, fine soils such as the bentonite in GCLs, have a larger fraction of unfrozen water than coarser soils, because water in large capillary pores undergoes a phase change, whereas water in small pores, or that absorbed to the surfaces of fine particles, persists in an unfrozen state. The exact temperature at which all the water in fine soils freezes is inconclusive and differs with mineralogy [14]. Anderson and Tice [15] considered that all water eventually freezes around −40 °C. In contrast, Kozlowski [16] argues that water remains unfrozen to −70 °C. Thus, for engineering applications, further studies are needed to corroborate the temperature at which all water in fine soils becomes frozen.

To predict the GCL behaviour in cold regions, the FCC of the GCL and the underneath subsoil must be defined as water-heat transport phenomena occur between both elements. When using numerical models, thermodynamic equilibrium during phase change must be simulated [5]. The aim is attained by using the freeze characteristic curve (FCC) as a relationship between unfrozen water content and cryo-suction. Such a relationship exists analogously for unsaturated soils. When soils dry, water is removed and is replaced by air, thus, producing an increase in the negative pore pressure (suction). Similarly, when unfrozen water becomes ice, the negative pore pressure (cryo-suction) increases while unfrozen water content decreases in the surroundings [13]. Hence, freezing soils act similarly to dry soils. The relationship between how much cryo-suction corresponds to any given freezing temperature is defined by the Clapeyron equation (Eq. 1) which was derived from the thermodynamic concept of Gibbs free energy [17].

$$P_{wf} = H_f \rho_w \ln\left(\frac{T + 273.15}{273.15}\right) \tag{1}$$

Where P_{wf} is cryo-suction (in MPa), H_f is the latent heat of fusion of water (in Joules), ρ_w is water density (in g/cm^3), T is freezing temperature (in Celsius). Use of Eq. 1 requires the assumptions that (a) pore ice pressure is equal to the atmospheric pressure and (b) there is a negligible influence of solutes. Several forms of the Clapeyron equation exist [5]; some versions do not assume atmospheric ice pressure and are more applicable for freezing-induced mechanical deformations. However, for water-heat transfer under non-deformable conditions, authors such as Kurylyk and Watanabe [5] consider a simplified Clapeyron equation, such as Eq. 1, to be appropriate. Also, it should be noted that the Clapeyron equation is only valid during thermal equilibrium. During disequilibrium (such as open systems or rapid early-stage freezing), the validity of any Clapeyron equation form has been contested [e.g. 18, 19] as it overestimates cryo-suction at the onset of freezing.

3 Experimental Methods Used to Obtain the FCC

Recently researchers have developed or enhanced techniques for measuring the unfrozen water content (UWC) in soils, which the most currently used are indicated in Table 1. Nuclear magnetic resonance (NMR) estimates the UWC by applying a radio frequency pulse into the sample that produces a time-delayed response, the strength and frequency of which is proportional to the liquid water fraction in the sample [20]. NMR is commonly considered as a calibration standard for other methods. Nevertheless, Kozlowski [e.g. 21] identified limitations in the technique including: difficulty in ensuring thermal equilibrium; uncertainty regarding the effect of quadrupolar nuclei (e.g., iron or manganese in the clay) on the results; and uncertainty in defining the increment in equilibration time in fine soils, among others.

Differential scanning calorimetry (DSC) estimates the UWC by determining the real heat flux function absorbed during the melting of frozen samples [22]. The aim is attained by a convolution analysis of the endothermic peak that occurs due to phase change from ice to water [22]. Hence, the application of DSC is limited to predict the amount of unfrozen water able to undergo a phase change, i.e., unbound water, which freezes close to the FP, between 0 and −10 °C [23]. Thus, bound water in clays (hygroscopic water) appears as "nonfreezable water" in DSC experiments. Despite their limitations, NMR and DSC have been used extensively and seem to be the most applicable techniques for the bentonite component of the GCL.

Time domain reflectometry (TDR) method measures the unfrozen water based on the relationship between the permittivity of the soil and its water content since water possesses a larger dielectric constant than other constituents in the soil (solids, air, ice) [24]. The TDR-measured permittivity can then be related to the unfrozen fraction through calibrations for saturated or unsaturated conditions [e.g. 13, 25]. While the method is practical and commonly used, the small thickness of the GCL challenges its applicability. Also, Watanabe and Wake [25] and Zhou et al. [24] argued that TDR might overestimate the ice content (underestimate UWC) near the FP. Thus, TDR is usually calibrated with NMR or Gamma Ray Attenuation.

Ultrasonic techniques used to estimate UWCs are based on measured velocities of compressional and shear waves. While the application of the method has been more limited than the other techniques, Christ and Park [26] consider that overestimation of the unfrozen water (underestimate ice content) is likely to occur near the FP.

Instead of measuring the UWC, manufactured pore water pressure gauges [19] and chilled mirror sensors [27] can measure the cryo-suction in the freezing soil directly which avoids the use of the Clapeyron equation for the conversion between temperature and cryo-suction when the FCC is used in simulation models. However, these techniques have limited use in laboratories to date.

Table 1. Methods utilized to measure unfrozen water content in porous media.

Method	Example of authors that applied the method
Differential scanning calorimeter (DSC)	Kozlowski [23], Kozlowski and Nartowska [28], among others
Nuclear magnetic resonance (NMR)	Tice et al. [20], Kruse and Darrow [14], among others
Time domain reflectometry (TDR)	Spaans and Baker [13], Zhou et al. [24], Watanabe and Wake [25], among others
Ultrasonic techniques	Christ and Park [26], Huang et al. [29], among others

4 Estimation of FCC by Prediction Methods

Several researchers developed methods to estimate the FCC from (a) considering empirical relationships between the unfrozen water and freezing temperatures and (b) evaluation of the similarities between the FCC and the water retention curve (WRC) which are listed in Table 2. The first group assumes that the FCC can be approximated by several functions such as: power curve [e.g. 15], exponential curve [e.g. 9, 30], simple linear piecewise functions [9] or a non-linear piecewise curve [31]. To produce such approximations, properties of the studied soil such as dry density, amount of clay content, surface area, porosity and hygroscopic water content are required. Thus, estimation of the FCC can be done based on elementary laboratory tests. The second group estimates the FCC from the WRC by combining the Clapeyron equation and WRC fitting parameters such as: (a) Brooks and Corey [32] [e.g. 33], (b) van Genuchten [34], [e.g. 35], and (c) Fredlund and Xing [36], [e.g. 37, 38], assuming that air-water interface energy and ice-water interface energy are equal.

Such assumption has been contested by Koopmans and Miller [39] and concluded that a proper FCC derived from WRC must correctly express a pressure discontinuity adjustment, for example as based on the capillary theory postulated by Miller [40], mainly for capillary-dominated soils (e.g., granular soils). As a result, some authors, such as Dall'Amico [35] have incorporated the pressure discontinuity adjustment derived from van Genuchten [34] fitting parameters in their FCC, as justified by modeling results. Kurylyk and Watanabe [5] recommended the usage of the pressure discontinuity adjustment but mentioned that the influence of freezing temperatures and types of soils on the adjustment factor need confirmation.

Table 2. Prediction methods to estimate the FCC.

Method	Strengths and limitations
ER: Power curve [e.g. 15]	Designed for soils governed by sorptive forces (e.g. clay soils)
ER: Exponential curve [e.g. 9, 30]	Good fit results for capillary soils (e.g. granular soils)
ER: Simple linear piecewise function [9]	Approximated fit results for capillary soils (e.g. granular soils)
ER: Non-linear piecewise curve [31]	Good fit results for both soil types including soils governed by sorptive forces (e.g. clay soils)
WRC: Brooks and Corey [e.g. 33]	Good fit results for capillary soils (e.g. granular soils)
WRC: van Genuchten [e.g. 35]	Good fit results for capillary soils (e.g. granular soils)
WRC: Fredlund and Xing [e.g. 37, 38]	Good fit results for both soil types

ER: FCC obtained by empirical relationship/equation.
WRC: FCC obtained by combining the Clapeyron equation and WRC fitting parameters.

A thorough comparison of empirical methods from both estimation groups has been carried out by Kurylyk and Watanabe [5] and Ren et al. [38]. The publications concluded that some empirical methods are more applicable to certain soils as a function of their grain size distribution. Hence, experimental data is required to corroborate which empirical method more accurately predicts the FCC for a specific soil type.

Regarding the remaining UFPFs, the thermal conductivity function in frozen soils is commonly greater than the one of unfrozen soils since the thermal conductivity of ice is greater than water. This function can be estimated from the FCC using empirical methods such as the one proposed by Johansen [41] by estimating the thermal conductivity as a function of the ice fraction in the soil at each freezing temperature. Such method is commonly used in the engineering practice with good results when consideration of the mineralogy (e.g., quartz in granular soils) is made. Differently, the hydraulic conductivity function in frozen soils is estimated from the WRC by using different approaches, as enumerated by Kurylyk and Watanabe [5]. In contrast to such prediction methods, more recently Watanabe and Osada [42] performed experimental measurements of hydraulic conductivity in a frozen soil, which aids in improving the accuracy of numerical simulations.

5 Conclusions

The FCC is a key component in the analysis of freezing and thawing processes that impact the GCL hydraulic performance. Currently, prediction models using the Clapeyron equation are not applicable to open systems or rapid early-stage freezing where equilibrium is unlikely to occur.

The acquisition of FCCs of GCLs by experimental methods is essential for more accurate prediction of GCL hydraulic performance in cold regions. The bentonite component of GCLs can be used to obtain the FCC experimentally using NMR or DSC. These experimental methods possess limitations, which need to be taken into consideration in GCL studies. An experimentally derived FCC of GCLs is required to validate existing empirical equations for GCLs. Once validated, or revised as needed, such an equation could be used for estimating GCL initial conditions/scenarios unattained in the experimental programs and for expanding the capabilities of existing prediction models.

Acknowledgments. This research was supported under the Australian Research Council's Linkage Projects funding scheme (project number LP140100516). The first author thanks the Peruvian National Program of Scholarships and Student Loans (PRONABEC) for funding his Ph.D. studies. The authors also acknowledge the funding provided to this project by Geofabrics Australasia Pty. Ltd.

References

1. Bouazza A (2002) Geosynthetic clay liners. Geotext Geomembr 20(1):3–17
2. Bouazza A, Singh RM, Rowe RK, Gassner F (2014) Heat and moisture migration in a geomembrane–GCL composite liner subjected to high temperatures and low vertical stresses. Geotext Geomembr 42(5):555–563
3. McWatters RS, Rowe RK, Wilkins D, Spedding T, Jones D, Wise L, Mets J, Terry D, Hince G, Gates WP, Di Battista V, Shoaib M, Bouazza A, Snape I (2016) Geosynthetics in Antarctica: performance of a composite barrier system to contain hydrocarbon-contaminated soil after three years in the field. Geotext Geomembr 44(5):673–685
4. Hosney MS, Rowe RK (2014) Performance of GCL after 10 years in service in the Arctic. J Geotech Geoenvironmental Eng 140(10):04014056
5. Kurylyk BL, Watanabe K (2013) The mathematical representation of freezing and thawing processes in variably-saturated, non-deformable soils. Adv Water Resour 60:160–177
6. Harlan RL (1973) Analysis of coupled heat-fluid transport in partially frozen soil. Water Resour Res 9(5):1314–1323
7. Guymon GL, Luthin JN (1974) A coupled heat and moisture transport model for Arctic soils. Water Resour Res 10(5):995–1001
8. Newman GP, Wilson GW (1997) Heat and mass transfer in unsaturated soils during freezing. Can Geotech J 34(1):63–70
9. McKenzie JM, Voss CI, Siegel DI (2007) Groundwater flow with energy transport and water–ice phase change: numerical simulations, benchmarks, and application to freezing in peat bogs. Adv Water Resour 30(4):966–983
10. Zhao Y, Si B, He H, Xu J, Peth S, Horn R (2016) Modeling of coupled water and heat transfer in freezing and thawing soils, Inner Mongolia. Water 8(10):424

11. Stähli M, Jansson P-E, Lundin L-C (1999) Soil moisture redistribution and infiltration in frozen sandy soils. Water Resour Res 35(1):95–103
12. Wen Z, Ma W, Feng W, Deng Y, Wang D, Fan Z, Zhou C (2012) Experimental study on unfrozen water content and soil matric potential of Qinghai-Tibetan silty clay. Environ Earth Sci 66(5):1467–1476
13. Spaans EJA, Baker JM (1996) The soil freezing characteristic: its measurement and similarity to the soil moisture characteristic. Soil Sci Soc Am J 60(1):13–19
14. Kruse AM, Darrow MM (2017) Adsorbed cation effects on unfrozen water in fine-grained frozen soil measured using pulsed nuclear magnetic resonance. Cold Reg Sci Technol 142:42–54
15. Anderson D, Tice A (1971) Low-temperature phases of interfacial water in clay-water systems. Proc Soil Sci Soc Am 35:47–54
16. Kozlowski T (2012) Modulated differential scanning calorimetry (MDSC) studies on low-temperature freezing of water adsorbed on clays, apparent specific heat of soil water and specific heat of dry soil. Cold Reg Sci Technol 78:89–96
17. Everett DH (1959) An introduction to the study of chemical thermodynamics. Longmans, London
18. Konrad JM, Duquennoi C (1993) A model for water transport and ice lensing in freezing soils. Water Resour Res 29(9):3109–3124
19. Zhang L, Ma W, Yang C, Yuan C (2014) Investigation of the pore water pressures of coarse-grained sandy soil during open-system step-freezing and thawing tests. Eng Geol 181:233–248
20. Tice AR, Anderson DM, Sterrett KF (1981) Unfrozen water contents of submarine permafrost determined by nuclear magnetic resonance. Eng Geol 18(1):135–146
21. Kozlowski T (2016) A simple method of obtaining the soil freezing point depression, the unfrozen water content and the pore size distribution curves from the DSC peak maximum temperature. Cold Reg Sci Technol 122:18–25
22. Kozlowski T (2003) A comprehensive method of determining the soil unfrozen water curves: 1. Application of the term of convolution. Cold Reg Sci Technol 36(1):71–79
23. Kozlowski T (2003) A comprehensive method of determining the soil unfrozen water curves: 2. Stages of the phase change process in frozen soil–water system. Cold Reg Sci Technol 36(1):81–92
24. Zhou X, Zhou J, Kinzelbach W, Stauffer F (2014) Simultaneous measurement of unfrozen water content and ice content in frozen soil using gamma ray attenuation and TDR. Water Resour Res 50(12):9630–9655
25. Watanabe K, Wake T (2009) Measurement of unfrozen water content and relative permittivity of frozen unsaturated soil using NMR and TDR. Cold Reg Sci Technol 59:34–41
26. Christ M, Park J-B (2009) Ultrasonic technique as tool for determining physical and mechanical properties of frozen soils. Cold Reg Sci Technol 58(3):136–142
27. Watanabe K, Takeuchi M, Osada Y, Ibata K (2012) Micro-chilled-mirror hygrometer for measuring water potential in relatively dry and partially frozen soils. Soil Sci Soc Am J 76 (6):1938–1945
28. Kozlowski T, Nartowska E (2013) Unfrozen water content in representative bentonites of different origin subjected to cyclic freezing and thawing. Vadose Zone J 12(1)
29. Huang X, Li D, Ming F, Bing H, Peng W (2015) Experimental study on acoustic characteristics and physico-mechanical properties of frozen silty clay. Yanshilixue Yu Gongcheng Xuebao/Chin J Rock Mech Eng 34(7):1489–1496
30. Bai R, Lai Y, Zhang M, Yu F (2018) Theory and application of a novel soil freezing characteristic curve. Appl Therm Eng 129:1106–1114

31. Kozlowski T (2007) A semi-empirical model for phase composition of water in clay–water systems. Cold Reg Sci Technol 49(3):226–236
32. Brooks RH, Corey AT (1964) Hydraulic properties of porous media. Colorado State University, Fort Collins
33. Sheshukov AY, Nieber JL (2011) One-dimensional freezing of nonheaving unsaturated soils: model formulation and similarity solution. Water Resour Res 47(11)
34. van Genuchten MT (1980) A closed-form equation for predicting the hydraulic conductivity of unsaturated soils. Soil Sci Soc Am J 44(5):892–898
35. Dall'Amico M (2010) Coupled water and heat transfer in permafrost modeling. Doctoral thesis, University of Trento, p 175
36. Fredlund DG, Xing A (1994) Equations for the soil-water characteristic curve. Can Geotech J 31(4):521–532
37. Azmatch T, Sego DC, Arenson LU, Biggar KW (2012) Using soil freezing characteristic curve to estimate the hydraulic conductivity function of partially frozen soils. Cold Reg Sci Technol 83–84:103–109
38. Ren J, Vanapalli SK, Han Z (2017) Soil freezing process and different expressions for the soil-freezing characteristic curve. Sci Cold Arid Reg 9(3):221–228
39. Koopmans RWR, Miller RD (1966) Soil freezing and soil water characteristic curves. Soil Sci Soc Am Proc 30:680–684
40. Miller RD (1980) Freezing phenomena in soils. In: Hillel D (ed) Applications of soil physics. Academic Press, New York, pp 254–299
41. Johansen O (1975) Thermal conductivity of soils. (CRREL Draft Translation 637, 1977), Trondheim, Norway
42. Watanabe K, Osada Y (2017) Simultaneous measurement of unfrozen water content and hydraulic conductivity of partially frozen soil near 0 °C. Cold Reg Sci Technol 142:79–84

Failure Mechanism of Two-Stage Mechanically Stabilized Earth Walls on Soft Ground

Xu Zhang[1], Jianfeng Chen[1(✉)], and Junxiu Liu[2]

[1] Tongji University, Shanghai 200092, China
jf_chen@tongji.edu.cn
[2] Anhui Jianzhu University, Hefei 230601, China

Abstract. A coupled of discrete-continuum numerical model was established for the two-stage Mechanically Stabilized Earth (TS-MSE) wall on soft ground based on centrifugal test. In the numerical model, the wall and soft ground were simulated using Particle Flow Code (PFC) and Fast Lagrangian Code (FLAC), respectively. The deformations of the TS-MSE wall on soft ground were analyzed under working loading. Then, the failure mechanism of the wall was further investigated. It was found that the numerical results were in good agreement with the measured results in the centrifugal test and that the TS-MSE wall can well adapt to the soft ground under the working loading. The TS-MSE wall on the soft ground generally suffers the deep seated failure mode, and the deep seated failure surface is composed of a Rankine failure surface in the unreinforced zone and an arc failure surface in the foundation. The strength of the reinforcements was gradually reduced to investigate the internal failure mechanism of the TS-MSE wall on the soft ground, and the internal failure mechanism of the wall exhibited that the reinforcements were fractured from the bottom to the top of the wall in sequence following the failure surface in the foundation, forming a Rankine failure surface passing through the end of the connection element at the bottom layer.

1 Introduction

The two-stage Mechanically Stabilized Earth (TS-MSE) wall is a new type of retaining structure for many applications on soft ground [1, 2]. The wall is constructed in two stages to adapt to the soft ground. The stage-one consists of building the flexible facing MSE wall and allows the consolidation of the soft ground. When the main consolidation is completed, a rigid plate is attached to the flexible facing MSE wall by a connection element in the stage-two. In New Jersey, eight field TS-MSE walls has been constructed on soft clay soil to support bridge abutment and demonstrated an excellent post-construction performance [2].

A few studies on TS-MSE wall on soft ground have been conducted using experiment and numerical modeling method. Chen et al. carried out centrifugal test and finite difference numerical modeling to investigate the performance of TS-MSE wall on soft ground under working loading [3, 4]. The test and simulation results showed a satisfactory performance despite the existence of the soft foundation. However, the failure mechanism of the TS-MSE wall on soft ground has not been investigated and

L. Zhan et al. (Eds.): ICEG 2018, ESE, pp. 624–632, 2019.
https://doi.org/10.1007/978-981-13-2224-2_77

understood. For the Geosynthetic-reinforced soil (GRS) wall on rigid foundation, the internal failure surface is assumed to be a Rankine failure surface passing through the toe of the wall [5]. However, for the TS-MSE wall with the rigid connection element, the internal failure surface cannot easily pass through the range from the rigid panel to the end of the connection element. The failure mechanism can also be affected by the large and differential settlement expected in soft ground, thus the internal failure surface of the TS-MSE wall on soft foundation will be different from that of the GRS wall on rigid foundation.

The coupled discrete-continuum method [6, 7] is an effective approach adopted for this study. In the method, the discrete element method (DEM) and the finite element or finite difference method are employed to model the granular media and continuum material, respectively. In this paper, a coupled discrete-continuum numerical model was established for the TS-MSE wall on soft clay foundation, based on the centrifugal test conducted by Chen et al. [3]. In this model, the TS-MSE wall and soft ground are simulated using the Particle Flow Code (PFC) and the Fast Lagrangian Code (FLAC), respectively. The numerical model was used to investigate deformation and the failure mechanism of the TS-MSE wall on soft ground.

2 Centrifugal Model Test [3]

The prototype TS-MSE wall with 3.0 m high was constructed on a 6.0 m thick soft ground, laying a 0.6 m thick sand cushion to compensate the settlement. The reinforcements with the length of 4.5 m were wrapped at the face to form the flexible facing. The rigid facing with the thickness of 40 mm was cast in place. The flexible and rigid facing were mechanically connected using three layers of connection elements with the vertical spacing of 1.2 m. The connection element is composed of the metal rods with the horizontal spacing of 1.2 m and a metal L-shape plate at a distance of 2.25 m to the wall facing. The TS-MSE wall model and the schematic of the connection element with a reduced size by ratio of 1/15 to the prototype wall is shown in Fig. 1.

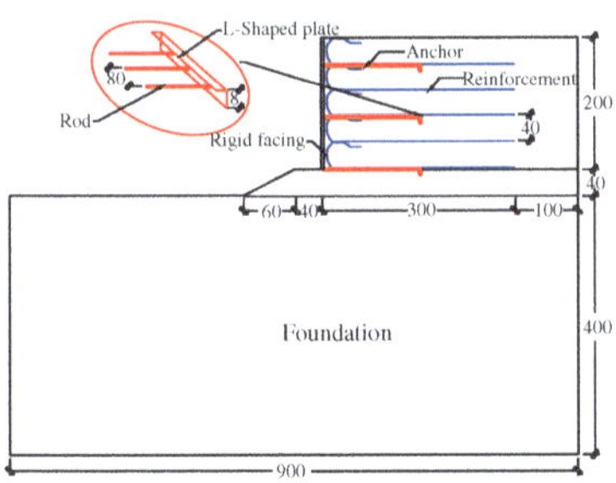

Fig. 1. Centrifugal model of the TS-MSE wall on soft ground (unit: mm)

The backfill used in the model test was Fujian Pingtan standard sand with unit weight of 16.2 kN/m^3 and internal friction angle of 27.3°. The foundation soil modeled by the 800-mesh kaolin clay with a unit weight of 17.8 kN/m^3. The reinforcements made of Polyamide mesh in the test had ultimate tensile strength of 13.5 kN/m based on the result of the wide-width test. A 2-mm-thick aluminum alloy plate was chosen to model the rigid panel based on the similarity criterion. The L-shape plate and rods of the connection element were modelled by 2-mm-thick L-shape aluminum alloy panel and 2-mm-diameter copper rods, respectively.

The processes of the centrifugal model test were divided into two stages:

(1) In stage-one, kaolin slurry at the water content of 100% was consolidated in the strong box of the geotechnical centrifuge under 15 g to prepare the soft foundation. Then, the flexible facing MSE wall with the connection elements installed at the designed elevations was constructed on the consolidated foundation. After the completion of the construction, the flexible facing MSE wall model was reloaded under 15 g until the completion of the settlement in the foundation.
(2) In stage-two, the rigid panel was mechanically attached to the flexible facing MSE wall, using the connection elements, meaning the completion of the construction of the TS-MSE wall. Then, two layers of steel ingot with the size of 5 cm × 5 cm × 2 cm (length × width × height) were laid on the TS-MSE wall to simulate the working surcharge loading, based on the model scale effect. Ultimately, the model was reloaded in the centrifuge under 15 g until the completion of the settlement in the foundation.

3 Numerical Model

3.1 Geometric Model

In the numerical model, the TS-MSE wall and the soft ground were simulated using PFC and FLAC, respectively. About 6000 particles with the diameter of 1 mm–2 mm were generated to simulate the backfill. Geotextile, anchors and rigid facing were generated with particles of 2 mm gain size. Particles with 4 mm gain size were generated on the top of wall to simulate the surcharge loading. The soft ground numerical model (900 mm × 400 mm) with standard boundary conditions was meshed with 40 × 20 grids, as shown in Fig. 2.

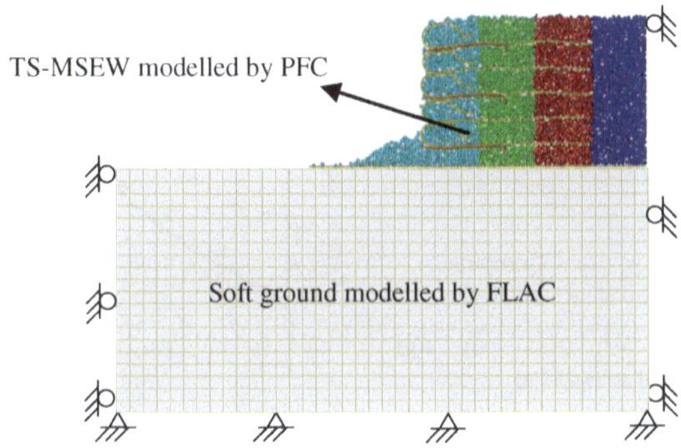

Fig. 2. The coupling of discrete-continuous model of the TS-MSE wall on soft ground

3.2 Numerical Parameters

The micro parameters of the materials used in the MSE wall are as shown in Table 1. The micro parameters of backfill soil were determined through the PFC simulation of direct shear test under the normal stresses of 50 kPa, 100 kPa, 150 kPa and 200 kPa, as shown in Fig. 3(a). The geotextile micro parameters were determined using PFC simulation of tensile test of geotextiles, as shown in Fig. 3(b). The parameters of rigid face were determined using PFC simulation of the theoretical three-point bending curve of the rigid panel, as shown in Fig. 3(c). The connection element micro parameters were determined using PFC simulation of tensile test, as shown in Fig. 3(d). The elastic modulus $E = 112$ GPa used in Fig. 3(d), which is quite close to the true value $E = 108$ GPa. Particles with gain size of 4 mm and density of 76 g/cm^3 were generated on the top of wall to simulate the surcharge load. The Mohr-Coulomb criterion was used to model the foundation soil with elastic modulus $E = 2.0$ MPa, passion radio $v = 0.3$ and consolidated undrained shear strength $c_{cu} = 0$ kPa, $\phi_{cu} = 27°$ based on triaxial shear test.

Table 1. Microscopic parameters of the reinforced retaining wall

	Fill	Geotextile	Rigid face	Connection element
Particles radius (mm)	1.0–2.0	1	2	2
Particle density (kg/m^3)	1600	20	2650	7500
Parallel bond normal stiffness (N/m)	–	4.0×10^4	5.0×10^{13}	9×10^{12}
Parallel bond shear stiffness (N/m)	–	4.0×10^4	5.0×10^{13}	9×10^{12}
Parallel bond normal strength (N)	–	1.4×10^4	5.0×10^{12}	9×10^{11}
Parallel bond shear strength (N)	–	1.4×10^4	5.0×10^{12}	9×10^{11}
Radius multiplier	–	1	1	1
Friction coefficient	0.5	0.45	0.45	0.3

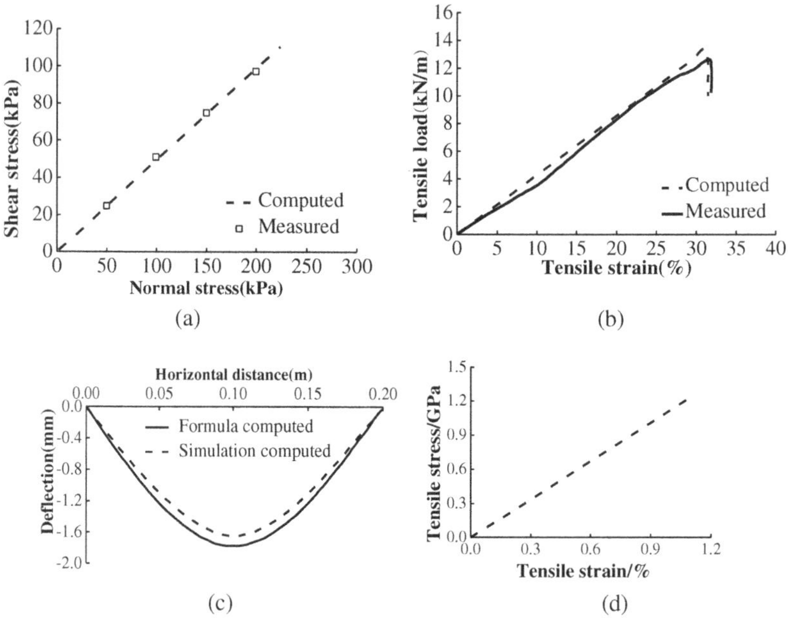

5

Fig. 3. PFC simulation of (a) direct shear test of fill, (b) tensile test of geotextiles, (c) the theoretical three-point bending curve of the panel, (d) tensile test of connection element

4 Results

Figure 4 shows the comparison of the measured and simulated deformation results in two stages, including the settlement in the wall base and in the wall top and the horizontal displacement of the facing. It can be seen that the numerical results agree well with the measured results in two stages. A large and differential settlement of the reinforced zone was greater than that of the unreinforced zone, and the maximum settlement was approximately located at the position of the L-shape panel. The settlement in the wall top was relatively uniform than that in wall base. In addition, the flexible facing and rigid facing basically maintained upright in two stages. These results demonstrate a satisfactory performance of the TS-MSE wall despite the large and differential settlement in the soft soil.

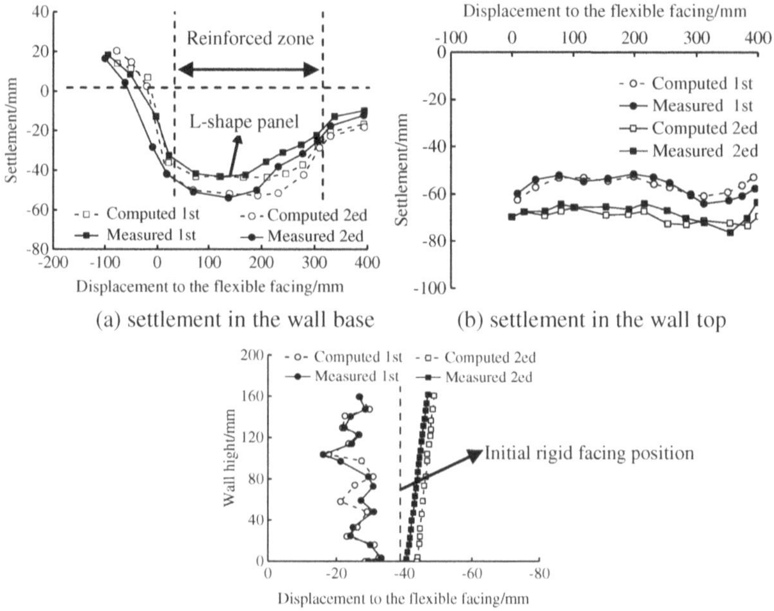

(a) settlement in the wall base (b) settlement in the wall top

(c) the wall facing displacement

Fig. 4. The comparison of the measured and simulated deformation results

5 Failure Mechanism of the TS-MSE Wall

Although a large and differential settlement was suffered for the wall under a 15 g acceleration, the overall structure of the wall was still stable. To investigate the failure mechanism of the TS-MSE wall on soft ground, a MSE wall model with the same geometry and parameters was constructed and was loaded by increasing acceleration. In the analysis, the TS-MSE wall on soft ground failed at 35 g acceleration, forming an obvious failure surface.

Figure 5(a) shows the deformation pattern of the wall and the contour of maximum shear strain increment of soft ground under 35 g. It can be seen that the TS-MSE wall on the soft ground suffers the deep seated failure mode, and the deep seated failure surface is a compound surface through the soft ground and unreinforced zone.

The shape of the failure surface in the unreinforced zone cannot be observed in Fig. 5(a) due to the difficulty to obtain the maximum shear strain increment for PFC, but the displacement can be available for both PFC and FLAC. So the displacement contour can be used to illustrate the failure surface of the TS-MSE wall on soft ground. As shown in Fig. 5(b), the failure surface is composed of an arc surface in the soft ground and a plane in the unreinforced zone, and the plane surface in unreinforced zone starts from the ending of the connected element at bottom layer and takes the angle of $45 + \varphi/2 = 59°$ (φ refers to the internal friction of the sand) on the side of horizontal as the Rankine failure surface.

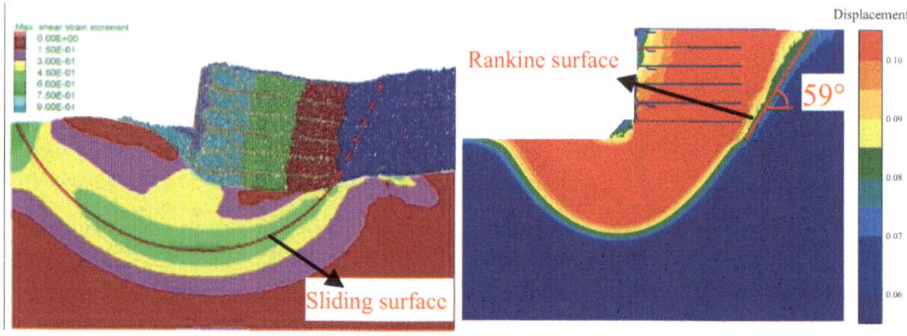

(a) (b)

Fig. 5. (a) Deformation pattern of TS-MSE wall and contour of maximum shear strain increment of foundation, (b) The displacement contour of the model in external failure (unit: m)

The simulated result, the deep seated failure in the TS-MSE wall on soft ground, is consistent with that obtained by Chen et al. [8, 9]. Chen et al. [8] conducted a field test to investigate the behaviors of a 7.6 m high MSE wall on a soft ground with 10 m length reinforcements. The field test results showed that a deep seated failure was observed when the wall achieved the finished level of 7.6 m. Extending reinforcement lengths is expected to increase the external stability of the MSE wall. Thus, Chen et al. [9] conducted finite element (FE) simulations on the above MSE wall with the doubled-length reinforcements and evaluated the stability of the MSE wall. However, the results showed that the wall still suffered a deep seated failure despite the increase of the global safety factor. These results indicated that the MSE wall on soft ground mainly suffers the deep seated failure mode. This can be explained by the fact that the entire stiffness of the reinforced zone is much higher than that of the soft foundation, thus the reinforced zone can deform entirely and is not easy to incur the internal failure surface in it.

However, the internal stability is one of the issues to be considered in the design of MSE wall and has not been investigated and understood. To investigate the internal stability of the TS-MSE wall on soft ground, the strength of the reinforcement was gradually reduced until the TS-MSE wall failed, with an obvious internal failure surface observed. In the analysis, the internal failure surface was observed in the TS-MSE wall on soft ground when the reinforcement strength was reduced to 1.5 kN m^{-1} under the gravity of 35 g. Figure 6 shows the deformation pattern of the wall and the contour of the max shear strain increment in the soft foundation under 35 g acceleration at 5000, 10000 and 40000 time steps. It can be seen that the reinforcements were initially fractured at the end of the connection element at bottom layer (Fig. 6(a)). Then, the residual reinforcements were fractured from bottom to top of the wall in sequence following with the arc failure surface in the soft ground, forming an internal failure surface in the reinforced zone, as shown in Fig. 6(c).

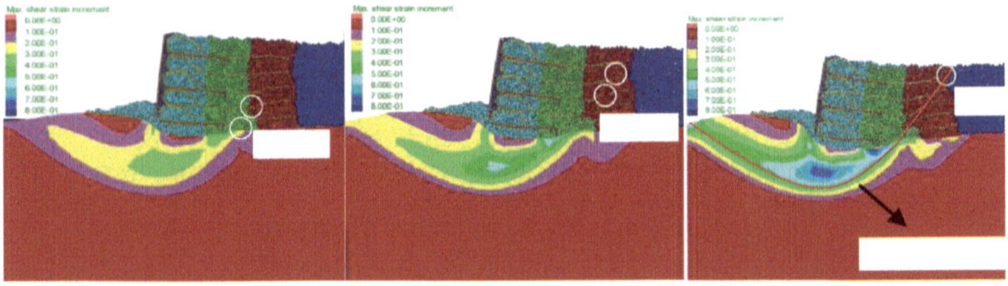

(a) 5000 steps (b) 10000 steps (c) 40000 steps

Fig. 6. Internal failure evolution of TS-MSE wall on soft foundation

To illustrate the shape of the internal failure surface, the displacement contour of the wall at 40000 steps is shown in Fig. 7. It can be seen that the internal failure surface was traced based on the shear zone and was plotted using a linear solid line. The linear solid line or the internal failure surface starts from the end of the connection element at the bottom and takes the angle of $45 + \varphi/2 = 59°$ on the side of horizontal as the Rankine failure surface.

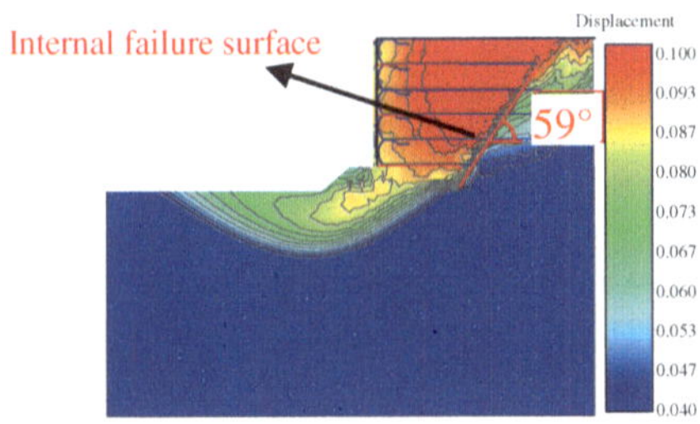

Fig. 7. The displacement contour of the model in internal failure (unit: m)

6 Conclusion

A coupled discrete-continuum numerical model was built to simulate the deformation of the TS-MSE wall on soft foundation, based on the centrifugal test. Then, the numerical model was further used to investigate the failure mechanism of the TS-MSE wall on soft ground. The external failure and internal failure were considered. Based on the test results, the following conclusions can be drawn:

(1) The numerical results are in good agreement with those measured in the centrifugal test in two stages. The wall suffered a large and differential settlement due to its own weight and the surcharge loading, but the overall structure of the wall was still stable, indicating that the TS-MSE wall can well adapt to the soft ground.
(2) The TS-MSE wall on the soft ground generally suffers the deep seated failure mode because the whole stiffness of the reinforced soil is much higher than that of the soft ground. The deep seated failure surface of the wall is composed of a Rankine failure surface in the unreinforced zone and an arc slip surface in the soft ground.

(3) The strength of the reinforcements was gradually reduced to investigate the internal failure mechanism of the TS-MSE wall on the soft ground, and the internal failure mechanism of the wall exhibited that the reinforcements were fractured from the bottom to the top of the wall in sequence following the failure surface in the foundation, forming a Rankine failure surface passing through the end of the connection element at the bottom.

Acknowledgement. The support from the National Natural Science Foundation of China under grants Nos. 41072200, 41572266 and 41772289 is gratefully acknowledged.

References

1. Tatsuoka F, Tateyama M, Uchimura T et al (1997) Geosynthetic reinforced soil retaining walls as important permanent structures. Geosynth Int 4(2):81–136
2. Bloomfield RA, Soliman AF, Abraham A (2001) Performance of mechanically stabilized earth walls over compressible soils. Balkema, Swets and Zeitlinger, pp 317–322
3. Chen JF, Liu JX, Xue JF (2014) Centrifugal test on a reinforced soil wall with flexible/rigid facings. J Tongji Univ (Nat Sci) 42(12):1805–1811 (in Chinese)
4. Chen JF, Liu JX, Shi ZM (2016) Numerical simulation of reinforced soil walls with flexible/rigid facings on yielding foundation. Chin J Rock Mech Eng 35(2):422–432 (in Chinese)
5. Berg RR, Christopher BR, Samtani NC (2009) Design of mechanically stabilized earth walls and reinforced soil slopes: volume I. Report No. FHWA-NHI-10-024. Federal Highway Administration (FHWA), Washington D C, USA
6. Cundall PA, Hart RD (1992) Numerical modelling of discontinua. Eng Comput 9(2):101–113
7. Itasca Consulting Group Inc. (1999) PFC2D (Particle flow code in 2 dimensions) online manual table of contents. Itasca Consulting Group Inc., Minnesota
8. Xue JF, Chen JF, Liu JX, Shi ZM (2014) Instability of a geogrid reinforced soil wall on thick soft Shanghai clay with prefabricated vertical drains: a case study. Geotext Geomembr 42 (2):302–311
9. Chen JF, Liu JX, Jian FX, Shi ZM (2014) Stability analyses of a reinforced soil wall on soft soils using strength reduction method. Eng Geol 177(10):83–92

Resilience of Australian Polymer-Modified Powdered Sodium Bentonite Geosynthetic Clay Liners to Downslope Bentonite Erosion

Will P. Gates[1(✉)], Daniel Gibbs[2], and Marc Amstberg[2]

[1] Australian Centre for Infrastructure Durability, Institute for Frontier Materials, Deakin University, Melbourne-Burwood, 221 Burwood Highway, Burwood, VIC 3125, Australia
will.gates@deakin.edu.au

[2] Geosynthetic Centre of Excellence, Geofabrics Australasia Pty Ltd., 11 Production Avenue, Molendinar, QLD 4214, Australia
{d.gibbs,m.amtsberg}@geofabrics.com.au

Abstract. Loss of bentonite from geosynthetic clay liners (GCLs), installed as a component of a composite liner beneath an exposed geomembrane (GM) on the side-slopes of landfills and exposed to thermal cycling, has become a topic of some importance recently as it raises questions on the long-term durability of modern landfill lining products. Laboratory drip testing was used to evaluate the resilience of four GCLs containing polymer-modified powdered bentonite against down-slope bentonite erosion. Onset erosion features were observed only after 60–140 drip cycles. In all probability the polymer-modified bentonite can be expected to self-heal if no further erosion takes place at these initial stages. A detailed assessment of the drip test and comparison with known field test results, indicates that the drip rate and duration of the laboratory test conducted at ≈22 °C is equivalent to in-field conditions where daytime temperatures of the GM/GCL may reach 65 to 75 °C. Thus, meaningful accelerated testing of polymer-modified powdered bentonite GCL products is possible. The results need to be validated with in-field tests under local conditions because recoverable service-lives of GCL products ultimately depend on the in-field conditions where these materials are deployed.

Keywords: Geosynthetic clay liners · Cyclical wetting and drying
Cyclical heating and cooling

1 Introduction

Downslope bentonite erosion within geosynthetic clay liners (GCLs) is of potential concern when composite liners (i.e. geomembranes (GMs) overlying GCLs), deployed on side-slopes, have been left exposed to diurnal temperature cycling for long periods of time, because cyclical heating and cooling of exposed composite liners can lead to bentonite migration coupled with water transfer within the composite liner (Brachman et al. 2015; Rowe et al. 2014, 2016). The mechanism is related to radiative heating of the exposed GM from solar radiation (Malik et al. 1982) which causes water vapour to

L. Zhan et al. (Eds.): ICEG 2018, ESE, pp. 633–640, 2019.
https://doi.org/10.1007/978-981-13-2224-2_78

move from the GCL and subgrade materials into the void space formed by wrinkles in the exposed GM (Chappel et al. 2012; Rowe et al. 2012). When ambient temperatures cool, condensation of this vapour results in excessive dripping and eventual down-slope migration of bentonite within some GCLs when installed on sides slopes beneath exposed GMs (Moreno et al. 2011). Recent laboratory and field research has revealed that down-slope erosion of bentonite from some types of GCLs exposed on slide slopes of landfills may lead to long-term impairment of their hydraulic function (Rowe et al. 2014; Take et al. 2015; Rowe et al. 2016).

The forms of bentonite (e.g., powdered vs granular, fine granular vs coarse granular, polymer-modified vs unmodified) and also the type of GCL (geotextiles, scrim reinforcement) have been observed to behave differently with respect to both field and laboratory assessments (Rowe et al. 2016). For example, polymer-modified granular bentonite GCLs were shown to experience significant erosion immediately on initiation of laboratory tests, but appeared to have experienced only minor (if any) evidence of downslope erosion after eighteen months at the QEUELTs experimental site. However, given the recentness and resulting lack of reporting by other groups worldwide, some questions remain as to the full scale of this issue with some GCL products, particularly those containing polymer-modified powdered bentonite. The purpose of this study was to determine the resilience of polymer-modified powdered GCL products manufactured in Australia – products which have had a history of excellent performance attributes in geotechnical and geoenvironmental engineering projects in the Austral-Asian region.

A series of laboratory drip tests were used to evaluate the resilience of Australian polymer-modified powdered bentonite GCL products against down-slope bentonite erosion. This paper reports on the most recent set of results.

2 Materials and Methods

The drip test by Ashe et al. (2014) has been adapted for this study. The X800, X1000, X2000 and X3000 GCLs used for this study were Australian manufactured, polymer-modified (anionic polymer), powdered GCL products. The X800/X1000 products are made with woven carrier and non-woven cover polypropylene geotextiles, while the X2000/X3000 products include scrim reinforced nonwoven carrier geotextiles. The X3000 contains the highest amount of bentonite of the four GCLs tested. 350 W $\times$ 550 L mm GCL coupons were cut from rolls (L in the roll direction), hydrated to ~100% gravimetric water content and then sealed in plastic to equilibrate for 24 h. Each coupon was then dried at 60 °C for >15 h to induce desiccation cracking of the bentonite within the GCL. The top end and both sides were taped, and the lower edge was heat-sealed to minimise slumping of bentonite and to ensure that any bentonite erosion was associated with the drip. The coupons were then affixed to a transparent Perspex® sheet at a 3H:1V slope.

Hydration cycles consisting of a wetting phase using deionised water (mostly around 1 μS/cm but generally $\leq$ 10 μS/cm), at a drip rate of 3L/hr maintained for 8 h followed by a drying phase of 16 h at 20 °C were established to mimic a summer diurnal cycle. Tests were not run over the weekend. Drip rate and water quality were confirmed daily. At the end of each cycle, bright lamps (6 $\times$ 800 lm LED) were used to illuminate the GCL sample so that backlit photos could be taken to gauge bentonite erosion (selected results shown in Table 1), and tests were stopped when at least two scores of 'E' or higher were measured. The magnitude of bentonite erosion was scored based on the approach taken by Brachman et al. (2015). Onset erosion was considered if any bentonite thinning was observed, and the width and length of any obvious erosion features within the GCL coupon were scored to gauge the progression of down slope erosion.

Table 1. Erosion scores for polymer-modified GCL products and X3000 erosion features

Cycles	X800	X1000	X2000	X3000	X3000 Erosion Features	
20	h	h	h	h	Day 10-h	Day 140-o
40	h	h	h	h	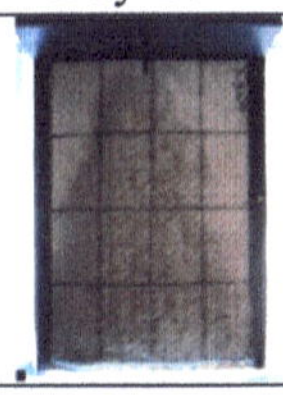	
50	h	o	h	h		
60	h	o	h	h		
70	o	EE	o	h		
80	EE	EEE	e	h		
90	EEE		E	h	Day 150-EE	Day 160-EEE
100			EE	h	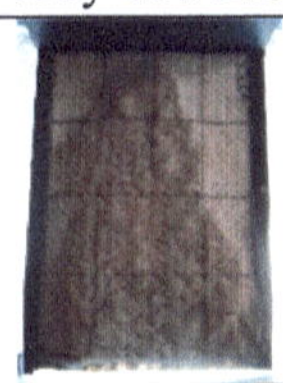	
110			EEE	h		
140				o		
150				EE		
160				EEE		

3 Results and Discussion

3.1 Drip Test Scores

Table 1 provides a summary of test scores on polymer-modified products. All products maintained scores of 'h' for up to 50 wet-dry cycles. Of the four products tested, the X1000 product, indicated onset of erosion ('o') after 50 drip cycles, irrecoverable erosion ('EE') after 70 cycles and irrecoverable extreme erosion ('EEE') thereafter. These erosion features were initially well away from the drip zone under which the bentonite continued to show good swollen gel stability. Previous tests carried out on X1000 indicated a hydrated ('h') state after 60 cycles. The X800 and X2000 products maintained 'h' status for at least 60 cycles, but a score of 'o' occurred within 70 cycles and 'e' (X2000) or 'EE' (X800) was noted within 80 cycles. The X3000 product maintained a hydrated state for up to 140 cycles, 'o' after 140 cycles, 'E' within 150 cycles and 'EEE' thereafter.

3.2 Field Relevance and Estimated Expected Field Lifetimes

A recoverable service-life, defined as the estimated in-field life-time where the ability of the bentonite to self-heal is retained, will be ultimately dependent on the in-field conditions where the composite liner is deployed. Thus, an evaluation of how relevant the laboratory test may be compared to field conditions is critical. The only in-field measurements to date come from the field studies led by the Rowe group, conducted at the Queens University Environmental Liner Test Site (QUELTS) in Ontario Canada, which have provided some important research-based validation of the laboratory test. It must be noted that accurate and valid correlation between the laboratory test for predicting the real field conditions in Australia (or elsewhere) carries a high degree of uncertainty.

With these limitations in mind it is possible to approximate the number of equivalent cycles for an average month of QUELTS field conditions. GCLs 5 and 6 in Rowe et al. (2016) were powdered bentonite GCLs with similar cover and carrier configurations to X3000 (GCL5) and X1000 (GCL6). These samples lasted at least 60 cycles in laboratory tests before minor 'e' erosion was observed. Both GCL samples were observed to only have scores of 'e' after 15 months in the QUELTS experiment (Rowe et al. 2016), but after 28 months, at least one 'EE' and one 'EEE' features were observed. Thus, as a rule of thumb, we can expect 70 cycles of the laboratory test to be equivalent to 28 months, or 2.5 cycles per month. The four polymer-modified powdered bentonite GCL samples tested (Table 1) were all observed to have 'e' scores or less after a low of 60 cycles (X1000) and a high of 150 cycles (X3000). The first recorded 'EE' score was after 70 cycles (X1000), 80 cycles (X800), 100 cycles (X2000) and 150 cycles (X3000). Thus, the X1000 product can be projected to maintain a score of 'e' or less for down slope bentonite erosion for at least 24 months of exposure to conditions equivalent to the QUELTS experiment while X800 and X2000 products may attain 32 months (or more for X2000). For the X3000 GCL, the projection may be as long as 60 months of QUELTS equivalent conditions.

It must be reiterated that any such projection of service life from laboratory tests would have a very high uncertainty if extrapolated to field conditions outside those typical of those experienced during the study at the QUELTS in Ontario, Canada. Factors such as the overall temperature ranges and relative humidity variation in the laboratories, as well as the purity of the water used, can have significant impacts on the number of cycles to which a particular GCL product may maintain a score of 'o' or less in lab tests, and this would then be expected to add to the error of the projection to expected field service life. For example, Rowe et al. (2016) discusses how for a sample that eroded within a few drip cycles with deionized water, just 35 ppm Ca^{2+} in the drip water nullified the effects of deionized water and no erosion was observed even after a considerable number of drip cycles.

Compared to various locations in Australia (Table 2), the biggest differences are related to the amount and time of precipitation (snow and rain in Ontario, no snow in most of Australia) and average annual and seasonal highs and lows (Canada much colder, with freezing temperatures for up to four months out of the year). In this regard, while total precipitation at Olympic Park in Sydney is similar to QUELTS, significant differences in temperatures, and expected thermal fluxes can be expected on exposed composite liners sited in similar climatic conditions. There might also be differences in the development and extent of wrinkles in the GM due to differences in base temperatures between sites rather than temperature differentials within sites. For example, Rowe et al. (2012) and Chapel et al. (2012) indicated that thermal expansion of exposed GMs resulted in several km of connected wrinkles and associated void space at QUELTS. While this has not been fully confirmed elsewhere in other field trials, it may be, for example, that in Australia many more km of connected wrinkles exist, and thus the amount of water per drip point may be largely reduced leading to less water per drip point impacting the GCL. Any such extrapolations would obviously carry excessive uncertainty until sufficient field data is collected.

Table 2. Climatic information for QUELTS from Government of Canada, (2017); for Australia forty-year averages. Source: Australian Bureau of Meteorology (2017)

	QUELTS*	VIC	QLD	NSW	WA	SA
Annual precipitation	952	709	1452	973	564	389
Annual daytime high (°C)	12.1	19.8	25.0	23.3	24.8	23.6
Annual night-time low (°C)	3.6	10.2	15.3	12.2	11.9	9.7
Annual day-night difference (°C)	8.5	9.6	9.7	11.1	12.9	13.9
Summer daytime high (°C)	25.9	26.1	28.5	28.5	32.0	31.8
Summer daytime low (°C)	16.3	14.4	20.3	17.6	16.7	14.9
Summer day-night difference (°C)	9.6	11.7	8.2	10.9	16.3	16.9
Winter daytime high (°C)	–2.6	13.7	20.6	17.4	17.9	15.3
Winter night-time low (°C)	–11.4	6.2	9.1	6.3	7.4	5.2
Winter day-night difference (°C)	–8.8	7.5	11.5	11.1	10.5	10.1

*QUELTS is ~40 km NNW of Kingston, Ontario, Canada; Vic: Moorabbin Airport, Melbourne, Victoria; QLD: Southport, Gold Coast, Queensland; NSW: Olympic Park, Sydney, New South Wales; WA: Guilford, Fremantle, Western Australia; SA: Roseworthy, Adelaide, South Australia.

We can still gain some appreciation for how informative these measurements may be and how realistically they may represent or estimate in-field down slope erosion. Firstly, it is obvious that 30 months exposure to QUELTS field conditions will be different from 30 months exposure to Australian field conditions. Due to the mild continental climate in Ontario, Canada (i.e., the high temperatures are less extreme and less desiccating compared to typical Australian conditions, Table 2) subgrade soils used there likely retain optimal moisture contents for longer times prior to installation of the GCL. The mean summer daytime high and mean summer daytime low temperatures given in Table 2 indicate that some Australian conditions are harsher than Canadian conditions in terms of total thermal flux that drives the solar still effect. While two of the four Australian sites (NSW and VIC) have similar annual and summer temperature differences as in Canada, two (SA and WA) have significantly higher overall temperatures as well as larger temperature differentials. The QLD site is overall warmer and wetter but has a lower summer temperature difference. Thus, while more water may be available for hydrating GCLs or moving into the void spaces of the composite liner, the temperature extremes within the composite liner are likely less important in moving water vapour at the QLD site. On the other hand, less water may be available in the subgrade in a SA or WA landfill at the time of GCL installation due to higher day time temperatures. Furthermore, the water that is present in a subgrade in SA or WA will be more susceptible to thermal induced migration due to the higher temperature differences between daytime highs and daytime lows (Table 2).

Additionally, there is some indication that the total amount of water available may be sufficiently lower in Australia to decrease the threat posed. For example, Adelaide, SA has less than half the rainfall as Kingston Ontario, and thus insufficient water might be available to contribute to the solar still moisture cycle. Due to the generally warmer climate, any day throughout the year can potentially create the solar still effect in NSW, and the VIC site probably has a similar number of days per year as QUELTS in which the solar still effect can take place. Both WA and SA will have limited solar still effect during the dry season. Finally, the subgrade soil type at the QUELTS site, reported as non-plastic with negligible clay (Rowe et al. 2016), will release more water far more readily than a soil containing a higher clay content, as are often used in Australia (Bouazza et al. 2017).

4 Conclusions

Based on the available evidence, exposed composite liners utilizing the polymer-modified powdered bentonite GCLs used in this study should maintain full integrity for at least 24 months (X1000) under conditions equivalent to the QUELTS field site. Lifetimes may vary significantly from the results reported here for products containing different polymer-modified bentonites. Only minor onset bentonite erosion that can still self-heal in a few locations of an exposed composite liner could be expected up to 24 month exposure of an X1000 GCL product. Depending on the factors listed above, erosion may be quite prevalent across the exposed composite liner surface after 30 months with X1000 products. Exposures longer than 30 months will result in increased risks of loss of X1000 liner integrity such that irrecoverable erosion should be expected in at least one location of the exposed composite.

The more robust GCL products (X2000, X3000), usually recommended for side slope constructions, can be projected to maintain integrity for more than 30 months if left exposed. The X2000 product should be expected to suffer only minor bentonite erosion in a few locations after 30 months exposure and that this will self-heal if the composite liner is covered. All available evidence suggests that the X3000 product is strongly resistant to down slope bentonite erosion and that it can be expected to last at least 60 months in an exposed condition before irrecoverable erosion could be expected to be a significant problem. Care should be taken, however, in extrapolating these results to other bentonites modified by different polymers for estimating service life of GCL products.

5 Recommendations

A number of factors need to be tested to clearly show how they influence a service life prediction from the laboratory drip test. These include drip rate, drip duration, desiccation temperature and duration, thermal regime of the site, aspect of the exposed composite liner, and the source of initial hydration water (i.e. the subgrade soil water retention properties). Clearly the drip rate (3L per hour) and dripping regime in the laboratory test is extreme in terms of the amount of water that drips over a period of time on a single point. In a real exposed composite liner, drip rates may sometimes exceed 3L/hour and therefore be a continuous flow, but only for a short time duration during the initial condensation phase of the solar still. Further experimental input is required to determine if short duration, continuous flow, which may occur in real exposed composite liners could result in different behaviour than observed in laboratory tests.

The dehydration regime of each wet-dry cycle of the laboratory drip test is probably too gentle compared to Australian climatic conditions, where ambient air temperatures routinely exceed 35 °C in the GCL. For example, an exposed liner in Melbourne on a southeast facing slope will undoubtedly experience cooling earlier in the afternoon than a northwest facing slope in Adelaide. The drip rate that would occur after the initial flush of condensate would be dependent on the on-going liquid-to-vapour phase transfer of water from the subgrade and GCL products. At this time, this is largely an unknown factor for exposed composite liners, and future research should focus on determining these variables.

Regardless of the results and reasoned arguments presented herein as to the resilience of polymer-modified powdered bentonite based GCLs, it is recommended that composite liners be covered as soon as possible to prevent issues related to exposure to solar radiation.

References

Ashe L, Rowe RK, Brachman RWI, Take WA (2014) Laboratory study of downslope erosion for ten different GCLs. J Geotech Geoenvironmental Eng. https://doi.org/10.1061/(ASCE)GT.1943-5606.0001191. 04014079

Bouazza A, Ali M, Gates WP, Rowe RK (2017) New insight on geosynthetic clay liner hydration: the key role of subsoils mineralogy. Geosynth Int 24:139–150

Brachman RWI, Rentz A, Rowe RK, Take WA (2015) Classification and quantification of downslope erosion from a GCL when covered only by a black geomembrane. Can Geotech J 52:395–412

Chappel MJ, Brachman RWI, Take WA, Rowe RK (2012) Large-scale quantification of wrinkles in a smooth, black, HDPE geomembrane. ASCE J Geotech Geoenvironmental Eng 137:671–679

Malik MAS, Tiwari G, Kumar A, Sodha MS (1982) Solar distillation: a practical study of a wide range of stills and their optimum design, construction, and performance. Pergamon Press, New York

Moreno L, Liu L, Neretnieks I (2011) Erosion of sodium bentonite by flow and colloid diffusion. Phys Chem Earth Parts A/B/C 36:1600–1606

Rowe RK, Ashe LE, Take WA, Brachman RWI (2014) Factors affecting the down-slope erosion of bentonite in a GCL. Geotext Geomembr 42:445–456

Rowe RK, Brachman RWI, Take AW, Rentz A, Ashe LE (2016) Field and laboratory observations of down-slope bentonite migration in exposed composite liners. Geotext Geomembr 44:686–706

Rowe RK, Chappel MJ, Brachman RWI, Take WA (2012) Field monitoring of geomembrane wrinkles at a composite liner test site. Can Geotech J 49:1196–1211

Take WA, Brachman RWI, Rowe RK (2015) Observations of bentonite erosion from solar-driven moisture migration in GGLs covered only by a black geomembrane. Geosynth Int 22:78–92

Experimental Study of Shear Strength of Geosynthetic Clay Liner for Monotonic Loading

Ji-Yun Chang(✉), Shi-Jin Feng, Yang Shen, Hao Shi, and Jia-Liang Shi

Department of Geotechnical Engineering, Tongji University, Shanghai 200092, China
1610148@tongji.edu.cn

Abstract. Internal shear strength of geosynthetic clay liner (GCL) is of vital importance for the stability of waste containment facilities where GCLs are widely used as hydraulic barriers. This paper presents an experimental investigation of the internal shear strength of hydrated needle-punched geosynthetic clay liners (NP GCLs) under monotonic loading. A new dynamic direct shear apparatus with the capability to cover a large range of normal stress and displacement rate is developed. Series of direct shear tests on NP GCL for different normal stress levels and displacement rates are performed. The failure mechanism of hydrated GCL is explored through inspection of tested specimens and analysis of test results. The intrinsic relationship between peak/residual shear strength of NP GCL and displacement rate is also studied. Experimental results indicated that displacement rate has obvious effect on the peak shear strength of NP GCL, while the influence of displacement rate on residual shear strength is of little significance for the same normal stress level.

Keywords: Geosynthetic clay liner · Direct shear test · Shear strength Landfill

1 Introduction

Geosynthetic clay liners (GCLs) are manufactured hydraulic barriers, which have been widely used in waste containment facilities. GCL is composed of bentonite clay and two layers of geosynthetic material, and generally there are geosynthetic reinforcements across the bentonite layer to improve the shear resistance. However, internal shear strength of GCL may decrease dramatically due to the hydration of bentonite (Fox and Stark 2015). The relatively low shear strength of GCL within liner system contributes to the formation of a potential slip interface, which is detrimental to stability of landfill (Feng et al. 2018; Kavazanjian et al. 2013). Hence, the internal shear strength of hydrated GCL has drawn extensive attention.

Various laboratory tests of GCLs have been reported in the literature. Initially, the internal shear strength of unreinforced GCL specimens were widely explored through direct shear tests (Daniel et al. 1993; Eid and Stark 1997). Effects of normal stress (σ_n = 27–139 kPa) and displacement rate (R = 0.0003–0.26 mm/min) were studied in

L. Zhan et al. (Eds.): ICEG 2018, ESE, pp. 641–648, 2019.
https://doi.org/10.1007/978-981-13-2224-2_79

their tests, and results indicated that: higher displacement rate, higher peak shear strength of GCL. However, application of unreinforced GCLs becomes limited and existence of reinforced fibres may result in more complicated shear mechanism of GCL.

For the internal shear strength of widely used thermal-bonded woven/nonwoven (W/NW) needle-punched (NP) GCL, numerous studies have been carried out (Stark and Eid 1996; Eid et al. 1999; McCartney et al. 2002; Zornberg et al. 2005). These experimental studies revealed that peak shear strength of GCL is closely related to displacement rate, while displacement rate has little influence on residual shear strength. Noteworthily, maximum shear displacement rate of abovementioned experiments was 36.5 mm/min, which was deemed in the quasi-static range. Fox et al. (2015) performed direct shear tests on GCLs using a large direct shear machine for normal stress levels ranging from 141 to 1382 kPa and shear displacement rates ranging from 0.1 to 30000 mm/min. Results of their tests displayed complex relationship between peak/residual shear strength of GCL and displacement rate. Generation of excess pore pressures, swift pullout or rupture of needle-punched fibres and drained or undrained frictional resistance of the bentonite all contribute to the complicated shear behaviour of GCL, which demonstrates the necessity of further investigation.

In this study, series of direct shear test on GCLs are conducted with newly developed large-scale dynamic direct shear test. GCL's failure mode is inspected and the relation between horizontal displacement and shear stress of the GCL specimen is systematically analysed. Effects of normal stress and displacement rate on peak and residual internal shear strength are studied, which facilitates the understanding of the shear mechanism of reinforced GCL.

2 Experimental Program

2.1 GCL Material

The tested materials are GCLs manufactured by CETCO (Jiangsu, China). As shown in Fig. 1, granular bentonite is sandwiched between a woven slit-film polypropylene geotextile and a nonwoven needle-punched polypropylene geotextile. Especially, the reinforcement is provided through needle-punching polypropylene fibers from the carrier geotextile through the bentonite and the cover geofabric layer. The thickness of GCL is around 5 mm with satisfactory uniformity.

2.2 Test Equipment

Direct shear tests on GCLs are performed with a large-scale dynamic shear apparatus for geosynthetic interface. The newly developed direct shear machine is powered by two high pressure oil pumps and controlled by a set of server systems. As shown in Fig. 2, this apparatus consists of three main subsystems: (1) the shear test system, (2) the control system and (3) the oil pump system. Two high-precision displacement sensors are connected with the horizontal and vertical actuators respectively. Therefore, when the bottom shear box moves during the shear tests, both horizontal and vertical displacement are detected by the displacement sensors and consequently test data are collected.

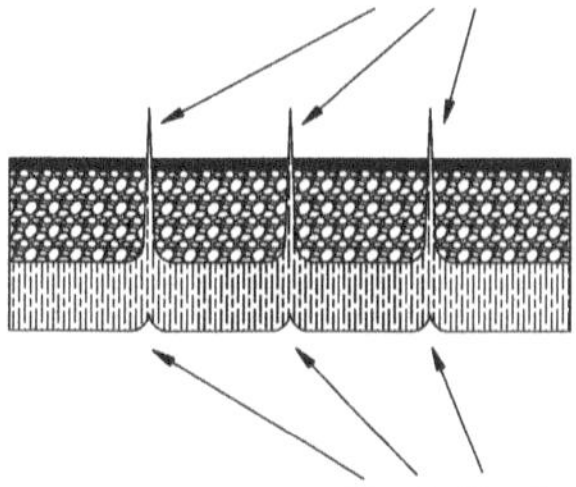

Fig. 1. Profile of the tested GCL.

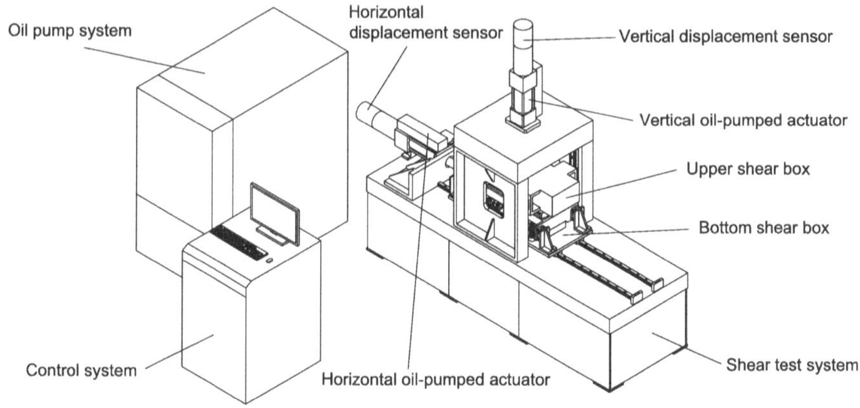

Fig. 2. Framework of the large-scale dynamic shear apparatus.

As presented in Fig. 3, The upper and bottom shear boxes are designed to model the shear process and two gripping plates are fixed on the shear boxes respectively. The gripping plates have a large number of steel sharp teeth (height = 1 mm; spacing distance = 2 cm) to grip the tested materials. The implementation of gripping plates not only permits the GCL specimen to fail along the weakest surface but also avoids possible progressive failure effects.

This large-scale dynamic shear apparatus enables both single-direction and bidirectional shear tests. The maximum displacement rate is 200 mm/min and the maximum frequency is 5 Hz for sinusoidal loading. Main technical specifications of this apparatus are summarized in Table 1.

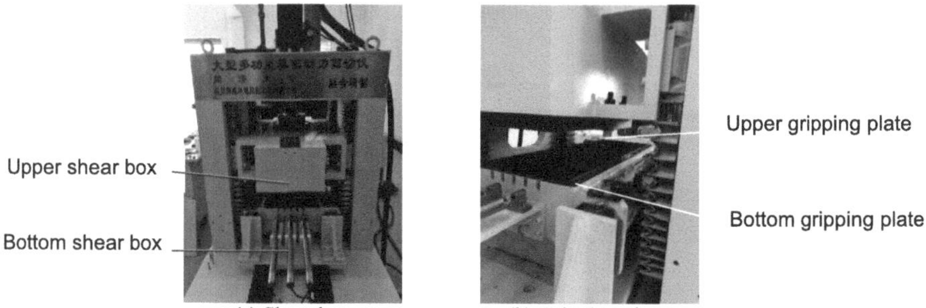

(a) Shear boxes (b) Gripping plates

Fig. 3. Pictures of crucial components of this apparatus.

Table 1. Technical specifications of dynamic shear apparatus.

Feature	Specification
Specimen size	600 mm × 200 mm
Maximum normal force	60 kN
Maximum shear force	35 kN
Maximum specimen thickness	100 mm
Maximum horizontal displacement	150 mm
Specimen gripping system	Teeth height: 1 mm; spacing distance: 2 mm
Minimum displacement rate	0.1 mm/min
Maximum displacement rate	200 mm/min

2.3 Test Procedure

Initially, the GCLs are wrapped with plastic film to keep certain water content and then GCL specimens are cut according to the size of shear apparatus. To perform the hydration of GCL promptly, the two-stage accelerated procedure proposed by Fox and Stark (2004) is applied. The tested GCL specimens are placed in a water tank for about two days with free access to tap water. During this stage, relatively low normal stress (10 kPa) is applied to specimens. Then, GCL specimen is placed on the gripping plate of the shear apparatus with the woven geotextile facing upward. Normal stress, which equals to the normal stress during shear process, is applied to GCL specimen for approximate one day to finish the second stage of hydration.

Once the two-stage hydration is finished, the bottom shear box moves according to the given displacement rate without any change of normal stress. After each test was completed, the GCL specimen is removed from the apparatus and the mode of failure is observed immediately. In this study, 6 displacement-controlled monotonic direct shear tests on GCL are conducted to a final displacement of 150 mm for two normal stress levels (σ_n = 250 and 400 kPa) and three displacement rates (R = 1, 10 and 100 mm/min).

3 Experimental Results

3.1 Failure Mode

Photographs of GCL specimens taken after direct shear tests are shown in Fig. 4. It is inferred from Fig. 4(a) that GCL's internal failure occurs at the interface between woven geotextile and bentonite under monotonic loading. Tested GCL specimens are peeled from woven geotextile-bentonite interface easily, and Fig. 4(b) indicates that rupture of the needle-punched fibers is the main failure mode. Besides, there displays certain amount of bentonite extrusion through the edges of the specimens, which may be due to the dissipation of excess pore pressure during the loading process. However, bentonite extrusion seems to have a minor effect on the strength loss according to other studies (Chen et al. 2010).

(a) GCL specimens after tests (b) Failure surface

Fig. 4. Visual inspection of GCL specimens after tests.

3.2 Stress-Displacement Relationship

Relationships for shear stress (τ) versus shear displacement (δ) at two normal stress levels are shown in Fig. 5. For different normal stress and shear rates, the stress-displacement behaviours are similar. During the loading process, there shows two stages (hardening and softening stage) with the peak shear strength (τ_p) being the division point. Displacement hardening effect is mainly attributed to the gradual tension of reinforcement fibres, while the post-peak strength reduction is the result of failure of the needle-punched reinforcement. Therefore, the residual shear strength (τ_r) of GCL at large shear displacement reflects the interface friction between woven geotextile and bentonite.

Noteworthily, there is an initial shear strength at very small displacement apart from the peak and residual shear strength in stress-displacement curves, which is similar to experimental results of other studies (Fox et al. 2015). Existence of the initial shear strength indicates that GCL specimens move against the gripping plates slightly before the gripping teeth begin to hold the GCL in the shear direction.

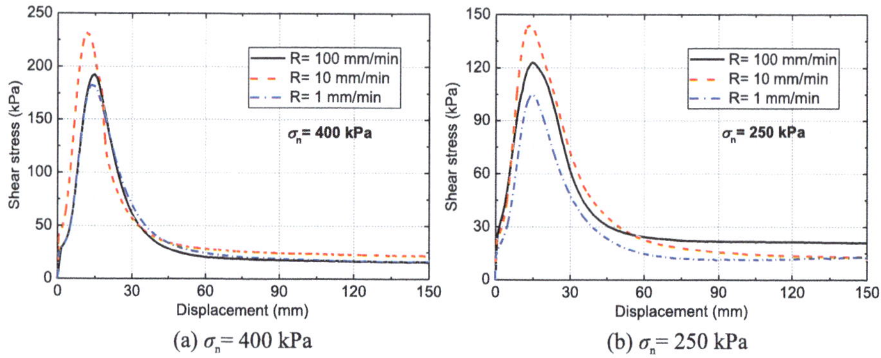

Fig. 5. Shear stress-displacement relationship of GCL specimens.

3.3 Shear Strength

Peak shear strengths of GCLs versus shear displacement rate are plotted in Fig. 6(a), where the tendencies are similar for different normal stress levels. Peak shear strength increases initially and then decreases with displacement rate, and the division is around 10 mm/min. This varying pattern is consistent with results of Fox et al. (2015), while the divisions show some discrepancy. This may be attributed to the difference of GCL materials and testing equipment. As for the mechanism of displacement rate's effect on peak shear strength, it is commonly recognized that variation of tensile resistance of needle-punched fibers plays an important part.

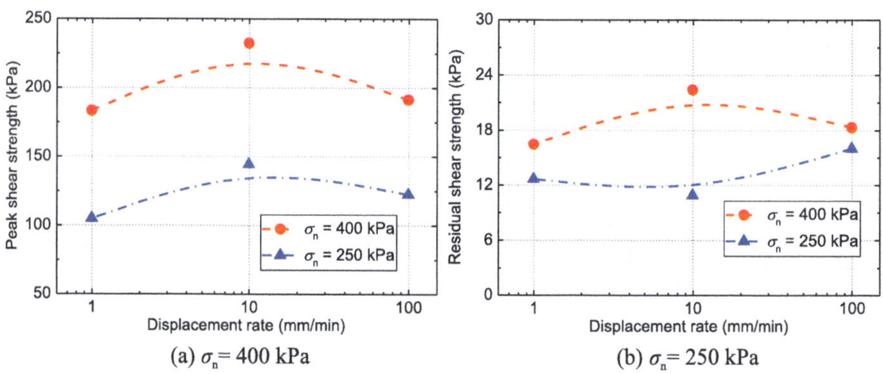

Fig. 6. Shear strength of GCL specimens for different displacement rate.

Residual shear strengths of GCLs, which represent the residual strengths of bentonite, are plotted versus shear displacement rate in Fig. 6(b). There displays no uniform law for the variation of residual shear strength with displacement rate. Theoretically, residual shear strength of GCL is closely related to the development of shear-induced excess pore pressure, which deserves deeper investigations.

3.4 Vertical Compression

Vertical compression of GCL specimens is shown in Fig. 7 where negative values indicate contraction. This figure reflects that GCL specimen generally suffers volume contraction during the direct shear process. It can also be inferred that lower displacement rate leads to larger magnitude of contraction. Moreover, GCL specimens experience smaller volume change when subjected to higher normal stress, which may be attributed to lower water content and higher stiffness of the bentonite after consolidation under corresponding normal stress.

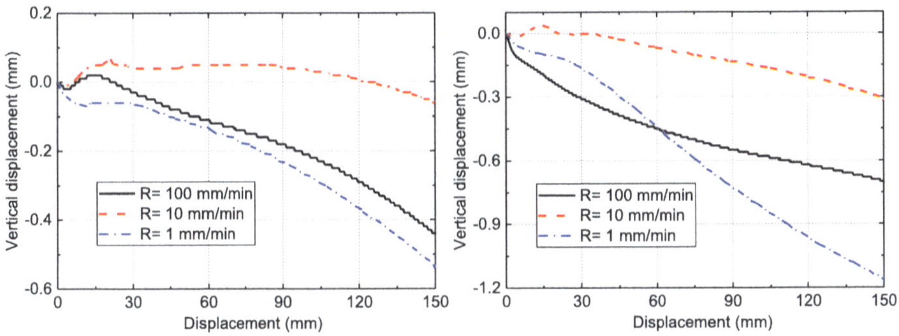

(a) σ_n= 400 kPa (b) σ_n= 250 kPa

Fig. 7. Thickness change of GCL specimens.

4 Conclusions

Experimental results of this study indicate that internal failure of hydrated GCL mainly occurs at the interface between woven geotextile and bentonite due to rupture of the needle-punched fibers for monotonic loading. Significant pre-peak hardening and post-peak softening are revealed in shear stress-displacement relationship. Displacement rate has an obvious effect on peak shear strength of GCL. Peak shear strengths increase and then decrease with rising displacement rate for each normal stress levels. However, displacement rate's influence on residual shear strength is of little significance. Besides, GCL specimens generally suffer volume contraction during shear process.

Nevertheless, only two normal stress levels and three displacement rates are considered herein, which is insufficient to get quantitative and universal rules. Therefore, shear tests considering larger range of normal stress and displacement rate, as well as influence of hydration and other factors, are still required.

References

Chen YM, Lin WA, Zhan TL (2010) Investigation of mechanisms of bentonite extrusion from GCL and related effects on the shear strength of GCL/GM interfaces. Geotext Geomembr 28 (1):63–71

Daniel DE, Shan HY, Anderson JD (1993) Effects of partial wetting on the performance of the bentonite component of a geosynthetic clay liner. In: Proceedings of the geosynthetics, Vancouver, Canada, vol 3, pp 1483–1496. Industrial Fabrics Association International, Roseville

Eid HT, Stark TD (1997) Shear behavior of an unreinforced geosynthetic clay liner. Geosynth Int 4(6):645–659

Eid HT, Stark TD, Doefler CK (1999) Effect of shear displacement rate on internal shear strength of a reinforced geosynthetic clay liner. Geosynth Int 6(3):219–239

Feng SJ, Chang JY, Chen HX (2018) Seismic analysis of landfill considering the effect of GM-GCL interface within liner. Soil Dyn Earthq Eng 107:152–163

Fox PJ, Stark TD (2004) State-of-the-art report: GCL shear strength and its measurement. Geosynth Int 11(3):141–175

Fox PJ, Sura JM, Nye CJ (2015) Dynamic shear strength of a needle-punched GCL for monotonic loading. J Geotech Geoenviron Eng 141(7):04015025

Fox PJ, Stark TD (2015) State-of-the-art report: GCL shear strength and its measurement–ten-year update. Geosynth Int 22(1):3–47

Kavazanjian E, Arab M, Matasovic N (2013) Performance of two geosynthetics-lined landfills in the northridge earthquake. In: 7th international conference on case histories in geotechnical engineering, Chicago

McCartney JS, Zornberg JG, Swan RH (2002) Internal and interface shear strength of geosynthetic clay liners (GCLs). Geotechnical Research Rep., Dept. of Civil, Environmental and Architectural Engineering, Univ. of Colorado, Boulder, CO, 450

Stark TD, Eid HT (1996) Shear behavior of reinforced geosynthetic clay liners. Geosynth Int 3 (6):771–786

Zornberg JG, McCartney JS, Swan RH (2005) Analysis of a large database of GCL internal shear strength results. J Geotech Geoenviron Eng 131(3):367–380

Failure Analysis of Reinforced Foundation Using Transparent Soils

J. F. Chen[1(✉)], X. P. Guo[1], J. F. Xue[2], and P. H. Guo[3]

[1] Tongji University, Shanghai, China
01016@tongji.edu.cn
[2] University of New South Wales, Canberra, Australia
[3] CCCC First Highway Consultants Co., Ltd., Xi'an, China

Abstract. A series of laboratory model loading tests were performed on strip foundations supported on unreinforced and geotextile-reinforced transparent soils. Fused silica sand and white oil were carefully mixed to prepare the transparent soil. The transparent soils were reinforced with geotextiles with the spacing of 0.25B and 0.5B, the length of 1B to 7B, and the depth of 0.5B, 1B and 1.5B (B is the width of the foundation). The deformations of the soils and geotextiles were monitored using a digital camera and a laser transmitter. The results show that for the soils with reinforcements of 0.5B spacing, the deformations of the geotextiles show that each peak point observed in the curves of the bearing capacity improvement versus foundation settlement is related to the failure of one layer of reinforcement, which indicates the progressive failure of different layers of reinforcements. However, for the soils with reinforcement of 0.25B spacing, failure of layers of reinforcements occurs from bottom to top. Moreover, most of the foundations show brittle failure behavior. It also reveals that the bearing capacity of reinforced soil foundation does not always increase with the increase in length of reinforcements when suitable reinforcement spacing is chosen.

Keywords: Reinforced foundation · Transparent soils · Failure mechanism Bearing capacity

1 Introduction

Reinforcement materials like metallic strips and geosynthetics have been widely used in geotechnical engineering to improve the bearing capacity and to reduce the settlement of shallow foundations. At present, most research focuses on factors that may affect the bearing capacity of reinforced soils, such as the spacing of the reinforcements, the embedment depth of the first layer, reinforcement depth and length, on the bearing capacity of reinforced soil foundation by using model test and/or numerical simulation [1–5]. Little research has been done to understand the failure mechanism of foundations on reinforced soils, which requires the understanding of the fracture of geosynthetics during loading. Since the geosynthetic members are buried in soil, the facture of the members during loading is not easy to monitor. Transparent soil, a

L. Zhan et al. (Eds.): ICEG 2018, ESE, pp. 649–657, 2019.
https://doi.org/10.1007/978-981-13-2224-2_80

transparent media allowing researchers to see through the soil, has been widely used by many researchers [6–9].

This paper is to reveal the soil displacement and fracture of geosynthetics under strip footing loading using geotextile reinforced transparent soils. Loading tests on strip footing resting on transparent soils reinforced with geotextiles of different layers and lengths were conducted in laboratory. The load settlement relationships of foundations were studied to investigate the failure mechanism of the foundations.

2 Laboratory Model Test

2.1 Test Set-Up

The test setup included a test box, a loading system, a laser transmitter and a digital camera (see Fig. 1). The box was made of 19 mm-thick tempered glass. Its inner dimensions were 500 mm (length) × 85 mm (width) × 500 mm (height). The strip footing was made of steel plate with the width of 40 mm (B) and length of 83 mm. A piece of sand paper was glued onto the base of footing to simulate the rough base condition. An electronic universal loading frame, WDW-600KN, was used to apply the static vertical loads. All the tests were conducted under the rate of 1.0 mm/min. The laser transmitter has a maximum output power of 150 mW and can generate laser light sheet. The tests were conducted in a dark room to minimize the effects of other light sources. A Canon EOS 500D, a digital camera with a resolution of 2592 × 1728 pixels, was used to capture images during the tests. The images were taken at 10 s intervals and stored in the computer during the test.

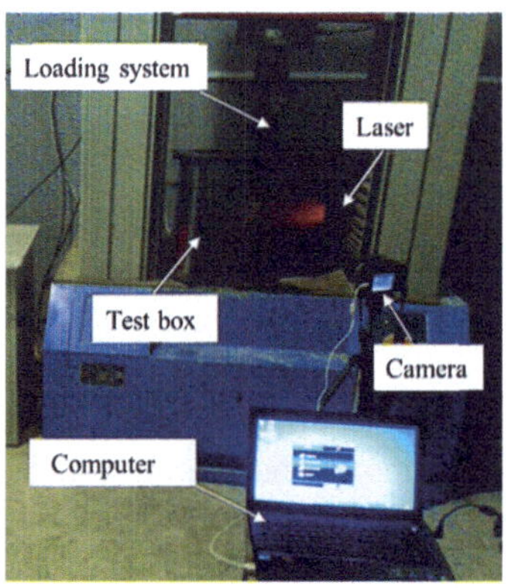

Fig. 1. Photo of the experimental set-up.

2.2 Materials

The transparent soil was prepared using the mixture of transparent particles and transparent fluid with a matching refractive index. High purity fused quartz consisting of 99.99% SiO_2, with specific gravity of 2.24, was used as the transparent material.

Its refractive index was 1.4585. The maximum and minimum values of dry unit weight were 1.48 and 1.08 g/cm^3. The average particle size (D_{50}) was 0.42 mm. The uniformity coefficient (C_u) and gradation coefficient (C_c) were 2.0 and 0.77, respectively. The colorless 3# and 15# mineral oil were mixed at the volume ratio of 1:3 to prepare the transparent fluid. The refractive index of the fluid was 1.4587 at 20 °C. The fused sand was poured into the oil mixture and stirred carefully to prepare the transparent soil samples. Direct shear tests on sand samples (dried, water saturated and oil saturated) at relative density of 85% showed that the friction of the sand is not much affected by the fluid. The friction angle of the sand is within the range of 43 ± 1°.

Polyamide net was used as the reinforcement in the tests. Its aperture size was 1.2 × 1.2 mm. The tensile strength of the net at 2% and 5% strain were 0.875 kN/m and 1.92 kN/m, respectively. The tensile strength of the net was 13.5 kN/m, with tensile modulus of 38.4 kN/m.

2.3 Testing Program

Figure 2 shows the schematic diagram of a typical model. The size of the transparent model bed used in the tests was 500 mm (width) × 240 mm (height), which ensures that the boundary effect could be neglected in this study according to Michalowski and Shi [10].

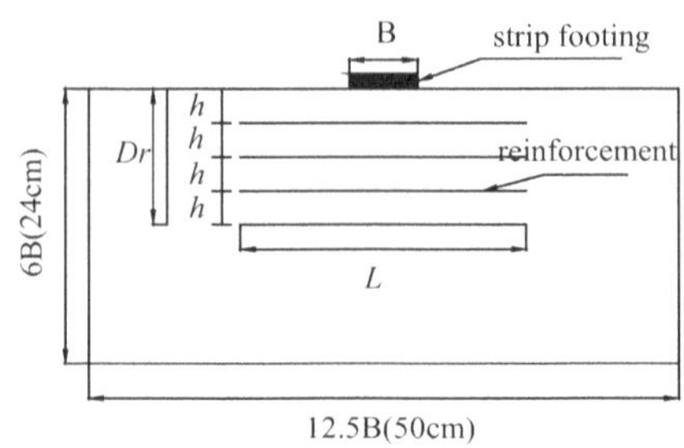

Fig. 2. Schematic diagram of model reinforced soil foundation.

To prepare the model foundations, the transparent fluid was poured into the test box. The dry fused quartz samples were dropped into the test box using a spoon, and stirred using a glass rod to eliminate air bubbles. The samples were compacted in layers using a shaking table to the desired relative density of 85%. On reaching the prescribed depth, the soil surface was leveled and a layer of reinforcement was laid on the transparent soil. The steps were repeated until the ground reaches the designed level. Table 1 presents the scheme of model tests.

Table 1. List of laboratory model tests.

Test no.	Reinforcement spacing (h/B)	Reinforcement length (L/B)	Reinforcement depth (D_r/B)
Unreinforced	–	–	–
1	0.25	1	1
2		3	0.5
3		3	1
4		3	1.5
5		7	1
6		7	1.5
7	0.5	3	1
8		3	2
9		5	1
10		7	2

3 Failure of the Foundations

3.1 Foundation on Soils with Reinforcements of 0.5B Spacing

Figures 3, 4 and 5 show the failure of the foundations on soils with reinforcements of 0.5B spacing. For soils with two layers of reinforcements, both layers of reinforcements fractured at the end of the test regardless of the reinforcement length. However, for the foundation on the shorter reinforcements, failure initiated in the top layer first and then migrated to the second layer, as shown in Fig. 3. For the longer reinforcements, the two layers of reinforcement failed almost simultaneously (see Fig. 4a). This may be because, the shorter reinforcements cannot provide enough confinement to the soils on the side of the foundation. So failure zone can be easily extended laterally and vertically, which resulted in the sequential failure of the two layers of reinforcements. But for the longer reinforcements, the movement of the side soils can be effectively restrained, thus the stress can be transferred effectively to the second layer of reinforcement. This resulted in the simultaneous failure of the reinforcements.

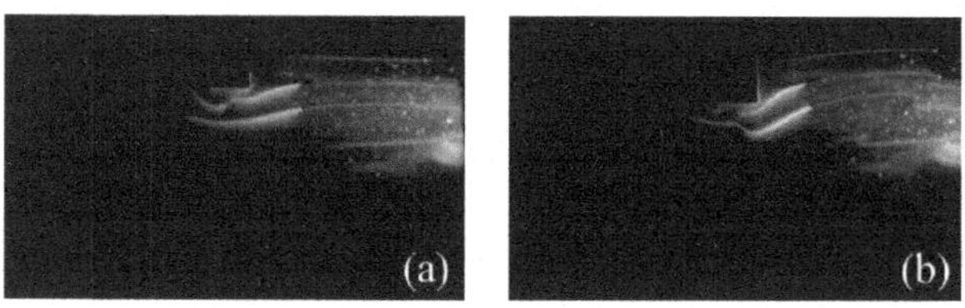

Fig. 3. Failure of the foundation with two layers of reinforcements of 0.5B spacing and 3B long at: (a) fracture of the top layer; (b) end of the test.

For soils with four layers of reinforcement, only the top layer of reinforcement failed for reinforcements with the length of 3B. For longer reinforcements, e.g. 7B, the

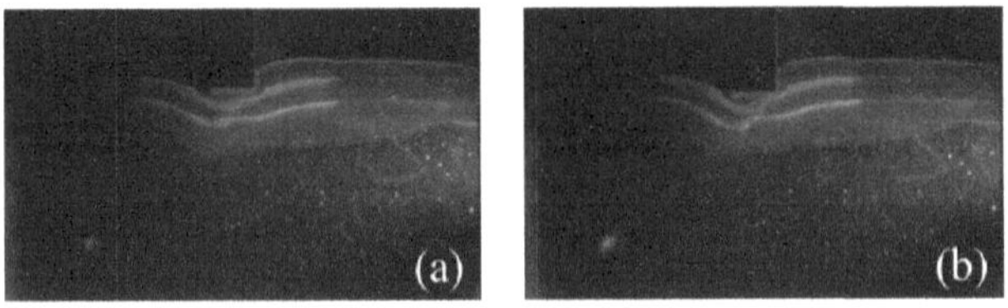

Fig. 4. Failure of the foundation with two layers of reinforcements of 0.5B spacing and 5B long at: (a) failure of the reinforcements; (b) end of the test.

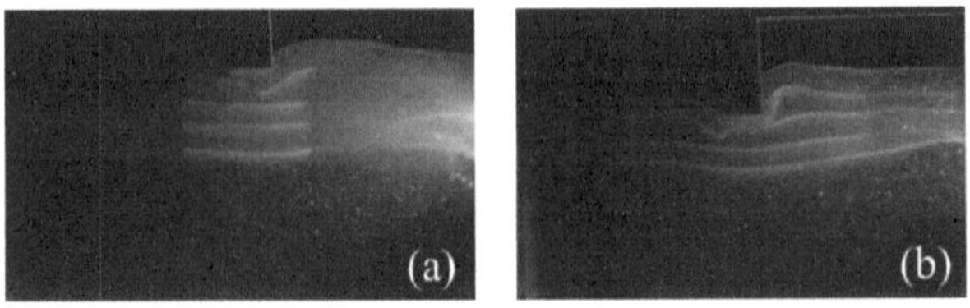

Fig. 5. Failure of the foundation with four layers of reinforcements at end of the tests for: (a) 0.5B spacing and 3B long; (b) 0.5B spacing and 7B long.

second layer of reinforcement fractured at the end of test. This may be because, for 4 layers of reinforcements, the failure stress of soil at greater depth was improved by including more layers of reinforcements, so is the stiffness of the reinforced soil. But for shorter reinforcements, e.g. 3B, due to limit effect of lateral reinforcement, large lateral displacement has been observed in the soil after the fracture of the top layer reinforcement. The second layer of reinforcement may have not been fully activated in providing support to the foundation (see Fig. 5a). For longer reinforcements, e.g. 7B, ground heave or soil shear has been effectively resisted by the extra length of reinforcements. So the second layer of reinforcement has been fully activated to provide extra support to the foundation (see Fig. 5b).

3.2 Foundation on Soils with Reinforcements of 0.25B Spacing

Figures 6, 7 and 8 show the failure of the foundations on soils with reinforcements of 0.25B spacing. For the soils with two layers of reinforcement, the reinforcements did not fracture at the end of the test (see Fig. 6a). This indicates that the reinforcement depth is too shallow and the effect of reinforcement is minimal. For the shorter reinforcement, e.g. 1B (see Fig. 6b), the reinforcements increased the influence depth of the foundation. The reinforcements and foundation formed a "deeper" embedded foundation, which resulted in a deeper and larger shear failure surface in soils.

Figure 7 shows the failure of the foundations with four layers of reinforcements of 0.25B spacing and 3B long. It can be observed that two bottom layers of reinforcements ruptured sequentially from the bottom to up. Thereafter all of the reinforcements failed gradually and a shear failure surface was formed in the end. Figure 8 presents the failure of the foundations with different layers of reinforcements and reinforcement lengths. The failure mechanisms of these foundations are similar to that of the

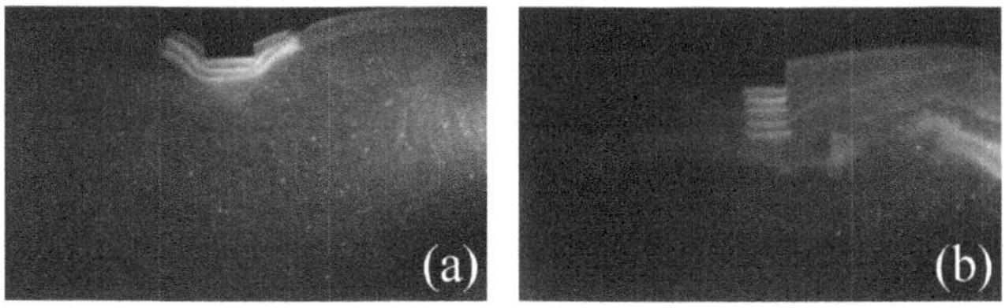

Fig. 6. Failure of the foundation at the end of the tests with: (a) two layers of reinforcements of 0.25B spacing and 3B long; (b) four layers of reinforcements of 0.25B spacing and 1B long.

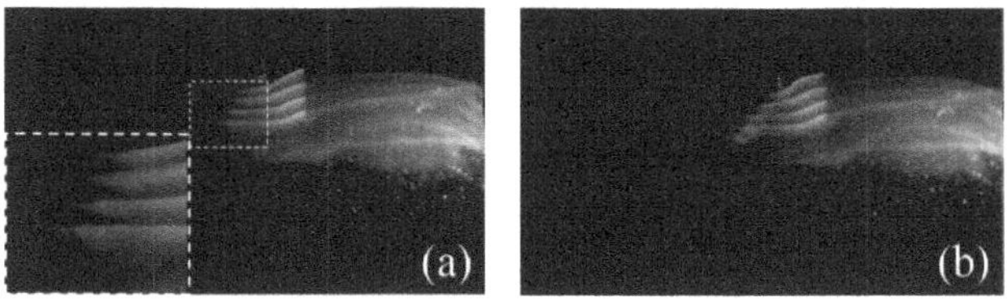

Fig. 7. Failure of the foundation with four layers of reinforcements of 0.25B spacing and 3B long at: (a) failure of two bottom layers of reinforcements; (b) end of the test.

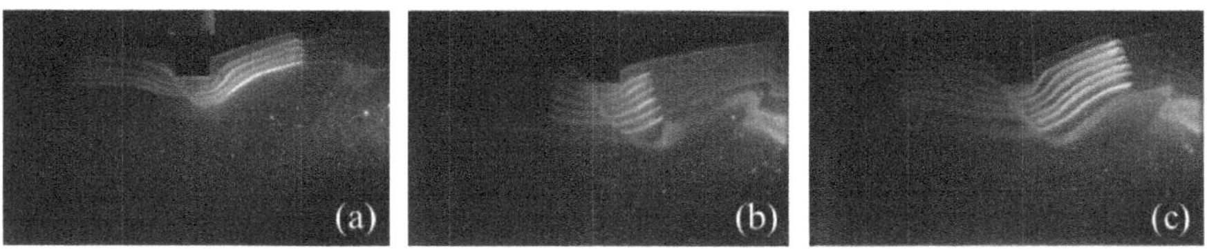

Fig. 8. Failure of the foundation with: (a) four layers of reinforcements of 0.25B spacing and 7B long; (b) six layers of reinforcements of 0.25B spacing and 3B long; (c) six layers of reinforcements of 0.25B spacing and 7B long at the end of the tests.

foundation shown in Fig. 7. This may be due to the fact that with denser reinforcements, the lateral displacement of soils in reinforced zone is restrained effectively, which causes that horizontal displacement first accumulates in the unreinforced zone near the lower right and left corner of the reinforced zone during loading, and then extends upward to reinforced zone, which results in the fracture of layers of the reinforcements successively from bottom to top and a shear failure surface in soils.

4 Load Settlement Behavior

To express and compare the results from reinforced and unreinforced soils, a dimensionless factor called bearing capacity improvement (*BCI*) is used to represent the performance improvement due to the provision of reinforcement. This factor is defined as the ratio of bearing pressure of reinforced soil or unreinforced soil at a given settlement (Q_R) to ultimate bearing pressure of unreinforced soil (Q_{ult}). The settlement

(S) is also expressed in dimensionless form in terms of the foundation width (B) as S/B (%). The relationships between bearing capacity improvement and foundation settlement for different reinforcement spacing are presented in Fig. 9a and b.

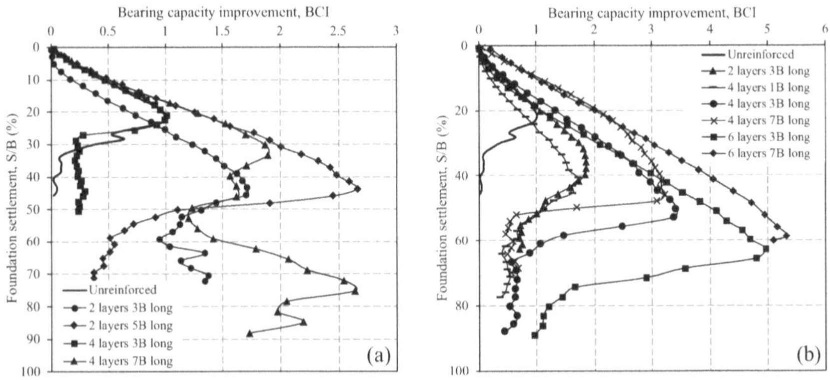

Fig. 9. Variation of bearing capacity improvement (BCI) with foundation settlement (S/B) for the reinforcements of: (a) 0.5B spacing; (b) 0.25B spacing.

Figure 9a shows the variations of bearing capacity improvement with foundation settlement for the reinforcements of 0.5B spacing. It shows that with two layers of reinforcements, the performance improvement shows slightly different behaviors for different reinforcement lengths. For the reinforcement with 3B length, the curve peaks and then drops, and then peaks again. The first peak is related to the failure of the first layer of the reinforcement, follows by the involvement of soil in between the reinforcement, and then the failure of the second layer of reinforcement. For the 5B long reinforcement, there is only one peak. This agrees with the simultaneous failure of the reinforcements. The results also show that the reinforcement greatly improves the bearing capacity of the foundations, except the ones with four layers of reinforcement with 3B long. This may be because, only one layer of the reinforcement contributes to the shear resistance, and the inclusion of the second layer of reinforcement at such a larger spacing may reduce the size of the failure zone, which results in little increment in bearing capacity.

Figure 9b shows the variations of bearing capacity improvement with foundation settlement for the reinforcements of 0.25B spacing. It can be seen that the behavior of the foundations on 2 layers of reinforcements of 3B long and 4 layers of reinforcement with 1B long is quite similar to that of foundation on unreinforced soils. The bearing pressure increases with settlement, and peaks after certain displacement, then followed by slowly decrease of bearing capacity. This is because the reinforcement only changed the shape or size of the failure zone, but did not contribute much to the total resistance. For soils reinforced with 4 layers and 6 layers of reinforcement of 3B and 7B length, the failure mechanism is quite different to that of the foundation on unreinforced soils. For these foundations, the pressure increases with displacement, and peaks, then followed by a sudden drop of bearing loading, which indicates the fractures of the reinforcements in all layers extend from bottom layer to top layer within a relatively short period. Unlike the foundations on 4 layers of reinforcements with 0.5B spacing and 7B length, the curves of the foundations on reinforcements of 0.25B only peak once, and decrease suddenly, which is almost like a brittle failure. Figure 9b also shows that, for same number of reinforcement layers, the reinforcement length has little effect on the bearing capacity of the foundations. This agrees with the finding by Huang [11] that the bearing capacity of foundations does not increase much when reinforcement length increases from 3B to 6B.

5 Conclusions

A series of laboratory model tests have been conducted to investigate the effect of reinforcement spacing, depth and length on the failure mechanism of foundations using transparent soils. The main conclusions are as follows:

1. For foundations on reinforced soils with 0.5B reinforcement spacing, ruptures of reinforcements extend from top layer to deeper layers, but only to the second layer the deepest. The failure of the reinforcements results in local peaks in the variation curves of bearing capacity improvement versus foundation settlement, with each peak related to the failure of one layer of reinforcement.
2. For foundations on reinforced soils with 0.25B reinforcement spacing, fracture of reinforcements occurs from bottom layer to top layer. The variations of bearing capacity improvement with foundation settlement show brittle failure mechanism of foundations, which indicating the failure of the reinforcements followed by the sudden reduction of resistance.
3. For foundations on reinforced soils with 0.25B reinforcement spacing, the bearing capacity is not greatly affected by reinforcement length once the reinforcement length is longer than 3B.

Acknowledgement. The support from the National Natural Science Foundation of China under grants Nos. 41772289 and 41572266 is gratefully acknowledged.

References

1. Huang CC, Tatsuoka F (1990) Bearing capacity of reinforced horizontal sandy ground. Geotext Geomembr 9(1):51–82
2. Latha GM, Somwanshi A (2009) Bearing capacity of square footings on geosynthetic reinforced sand. Geotext Geomembr 27(4):281–294
3. Omar MT, Das BM, Puri VK, Yen SC (1993) Ultimate bearing capacity of shallow foundations on sand with geogrid reinforcement. Can Geotech J 30(3):545–549
4. Shin EC, Das BM, Lee ES, Atalar C (2002) Bearing capacity of strip foundation on geogrid-reinforced sand. Geotech Geol Eng 20(2):169–180
5. Tafreshi SNM, Dawson AR (2010) Comparison of bearing capacity of a strip footing on sand with geocell and with planar forms of geotextile reinforcement. Geotext Geomembr 28:72–84

6. Ni Q, Hird CC, Guymer I (2010) Physical modelling of pile penetration in clay using transparent soil and particle image velocimetry. Geotechnique 60(2):121–132
7. Bathurst RJ, Ezzein FM (2015) Geogrid and soil displacement observations during pullout using a transparent granular soil. Geotech Test J 38(5):673–685
8. Kong GQ, Cao ZH, Zhou H, Sun XJ (2015) Analysis of piles under oblique pullout load using transparent-soil models. Geotech Test J 38(5):725–738
9. De Cataldo D, Chen K, Airey D (2017) Three-dimensional deformations in transparent soil using fluorescent markers. Int J Phys Model Geotech 17:1–13
10. Michalowski RL, Shi L (2003) Deformation patterns of reinforced foundation sand at failure. J Geotech Geoenviron Eng 129(5):439–449
11. Huang CC (2017) Model tests on the bearing capacity of reinforced saturated sand ground. Geosynth Int 24(2):114–124

Effect of Freeze-Thaw Cycling Concurrent with MSW Landfill Leachate on the Hydraulic Conductivity of Geosynthetic Clay Liner

Bao Wang[1(✉)], Bin Chen[2], Dou Tongtong[1], and Lou Wang[3]

[1] Department of Environmental and Municipal Engineering, Xi'an University of Architecture and Technology, Xi'an 710055, Shaanxi, China
wangbao@xauat.edu.cn

[2] Department of Civil Engineering, Xi'an University of Architecture and Technology, Xi'an 710055, Shaanxi, China

[3] Xi'an Branch of China Urban Construction Design and Research Institute Co., Ltd., Xi'an 710065, Shaanxi, China

Abstract. A study was conducted to determine the effect of freeze-thaw cycling concurrent with municipal solid waste (MSW) landfill leachate on the hydraulic conductivity of a geosynthetic clay liner (GCL). The GCL was hydrated with MSW landfill leachate and subsequently subjected to 5, 10, 15 and 20 freeze-thaw cycles. After the prescribed numbers of freeze-thaw cycles was met, the hydraulic conductivity of the GCL was tested in a flexible wall permeameter. Hydraulic conductivity of GCL experienced freeze-thaw cycles was slightly higher than those without frozen, with the hydraulic conductivity increased with increasing of number of freeze-thaw cycles. The increase in the hydraulic conductivity of GCL could be a result of the contraction of double-layer, resulting from the increases in the concentrations of cations in the pore water (i.e., MSW landfill leachate) induced by freeze.

Keywords: Geosynthetic clay liners · Freeze-thaw · MSW landfill leachate Hydraulic conductivity

1 Introduction

Geosynthetic clay liners (GCLs) are thin manufactured hydraulic product that consisted of a thin layer of bentonite sandwiched between two geotextiles or mixed with an adhesive and bonded to a geomembrane. Due to their low hydraulic conductivity to water, strong attenuation capacity, ease of installation, perceived resistance to environmental distress, and relatively low cost, GCLs are now widely used as an alternative to traditional compacted clay liners (CCLs) as the composite liner for municipal solid waste (MSW) landfills. When there is no geomembrane, the hydraulic performance of a GCL is controlled by the hydraulic conductivity of the sodium-bentonite, which is composed primarily of the mineral montmorillonite. The large specific surface, strong negative charge and interlayer separation result in bentonite maintain low hydraulic conductivity [1, 2]. However, these same factors also make bentonite vulnerable to chemical interactions that can cause increase in hydraulic conductivity.

L. Zhan et al. (Eds.): ICEG 2018, ESE, pp. 658–663, 2019.
https://doi.org/10.1007/978-981-13-2224-2_81

Several studies have addressed how the hydraulic conductivity of GCLs is affected by permeation with actual MSW landfill leachate [3, 4]. These studies have shown that permeation with real MSW landfill leachate appears to have no detrimental effect on the hydraulic conductivity of GCLs. These results are encouraging and provide strong evidence that GCL can perform well in the applications of MSW landfill. However, up to the present time, there is no available information about the effect of free-thaw cycling concurrent with MSW landfill leachate on hydraulic conductivity of GCLs.

The scenario of freeze-thaw cycling concurrent with MSW landfill leachate permeation generally occurs in the small landfill constructed in developing countries, where, due to poor management and terrible design for the leachate drainage system, the leachate generated in this kind of landfills can not be collected and discharged out immediately, then built up and start going out of the working cell, and ultimately flow over the bottom systems (consist of GCLs and GM) in the unused cell. Because there is no enough MSW covered on, the GCLs that have contacted with MSW landfill leachate in the unused cell will be subjected to freeze-thaw cycling. The question raised is whether the freeze-thaw cycling coincident with MSW landfill leachate hydration has damage the hydraulic performance of the GCL.

The purpose of this paper is to investigate the combined effect of freeze-thaw cycling and MSW landfill leachate on the hydraulic conductivity of GCL.

2 Material

2.1 Geosynthetic Clay Liner (GCL)

One GCL product containing natural sodium bentonite was used in present study. The GCL consists of a thin layer of granular sodium bentonite held together by needle-punched fibres. The initial unconfined thickness of the air-dry GCL specimen ranged from 5.1 to 5.8 mm. The mass per unit area of air-dried bentonite in the GCL was 5.6 kg/m^2. X-ray diffraction showed that the bentonite component of the GCL contained 52% montmorillonite, 38% quartz, 3% boehmite, 3% cristobalite, 2% feldspar and 2% other minerals.

2.2 Permeant Liquids

The liquids used in this study consisted of distilled water (DW) and synthetic MSW landfill leachate. Properties of the MSW landfill leachate are summarized in Table 1. The leachate composition was based on a MSW landfill leachate used by Rauen and Benson [5]. The parameter RMD in Table 1 represents a ratio of the concentrations of monovalent and divalent cations in the permeant solution.

Table 1. Chemical composition of synthetic MSW landfill leachate

Ions concentration (mg/L)					pH	EC (μS/cm)	RMD
Na^+	K^+	Mg^{2+}	Ca^{2+}	Cl^-			
3000	763	538	4.5	5724	6.34	17.02	0.96

3 Methods

3.1 Hydration (or Hydraulic Conductivity Test)

GCL will be hydrated by the MSW landfill leachate overflowed above it. To simulate this hydration process, GCL specimen was permeated with MSW leachate using flexible wall permeameter. To simulate the extreme condition, the hydration process was conducted until the chemical equilibrium between GCL and the leachate was archived. The reason to use this method to hydrate GCL is that the hydraulic conductivity of the GCL to MSW landfill leachate and without frozen can be obtained during the hydration process.

GCL specimens were trimmed to a nominal diameter of 101 mm following the procedures described in ASTM D 6766. During the process, a small amount of DW or MSW landfill leachate was applied along the cutting edge of the GCL specimens in order to prevent loss of bentonite.

Hydration was conducted following the procedure described in ASTM D 6766. After setting on the pedestal of the flexible wall permeameter, the GCL specimen was prehydrated directly with MSW landfill leachate for 48 h under an effective stress of 30 kPa by closing the effluent port. The effective stress of 30 kPa was selected to simulate the stress on GCL in a MSW landfill cell with leachate collection system. Tests were conducted under a hydraulic gradient of 225. No backpressure was used in test so that effluent solution could be collected conveniently for chemical analysis in bag.

Hydration then was initiated by opening the effluent port and continued until the termination condition stipulated in ASTM D 6766 had been satisfied. After termination, the GCL specimen was carefully removed from flexible wall permeameter and sealed with several layers of plastic warp and placed in a plastic bag to prevent desiccation. A control test was conducted following the method described previously except using DW instead of MSW landfill leachate.

3.2 Freeze-Thaw Cycling

The freeze-thaw cycling test was conducted in general accordance with the method described in ASTM D 6035. GCL specimen was subjected to 0, 5, 10, 15 and 20 freeze-thaw cycles. After sealing in plastic warp and plastics, the GCL specimen was placed in a refrigerated and heating circulator. The freezing temperature and thawing temperature were set at –20 °C and 20 °C, respectively. In order to complete one cycle in 24 h period, a 12 h period was used for both freezing and thawing. Previous studies conducted by Makusa et al. [6] show that 12 h freezing time were sufficient to freeze GCL. The freezing and thawing cycle was replicated for 5 times for one set of treatment. After that, the GCL was removed from the refrigerated and heating circulator and was set in the flexible wall permeameter to test the hydraulic conductivity using the same method aforementioned in hydration process. Once the hydraulic conductivity was obtained, the specimen was replaced in the refrigerated and heating circulator and subjected to another 5 freezing and thawing cycles. The procedure was repeated until the hydraulic conductivity of the specimen was obtained after 20 freeze-thaw cycles.

4 Results and Discussion

The hydraulic conductivity versus pore volumes of flow (PVF) for the specimens before freezing is shown in Fig. 1 and is summarized in Table 2. The hydraulic conductivity for the specimen permeated with DW decreased quickly during the initial 0.2 PVF as the bentonite was hydrated. After that, the hydraulic conductivity levelled off and stabilized at 1.50×10^{-11} m/s. Permeation with MSW landfill leachate resulted in similar behaviour for GCL, except for the slightly higher equilibrium hydraulic conductivity (2.07×10^{-11} m/s). The slightly higher hydraulic conductivity to MSW landfill leachate was the result of contraction of double layer induced by the higher concentration of cations in the MSW landfill leachate (Table 1).

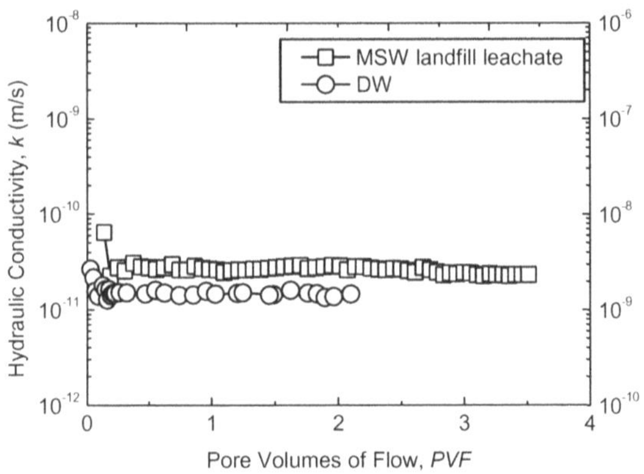

Fig. 1. Hydraulic conductivity versus pore volumes of flow for geosynthetic clay liner permeated with MSW landfill leachate or DW and without freeze-thaw cycles

Results of hydraulic conductivity test conducted on GCL specimens subjected to freeze and thaw cycling are summarized in Table 2 and Fig. 2. The data from other studies where GCLs were subjected to freeze-thaw cycling and permeated with tap water, deionized water or cover pore water under relatively low stress condition (14–20 kPa) are also shown for comparison in Fig. 2. As shown in Fig. 2, there is a trend of increasing hydraulic conductivity with increasing number of freeze-thaw cycles for the GCL tested in present study (i.e., hydrated with MSW landfill leachate). At the end of the test, the ratio of the hydraulic conductivity obtained after 20 freeze-thaw cycles relative to the hydraulic conductivity prior to freezing, or k_{20}/k_0, was of 2.14.

Table 2. Summary of hydraulic conductivities of GCL specimens before and after freeze-thaw cycles

Permeant liquid	k_0^a ($\times 10^{-11}$ m/s)	k_n^b ($\times 10^{-11}$ m/s)				k_{20}/k_0
		k_5	k_{10}	k_{15}	k_{20}	
MSW landfill leachate	2.07	3.52	4.43	5.58	7.54	2.14
DW	1.50	NA[c]	NA	NA	NA	NA

Note: a Hydraulic conductivity before freeze-thaw cycles. b Hydraulic conductivity after n freeze-thaw cycles. c Test are still being conducted.

The observed increase in the hydraulic conductivity of GCL should be a result of a contraction of double-layer because of increases in the concentrations of cations in the pore water (i.e., MSW landfill leachate) induced by freeze. During the freezing, dissolved compounds in the water phase are rejected into the temporarily unfrozen water [7]. This process increased the concentrations of cations in the unfrozen pore water and then resulted in decrease in thickness of double layers of bentonite. In addition, some cracks (or large pores) developed in the bentonite layer of the GCL due to the formation of ice lenses also was responsible for the increase in hydraulic conductivity [8], despite some of them could be closed by "thaw consolidation".

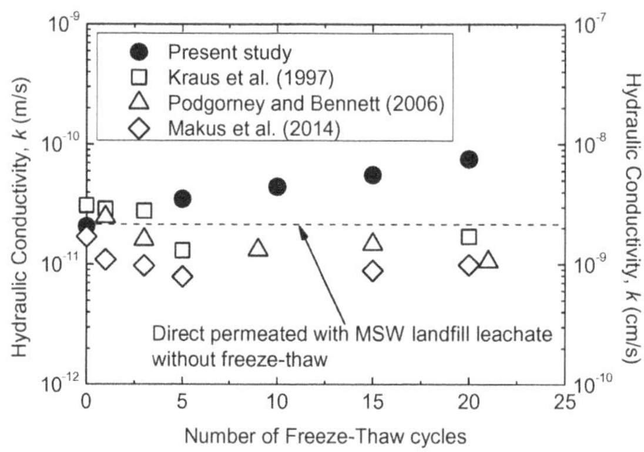

Fig. 2. Hydraulic conductivity versus freeze-thaw cycles for GCL permeated with MSW landfill leachate. GCLs from other studies were hydrated and permeated with Tap water, DI water or cover soil pore water.

The results presented here appeared to counter those obtained by Hewitt and Daniel [9], Kraus et al. [10], Podgorney and Bennett [11] and Makusa et al. [6], who found that freeze-thaw cycles did not produce any or appreciable detrimental effect on the hydraulic conductivity of GCLs. In their studies, GCL specimens were generally hydrated and permeated with dilute solution (i.e., deionized water, DW or local soil pore water). The test results obtained by Rowe et al. [12] on GCLs exhumed from the field at Brevoort Island also show that freeze-thaw cycling did not impact the hydraulic performance of GCL to water. The recovered GCLs experienced 3 years of natural freeze-thaw cycling and chemical interaction with local soil pore water. The difference in effect of freeze-thaw cycling between present study and previous investigations could be explained by the significantly low cations concentration in the pore water within the GCL specimens used in previous investigations. Due to the initial concentrations of the cations are significantly low, even though the freeze increases them to higher values, which still can not be high enough to contract the double-layer of the clay particles. So, there is no increase in hydraulic conductivity can be observed in previous studies.

5 Conclusions

The objective of this study was to assess the combined effect of freeze-thaw cycling and hydration of MSW landfill leachate on the hydraulic conductivity of GCL. Based on the finding of this study, the following conclusions can be made:

- The hydraulic conductivity for the GCL was 2.07×10^{-11} m/s based on the permeation of MSW landfill leachate before freeze-thaw cycling.
- The combined effect of freeze-thaw cycling and hydration with MSW landfill leachate resulted in minor increase in the hydraulic conductivity of GCL. The GCL exposed to as many as 20 freeze-thaw cycles had a hydraulic conductivity 2.7 times higher than that of not exposed to freeze-thaw.

Acknowledgment. Financial support for this study was provided by National Natural Science Foundation of China (NSFC), under Grant No. 41602291.

References

1. Shackelford CD, Benson CH, Takeshi K et al (2000) Evaluating the hydraulic conductivity of GCLs permeated with non-standard liquids. Geotext Geomembr 18(2–4):133–161
2. Young JH, Takeshi K, Benson CH et al (2001) Hydraulic conductivity and swelling of nonprehydrated GCLs permeated with single-species salt solutions. J Geotech Geoenviron Eng 127(7):557–567
3. Rubl JL, Daniel DE (1997) Geosynthetic clay liners permeated with chemical solutions and leachates. J Geotech Geoenviron Eng 123(4):369–381
4. Bradshaw SL, Benson CH, Rauen TL (2016) Hydraulic conductivity of geosynthetic clay liners to recirculated municipal solid waste leachates. J Geotech Geoenviron Eng 142 (2):4015074
5. Rauen TL, Benson CH (2008) Hydraulic conductivity of a geosynthetic clay liner permeated with leachate from a landfill with leachate recirculation. In: GeoAmericas Conference Proceedings, pp 76–83
6. Makusa GP, Bradshaw SL, Erin B et al (2014) Freeze–thaw cycling concurrent with cation exchange and the hydraulic conductivity of geosynthetic clay liners. Can Geotech J 51 (6):591–598
7. Sorab P, Yavuz CM (1991) Solute rejection in freezing soils. Water Resour Res 27(1):99–108
8. Kerry RR, Toshifumi M, Bathurst RJ (2006) Compatibility with Jet A-1 of a GCL subjected to freeze-thaw cycles. J Geotech Geoenviron Eng 132(12):1526–1537
9. Hewitt RD, Daniel DE (1997) Hydraulic conductivity of geosynthetic clay liners after freeze-thaw. J Geotech Geoenviron Eng 123(4):305–313
10. Kraus JF, Benson CH, Erickson AE et al (1997) Freeze-thaw cycling and hydraulic conductivity of bentonitic barriers. J Geotech Geoenviron Eng 123(3):229–238
11. Podgorney RK, Bennett JE (2006) Evaluating the long-term performance of geosynthetic clay liners exposed to freeze-thaw. J Geotech Geoenviron 132(2):265–268
12. Rowe RK, Mukunoki T, Bathurst RJ (2008) Hydraulic conductivity to Jet-A1 of GCLs after up to 100 freeze–thaw cycles. Géotechnique 58(6):503–511

Hydraulic Conductivity of Bentonite-Polymer Geosynthetic Clay Liners to Coal Combustion Product Leachates

Jian-Nan Chen[1(✉)], Kuo Tian[2], and Craig H. Benson[3]

[1] Southwest Jiaotong University, Chengdu 611756, China
jchen@swjtu.edu.cn
[2] George Mason University, Fairfax, VA 22030, USA
ktian@gmu.edu
[3] University of Virginia, Charlottesville, VA 22904, USA
chbenson@virginia.edu

Abstract. Experiments were conducted to evaluate the compatibility of bentonite-polymer (B-P) GCLs to Coal Combustion Product (CCP) leachates. The B-P GCLs were manufactured by dry-mixing polymer and sodium bentonite (NaB). Six synthetic leachates were selected based on database of coal ash leachates. Hydration characteristics (i.e., free swell index) and hydraulic conductivity of B-P GCLs were investigated using deionized water and synthetic leachates. The hydraulic conductivity of B-P GCL with low polymer loading (1.2%) is strongly affected by the ionic strength of leachates and controlled by swelling of bentonite. However, the hydraulic conductivity of B-P GCL with high polymer loading ($\geq 5.1\%$) is controlled primarily by hydrogel blockage in the pore space. Polymer hydrogel forms reticular structure and bonds free water when hydrated with DI water. However, under 20 mM CaCl2, the reticular structure of polymer hydrogel is inhibited by self-entanglement due to divalent cations and high ionic strength.

Keywords: Bentonite · Polymer · Coal combustion product · Leachate Hydraulic conductivity · Swell index · Hydration

1 Introduction

Coal combustion products (CCPs) are residuals from coal-fired boilers that are disposed in waste containment facilities when they cannot be used beneficially in other applications [1]. In United States, the landfill of coal ash is subject to the requirements of Subtitle D in the Resource Protection and Recycling Act (RCRA), where the disposal method of coal ash is equivalent to that of municipal solid waste (MSW) [2]. Geosynthetic clay liners (GCLs) are used as hydraulic barriers in disposal facilities for CCPs. GCLs consist of a thin layer of bentonite clay (5–10 mm) sandwiched between two geotextiles bonded by stitching or needle punching. In some cases, a geomembrane is laminated to the GCL. The effectiveness of a GCL as a barrier to fluid flow is controlled by the hydraulic conductivity of the bentonite, which is comprised primarily of the clay mineral montmorillonite [3, 4]. When contacted with water, CCPs can

L. Zhan et al. (Eds.): ICEG 2018, ESE, pp. 664–671, 2019.
https://doi.org/10.1007/978-981-13-2224-2_82

release a variety of cations and anions into solution, which could negatively impact the hydraulic conductivity of GCLs [5].

Jo et al. [4] and Xu et al. [6] studied the effects of ionic strength, ionic valence and pH on the swelling and hydraulic conductivity of sodium bentonite (NaB) GCL using single-species solutions. The results showed the hydraulic conductivity of GCL increased with the increase of ionic strength and ionic valence (e.g. calcium, magnesium, aluminum) or under extreme pH (pH > 13 or pH < 2). Kolstad et al. [7] studied the effects of multi-species solutions on the swelling and hydraulic conductivity of Na-bentonite GCLs. The relative content of monovalent cations and polyvalent cations in leachate (relative abundance of monovalent and polyvalent cation, referred to as "RMD") is an important factor influencing the swelling and hydraulic conductivity of NaB GCLs, and can be expressed as follow:

$$RMD = M_M/\sqrt{M_D} \tag{1.1}$$

where Mm is the total molar concentration of monovalent cations; M_D is the total molar concentration of the polyvalent cations. Therefore, low RMD value indicates that the solution has relatively higher concentration of polyvalent cations.

Since the chemical composition of leachate is much more complex than the aforementioned solutions, many studies have investigated the compatibility of GCL to on-site leachate or simulated leachate [8–10]. Chen et al. [5] tested the compatibility of NaB GCL to CCP leachate. The study found that when the ionic strength of leachate increased from 39.5 to 755 mM, the hydraulic conductivity of GCL increased by four orders of magnitude (from 10^{-10} to 10^{-6} m/s).

Considering the incompatibility between bentonite and leachate, bentonite with improved chemical resistivity by adding polymers has been studied and investigated under harsh chemical conditions [11–14]. Katsumi et al. [14] evaluated the chemical compatibility of multi-swellable bentonite (MSB) and dense-prehydrated GCL (DPH-GCL) to NaCl and $CaCl_2$ solutions. The results show that MSB and DPH-GCL can maintain osmotic swell and lower hydraulic conductivity than that of NaB GCL in solution with ion concentration $\leq$ 1.0 M. Scalia [13] studied the hydraulic conductivity of bentonite-polymer composite (BPC) under various concentrations of CaC12. The hydraulic conductivity of BPC is up to four orders of magnitude lower than that of NaB GCL to the same $CaCl_2$ solution. However, there are still uncertainties in the use of bentonite-polymer (B-P) GCLs. Shackelford et al. [15] studied the compatibility of a B-P GCL, manufactured by dry-mixing polymer and sodium bentonite, to simulated leachates. The results show that the hydraulic conductivity of B-P GCL is 9.1 to 21 times higher than that of NaB GCL. Therefore, chemical compatibility study is still necessary before using B-P GCL as hydraulic barrier.

In this study, the chemical compatibility of B-P GCLs to CCP leachates was studied. The hydration behavior of B-P GCLs to CCP leachates were investigated by index tests (i.e., free swell tests) and micro-scale image analysis to understand the mechanism controlling the hydraulic conductivity of B-P GCLs.

2 Materials and Methods

2.1 Geosynthetic Clay Liners

Two types of B-P GCL (i.e. CP1.2 and CP5.1, numbers refer to the mass percentage of polymers) and one type of NaB GCL were used in this study. All three GCLs are manufactured by the same company, and widely available in the North America. The B-P GCLs are manufactured by dry mixing sodium bentonite (same bentonite as NaB GCL) and same polymer between needle stitched woven (mass per unit area = 109 g/m^2) and non-woven geotextiles (mass per unit area = 204 g/m^2). Table 1 summarizes the unit mass, initial water content, initial thickness, bound cations and cation exchange capacity (CEC) of NaB and B-P GCLs. All three GCLs contain granular bentonite with $D50 > 0.7$ mm. Quantitative X-ray diffraction analysis indicates that the GCLs mainly contain montmorillonite (79% to 86%), with a small amount of quartz, plagioclase, feldspar, illite, mica and calcite.

Table 1. Physical and chemical properties of GCLs used in current study.

GCL		NaB	CP1.2	CP5.1
Mass per Unit Area (kg/m^2)		3.7	3.6	3.6
Initial Thickness (mm)		7.9–9.8	7.5–8.8	7.1–8.3
Initial Water Content (%)		10.1	11.2	4.4
Montmorillonite Content (%)		84	86	79
Bound Cations (Mole Fraction)	Na	0.47	0.53	0.14
	K	0.04	0.03	0.01
	Ca	0.30	0.25	0.13
	Mg	0.13	0.11	0.04
CEC (cmol + /kg)		71.3	71.4	89.6
Loss on Ignition (%)		1.6 ± 0.01	2.8 ± 0.02	6.6 ± 1.9
Polymer Content (%)		0	1.2	5.1

The CEC and bound cations [i.e., calcium (Ca), sodium (Na), magnesium (Mg), potassium (K)] of NaB and B-P GCLs, are measured per ASTM D7503. Average values of five tests are reported in Table 1. The loss on ignition (LOI) of B-P GCLs is measured in accordance with ASTM D7348, which is used to measure the polymer loading in B-P GCLs. The specific method for calculating the polymer contents is described in Scalia [16].

2.2 Synthetic CCP Leachate

Synthetic CCP leachates were created after analyzing the ionic strength and RMD of leachate data from Electric Power Research Institute (EPRI) [1, 17]. Five CCP leachates were selected for testing that represent different conditions encountered in CCP disposal facilities (solid symbols in Fig. 1): (1) typical CCP leachate (I = 39.5 mM and RMD = 0.16 M$^{1/2}$), (2) strongly divalent ("low RMD") cation ash leachate

(I = 48.0 mM, RMD = 0.014 $M^{1/2}$), (3) flue gas desulfurization (FGD) residue leachate (I = 96.8 mM, RMD = 0.39 $M^{1/2}$), (4) high ionic strength ash leachate (I = 177 mM, RMD = 1.0 $M^{1/2}$), and (5) trona ash leachate (I = 755 mM, RMD = 4.5 $M^{1/2}$). Synthetic leachates representing these conditions were created by dissolving reagent-grade $CaSO_4$, Na_2SO_4, $MgSO_4$, K_2SO_4, NaCl, and $CaCl_2$ in deionized (DI) water [5]. One additional site-specific leachate (CCP1: I = 975 mM, RMD = 1.86 $M^{1/2}$) was also used in the study, which represents the chemical composition of CCP leachate in mid-west US.

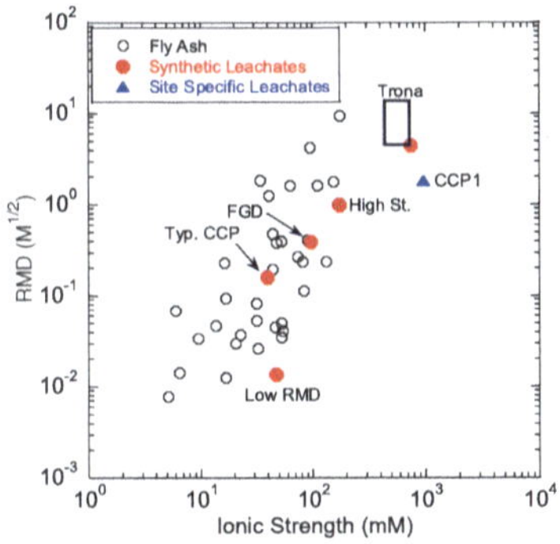

Fig. 1. RMD vs. ionic strength of CCP leachates.

2.3 Hydraulic Conductivity and Swell Index

Hydraulic conductivity tests were conducted on the 100-mm-diameter GCL specimens using flexible-wall permeameters in accordance with ASTM D5084 and ASTM D6766. The falling headwater-constant tailwater method was used. A 20 kPa effective stress was applied initially, and the average hydraulic gradient was 190. GCL specimens were prepared in accordance with the procedure in Chen et al. [5] and hydrated with leachate in the permeameter for 48 h with the effluent valve closed. Permeation followed immediately afterwards and was conducted until the hydraulic conductivity was steady, the incremental inflow and outflow were equal, and the pH, major cation concentrations (Na, Ca, Mg, K), and electrical conductivity of the effluent were within 10% of those of the influent.

Swell index (SI) tests were conducted on Na-bentonite with each leachate and DI water (control) in accordance with ASTM D5890. Bentonite from the GCL was ground to fine powder by a mortar and pestle until at least 65% passed the No. 200 US standard sieve. The powdered sample was oven-dried at 105 °C for 24 h prior to testing. A 100-mL graduated cylinder was filled to 90 mL with testing liquid, and 2 g of ground bentonite was added to the graduated cylinder in 0.1 g increments. The graduated cylinder was then filled to 100 mL with leachate. Bentonite volumes were recorded as the SI after 24 h of hydration.

3 Results and Discussion

3.1 Hydraulic Conductivity of B-P GCL to CCP Leachates

Hydraulic conductivities of the GCLs to the CCP leachates are shown in Fig. 2a as a function of leachate ionic strength. The hydraulic conductivity of CP1.2 and NaB GCLs increases up to 5 order of magnitude (10^{-11} to 10^{-6} m/s) with the increase of the ionic strength (from 0.001 to 755 mM). Although the hydraulic conductivity is dominated by the influence of ionic strength, the effect of RMD of the leachate is evident at lower ionic strengths (< 50 mM). The hydraulic conductivity of NaB and CP1.2 GCLs to Low RMD leachate are 1.3 to 200 times higher than that to Typ.CCP leachate, which has identical ionic strength but lower ratio of polyvalent cations. Compared to NaB and CP1.2 GCLs, CP5.1 GCL has relatively better chemical compatibility, and maintains low hydraulic conductivity ($< 2 \times 10^{11}$ m/s) when ionic strength $\leq$ 755 mM. However, as ionic strength reaches 975 mM, the hydraulic conductivity of CP5.1 GCL increases 3 orders of magnitude to 10^{-8} m/s, which is also 4 orders of magnitude higher than that to DI water.

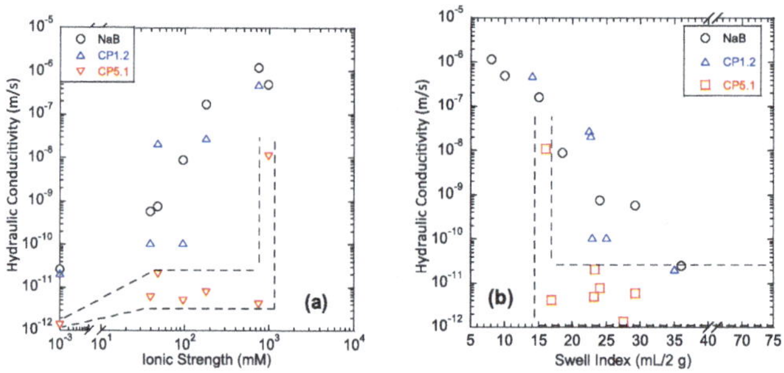

Fig. 2. Hydraulic conductivity of GCLs as a function of (a) ionic strength of leachate and (b) swell index of bentonite.

3.2 Free Swell and Hydraulic Conductivity

Hydraulic conductivity of the GCLs as a function of SI is shown in Fig. 2b (average swell index from two replicate tests are reported). The hydraulic conductivity and swelling index of NaB and CP1.2 GCLs showed a strong positive correlation, which indicates that swelling of bentonite is the main controlling mechanism of hydraulic conductivity. However, no obvious relationship is evident between hydraulic conductivity of CP5.1 and free swell. Although the swell index of bentonite in CP5.1 GCL decreases with the increase of ionic strength, low hydraulic conductivity (10^{-11} m/s) is still observed even at low swell state (17 mL/2 g), which is three orders of magnitude lower than that of NaB GCL (10^{-8} m/s). This phenomenon indicates as the increase of polymer content, the controlling mechanism of hydraulic conductivity changes from swell control to a more complicated controlling mechanism.

3.3 Mechanism Controlling the Hydraulic Conductivity of B-P GCLs

Testing results show that the mechanism controlling the hydraulic conductivity of B-P GCLs changes with the increase of polymer content. The common polymers used in B-P GCLs are usually hydrogels, such as sodium carboxymethyl cellulose (Na-carboxymethylcellulose) and sodium polyacrylate (Na-polyacrylate) [11, 13]. Hydrogel is a polymer network structure formed by cross-linking of hydrophilic functional groups, which can bond free water to produce swelling. The microstructure of bentonite-polymer in CP5.1 GCL hydrated in DI water and 20 mM $CaCl_2$ was analysed by scanning electron microscope (SEM) (Fig. 3).

Fig. 3. Micro structure (by SEM) of bentonite-polymer composite hydrated in (a) DI water and 20 mM $CaCl_2$

When hydrated by DI water, the polymer forms a densely reticular structure, and effectively fills the pore space of bentonite. Since the polymer has the very strong hydrophilicity, free water is bonded by the function group forming high viscosity polymer hydrogel. The hydrogel further restricts the flow of free water, and reduces the hydraulic conductivity of B-P GCL by blocking the pore space. However, when hydrated by 20 mM $CaCl_2$, the polymer hydrogel presents a loosely reticular structure. The reticular structure, on the contrary, has large porosity, and not able to fill the pore space within the bentonite. This phenomenon is likely due to the bridging effect by the divalent cations in the solution [18]. Since Na-carboxymethylcellulose and Na-polyacrylate are anionic polymers, the presence of divalent cations may lead the combination (bridge) of the function groups on the same polymer main chain by ionic bond. The combination further generates self-entanglement and reduces the cross-linking between polymer main chains [19]. Additionally, since polymer and ions in the solution are both hydrophilic, the competition on water bonding exists and further hinders the formation of the reticular structure.

4 Conclusions

The hydraulic conductivity of GCL can be effectively reduced by adding polymers to the traditional sodium bentonite. As the polymer loading increases (from 2.1% to 5.1%), the hydraulic conductivity of GCL reduces by up to 4 orders of magnitude. Although leachate chemistry still inhibits the swelling of the original sodium bentonite, the bentonite-polymer GCL can still maintain the low hydraulic conductivity. The polymer hydrogel fills the pore space of the sodium bentonite, and the mechanism controlling the hydraulic conductivity of GCLs changes from swelling control mechanism to the blocking control mechanism.

However, leachates with high ionic (975 mM in this study) could restrict the formation of reticular structure in polymer hydrogel. The ions in the leachates induce self-entanglement of function groups and reducing the hydration ability of polymer. Consequently, the hydrogel is not able to form enough blockage of the flow, which further results in higher hydraulic conductivity.

Acknowledgements. The Electric Power Research Institute (EPRI) provided financial support for this study. National Science Foundation of China (41701347) supported the partial travel expenses of Dr. Chen.

References

1. EPRI (2009) Coal ash: characteristics, management, and environmental issues. Electric Power Research Institute, Palo Alto, CA. 1019022
2. USEPA 40 CFR Parts 257 and 261 (2015) Hazardous and solid waste management system; disposal of coal combustion residuals from electric utilities; final rule. USEPA
3. Shackelford C, Benson C, Katsumi T, Edil T, Lin L (2000) Evaluating the hydraulic conductivity of GCLs permeated with non-standard liquids. Geotext Geomembr 18(2–4):133–162
4. Jo H, Katsumi T, Benson C, Edil T (2001) Hydraulic conductivity and swelling of non-prehydrated GCLs permeated with single species salt solutions. J Geotech Geoenviron Eng 127(7):557–567
5. Chen J, Benson CH, Edil TB (2017) Hydraulic conductivity of geosynthetic clay liners with sodium bentonite to coal combustion product leachates. J Geotech Geoenviron Eng 144 (3):04018008
6. Xu C, Li Z, Gao Y (2009) Influence of solution characteristics on swelling and hydration performance of geosynthetic clay liners. J Tonji Univ (Nat Sci) 37(1):36–40
7. Kolstad D, Benson C, Edil T (2004) Hydraulic conductivity and swell of non-prehydrated GCLs permeated with multi-species inorganic solutions. J Geotech Geoenviron Eng 130 (12):1236–1249
8. Ruhl J, Daniel D (1997) Geosynthetic clay liners permeated with chemical solutions and leachates. J Geotech Geoenviron Eng 123(4):369–381
9. Bradshaw S, Benson C (2013) Effect of municipal solid waste leachate on hydraulic conductivity and exchange complex of geosynthetic clay liners. J Geotech Geoenviron Eng 140(4):04013038

10. Katsumi T, Ishimori H, Ogawa A et al (2007) Hydraulic conductivity of nonprehydrated geosynthetic clay liners permeated with inorganic solutions and waste leachates. Soils Found JGS 1:79–96
11. Di Emedio G, Van Impe W, Flores V (2011) Advances in geosynthetic clay liners: polymer enhanced clays. In: GeoFrontiers 2011. Advances in geotechnical engineering, American society of civil engineers, Reston, pp 1931–1940
12. Mazzieri F, Emidio G, Van Impe P (2010) Diffusion of calcium chloride in a modified bentonite: Impact on osmotic efficiency and hydraulic conductivity. Clays Clay Miner 58 (3):351–363
13. Scalia J, Benson C, Bohnhoff G, Edil T, Shackelford C (2014) Long-term hydraulic conductivity of a bentonite-polymer composite permeated with aggressive inorganic solutions. J Geotechnol Geoenviron Eng 140:040130251
14. Katsumi T, Ishimori H, Onikata M, Fukagawa R (2008) Long-term barrier performance of modified bentonite materials against sodium and calcium permeant solutions. Geotext Geomembr 26(1):14–30
15. Shackelford CD, Sevick GW, Eykholt GR (2010) Hydraulic conductivity of geo-synthetic clay liners to tailings impoundment solutions. Geotext Geomembr 28(2):149–162
16. Scalia J (2012) Bentonite-polymer composites for containment applications. Ph.D. Dissertation, University of Wisconsin, Madison, WI
17. EPRI (2006) Characterization of field leachates at coal combustion product management sites. Electric Power Research Institute, Palo Alto, p 1012578
18. Tian K, Benson C, Likos W (2016) Hydraulic conductivity of geosynthetic clay liners to low-level radioactive waste leachate. J Geotech Geoenviron Eng 04016037:1–12
19. Bulo RE, Donadio D, Laio A, Molnar F, Rieger J et al (2007) Site binding of Ca2+ ions to polyacrylates in water: a molecular dynamics study of coiling and aggregation. Macromolecules 40:3437–3442

Containing Bauxite Liquor Using Bentonite-Polymer Composite Geosynthetic Clay Liners

Kuo Tian[1(✉)] and Craig H. Benson[2]

[1] George Mason University, Fairfax, VA 22020, USA
ktian@gmu.edu
[2] University of Virginia, Charlottesville, VA 22904, USA

Abstract. Experiments were conducted to evaluate the hydraulic conductivity of geosynthetic clay liners (GCLs) to bauxite liquor from aluminum refining. Tests were conducted with a GCL containing conventional sodium-bentonite (Na-B) and one GCLs containing a bentonite-polymer composite. B-P has polymer loading of 6.8%. Hydraulic conductivity tests were conducted on the GCLs with bauxite liquor in flexible-wall permeameters following procedures in ASTM D6766. The bauxite liquor used in the experiments has pH 13 and ionic strength = 700 mM. Hydraulic conductivity of the Na-B GCL increased to approximately 10^{-7} m/s during permeation, whereas the B-P GCL maintained low hydraulic conductivity ($\sim 4.3 \times 10^{-12}$ m/s). Suppression of bentonite swelling by the bauxite liquor is the primary factor responsible for the higher hydraulic conductivity of the Na-B GCL. Low hydraulic conductivity of the B-P GCL to bauxite liquor is attributed to polymer clogging intergranular pores controlling flow of bauxite liquor through the GCL. The findings indicate that B-P GCLs with sufficient polymer loading can be used effectively in composite liners for disposal facilities containing bauxite liquor.

Keywords: Bauxite liquor · Bentonite-polymer composite
Geosynthetic clay liner · Hydraulic conductivity

1 Introduction

Geosynthetic clay liners (GCLs) are factory-manufactured barriers comprised of a layer on sodium-bentonite (Na-B) sandwiched between two geotextiles (Bradshaw and Benson 2014; Scalia et al. 2014; Tian et al. 2016). GCLs are common elements in waste containment facilities due to their low hydraulic conductivity to water (typically < 10^{-10} m/s) and ease of installation (Shackelford et al. 2000). The low hydraulic conductivity of GCLs is primarily due to the osmotic swell of Na-B, which creates narrow and tortuous flow paths for liquids (Bradshaw and Benson et al. (2014). The hydraulic conductivity of GCLs can be affected by chemical interactions between the Na-B and liquid to be contained (Jo et al. 2001, 2005; Kolstad et al. 2004). GCLs permeated with more aggressive leachates having higher ionic strength and/or a predominance of polyvalent cations can be orders of magnitude more permeable than GCLs permeated with deionized (DI) or tap water due to the reduction of swelling of

L. Zhan et al. (Eds.): ICEG 2018, ESE, pp. 672–678, 2019.
https://doi.org/10.1007/978-981-13-2224-2_83

Na-B (Jo et al. 2001, 2005; Kolstad et al. 2004; Shackelford et al. 2010; Scalia et al. 2014; Tian et al. 2015, 2016).

Bauxite liquor can be particularly aggressive with GCLs. Benson et al. (2010) evaluated the hydraulic conductivity of two conventional Na-B GCLs to a bauxite liquor from an aluminum refining operation. One GCL contained granular bentonite and the other powdered bentonite. The leachate had pH 12.2 and an ionic strength of 774 mM, with Na (409 mM) and Al (129 mM) being the predominant cations. The hydraulic conductivity of one GCL containing granular bentonite increased to as high as 1.8×10^{-8} m/s when permeated with bauxite liquor, whereas the hydraulic conductivity of the GCL with powdered bentonite decreased to approximately 2.5×10^{-12} m/s. The increase in hydraulic conductivity was attributed to reduction in swelling of bentonite granules in the bauxite liquor. Precipitation of aluminum complexes in the fine pores of the powdered bentonite was hypothesized to clog flow channels, resulting in lower hydraulic conductivity.

Recent studies have focused on using polymer-modified bentonite to improve the chemical compatibility of GCLs with aggressive liquids like bauxite liquor (Katsumi et al. 2008, Scalia et al. 2014, De Camillis et al. 2016, Tian et al. 2016, 2017). GCLs containing bentonite-polymer (B-P) composites have been evaluated for containment of mine leachate (Shackelford et al. 2010), coal combustion product (Salihoglu et al. 2016), and low-level radioactive waste and mixed waste (Tian et al. 2016). Tian et al. (2016) proposed that a polymer clogging mechanism controls hydraulic conductivity of B-P GCL to aggressive leachates.

The objective of this study was to evaluate the effect of bauxite liquor on the hydraulic conductivity of geosynthetic clay liners (GCLs) used in composite liner systems. Tests were conducted on a GCL containing conventional Na-B and one GCL containing bentonite-polymer (B-P) composites. The bauxite liquor used in the experiments has pH 13 and ionic strength ~700 mM. Hydraulic conductivity tests were conducted in flexible-wall permeameters at an effective confining stress of 20 kPa.

2 Method

Two GCLs were evaluated in this study. One GCL contained conventional Na-B and the other contained a bentonite-polymer composite (labeled as B-P). Each GCL consists of granular material (bentonite granules or granules of bentonite-polymer composite) sandwiched between a non-woven geotextile (top) and a woven geotextile (bottom). For all GCLs, polypropylene geotextiles bonded by needlepunching were employed.

Physical properties of the GCLs are summarized in Table 1. Polymer loading was determined using the loss on ignition method per ASTM D7348 (2013). Swell Index value was conducted based on ASTM D5890 (2011).

The bauxite liquor used in the testing program was obtained from an aluminum refinery and analyzed for elemental concentrations by inductively coupled plasma-optical emission spectroscopy (ICP-OES). The bauxite liquor has pH 13.0, ionic

Table 1. Physical properties of the GCLs used in the study

GCL	Mass per unit area (kg/m^2)	Polymer loading (%)	Swell index in DI water (mL/2.0 g)	Swell index in bauxite liquor (mL/2.0 g)
Na-B	3.7	–	25.0	9.0
B-P	3.6	6.8	–	12.0

strength of approximately 700 mM, and the cations are predominantly Al (—119 mM) and Na (~ 456 mM).

Hydraulic conductivity tests on the GCL specimens were conducted in flexible-wall permeameters using the falling headwater and constant tailwater method described in ASTM D6766 (2012). Backpressure saturation was not applied to simulate the *in situ* condition and to avoid unrealistic changes in geochemistry. After prehydration, the effective confining stress was set at 20 kPa and the hydraulic gradient was set at 125. Influent for the specimens was contained in 50 mL burettes sealed with parafilm to prevent evaporation and carbonation. Effluent was collected in 60-mL polyethylene bottles sealed with parafilm.

Equilibrium was defined using the hydraulic and chemical equilibrium criteria in ASTM D6766 (2012) along with an additional criterion for influent and effluent concentrations. The criteria in D6766 require no temporal trend in the hydraulic conductivity measurements, hydraulic conductivity falling within 25% of the mean for three consecutive measurements, incremental effluent volume (Q_{out}) within 25% of the incremental influent volume (Q_{in}) for at least 3 measurements, and the ratio Q_{out}/Q_{in} exhibiting no temporal trend. Chemical equilibrium is defined as the electrical conductivity and pH of the effluent (EC_{out} and pH_{out}) showing no temporal trend and falling within 10% of the electrical conductivity and pH of the influent (EC_m and pH_m).

3 Results and Discussion

One specimen of each GCL was permeated with bauxite liquor or DI water. The tests were conducted for 160–280 d. Na-B GCLs permeated with bauxite liquor met the termination criteria in ASTM D6766, whereas the B-P GCLs had not reached chemical equilibrium at termination. Pore volumes of flow (PVF) were calculated based on the final properties of each GCL specimen (e.g., thickness, diameter, and mass).

Hydraulic conductivity of the Na-B and B-P GCLs permeated with bauxite liquor as a function of PVF is shown in Fig. 1. Hydraulic conductivity of the Na-B GCLs increased to 1.4×10^{-7} m/s immediately, which is approximately four orders of magnitude higher than the hydraulic conductivity to DI water (1.5×10^{-7} m/s vs. 3.4×10^{-11} m/s) (Fig. 2) and much higher than normally required for a composite liner in a containment facility for bauxite liquor (usually $< 10^{-10}$ m/s). In contrast, the hydraulic conductivity of B-P slowly decreased to approximately 4.3×10^{-12} m/s when permeated with bauxite liquor.

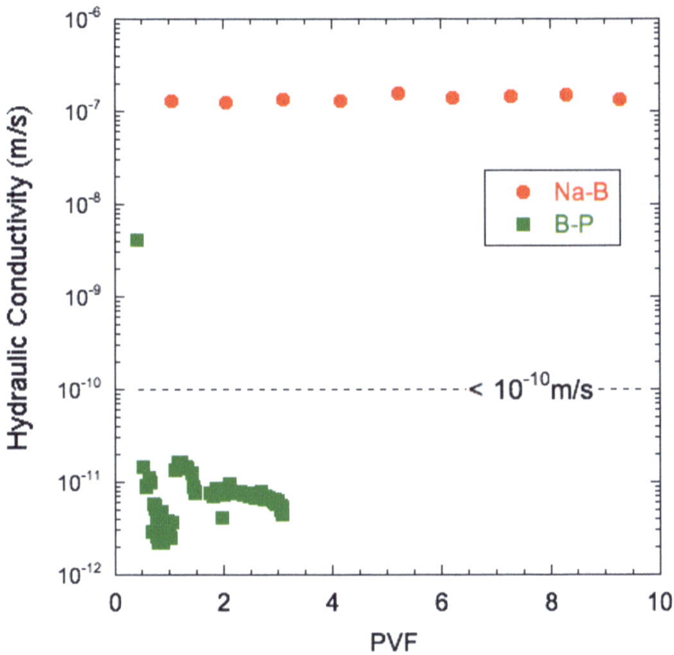

Fig. 1. Hydraulic conductivity vs. pore volumes of flow for Na-B and B-P GCL permeated with bauxite liquor.

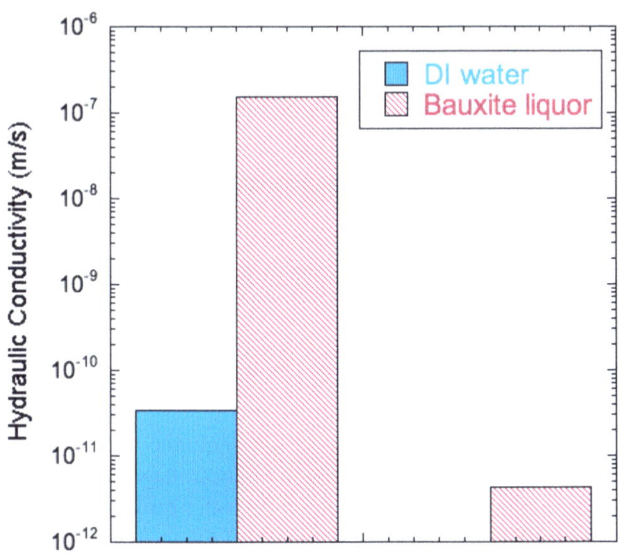

Fig. 2. Hydraulic conductivity of Na-B and B-P GCL permeated with DI water and bauxite liquor.

Hydraulic conductivity of the GCLs to bauxite liquor as a function of swell index is show in Fig. 3 along with data from Jo et al. (2001). The data for the tests with bauxite liquor are consistent with the data from Jo et al. (2001), and confirm that suppression of swelling of Na-B results in larger pore spaces between bentonite granules and higher hydraulic conductivity. Jo et al. (2001) report that Na-B having a swell index > 16 mg/L can maintain low hydraulic conductivity ($<10^{-10}$ m/s). In this study, the NaB in the GCL had a swell index of 25.0 mL/2.0 g in DI water and decreased to 9.0 mL/2.0 g in bauxite liquor. Reduction in swelling of the bentonite is the primary factor responsible for the higher hydraulic conductivity of the conventional Na-B GCL to bauxite liquor relative to DI water.

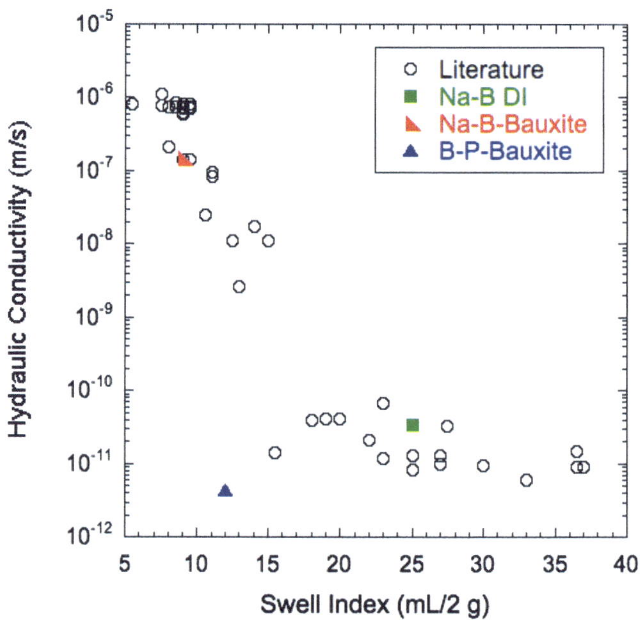

Fig. 3. Hydraulic conductivity of Na-B and B-P GCL permeated with bauxite liquor and DI water as a function of swell index. Literature data from Jo et al. (2001).

Low hydraulic conductivity of the B-P GCL is attributed to polymer clogging intergranular pores controlling flow of bauxite liquor through the GCL, as described in Tian et al. (2016, 2017). The polymer forms a three-dimensional hydrogel structure that occupies the intergranular pores space when the bentonite swelling is suppressed, as shown in Fig. 4 (Tian et al. 2016). Under hydration stage, the polymer chain will bound lots of water molecules, and the immobile water molecules will block the flow channels, resulting in low hydraulic conductivity of GCL. These findings suggest that B-P GCLs can be used effectively in composite liners for disposal of residuals from aluminum refining.

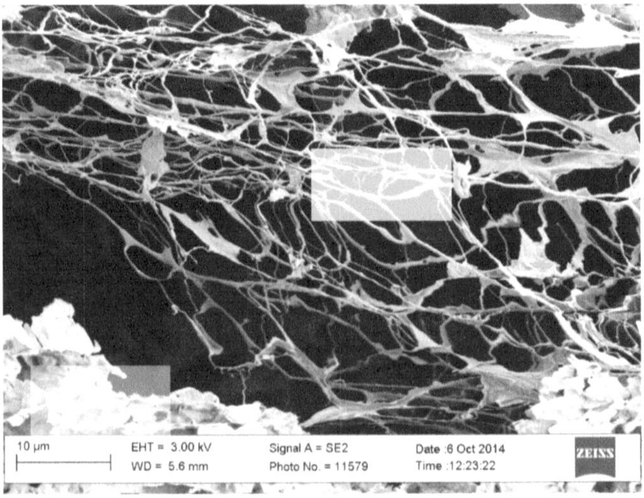

Fig. 4. SEM image of B-P from GCL in DI water after freeze-drying

4 Conclusion

This study evaluated how the hydraulic conductivity of GCLs is affected by permeation with bauxite liquor. Two commercially available GCLs were evaluated - one with conventional sodium-bentonite (Na-B) and the other with a bentonite-polymer composite (B-P).

Hydraulic conductivity of the GCL containing conventional Na-B increased immediately when permeated with bauxite liquor. The final hydraulic conductivity was approximately four orders of magnitude higher than the hydraulic conductivity to DI water (1.5×10^{-7} m/s vs. 3.4×10^{-11} m/s). The high hydraulic conductivity of Na-B GCL to bauxite liquor is attributed to bauxite liquor suppressing the swelling of Na-B.

Hydraulic conductivity of the prehydrated B-P GCL to bauxite liquor was very low (4.3×10^{-11} m/s). The low hydraulic conductivity of the B-P GCL is attributed to polymer hydrogel clogging intergranular pores that conduct flow in the Na-B GCL. This finding suggests that B-P GCL can be used to line disposal facilities containing residuals from aluminum refining, including bauxite liquor.

References

ASTM D5890 (2011) Standard Test Method for Swell Index of Clay Mineral Component of Geosynthetic Clay Liners, ASTM International, West Con- shohocken, PA

ASTM D6766 (2012) Standard Test Method for Evaluation of Hydraulic Properties of Geosynthetic Clay Liners Permeated with Potentially Incompatible Aqueous Solutions, ASTM International, West Conshohocken, PA

ASTM D7348 (2013) Standard Test Methods for Loss on Ignition (LOI) of Solid Combustion Residues, ASTM International, West Conshohocken, PA

Benson C, Oren A, Gates W (2010) Hydraulic conductivity of two geosynthetic clay liners permeated with a hyperalkaline solution. Geotext Geomembr 28(2):206–218

Bradshaw S, Benson C (2014) Effect of municipal solid waste leachate on hydraulic conductivity and exchange complex of geosynthetic clay liners. J Geotech Geoenviron Eng 140 (4):04013038

De Camillis M, Di Emidio G, Bezuijen A, Verastegui-Flores RD (2016) Hydraulic conductivity and swelling ability of a polymer modified bentonite subjected to wet-dry cycles in seawater. Geotext Geomembr 44(5):739–747

Jo H, Katsumi T, Benson C, Edil T (2001) Hydraulic conductivity and swelling of non-prehydrated GCLs permeated with single species salt solutions. J Geotech Geoenviron Eng 127(7):557–567

Jo H, Benson C, Lee J, Shackelford C, Edil T (2005) Long-term hydraulic conductivity of a non-prehydrated geosynthetic clay liner permeated with inorganic salt solutions. J Geotech Geoenviron Eng 131(4):405–417

Katsumi T, Ishimori H, Onikata M, Fukagawa R (2008) Long-term barrier performance of modified bentonite materials against sodium and calcium permeant solutions. Geotext Geomembr 26(1):14–30

Kolstad D, Benson C, Edil T (2004) Hydraulic conductivity and swell of nonprehydrated GCLs permeated with multispecies inorganic solutions. J Geotech Geoenviron Eng 130(12):1236–1249

Salihoglu, H., Chen, J., Likos, W., Benson, C.: Hydraulic conductivity of bentonite-polymer geosynthetic clay liners in coal combustion product leachates. In: Proceeding of Geo-Chicago, Chicago, pp 438–447 (2016)

Shackelford C, Benson C, Katsumi T, Edil T, Lin L (2000) Evaluating the hydraulic conductivity of GCLs permeated with non-standard liquids. Geotext Geomembr 18:133–161

Shackelford C, Sevick G, Eykholt G (2010) Hydraulic conductivity of geosynthetic clay liners to tailings impoundment solutions. Geotext Geomembr 28(2):206–218

Scalia J, Benson C, Bohnhoff G, Edil T, Shackelford C (2014) Long-term hydraulic conductivity of a bentonite-polymer composite permeated with aggressive inorganic solutions. J Geotech Geoenviron Eng 140(3):04013025

Tian K, Benson C (2015) Effect of low-level radioactive waste leachate on hydraulic conductivity of a geosynthetic clay liner. Geosynthetics 2015, IFAI, St. Paul, MN, pp 236–244

Tian K, Benson C, Likos W (2016) Hydraulic conductivity of geosynthetic clay liners to low-level radioactive waste leachate. J Geotech Geoenviron Eng 142(8):04016037

Tian K, Benson C, Likos W (2017) Effect of anion ratio on hydraulic conductivity of a bentonite-polymer geosynthetic clay liner. In: Proceeding of geotechnical frontiers 2017, ASCE, Orlando, FL. https://doi.org/10.1061/9780784480434.018

Using Fluid Loss to Evaluate the Hydraulic Conductivity of Geosynthetic Clay Liners Under Mining Leachates

Yang Liu[1,2(✉)], Yu Hao[1], and Lizhen Wang[3]

[1] Department of Geology, Hunan University of Science and Technology, Xiangtan 411201, Hunan, China
liuyang2585899@163.com
[2] Hunan Provincial Key Laboratory of Shale Gas Resource Utilization, Hunan University of Science and Technology, Xiangtan 411201, Hunan, China
[3] Changsha Research Institute of Mining and Metallurgy Co., Ltd., Changsha 410012, Hunan, China

Abstract. Geosynthetic clay liners (GCLs) are increasingly used in mining applications, their hydraulic performance need to be further assessed. In this study, the fluid loss (FL) tests of three bentonites from different GCLs were conducted under different mining leachates. The results suggest that the FL values in mining leachates greatly increased compared to those in water. The greater ionic strength of the leachates led to much higher fluid loss. The triaxial hydraulic conductivity tests using mining leachates as permeant solutions were also performed, the hydraulic conductivity (*k*) value increased with the increased ionic strength of the mining leachates. The relationship between fluid loss and hydraulic conductivity was also established. The *k* calculated from the *FL* test (k_{FL}) is in good agreement with that from triaxial hydraulic conductivity test (k_{Tri}). Therefore, the *FL* test could be considered as a quick method to evaluate the hydraulic conductivity under the studied mining leachates. This study would provide useful reference to the mining application of GCLs.

Keywords: Geosynthetic Clay Liners · Mining leachates · Fluid loss Hydraulic conductivity

1 Introduction

Geosynthetic clay liners (GCLs) are usually employed as hydraulic barriers because of their low hydraulic conductivity to water (Bouazza 2002). Now they are gradually applied in mining facilities (e.g., heap leach pads, brine evaporation ponds, waste rock dumps, etc.) (Hornsey et al. 2010). Therefore, GCLs would be undergone the new environmental loads (Gates and Buckley 2009; Shackelford et al. 2010; Liu et al. 2013) and the hydraulic performance of a GCL would be a concern.

Many studies have been conducted on the k value of GCLs with non-standard liquids, such as saline solutions (Shackelford et al. 2000; Kolstad et al. 2004) and acidic or alkaline solutions (Benson et al. 2010; Gates and Bouazza 2010). The results of these investigations suggest that these non-standard solutions result in a increased *k* in GCLs relative to water.

L. Zhan et al. (Eds.): ICEG 2018, ESE, pp. 679–685, 2019.
https://doi.org/10.1007/978-981-13-2224-2_84

The flexible-wall hydraulic conductivity tests are time-consuming, therefore a quick method to evaluate the *k* value is greatly needed. The *FL* test has been used to quickly evaluate *k*. A low *FL* means low flow passing through the filter cake within a given time and therefore implies a low *k* value. Olsta et al. (2004) correlated the fluid loss values with *k* for DI water, distilled water, and tap water, they found that the *k* values usually raised with the increased fluid loss values, particularly when the *FL* values are greater than 17 mL. The *FL* went up from 13 to 20 mL result in the increase of *k* from 5×10^{-11} to 1×10^{-9} m/s. Chung and Daniel (2008) suggested the chance of backcalculating *k* from the fluid loss test, showing the potential of using fluid loss test to evaluate *k* value. Liu et al. (2014) calculated the k values with *FL* methods and correlated with the *k* values obtained from the flexible-wall hydraulic tests. They found that the *FL* test would be used to rapidly estimate the *k* values under DI water and as well as low acidic concentration (≤ 0.015 mol/L) conditions. Other methods including the filter press tests have been performed to quickly estimate the *k* of bentonite cakes (Rosin-Paumier et al. 2010; Rosin-Paumier and Touze-Foltz 2012).

The purpose of this study was to identify the hydraulic performance of two commercial GCLs under various concentrations of mining leachates and to quickly estimate the *k* by backcalculating from the *FL* test.

2 Materials and Methods

2.1 Materials

GCL1 and GCL2, both needle-punched GCLs from China, were used in this study. GCL1 has an unhydrated thickness at ~6.5 mm. The content of powdered sodium bentonite in GCL1 is 4.8 kg/m^2. Its *k* value under water condition was $\sim 1.8 \times 10^{-11}$ m/s with a 35 kPa effective stress. The bentonite component (B1) had an swell index (*SI*) of 24 mL/2 g. GCL2 has an unhydrated thickness at ~7.0 mm. The content of powdered sodium bentonite is 5.0 kg/m^2. Its *k* value under water condition was $\sim 1.0 \times 10^{-11}$m/s with a 35 kPa effective stress. The bentonite component (B2) had an *SI* of 30 mL/2 g. Two bentonite cakes having mass per unit area of ~4.8 kg/m^2 and ~5.0 kg/m^2 were made from GCL1 and GCL2, respectively. The detailed bentonite cakes preparation process was given in Liu et al. (2015). The information of GCL1 and GCL2 is shown in Table 1.

Table 1. The information of the GCLs used in this study

Specimens	GCL1	GCL2
Type of bentonite	Sodium/powdered	Sodium/powdered
Mass per unit area of bentonite component (kg/m^2)	4.8	5.0
Initial un-hydrated thickness (mm)	6.5	7.0
CEC (cmol/kg)*	94	102
Exchange cations (of total CEC)		
Na	82	89
K	1	2

(continued)

Table 1. (*continued*)

Specimens	GCL1	GCL2
Ca	6	5
Mg	5	6
Swell index (mL/2 g)	24	30
Liquid limit (%)	434	487
Plastic index (%)	392	448

Note:* by methylene blue adsorption;

2.2 Mining Leachates

Two mining leachates with different ionic strength from Hunan Province were applied in this research. The *pH* and electrical conductivity (*EC*) were measured respectively. The *pH, EC* and ionic strength of these two mining leachates are shown in Table 2.

Table 2. Properties of the mining leachates

Leachates	*pH*	*EC* (mS/cm)	Ionic strength (mol/L)
Mining leachate 1 (M1)	6.8	126.9	0.9
Mining leachate 2 (M2)	7.2	203.4	1.6

2.3 Fluid Loss Tests

The *FL* test was conducted according to a modified process as described in Liu et al. (2014). The *k* of the filter cake calculated from the *FL* tests (k_c) based on the study of Chung and Daniel (2008) and the filtration model presented by Rushton et al. (2000) based on Darcy's law:

$$k_c = \frac{\beta \gamma_w}{2A^2} \frac{V_f^2}{P_0 t} \tag{1}$$

where P_0 is the applied stress (kPa), γ_w is the unit weight of liquids (kN/m^3), V_f is the filtrate volume (m^3), k_c is in m/s, t is the filtration time (s) and A is the filter cake area (m^2), β is the constant of proportionality when the volume of the filter cake was expected to increase linearly for each unit of the suspension filtered.

2.4 Hydraulic Conductivity Tests

The flexible-wall hydraulic conductivity test were conducted in the study. The detailed information was described in Liu et al. (2015).

The termination criteria of the hydraulic conductivity test followed ASTM D5084 and as recommended by Shackelford et al. (1999). The hydraulic conductivity test would be terminated until the quantity of flow, *pH* and *EC* all tended toward a steady state. Namely, the inflow and outflow (Q_{in}/Q_{out}) and the ratios of *pH* and *EC* in inflow and outflow, pH_{in}/pH_{out}, EC_{in}/EC_{out} were within 1.00 ± 0.15.

3 Results and Discussion

3.1 Fluid Loss Tests

Figure 1 shows the *FL* test results of two GCLs under different mining leachates. The results suggest that the *FL* of both bentonites went up with increased ionic strength.

B2 always has the relatively higher fluid loss values than those of B1 over the range of mining leachates due to the lower initial swell index as shown in Table 1. The total increases of fluid loss were 228% and 277% for B1 and B2, respectively. For both two bentonites, around 70% increase of total occurred when using lower ionic strength mining leachate M1, which indicates that the bentonite is sensitive to the effect of ionic strength (Liu et al. 2013), meanwhile, M1 is strong enough to cause the collapse of the clay layer and to lead to a high fluid loss since the *FL* values for each bentonite under both M1 and M2 are not in great differences (Liu et al. 2014).

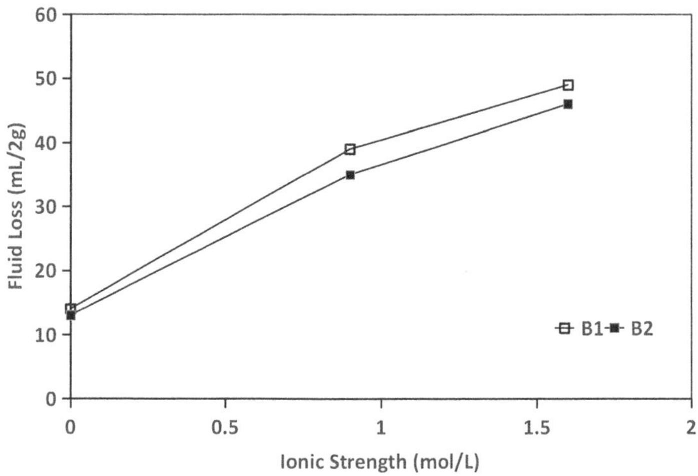

Fig. 1. The *FL* values of two bentonites with different ionic strengths

3.2 Hydraulic Conductivity Tests

The k_{Tri} values tested from the triaxial hydraulic conductivity tests, for different ionic strengths of mining leachates are shown in Fig. 2. Generally, higher *k* values were obtained when the ionic strength increased. However, GCL1 showed the relatively smaller increase ($\sim$200 times) in hydraulic conductivity than GCL2 ($\sim$400 times), which is in general agreement with the observation for the fluid loss tests as shown in Fig. 1.

Figure 3 shows the *k* values back-calculated from the *FL* tests (k_{FL}) based on Eq. 1. Similarly, k_{FL} increases with growing ionic strength since they were calculated based on the *FL* test. GCL1 always had lower k_{FL} values when permeated with both two mining leachates although GCL has an initial *k* in water. This would be attributed to the higher exchangeable cations.

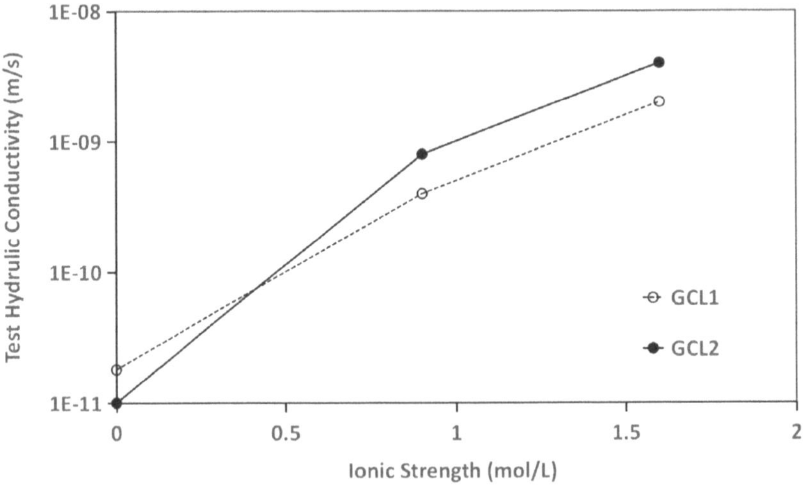

Fig. 2. The tested hydraulic conductivity values from triaxial flexible-wall permeameters with different ionic strengths

The k values back-calculated from the FL tests (k_{FL}) and triaxial hydraulic conductivity tests (k_{Tri}) are shown in Table 3. As shown, the ratio of k_{FL} values were to k_{Tri} is 2.3 to 15, i.e., k_{FL}/k_{Tri} = 2.3–15, under both DI water and mining leachates conditions. Namely, k_{FL} was relatively conservative (i.e., $k_{FL}/k_{Tri} > 1$). It is worth to mention that the back-calculated k_{FL} assumed to be performed under the condition of maximum amount of leachate as described in the fluid loss test. Furthermore, the effect of geotextiles and bonding process should also be considered since these are proved to affect the hydraulic conductivity of GCLs (Petrov et al. 1997). However, the discussions in this study still support the validity of a rapid and effective estimation of hydraulic conductivity by using the FL tests with the mining leachates used in this investigation.

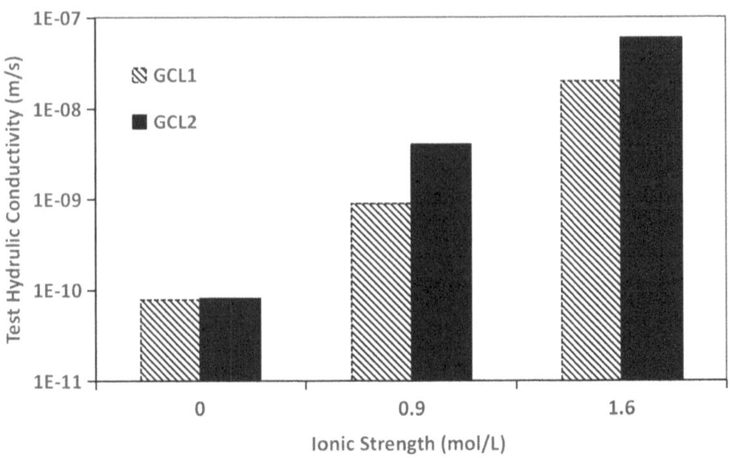

Fig. 3. The k values back-calculated from the FL test with different ionic strengths

Table 3. Properties of mining leachates used in this study

Ionic strength (mol/L)		0 (water)	0.9 (M1)	1.6 (M2)
GCL1	k_{FL} (m/s)	7.8×10^{-11}	9×10^{-10}	2×10^{-8}
	k_{Tri} (m/s)	1.8×10^{-11}	4×10^{-10}	2×10^{-9}
	k_{FL}/k_{Tri}	4.3	2.3	10
GCL2	k_{FL} (m/s)	8.1×10^{-11}	4×10^{-9}	6×10^{-8}
	k_{Tri} (m/s)	1.0×10^{-11}	8×10^{-10}	4×10^{-9}
	k_{FL}/k_{Tri}	8.1	5	15

4 Conclusions

The *FL* of both bentonites increased with the increasing ionic strength. B2 always has greater *FL* values than those of B1 under both two mining leachates due to the lower initial swell index as shown in Table 1. Around 70% increase of total occurred in permeation with the lower ionic strength mining leachate M1, which indicates the sensitivity of the bentonite to the effect of ionic strength.

Greater hydraulic conductivity values were obtained as the ionic strength increased. However, GCL1 showed the relatively smaller increase ($\sim$200 times) in hydraulic conductivity than GCL2 ($\sim$400 times), which is in general agreement with the observation for the *FL* tests. The calculated k_{FL} increases with increased ionic strength since they were calculated based on the *FL* test. k_{FL} was conservative (i.e., $k_{FL}/k_{Tri} > 1$) under both DI water and mining leachates conditions. The ratio of k_{FL}/k_{Tri} is 2.3–15. Thus, the *FL* test could still be considered as a rapid method to evaluate the hydraulic conductivity under both two mining leachates used in this study, even if the effect of geotextiles and bonding process, etc. were not considered. Further tests are still encouraged to improve the evaluation of the *k* value based on the *FL* tests under different non-standard solutions.

Acknowledgments. The first author was funded by the Open Research Fund Program of Hunan Provincial Key Laboratory of Shale Gas Resource Utilization, Hunan University of Science and Technology (Grant No. E21818); the National Natural Science Foundation of China (Project No. 51608192); the Natural Science Foundation of Hunan Province (Grant No. 2016JJ4031).

References

Bouazza A (2002) Geosynthetic clay liners. Geotext Geomembr 20(1):3–17

Hornsey WP, Scheirs J, Gates WP, Bouazza A (2010) The impact of mining solutions/liquors on geosynthetics. Geotext Geomembr 28(2):191–198

Gates WP, Buckley J (2009) Geosynthetic clay liners–is the key component being overlooked? In: GIGSA GeoAfrica 2009 Conference, Cape Town, 2–5 September, pp 1–10

Shackelford CD, Sevick GW, Eykholt GR (2010) Hydraulic conductivity of geosynthetic clay liners to tailings impoundment solutions. Geotext Geomembr 28(2):149–162

Liu Y, Gates WP, Bouazza A (2013) Acid induced degradation of the bentonite component used in geosynthetic clay liners. Geotext Geomembr 36(2–4):71–80

Petrov RJ, Rowe RK, Quigley RM (1997) Selected factors influencing GCL hydraulic conductivity. J Geotech Geoenviron Eng 123(8):683–695

Shackelford CD, Benson CH, Katsumi T, Edil TB, Lin L (2000) Evaluating the hydraulic conductivity of GCLs permeated with non-standard liquids. Geotext Geomembr 18(2–4):133–161

Kolstad DC, Benson CH, Edil TB, Jo HY (2004) Hydraulic conductivity of a dense prehydrated GCL permeated with aggressive inorganic solutions. Geosynth Int 11(3):233–241

Benson C, Ören A, Gates WP (2010) Hydraulic conductivity of two geosynthetic clay liners permeated with a hyperalkaline solution. Geotext Geomembr 28(2):206–218

Gates WP, Bouazza A (2010) Bentonite transformations in strongly alkaline solutions. Geotext Geomembr 28(2):219–225

Olsta J, Daniel D, Chung J (2004) Various aspects of sodium bentonite testing. In: Advances in geosynthetic clay liner technology: 2nd symposium. ASTM STP 1456. ASTM International, West Conshohocken, PA, pp 3–10

Chung J, Daniel D (2008) Modified fluid loss test as an improved measure of hydraulic conductivity for bentonite. Geotech Test J 31(3):243–251

Liu Y, Gates WP, Bouazza A, Rowe RK (2014) Fluid loss as a quick method to evaluate the hydraulic conductivity of geosynthetic clay liners under acidic conditions. Can Geotech J 51 (2):158–163

Rosin-Paumier S, Touze-Foltz N, Pantet A, Monnet P, Didier G, Guyonnet D (2010) Swell index, oedopermeametric, filter press and rheometric tests for identifying the qualification of bentonites used in GCLs. Geosynth Int 17(1):1–11

Rosin-Paumier S, Touze-Foltz N (2012) Hydraulic and chemical evolution of GCLs during filter press and oedopermeametric tests performed with real leachate. Geotext Geomembr 33:15–24

Rushton A, Ward AS, Holdich RG (2000) Solid-liquid filtration and separation technology, 2nd edn. Wiley-VCH Verlag GmbH, Weinbeim

Liu Y, Bouazza A, Gates WP, Rowe RK (2015) Hydraulic performance of geosynthetic clay liners to sulfuric acid solutions. Geotext Geomembr 43(2):14–23

Shackelford CD, Malusis MA, Majeski MJ, Stern RT (1999) Electrical conductivity breakthrough curves. J Geotech Geoenviron Eng 125(4):260–270

Laboratory Tests of the Influence of Clogging on the Hydraulic Properties of Nonwoven Geotextiles

Anna Miszkowska(✉), Eugeniusz Koda, Anna Sieczka, and Piotr Osiński

Faculty of Civil and Environmental Engineering, Warsaw University of Life Sciences, Nowoursynowska 159, 02-776 Warsaw, Poland
anna_miszkowska@sggw.pl

Abstract. Since decades the use of nonwoven geotextile filters is a common application in geo-environmental engineering. The main functions of the filter are to prevent the movement of the base soil fine particles allowing the liquid to flow as freely as possible. Geotextiles are easy to transport and install, their main advantage is to help to minimize clogging effect of drainage and the proper selection of geotextile filter layers is essential to ensure durability of drainage systems. However, the design of a geosynthetics filter is a very complex process because of the large number of parameters involved. In this paper the current geotextile filter design criteria: retention, permeability and anti-clogging were showed. What is more, laboratory tests of water permeability normal to the plane of worn nonwoven geotextile samples taken from Białobrzegi earthfill dam after more than 20 years of exploitation were presented. Therefore the factors affecting the geotextile filters clogging were illustrated. The results show a marked influence of clogging on hydraulic properties of tested nonwoven geotextiles and allowed to verify the effectiveness of applied geofilters in earthfill dam.

Keywords: Nonwoven geotextile · Water permeability · Clogging Earthfill dam

1 Introduction

Geotextiles manufactured from polyester, polypropylene, polyester, polyethylene or polyamides have been used for 40 years now, especially as filters to prevent the migration of fine particles and allow adequate seepage to flow through the geotextile plane. These geosynthetics are easy to install, more permeable and much thinner than granular filters (Koda et al. 2016; Palmeira and Trejos Galvis 2017; Miszkowska et al. 2017).

The main mechanism affecting the durability of geotextile filters is clogging which means the reduction of the geotextile's permeability to a point where flow through it results in the hydraulic system's non-performance. The clogging can be caused by physical, chemical and biological processes (Junqueira et al. 2006; McIssac and Rowe 2006; Veylon et al. 2016; Correia et al. 2017). The mechanism behind physical

L. Zhan et al. (Eds.): ICEG 2018, ESE, pp. 686–691, 2019.
https://doi.org/10.1007/978-981-13-2224-2_85

clogging is the accumulation of fine particles upstream of the geotextile or within the geotextile. For that reason, the knowledge of the interaction between the geotextile filter and the base soil is required for the properly design of filter.

In particularly, the retention, permeability and anti-clogging capability of geotextiles are important design concerns in filtration applications (Giroud 1982; Wu et al. 2008; Giroud 2010; Moraci 2010: Cazzuffi et al. 2016).

The retention criterion is usually expressed, as below:

$$O_F \leq R_R D_n \tag{1.1}$$

where: O_F is the characteristic opening size of geotextile (for example O_{90}), R_R is a Retention ratio dependent on the criterion and D_n is the indicative diameter of the base soil particles (usually D_{85}).

The permeability criterion is generally expressed, as follows:

$$k_{GT} \geq \lambda k_s \tag{1.2}$$

where: k_{GT} is the permeability of geotextile normal to the plane, λ is a constant depending on the criterion and k_s is the soil permeability.

In case of anti-clogging capabilities, the gradient ratio test is a popular method to evaluate the filtration performance of geotextiles in contact with soils. The gradient ratio is defined as:

$$GR = i_{LG}/i_S \tag{1.3}$$

where: i_{LG} is the hydraulic gradient along the soil thickness and i_S is the hydraulic gradient at soil-geotextile interface.

A GR value of 3 has been adopted as an upper limit for the acceptance of soil-geotextile filter compatibility.

Apart from value of GR, the number of constrictions m, the number of "windows" delimited by three or more fibers (ASTM D7178-16) in which soil particles could migrate, should be also determined for nonwoven geotextiles:

$$m = t/d_c \tag{1.4}$$

where: t is the geotextile thickness and d_c is the average distance between two constrictions.

The number of constrictions can be used to differentiate nonwoven geotextiles with different structures but similar opening sizes. Optimal constrictions numbers should range between 25 and 45 (Giroud 2010). However more research is needed to define the implication of number of constrictions on the prevalence of the clogging mechanism.

In this paper the permeability criterion was checked for nonwoven geotextile after exploitation.

2 Materials and Methods

The main aim of this paper is to examine the water permeability normal to the plane under load of 2, 5, 20 and 200 kPa of PP, PET nonwoven geotextile samples (Table 1) taken from the earthfill dam Białobrzegi in Poland after 22 years of exploitation in drainage system (Fig. 1).

Table 1. The properties of tested materials before exploitation

Parameter	Unit	Nonwoven geotextile
Mass per unit	g/m^2	460
Thickness under 2 kPa	mm	5.10
Tensile strength - MD	N	500
Tensile strength - CMD	N	650
Permeability coefficient	m/s	0.00268

The water permeability normal to the plane under load was determined according to EN ISO 10776:2012 in the Laboratory Water Center WULS (Fig. 2).

At the beginning the geotextiles samples were placed under water containing a wetting agent and were left to saturate for 24 h. The surface of geotextile specimen was 0.001963 m^2.

Then the specimen was placed in the cylinder with glass aggregate and supporting steel mesh to reproduce the soil natural conditions and to avoid deformation of geotextile by the pressure of water flowing through the specimen. The value of hydraulic gradient was subsequently: 3 and 5 mm.

Testing the water permeability coefficient k_g was carried out with the loads of subsequently: 2, 5, 20 and 200 kPa by using a special piston (Fig. 3) and calculated using the following formula:

$$k_{GT} = (V \times g)/(A \times t \times \Delta h) \tag{1.5}$$

where: V is the water volume measured (m^3), g is the average thickness of 10 samples of tested material under given load (m), A is the exposed specimen area (m^2), t is the time measured to achieve the volume V (s) and Δh is the pressure differential under and over the specimen, expressed as the height of water column (m).

3 Results and Discussion

Table 2 presents the thickness of tested nonwoven geotextile specimens under different load.

Figure 4 shows the change of the mean values of water permeability coefficient under load of tested samples.

The obtained results confirm the reduction of water permeability characteristics under load. The permeability coefficient decreased approximately 1.42, 1.64, 1.90 and

Fig. 1. Nonwoven geotextile samples after exploitation

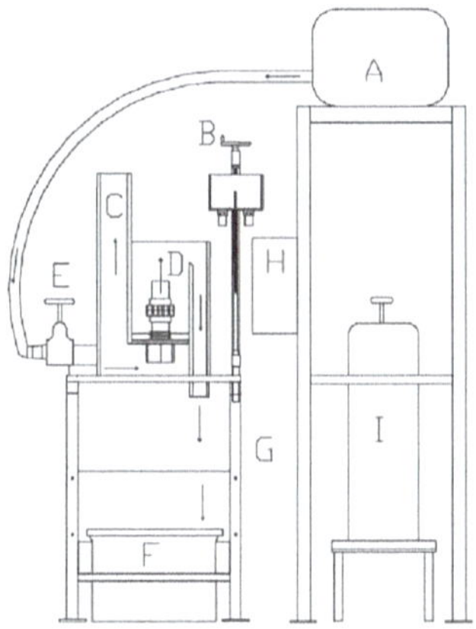

Fig. 2. Scheme of laboratory equipment: A - water tank, B - adjustment of head loss, C - glass casing, D - sample holder, E - throttle walve, F - water collection tank, G - fastening, H - recording equipment, I - water deaeration tank (Stępień et al. 2012)

Fig. 3. Piston and sample holder with glass aggregate and steel mesh

Table 2. Thickness characteristics of tested materials

Load (kPa)	Thickness (mm)
0	5.05
2	4.62
5	4.24
20	3.39
200	2.14

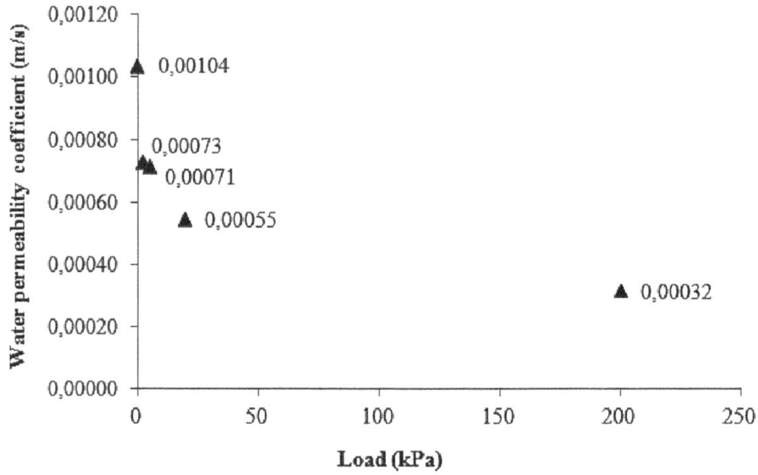

Fig. 4. Relationship between load and water permeability coefficient for tested nonwoven geotextile

3.27 times for the tested nonwoven geotextile samples subjected to respective loads of 2, 5, 20 and 200 kPa.

According to formula 1.2 the permeability of geotextile normal to the plane should be greater than the soil permeability. To presuppose $\lambda = 10$ (Giroud 1982) and to know that at the Bialobrzegi earthfill dam the permeability coefficient of the soil (clSa) is equal 7.65×10^{-5} m/s, the condition is not met for the geotextile under load. What is more the water permeability coefficient of tested geotextile before exploitation was 0.00268 m/s so after 22 years this permeability characteristic decreased 2.6 times.

Decreasing of water permeability coefficient strongly depends on clogging. What is more, it is necessary to take sample from different place of drainage system of earthfill dam Białobrzegi to compare test results and to summarise unmistakable conclusions. Similar tests were performed for another geotextile sample taken from Białobrzegi earthfill dam and published by Miszkowska and Koda (2017). In mentioned paper, the permeability criterion was met even for the most unfavourable case (under load of 200 kPa). It shows that clogging can occur to varying degrees depending on factors such as physical parameters of geotextile after exploitation (especially on thickness and opening size) and variable soil and water conditions.

4 Conclusions

The present research shows that clogging has influence on the decrease of permeability of nonwoven geotextile. After 22 years of exploitation water permeability coefficient of tested samples decreased 2.6 times. What is more, the hydraulic properties of geotextile decrease under load, what should be included in the criteria for geotextile filter layers. Because of the variety of results obtained, more research would be beneficial for a better understanding of the effects of clogging on the filtration properties of nonwoven geotextile exploited in earthfill dam Białobrzegi.

References

ASTM D7178-16 standard practise for determining the number of constrictions "m" of non-woven geotextiles as a complementary filtration property

Cazzuffi D, Moraci N, Mandaglio MC, Ielo (2016) Evolution in design of geotextile filters. In: Proceedings of the geo-Chicago 2016, Chicago, USA, pp 40–63

Correia LGCS, Ehrlich M, Mendoca MB (2017) The effect of submersion in the ochre formation in geotextile filters. J Geotext Geomembr 45:1–7

EN ISO 10776:2012 Geotextiles and geotextile-related products. Determination of water permeability characteristics normal to the plane, under load

Giroud JP (1982) Filter criteria for geotextiles. In: Proceedings of the 2nd international conference on geotextiles, Las Vegas, USA, vol 1, pp 103–108

Giroud JP (2010) Development of citeria for geotextile and granular filters. In: Proceedings of the 9th international conference on geosynthetics, Guaruja, Brazil, pp 45–64

Junqueira FF, Silva ARL, Palmeira EM (2006) Performance of drainage systems incorporating geosynthetics and their effect on leachate properties. J Geotext Geomembr 24:311–324

Koda E, Miszkowska A, Stępień S (2016) Quality control of non-woven geotextiles used in a drainage system in an old remedial landfill. In: Proceedings of the geo-Chicago 2016, Chicago, USA, pp 254–263

McIssac R, Rowe RK (2006) Effect of filter-separators on the clogging of leachate collection systems. Can Geotech J 43:674–693

Miszkowska A, Koda E (2017) Change of water permeability of nonwoven geotextile exploited in earthfill dam. In: Proceedings of the 24th international PhD students conference, Brno, Czech Republic, pp 790–795

Miszkowska A, Lenart S, Koda E (2017) Changes of permeability of nonwoven geotextiles due to clogging and cyclic water flow in laboratory conditions. Water 9:660

Moraci N (2010) Geotextile filter: design, characterization and factors affecting clogging and blinding limit states. In: Proceedings of the 9th international conference on geosynthetics, Guaruja, Brazil, pp 413–435

Palmeira EM, Trejos Galvis HL (2017) Opening sizes and filtration behaviour of nonwoven geotextiles under confined and partial clogging conditions. Geosynt Int 24(2):125–138

Stępień S, Jędryszek M, Koda E (2012) Ocena zmian przepuszczalnosci geowlokniny zastosowanej jako filtr w systemie ujęcia odciekow ze skladowiska odpadow komunalnych. Prz Nauk Inz Kszt Srod 57:159–170

Veylon G, Stoltz G, Meriaux P, Faure Y-H, Touze-Foltz N (2016) Performance of geotextile after 18 years' service in drainage trenches. J Geotext Geomembr 44:515–533

Wu Ch-S, Hong Y-S, Wang R-H (2008) The influence of uniaxial tensile strain on the pore size and filtration characteristics of geotextiles. J Geotext Geomembr 26:250–262

Mechanical Damage of Geosynthetics Used in the Base Liner System of Landfills Caused by Different Aggregates

José Ricardo Carneiro(✉), Ana Cláudia Brás, and Maria de Lurdes Lopes

Construct-Geo, Faculty of Engineering, University of Porto, Rua Dr. Roberto Frias S/N, 4200-465 Porto, Portugal
rcarneir@fe.up.pt

Abstract. The installation process may cause unwanted changes in the mechanical behaviour of the geosynthetics. In this work, some geosynthetics used in the base liner system of landfills (a combination of a geotextile, a geomembrane and a geosynthetic clay liner) were submitted to mechanical damage under repeated loading tests (procedure adapted from EN ISO 10722), being monitored the changes occurred in their short-term behaviour. The mechanical damage under repeated loading tests consisted in placing the geosynthetics between a flexible base (granite residual soil with a relative density of 80%) and an aggregate and submitting them to cyclic dynamic loading. The geosynthetics were damaged with a natural aggregate (gravel 14/20), a recycled ceramic aggregate and *corundum* (synthetic aggregate used in EN ISO 10722). The damage suffered by the geosynthetics was evaluated by monitoring changes in their tensile, tearing (only for the geomembrane) and static puncture properties.

Keywords: Geosynthetics · Landfills · Base liner systems
Mechanical damage

1 Introduction

The landfill base liner systems are designed to protect soils and groundwater from the contaminants present in waste. Without proper containment, the waste (and resulting leachates) would be in contact with the surrounding environment, which could have a negative impact in human health. The base liner systems of municipal solid waste landfills are usually formed by a combination of geosynthetics (a geosynthetic clay liner, a geomembrane and a geotextile) installed over a layer of compacted clay (a granular drainage layer is placed over the geosynthetics).

The geosynthetic clay liners (GCL) are composite materials (typically formed by a layer of bentonite supported by geotextiles) that act as mineral sealing barriers, while geomembranes are impermeable sheets used for fluid containment (liquids or gases). The geotextiles are permeable fabrics normally used to drain, filter, separate, protect or reinforce. However, their main function when part of a base liner system is to protect

L. Zhan et al. (Eds.): ICEG 2018, ESE, pp. 692–699, 2019.
https://doi.org/10.1007/978-981-13-2224-2_86

the geomembrane from punctures or tears induced by the particles of the overlying drainage layer (often from natural origin – gravel).

The installation of the drainage layer may induce damage to the geosynthetics of the base liner system, causing unwanted changes in their mechanical and/or hydraulic properties (these changes must be known and accounted in design). The evaluation of installation damage can be performed by field tests (installation under real conditions and immediate recovery of the materials for analysis) or by laboratory tests that try to simulate the damaging actions (for example, the test described in EN ISO 10722 [1]).

This work studies the effect of mechanical damage under repeated loading (for simplification purposes, hereinafter designated only by mechanical damage) in the short-term mechanical behaviour of geosynthetics used in the base liner system of landfills. For that purpose, the geosynthetics (a combination of a geotextile, a geomembrane and a GCL) were submitted to mechanical damage tests with different aggregates (tests adapted from EN ISO 10722 [1]). The geosynthetics were damaged with a natural aggregate (gravel 14/20 – crushed granite), with a recycled ceramic aggregate (construction and demolition waste) and with *corundum* (synthetic aggregate used in EN ISO 10722 [1]). The main goals of the work included: (1) evaluation of the effect of different aggregates in the damage suffered by the geosynthetics, (2) comparison of the damage caused by the natural and recycled aggregates with the damage induced by *corundum* and (3) evaluation if the recycled ceramic aggregate (in terms of mechanical damage induced to the geosynthetics) can be used as a viable alternative for the granular drainage layer (thereby reducing the use of natural resources).

2 Materials and Methods

2.1 Geosynthetics

The geosynthetics studied in this work included a nonwoven polypropylene (PP) geotextile, a high-density polyethylene (HDPE) geomembrane and a GCL (formed by a layer of sodium bentonite supported by two PP geotextiles). The main characteristics of the geosynthetics can be found in Table 1. The sampling and preparation of test-specimens (for the characterisation and mechanical damage tests) was carried out according to EN ISO 9862 [2].

Table 1. Main characteristics of the geosynthetics

Geosynthetic	Composition	μ_A (g.m^{-2})	t (mm)
Geotextile	PP	411 (17)	2.81 (0.10)
Geomembrane	HDPE	1773 (18)	1.96 (0.03)
GCL	PP + Bentonite	6858 (171)	8.44 (0.48)

(standard deviations in brackets)
μ_A - Mass per unit area; t - Thickness

2.2 Mechanical Damage Tests

The mechanical damage tests were performed in a prototype equipment formed by a lower box (where the test base was placed), an upper box (where the aggregates were placed), a loading plate and a compression machine (the geosynthetics were placed between the lower box and upper box) (Fig. 1). The full description of the equipment can be found in Lopes and Lopes [3].

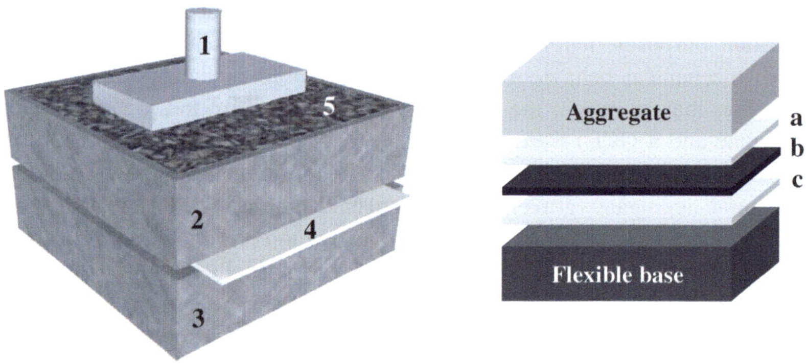

Fig. 1. Schematic representation of the equipment used in the mechanical damage tests: (1) loading plate; (2) upper box (aggregate inside); (3) lower box (flexible base); (4) geosynthetics; (5) aggregate; (a) geotextile; (b) geomembrane; (c) GCL

The geosynthetics (simultaneous testing of a combination of a geotextile, a geomembrane and a GCL) were placed between a flexible smooth base (residual soil of granite with a relative density of 80%) and an aggregate and submitted to cyclic dynamic loading between 5 and 500 kPa at the frequency of 1 Hz during 200 cycles (procedure adapted from EN ISO 10722 [1]) (Fig. 1). The aggregates used in the mechanical damage tests included gravel 14/20 (crushed granite), a recycled ceramic aggregate (construction and demolition waste – C&DW) and *corundum* (standard aggregate) (Fig. 2). The recycled ceramic aggregate was obtained from a recycling plant (RCD, SA – Figueira da Foz, Portugal) and came from the demolition of buildings and cleaning of lands with illegal deposition of construction and demolition residues (its main components included tiles, bricks, mosaics and masonry units). Some parameters for characterising the grain size distribution of the aggregates can be found in Table 2.

2.3 Evaluation of the Damage Suffered by the Geosynthetics

The degradation occurred in the geosynthetics during the mechanical damage tests was evaluated by monitoring changes in their mechanical properties. The mechanical characterisation of the geosynthetics included tensile tests (ASTM D638 [4] for the geomembrane and EN ISO 10319 [5] for the other geosynthetics), tearing tests (ASTM D1004 [6], only for the geomembrane) and static puncture tests (EN ISO 12236 [7]).

Fig. 2. Aggregates used in the mechanical damage tests: (a) gravel 14/20; (b) recycled ceramic aggregate (C&DW); (c) corundum

Table 2. Characterisation of the grain size distribution of the aggregates

Aggregate	D_{10} (mm)	D_{30} (mm)	D_{50} (mm)	D_{60} (mm)	D_{Max} (mm)
Gravel 14/20	9.4	13.0	16.3	17.1	20.0
C&DW	0.7	8.4	12.9	14.6	22.4
Corundum	5.8	7.1	7.9	8.4	10.0

D_X – x% grain size; D_{Max} – Maximum grain size

The tensile and tearing tests were performed with specimens prepared in the machine direction of production. Each tensile and static puncture test included the analysis of 5 specimens, while each tearing test included the analysis of 10 specimens.

The mechanical parameters determined in the tensile tests of the geotextile and GCL were tensile strength (T, in $kN.m^{-1}$) and elongation at maximum load (E_{ML}, in %). The tensile tests of the geomembrane included the determination of tension at yield (T_{Yield}, in MPa) and elongation at yield (E_{Yield}, in %). Tearing strength (F_T, in N) and puncture strength (F_P, in kN) were the parameters obtained in the tearing and static puncture tests, respectively. Some results are expressed in terms of retained strength (in %), obtained by dividing the strength (tensile, tearing or puncture) of the damaged samples by the respective strength of undamaged samples.

3 Results and Discussion

3.1 Geotextile

The mechanical damage tests had different effects on the geotextile, depending on the aggregate used. The tests with *corundum* provoked some visible damage in the geotextile, like cuts in fibres or abrasion. By contrast, the geotextile had no visible damage after the tests with the other aggregates.

The tensile properties of the geotextile suffered some changes (reductions) after the mechanical damage tests (Table 3). The test with *corundum* caused the highest reduction in the tensile strength (retained tensile strength of 76.1%) and elongation at maximum load (decrease from 82.7% to 50.9%) of the geotextile. The reduction in tensile strength induced by the recycled ceramic aggregate was not much different (slightly lower) from the reduction provoked by the natural aggregate. Indeed, the geotextile had retained tensile strengths of 93.9% and 91.1% after the mechanical damage tests with the recycled ceramic aggregate and with gravel 14/20, respectively. The reduction occurred in elongation at maximum load was also slightly less pronounced after the test with the recycled ceramic aggregate than after the test with gravel 14/20.

Table 3. Properties of the geotextile before and after the mechanical damage tests

Test	T ($kN.m^{-1}$)	E_{ML} (%)	F_P (kN)
Undamaged	29.49 (2.57)	82.7 (4.2)	5.39 (0.15)
MD test: gravel 14/20	26.87 (2.12)	61.2 (5.4)	4.70 (0.29)
MD test: C&DW	27.70 (1.88)	65.7 (2.9)	5.06 (0.21)
MD test: *corundum*	22.45 (1.21)	50.9 (2.5)	3.56 (0.44)

(standard deviations in brackets)

The mean curves "tensile force-elongation" obtained before and after the mechanical damage tests can be found in Fig. 3. In addition to the described changes in tensile strength and elongation at maximum load, Fig. 3 also shows some differences (small increases) in the rigidity of the geotextile (between about 20% and 60% elongation) after the tests with the recycled ceramic aggregate and with gravel 14/20. These small differences do not necessarily mean a change in the tensile behaviour of the geotextile after the mechanical damage tests. In fact, they may be due to the heterogeneity that typically exists in nonwoven geotextiles.

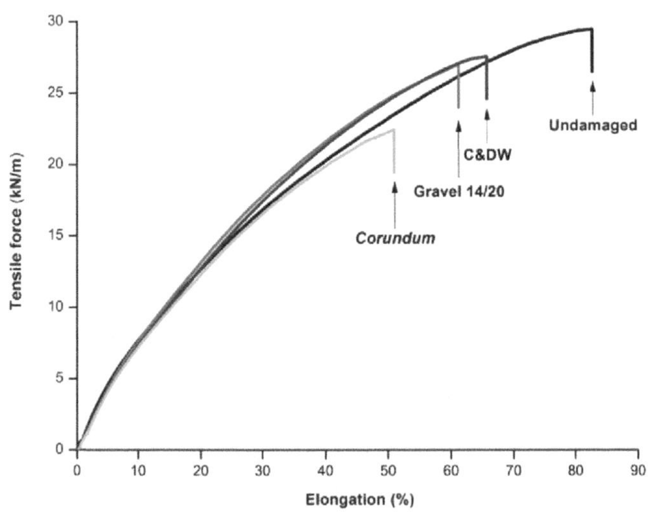

Fig. 3. Mean curves "tensile force-elongation" obtained for the geotextile before and after the mechanical damage tests

Similarly to what happened for tensile strength, the puncture strength of the geotextile also suffered the highest reduction after the mechanical damage test with *corundum* (Table 3). This reduction (decrease of 34.0%) was even higher than the reduction occurred in tensile strength (decrease of 23.9%). The reduction in puncture strength caused by the natural aggregate (retained puncture strength of 87.2%) was more pronounced than the reduction induced by the recycled ceramic aggregate (retained puncture strength of 93.9%).

The damage (deterioration of mechanical properties) caused by the natural aggregate and by the recycled ceramic aggregate was less pronounced than the damage provoked by *corundum*. This can be explained by the fact that *corundum* (despite having particles of lower dimension when compared to the other aggregates) had rough and angular particles, inducing cuts in fibres and having an abrasive effect in the nonwoven structure of the geotextile (as observed by naked eye).

3.2 Geomembrane

Contrarily to what happened for the geotextile (in direct contact with the different aggregates), the mechanical damage tests did not induce much relevant changes in the short-term mechanical behaviour of the geomembrane (tested between the geotextile and the GCL) (Table 4). Indeed, the geomembrane had high retained tensions at yield (between 103.5% and 107.0%), high retained tearing strengths (between 101.7% and 106.4%) and high retained puncture strengths (between 94.1% and 94.9%) after the mechanical damage tests. In addition, elongation at yield remained practically unchanged. The absence of considerable damage (reflected by the existence of only minor changes in the mechanical behaviour of the geomembrane, which may be due to variability provoked by some heterogeneity) indicated that the geotextile was able to protect properly the geomembrane against mechanical damage (independently of the aggregate used), fulfilling properly its protective function.

Table 4. Properties of the geomembrane before and after the mechanical damage tests

Test	T_{Yield} (MPa)	E_{Yield} (%)	F_T (N)	F_P (kN)
Undamaged	20.0 (0.6)	20.5 (1.2)	345 (6)	5.93 (0.09)
MD test: gravel 14/20	21.4 (0.3)	20.5 (0.9)	364 (10)	5.58 (0.12)
MD test: C&DW	21.2 (0.2)	20.9 (1.0)	351 (11)	5.60 (0.05)
MD test: *corundum*	20.7 (0.4)	20.5 (0.8)	367 (19)	5.63 (0.09)

(standard deviations in brackets)

3.3 Geosynthetic Clay Liner

Similarly to what happened for the geomembrane, the mechanical properties of the GCL also remained practically unchanged after the mechanical damage tests (Table 5). Indeed, the GCL had retained tensile strengths between 96.6% and 101.5% and retained puncture strengths between 102.8% and 104.8% after the mechanical damage tests. Elongation at maximum load also suffered only minor changes. The geotextile and geomembrane (placed above the GCL) protected the GCL against the damage provoked by the aggregates. In addition, the flexible base (placed below the GCL) was formed by fine particles (D_{50} of 1.0 mm and D_{Max} of 4.8 m), had a smooth surface and did not induce damage to the GCL.

Table 5. Properties of the GCL before and after the mechanical damage tests

Test	T ($kN.m^{-1}$)	E_{ML} (%)	F_P (kN)
Undamaged	19.15 (0.90)	28.7 (2.7)	2.90 (0.17)
MD test: gravel 14/20	19.39 (1.36)	29.3 (2.5)	3.04 (0.36)
MD test: C&DW	18.50 (0.92)	27.4 (1.7)	2.98 (0.34)
MD test: *corundum*	19.44 (1.71)	28.6 (1.3)	2.99 (0.31)

(standard deviations in brackets)

4 Conclusions

The mechanical damage tests provoked some relevant changes in the short-term mechanical behaviour of the geotextile. The highest deterioration of its mechanical properties was induced by *corundum* (aggregate with rough and angular particles, which caused cuts in fibres and abrasion). The damage provoked by the natural aggregate (gravel 14/20) was not much different from the damage induced by the recycled ceramic aggregate. Still, the deterioration of the mechanical properties of the geotextile was slightly more pronounced in the mechanical damage tests with the natural aggregate (probably due to the fact that gravel 14/20 was formed by particles with higher grain size than the recycled ceramic aggregate).

Contrarily to what happened for the geotextile, the geomembrane and the GCL had no relevant changes in their mechanical properties after the mechanical damage tests (independently of the aggregate used). This way, the geotextile was able to protect the geomembrane from damage.

Regarding the possibility of using the recycled ceramic aggregate in the drainage layer of landfills, this work does not allow definitive conclusions since it only studied the mechanical damage induced by the aggregate to the geosynthetics present in the base liner system (further studies, like the determination of the drainage capabilities of the aggregate, must be performed). Nevertheless, the damage provoked by the recycled ceramic aggregate in the short-term mechanical behaviour of the geotextile was lower than the damage induced by *corundum* or by the natural aggregate, which opens good perspectives for the use of this type of aggregates in the drainage layer of landfills (reducing the exploitation of natural resources).

Acknowledgements. This work was financially supported by projects POCI-01-0145-FEDER-028862 and POCI-01-0145-FEDER-007457, funded by FEDER funds through COMPETE 2020 – "Programa Operacional Competitividade e Internacionalização" (POCI) and by national funds (PIDDAC) through FCT/MCTES. J.R. Carneiro would like to thank FCT for the research grant SFRH/BPD/88730/2012 (grant supported by POPH/POCH/FSE funding).

References

1. EN ISO 10722 (2007) Geosynthetics – Index test procedure for the evaluation of mechanical damage under repeated loading. Damage caused by granular material
2. EN ISO 9862 (2005) Geosynthetics – Sampling and preparation of test specimens
3. Lopes MP, Lopes ML (2003) Equipment to carry out laboratory damage during installation tests on geosynthetics. Geotech J Port Geotech Soc 98:7–24 (in Portuguese)
4. ASTM D638 (2014) Standard test method for tensile properties of plastics
5. EN ISO 10319 (2008) Geosynthetics – Wide-width tensile test
6. ASTM D1004 (2013) Standard test method for tear resistance (graves tear) of plastic film and sheeting
7. EN ISO 12236 (2006) Geosynthetics – Static puncture test (CBR test)

Resistance of Geotextiles Against the Isolated and Combined Effect of Mechanical Damage Under Repeated Loading and Abrasion

José Ricardo Carneiro(✉), Filipa Escórcio, and Maria de Lurdes Lopes

Faculty of Engineering, Construct-Geo, University of Porto, Rua Dr. Roberto Frias s/n, 4200-465 Porto, Portugal
rcarneir@fe.up.pt

Abstract. The installation procedures (which may cause mechanical damage) and abrasion may provoke unwanted changes in the properties of the geosynthetics, affecting their performance. This work evaluated the resistance of two geotextiles (with different structures) against two degradation mechanisms: mechanical damage under repeated loading and abrasion. The geotextiles were (1) exposed in isolation to each degradation mechanism and (2) exposed consecutively to both degradation mechanisms. The damage suffered by the geotextiles (in the degradation tests) was evaluated by monitoring changes in their tensile and static puncture behaviour. Based on the changes occurred in tensile and puncture strength, reduction factors were determined. The reduction factors obtained in the successive exposure to mechanical damage under repeated loading and abrasion were compared with the reduction factors determined by the traditional methodology for the combined effect of those degradation mechanisms.

Keywords: Geotextiles · Mechanical damage · Abrasion · Reduction factors

1 Introduction

In their applications, the geosynthetics can be exposed to many degradation agents capable of having a negative effect on their short and long-term behaviour. The most common degradation agents include: chemical substances (like acids or alkalis), high temperatures, oxygen, solar radiation and other weathering agents, creep or abrasion. The installation process can also induce damage to the geosynthetics.

In order to account for the degradation that geosynthetics suffer over time, reduction factors are often used in design. Each reduction factor usually represents a decrease in resistance (known or estimated) induced by one, or more, degradation mechanisms. For example, for reinforcement applications, the tensile strength of geosynthetics is typically affected by a set of reduction factors accounting for the effects of installation damage, creep, atmospheric agents and chemical and biological agents.

L. Zhan et al. (Eds.): ICEG 2018, ESE, pp. 700–707, 2019.
https://doi.org/10.1007/978-981-13-2224-2_87

The actual design methods, standard degradation tests and most studies found in literature consider the isolated action of the degradation agents, not contemplating possible interactions between them. However, in real cases, the damage suffered by the geosynthetics will always be due to the combined effect of many degradation agents, being possible the occurrence of interactions between them. The combined effect of two degradation agents can be different (more severe) from the sum of their individual actions. Examples of interactions can be found in Carneiro *et al.* [1], Dias *et al.* [2] and Carneiro *et al.* [3].

In this work, two geotextiles (with different structures) were submitted to mechanical damage under repeated loading, abrasion and mechanical damage under repeated loading followed by abrasion (combined action). The main goals of the work included: (1) determination of the effect of the degradation mechanisms in the mechanical properties of the geotextiles, (2) evaluation of possible interactions between the degradation mechanisms and (3) comparison of the reduction factors obtained for the combined effect of mechanical damage under repeated loading and abrasion by the traditional methodology (determination of reduction factors in isolation for each degradation mechanism and further multiplication) and by the successive exposure to both degradation mechanisms.

2 Materials and Methods

2.1 Geotextiles

This work studied two geotextiles: a woven (designated by W200) and a nonwoven (designated by NW350). The designations attributed to the geotextiles contain information about their type ("W" for woven and "NW" for nonwoven) and about their mass per unit area (200 $g.m^{-2}$ and 350 $g.m^{-2}$; values defined by the producer). The main characteristics of the geotextiles can be found in Table 1. The sampling process (for the characterisation and degradation tests) was carried out according to EN ISO 9862 [4].

Table 1. Main characteristics of the geotextiles

Geotextile	Polymer	μ_A ($g.m^{-2}$)	t (mm)
W200	HDPE	204 (± 1)	0.85 (± 0.01)
NW350	PP	349 (± 9)	2.94 (± 0.06)

(95% confidence intervals in brackets)
μ_A - Mass per unit area; t - Thickness, HDPE - High-density polyethylene; PP - Polypropylene

2.2 Degradation Tests

Initially, the geotextiles were exposed in isolation (single exposure) to mechanical damage under repeated loading and abrasion. Then, the materials were successively exposed to both degradation mechanisms (multiple exposure).

The mechanical damage under repeating loading tests (for simplification, hereinafter designated only by mechanical damage tests) were carried out according to EN ISO 10722 [5]. The geotextiles were placed between two layers of a synthetic aggregate (*corundum*) and submitted to dynamic loading between (5.0 ± 0.5) kPa and (500 ± 10) kPa at the frequency of 1 Hz for 200 cycles. The aggregate (grain size between 5 and 10 mm) was sieved (5 mm aperture sieve) after every 3 uses (the particles retained in the sieve were reused) and totally discarded after 20 uses.

The abrasion tests (EN ISO 13427 [6]) consisted in placing the geotextiles in a stationary platform where they were rubbed by a P100 abrasive. The abrasive (installed in a sliding platform) was moved under controlled pressure (6 kPa) along a horizontal axis during 750 cycles (each cycle corresponded to a double passage of the abrasive by the geotextiles).

2.3 Evaluation of the Damage Suffered by the Geotextiles

The damage suffered by the geotextiles during the degradation tests was evaluated by monitoring changes in their tensile and static puncture behaviour (tests according to EN ISO 10319 [7] and EN ISO 12236 [8], respectively). The mechanical parameters determined in the tensile tests (values in the machine direction of production) included tensile strength (T, in $kN.m^{-1}$) and elongation at maximum load (E_{ML}, in %). Puncture strength (F_P, in kN) was the mechanical parameter determined in the static puncture tests. Each tensile and static puncture test included the analysis of 5 specimens. The results are presented with 95% confidence intervals determined according to Montgomery and Runger [9]. Some results are expressed in terms of retained strength (in %), obtained by dividing the strength (tensile or puncture) of the damaged samples by the respective strength of reference samples (undamaged).

The reduction factors (RF) for the effects of mechanical damage (RF_{MD}), abrasion (RF_A) and mechanical damage followed by abrasion (RF_{MD+A}) were obtained by the following equation:

$$RF = R_{Undamaged}/R_{Damaged} \quad (1)$$

Where, $R_{Undamaged}$ and $R_{Damaged}$ represent, respectively, the resistance (tensile or puncture strength) of the geotextiles before and after the degradation tests. The reduction factor obtained by the traditional methodology for the combined effect of mechanical damage and abrasion ($RF_{MD+A\ Trad}$) was obtained by Eq. 2.

$$RF_{MD+A\,Trad} = RF_{MD} \text{x } RF_A \quad (2)$$

The reduction factors determined in this work correspond to specific conditions and cannot be generalized nor applied directly in the design. For being used in the design, the reduction factors must be analysed case by case, taking into consideration the particular conditions of the construction.

3 Results and Discussion

3.1 Woven Geotextile W200

The mechanical damage tests caused cuts in filaments and some holes in geotextile W200. By contrast, no visible defects were found in geotextile W200 after the abrasion test. The successive exposure to both degradation tests provoked similar damage to that caused by the mechanical damage test in isolation.

The mechanical damage test induced a pronounced reduction in the mechanical strength of geotextile W200 (retained tensile strength of 47.4% and retained puncture strength of 36.5%) (Table 2). The reduction in tensile strength was accompanied by a decrease in elongation at maximum load (from 52.1% to 20.0%).

Table 2. Mechanical properties of geotextile W200 before and after the degradation tests

Test	T ($kN.m^{-1}$)	E_{ML} (%)	F_P (kN)
Undamaged	26.59 (± 0.92)	52.1 (± 6.1)	4.22 (± 0.11)
Mechanical damage (MD)	12.60 (± 0.67)	20.0 (± 3.2)	1.54 (± 0.22)
Abrasion	23.21 (± 1.43)	33.0 (± 5.7)	3.96 (± 0.26)
MD + Abrasion	10.62 (± 1.18)	16.0 (± 3.1)	1.04 (± 0.24)

(95% confidence intervals in brackets)

The changes induced by the abrasion test in the mechanical properties of geotextile W200 were less pronounced than the changes provoked by the mechanical damage test (Table 2). Indeed, the geotextile W200 still had relatively high tensile and puncture strengths (retained strengths of 87.3% and 93.8%, respectively) after abrasion. The reduction occurred in elongation at maximum load (from 52.1% to 33.0%) was also lower. The successive exposure to both degradation mechanisms caused the highest reductions in the tensile properties and puncture strength of geotextile W200 (retained tensile strength of 39.9% and retained puncture strength of 24.6%; decrease of elongation at maximum load from 52.1% to 16.0%).

3.2 Nonwoven Geotextile NW350

The degradation tests had different effects in geotextile NW350 (different types of visible damage). The mechanical damage test caused cuts in fibres and imprisonment of fine particles in the nonwoven structure (Fig. 1b) (particles resulting from the fragmentation of *corundum*). The abrasion test provoked the disintegration of the surface of geotextile NW350 (in contact with the abrasive) – the change in texture is perceptible in Fig. 1c. The combined action of both degradation tests led to cuts in fibres, which lined up and formed clusters perpendicularly to the direction of motion of the sliding platform (where the abrasive was installed) (Fig. 1d).

The changes occurred in the tensile properties (caused by the degradation tests) were much less pronounced in geotextile NW350 than in geotextile W200 (materials with different structures and masses per unit area) (Table 3). Indeed, geotextile NW350

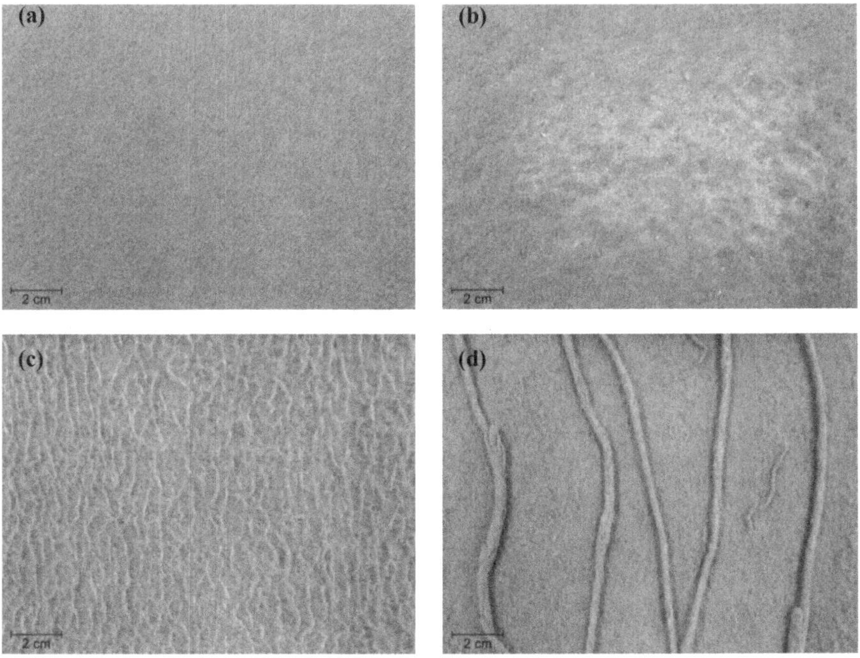

Fig. 1. Geotextile NW350 (a) before the degradation tests (undamaged), (b) after mechanical damage, (c) after abrasion and (d) after mechanical damage followed by abrasion

Table 3. Mechanical properties of geotextile NW350 before and after the degradation tests

Test	T ($kN.m^{-1}$)	E_{ML} (%)	F_P (kN)
Undamaged	21.96 (± 0.29)	84.5 (± 8.4)	4.45 (± 0.31)
Mechanical damage (MD)	19.12 (± 1.04)	65.7 (± 5.7)	3.54 (± 0.20)
Abrasion	20.46 (± 0.51)	74.3 (± 10.8)	4.22 (± 0.14)
MD + Abrasion	16.10 (± 2.95)	57.6 (± 12.2)	3.52 (± 0.12)

(95% confidence intervals in brackets)

still had a relatively high tensile strength after the mechanical damage test (retained tensile strength of 87.1% for geotextile NW350 and retained tensile strength of 47.4% for geotextile W200). For geotextile NW350 (and contrarily to what happened for geotextile W200), the effects of the isolated actions of the mechanical damage and abrasion tests were relatively close (retained tensile strengths of 87.1% and 93.2%, respectively). The highest reduction in tensile strength (decrease of 26.7%) occurred, once again, after the successive exposure to mechanical damage and abrasion. The reductions observed in tensile strength were accompanied by reductions in elongation at maximum load (highest decrease induced by the combined action of both degradation mechanisms).

Similarly to what was observed for tensile strength, the reductions occurred in puncture strength (caused by the degradation tests) were much less pronounced in geotextile NW350 than in geotextile W200 (retained puncture strengths of 79.1% and 24.6% for geotextiles NW350 and W200, respectively, after the combined action of both degradation mechanisms) (Table 3). For geotextile NW350, and contrarily to what happened for tensile strength, the reduction in puncture strength induced by the isolated action of mechanical damage (retained puncture strength of 79.6%) was identical to the reduction caused by the combined action of mechanical damage and abrasion (retained puncture strength of 79.1%). The isolated effect of abrasion led to the lowest variation (a slight reduction) in the puncture strength of geotextile NW350 (retained puncture strength of 94.8%).

Taking into account the results obtained for both geotextiles, the isolated action of abrasion was less damaging (lower deterioration of mechanical properties) than the isolated action of mechanical damage. The most pronounced reductions in mechanical strength (which leads to higher reduction factors) occurred in the successive exposure of the geotextiles to both degradation mechanisms.

3.3 Comparison of Reduction Factors: Successive Exposure vs. Traditional Methodology

Due to the highest reductions occurred in mechanical strength, the highest reduction factors were obtained in the successive exposure of the geotextiles to mechanical damage and abrasion (Table 4). The isolated effect of abrasion led to the lowest reduction factors (close to 1), reflecting the inexistence of pronounced changes in the mechanical strength of the geotextiles.

Table 4. Reduction factors obtained for tensile strength and puncture strength

Geotextile	Reduction factor	Tensile strength	Puncture strength
W200	RF_{MD}	2.11	2.74
	RF_{A}	1.15	1.07
	RF_{MD+A}	2.50	4.06
	$RF_{MD+A\ Trad}$	2.43	2.93
NW350	RF_{MD}	1.15	1.26
	RF_{A}	1.07	1.05
	RF_{MD+A}	1.36	1.26
	$RF_{MD+A\ Trad}$	1.23	1.32

The reduction factors obtained in the successive exposure of the geotextiles to mechanical damage and abrasion were compared with the reduction factors determined by the traditional methodology for the combined effect of those degradation mechanisms (determination of the reduction factors in isolation for each degradation mechanism and further multiplication) (Table 4).

For geotextile W200, the reduction factors obtained in the successive exposure to mechanical damage and abrasion were higher than those obtained by the traditional methodology for the combined effect of both degradation mechanisms. The difference was minimal for the reduction factors obtained for tensile strength (2.50 and 2.43, respectively) but pronounced for the reduction factors obtained for puncture strength (4.06 and 2.93, respectively). These results indicate that the reduction factors determined by the traditional methodology may not be representing accurately (by underestimating) the combined effect of mechanical damage and abrasion.

Regarding geotextile NW350, the reduction factor obtained for tensile strength by the traditional methodology for the combined effect of mechanical damage and abrasion (reduction factor of 1.23) was slightly lower than the reduction factor obtained in the successive exposure to both degradation mechanisms (reduction factor of 1.36). By contrast, for puncture strength the traditional methodology led to a slightly higher reduction factor.

4 Conclusions

The degradation tests (isolated and successive exposure to mechanical damage and abrasion) caused some relevant reductions in the mechanical properties of the geotextiles. The isolated action of mechanical damage led to a higher deterioration of mechanical properties than the isolated action of abrasion. The combined action of mechanical damage and abrasion induced the highest deterioration of the mechanical properties of the geotextiles.

The reduction factors obtained in the successive exposure of the geotextiles to mechanical damage and abrasion tended to be higher than the reduction factors determined by the traditional methodology for the combined action of those degradation mechanisms. Therefore, the traditional methodology may not always be able to represent correctly (by underestimating) the resistance changes provoked by the combined action of multiple degradation mechanisms (due to the occurrence of interactions between them).

Finally, the definition of accurate reduction factors (taking into account the interactions that occur between the different degradation mechanisms) may contribute to improve the application of geosynthetics in Civil Engineering. Indeed, the definition of reliable reduction factors will allow a better design.

Acknowledgements. This work was financially supported by projects POCI-01-0145-FEDER-028862 and POCI-01-0145-FEDER-007457, funded by FEDER funds through COMPETE 2020 – "Programa Operacional Competitividade e Internacionalização" (POCI) and by national funds (PIDDAC) through FCT/MCTES. J.R. Carneiro would like to thank FCT for the research grant SFRH/BPD/88730/2012 (grant supported by POPH/POCH/FSE funding).

References

1. Carneiro JR, Almeida PJ, Lopes ML (2014) Some synergisms in the laboratory degradation of a polypropylene geotextile. Constr Build Mater 73:586–591
2. Dias M, Carneiro JR, Lopes ML (2017) Resistance of a nonwoven geotextile against mechanical damage and abrasion. Ciência e Tecnologia dos Materiais 29:177–181
3. Carneiro JR, Almeida PJ, Lopes ML (2018) Laboratory evaluation of interactions in the degradation of a polypropylene geotextile in marine environments. Advances in Materials Science and Engineering, article ID 9182658, 10p
4. EN ISO 9862 (2005) Geosynthetics – Sampling and preparation of test specimens
5. EN ISO 10722 (2007) Geosynthetics – Index test procedure for the evaluation of mechanical damage under repeated loading. Damage caused by granular material
6. EN ISO 13427 (1998) Geotextiles and geotextile-related products – Abrasion damage simulation (sliding block test)
7. EN ISO 10319 (2008) Geosynthetics – Wide-width tensile test
8. EN ISO 12236 (2006) Geosynthetics – Static puncture test (CBR test)
9. Montgomery DC, Runger GC (2010) Applied statistics and probability for engineers. Wiley, Hoboken

The Effect of Pore Size Deformation on Multi-pore Model of Woven Geotextiles in Tension Test

Wenfang Zhao and Xiaowu Tang(✉)

Zhejiang University, Hangzhou 310058, China
tangxiaowu@zju.edu.cn

Abstract. The filtration performance of woven geotextiles is a constant issue in water proof and flood prevention site. Woven geotextiles have been characterized by a two-dimensional single-pore model and linear change laws of characteristic pore size and percent open area (POA) with increasing of tensile strain have been applied by people before. However, percent open area (POA) which is decided by the ratio of pores area and geotextiles area is proved to be non-linear as tensile strain increases. In this paper, change laws of percent open area (POA) and characteristic pore size in uniaxial tension test are explored by using imagine analysis. Considering the effect of "pore size deformation coefficient" and "textile shrinkage coefficient", "multi-pore model" of woven geotextiles is come up with in this paper in order to improve change laws of percent open area (POA) and characteristic pore size in uniaxial tension test.

Keywords: Woven geotextiles · Percent open area · Characteristic pore size
Uniaxial tension strain · Imagine analysis · Pore size deformation
Textile shrinkage · Multi-pore model

1 Introduction

Geotextiles have been widely used in fields for water proof and urban construction as important filtration materials. The characteristic pore size and percent open area (POA) are important parameters reflecting the filtration performance of geotextiles [5, 7].

A test that exploring the relationship of tension, characteristic pore size and flow rate has been carried out by Wu [8] for needle-punched non-woven geotextiles. It is pointed out that as tensile strain increases, the flow rate increased at first, then it decreased, at the later stage, flow rate increased again in test. However, there are contradictions in the results of changing law of pore size and percent open area in tensile tests under different experimental conditions. By changing the number of weights, Fourie [10] explored the relationship between tensile force and pore size for different thicknesses of woven geotextiles without controlling the changing of tensile strain. It is pointed out that the pore size of thinner woven geotextiles increases as tensile force increasing while the pore size decreases for thicker woven geotextiles. Wu [9] found that the pore size increased with tension increasing. She [4] pointed out that

L. Zhan et al. (Eds.): ICEG 2018, ESE, pp. 708–714, 2019.
https://doi.org/10.1007/978-981-13-2224-2_88

the percent open area of woven geotextiles increased linearly with tensile strain increasing.

The change law of percent open area and tension for woven geotextiles has been studied, however, it did not take into account changes in the shape of the pores under tension as well as textiles area shrinkage. In this paper, by tracking the change in pore size and textile area, change laws of percent open area with tensile strain increasing was explored under strictly strain controlled conditions for test sample W_{150}. The multi-pore model considering not only fabric shrinkage but also the pore deformation was come up with in order to improve change laws of pore size and percent open area during the whole stage of tensile strain increasing in practical filtration applications.

2 Test Materials and Methods

2.1 Test Materials

Tested woven geotextile sample W_{150} is commonly used with the approximate mass per unit area of 150 g/m^2. Main physical and mechanical parameters are shown in Table 1.

Table 1. Physical and mechanical properties of W150 geotextiles.

W_{150} physical and mechanical properties of fabrics	Physical parameters
Thickness (mm)	0.22
Unit mass (g/m^2)	159
Vertical tensile strength (kN/m)	25.3
Zonal tensile strength (kN/m)	22.5
Film width (mm)	2.10
Film thickness (mm)	0.10
Material density (kg/m^3)	900

Ten W_{150} geotextiles with the same length of 20 cm and width of 10 cm were used to perform the longitudinal tensioning. Lengths of two ends were 5 cm and the area of tension for studying was 10 cm $\times$ 10 cm. The tension rate during test was 10 mm/min and peak strain was found to be 12.9% after which the films broke. Geotextiles were photographed with uniaxial tensile strains of 0, 3%, 6%, 9%, and 12%.

2.2 Digital Image Method

A uniform light source was placed behind the tested woven geotextiles and the focal length of SLR camera was set to be 50 mm while the number of pixel was set to be 2 million. The tripod was set to be self-timer mode and was corrected horizontally before the test started. The captured color image was converted into a binary image by using Matlab and each aperture pixel change was tracked and converted into actual pore size. Two converted binary images were shown in Fig. 1.

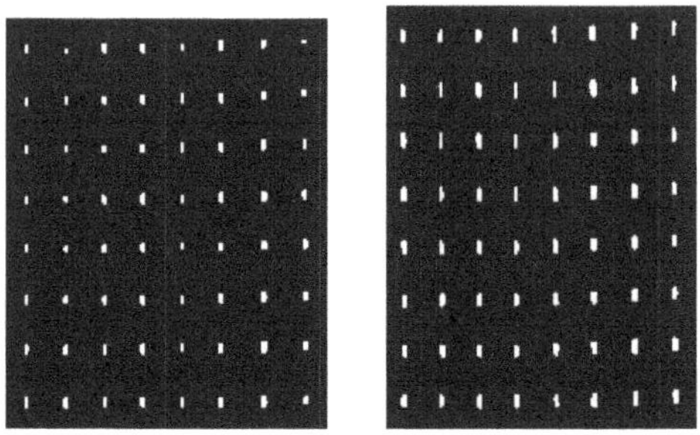

(a) Tensile strainε=3% (b) Tensile strainε=12%

Fig. 1. Binary images for tensile strains of 3% and 12%.

3 Analytical Theoretical Model

3.1 Single-Pore Model

The theoretical single-pore model made by She [4] was supposed to be made up of same squares. With tensile strain increasing, the percent open area of pores increases linearly as well. The theoretical characteristic pore size value and percent open area [4] are shown in (1) and (2).

$$O^{un} = \sqrt{A_0^{un}} = a\sqrt{\left(\frac{2t\rho}{\mu} - \frac{1}{\sqrt{1+\varepsilon}}\right)\left[\frac{2t\rho}{\mu}(1+\varepsilon) - 1\right]} \tag{1}$$

$$POA_{\varepsilon}^{un} = \frac{A_0^{un}}{A_s^{un}} = \frac{\left(2t\rho - \frac{\mu}{\sqrt{1+\varepsilon}}\right)[2t\rho(1+\varepsilon) - \mu]}{(2t\rho)^2(1+\varepsilon)} \tag{2}$$

3.2 Multi-pore Model and Percent Open Area

The zonal films are horizontally arranged while the longitude films are inclined with the same angle of α as shown in Fig. 2. Relationship of the minimum pore size b and film width a is shown in (3) and percent open area at the beginning is shown in (4).

$$b = \frac{\left(2a + 3acos\alpha - \frac{2a\mu}{\rho t} - \frac{3acos\alpha\mu}{\rho t}\right)}{\frac{8cos\alpha\mu}{\rho t}} + \frac{\sqrt{\left(2a + 3acos\alpha - \frac{2a\mu}{\rho t} - \frac{3acos\alpha\mu}{\rho t}\right)^2 - \frac{24cos\alpha\mu}{\rho t}\left(\frac{a^2\mu}{\rho t} - 2a^2\right)}}{\frac{8cos\alpha\mu}{\rho t}} \tag{3}$$

Where pore width $b_{1i} = i * b$, b_{21} and b_{22} are equal to b, b_{23} are equal to $2b$. Cell width M_{1i} is equal to $a/\cos\alpha + b_{1i}$, M_{2i} is equal to $a + b_{2i}$. (i = 1, 2, 3)

$$POA_0 = \frac{8b^2}{\left(\frac{a}{\cos\alpha} + 2b\right)(3a + 4b)} \tag{4}$$

With the increasing of the tensile strain in the warp direction, the duration of pore size deformation was divided into four stages.

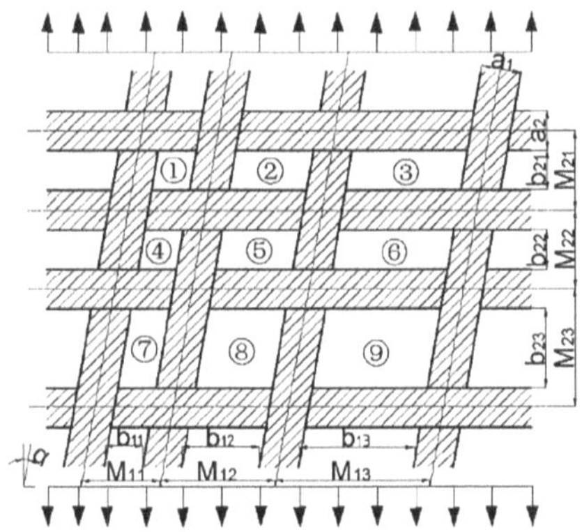

Fig. 2. The theoretical multi-pore model at 0% tensile strain.

At the first stage of tensile strain varying from 0% to 3%, small pore sizes increase rapidly as zonal films move vertically until the distance between adjacent latitudinal strips becomes 5/3b. The warp strip film rotates gradually until the angle of α reduces to 0 as shown in Fig. 3. As a result, the aperture area increases and percent open area increases as well, which is shown in (5).

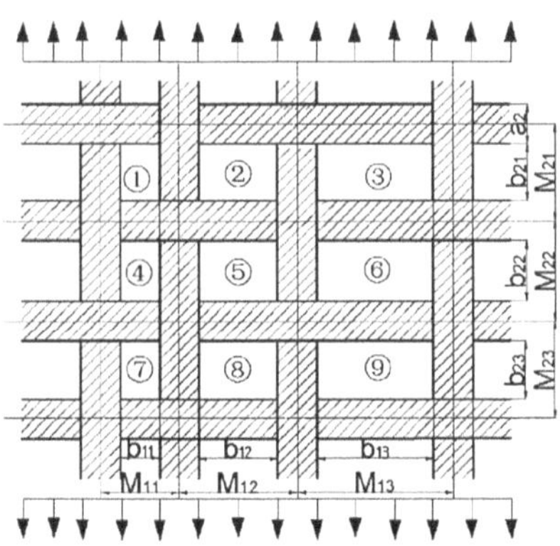

Fig. 3. The theoretical multi-pore model at 3% tensile strain.

Where pore width $b_{1i} = i * b$, $b_{2i} = 5/3b$. Cell width $M_{1i} = a + b_{1i}$, $M_{2i} = a + 5/3b$. (i = 1, 2, 3)

$$POA_1 = \frac{10b^2}{(3a+5b)(a+2b)} \tag{5}$$

In the second stage of tensile strain from 3% to 6%, the warp strip film becomes thinner with tensile strain increasing. The latitudinal strip films do not move and the percent open area at tensile strain of 6% is shown in (6).

$$POA_2 = \frac{(2b+av(\varepsilon-3\%))\left[\left(\frac{5}{3}b+a\right)(1+(\varepsilon-3\%))-a\right]}{(2b+a)\left(\frac{5}{3}b+a\right)(1+(\varepsilon-3\%))} \tag{6}$$

Where v is equal to $\frac{1}{\varepsilon-3\%}\left(1-\frac{1}{\sqrt{1+\varepsilon-3\%}}\right)$.

In the third stage, tensile strain varies from 6% to 9%. Warp films continue to be thinner and laterally move inside until the distance between any two adjacent warp films becomes 5/3b. The percent open area is shown in (7).

$$POA_3 = \frac{(x+av(\varepsilon-3\%))[(x+a)(1+(\varepsilon-3\%))-a]}{(x+a)^2(1+(\varepsilon-3\%))} \tag{7}$$

Where x is the equivalent pore size during tensile strain varying from 6% to 9%, which is shown in (8).

$$x = \frac{100b}{9}(\varepsilon-3\%)+b \tag{8}$$

The tensile strain of the fourth stage varies from 9% to 12%. Zonal films become thinner uniformly as tensile strain increases and POA is shown in (9).

$$POA_4 = \frac{\left(\frac{5}{3}b+av(\varepsilon-3\%)\right)\left[\left(\frac{5}{3}b+a\right)(1+(\varepsilon-3\%))-a\right]}{\left(\frac{5}{3}b+a\right)^2(1+(\varepsilon-3\%))} \tag{9}$$

4 Results

4.1 Pore Size

Sixty-four pores were selected in test with eight pores being arranged vertically and horizontally respectively. As a result of average tangential value of inclination of warp films, the inclination angle is supposed to be 20° for W_{150} in this test.

The results of experimental pore areas, experimental geotextile areas and percent open areas during the uniaxial tensile test are shown in Table 2.

Table 2. Total pore areas, geotextile areas and POA in tension test.

Tensile strain (%)	Total pore area (mm^2)	Geotextile area (mm^2)	Experimental POA (%)	Theoretical POA (%)
0	2.47	144	1.72	1.71
3	4.77	225	2.12	2.08
6	5.01	197	2.55	2.66
9	6.98	249	2.80	2.88
12	9.84	295	3.33	3.48

Table 2 shows that as tensile strain increases, the total pore area and percent open area increase as well. At the initial stage of uniaxial tension test, the tensile force acting on the woven geotextile sample has not overcome the frictional force between slats, shapes of pores are quite irregular and pore areas distribute non-uniformly. However, as tensile strain increases, geotextile area becomes smaller since warp and weft films have overcome the frictional resistance between slats. As films become thinner and thinner, pores gradually become more and more uniform. The width of films continue to decrease while the length of geotextiles increases and the pore size increases as well.

4.2 Percent Open Area

POA is the ratio of pore area to geotextile area. Figure 4 indicates three change laws of theoretical percent open area of the single-pore model, theoretical percent open area of the multi-pore model and the experimental percent open area with tensile strain increasing. It is noted that there are large errors between results of theoretical single-pore model and experimental percent open area since pores deform non-uniformly in practice. In the initial stage of stretching, the experimental POA increases at a slower speed than the theoretical single-pore value since the deformation of pore shape is the main change in this stage, however, the speed becomes quite slower at the middle stage since both geotextile areas and pore areas become smaller at this stage. The change of percent open areas then increases at the same speed as theoretical single-pore model as the result of uniform distribution of pore areas. Figure 4 shows that the result of multi-pore model is better than that of single-pore model.

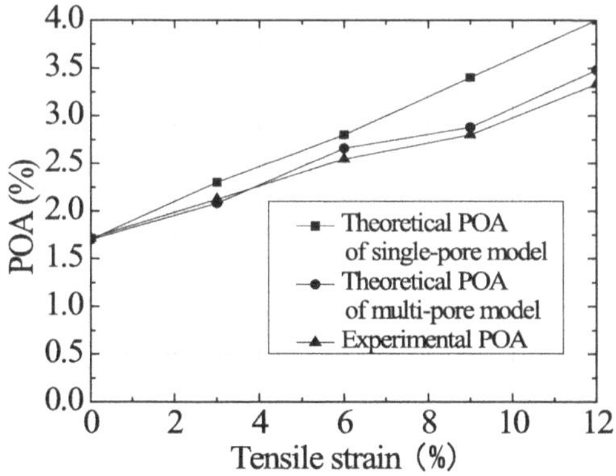

Fig. 4. Change laws of theoretical POA and experimental POA with tension strain increasing.

5 Conclusion

5.1 Movement of Strips During Tension Stage Due to Friction Resistance

It has been found that the friction resistance between strips prevents the movement between strips during the early stage of geotextile stretching, however, the tensile force of the woven geotextile in the middle stage overcomes the friction resistance between these strips and the pore deformation is obvious, then films move laterally and geotextile area shrinks until pores approximately become squares and pore size and percent open area change linearly.

5.2 Comparison of Experimental Results and Theoretical Values

Experimental results show that the theoretical percent open area of the multi-pore model is more similar to practical value than that of a single-pore model since the model combined the non-uniform distribution of 64 selected pore areas and irregular pore shapes which are not considered in uniform square model people applied before. In addition, the multi-pore model improved not only the specific change process of tension test including initial stage, middle stage and later stage but also the filtration measurement methods in construction sites.

Acknowledgement. The authors would like to acknowledge the financial support by the National Natural Science Foundation of China (Grant No. 51779218) and the Heritage Conservation Science and Technology Project of Zhejiang Province (Grant No. 2013010).

References

1. Xiaowu T, Lin T (2013) Prediction of pore size characteristics of woven slit-film geotextiles problem to tensile strains. Geotext Geomembr 38:43–50
2. Lin T (2012) Opening size of woven geotextiles varied with tensile loads. In: the 5th Asian society of geosynthetics, IGS Student Award, Thailand, pp 1073–1076
3. Lin T, Xiaowu T, Wei S, Bosong G (2013) Influence of uniaxial tensile strain on filtration characteristics of geotextiles. Chin J Geotech Eng 35(4):785–788
4. Wei S, Xiaowu T (2012) Change of pore size of woven geotextiles affected by uniaxial tension using image analysis method. Chin J Geotech Eng J 34(8):1522–1526
5. Bin B, Xiaowu T, Lin T, Shaoxing Q (2015) Influence of equal biaxial tension on opening sizes of nonwoven geotextiles. Rock Soil Mech 36(6):1615–1621, 1626
6. Lin T, Xiaowu T, Shaoxing Q (2016) Influence of biaxial tensile strains on pore size parameters of woven geotextiles. Chin J Geotech Eng 38(Suppl. 1):134–140
7. Wenfang Z, Xiaowu T, Bin B (2016) Effect of unsymmetrically biaxial tension on percent open area of geotextiles. Chin J Geotech Eng 38(Suppl 1):109-103
8. Wu CS, Hong YS (2016) The influence of tensile strain on the pore size and flow capability of needle-punched nonwoven geotextiles. Geosynth Int 23(6):422–434
9. Wu CS, Hong YS, Wang RH (2008) The influence of uniaxial tensile strain on the pore size and filtration characteristics of geotextiles. Geotext Geomembr 26(2):250–262
10. Fourie AB, Addis PC (1999) Changes in filtration opening size of woven geotextiles subjected to tensile loads. Geotext Geomembr 17(5):331–340

Experimental Investigation of Volume Change and Hydraulic Conductivity on Geosynthetic Clay Liner

Guangwei Zhang[1,2(✉)], Hhuyuan Zhang[1(✉)], Jinwen Liu[1,3], Lang Zhou[1], Ming Yan[1], and Jinfang Wang[4]

[1] Key Laboratory of Mechanics on Disaster and Environment in Western China, Lanzhou University, Ministry of Education, Lanzhou, China
602073909@qq.com
[2] PetroChina Research Institute of Exploration & Development-Northwest, Lanzhou, China
[3] Geo-Environment Monitoring Institute of Gansu Province, Lanzhou, China
[4] School of Resource and Environmental Science, Lanzhou University, Lanzhou, China

Abstract. The objective of this paper is to evaluate the hydraulic conductivity of a commercially-available geosynthetic clay liner (GCL) by Flexible-wall permeameter. The GCLs are produced by the same factory, but the content of bentonite is different. And hydraulic conductivity (HC) tests are performed on GCL specimens with different HC results (i.e., 10^{-9} cm/s and 10^{-6} cm/s). After 18 days, specimens were complete saturated under backpressure 20 kPa. The results of these tests indicate that there is a decrease in hydraulic conductivity of the GCL may be expected due to the consolidation of GCL. Measured hydraulic conductivities of GCL specimens with high HC permeated with distilled water under 20 kPa stress conditions that there is a slightly decrease and the HC keep in one order of magnitude, whereas measured hydraulic conductivities of low HC GCL specimens tested under similar conditions the HC keep in one order of magnitude. The test results show that similar hydraulic conductivity values are obtained for GCL specimens after complete saturated. In addition, the flow rate decreases with the time of test. The research findings may be of interest to researchers and engineers who design liners for landfills and other liquid containment facilities.

Keywords: Landfills · Geosynthetic clay liner · Hydraulic conductivity Volume change · Consolidation

1 Introduction

Geosynthetic clay liners (GCLs) consisting of bentonite placed between two geotextiles are used in various barrier systems to provide containment of contaminated soil and water [1]. Use of GCLs in such applications is desirable due to their extremely low hydraulic conductivity to water, simple and rapid installation, and easier quality assurance relative to CCLs [2]. Bentonite refers to any material that is primarily smectite group of minerals

L. Zhan et al. (Eds.): ICEG 2018, ESE, pp. 715–723, 2019.
https://doi.org/10.1007/978-981-13-2224-2_89

and whose physical properties are dictated by the smectite minerals. Characteristics of smectite minerals include large cation-exchange capacity, large specific surface area, high swelling potential, and low hydraulic conductivity to water [3].

The quality of bentonite has a significant role in the performance of GCLs. It found that the GCL with the higher quality bentonite is more susceptible to chemical attack than the GCL with the lower quality bentonite [4, 5]. In general, it showed that GCLs can withstand at least three freeze-thaw cycles without significant changes in hydraulic conductivity [6–8]. And bentonites are sensitive to chemical composition of the pore liquid that influences the thickness of the adsorbed layer, liquids that can cause the adsorbed layer to collapse also can cause the hydraulic conductivity to increase. The number of pore volumes of permeant flow and consequently the level of chemical equilibrium is shown to have a significant effect on the hydraulic conductivity. Several studies have been conducted to assess the performance of GCLs permeated with water and some chemical solution [8–10].

Effective stress is a significant variable that controls the behavior of bentonite, decreasing both hydraulic conductivity and the susceptibility of bentonite to chemical alterations. Increasing the effective stress on a GCL decreases the void ratio within the bentonite layer, which tends to lower its hydraulic conductivity [11–14].

However, it cost a lot of time to determine the hydraulic conductivity by permeant solutions in the lab. And few of studies focus on the volume changes of GCL in pilot process. This paper is to evaluate the hydraulic conductivity of GCLs were completely saturated under a low effective stress. Meanwhile, the influence of consolidation on hydraulic performance is also concerned in the test.

2 Materials and Methods

2.1 Geosynthetic Clay Liner

The GCL tested in this study is sold commercially and is a reinforced (needle-punched) GCL consisting of a layer of granular sodium bentonite between two nonwoven polypropylene geotextiles. The GCL bentonite is classified as a highplasticity clay (CH) according to the Unified Soil Classification System (ASTM D2487). For the sample roll utilized in this study, the average oven-dried bentonite mass per unit area was 5.00 kg/m^2 based on three bentonite samples collected from the sample roll. This value was used as the baseline bentonite mass per unit area for the GCL in this study. The natural moisture content of the GCL bentonite was about 12.00% at the time of specimen preparation. The ovendried GCL and the physical property of the GCL can be seen from Fig. 1 and Table 1. The mass per unit area and moisture content were determined in lab accordance with the standard for soil test method GB/T 50123-1999 of China.

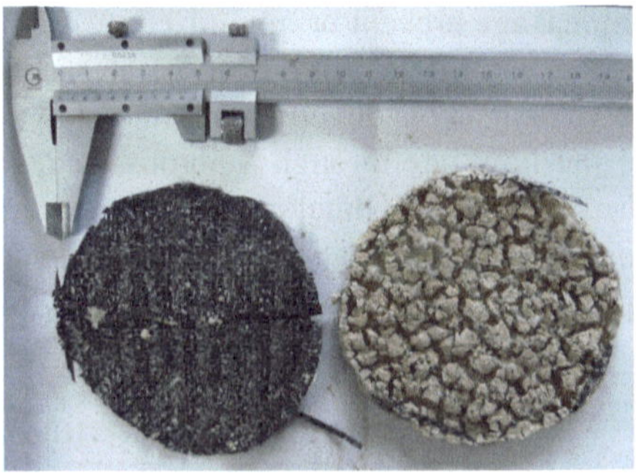

Fig. 1. Ovendried GCL after the test.

Table 1. Physical property of the GCL used.

Characteristic	Thickness after test (cm)	Mass per unit area (kg/m^2)	Moisture content (%)	
			Before test	Final test
GCL1	1.60	12.80	207	207
GCL2	1.00	7.50	229	238
GCL3	1.10	8.60	184	218
Standard of experiment	GB/T50123-1999			

2.2 Hydraulic Conductivity Tests

Specimens of the GCLs with nominal diameters of 103 mm were permeated using the falling-head procedure and flexible-wall permeameters as described in ASTM D5084. The test method may be utilized on all specimen types (undisturbed, reconstituted, remolded, compacted, etc.) that have a hydraulic conductivity less than 10^{-4} cm/s. The specimens were prehydrated by distilled water (DW), and then be put in a humidor for 48 h. Distilled water is also as the permeant liquid. Before the test, Specimens were exposed to the distilled water for at least 7 days by opening the influent port keeping effluent port closed under 20 kPa. Backpressure was also keep in 20 kPa in the whole test. The thickness of specimens was measured before and after the hydraulic conductivity tests using a calliper and a ruler. All the specimens were permeated at an average hydraulic gradient of 500 such that the average effective stresses at the cell, bottom (inflow), and top (outflow) of the specimens were 35 kPa, 15 kPa, and 0 kPa, respectively. While this hydraulic gradient is higher than the maximum gradient (i.e., 30) stipulated in ASTM D5084 for soils with low hydraulic conductivity 10^{-7} cm/s, hydraulic gradients ranging from 50 to 600 typically are used for measuring the hydraulic conductivity of GCLs. As shown by same scholar, the hydraulic conductivity of GCLs is affected to a greater extent by average effective stress than by the magnitude of hydraulic gradient. In summary, flexible-wall permeameter has the following advantages: (1) side-wall leakage prevent or minimized; (2) stress on specimen can be controlled; (3) volume changes or deformation can be measured.

In this tests, the flow velocity (v) is directly proportional to the hydraulic gradient I. Darcy's law, describing the flow rate through a porous medium, could be written as follows: $v = k \times I$, where k is the hydraulic conductivity (cm/s). The head of the influent at certain time intervals are recorded to calculate the hydraulic conductivities of the specimen according to Darcy's law. The hydraulic conductivity of the GCL can be defined as follows:

$$k = 2.3\frac{aL}{A(t_0 - t_s)}\lg\frac{(h_{in} - h_{out})_{initial} + 10p_{initial}}{(h_{in} - h_{out})_{final} + 10p_{final}} \quad (1)$$

Where k is hydraulic conductivities (cm/s); a is basal area of intake pipe or outlet pipe (cm^2); L is the height of the specimen (cm); A is the basal area of the specimen (cm^2); t_0 and t_s the times for the test (s); h_{in} and h_{out} are hydraulic heads in the intake or outlet pipe (cm), respectively; and p_{in} and p_{out} are the gas pressure on the hydraulic head (kPa).

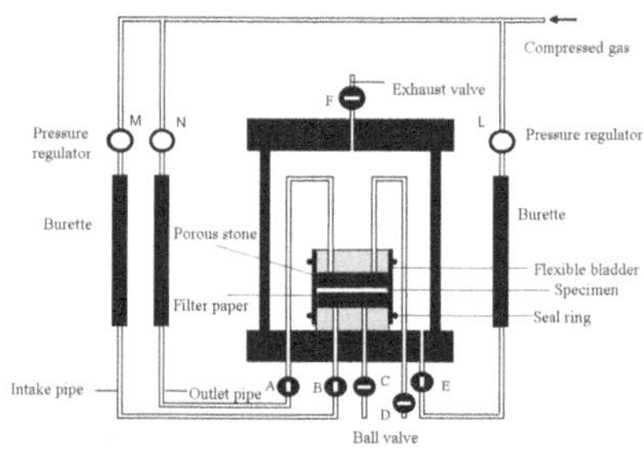

Fig. 2. Schematic diagram of flexible-wall permeameter.

3 Results and Discussion

3.1 Volume Changes of Saturation Processes

The volume changes for specimens of saturation under backpressure are shown in Figs. 3(a), 3(b), and 3(c). Figure 3(a) showed the GCL1 volume changes of exposed to the DW under 20 kPa backpressure. The cell total reduction for the specimen of the GCL1 was approximately 4.7 ml in the first day, and subsequently decreased to about 0 ml after 6 days. Thereafter, the volume of GCL1 remains essentially constant after one week. For the volume of the inflow pipe, there was a rise in the first day, because the specimen of the GCL1 was compressed under the backpressure 20 kPa. After that, there was a slightly increase on liquid level of the inflow of GCL1 in the 17 days. In the last day, the liquid level stayed the same. The volume of the outflow pipe, there is a drop in the first day, and then there was the slightly increase on liquid level of the outflow of GCL1 in the next 17 days. In the 18 days, the volume of the inflow and outflow were the same, specimens were complete saturated.

Figure 3(b) showed the GCL2 volume changes of exposed to the DI under 20 kPa backpressure. The cell total reduction for the specimen of the GCL2 was approximately 6.7 ml in the first day, and subsequently decreased to about 0 ml after 3 days. Thereafter, the volume of GCL1 remains essentially constant after 3 days. For the volume of the inflow pipe, there was a rise in the first day, because the specimen of the GCL2 was compressed under the backpressure 20 kPa. After that, there was a slightly increase on liquid level of the inflow of GCL2 in the 17 days. In the last day, the liquid level stayed the same. The volume of the outflow pipe, there is a drop in the first day, and then there was the slightly increase on liquid level of the outflow of GCL2 in the next 17 days. In the 18 days, the volume of the inflow and outflow were the same, specimens were complete saturated.

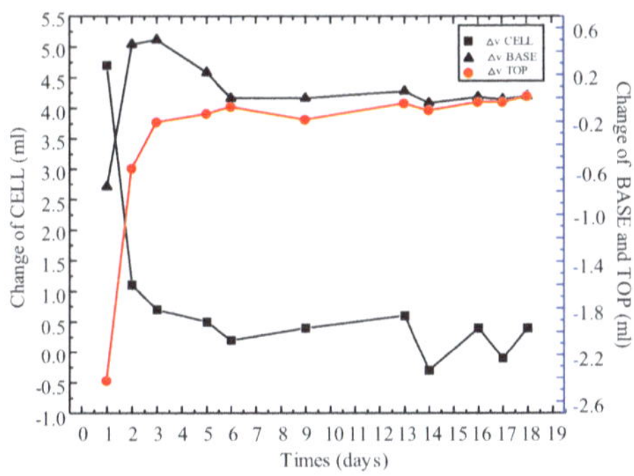

Fig. 3a. GCL1 volume changes of exposed to the DI under 20 kPa backpressure.

Figure 3(c) showed the GCL3 volume changes of exposed to the DI under 20 kPa backpressure. The cell total reduction for the specimen of the GCL3 was approximately 0.7 ml in the first day, and also stayed in the level of about 0 ml. For the volume of DW in the inflow pipe, there was a rise in the first day, because the specimen of the GCL3 was compressed under the backpressure 20 kPa. After that, there was a slightly increase on liquid level of the inflow of GCL3 in the next 17 days. In the last day, the liquid level stayed the same. The volume of the outflow pipe, there is a rise in the 18th day. In the last day, the volume of the inflow and outflow were the same, specimens were complete saturated.

As shown in Figs. 3(a), 3(b), and 3(c), the GCL was normally consolidated for values of backpressure 20 kPa, the volume of GCL decrease in the saturation before the HC test. Because of the backpressure, the specimens were consolidated and drained in the beginning, and then to saturate by DW. In saturation process under backpressure, the level of DW in the inflow pipe decreased first and then increased. By contrast, the level of DW in the inflow pipe was slowly rising. Under backpressure 20 kPa, the specimens were completely saturated after 18 days. An that time, volume of DW inflow and outflow were same.

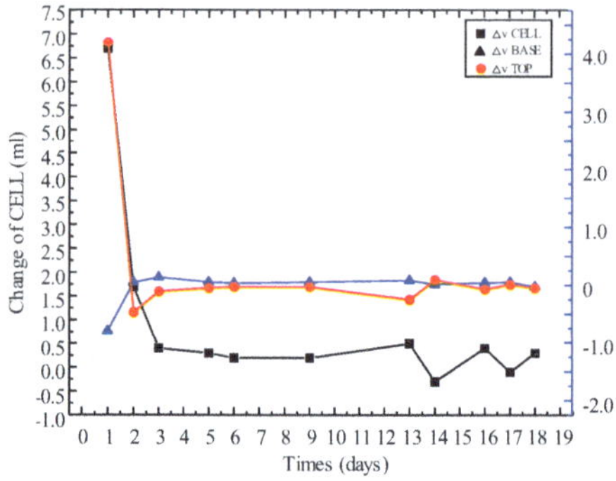

Fig. 3b. GCL2 volume changes of exposed to the DI under 20 kPa backpressure.

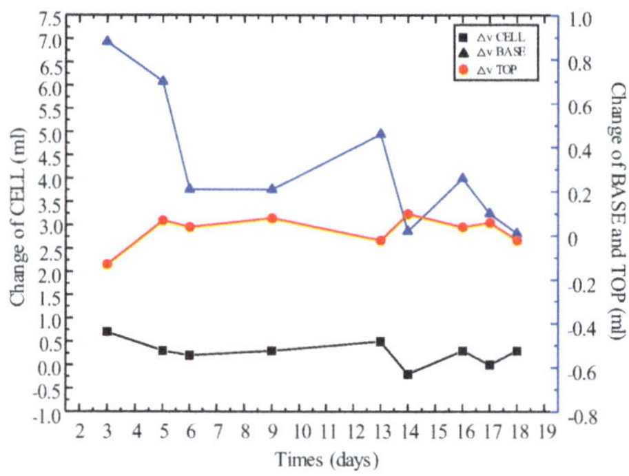

Fig. 3c. GCL3 volume changes of exposed to the DI under 20 kPa backpressure.

3.2 Hydraulic Conductivity Tests

The results of the hydraulic conductivity tests for specimens GCL1 are shown in Figs. 4(a), 4(b), and 4(c). All specimen were under the same experimental condition, such that bottom (inflow), and top (outflow) of the specimens were 35 kPa, 15 kPa, and 0 kPa, respectively.

Figure 4(a) showed the HC of GCL1 change with time and the volume changes in the cell. It can be seen from Fig. 4(a) that the HC of GCL1 decreased with the time in period of tests. In the first day, the HC of GCL1 range from to 1.2×10^{-9} cm/s to 2.0×10^{-9} cm/s. And after 60 days later, the HC of GCL1 decreased under 1.0×10^{-9} cm/s. Compare to the 60 days, the there was only a slight decrease about HC of GCL1, keeping under 0.8×10^{-9} cm/s. In the test process, HC of the GCL1 was a slightly decrease under backpressure 20 kPa, and the HC keep in one order of magnitude. As for the volume of GCL1, it was also a decrease process, the decrease of volume of GCL1 and the change of HC were synchronous. Seepage consolidation happened in the HC tests. So confining stress contributed to lower bulk GCL void ratios and hence lower hydraulic conductivity values. So confining stress contributed to lower bulk GCL void ratios and hence significantly lower hydraulic conductivity values.

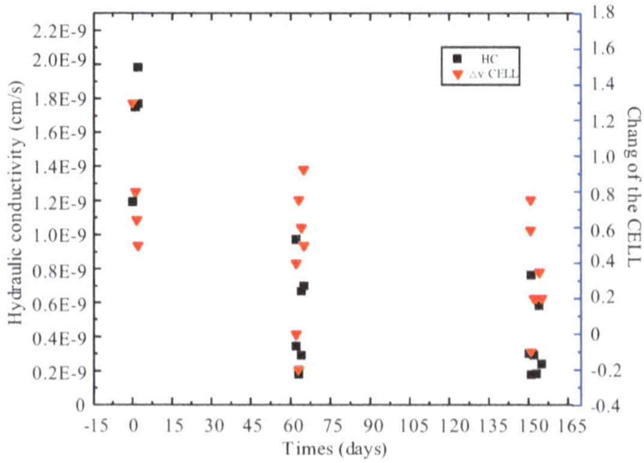

Fig. 4a. GCL1 variation of hydraulic conductivity and cell volume.

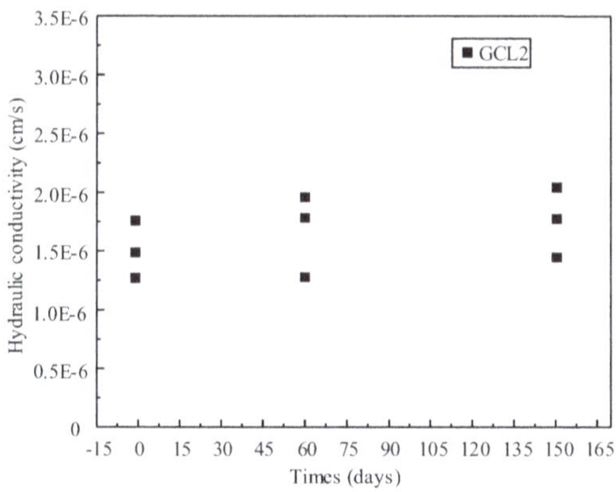

Fig. 4b. GCL2 variation of hydraulic conductivity.

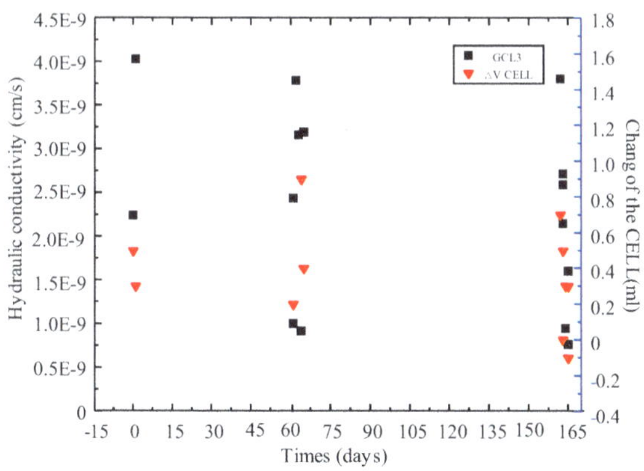

Fig. 4c. GCL1 variation of hydraulic conductivity and cell volume.

Figure 4(b) showed the HC of GCL2 change with time. It can be seen from Fig. 4 (b) that the HC of GCL2 keep in the same level in whole period of test. The HC of GCL2 range from to 1.2×10^{-9} cm/s to 2.2×10^{-9} cm/s. Because of the GCL2 was highly penetrable, there was scarcely any consolidation, so the HC keep in the same high level, more than 1.2×10^{-9} cm/s.

Figure 4(c) showed the HC of GCL3 change with time and the volume changes in the cell. It can be seen from Fig. 4(c) that the HC of GCL3 decreased with the time in period of tests. In the first day, the HC of GCL3 range from to 2.2×10^{-9} cm/s to 4.1×10^{-9} cm/s. And after 60 days later, the HC of GCL3 decreased under 4.0×10^{-9} cm/s. Compare to the 60 days, the there was only decrease about HC of GCL3, keeping under 3.0×10^{-9} cm/s. Similarly, HC of the GCL3 was a slightly decrease under backpressure 20 kPa, and the HC keep in one order of magnitude. As for the volume of GCL3, it was also a decrease process, the decrease of volume of GCL3 and the change of HC were synchronous. Seepage consolidation happened in the HC tests. So confining stress contributed to lower bulk GCL void ratios and hence lower hydraulic conductivity values. Some authors have studied on the GCL void ratio effect the HC of GCL. To normalize the effect of variable mass of bentonite, and its effect on the final GCL height, in the discussion it was prove useful to describe the results in terms of the bulk GCL void ratio, eB defined as the ratio of the volume of voids (geotextiles and bentonite) to the volume of solids (geotextiles and bentonite) in the GCL.

$$e_B = \frac{Vt - Vs}{Vs} = \frac{H_{GCL} - H_s}{H_s} \tag{2}$$

where Vs = total volume of solids (geotextiles and bentonite); Vt = total *volume* of the GCL; H_{GCL} = GCL height; and Hs = height of solids in the GCL (geotextiles and bentonite). As the results showed, the leakage rate through GCL under 20 kPa were significantly different from the leakage of GCL, which HC is stable. There is no linear relation between the flow rate to the confining stress. The flow rate in a different way according to the stress applied.

4 Conclusion

The difference between the hydraulic performances of complete saturated GCLs at penetration Levels is evaluated by permeation DW. The GCL with the low HC is about 10^{-9} cm/s and a HC of GCL is 10^{-6} cm/s

The volume changes for specimens under backpressure in the saturation time. For the volume of the inflow pipe, there is a rise in the first day, because the specimen of the GCL under the backpressure 20 kPa. There is a slightly decrease on liquid level of the inflow of GCL in the 17 days. Similarly, the volume of the outflow pipe, there is a drop in the first day, and then there is the slightly increase on liquid level of the outflow of GCL in the next 17 days. After 18 days, the volume of the inflow and outflow were the same, specimens were complete saturated. So it is very useful to saturate GCL under backpressure 20 kPa.

Seepage consolidation happened in the HC tests. So confining stress contributed to lower bulk GCL void ratios and hence lower hydraulic conductivity values, even the backpressure is 20 kPa.

Acknowledgements. The research was supported by the Key Laboratory of Mechanics on Disaster and Environment in Western China, Ministry of Education, (Lanzhou University), China (No.201603), the Doctoral Program of Higher Education of China (grant no. 20110211110025, 20120211120008) and the Gansu Province Science Foundation for Youths (No.1208RJYA040). Sincere thanks to the reviewers for their time and courtesy, for assisting the authors to maintain the quality of the paper.

References

1. Rowe R, Mukunoki T, Sangam H (2005) Benzene: toluene, ethylbenzene, m&p-xylene, o-xylene diffusion and sorption for a geosynthetic clay liner at two temperatures. J Geotech Geoenviron Eng 131(10):1211–1221
2. Malusis MA, Scalia J (2007) Hydraulic conductivity of an activated carbon-amended geosynthetic clay liner. Geoenviron Eng 163:1–13
3. Gleason M, Daniel D, Eykholt G (1997) Calcium and sodium bentonite for hydraulic containment applications. J Geotech Geoenviron Eng 123(5):438–445
4. Lee J, Shackelford C (2005) Impact of bentonite quality on hydraulic conductivity of geosynthetic clay liners. J Geotech Geoenviron Eng 131(1):64–77
5. Lorenzetti R, Bartelt-Hunt S, Burns S, Smith J (2005) Hydraulic conductivities and effective diffusion coefficients of geosynthetic clay liners with organobentonite amendments. Geotext Geomembr 23(5):385–400
6. Hewitt R, Daniel D (1997) Hydraulic conductivity of geosynthetic clay liners after freeze-thaw. J Geotech Geoenviron Eng 123(4):305–313
7. Kraus J, Benson C, Erickson A, Chamberlain E (1997) Freeze-thaw cycling and hydraulic conductivity of bentonitic barriers. J Geotech Geoenviron Eng 123(3):229–238
8. Petrov R, Rowe R, Quigley R (1997) Selected factors influencing GCL hydraulic conductivity. J Geotech Geoenviron Eng 123(8):683–695
9. Jo H, Benson C, Shackelford C, Lee J, Edil T (2005) Long-term hydraulic conductivity of a geosynthetic clay liner permeated with inorganic salt solutions. J Geotech Geoenviron Eng 131(4):405–417
10. Kolstad D, Benson C, Edil T (2004) Hydraulic conductivity and swell of nonprehydrated geosynthetic clay liners permeated with multispecies inorganic solutions. J Geotech Geoenviron Eng 130(12):1236–1249
11. Shackelford C, Benson C, Katsumi T, Edil T, Lin L (2000) Evaluation of the hydraulic conductivity of GCLs permeated with non-standard liquids. Geotext Geomembr 18(2–4):133–161
12. Benson C, Meer S (2009) Relative abundance of monovalent and divalent cations and the impact of desiccation on geosynthetic clay liners. J Geotech Geoenviron Eng 135(3):349–358
13. Malusis M, Barben E, Evans J (2009) Hydraulic conductivity and compressibility of soil-bentonite backfill amended with activated carbon. J Geotech Geoenviron Eng 135(5):664–672
14. Kang J, Shackelford C (2010) Consolidation of a geosynthetic clay liner under isotropic states of stress. J Geotech Geoenviron Eng 136(1):253–259

Field Monitoring of a Geogrid Reinforced MSW Slope

Pengcheng Ma(✉), Jiwu Lan, and Han Ke

Zhejiang University, Hangzhou Yuhangtang Road 866#, Hangzhou, China
{mapengcheng, lanjiwu}@zju.edu.cn, boske@126.com

Abstract. Geogrids are used to reinforce the MSW slope in some landfills, but the interactions between MSW and the geogrid reinforcement are not clear at present. Xingfeng landfill used geogrids to strengthen the MSW slope during the project of vertical expansion. To observe the behavior of the reinforced MSW slope, systematic field monitoring was carried out. The vertical earth pressures in the reinforced MSW and the tensile strains of the geogrid reinforcement were measured by earth pressure cells and flexible strain sensors, respectively. The distribution of the vertical earth pressures and the reinforcement strains was nonlinear along the length of reinforcement. The vertical earth pressures were adjusted to not equal the theoretical values due to the deformation of the geogrid reinforcement, indicating the interactions between MSW and geogrid reinforcement were relatively effective. However, due to the geogrid reinforcement is relatively short and doesn't traverse the potential slip surface, the risks of slide still cannot be ignored for this reinforced MSW slope.

Keywords: Geogrid reinforcement · MSW slope · Field behavior

1 Introduction

Geosynthetics reinforced soil is used widely to strengthen steep slopes, build high retaining walls and dispose weak foundations [1–3]. Through the interlockings and frictions between soils and reinforcement, the geosynthetics play an important role to transfer and sustain the tensile forces, as well as restrain the soil deformations. Then the strength and stiffness of the soils can be improved. Geogrid is a common geosynthetic material. Due to the grid structure, the interactions between soils and geogrid reinforcement are usually stronger than other geosynthetics. These interactions include the frictions between soils and geogrids surfaces, the resistance of the transverse ribs to soil particles and the interlocking between soil particles and geogrids. Therefore, geogrids are frequently-used in reinforced soil structures.

Landfilling is the main way to dispose municipal solid waste (MSW). The stability of MSW slope is a critical problem in large-scale landfills. As the height of the slope increased, the risks of landslide grow significantly, especially in some special circumstances such as vertical expansion. To increase the stability, geosynthetics were used to reinforce the MSW slope in some landfills. For example, the landfill in Sestri Levante, Italy used geogrids to reinforce the MSW slope during vertical expansion. This project achieved good results finally.

L. Zhan et al. (Eds.): ICEG 2018, ESE, pp. 724–731, 2019.
https://doi.org/10.1007/978-981-13-2224-2_90

Although the geogrid reinforcement has already used in several landfills worldwide, the interactions between MSW and geogrids are not quite clear at present. The observations of field behaviors and achievements of the theoretic study are also absent, which is adverse for the engineering design.

The expansion project in Xingfeng landfill was carried out in 2017. To improve the stability of its eastern slope, geogrid reinforcement was used. This paper presents systematic field monitoring to the behavior of the reinforced MSW slope in Xingfeng landfill. The monitoring items include the vertical earth pressures of reinforced MSW and the tensile strains of the geogrid reinforcement. The vertical earth pressures were measured by the earth pressure cells and the reinforcement strains were measured by the flexible strain sensors. On the basis the monitoring data, the interactions between MSW and geogrid reinforcement were analyzed.

2 Landfill Site

Xingfeng landfill is one of the biggest landfills in China. It came into service in 2002 and the designed service life is 25 years. However, due to the fast increase of MSW, Xingfeng landfill is in a state of overload operation for a long time. The long-term overload operation ran out the designed capacity in advance and promoted the project of expansion. As shown in Fig. 1, the region of vertical expansion (the green region in Fig. 1) is in the middle of the whole waste body and its elevation was 160–170 m before expansion. The target of the expansion project is to increase the elevation to 180–190 m, meaning some layers of MSW with the total thickness of 20 m were going

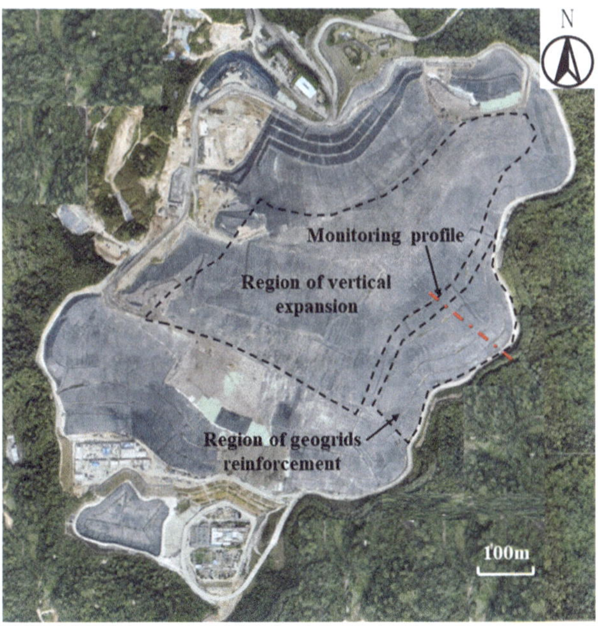

Fig. 1. Schematic plan of Xingfeng landfill (aerial photograph).

to be dumped on the top of the existing waste body. To ensure the safety of the eastern slope (the blue region in Fig. 1), geogrids were chosen to reinforce the MSW slope by a designing institute.

Figure 2 shows the monitoring profile of the reinforced MSW slope and some construction details. The isometric geogrids with the length of 33.0 m were used from the toe to the top of the slope. The vertical spacings of the geogrid reinforcement were all 1.0 m. On the slope surface, each geogrid was wrapped around the geotextile gravel bags. The width of gravel bags wall is about 0.5 m. The geogrids were connected each other vertically by the connecting rods. During the construction, the excavation of the aged MSW were needed. Then the aged MSW excavated were backfilled hierarchically and the geogrids were paved according to the design length and spacing. The geogrids used in this project are high density polyethylene (HDPE) uniaxial geogrids. Figure 3 is a site photograph of the reinforced MSW slope.

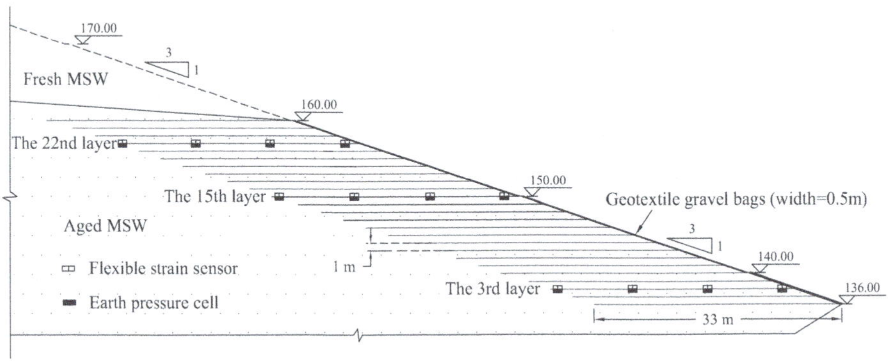

Fig. 2. Monitoring profile of the reinforced MSW slope and the arrangement of transducers.

Fig. 3. The finished reinforced MSW slope

3 Scheme of the Field Monitoring

The arrangement of the transducers along the monitoring profile is shown in Fig. 2 and Table 1. The flexible strain sensors and the earth pressure cells were installed on the 3rd layer, the 15th layer and the 22nd layer, to measure the strains of geogrid reinforcement and the vertical earth pressures. For the convenience to compare different kind of data, the earth pressure cells and the flexible strain sensors were installed adjacently on the same layers and the same positions.

Table 1. Arrangement of the transducers along the monitoring profile

Transducer	Layer number and distance to the slope surface (m)
Flexible strain sensor	The 3rd layer, 2.0, 12.0, 22.0 and 32.0
	The 15th layer, 2.0, 12.0, 22.0 and 32.0
	The 22nd layer, 2.0, 12.0, 22.0 and 32.0
Earth pressure cell	The 3rd layer, 2.0, 12.0, 22.0 and 32.0
	The 15th layer, 2.0, 12.0, 22.0 and 32.0
	The 22nd layer, 2.0, 12.0, 22.0 and 32.0

4 Results of Field Monitoring

The field monitoring began from June 2017, when the reinforced MSW slope was finished. However, a small-scale slide happened in June to July 2017. The possible reason for this slide may be the high pore pressure in the slope, as the result of the rainfall penetration during the excavation. The behaviors of the reinforced MSW slope under large deformations were observed during this slide. In November 2017, the fresh MSW began to be dumping on the top of the reinforced slope, providing an opportunity to observe its behaviors under the top overload. This paper mainly presents the monitoring results measured by the transducers installed on the 15th layer.

4.1 Vertical Earth Pressures

As shown in Fig. 4, the distribution of the vertical earth pressures along the length of geogrid reinforcement were obvious nonlinear and showed a single-peak appearance. One curve represents the monitoring results of the vertical earth pressures of one measurement. During the monitoring period, although the magnitudes were changed, the distribution laws of the vertical earth pressures were relatively consistent. In this reinforced MSW slope, the maximum earth pressures appeared in the end of geogrid reinforcement while the earth pressure near the slope surface was relatively lower. This behavior observed is similar to the monitoring data from a geogrid reinforced soil retaining wall [3, 4]. The nonlinear distribution of the vertical earth pressures may attribute to the interactions among MSW, geogrid reinforcement and geotextile gravel bags. The vertical earth pressures near the slope surface were smaller than the theoretical values (the theoretical value at one point was the product of the unit weight of

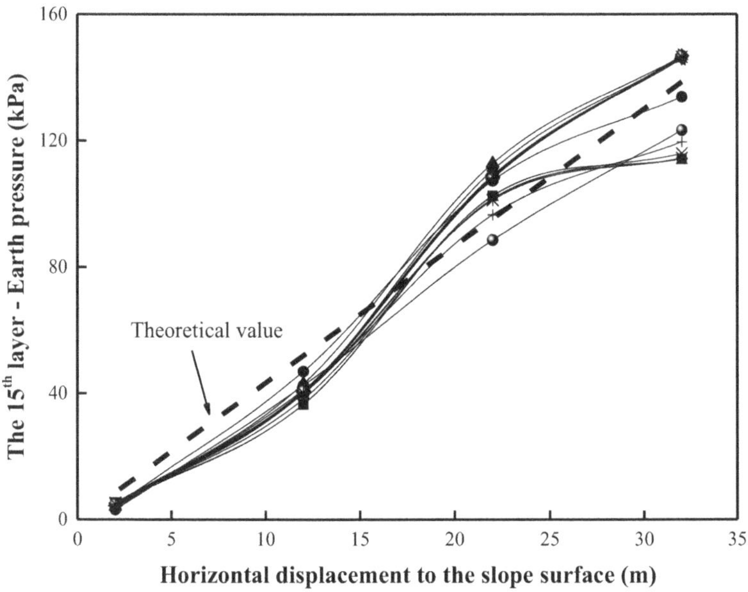

Fig. 4. The distribution of the vertical earth pressures along the length of geogrid reinforcement on the 15th layer from August 2017 to May 2018 (the period of slide is not included)

MSW and the vertical distance form this point to the slope surface) while the earth pressures in the rear part of the reinforcement were slightly larger than the theoretical values. This indicates that the geogrid reinforcement had obvious influences on the earth pressures in reinforced MSW.

Figure 5 shows the changes of the vertical earth pressures over time from August 2017 to May 2018. When beginning to dump the fresh MSW on the top of the reinforced slope, the vertical earth pressures in the middle and the end of the geogrid reinforcement appeared the obvious increase while the vertical earth pressures near the slope surface were influenced much less.

The vertical earth pressures near the slope surface were relatively small and were influenced less by the slide and the dumping may be due to the existence of the gravel bags. The gravel bags may move outwards to release some stresses and may also withstand some vertical stresses for the reinforced MSW, then the vertical earth pressures near the slope surface were reduced.

4.2 Strains of Geogrid Reinforcement

The distribution the reinforcement strains along the length of geogrid reinforcement are shown in Fig. 6. The single-peak nonlinear distribution was also observed apparently. Being different from the vertical earth pressures, the peak of the reinforcement strains was at the distance of 10–15 m to the slope surface, while vertical earth pressures in this region were relatively small. From the peak to the end of the geogrid reinforcement, the strains decreased gradually.

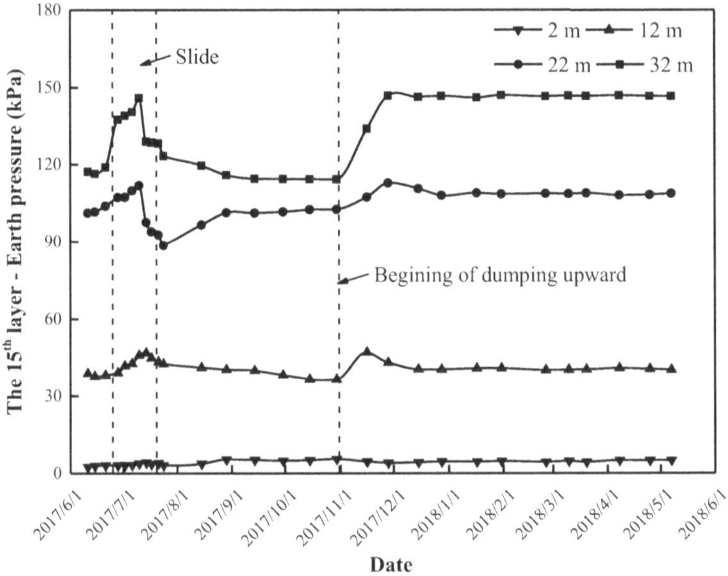

Fig. 5. The variation of the vertical earth pressures on the 15th layer from June 2017 to May 2018

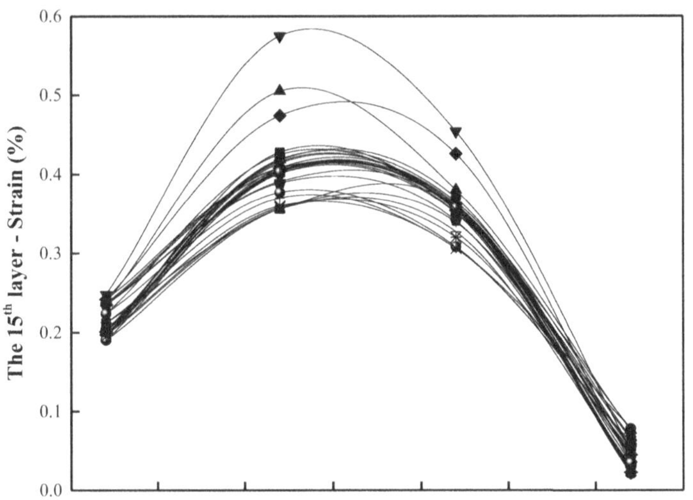

Fig. 6. The distribution of the reinforcement strains along the length of geogrid reinforcement on the 15th layer from June 2017 to May 2018

Figure 7 shows the changes of the reinforcement strains over time from June 2017 to May 2018. Similar to the monitoring results of the vertical earth pressures, the reinforcement strains increased obviously during the waste slide in July 2017. After the slide, the strains decreased and became stable. The reinforcement strains also increased when the fresh MSW was dumped on the top of the reinforced slope, except the strains in front of the gravel bags and at the end of the reinforcement. From December 2017 to May 2018, the strains didn't change significantly, indicating the reinforced MSW slope didn't produce large deformations during this period.

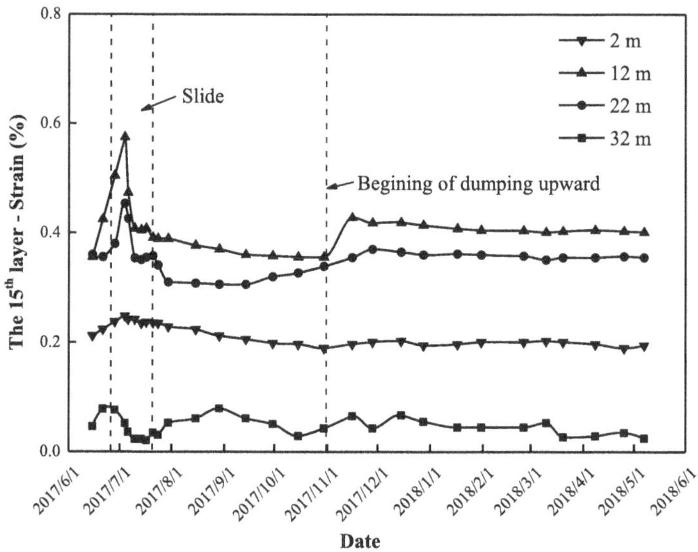

Fig. 7. The variation of the reinforcement strains on the 15th layer from June 2017 to May 2018

5 Discussion

The magnitudes and distribution of the vertical earth pressures and the reinforcement strains depend on the interactions between MSW and geogrid reinforcement. Based the monitoring data, the interactions in this reinforced MSW slope are analyzed primarily. As mentioned above, the vertical earth pressures in the slope were not equal to the theoretical values. The phenomena were related to the influences of the reinforcement. In the region at the distance of 5–10 m to the slope surface, the vertical earth pressures were smaller than the theoretical values. The stress differences between them were sustained by the reinforcement. Correspondingly, to transfer the stresses and withstand the stress differences, the reinforcement there also produced larger strains. Another example which can reflect the interactions is the tensile strains at the end of the reinforcement. Due to the fact that the anchorage length was short at the end of the reinforcement, the anchorage effect was weaker, then the tensile strains there were smaller correspondingly. The monitoring results show that the interactions between MSW and geogrid reinforcement were relatively effective in this reinforced MSW slope.

However, the stability of this reinforced MSW slope in Xingfeng landfill is not optimistic. The waste slide happened in June to July 2017 indicating security risks still are the critical problem we should consider. The reason why the geogrid reinforcement didn't stop the slide may be that the reinforcement is relatively short and doesn't traverse the potential slip surface. Therefore, when designing the reinforced slope, the evaluation for the location of potential slip surface is necessary.

6 Conclusion

Systematic field monitoring to a reinforced MSW slope was carried out in Xingfeng landfill, the following conclusions can be drawn:

1. The vertical earth pressures appeared single-peak nonlinear distribution along the length of geogrid reinforcement. The maximum value was at the end of the reinforcement.
2. The reinforcement strains also appeared single-peak nonlinear distribution along the length of geogrid reinforcement. The maximum value was at the distance of 10–15 m to the slope surface, which was different from the vertical earth pressures.
3. The adjustments of the vertical earth pressures were achieved through the deformations of geogrid reinforcement, indicating the interactions between MSW and geogrid reinforcement were relatively effective.
4. To increase the stability of the reinforced slope, the reinforcement should traverse the potential slip surface.

References

1. Runser DJ, Fox PJ, Bourdeau PL (2001) Field performance of a 17 m-high reinforced soil retaining wall. Geosynth Int 8(5):367–391
2. Cao WZ, Zheng JJ, Zhang J, Zhang RJ (2016) Field test of a geogrid-reinforced and floating pile-supported embankment. Geosynth Int 23(5):348–361
3. Yang G, Zhang B, Lv P, Zhou Q (2009) Behaviour of geogrid reinforced soil retaining wall with concrete-rigid facing. Geotext Geomembr 27(5):350–356
4. Yoo C, Kim S (2008) Performance of a two-tier geosynthetic reinforced segmental retaining wall under a surcharge load: Full-scale load test and 3D finite element analysis. Geotext Geomembr 26(6):460–472

Analytical Study of the Variation of Pore Size Characteristics with Biaxial Tensile Strains on Woven Slit-Film Geotextiles

Lin Tang[1(✉)], Xiao-Wu Tang[2], and Shao-xing Qu[3]

[1] Harbin Institute of Technology at Weihai, Weihai 264209, China
tanya_tang3@hit.edu.cn
[2] Research Center of Costal and Urban Geotechnical Engineering, College of Civil Engineering and Architecture, Zhejiang University, Hangzhou 310058, China
[3] Department of Engineering Mechanics, Zhejiang University, Hangzhou 310027, China

Abstract. Woven geotextiles are typically subjected to complex unequal biaxial tensile strains in filtration applications. Analytical solutions have been proposed to predict the per cent open area (POA) and analytical pore size of woven slit-film geotextiles under certain biaxial tensile strains. The influence of unequal biaxial tensile strains on pore size characteristics of one woven slit-film polypropylene geotextiles was evaluated by image analysis. The ratios of warp to weft strains are set to be 2:1, 3:1 and 4:1, respectively. Comparisons of experiments and predicted results suggest that POA and pore size increase with the biaxial tensile strain. The increasing trend of POA and O_{95} can be predicted by the analytical solutions.

Keywords: Woven slit-film geotextile · Pore size · Biaxial tensile strain
Analytical solution

1 Introduction

Geotextiles are used in filtration applications to prevent soil erosion without impending the flow of seepage water though the soil. To ensure effective filtration, the pore sizes of geotextiles typically need to meet several criteria, including retention, permeability, and anti-clogging capabilities (Koerner 1998). These criteria are often based on the relationship between the characteristic pore sizes and soil grain sizes. The filtration applications of woven slit-film geotextiles are typically subjected to tensile strains, which can cause pore size variations. For example, interface friction between the fill and geotextile tube cause a significant increase in the strain of geotextiles (Plaut and Stephens 2012). Furthermore, the woven slit-film geotextile filters used beneath the erosion control structures are often working under tensile strains. Rowe and Myleville (1990) demonstrated that the strain imposed on geotextiles in the direction perpendicular to the long axis of the embankment can reach 10%. Izadi et al. (2018) evaluated the strains in the geotextile caused by impact loadings ranged from 3.5% up to 5%.

L. Zhan et al. (Eds.): ICEG 2018, ESE, pp. 732–737, 2019.
https://doi.org/10.1007/978-981-13-2224-2_91

However, the characteristic pore sizes are commonly tested in the unstrained condition, which results in soil clogging or leaking despite the fact that the design criteria were satisfied (Fourie and Addis 1999). Therefore, accurate determination of pore size parameters under certain tensile strains is important in the filtration design of woven slit-film geotextiles. The effect of tensile strains on the pores has been investigated, and an analytical solution has been derived to calculate the values of pore size characteristics subjected to designated biaxial tensile strains.

2 Analytical Solution

2.1 Model of the Change in Pore Size

In the past, we have been able to predict the pore size characteristics of woven slit-film geotextiles under the uniaxial tensile strains using the analytical pore model (Tang et al. 2013). The pore model assumed that a woven slit-film geotextile is a two-dimensional structure consisting of warp and weft slit films that are perpendicular to each other. In this research work, we extend the previous pore model for unequal biaxial tensile loading condition. The shape of an unstrained pore is assumed to be square, and a strained pore is considered as a rectangle.

Then, the plane structure of a woven geotextile and the loading direction are illustrated in Fig. 1, where b is the length of one side of a pore square, a is the width of the slit film, t and T are the thicknesses of the silt film and geotextile, respectively. μ is the mass per unit area of the geotextile, and ρ is the density of the slit film. The warp and weft films are in contact at the four points P, Q, R and S. Taking the square PQRS as a unit, the unstrained pore size b is expressed by the measurable physical properties a, t, ρ and μ in Eq. (1). In the following equations, all instances of b are replaced by Eq. (1), which is a convenient assumption for calculation.

$$b = \frac{2at\rho}{\mu} - a \tag{1}$$

The unequal biaxial tensile strains ε_x and ε_y are loaded along the warp and weft direction, respectively, as shown in Fig. 1. The variation of silt film is simplified as a plane stress problem. Thus, the change in the plane pore size is attributed mainly to the tensile axial strain and lateral strain of the silt films. The ratio of the two strains ε_y and ε_x is expressed as follows:

$$\varepsilon_y = n\varepsilon_x \tag{2}$$

Thus, the distance between the axis of two adjacent warp slit films is $(a+b)(1+\varepsilon_x)$, and the distance between the axis of two adjacent weft slit films increases to $(a+b)(1+\varepsilon_y)$, as shown in Fig. 1.

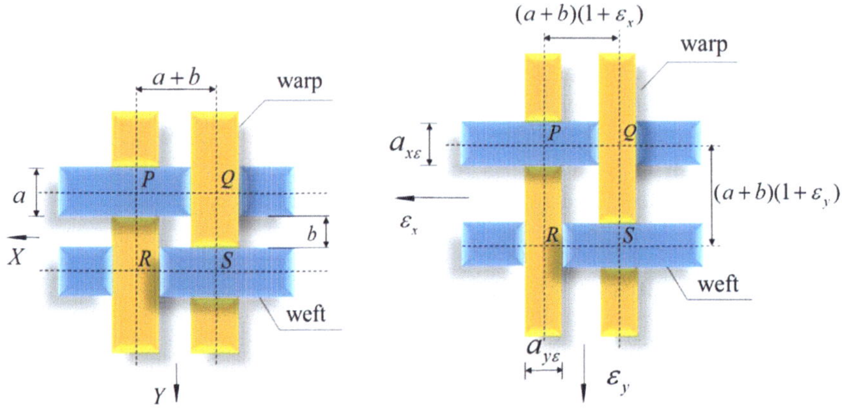

Fig. 1. Model of the change in pore size

2.2 Analytical Solution of the Per cent Open Area

The per cent open area (POA) is the ratio of open area to total area of woven geotextiles, which has applicability only for monofilament woven geotextiles. The strained POA_{ε}^{bi} can be expressed as a function of ε_x as follows:

$$POA_{\varepsilon}^{bi} = \left[1 - \frac{\mu(1 - \upsilon n\varepsilon_x)}{2t\rho(1 + \varepsilon_x)}\right]\left[1 - \frac{\mu(1 - \upsilon\varepsilon_x)}{2t\rho(1 + n\varepsilon_x)}\right] \tag{3}$$

υ is the Poisson's ratio of the slit film. In the uniaxial equation (Tang et al. 2013), the slit film was taken as an incompressible material, and the value of υ was presented as a function of the strain in Eq. (9) (Giroud 2004). Whereas, the variation of pore size parameters seems to be sensitive to υ. The precise measurements of υ can improve the accuracy of the analytical solution. Furthermore, the value of analytical POA_{ε}^{bi} is very sensitive to the thickness of slit film t.

$$\upsilon = \frac{1}{\varepsilon}\left(1 - \frac{1}{\sqrt{1 + \varepsilon}}\right) \tag{4}$$

2.3 Analytical Solution of Pore Size

The analytical pore size is taken as the length of the size of an equivalent square, which has the same area as the actual irregularly shaped pore (Dierickx 1999). When a pore is square, the analytical pore size of the pore is equal to the side dimension of the square. And when different shapes of pores have the same area, circular pores have the shortest perimeter. Hence, the perimeter of a non-square pore is long enough for a spherical particle with the analytical pore size to pass through under hydrodynamic or other conditions. When the biaxial strain of a geotextile reaches ε_x and ε_y, the strained analytical pore size O^{bi} is expressed as a function of ε_x as follows:

$$O^{bi} = a\sqrt{\left[\frac{2t\rho}{\mu}(1+\varepsilon_x) - 1 + vn\varepsilon_x\right]\left[\frac{2t\rho}{\mu}(1+n\varepsilon_x) - 1 + v\varepsilon_x\right]} \tag{5}$$

The biaxial strained pore sizes obtained from the equations are defined as "analytical pore size (O^{bi})". The analytical pore size is derived from the average of each physical parameter, but it does not mean the average of all pore sizes. There is no direct correlation between the analytical pore size and O_{50}, O_{95} or any other characteristic opening sizes. The values of O^{bi} under different strains predict the variations of pore sizes with strains.

3 Experimental Procedure

The influences of unequal biaxial tensile strains on POA and O_{95} of one woven slit-film polypropylene geotextile were evaluated by image analysis. Details of the geotextile used are described in Table 1. The ratios of warp to weft strains were set to be 2:1, 3:1 and 4:1, respectively. Comparisons of experimental results of POA and O_{95} and analytical curves are shown in Figs. 2 and 3. The Experimental results suggest that POA and O_{95} increase approximately linearly with the unequal biaxial tensile strain. With the same strain in the weft direction, the POA and O_{95} increase faster with the larger ratios of warp to weft strains. The slope of the analytical curve increases with the increase of the Poisson's ratio of the slit film υ. The υ of the slit film on W_{150} were 0.36, which was tested by the image analysis. The analytical curves of POA are in good accordance with the variations in the experimental values. The slopes of the analytical curves and experimental results were linearly fitted, as shown in Table 2. The slopes of the analytical pore sizes correlate well with that of the experimental O_{95}. Whereas, the values of O_{95} do not agree well with the analytical pore size O^{bi}.

Table 1. Physical and mechanical parameters of woven geotextiles

Geotextile	Thickness (mm)	Mass per unit area (g/m^2)	Warp tensile strength (kN/m)	Weft tensile strength (kN/m)	Density of slit-film (kg/m^3)	a (mm)	t (mm)
W_{150}	0.27	159.4	25.3	22.5	900	2.060	0.102

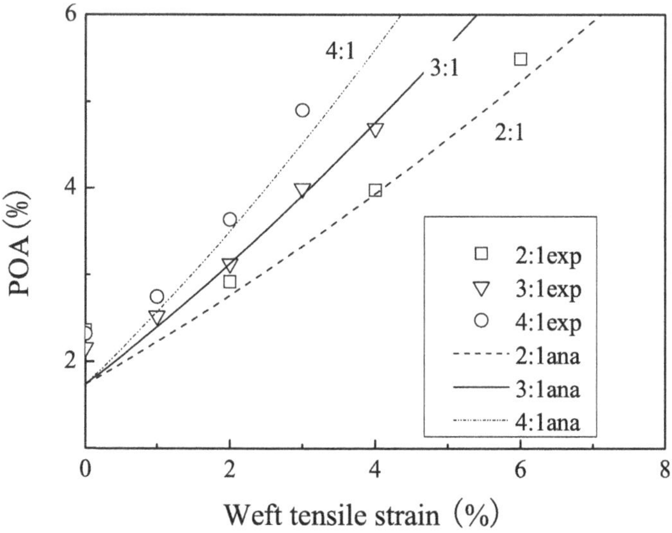

Fig. 2. Analytical and experimental variation of POA with biaxial tensile strain

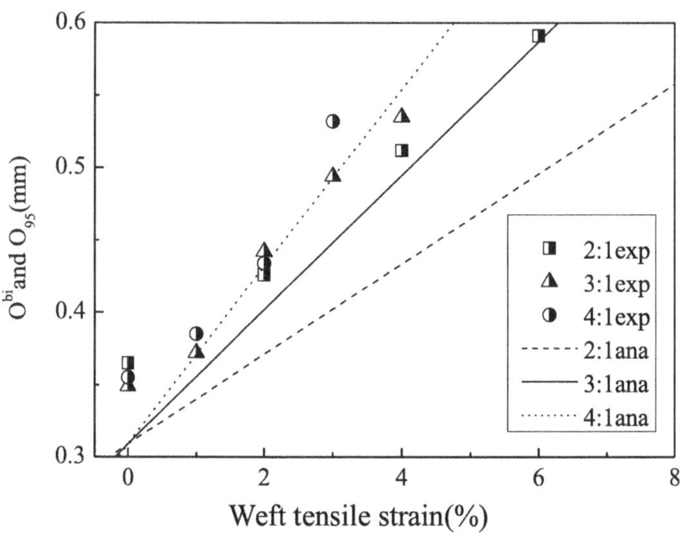

Fig. 3 Variation of the O^{bi} and O_{95} with biaxial tensile strain

Table 2. Slopes of the analytical curves and experimental results

2:1exp	2:1ana	3:1exp	3:1ana	4:1exp	4:1ana
0.038	0.031	0.049	0.046	0.058	0.061

4 Conclusions

Analytical solutions of the pore size characteristics of woven slit-film geotextiles subjected to unequal biaxial tensile strains have been proposed. The analytical model is based on the two-dimensional structure of woven geotextiles. The variation in the pore size of the geotextiles under biaxial tensile strain is mainly attributed to the axial and lateral strains of the slit films in the loading direction. The strained per cent open area (POA) and analytical pore size can be calculated based on the physical properties of the geotextiles and the tensile strains. The experimental values obtained via image analysis correspond with the analytical predictions. The POA and O_{95} both increase with increasing tensile strain. The slopes of the analytical curves of POA and O_{95} correlate well with that of the experimental results.

Acknowledgements. This study was supported by National Natural Science Foundation of China (Grant 51708160), the Shandong Provincial Natural Science Foundation, China (ZR2015PE006), the Fundamental Research Funds for the Central Universities (HIT. NSRIF.2017018), and Research Innovation Foundation in Harbin Institute of Technology (2018).

References

Koerner RM (1998) Designing with geosynthetics. Prentice Hall, Upper Saddle River

Plaut RH, Stephens TC (2012) Analysis of geotextile tubes containing slurry and consolidated material with frictional interface. Geotext Geomembr 32:38–43

Rowe RK, Myleville BLJ (1990) Implications of adopting an allowable geosynthetic strain in estimating stability. In: Proceedings of the fourth international conference on geotextiles, geomembranes and related products, vol 1, pp 131–136 (1990)

Izadi E et al (2018) A laboratory investigation on the impact resistance of a woven geotextile. Geotext Geomembr 46:91–100

Fourie AB, Addis PC (1999) Changes in filtration opening size of woven geotextiles subjected to tensile loads. Geotext Geomembr 17(5):331–340

Tang XW, Tang L, She W, Gao BS (2013) Prediction of pore size characteristics of woven slit-film geotextiles subjected to tensile strains. Geotext Geomembr 38:43–80

Giroud JP (2004) Poisson's ratio of unreinforced geomembranes and nonwoven geotextiles subjected to large strains. Geotext Geomembr 22(4):297–305

Dierickx W (1999) Opening size determination of technical textiles used in agricultural applications. Geotext Geomembr 17:231–245

Correction to: Proceedings of the 8th International Congress on Environmental Geotechnics Volume 2

Liangtong Zhan, Yunmin Chen, and Abdelmalek Bouazza

Correction to:
L. Zhan et al. (Eds.):
***Proceedings of the 8th International Congress on Environmental Geotechnics Volume 2*, ESE,**
https://doi.org/10.1007/978-981-13-2224-2

The original version of the book was inadvertently published without incorporating the author corrections. The book has been updated with the changes.

The updated version of the book can be found at
https://doi.org/10.1007/978-981-13-2224-2

L. Zhan et al. (Eds.): ICEG 2018, ESE, p. C1, 2019.
https://doi.org/10.1007/978-981-13-2224-2_92

Author Index

L. Zhan et al. (Eds.): ICEG 2018, ESE, pp. 739–742, 2019.
https://doi.org/10.1007/978-981-13-2224-2

The manufacturer's authorised representative in the EU is Springer Nature Customer Service Centre GmbH, Europaplatz 3, 69115 Heidelberg, Germany. If you have any concerns regarding our products, please contact ProductSafety@springernature.com

Printed and bound by CPI Group (UK) Ltd, Croydon, CR0 4YY
25/09/2026
02222884-0019